Optical Engineering Science

Optical Engineering Science

Stephen Rolt
University of Durham
Sedgefield, United Kingdom

WILEY

This edition first published 2020

Registered Offices
John Wiley & Sons, Inc., 111 River Street, Hoboken, NJ 07030, USA
John Wiley & Sons Ltd, The Atrium, Southern Gate, Chichester, West Sussex, PO19 8SQ, UK

Editorial Office
The Atrium, Southern Gate, Chichester, West Sussex, PO19 8SQ, UK

For details of our global editorial offices, customer services, and more information about Wiley products visit us at www.wiley.com.

Wiley also publishes its books in a variety of electronic formats and by print-on-demand. Some content that appears in standard print versions of this book may not be available in other formats.

Library of Congress Cataloging-in-Publication Data

Names: Rolt, Stephen, 1956- author.
Title: Optical engineering science / Stephen Rolt, University of Durham, Sedgefield, United Kingdom.
Description: First edition. | Hoboken, NJ : John Wiley & Sons, 2020. | Includes bibliographical references and index.
Identifiers: LCCN 2019032028 (print) | LCCN 2019032029 (ebook) | ISBN 9781119302803 (hardback) | ISBN 9781119302797 (adobe pdf) | ISBN 9781119302810 (epub)
Subjects: LCSH: Optical engineering. | Optics.
Classification: LCC TA1520 .R65 2019 (print) | LCC TA1520 (ebook) | DDC 621.36–dc23
LC record available at https://lccn.loc.gov/2019032028
LC ebook record available at https://lccn.loc.gov/2019032029

Cover Design: Wiley
Cover Images: Line drawing cover image courtesy of Stephen Rolt, Background: © AF-studio/Getty Images

Set in 10/12pt Warnock by SPi Global, Chennai, India
Printed and bound in Singapore by Markono Print Media Pte Ltd

10 9 8 7 6 5 4 3 2 1

Contents

Preface

The book is intended as a useful reference source in optical engineering for both advanced students and engineering professionals. Whilst grounded in the underlying principles of optical physics, the book ultimately looks toward the practical application of optics in the laboratory and in the wider world. As such, examples are provided in the book that will enable the reader to understand and to apply. Useful exercises and problems are also included in the text. Knowledge of basic engineering mathematics is assumed, but an overall understanding of the underlying principles should be to the fore.

Although the text is wide ranging, the author is keenly aware of its omissions. In compiling a text of this scope, there is a constant pre-occupation of what can be omitted, rather than what is to be included. This tyranny is imposed by the manifest requirement of brevity. With this limitation in mind, choice of material is dictated by the author's experience and taste; the author fully accepts that the reader's taste may vary somewhat.

The evolution of optical science through the ages is generally seen as a progression of ideas, an intellectual journey culminating in the development of modern quantum optics. Although some in the ancient classical world thought that the sensation of vision actually originates in the eye, it was quickly accepted that vision arises, in some sense, from an external agency. From this point, it was easy to visualise light as beams, rays, or even particles that have a tendency to move from one point to another in a straight line before entering the eye. Indeed, it is this perspective that dominates geometric optics today and drives the design of modern optical systems.

The development of ideas underpinning modern optics is, to a large extent, attributed to the early modern age, most particularly the classical renaissance of the seventeenth century. However, many of these ideas have their origin much earlier in history. For instance, Euclid postulated laws of rectilinear propagation of light, as early as 300 BCE. Some understanding of the laws of propagation of light might have underpinned Archimedes' famous solar concentrator that (according to legend) destroyed the Roman fleet at the siege of Syracuse in 212 BCE. Whilst the law governing the refraction of light is famously attributed to Willebrord Snellius in the seventeenth century, many aspects of the phenomenon were understood much earlier. Refraction of light by water and glass was well understood by Ptolemy in the second century CE and, in the tenth century, Ibn Sahl and Ibn Al-Haytham (Alhazen) analysed the phenomenon in some detail.

From the early modern era, the intellectual progression in optics revolved around a battle between particle (corpuscular) or ray theory, as proposed by Newton, and wave theory, as proposed by Huygens. For a time, in the nineteenth century, the journey seemed to be at an end, culminating in the all-embracing description provided by Maxwell's wave equations. The link between wave and ray optics was provided by Fermat's theorem which dictated the light travels between two points by the path that takes the least time and this could be clearly derived from Maxwell's equations. However, this clarity was removed in the twentieth century when the ambiguity between the wave and corpuscular (particle) properties of light was restored by the advent of quantum mechanics.

This progression provides an understanding of the history of optics in terms of an intellectual journey. This is the way the history of optics is often portrayed. However, there is another strand to the development of optics that is often ignored. When Isaac Newton famously procured his prism at the Stourbridge Fair in Cambridge in 1665, it is clear that the fabrication of optical components was a well-developed skill at the time. Indeed, the construction of the first telescope (attributed to Hans Lippershey) would not have been possible without the technology to grind lenses, previously mastered by skilled spectacle makers. The manufacture of lenses for spectacles had been carried out in Europe (Italy) from at least the late thirteenth century CE. However, the origins of this skill are shrouded in mystery. For instance, Marco Polo reported the use of spectacles in China in 1270 and these were said to have originated from Arabia in the eleventh century.

So, in parallel to the more intellectual journey in optics, people were exercising their practical curiosity in developing novel optical technologies. In many early cultures, polished mirrors feature as grave goods in the burials of high-status individuals. One example of this is a mirror found in the pyramid build for Sesostris II in Egypt in around 1900 BCE. The earliest known lens in existence is the Nimrud or Layard lens attributed to the Assyrian culture (750–710 BCE). Nero is said to have watched gladiatorial contests through a shaped emerald, presumably to correct his myopic vision. Abbas Ibn Firnas, working in Andalucia in the ninth century CE developed magnifying lenses or 'reading stones'.

These two separate histories lie at the heart of the science of optical engineering. On the one hand, there is a desire to understand or analyse and on the other hand there is a desire to create or synthesise. An optical engineer must acquire a portfolio of fundamental knowledge and understanding to enable the creation of new optical systems. However, ultimately, optical engineering is a practical discipline and the motivation for acquiring this knowledge is to enable the design, manufacture, and assembly of better optical systems. For this knowledge to be fruitful, it must be applied to specific tasks. As such, this book focuses, initially, on the fundamental optics underlying optical design and fabrication. Notwithstanding the advent of powerful software and computational tools, a sound understanding and application of the underlying principles of optics is an essential part of the design and manufacturing process. An intuitive understanding greatly aids the use of these sophisticated tools.

Ultimately, preparation of an extensive text, such as this, cannot be a solitary undertaking. The author is profoundly grateful to a host of generous colleagues who have helped him in his long journey through optics. Naturally, space can only permit the mention of a few of these. Firstly, for a thorough introduction and grounding in optics and lasers, I am particularly indebted to my former DPhil Supervisor at Oxford, Professor Colin Webb. Thereafter, I was very fortunate to spend 20 years at Standard Telecommunication Laboratories in Harlow, UK (later Nortel Networks), home of optical fibre communications. I would especially like to acknowledge the help and support of my colleagues, Dr Ken Snowdon and Mr Gordon Henshall during this creative period. Ultimately, the seed for this text was created by a series of Optical Engineering lectures delivered at Nortel's manufacturing site in Paignton, UK. In this enterprise, I was greatly encouraged by the facility's Chief Technologist, Dr Adrian Janssen.

In later years, I have worked at the Centre for Advanced Instrumentation at Durham University, involved in a range of Astronomical and Satellite instrumentation programmes. By this time, the original seed had grown into a series of Optical Engineering graduate lectures and a wide-ranging Optical Engineering Course delivered at the European Space Agency research facility in Noordwijk, Netherlands. This book itself was conceived, during this time, with the encouragement and support of my Durham colleague, Professor Ray Sharples. For this, I am profoundly grateful. In preparing the text, I would like to thank the publishers, Wiley and, in this endeavour, for the patience and support of Mr Louis Manoharan and Ms Preethi Belkese and for the efforts of Ms Sandra Grayson in coordinating the project. Most particularly, I would like to acknowledge the contribution of the copy-editor, Ms Carol Thomas, in translating my occasionally wayward thoughts into intelligible text.

This project could not have been undertaken without the support of my family. My wife Sue and sons Henry and William have, with patience, endured the interruption of many family holidays in the preparation of the manuscript. Most particularly, however, I would like to thank my parents, Jeff and Molly Rolt. Although their early lives were characterised by adversity, they unflinchingly strove to provide their three sons with the security and stability that enabled them to flourish. The fruits of their labours are to be seen in these pages.

Finally, it remains to acknowledge the contributions of those giants who have preceded the author in the great endeavour of optics. In humility, the author recognises it is their labours that populate the pages of this book. On the other hand, errors and omissions remain the sole responsibility of the author. The petty done, the vast undone…

Glossary

AC	Alternating current
AFM	Atomic force microscope
AM0	Air mass zero
AM1	Air mass one (atmospheric transmission)
ANSI	American national standards institute
APD	Avalanche photodiode
AR	Antireflection (coating)
AS	Astigmatism
ASD	Acceleration spectral density
ASME	American society of mechanical engineers
BBO	Barium borate
BRDF	Bi-directional reflection distribution function
BS	Beamsplitter
BSDF	Bi-directional scattering distribution function
CAD	Computer aided design
CCD	Charge coupled device
CD	Compact disc
CGH	Computer generated hologram
CIE	Commission Internationale de l'Eclairage
CLA	Confocal length aberration
CMM	Co-ordinate measuring machine
CMOS	Complementary metal oxide semiconductor
CMP	Chemical mechanical planarisation
CNC	Computer numerical control
CO	Coma
COTS	Commerical off-the-shelf
CTE	Coefficient of thermal expansion
dB	Decibel
DC	Direct current
DFB	Distributed feedback (laser)
DI	Distortion
E-ELT	European extremely large telescope
EMCCD	Electron multiplying charge coupled device
ESA	European space agency
f#	F number (ratio of diameter to focal distance)
FAT	Factory acceptance test
FC	Field curvature

FEA	Finite element analysis
FEL	Filament emission lamp
FEL	Free electron laser
FFT	Fast Fourier transform
FRD	Focal ratio degradation
FSR	Free spectral range
FT	Fourier transform
FTIR	Fourier transform infra-red (spectrometer)
FTR	Fourier transform (spectrometer)
FWHM	Full width half maximum
GRIN	Graded index (lens or fibre)
HEPA	High- efficiency particulate air (filter)
HST	Hubble space telescope
HWP	Half waveplate
IEST	Institute of environmental sciences and technology
IFU	Integral field unit
IICCD	Image intensifying charge coupled device
IR	Infrared
ISO	International standards organisation
JWST	James Webb space telescope
KDP	Potassium dihydrogen phosphate
KMOS	K-band multi-object spectrometer
LA	Longitudinal aberration
LCD	Liquid crystal display
LED	Light emitting diode
LIDAR	Light detection and ranging
MTF	Modulation transfer function
NA	Numerical aperture
NASA	National Aeronautics and Space Administration
NEP	Noise equivalent power
NIRSPEC	Near infrared spectrometer
NIST	National institute of standards and technology (USA)
NMI	National measurement institute
NPL	National physical laboratory (UK)
NURBS	Non-uniform rational basis spline
OPD	Optical path difference
OSA	Optical society of America
OTF	Optical transfer function
PD	Photodiode
PMT	Photomultiplier tube
PPLN	Periodically poled lithium niobate
PSD	Power spectral density
PSF	Point spread function
PTFE	Polytetrafluoroethylene
PV	Peak to valley
PVA	Polyvinyl alcohol
PVr	Peak to valley (robust)
QMA	Quad mirror anastigmat
QTH	Quartz tungsten halogen (lamp)

QWP	Quarter waveplate
RMS	Root mean square
RSS	Root sum square
SA	Spherical aberration
SI	Système Internationale
SLM	Spatial light modulator
SNR	Signal to noise ratio
TA	Transverse aberration
TE	Transverse electric (polarisation)
TGG	Terbium gallium garnet
TM	Transverse magnetic (polarisation)
TMA	Three mirror anastigmat
TMT	Thirty metre telescope
USAF	United States Airforce
UV	Ultraviolet
VCSEL	Vertical cavity surface emitting laser
VPH	Volume phase hologram
WDM	Wavelength division multiplexing
WFE	Wavefront error
YAG	Yttrium aluminium garnet
YIG	Yttrium iron garnet
YLF	Yttrium lithium fluoride

About the Companion Website

This book is accompanied by a companion website:

www.wiley.com/go/Rolt/opt-eng-sci

The website includes:

- Problem Solutions
- Spreadsheet tools

Scan this QR code to visit the companion website.

1

Geometrical Optics

1.1 Geometrical Optics – Ray and Wave Optics

In describing *optical* systems, in the narrow definition of the term, we might only consider systems that manipulate visible light. However, for the optical engineer, the application of the science of optics extends well beyond the narrow boundaries of human vision. This is particularly true for modern instruments, where reliance on the human eye as the final detector is much diminished. In practice, the term optical might also be applied to radiation that is manipulated in the same way as visible light, using components such as lenses, mirrors, and prisms. Therefore, the word 'optical', in this context might describe electromagnetic radiation extending from the vacuum ultraviolet to the mid-infrared (*wavelengths from ~120 to ~10 000 nm*) and perhaps beyond these limits. It certainly need not be constrained to the narrow band of visible light between about 430 and 680 nm. Figure 1.1 illustrates the electromagnetic spectrum.

Geometrical optics is a framework for understanding the behaviour of light in terms of the propagation of light as highly directional, narrow bundles of energy, or *rays*, with 'arrow like' properties. Although this is an incomplete description from a theoretical perspective, the use of ray optics lies at the heart of much of practical optical design. It forms the basis of optical design software for designing complex optical instruments and geometrical optics and, therefore, underpins much of modern optical engineering.

Geometrical optics models light entirely in terms of infinitesimally narrow beams of light or rays. It would be useful, at this point, to provide a more complete conceptual description of a ray. Excluding, for the purposes of this discussion, quantum effects, light may be satisfactorily described as an electromagnetic wave. These waves propagate through free space (vacuum) or some optical medium such as water and glass and are described by a wave equation, as derived from Maxwell's equations:

$$\frac{\partial^2 E}{\partial x^2} + \frac{\partial^2 E}{\partial y^2} + \frac{\partial^2 E}{\partial z^2} = \frac{n^2}{c^2}\frac{\partial^2 E}{\partial z^2} \tag{1.1}$$

E is a *scalar* representation of the local electric field; **c** is the velocity of light in free space, and **n** is the refractive index of the medium.

Of course, in reality, the local electric field is a ***vector*** quantity and the ***scalar theory*** presented here is a useful initial simplification. Breakdown of this approximation will be considered later when we consider polarisation effects in light propagation. If one imagines waves propagating from a central point, the wave equation offers solutions of the following form:

$$E = \frac{E_0}{r} e^{i(kr-\omega t)} \tag{1.2}$$

Equation (1.2) represents a spherical wave of angular frequency, ω, and spatial frequency, or wavevector, k. The velocity that the wave disturbance propagates with is ω/k or c/n. In free space, light propagates at the speed of light, c, a fundamental and ***defined*** constant in the SI system of units. Thus, the refractive index, n, is the ratio of the speed of light in free space to that in the specified medium. All points lying at the same distance,

Optical Engineering Science, First Edition. Stephen Rolt.

Companion website: www.wiley.com/go/Rolt/opt-eng-sci

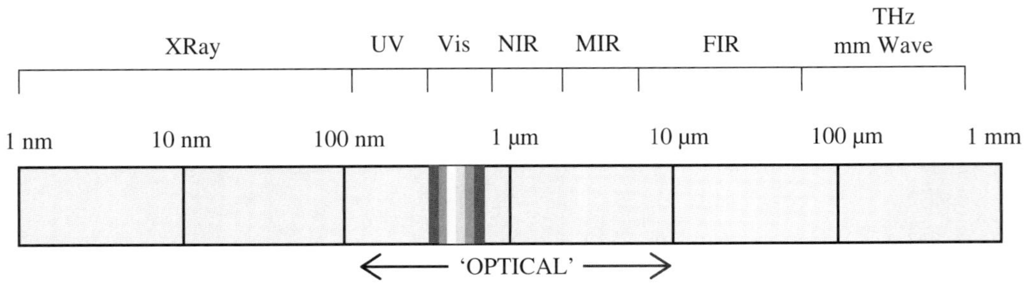

Figure 1.1 The electromagnetic spectrum.

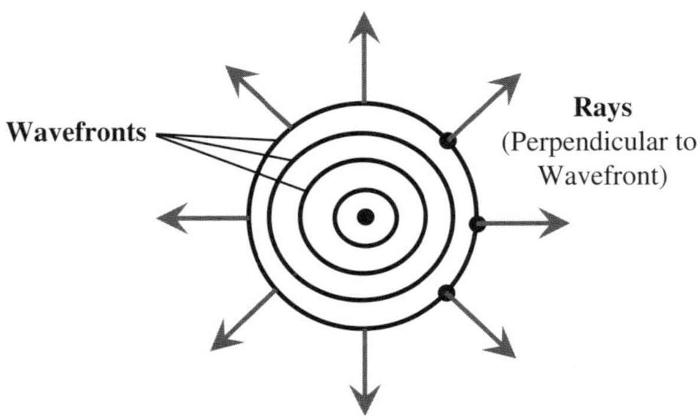

Figure 1.2 Relationship between rays and wavefronts.

r, from the source, will oscillate at an angular frequency, ω, and in the same phase. Successive surfaces, where all points are oscillating entirely in phase are referred to as wavefronts and can be viewed at the crests of ripples emanating from a point disturbance. This is illustrated in Figure 1.2. This picture provides us with a more coherent definition of a ray. *A ray is represented by the vector normal to the wavefront surface in the direction of propagation.* Of course, Figure 1.2 represents a simple spherical wave, with waves spreading from a single point. However, in practice, wavefront surfaces may be much more complex than this. Nevertheless, the precise definition of a ray remains clear:

At any point in space in an optical field, a ray may be defined as the unit vector perpendicular to the surface of constant phase at that point with its sense lying in the same direction as that of the energy propagation.

1.2 Fermat's Principle and the Eikonal Equation

Intuition tells us that light 'travels in straight lines'. That is to say, light propagates between two points in such a way as to minimise the distance travelled. More generally, in fact, all geometric optics is governed by a very simple principle along similar lines. Light always propagates between two points in space in such a way as to minimise the *time* taken. If we consider two points, A and B, and a ray propagating between them within a medium whose refractive index is some arbitrary function, $n(r)$, of position then the time taken is given by:

$$\tau = \frac{1}{c}\int_A^B n(r)ds \tag{1.3}$$

c is the speed of light in vacuo *and ds is an element of path between A and B*

This is illustrated in Figure 1.3.

Figure 1.3 Arbitrary ray path between two points.

Fermat's principle may then be stated as follows:

Light will travel between two points A and B such that the path taken represents a local minimum in the total optical path between these points.

Fermat's principle underlies all ray optics. All laws governing refraction and reflection of rays may be derived from Fermat's principle. Most importantly, to demonstrate the theoretical foundation of ray optics and its connection with physical or wave optics, Fermat's principle may be directly derived from the wave equation. This proof demonstrates that the path taken represents, in fact, a ***stationary*** solution with respect to other possible paths. That is to say, technically, the optical path taken ***could*** represent a local ***maximum*** or ***inflexion*** point rather than a minimum. However, for most practical purposes it is correct to say that the path taken represents the minimum possible optical path.

Fermat's principle is more formally set out in the Eikonal equation. Referring to Figure 1.2, if instead of describing the light in terms of rays it is described by the wavefront surfaces themselves. The function $S(r)$ describes the *phase* of the wave at any point and the Eikonal equation, which is derived from the wave equation, is set out thus:

$$\frac{\partial^2 S(r)}{\partial x^2} + \frac{\partial^2 S(r)}{\partial y^2} + \frac{\partial^2 S(r)}{\partial z^2} = n^2 \tag{1.4}$$

The important point about the Eikonal equation is not the equation itself, but the assumptions underlying it. Derivation of the Eikonal equation assumes that the rate of change in phase is small compared to the wavelength of light. That is to say, the radius of curvature of the wavefronts should be significantly larger than the wavelength of light. Outside this regime the assumptions underlying ray optics are not justified. This is where the effects of the wave nature of light (i.e. diffraction) must be considered and we enter the realm of physical optics. But for the time being, in the succeeding chapters we may consider that all optical systems are adequately described by geometrical optics.

So, for the purposes of this discussion, it is one simple principle, Fermat's principle, that provides the foundation for all ray optics. For the time being, we will leave behind specific consideration of the detailed behaviour of individual optical surfaces. In the meantime, we will develop a very generalised description of an idealised optical system that does not attribute specific behaviours to individual components. Later on, this 'black box model' will be used, in conjunction with Gaussian optics to provide a complete first order description of complex optical systems.

1.3 Sequential Geometrical Optics – A Generalised Description

In applying geometrical optics to a real system, we are attempting to determine the path of a ray(s) through the system. There are a few underlying characteristics that underpin most optical systems and help to simplify analysis. First, *most* optical systems are sequential. An optical system might comprise a number of different elements or surfaces, e.g. lenses, mirrors, or prisms. In a sequential optical system, the order in which light propagates through these components is unique and pre-determined. Second, in most practical systems, light is constrained with respect to a mechanical or optical axis of symmetry, the optical axis, as illustrated in Figure 1.4. In real optical systems, light is constrained by the use of physical apertures or 'stops'; this will be discussed in more detail later.

Of course, in practice, the optical axis need not be a continuous, straight line through an optical system. It may be bent, or folded by mirrors or prisms. Nevertheless, there exists an axis throughout the system with respect to which the rays are constrained.

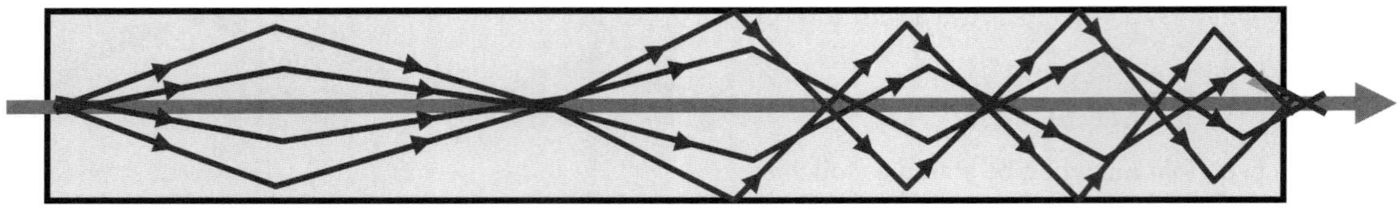

Figure 1.4 Constraint of rays with respect to optical axis.

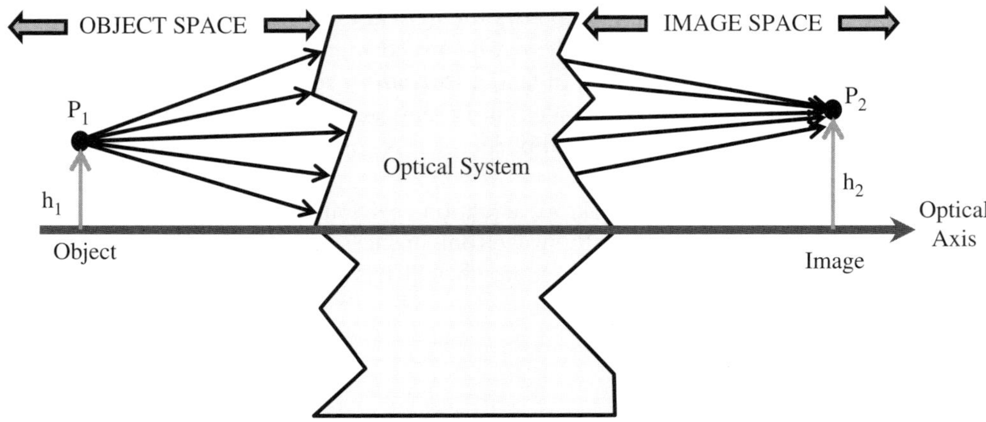

Figure 1.5 Generalised optical system and conjugate points.

1.3.1 Conjugate Points and Perfect Image Formation

We consider an ideal optical system which consists of a point source of light, the ***object***, and an optical system that collects the light and re-directs *all rays* emanating from this point source or object, such that the rays converge onto a single point, the ***image*** point. At this stage, the interior workings of the optical system are undefined; the system behaves as a 'black box'. The object is said to be located in object space and the image in image space and the pair of points are said to be ***conjugate points***. This is illustrated in Figure 1.5.

In Figure 1.5, the two points P_1 and P_2 are conjugate. The optical system can be simple, for example a single lens, or it can be complex, containing many optical elements. The description above is entirely generalised. Where the object point lies on the optical axis, its image or conjugate point also lies on the optical axis. In Figure 1.5, the object point has a height of h_1 with respect to the optical axis and its corresponding image point has a height of h_2 with respect to the same axis. The ratio of these two heights gives the system (transverse) magnification, M:

$$M = h_2/h_1 \tag{1.5}$$

Points occupying a plane perpendicular to the optical axis are conjugate to points lying on another plane perpendicular to the optical axis. These planes are known as conjugate planes.

1.3.2 Infinite Conjugate and Focal Points

Where an image or object is located at infinity, all rays emerging from or travelling to these locations will be parallel with respect to each other. In this instance, the point located at infinity is said to be at an ***infinite conjugate***. The corresponding conjugate point to the infinite conjugate is known as a ***focal point***. There are two focal points. The ***first focal point*** is located in the object space with the corresponding image located at the infinite conjugate. The ***second focal point*** is located in the image space with the object placed at the infinite conjugate. Figure 1.6 depicts the first focal point:

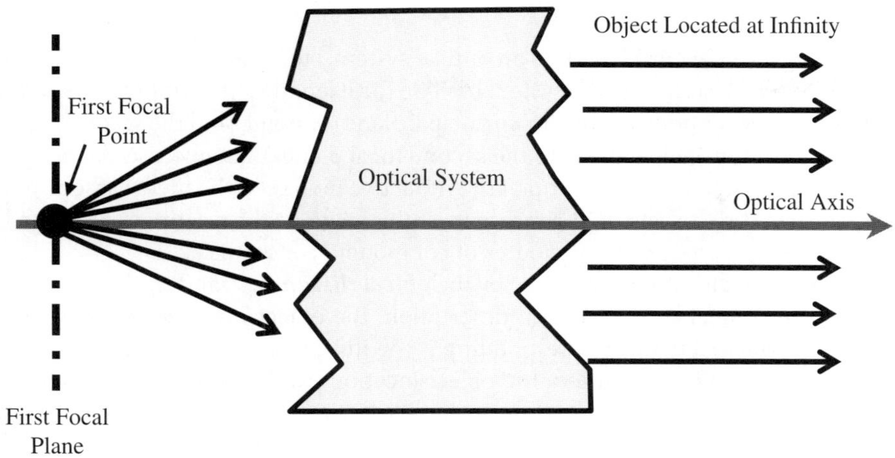

Figure 1.6 Location of first focal point.

As well as focal points, there are two corresponding ***focal planes***. The two focal planes are planes perpendicular to the optical axis that contain the relevant focal point. For all points lying on the relevant focal plane, the conjugate point will lie at the infinite conjugate. In other words, all rays will be parallel with respect to each other. In general, the rays will not be parallel to the optic axis. This would only be the case for a conjugate point lying on the optical axis.

1.3.3 Principal Points and Planes

All points lying on a particular conjugate plane are associated with a specific transverse magnification, M, which is equal to the ratio of the image and object heights. For an ideal system, there exist two conjugate planes where the magnification is unity. These are known as the ***principal planes***. Thus, there are two principal planes and the points where the optical axis intersects the principal planes are known as principal points. The first principal point (plane) is located in object space and the second principal point (plane) is located in image space. The arrangement is illustrated schematically in Figure 1.7.

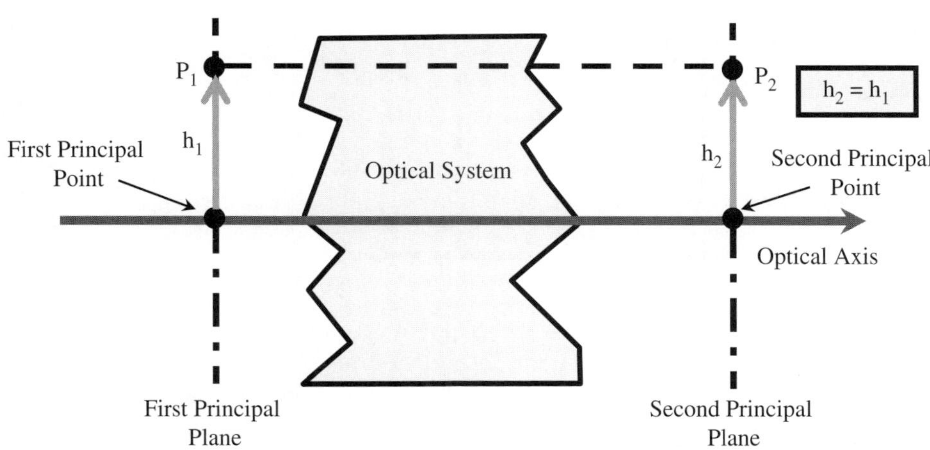

Figure 1.7 Principal points and principal planes.

1.3.4 System Focal Lengths

The reader might be used to ascribing a single focal length to an optical system, such as for a magnifying lens or a camera lens. However, in this general description, the system has two focal lengths. The ***first focal length***, f_1, is the distance from the first focal plane (or point) to the first principal plane (or point) and the ***second focal length***, f_2, is the distance from the second principal plane to the second focal plane. In many cases, f_1 and f_2 are identical. In fact, the ratio f_1/f_2 is equal to n_1/n_2, the ratio of the refractive indices of the media associated with the object and image spaces. However, this need not concern us at this stage, as the treatment presented here is entirely general and independent of the specific attributes of components or media.

In classical geometrical optics, the object location is denoted by the ***object distance***, u, and the image location by the ***image distance***, v. In the context of this general description, the object distance is simply the distance from the object to the first principal plane. Correspondingly, the image distance, v, is the distance from the second principal plane to the image. In addition, the object location can be described by the distance, x_1, separating the object from the corresponding focal plane. Similarly, x_2 represents the distance from the image to the second focal plane. This is illustrated in Figure 1.8.

1.3.5 Generalised Ray Tracing

This general description of an optical system is very economical in that the definition of conjugate points, focal planes, and principal planes provides sufficient information to determine the path of a ray in the image space, given the path of the ray in the object space. No assumptions are made about the internal workings of the optical system; it is merely a 'black box'.

We see how input rays originating in the object space are mapped onto the image space for specific scenarios where the object is located at the input focal plan, the infinite conjugate, or the first principal plane. How can this be extended to determine the output path of any input ray? The general principle is set out in Figure 1.9. First, the input ray is traced from point P_1 as far as its intersection with the (first) principal plane at A_1. We know that this point, A_1, is conjugated with point A_2, lying at the same height at the second principal plane. This follows directly from the definition of principal planes. Second, we draw a dummy ray originating from the first focal point, f_1, but parallel to the input ray and trace it to where it intersects the first principal plane at B_1. We know that B_1 is conjugated with point B_2, lying at the same height on the second principal plane.

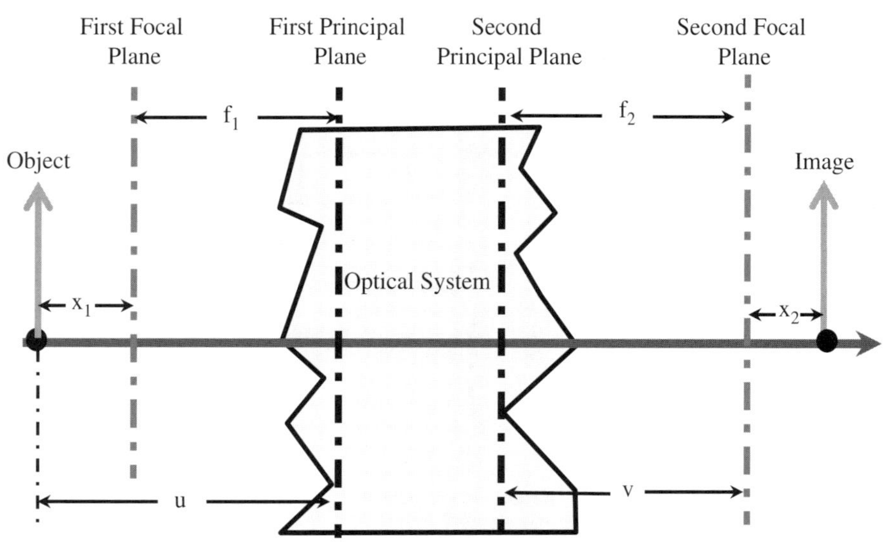

Figure 1.8 System focal lengths.

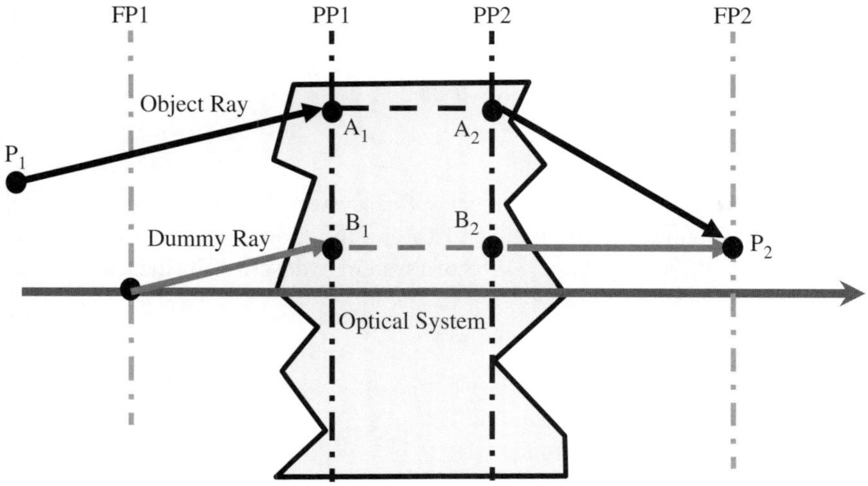

Figure 1.9 Tracing of arbitrary ray.

Since this ray originated from the first focal point, its path must be parallel to the optical axis in image space and thus we can trace it as far as the second focal plane at P_2. Finally, since the object ray and dummy rays are parallel in object space, they must meet at the second focal plane in the image space. Therefore, we can trace the image ray to point P_2, providing a complete definition of the path of the ray in image space.

1.3.6 Angular Magnification and Nodal Points

The ***angular magnification*** of an optical system is the ratio of the angle (with respect to the optical axis) of a ray in image space and that of its conjugate in object space. There exists a pair of conjugate points lying on the optical axis where, for all possible rays, the angular magnification is unity. These are the ***nodal points***. The ***first nodal point*** is located in object space and the ***second nodal point*** is located in image space. This is set out in Figure 1.10, where for a general conjugate pair, the angular magnification, α, is equal to θ_2/θ_1. For

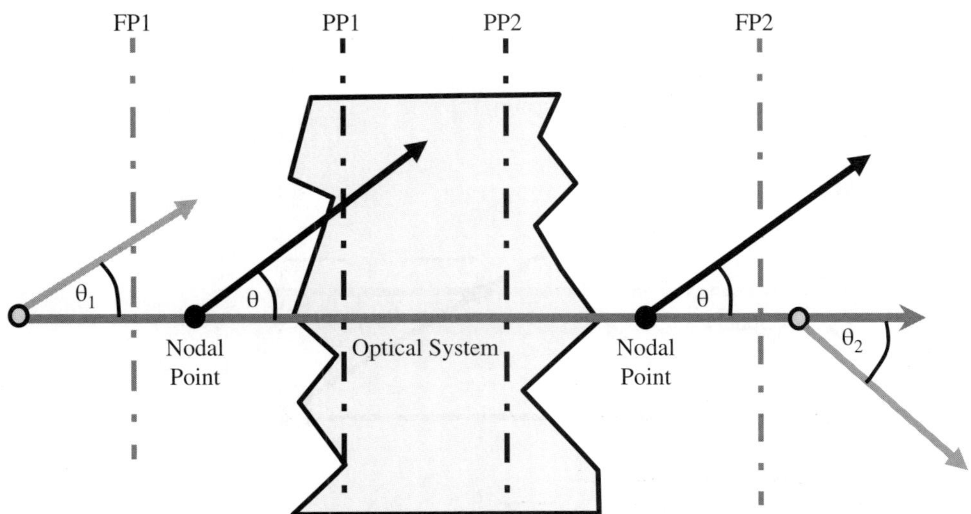

Figure 1.10 Angular magnification and nodal points.

the nodal points, $\theta_2 = \theta_1$; that is to say, the angular magnification is unity. Where the two focal lengths are identical, or the object and image spaces are within media of the same refractive index, the nodal points are co-located with the principal points.

1.3.7 Cardinal Points

This brief description has provided a complete definition of an ideal optical system. No matter how complex (or simple) the optical system, this analysis defines the complete end-to-end functionality of an ideal system. On this basis, an optical designer will specify the six ***cardinal points*** of a system to describe the ideal behaviour of a design. These six cardinal points are:

First Focal Point
Second Focal Point
First Principal Point
Second Principal Point
First Nodal Point
Second Nodal Point

The principal and nodal points are co-located if the two system focal lengths are identical.

1.3.8 Object and Image Locations - Newton's Equation

The location of the cardinal points has given us a complete description of a generalised optical system. Given that the function of an optical system might be to produce an image of an object located at a specific point, we might want to know the location of that image. Figure 1.11 shows the relationship between a generalised object and image.

Referring to Figure 1.11 and by using similar triangles it is possible to derive two separate relations for the magnification h_2/h_1:

$$M = \frac{h_2}{h_1} = -\frac{f_1}{x_1} = -\frac{x_2}{f_2}$$

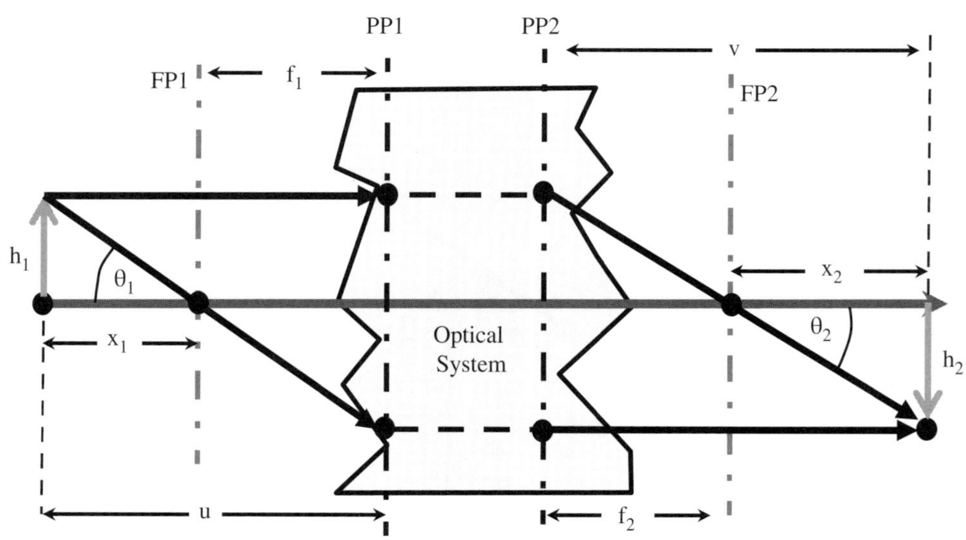

Figure 1.11 Generalised object and image.

And:

$$\text{Newton's Equation:}\quad x_1x_2 = f_1f_2 \tag{1.6}$$

The above equation is ***Newton's Equation*** and may be re-cast into a more familiar form using the definitions of object and image distances, u and v, as previously set out.

$$\frac{1}{u} + \left(\frac{f_2}{f_1}\right)\frac{1}{v} = \frac{1}{f_1} \tag{1.7}$$

If $f_1 = f_2 = f$, we are left with the more familiar lens equation. However, Eq. (1.7) is generally applicable to all optical systems. Most importantly, Eq. (1.7) will give the locations of the object and image in systems of arbitrary complexity. Many readers might have encountered Eq. (1.7) in the context of a simple lens where object and image distances are obvious and easy to determine. For a more complex system, one has to know the location of the principal planes as well in order to determine the object and image distances.

1.3.9 Conditions for Perfect Image Formation – Helmholtz Equation

Thus far, we have presented a description of an idealised optical system. Is there a simple condition that needs to be fulfilled in order to generate such an ideal image? It is easy to see from Figure 1.11 that the following relations apply:

$$f_1 \tan\theta_1 = h_2 \, and f_2 \tan\theta_2 = h_1$$

Therefore:

$$h_1f_1 \tan\theta_1 = h_2f_2 \tan\theta_2$$

As we will be able to show later, the ratio f_2/f_1 is equal to the ratio of the refractive indices, n_2/n_1, in the two media (object and image space). Therefore it is possible to cast the above equation in its more usual form, the ***Helmholtz equation***:

$$\text{Helmholtz equation:}\quad h_1n_1 \tan\theta_1 = h_2n_2 \tan\theta_2. \tag{1.8}$$

One important consequence of the Helmholtz equation is that there is a clear, inextricable linkage between transverse and angular magnification. Angular magnification is inversely proportional to transverse magnification. For small θ, tan θ and θ are approximately equal. So in the small signal approximation, the angular magnification, α is given by:

$$\alpha = \frac{\theta_2}{\theta_1} \approx \frac{h_1n_1}{h_2n_2}$$

Hence:

$$\alpha \approx \left(\frac{n_1}{n_2}\right)\frac{1}{M} \tag{1.9}$$

We have, thus far, introduced two different types of optical magnification – transverse and angular. There is a third type of magnification that we need to consider, ***longitudinal magnification***. Longitudinal magnitude, L, is defined as the shift in the axial image position for a unit shift in the object position, i.e.:

$$L = \frac{dx_2}{dx_1} \tag{1.10}$$

From Newton's Eq. (1.6):

$$\frac{dx_2}{dx_1} = \frac{f_1f_2}{x_1^2} = -\left(\frac{f_2}{f_1}\right)M^2$$

And:

$$L = -\left(\frac{f_2}{f_1}\right)M^2 \tag{1.11}$$

Thus, the longitudinal magnification is proportional to the square of the transverse magnification.

1.4 Behaviour of Simple Optical Components and Surfaces

1.4.1 General

The analysis presented thus far is entirely independent of the optical components that might populate the idealised optical system. In this section we will begin to consider, from the perspective of ray optics, the behaviour of real elements that make up this generalised system. At a basic level, only a few behaviours need to be considered in order to understand the propagation of rays through a real optical system. These are:

Propagation through a homogeneous medium
Refraction at a planar surface
Refraction at a curved (spherical) surface
Refraction through lenses
Reflection at a planar surface
Reflection at a curved (spherical) surface

As previously set out, the path of rays through a system is governed entirely by Fermat's principle. From this point, we will apply the simplest definition of Fermat's principle and assume that the time or optical path of rays is ***minimised***. As far as ***propagation through a homogeneous medium*** is concerned, this leads to a perhaps obvious and trivial conclusion that ***light travels in straight lines***. In fact, this describes a specific application of Fermat's principal, known as ***Hero's principle***, namely that light follows the path of minimum distance between two points within a homogeneous medium.

1.4.2 Refraction at a Plane Surface and Snell's Law

The law governing refraction at a planar surface is universally attributed to Willebrord Snellius and referred to as ***Snell's law***. This states that both incident and refracted rays lie in the same plane and their angles of incidence and refraction (*with respect to surface normal*) are given by:

$$\text{n}_1 \text{ and } \text{n}_2 \text{ are the refractive indices of the two media}: \quad n_1 \sin\theta_1 = n_2 \sin\theta_2. \tag{1.12}$$

This is illustrated in Figure 1.12.

The refractive indices of some optical materials (at 550 nm) are listed below:

Glass (BK7): 1.52
Plastic (Acrylic): 1.48
Water: 1.33
Air: 1.00027

Snell's law is, in fact, a direct consequence of Fermat's principle. The reader may wish to derive this through the application of differential calculus. In finding the optimum path from a point in one medium to a point in another medium, the ray will attempt, as far as possible, to minimise its path through the higher index medium. Snell's law thus represents the minimum optical path condition in this instance. Where the ray passes from a high index material to a low index material, there exists an angle of incidence where the angle of refraction

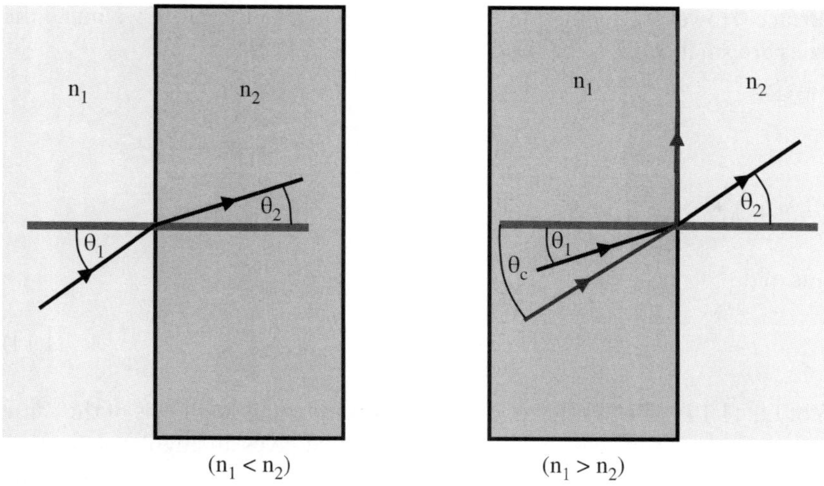

Figure 1.12 Refraction at a plane surface.

is 90°. This angle is known as the ***critical angle*** and, for angles of incidence beyond this, the ray is ***totally internally reflected***. The critical angle, θ_c, is given by:

$$n_2 < n_1 \quad n_1 \sin\theta_c = \frac{n_2}{n_1} \tag{1.13}$$

A single refractive surface is an example of an ***afocal system***, where both focal lengths are infinite. Although it does not bring a parallel beam of light to a focus, it does form an image that is a geometrically true representation of the object.

1.4.3 Refraction at a Curved (Spherical) Surface

Most, if not all, curved optical surfaces are at least approximately spherical and are widely employed in the fabrication of lens components. Figure 1.13 illustrates refraction at a spherical surface.

As before, the special case of refraction at a spherical surface may be described by Snell's law:

$$n_1 \sin\theta_1 = n_2 \sin\theta_2$$

If we now wish to calculate the angle ϕ in terms of θ, this process is, in principle, straightforward. We need also to take into account the angle the surface normal makes with the optical axis, Δ, and the radius

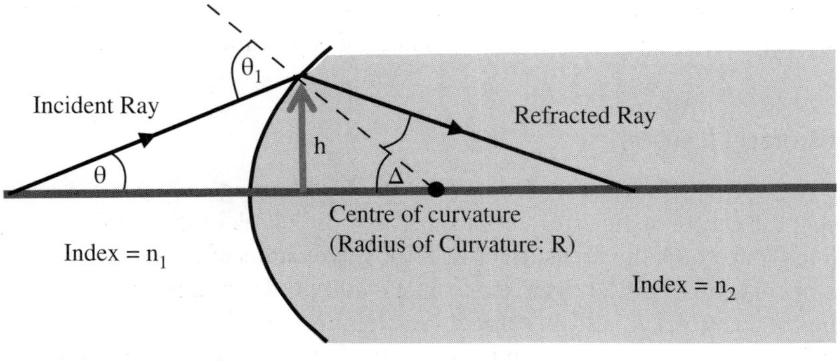

Figure 1.13 Refraction at a spherical surface.

of curvature, R, of the spherical surface. However, calculation is a little unwieldy, so therefore we make the simplifying assumption that all angles are small and:

$$\sin\theta \approx \theta$$

Hence:

$$\theta_2 \approx \frac{n_1}{n_2}\theta_1; \quad \Delta \approx \frac{h}{R} \text{ and } \theta_1 = \theta + \Delta; \quad \varphi = \theta_2 - \Delta$$

We can finally calculate ϕ in terms of θ:

$$\phi \approx \frac{n_1}{n_2}\theta - \left(\frac{n_2 - n_1}{n_2}\right)\frac{h}{R} \tag{1.14}$$

There are two terms on the RHS of Eq. (1.14). The first term, depending on the input angle θ is of the same form as Snell's law (for small angles) for a plane surface. The second term, which gives an angular deflection proportional to the height, h, and inversely proportional to the radius of curvature R, provides a focusing effect. That is to say, rays further from the optic axis are bent inward to a greater extent and have a tendency to converge on a common point. The sign convention used here assumes that positive height is vertically upward, as displayed in Figure 1.13 and a positive spherical radius corresponds to a scenario in which the centre of the sphere lies to the right of the point where the surface intersects the optical axis. Finally, a positive angle is consistent with an increase in ray height as it propagates from left to right in (1.13).

Equation (1.14) can be used to trace any ray that is incident upon a spherical refractive surface. If this surface is deemed to comprise 'the optical system' in its entirety, then one can use Eq. (1.14) to calculate the location of all Cardinal Points, expressed as a displacement, z along the optical axis. Positive z is to the right and the origin lies at the intersection of the optical axis and the surface. The Cardinal points are listed below.

Cardinal points for a spherical refractive surface

First Focal Point: $z = -\left(\frac{n_1}{n_2 - n_1}\right)R$ *First Focal Length*: $\left(\frac{n_1}{n_2 - n_1}\right)R$

Second Focal Point: $z = \left(\frac{n_2}{n_2 - n_1}\right)R$ *Second Focal Length*: $\left(\frac{n_2}{n_2 - n_1}\right)R$

Both Principal Points: $z = 0$

Both Nodal Points: $z = R$

In this instance, the two focal lengths, f_1 and f_2 are different since the object and image spaces are in different media. If we take the first focal length as the distance from the first focal point to the first principal point, then the first focal length is positive. Similarly, the second focal length, the distance from the second principal point to the second focal point, is also positive. The principal points are both located at the surface vertex and the nodal points at the centre of curvature of the sphere. It is important to note that, in this instance, the principal and nodal points do not coincide. Again, this is because the refractive indices of object and image space differ.

1.4.4 Refraction at Two Spherical Surfaces (Lenses)

Figure 1.14 shows a lens made up of two spherical surfaces, of radius, R_1 and R_2. Once again, the convention is that the spherical radius is positive if the centre of curvature lies to the right of the relevant vertex.

So, in the biconvex lens illustrated in Figure 1.14, the first surface has a positive radius of curvature and the second surface has a negative radius of curvature. The lens is made from a material of refractive index n_2 and is bounded by two surfaces with radius of curvature R_1 and R_2 respectively. It is immersed totally in a medium of refractive index, n_1 (e.g. air). In addition, it is assumed that the lens has negligible thickness (***the thin lens approximation***). Of course, as for the treatment of the single curved surface, we assume all angles are small

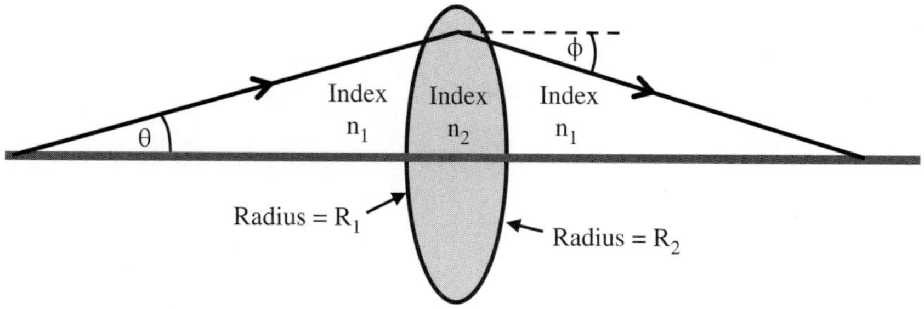

Figure 1.14 Refraction by two spherical surfaces (lens).

and $\theta \sim \sin\theta$. First, we might calculate the angle of refraction, ϕ_1, produced by the first curved surface, R_1. This can be calculated using Eq. (1.14):

$$\phi_1 \approx \frac{n_1}{n_2}\theta - \left(\frac{n_2 - n_1}{n_2}\right)\frac{h}{R_1}$$

Of course, the final angle, ϕ, can be calculated from ϕ_1 by another application of Eq. (1.14):

$$\phi \approx \frac{n_2}{n_1}\phi_1 - \left(\frac{n_1 - n_2}{n_1}\right)\frac{h}{R_2}$$

Substituting for ϕ_1 we get:

$$\phi \approx \theta - \left(\frac{n_2 - n_1}{n_1}\right)\left[\frac{h}{R_1} - \frac{h}{R_2}\right] \tag{1.15}$$

As for Eq. (1.14) there are two parts to Eq. (1.15). First, there is an angular term that is equal to the incident angle. Second, there is a focusing contribution that produces a deflection proportional to ray height. Equation (1.15) allows the tracing of all rays in a system containing the single lens and it is straightforward to calculate the Cardinal points of the thin lens:

Cardinal points for a thin lens

First Focal Point: $z = -\left(\frac{n_1}{n_2 - n_1}\right)\frac{R_1R_2}{R_1 - R_2}$ *First Focal Length*: $\left(\frac{n_1}{n_2 - n_1}\right)\frac{R_1R_2}{R_1 - R_2}$

Second Focal Point: $z = \left(\frac{n_1}{n_2 - n_1}\right)\frac{R_1R_2}{R_1 - R_2}$ *Second Focal Length*: $\left(\frac{n_1}{n_2 - n_1}\right)\frac{R_1R_2}{R_1 - R_2}$

Both Principal Points: At centre of lens

Both Nodal Points: At centre of lens

Since both object and image spaces are in the same media, then both focal lengths are equal and the principal and nodal points are co-located. One can take the above expressions for focal length and cast it in a more conventional form as a ***single*** focal length, f. This gives the so-called ***Lensmaker's Equation***, where it is assumed that the surrounding medium (air) has a refractive index of one (i.e. $n_1 = 1$) and we substitute n for n_2.

$$\frac{1}{f} = (n-1)\left[\frac{1}{R_1} - \frac{1}{R_2}\right] \tag{1.16}$$

1.4.5 Reflection by a Plane Surface

Figure 1.15 shows the process of reflection at a plane surface. As in the previous case of refraction, the reflected ray lies in the same plane as the incident ray and the angle of reflection is equal and opposite to the angle of incidence.

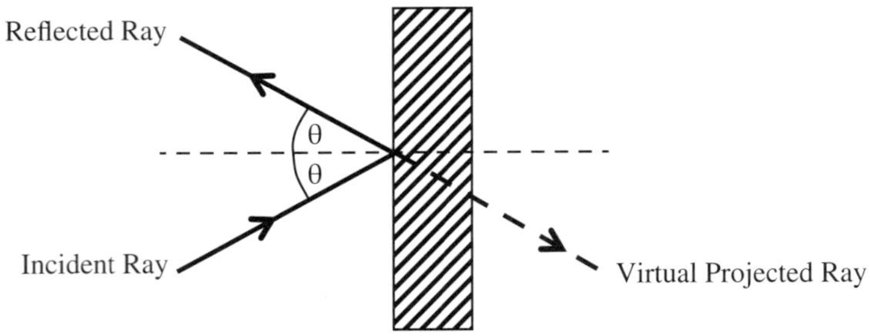

Figure 1.15 Reflection at a plane surface.

The virtual projected ray shown in Figure 1.15 illustrates an important point about reflection. If one considers the process as analogous to refraction, then a mirror behaves as a refractive material with an index of −1. This, in itself has an important consequence. The image produced is inverted in space. As such, there is no combination of positive magnification and pure rotation that will map the image onto the object. That is to say, a right handed object will be converted into a left handed image. More generally, if an optical system contains an ***odd number of reflective elements***, the parity of the image will be reversed. So, for example, if a complex optical system were to contain nine reflective elements in the optical path, then the resultant image could not be generated from the object by rotation alone. Conversely, if the optical system were to contain an even number of reflective surfaces, then the parity between the object and image geometries would be conserved.

Another way in which a plane mirror is different from a plane refractive surface is that a plane mirror is the one (and perhaps only) example of a ***perfect imaging system***. Regardless of any approximation with regard to small angles discussed previously, following reflection at a planar surface, ***all rays*** diverging from a single image point would, when projected as in Figure 1.15, be seen to emerge ***exactly*** from a ***single*** object point.

1.4.6 Reflection from a Curved (Spherical) Surface

Figure 1.16 illustrates the reflection of a ray from a curved surface.

The incident ray is at an angle, θ, with respect to the optical axis and the reflected ray is at an angle, φ to the optical axis. If we designate the incident angle as θ_1 and the reflected angle as θ_2 (with respect to the local surface normal), then the following apply, ***assuming all relevant angles are small***:

$$\theta_1 = \theta + \Delta; \quad \theta_2 = -\theta_1; \quad \varphi = \theta_2 - \Delta \text{ and } \Delta \approx \frac{h}{R}$$

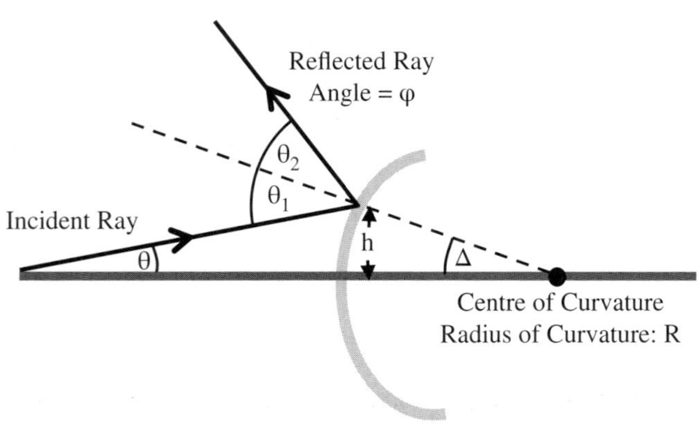

Figure 1.16 Reflection from a curved surface.

We now need to calculate the angle, φ, the refracted ray makes to the optical axis:

$$\phi = -\theta - \frac{2h}{R} \tag{1.17}$$

In form, Eq. (1.17) is similar to Eq. (1.14) with a linear dependence of the reflected ray angle on both incident ray angle and height. The two equations may be made to correspond exactly if we make the substitution, $n_1 = 1$, $n_2 = -1$. This runs in accord with the empirical observation made previously that a reflective surface acts like a medium with a refractive index of −1. Once more, the sign convention observed dictates that positive axial displacement, z, is in the direction from left to right and positive height is vertically ***upwards***. A ray with a positive angle, θ, has a positive gradient in h with respect to z.

As with the curved refractive surface, a curved mirror is image forming. It is therefore possible to set out the Cardinal Points, as before:

Cardinal points for a spherical mirror

First Focal Point: $z = \frac{R}{2}$ *First Focal Length*: $-\frac{R}{2}$

Second Focal Point: $z = \frac{R}{2}$ *Second Focal Length*: $\frac{R}{2}$

Both Principal Points: At vertex

Both Nodal Points: At centre of sphere

The focal length of a curved mirror is half the base radius, with both focal points co-located. In fact, the two focal lengths are of opposite sign. Again, this fits in with the notion that reflective surfaces act as media with a refractive index of −1. Both nodal points are co-located at the centre of curvature and the principal points are also co-located at the surface vertex.

1.5 Paraxial Approximation and Gaussian Optics

Earlier, in order to make our lens and mirror calculations simple and tractable, we introduced the following approximation:

$$\theta << 1 \text{ and } \theta \approx \sin\theta \approx \tan\theta$$

That is to say, all rays make a sufficiently small angle to the optical axis to make the above approximation acceptable in practice. When this approximation is applied more generally to an entire optical system, it is referred to as the ***paraxial approximation*** (i.e. *'almost axial'*). If the same consideration is applied to ray heights as well as angles, the paraxial approximations lead to a series of equations describing the transformation of ray heights and angles that are linear in both ray height and angle. This first order theory is generally referred to as ***Gaussian optics***, named after Carl Friedrich Gauss.

If we now assume that all rays are confined to a single plane containing the optical axis, then we can describe all rays by two parameters: θ – the angle the ray make to the optical axis and h – the height above the optical axis. If, after transformation by an optical surface, these parameters change to θ' and h′, it is possible to write down a series of linear equations describing all transformations. These are set out in Eqs. 1.18–1.21:

Refraction at a plane surface: $$\theta' = \frac{n_1}{n_2}\theta \quad (h' = h) \tag{1.18}$$

Refraction at a curved surface: $$\theta' = \frac{n_1}{n_2}\theta - \left(\frac{n_2 - n_1}{n_2}\right)\frac{h}{R} \quad (h' = h) \tag{1.19}$$

Reflection at plane mirror: $$\theta' = -\theta \quad (h' = h) \tag{1.20}$$

Reflection at curved surface: $$\theta' = -\theta - \frac{2h}{R} \quad (h' = h) \tag{1.21}$$

Even the most complex optical system may be described as a combination of all the above elements. At first sight, therefore, it would seem that this provides a complete description of the first order behaviour of an optical system. However, there is one important, but seemingly trivial, aspect that is not considered here. This is the case of ray propagation through space. The equations are, of course simple and obvious, but we include them for completeness.

$$\textbf{\textit{Propagation through space}}: \qquad \theta' = \theta \qquad h' = h + d\theta \qquad (d \text{ is propagation distance}) \tag{1.22}$$

Equation (1.8) introduced the Helmholtz equation, a necessary condition for perfect image formation for an ideal system. It is clear that Gaussian optics represents a mere approximation to the ideal of the Helmholtz equation. The contradiction between the two suggests that there may be imperfections in the ideal treatment of Gaussian optics. This will be considered later when we will look at optical imperfections or aberrations. In the meantime, we will consider a very powerful realisation of Gaussian optics that takes the basic linear equations previously set out and expresses them in terms of matrix algebra. This is the so-called Matrix Ray Tracing technique.

1.6 Matrix Ray Tracing

1.6.1 General

In Section 1.4 we looked at the behaviour of some very simple components, mirrors and lenses, deriving the locations of the Cardinal Points. As discussed previously, the Cardinal Points provide a complete description of the first order properties of an optical system, ***no matter how complex***.

The question then is how do we calculate the properties of a more complex optical system, such as the camera lens depicted in Figure 1.17? It is not immediately obvious where the Cardinal points lie or what the focal length is. However, we can combine the generalised description of an optical system with the treatment of Gaussian optics to produce a model that describes the entire system as a black box acting on rays with a simple linear transformation. The black box may be visualised as below in Figure 1.18.

Following the basic premise of Gaussian optics, we can relate the input and output rays using a set of linear equations:

$$h_{out} = Ah_{in} + B\theta_{in} \tag{1.23}$$

$$\theta_{out} = Ch_{in} + D\theta_{in} \tag{1.24}$$

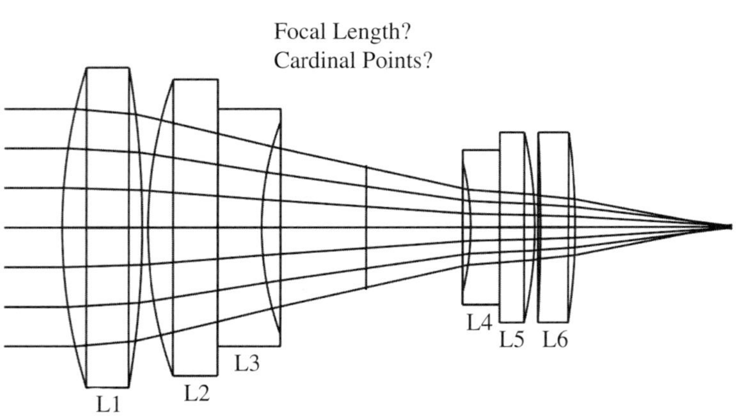

Figure 1.17 Complex optical system.

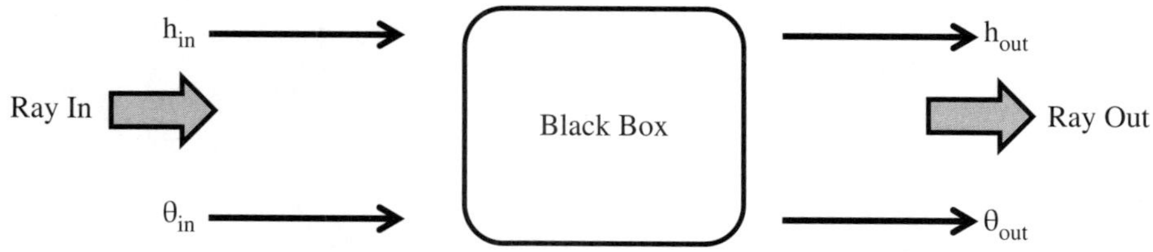

Figure 1.18 Modelling of complex systems.

Equations (1.23) and (1.24) may be combined in a matrix representation:

$$\begin{bmatrix} h_{out} \\ \theta_{out} \end{bmatrix} = \begin{bmatrix} A & B \\ C & D \end{bmatrix} \begin{bmatrix} h_{in} \\ \theta_{in} \end{bmatrix} \tag{1.25}$$

Equation (1.25) sets out the ***Matrix Ray Tracing*** convention used in this book. The reader should be aware that other conventions are used, but this is the most widely used. Equation (1.25) can be used to describe the overall system matrix or that of individual components. The question is how to build up a complex system from a large number of optical elements. The camera lens shown in Figure 1.17 has six lenses and we might represent each lens as a single matrix, i.e. $\mathbf{M_1}, \mathbf{M_2}, \ldots, \mathbf{M_6}$. Each matrix describes the relationship between rays incident upon the lens and those leaving. The impact of successive optical elements is determined by successive matrix multiplication. So the system matrix for the lens as a whole is given by the matrix product of all elements:

$$\mathbf{M_{system}} = \mathbf{M_6} \times \mathbf{M_5} \times \mathbf{M_4} \times \mathbf{M_3} \times \mathbf{M_2} \times \mathbf{M_1} \tag{1.26}$$

Note the order of the multiplication; this is important. M_1 represents the first optical element seen by rays incident upon the system and the multiplication procedure then works through elements 2–6 successively. For purposes of illustration, each lens has been treated as being represented by a single matrix element. In practice, it is likely that the lens would be reduced to its basic building blocks, namely the two curved surfaces ***plus the propagation (thickness) between the two surfaces.*** We also must not forget the propagation through the air between the lens elements.

Representation of the key optical surfaces can be determined by casting Eqs. (1.18)–(1.22) in matrix format.

$$\textbf{Refraction at a plane surface}: \quad \begin{bmatrix} 1 & 0 \\ 0 & n_1/n_2 \end{bmatrix} \tag{1.27a}$$

$$\textbf{Refraction at a curved surface (Radius R)}: \quad \begin{bmatrix} 1 & 0 \\ \left[\dfrac{n_1 - n_2}{n_2}\right]\dfrac{1}{R} & \dfrac{n_1}{n_2} \end{bmatrix} \tag{1.27b}$$

$$\textbf{Reflection by plane mirror}: \quad \begin{bmatrix} 1 & 0 \\ 0 & -1 \end{bmatrix} \tag{1.27c}$$

$$\textbf{Reflection by curved mirror (Radius R)}: \quad \begin{bmatrix} 1 & 0 \\ -\frac{2}{R} & -1 \end{bmatrix} \tag{1.27d}$$

$$\textbf{Propagation through space (distance d)}: \quad \begin{bmatrix} 1 & d \\ 0 & 1 \end{bmatrix} \tag{1.27e}$$

$$\textbf{Effect of lens (focal length f)}: \quad \begin{bmatrix} 1 & 0 \\ -1/f & 1 \end{bmatrix} \tag{1.27f}$$

n_1 and n_2 represent the refractive index of first and second media respectively.

1.6.2 Determination of Cardinal Points

It is very straightforward to calculate the Cardinal Points of a system from the system matrix:

$$\begin{bmatrix} A & B \\ C & D \end{bmatrix}$$

The matrix above represents the system matrix after propagating through all optical elements as shown in Figure 1.17. However, the convention adopted here is that an additional transformation is added after the final surface. This additional transformation is free space propagation to the original starting point. It must be emphasised that, this is merely a ***convention***, and that the final step traces a ***dummy ray*** as opposed to a real ray. That is to say, in reality, the light does not propagate backwards to this point. In fact, this step is a virtual back-projection of the real ray which preserves the original ray geometry. The logic of this, as will be seen, is that in any subsequent analysis, the location of all cardinal points is referenced with respect to a common starting point. If this step were dispensed with, then the three first Cardinal Points would be referenced to the start point and the three second Cardinal Points to the end point. With this in mind, the Cardinal Points, as referenced to the common start point are set out below; the reader might wish to confirm this.

$$\text{First FocalPoint}: \quad z = \frac{D}{C} \qquad \textit{First Focal Length}: \; -\frac{(AD-BC)}{C}$$

$$\textit{Second Focal Point}: \quad z = -\frac{A}{C} \qquad \textit{Second Focal Length}: \; -\frac{1}{C}$$

$$\textit{First Principal Point}: \quad z = B + D\frac{(1-A)}{C} \qquad \textit{Second Principal Point}: \quad z = \frac{(1-A)}{C}$$

$$\textit{First Nodal Point}: \quad z = \frac{(D-1)}{C} \qquad \textit{Second Nodal Point}: \quad z = \frac{A(D-1)-BC}{C}$$

The ***determinant*** of the matrix, (**AD–BC**), is a key parameter. The ratio of the two focal lengths of the system is simply given by the determinant. That is to say the ratio of the two focal lengths is given by:

$$\frac{f_1}{f_2} = (AD - BC) = Determinant \tag{1.28}$$

Inspecting all matrix expressions in Eqs. (1.27a–1.27f), the determinant of the matrix is simply n_1/n_2, the ratio of the indices in the two media, ***for all possible scenarios***. Since the determinant of a matrix product is simply the product of the individual determinants, then the determinant of the overall system matrix is simply the ratio of the refractive indices in image and object space. Thus:

$$\frac{f_1}{f_2} = \frac{n_{image}}{n_{object}} \tag{1.29}$$

This relationship was anticipated in the more generalised discussion in 1.3.9. Looking at the relationships for the principal and nodal points, it is clear when the determinant of the system matrix is unity, i.e. object and image space indices are the same, then the principal and nodal points are co-located.

In addition to the principal and nodal points, ***anti-principal points*** and ***anti-nodal points*** are sometimes (rarely) specified. Anti-principal points are conjugate points where the magnification is −1. Similarly, anti-nodal points are conjugate points where the angular magnification is −1.

1.6.3 Worked Examples

We can now use the foregoing analysis to see how matrix ray tracing might be used in practice. Here we focus on a number of useful practical examples.

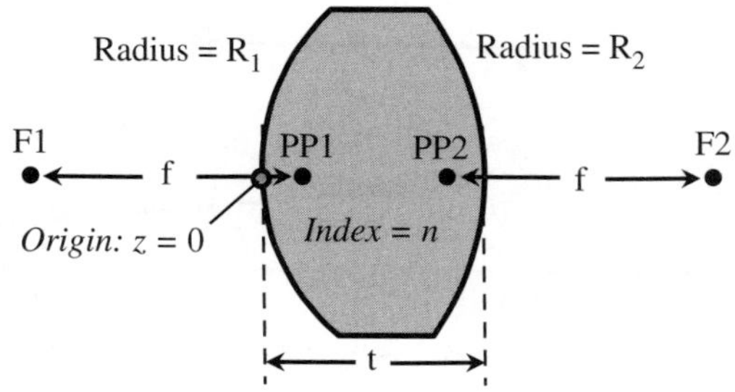

Figure 1.19 Thick lens.

Worked Example 1.1 ***Thick Lens***

The matrix for the system is simply as below – ***note the order***:

$$M = \underbrace{\begin{bmatrix} 1 & -t \\ 0 & 1 \end{bmatrix}}_{\textit{Translate to Origin}} \underbrace{\begin{bmatrix} 1 & 0 \\ \dfrac{(n-1)}{R_2} & 1/n \end{bmatrix}}_{\text{Second Surface}} \underbrace{\begin{bmatrix} 1 & t \\ 0 & 1 \end{bmatrix}}_{\text{Lens Thickness}} \underbrace{\begin{bmatrix} 1 & 0 \\ \dfrac{1-n}{n}\dfrac{1}{R_1} & 1/n \end{bmatrix}}_{\text{First Surface}}$$

We have two translations. The first translation represents the thickness of the lens and the second translation, ***by convention***, traces the refracted rays back to the origin in z. This is so that, in interpreting the formulae for Cardinal points, we can be sure that they are all referenced to a common origin, located as in Figure 1.19. Positive axial displacement (z) is to the right and a positive radius, R, is where the centre of curvature lies to the right of the vertex. The final matrix is as below:

$$M = \begin{bmatrix} \left[1 + t\left((n-1)\left(\dfrac{1}{R_1} - \dfrac{1}{nR_1} - \dfrac{1}{R_2}\right) + \dfrac{t(n-1)^2}{nR_1R_2}\right)\right] & \left[\dfrac{t^2(1-n)}{nR_2}\right] \\ \left[-(n-1)\left(\dfrac{1}{R_1} - \dfrac{1}{R_2}\right) - \dfrac{t(n-1)^2}{nR_1R_2}\right] & \left[1 + \dfrac{t(n-1)}{nR_2}\right] \end{bmatrix}$$

As both object and image space are in the same media, there is a common focal length, f, i.e. $f_1 = f_2 = f$. All relevant parameters are calculated from the above matrix using the formulae tabulated in Section 1.6.2.

The focal length, f, is given by:

$$\frac{1}{f} = \underbrace{(n-1)\left(\frac{1}{R_1} - \frac{1}{R_2}\right)}_{\text{Lensmaker}} + \underbrace{\frac{t(n-1)^2}{nR_1R_2}}_{\text{'Thickness Term'}}$$

The formula above is similar to the simple, 'Lensmaker' formula for a thin lens. In addition there is another term, linear in thickness, t, which accounts for the lens thickness.

The focal positions are as follows:

$$F_1 = -f\left(1 + \frac{t(n-1)}{nR_2}\right) \qquad F_2 = f + t - \frac{(n-1)}{nR_1}tf$$

The principal points are as follows:

$$P_1 = -f\frac{t(n-1)}{nR_2} \qquad P_2 = t - \frac{(n-1)}{nR_1}tf$$

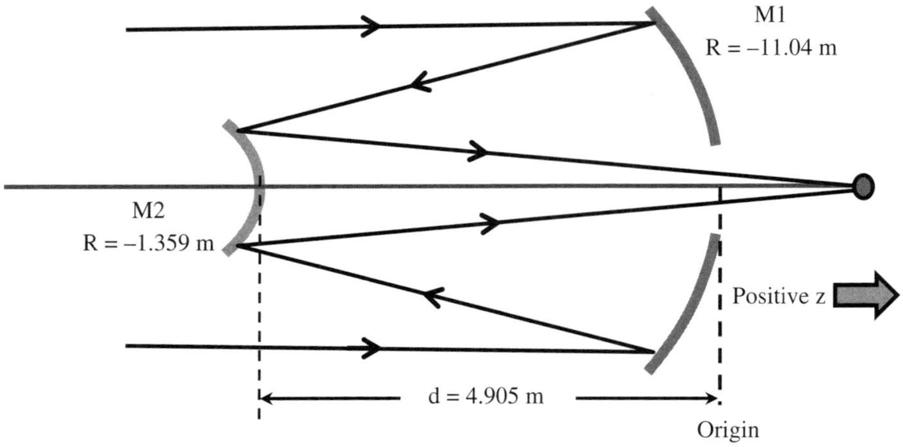

Figure 1.20 Hubble space telescope schematic.

Of course, since the refractive indices of the object and image spaces are identical, the nodal points are located in the same place as the principal points. If we take the example of a biconvex lens where $R_2 = -R_1$, then:

$$P_1 = \left[\frac{t}{2n}\right]\left[\frac{1}{1 + (t/2n(n-1)R_1)}\right]$$

So, for a biconvex lens with a refractive index of (1.5), then the principal points lie about one third of the thickness from their respective vertices.

Worked Example 1.2 ***Hubble Space Telescope***

The telescope part of the Hubble Space Telescope instrument is made up of two mirrors, a primary and a secondary. Characteristics of the telescope are shown in Figure 1.20. *Data is courtesy of the National Aeronautics and Space Administration.*

There are four matrix elements to consider here. First, there is a mirror with a radius of −11.04 m (note sign), followed by a translation of −4.905 m (again note sign). The third matrix element is a mirror (M2) of radius − 1.359 m. Finally, we translate by +4.905 m, so that both the input and output co-ordinates are referenced with respect to the same origin. The matrices are as below:

$$M = \begin{bmatrix} 1 & 4.905 \\ 0 & 1 \end{bmatrix}\begin{bmatrix} 1 & 0 \\ -\dfrac{2}{-1.359} & -1 \end{bmatrix}\begin{bmatrix} 1 & -4.905 \\ 0 & 1 \end{bmatrix}\begin{bmatrix} 1 & 0 \\ -\dfrac{2}{-11.04} & -1 \end{bmatrix} = \begin{bmatrix} .0271 & 45.217 \\ -.0172 & 8.219 \end{bmatrix}$$

The focal positions are:

F1: −477.93 m **F2:** 1.574 m

The principal points are at:

P1: −419.777 m **P2:** −56.579 m

$f_1 = f_2 = 58.153$ m

Since object and image space are in the same media, then the two focal lengths are the same. In addition, the nodal and principal points are co-located. However, when dealing with mirrors, one must be a little cautious. Each reflection is equivalent to a medium with a refractive index of −1, so that the matrix of a reflective surface will always have a determinant of −1. Therefore, any system having an ***even number*** of reflective surfaces, as in this example, then its matrix will have a determinant of 1. As such, the two focal lengths will be the same and principal and nodal points co-located. However, where there are an ***odd number*** of reflective surfaces, assuming object and image spaces are surrounded by the same media, then $f_2 = -f_1$. In this instance, principal and nodal points are separated by twice the focal length.

Although, in terms of overall length, the telescope is compact, ~5 m primary–secondary separation, the focal length, at 58 m, is long. The focal length of the instrument is fundamental in determining the 'plate scale' the separation of imaged objects (stars, galaxies) at the (second) focal plane as a function of their ***angular*** separation. As such, a long focal length, of the order of 60 m, may have been a ***requirement*** at the outset. At the same time, for practical reasons, a compact design may also have been desired. One may begin to glance, therefore, at the significance, at the very outset of these very basic calculations in the design of complex optical instruments.

1.6.4 Spreadsheet Analysis

For the examples previously set out, matrix multiplication is a quick and convenient method for calculating the first order parameters of an optical system. Nonetheless, it must be recognised that as systems become more complex, with more optical surfaces, these calculations can become quite tedious. However, these matrix calculations are easy to embed with spreadsheet tools enabling the automatic computation of all cardinal points. By way of example, the previous calculation is set out and automated using a simple spreadsheet tool.

Translate		2nd Mirror		Translate		1st Mirror			
d	4.905	R	−1.359	d	−4.905	R	−11.04		
1	4.905	1	0	1	−4.905	1	0		
0	1	1.472	−1	0	1	0.181	−1		
0.027	45.217	0.111	4.905	0.111	4.905	1	0	1	0
−0.017	8.219	−0.017	8.219	0.181	−1	0.181	−1	0	1
1st Focal Point	−477.93 m			**2nd Focal Point**		1.574 m			
1st Principal Point	−419.78 m			**2nd Principal Point**		−56.579 m			
1st Nodal Point	−419.78 m			**2nd Nodal Point**		−56.579 m			
Focal Length 1	58.153 m			**Focal Length 2**		58.153 m			

In the exercises that follow, the reader may choose to use this method to simplify calculations.

Further Reading

Born, M. and Wolf, E. (1999). *Principles of Optics*, 7e. Cambridge: Cambridge University Press. ISBN: 0-521-642221.

Haija, A.I., Numan, M.Z., and Freeman, W.L. (2018). *Concise Optics: Concepts, Examples and Problems.* Boca Raton: CRC Press. ISBN: 978-1-1381-0702-1.

Hecht, E. (2017). *Optics*, 5e. Harlow: Pearson Education. ISBN: 978-0-1339-7722-6.

Keating, M.P. (1988). *Geometric, Physical, and Visual Optics.* Boston: Butterworths. ISBN: 978-0-7506-7262-7.

Kidger, M.J. (2001). *Fundamental Optical Design.* Bellingham: SPIE. ISBN: 0-81943915-0.

Kloos, G. (2007). *Matrix Methods for Optical Layout.* Bellingham: SPIE. ISBN: 978-0-8194-6780-5.

Longhurst, R.S. (1973). *Geometrical and Physical Optics*, 3e. London: Longmans. ISBN: 0-582-44099-8.

Riedl, M.J. (2009). *Optical Design: Applying the Fundamentals*. Bellingham: SPIE. ISBN: 978-0-8194-7799-6.
Saleh, B.E.A. and Teich, M.C. (2007). *Fundamentals of Photonics*, 2e. New York: Wiley. ISBN: 978-0-471-35832-9.
Smith, F.G. and Thompson, J.H. (1989). *Optics*, 2e. New York: Wiley. ISBN: 0-471-91538-1.
Smith, W.J. (2007). *Modern Optical Engineering*. Bellingham: SPIE. ISBN: 978-0-8194-7096-6.
Walker, B.H. (2009). *Optical Engineering Fundamentals*, 2e. Bellingham: SPIE. ISBN: 978-0-8194-7540-4.

2

Apertures Stops and Simple Instruments

2.1 Function of Apertures and Stops

In the previous chapter, we were introduced to sequential geometric optics. The simple analysis presented there is contingent upon the paraxial approximation. It is assumed that all rays in their sequential progress through the optical system always subtend a negligibly small angle with respect to the optical axis. In this scenario, the effect of all optical elements may be described in terms of a simple set of linear (in ray height and angle) equations leading to perfect image formation. This analysis, as previously outlined, is referred to as Gaussian optics.

Of course, for real, non-ideal imaging systems, the assumptions underlying the paraxial approximation break down. An inevitable consequence of this is the creation of imperfections or aberrations in the formation of images. A full treatment of these optical aberrations forms the subject of succeeding chapters. In the meantime, consideration of the paraxial approximation might suggest that these imperfections or aberrations would be enhanced for rays that make a large angle with respect to the optical axis. It seems sensible, therefore, to restrict rays emanating from an object to a specific, restricted range of angles. In practice, for most systems, this is done by inserting an opaque obstruction with a circular aperture. This circular aperture is centred on the optical axis and is known as an ***aperture stop*** and restricts rays emanating from an object. To further control scattered light, the aperture stop is usually blackened in some manner.

In addition to selecting rays close to the optical axis and thus reducing imperfections, aperture stops also control and define the amount of light entering an optical system. This will be explored in more detail in the chapters relating to radiometry or the study of the analysis and measurement of optical flux. Naturally, the larger the aperture, then the more light is passed through the system. Most usually, the system aperture is formed by a purpose made mechanical aperture that is distinct from the optical elements themselves. However, on occasion, the system aperture may be formed by the physical boundary of an optical component, such as a lens or a mirror. This is true, for example, for a reflecting or refracting telescope, where the boundary of the first, or primary mirror, forms the aperture stop.

2.2 Aperture Stops, Chief, and Marginal Rays

This principle is illustrated in Figure 2.1 which shows an object together with a corresponding aperture stop. Note that the centre of the aperture stop corresponds to the intersection of its plane with the optical axis.

The aperture stop plays an important role in image formation and the analysis of optical systems. There are a number of important definitions relating to the aperture stop and its location. Of key significance is the ***chief ray*** which is a ray that that emanates from the object and intersects the plane of the aperture stop at its centre located at the optical axis. The angle, θ, that this ray makes with respect to the optical axis is known as the ***field angle***. Another ray of critical importance is the ***marginal ray*** that emanates from the point where the object plane intersects the optic axis and strikes the edge of the aperture. The angle, Δ, the marginal ray makes with

Optical Engineering Science, First Edition. Stephen Rolt.

Companion website: www.wiley.com/go/Rolt/opt-eng-sci

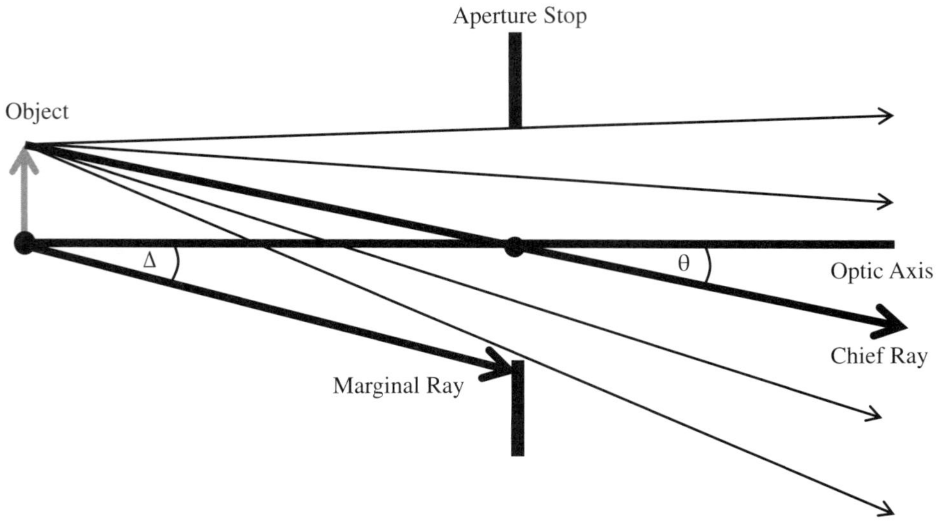

Figure 2.1 Aperture stop.

the axis effectively defines the size of the half angle of the cone of light emerging from a single on-axis point at the object plane and admitted by the aperture stop. The size of the aperture stop may be described either by its physical size or by the angle subtended. In the latter case, one of the most common ways of describing the aperture of an optical system is in terms of the ***numerical aperture*** (NA). The numerical aperture, is the product of the local refractive index, n, and the sine of the marginal ray angle, Δ.

$$NA = n \sin \Delta \tag{2.1}$$

A system with a large numerical aperture, allows more light to be collected. Such a system, with a high numerical aperture is said to be ***'fast'***. This terminology has its origins in photography, where the efficient collection of light using wide apertures enabled the use of short exposure times. An alternative convention exists for describing the relative size of the aperture, namely the ***f-number***. For a lens system, the f-number, N, is given as the ratio of the lens focal length to the aperture diameter:

$$N = \frac{f_{lens}}{D_{aperture}} \tag{2.2}$$

This f-number is actually written as f/N. That is to say, a lens with a focal ratio of 10 is written as f/10. The f-number has an inverse relationship to the numerical aperture and is based on the stop diameter rather than its radius. For small angles, where sinΔ = Δ, then the following relationship between the f-number and numerical aperture applies:

$$N = \frac{1}{2 \times NA} \tag{2.3}$$

In this narrative, it is assumed that the aperture is a circular aperture, with an entire, unobstructed circular area providing access for the rays. In the majority of cases, this description is entirely accurate. However, in certain cases, this circular aperture may be partly obscured by physical or mechanical hardware supporting the optics or by holes in reflective optics. Such features are referred to as ***obscurations***.

At this stage, it is important to emphasise the tension between fulfilment of the paraxial approximation and collection of more light. A 'fast' lens design naturally collects more light, but compromises the paraxial approximation and adds to the burden of complexity in lens and optical design. This inherent contradiction is explored in more detail in subsequent chapters.

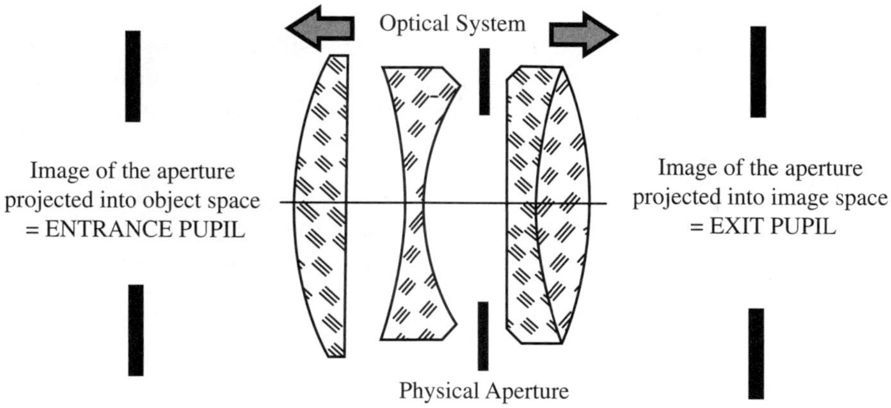

Figure 2.2 Location of entrance and exit pupils.

2.3 Entrance Pupil and Exit Pupil

The physical aperture stop may not actually be located conveniently in object space as shown in Figure 2.1. On the other hand, it may be located anywhere within the sequential train of optical components that make up the optical system. An example of this is shown in Figure 2.2, a situation that is true of many camera lenses, where the physical stop is located between lenses.

In the situation described, the ***entrance pupil*** is the image of the physical aperture as projected into object space. Correspondingly, the ***exit pupil*** is the image of the physical aperture as projected into image space. The exit pupil is located in the conjugate plane to the entrance pupil and may be regarded as the image of the entrance pupil. Along with the cardinal points of a system, the location of the entrance and exit pupils are key parameters that describe an optical system. Most particularly, the numerical aperture in object space is defined by the angle of the marginal ray that intersects the edge of the ***entrance pupil.***

Worked Example 2.1 ***Cooke Triplet***

Figure 2.3 shows a simplified illustration of an early type of camera lens, the Cooke triplet.

By convention, image space is assumed to be on the left-hand side of the illustration. All lenses are assumed to have no tangible thickness (thin lens approximation) and the axial origin lies at the first lens. Positive axial displacement is to the right.

Figure 2.3 Cooke triplet.

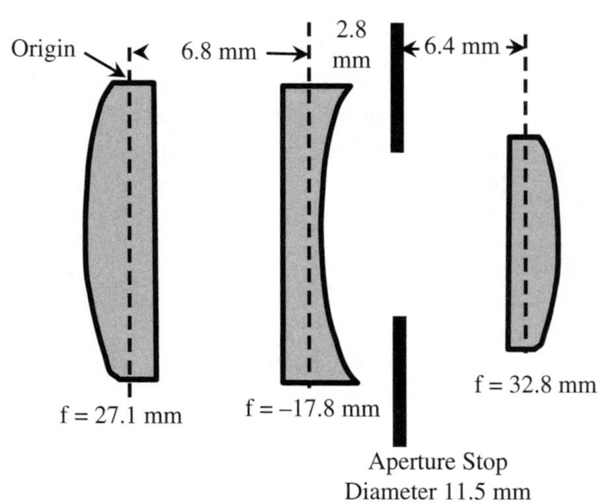

i) *Position and Size of Exit Pupil*

It is easiest, first of all, to calculate the position of the exit pupil, as this is the stop imaged by a single lens (the third lens) of focal length 32.8 mm. The position of the aperture stop, the object in this instance, is 6.4 mm to the left of this lens. The distance, v, of the exit pupil from the third lens is therefore given by:

$$\frac{1}{v}=\frac{1}{f}-\frac{1}{u}=\frac{1}{32.8}-\frac{1}{6.4} \qquad v=-7.95$$

Thus, the exit pupil is 7.95 mm to the left of the third lens and 8.05 mm from the origin. The magnification is given by (minus) the ratio of image and object distances and so it is easy to calculate the size of the exit pupil:

$$d_{exit}=M\times d_0=\frac{7.95}{6.4}\times 11.5 \qquad d_{\mathrm{exit}}=14.3\ \mathrm{mm}$$

ii) *Cardinal Points of the Lens*

The distance between the first and second lenses is 6.8 mm and between the second and third lenses is 9.2 mm. By convention, we retrace dummy rays −16 mm back to the origin at the first lens, so that all matrix ray tracing formulae are referred to a common origin. The matrix for the system is given below:

$$M=\begin{bmatrix}1 & -16\\ 0 & 1\end{bmatrix}\begin{bmatrix}1 & 9.2\\ 0 & 1\end{bmatrix}\begin{bmatrix}1 & 0\\ -\frac{1}{32.8} & 1\end{bmatrix}\begin{bmatrix}1 & 0\\ \frac{1}{17.8} & 1\end{bmatrix}\begin{bmatrix}1 & 6.8\\ 0 & 1\end{bmatrix}\begin{bmatrix}1 & 0\\ -\frac{1}{27.1} & 1\end{bmatrix}=\begin{bmatrix}1.1025 & 6.9216\\ -0.01911 & 0.7871\end{bmatrix}$$

To calculate the position of the exit pupil we need to know the focal length of the system and the positions of the two focal points. Following the matrix relations set out in Chapter 1, we can calculate the following:

Focal length: 52.3 mm
Location of First Focal Point: −41.2 mm
Location of Second Focal Point: 57.7 mm

All distances are referenced to the axial origin at the first lens. There is, of course, a single effective focal length as both object and image spaces are considered to lie within media of the same refractive index.

iii) *Position and Size of the Entrance Pupil*

The imaged pupil or exit pupil lies in image space, 8.05 mm from the origin. This is 49.65 mm to the ***left*** of the second focal point. In applying Newton's formula, the second focal distance, x_2 is then equal to −49.65. We can now calculate the first focal distance to determine the position of the entrance pupil.

$$x_1=\frac{f^2}{x_2}=\frac{52.3^2}{-49.65}=-55.1$$

The object or entrance pupil therefore lies 55.1 mm to the right of the first focal point and 13.9 mm (−41.2 + 55.1) to the right of the first lens.

The location of the entrance pupil expressed as an object distance is 52.3 – 55.1 or −2.8 mm. Similarly the location of the exit pupil expressed as an image distance is equal to −49.65 + 52.3 or +2.65 mm. The magnification (image/object) is, in this instance equal to 2.65/2.8 or 0.946. Therefore we have:

$$\frac{d_{exit}}{d_{entrance}}=0.946 \qquad d_{entrance}=\frac{d_{exit}}{0.946}=\frac{14.3}{0.946}=15.1$$

The diameter of the entrance pupil is, therefore, 15.1 mm

So, in summary we have:

Entrance Pupil Axial Location: 13.9 mm	Entrance Pupil Diameter: 15.1 mm
Exit Pupil Axial Location: 8.05 mm	Exit Pupil Diameter: 14.3 mm

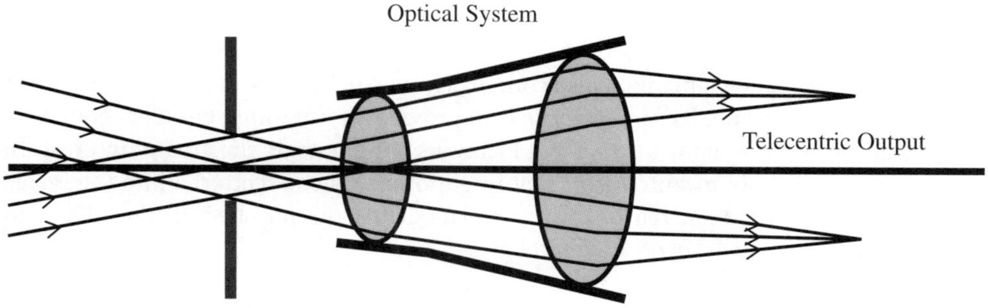

Figure 2.4 Optical system with a telecentric output.

2.4 Telecentricity

In the previous example, both entrance and exit pupils were located at finite conjugates. However, a system is said to be ***telecentric*** if the exit pupil (or entrance pupil) is located at infinity. In the case of a telecentric output, this will occur where the entrance pupil is located at the first focal point. In this instance, all chief rays will, in image space, be parallel. This is shown in Figure 2.4 which illustrates a telecentric output for two different field positions.

A telecentric output, as represented in Figure 2.4 is characterised by a number of converging ray bundles, each emanating from a specific field location, whose central or chief rays are parallel. There are a number of instances where optical systems are specifically designed to be telecentric. Telecentric lenses, for instance, have application in machine vision and metrology where non-telecentric output can lead to measurement errors for varying (object) axial positions.

2.5 Vignetting

The aperture stop is the principal means for controlling the passage of rays through an optical system. Ideally, this would be the only component that controls the admission of light to the optical system. In practice, other optical surfaces located away from the aperture stop may also have an impact on the admission of light into the system. This is because these optical components, for reasons of economy and other optical design factors, have a finite aperture. As a consequence, some rays, particularly those for larger field angles, may miss the lens or component aperture altogether. So, in this case, for field positions furthest from the optical axis, some of the rays will be clipped. This process is known as ***vignetting***. This is shown in Figure 2.5.

Vignetting tends to darken the image for objects further away from the optical axis. As such, it is an undesirable effect. At the same time, it can be used to control optical imperfections or aberrations by deliberately removing more marginal rays.

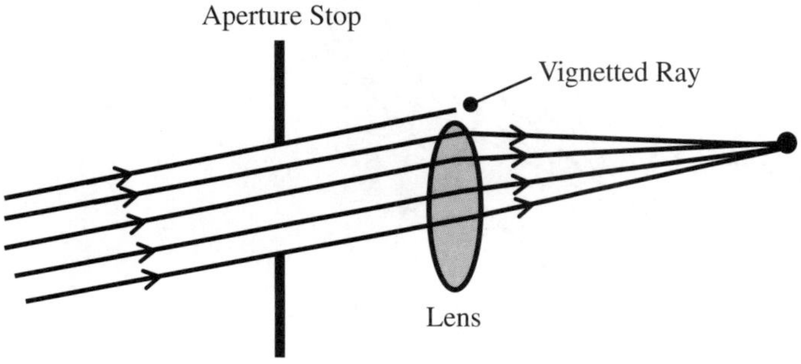

Figure 2.5 Vignetting.

2.6 Field Stops and Other Stops

In addition to the aperture stop, an optical system might also contain a ***field stop***. This is an aperture located in a plane that is conjugate with the image plane. Its first purpose is to provide a crisp (often circular) boundary to the viewable image. Secondly, it excludes light from object locations lying outside the area of interest. In so doing, the field stop reduces the level of unwanted light that might otherwise be scattered into the image plane and so reduce image contrast. For the same reason, other, intermediate stops may be introduced into an optical design in order to further reduce the level of scattered light.

2.7 Tangential and Sagittal Ray Fans

The analysis pursued hitherto has considered the propagation of rays in a single plane. From an analytical perspective, for ray tracing in an ideal system and determining the cardinal points of that system, this is a perfectly acceptable approach. However, in reality, rays are not necessarily confined to the plane containing the object and the optical axis. With the selection of rays delineated by a two-dimensional, ***circular*** aperture, we must expect some rays to be out of this plane. A group of co-planar rays, emanating from a single object point and bounded by the entrance pupil is referred to as a ***ray fan***. A ray fan that lies in the plane defined by the object and optical axis is known as the tangential ray fan. The sagittal ray fan emanates from the same object point and lies in a plane that is perpendicular to that of the tangential ray fan. This is illustrated in Figure 2.6.

The tangential ray fan is also referred to as the ***meridional ray fan***; the two terms are equivalent. In general any ray that is not in the tangential plane, i.e. not a tangential ray, is referred to as a ***skew ray***. A skew ray will never cross the optic axis.

2.8 Two Dimensional Ray Fans and Anamorphic Optics

The introduction of two distinct sets of ray fans, tangential and sagittal, together with the inclusion of skew rays confirms that sequential ray propagation in an axial geometry is essentially a two-dimensional problem.

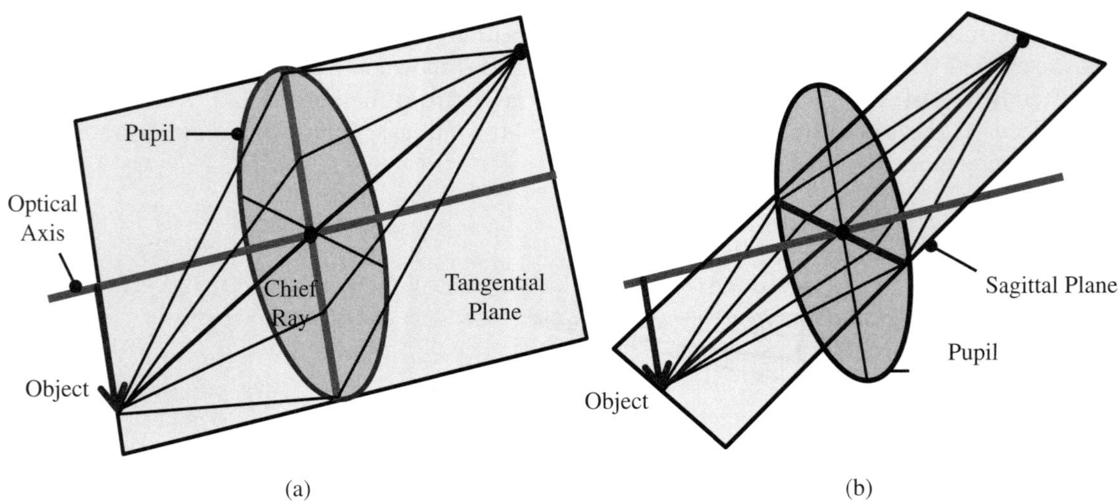

Figure 2.6 (a) Tangential ray fan; (b) Sagittal ray fan.

Hitherto, all discussion and, in particular, the matrix analysis, has been presented in a strictly one-dimensional form. However, the strict description of a ray in two dimensions requires the definition of four parameters, two spatial and two angular. In this more complete description, a ray vector would be written as:

$$Ray = \begin{bmatrix} h_x \\ \theta_x \\ h_y \\ \theta_y \end{bmatrix}$$

h_x is the x component of the distance of the ray from the optical axis
θ_x is the x component of the angle of the ray to the optical axis
h_y is the y component of the distance of the ray from the optical axis
θ_y is the y component of the angle of the ray to the optical axis

In this two dimensional representation, the matrix element representing each optical element would be a 4×4 matrix instead of a 2×2 matrix. However, the matrix is not fully populated in any realistic scenario. For a rotationally symmetric optical system, as we have been considering thus far, there can only be four elements:

$$M = \begin{bmatrix} A & B & 0 & 0 \\ C & D & 0 & 0 \\ 0 & 0 & A & B \\ 0 & 0 & C & D \end{bmatrix}$$

That is to say, the impact of each optical surface is identical in both the x and y directions in this instance. However, there are optical components where the behaviour is different in the x and y directions. An example of this might be a ***cylindrical lens,*** whose curvature in just one dimension produces focusing only in one direction. The two dimensional matrix for a cylindrical lens would look as follows:

$$M_{cyl} = \begin{bmatrix} 1 & 0 & 0 & 0 \\ -1/f & 1 & 0 & 0 \\ 0 & 0 & 1 & 0 \\ 0 & 0 & 0 & 1 \end{bmatrix}$$

A component that possesses different paraxial properties in the two dimensions is said to be ***anamorphic***. A more general description of an anamorphic element is illustrated next:

$$M_{anamorph} = \begin{bmatrix} A_x & B_x & 0 & 0 \\ C_x & D_x & 0 & 0 \\ 0 & 0 & A_y & B_y \\ 0 & 0 & C_y & D_y \end{bmatrix}$$

Note there are no non-zero elements connecting ray properties in different dimensions, x and y. This would require the surfaces produce some form of skew behaviour and this is not consistent with ideal paraxial behaviour. Since this is the case, the two orthogonal components, x and y, can be separated out and presented as two sets of 2×2 matrices and analysed as previously set out. All relevant optical properties, cardinal points are then calculated separately for x and y components. Even if focal points are identical for the two dimensions, the principal planes may not be co-located. This gives rise to different focal lengths for the x and y dimension and potentially differential image magnification. This differential magnification is referred to as ***anamorphic magnification***. Significantly, in a system possessing anamorphic optical properties, the exit pupil may not be co-located in the two dimensions.

2.9 Optical Invariant and Lagrange Invariant

The field angle, i.e. the angle of the chief ray and the marginal ray angles, will change as the rays propagate through an optical system. The relationship between these angles is inherently constrained by the magnification properties of the optical system in the paraxial approximation. The ***optical invariant*** is a parameter that, in the paraxial approximation, constrains the relationship between any two rays that propagate through an optical system. We now have two general rays as described by their ray vectors:

$$Ray_1 = \begin{bmatrix} h_1 \\ \theta_1 \end{bmatrix} \qquad Ray_2 = \begin{bmatrix} h_2 \\ \theta_2 \end{bmatrix}$$

The optical invariant, O, is given by:

$$O = n(h_1\theta_2 - h_2\theta_1) \qquad n \text{ is the local refractive index} \tag{2.4}$$

The optical invariant is, in the paraxial approximation, preserved on passage through an optical system. That is to say:

$$n'(h_1'\theta_2' - h_2'\theta_1') = n(h_1\theta_2 - h_2\theta_1) \tag{2.5}$$

n', h', θ', etc. *are ray parameters following propagation.*

Derivation of the above invariant is straightforward using matrix analysis.

$$\begin{bmatrix} h_1' \\ \theta_1' \end{bmatrix} = \begin{bmatrix} A & B \\ C & D \end{bmatrix} \begin{bmatrix} h_1 \\ \theta_1 \end{bmatrix} \text{ and } \begin{bmatrix} h_2' \\ \theta_2' \end{bmatrix} = \begin{bmatrix} A & B \\ C & D \end{bmatrix} \begin{bmatrix} h_2 \\ \theta_2 \end{bmatrix}$$

Hence:

$$(h_1'\theta_2' - h_2'\theta_1') = (AD - BC)(h_1\theta_2 - h_2\theta_1) = \det(M)(h_1\theta_2 - h_2\theta_1)$$

From (1.23) we know that the determinant of the matrix is given by the ratio of the refractive indices in the relevant media, so:

$$(h_1'\theta_2' - h_2'\theta_1') = \frac{n}{n'}(h_1\theta_2 - h_2\theta_1)$$

Finally we arrive at Eq. (2.5)

$$n'(h_1'\theta_2' - h_2'\theta_1') = n(h_1\theta_2 - h_2\theta_1)$$

The optical invariant is a generalised constraint that relates system lateral and angular magnification and applies to any arbitrary pair of rays. A very specific descriptor is created when the ray pair consists of the chief ray and the marginal ray. This special case of the optical invariant is known as the ***Lagrange invariant***. The Lagrange invariant, H is given by:

$$H = n(h_{marginal}\theta_{chief} - h_{chief}\theta_{marginal}) \tag{2.6}$$

If we now simply evaluate H at the entrance and exit pupils where, by definition, h_{chief} is zero, then the product $nh_{marginal}\theta_{chief}$ is constant. The Lagrange invariant then simply articulates the fact that the angular and lateral magnifications are inversely related. In fact, the Lagrange invariant captures a more fundamental constraint to an optical system. If the object plane is uniformly illuminated, then the total light flux emanating from the plane is proportional to the square of the maximum field angle. The proportion of that flux that is admitted by the entrance pupil is itself proportional to the square of the marginal ray height. Therefore, the total flux passing through an optical system is proportional to the square of the Lagrange invariant, H^2. Thus the Lagrange invariant is an expression of the conservation of energy as light propagates through an optical system. This will become of paramount significance when, in later chapters, we consider source brightness or radiance and the impact of the optical system on optical flux flowing through it.

2.10 Eccentricity Variable

The ***eccentricity variable***, E, is a measure of how far an axial location in an optical system is from the stop. It is expressed in terms of the chief ray to marginal ray height at that particular location. Of course, at the pupil (entrance or exit) itself, the eccentricity variable will be zero. The eccentricity variable is defined as:

$$E = \frac{h_{chief}}{h_{marginal}H} \qquad H \text{ is the Lagrange invariant.} \tag{2.7}$$

E is of course infinite at the focal point of a system. The variable is of great significance in the analysis of optical imperfections or aberrations where the distance of a component from the aperture stop is of critical importance.

2.11 Image Formation in Simple Optical Systems

These introductory chapters provide a complete description of ideal optical systems. That is to say, in the paraxial approximation, where imaging imperfections, or aberrations may be ignored, the analysis presented is substantially complete. Some very simple optical instruments are introduced at this point; their deficiencies are discussed later.

2.11.1 Magnifying Glass or Eye Loupe

The magnifying glass or eye loupe is perhaps the simplest optical system conceivable, in that is consists of a single lens that is intended to be used with the eye to magnify close objects. Our ability to resolve small, close objects is limited by our ability to focus at close quarters. Typically, although this varies with age and other factors, a comfortable distance for viewing near objects is about 250 mm. If the eye can resolve an angle of 1 arcminute, then this corresponds to a resolution of somewhat under 0.1 mm. Addition of a simple lens allows the eye to view objects at a much shorter distance. This is shown in Figure 2.7.

For the two cases illustrated in Figure 2.7, the eye's focussing power remains the same. Therefore, addition of a lens of focal length f will change the closest approach distance, d_0, to:

$$\frac{1}{d} = \frac{1}{d_0} + \frac{1}{f}$$

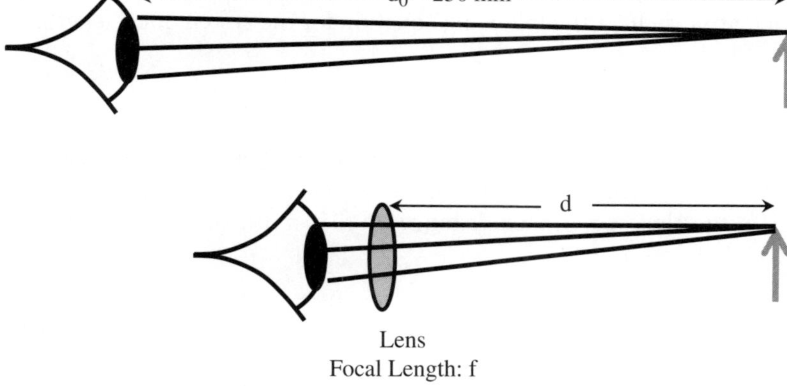

Figure 2.7 Simple magnifying lens.

If the magnification, M, provided by the lens is defined as the ratio of the final image sizes in the two scenarios, the magnification is given by:

$$M = 1 + \frac{d_0}{f} \tag{2.8}$$

In describing magnifying lenses, as suggested earlier, d_0 is defined to be 250 mm. Thus, a lens with a focal length of 250 mm would have a magnification of ×2 and a lens with a focal length of 50 mm would have a magnification of ×6. In practice, simple lenses are only useful up to a magnification of ×10. This is partly because of the introduction of unacceptable aberrations, but also because of the impractical short working distances introduced by lenses with a focal length of a few mm. For higher magnifications, the compound microscope must be used.

Naturally, the pupil of this simple system is defined by the pupil of the eye itself. The size of the eye's pupil varies from about 3 mm in bright light, to about 7 mm under dim lighting conditions, although this varies with individuals.

2.11.2 The Compound Microscope

In the preceding subsection, the limitations of a simple magnifying lens were made clear. Overall, its functionality, in delivering a high magnification, is to convey an intermediate image, located at the infinite conjugate, to the human eye. Furthermore, to provide maximum magnification, the focal length of this lens must be as small as possible. For practical reasons, a magnification of greater than ×10 cannot be delivered. This difficulty is solved by the compound microscope where a two-lens system is used to provide a system focal length that is considerably shorter than would be afforded by a single lens. In essence, a compound microscope consists of two lenses, or lens groups. The first lens is the objective lens that lies close to the object and the second lens, in the traditional microscope, is the eyepiece. Of course, in many modern instruments, the eye is replaced by a pixellated detector chip. Nonetheless the logic followed here still applies. Figure 2.8 shows the general set up.

The two lenses are separated by the nominal tube length, d, and an intermediate image is formed by the objective lens within the tube. The eyepiece then presents the final image to the eye at the infinite conjugate. In other words, the intermediate image is designed to be located at a distance f_2 (the eyepiece focal length) from the eyepiece. If the objective lens focal length is f_1, then the matrix of the system is:

$$M = \begin{bmatrix} 1 & -s \\ 0 & 1 \end{bmatrix} \begin{bmatrix} 1 & 0 \\ -1/f_2 & 1 \end{bmatrix} \begin{bmatrix} 1 & s \\ 0 & 1 \end{bmatrix} \begin{bmatrix} 1 & 0 \\ -1/f_1 & 1 \end{bmatrix} = \begin{bmatrix} 1 + s/f_2 - s^2/f_1 f_2 & s^2/f_2 \\ -1/f_1 - 1/f_2 + s/f_1 f_2 & 1 - s/f_2 \end{bmatrix}$$

The entire co-ordinate system is referenced to the position of the objective lens. Of particular relevance here is the first focal length. From the above matrix we have the following equation for the system focal length:

$$\frac{1}{f_{system}} = \frac{1}{f_1} + \frac{1}{f_2} - \frac{s}{f_1 f_2} \tag{2.9}$$

The logic of Eq. (2.9) is that a shorter system focal length can be created than would be reasonably practical with a single lens. Using the same definition as used for the simple magnifying lens, the effective system magnification, M_{system}, is given by the ratio of the closest approach distance d_0, (250 mm), and the system focal length. The system magnification, is given by:

$$M_{system} = \frac{d_0 s}{f_1 f_2} - \frac{d_0}{f_1} - \frac{d_0}{f_2} = \frac{d_0(s - f_1 - f_2)}{f_1 f_2} \tag{2.10}$$

The bracketed quantity, $(s - f_1 - f_2)$, i.e. the lens separation minus the sum of the lens focal lengths is known as the ***optical tube length*** of the microscope, and this will be denoted as d. Generally, for optical microscopes,

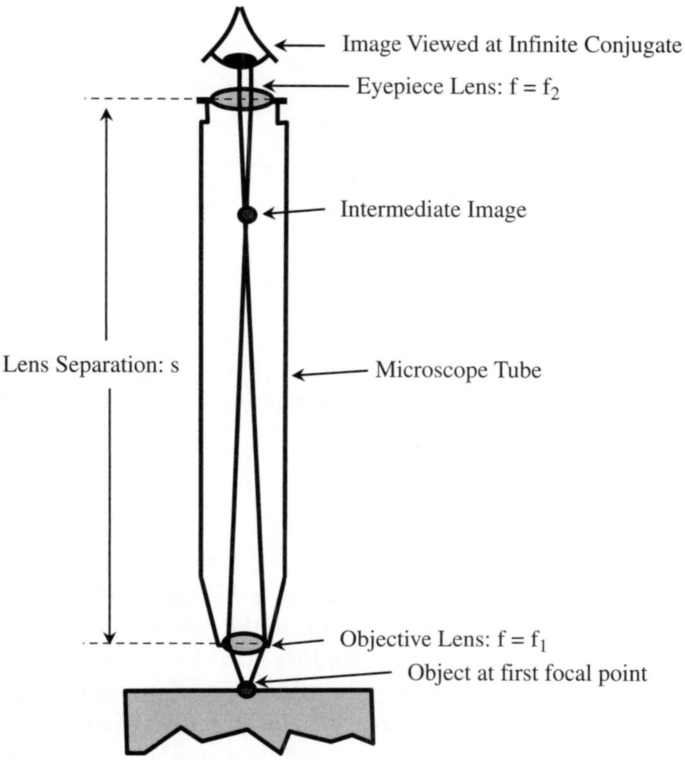

Figure 2.8 Compound microscope.

this tube length is standardised across many commercial instruments with the standard values being 160 or 200 mm. Equation (2.10) may be rewritten as:

$$M_{system} = \frac{dd_0}{f_1 f_2} = M_{objective} \times M_{eyepiece} \tag{2.11}$$

The above formula gives the total magnification of the instrument as the product of the individual magnifications of the objective lens and eyepiece. In this context, these individual magnifications are defined as in Eqs. (2.12a) and (2.12b):

$$M_{objective} = \frac{d}{f_1} \tag{2.12a}$$

$$M_{eyepiece} = \frac{d_0}{f_1} \tag{2.12b}$$

The equations above establish the standard definitions for microscope lens powers. For example, the magnification of microscope objectives is usually in the range of ×10 to ×100. For a standard tube length, d, of 160 mm, this corresponds to an objective focal length ranging from 16 to 1.6 mm. A typical eyepiece, with a magnification of ×10 has a focal length of 25 mm ($d_0 = 250$ mm). By combining a ×100 objective lens with a ×10 eyepiece, a magnification of ×1000 can be achieved. This illustrates the power of the compound microscope.

The entrance pupil is defined by the aperture of the objective lens. This entrance pupil is re-imaged by the eyepiece to create an exit pupil that is close to the eyepiece. Ideally, this should be co-incident with the pupil of the eye. The distance of the exit pupil from final mechanical surface of the eyepiece is known as the ***eye relief***. Placing the exit pupil further away from the physical eyepiece provides greater comfort for the user, hence the

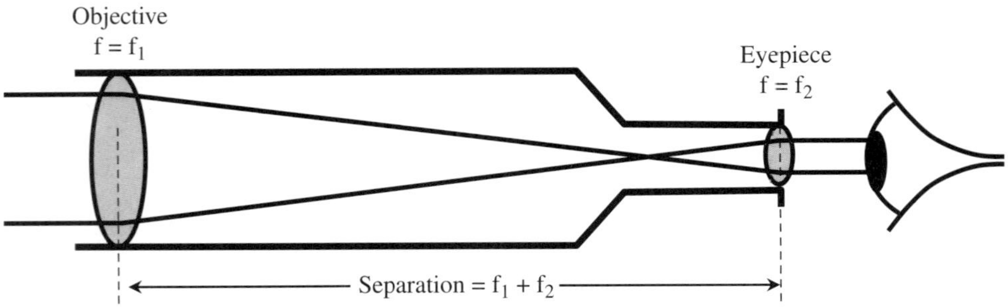

Figure 2.9 Basic optical telescope.

term 'eye relief'. Objective lens aperture tends to be defined by numerical aperture, rather than f-number and range from 0.1 to 1.3 (for oil immersion microscopes).

2.11.3 Simple Telescope

A classical optical telescope is an example of an ***afocal system***. That is to say, no clearly defined focus is presented either in object or image space. As the name suggests, the telescope views distant objects, nominally at the infinite conjugate and provides a collimated output for ocular viewing in the case of a traditional instrument. As far as the instrument is concerned, both object and image are located at the infinite conjugate. Of course, this narrative does assume that the instrument is designed for ocular viewing as opposed to image formation at a detector or photographic plate. In any case, the design principles are similar. Fundamentally, the telescope provides angular magnification of a distant object, and this angular magnification is a key performance attribute.

The basic layout of a simple telescope is shown in Figure 2.9. Light from the distant object is collected by an objective lens whose focal length is f_1 and then collimated by an eyepiece with a focal length of f_2. These two lenses are separated by the sum of their focal lengths, thus creating an afocal system with an angular magnification given by the ratio of the lens focal lengths.

The matrix of the telescope is similar to that of the compound microscope, with an objective lens and eyepiece separated by some fixed distance.

$$M = \begin{bmatrix} 1 + s/f_2 - s^2/f_1 f_2 & s^2/f_2 \\ -1/f_1 - 1/f_2 + s/f_1 f_2 & 1 - s/f_2 \end{bmatrix}$$

The separation, s, is simply the sum of the two focal lengths and the system matrix is given by:

$$M = \begin{bmatrix} -f_2/f_1 & f_1^2/f_2 + 2f_1 + f_2 \\ 0 & -f_1/f_2 \end{bmatrix} \tag{2.13}$$

The angular magnification (the D value of the matrix) is simply $-f_1/f_2$. It is important to note the sign of the magnification, so that for two positive lenses, then the magnification is negative. In line with the previous discussion with regard to the optical invariant, the linear magnification (given by matrix element A) is the inverse of the angular magnification. Also, the C element of the matrix, attesting to the focal power of the system, is actually zero and is characteristic of an afocal system.

As in the case of the microscope, the objective lens forms the system entrance pupil. The exit pupil is formed by the eyepiece imaging the objective lens. This is located a short distance, approximately f_1 from the eyepiece, this distance determining the 'eye relief'. Ideally, for ocular viewing, the pupil of the eye should be co-incident with the exit pupil. Unlike the compound microscope, the exit pupil of a simple (ocular) telescope is relatively large, about the size of the pupil of the eye. Clearly, if the exit pupil were significantly larger than the pupil of the eye, then any light falling outside the ocular pupil would be wasted. In fact, in a typical telescope, where f1 ≫ f2, the size of the exit pupil is approximately given by the diameter of the objective lens multiplied by the ratio of the focal lengths.

As an example, a small astronomical refracting telescope might comprise a 75 mm diameter objective lens with a focal length of 750 mm (f/10) and might use a ×10 eyepiece. Eyepiece magnification is classified in the same way as for microscope eyepieces and so the focal length of this eyepiece would be 25 mm, as derived from Eq. (2.12b). The angular magnification (f_1/f_2) would be ×30 and the size of the pupil about 3 mm, which is smaller than the pupil of the eye.

In the preceding discussion, the basic description of the instrument function assumes ocular viewing, i.e. viewing through an eyepiece. However, increasingly, across a range of optical instruments, the eye is being replaced by a detector chip. This is true of microscope, telescope, and camera instruments.

2.11.4 Camera

In essence, the function of a camera is to image an object located at the infinite conjugate and to form an image on a light sensitive planar surface. Of course, traditionally, this light sensitive surface consisted of a film or a plate upon which a silver halide emulsion had been deposited. This allowed the recording of a latent image which could be chemically developed at a later stage. Depending upon the grain size of the silver halide emulsion, feature sizes of around 10–20 μm or so could be resolved. That is to say, the ultimate system resolution is limited by the recording media as well as the optics. For the most part, this photographic film has now been superseded by pixelated silicon detectors, allowing the rapid and automatic capture and processing of images. These detectors are composed of a rectangular array of independent sensor areas (usually themselves rectangular) that each produce a charge in proportion to the amount of light collected. Resolution of these detectors is limited by the pixel size which is analogous to the grain size in photographic film. Pixel size ranges from a one micron to a few microns.

Optically from a paraxial perspective, the camera is an exceptionally simple instrument. Its purpose is simply to image light from an object located at the infinite conjugate onto the focal plane, where the sensor is located. As such, from a system perspective one might regard the camera as a single lens with the sensor located at the second focal point. This is illustrated in Figure 2.10.

If this system is the essence of simplicity, then the ***Pinhole Camera***, a very early form of camera, takes this further by dispensing with the lens altogether! A pinhole camera relies on a very small system aperture (a pinhole) defining the image quality. In this embodiment of the camera, all rays admitted by the entrance pupil follow closely the chief ray. However, light collection efficiency is low. Whilst in the paraxial approximation, the camera presents itself as a very simple instrument, as indeed early cameras were, the demands of light collection efficiency require the use of a large aperture which results in the breakdown of the paraxial

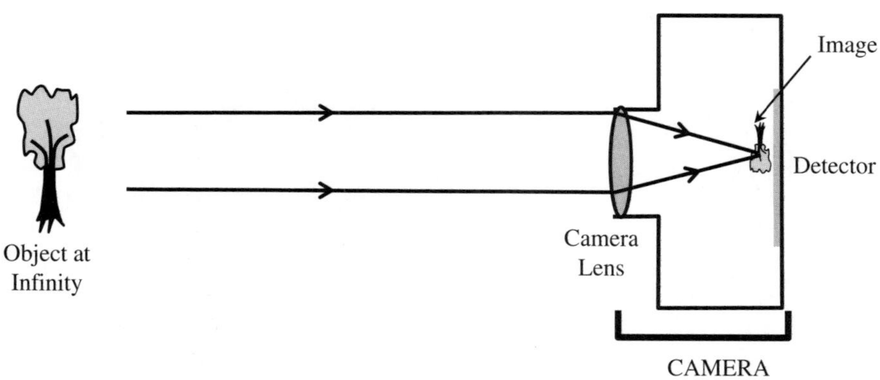

Figure 2.10 Basic camera.

approximation. As we shall see in later chapters, this leads to the creation of significant imperfections, or aberrations, in image formation which can only be combatted by complex multi-element lens designs. Thus, in practice, a modern camera, i.e. its lens, is a relatively complex optical instrument.

In defining the function of the camera, we spoke of the imaging of an object located at infinity. In this context, 'infinity' means a substantially greater object distance than the lens focal length. For the traditional 35 mm format photographic camera, a typical standard lens focal length would be 50 mm. The '35 mm' format refers to the film frame size which was 36 mm × 24 mm (horizontal × vertical). As mentioned in Chapter 1, the focal length of the camera lens determines the 'plate scale' of the detector, or the field angle subtended per unit displacement of the detector. Overall, for this example, plate scale is 1.15° mm^{-1}. The total field covered by the frame size is ±20° (Horizontal) × ±13.5° (Vertical). 'Wide angle' lenses with a shorter focal length lens (e.g. 28 mm) have a larger plate scale and, naturally a wider field angle. By contrast, telephoto lenses with longer focal lengths (e.g. 200 mm), have a smaller plate scale, thus producing a greater magnification, but a smaller field of view.

Modern cameras with silicon detector technology are generally significantly more compact instruments than traditional cameras. For example, a typical digital camera lens might have a focal length of about 8 mm, whereas a mobile phone camera lens might have a focal length of about half of this. The plate scale of a digital camera is thus considerably larger than that of the traditional camera. Overall, as dictated by the imaging requirements, the field of view of a digital camera is similar to its traditional counterpart, although, in practice, equivalent to that of a wide field lens. Therefore, in view of the shorter focal length, the detector size in a digital camera is considerably smaller than that of a traditional film camera, typically a few mm. Ultimately, the miniaturisation of the digital camera is fundamentally driven by the resolution of the detector, with the pixel size of a mobile phone camera being around 1 μm. This is over an order of magnitude superior to the resolution, or 'grain size' of a high specification photographic film.

Further Reading

Haija, A.I., Numan, M.Z., and Freeman, W.L. (2018). *Concise Optics: Concepts, Examples and Problems*. Boca Raton: CRC Press. ISBN: 978-1-1381-0702-1.

Hecht, E. (2017). *Optics*, 5e. Harlow: Pearson Education. ISBN: 978-0-1339-7722-6.

Keating, M.P. (1988). *Geometric, Physical, and Visual Optics*. Boston: Butterworths. ISBN: 978-0-7506-7262-7.

Kidger, M.J. (2001). *Fundamental Optical Design*. Bellingham: SPIE. ISBN: 0-81943915-0.

Kloos, G. (2007). *Matrix Methods for Optical Layout*. Bellingham: SPIE. ISBN: 978-0-8194-6780-5.

Longhurst, R.S. (1973). *Geometrical and Physical Optics*, 3e. London: Longmans. ISBN: 0-582-44099-8.

Smith, F.G. and Thompson, J.H. (1989). *Optics*, 2e. New York: Wiley. ISBN: 0-471-91538-1.

3

Monochromatic Aberrations

3.1 Introduction

In the first two chapters, we have been primarily concerned with an idealised representation of geometrical optics involving perfect or Gaussian imaging. This treatment relies upon the paraxial approximation where all rays present a negligible angle with respect to the optical axis. In this situation, all primary optical ray behaviour, such as refraction, reflection, and beam propagation, can be represented in terms of a series of linear relationships involving ray heights and angles. The inevitable consequence of this paraxial approximation and the resultant linear algebra is apparently perfect image formation. However, for significant ray angles, this approximation breaks down and imperfect image formation, or ***aberration***, results. That is to say, a bundle of rays emanating from a single point in object space does not uniquely converge on a single point in image space.

This chapter will focus on ***monochromatic aberrations*** only. These aberrations occur where there is departure from ideal paraxial behaviour at a single wavelength. In addition, ***chromatic aberration*** can also occur where first order paraxial properties of a system, such as focal length and cardinal point locations, vary with wavelength. This is generally caused by dispersion, or the variation in the refractive index of a material with wavelength. Chromatic aberration will be considered in the next chapter.

A simple scenario is illustrated in Figure 3.1 where a bundle of rays originating from an object located at the infinite conjugate is imaged by a lens. Figure 3.1a presents the situation for perfect imaging and Figure 3.1b illustrates the impact of aberration.

In Figure 3.1b, those rays that are close to the axis are brought to a focus at the ***paraxial focus***. This is the ideal focus. However, those rays that are further from the axis are brought to a focus at a point closer to the lens than the paraxial focus. In fact, the behaviour illustrated in Figure 3.1b is representative of a simple lens; marginal rays are brought to a focus closer to the lens than the chief ray. However, in general terms, the sense of the aberration could be either positive or negative, with the marginal rays coming to a focus either before or after the paraxial focus.

3.2 Breakdown of the Paraxial Approximation and Third Order Aberrations

In formulating perfect or Gaussian imaging we assumed all relationships are linear. For example, Snell's law of refraction was reduced in the following way:

$$n_1 \sin\theta_1 = n_2 \sin\theta_2 \rightarrow n_1\theta_1 \approx n_2\theta_2 \tag{3.1}$$

In making the paraxial approximation, we are considering just the first or linear term in the Taylor series. The next logical stage in the process is to consider higher order terms in the Taylor series.

$$\sin\theta = \theta - \frac{\theta^3}{3!} + \frac{\theta^5}{5!} - \frac{\theta^7}{7!} + \frac{\theta^9}{9!} - \frac{\theta^{11}}{11!} + \ldots\ldots \tag{3.2}$$

Optical Engineering Science, First Edition. Stephen Rolt.

Companion website: www.wiley.com/go/Rolt/opt-eng-sci

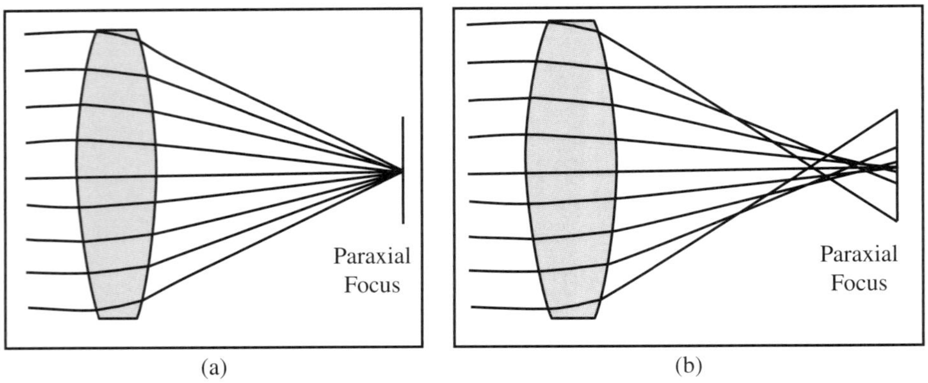

Figure 3.1 (a) Gaussian imaging. (b) Impact of aberration.

Following the term that is linear in θ, we have terms that are cubic or third order in θ. Of course, these third order terms are followed by fifth and seventh order terms etc. in succession. ***Third order aberration theory*** deals exclusively with those imperfections associated with the third order departure from ideal behaviour, as illustrated in Eq. (3.2). Much of classical aberration theory is restricted to consideration of these third order terms and is, in effect a refinement or successive approximation to paraxial theory. Higher order ($\geq$5) terms can be important in practical design scenarios. However, these are generally dealt with by numerical computation, rather than by a simple generically applicable theory.

Third order aberration theory forms the basis of the classical treatment of monochromatic aberrations. Unless specific steps are taken to correct third order aberrations in optical systems, then third order behaviour dominates. That is to say, error terms in the ray height or angle (compared to the paraxial) have a cubic dependence upon the angle or height. As a simple illustration of this, Figure 3.1b shows rays originating from a single object (at the infinite conjugate). For perfect image formation, the height of all rays at the paraxial focus should be zero, as in Figure 3.1a. However, the consequence of third order aberration is that the ray height at the paraxial focus is proportional to the third power of the original ray height (at the lens).

In dealing with third order aberrations, the location of the entrance pupil is important. Let us assume, in the example set out in Figure 3.1b, that the pupil is at the lens. If the radius of the entrance pupil is r_0 and the height a specific ray at this point is h, then we may define a new parameter, the ***normalised pupil co-ordinate***, p, in the following way:

$$p = \frac{h}{r_0} \tag{3.3}$$

The normalised pupil co-ordinate can have values ranging from −1 to +1, with the extremes representing the marginal ray. The chief ray corresponds to $p = 0$. At this stage, it is useful to provide a specific and quantifiable definition of aberration. The quantity, ***transverse aberration***, is defined as the difference in height of a specific ray and the corresponding chief ray as measured at the paraxial focus. The 'corresponding chief ray' emanates from the same object point as the ray under consideration. In addition, the term ***longitudinal aberration*** is also used to describe aberration. Longitudinal aberration (LA) is the axial distance from the point at which the ray in question intersects the chief ray and the location of the paraxial focus. The transverse aberration (TA) and longitudinal aberration definitions are illustrated in Figure 3.2.

In keeping with the previous arguments, the TA has a third order dependence upon the pupil function. This is illustrated in Eq. (3.4):

$$TA \propto p^3 \tag{3.4}$$

Transverse aberration has dimensions of length, whereas the pupil function is a dimensionless ratio. Geometrically, the LA is approximately equal to the transverse aberration divided by the ray angle which itself is

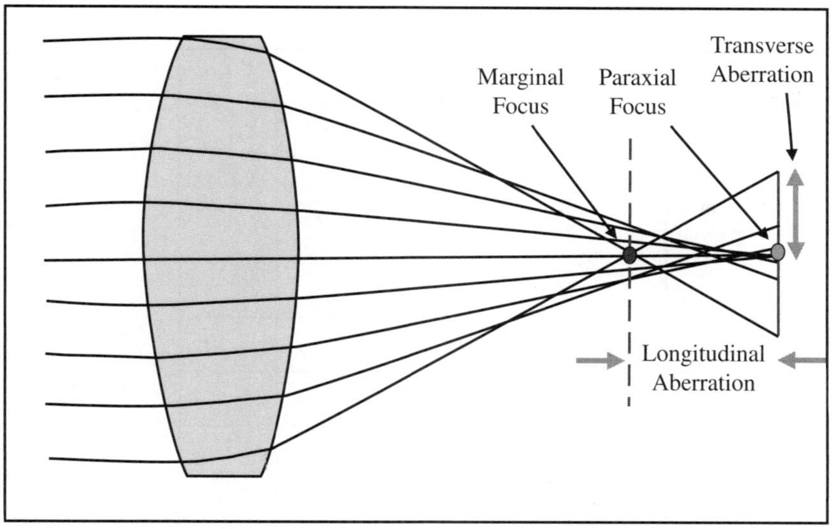

Figure 3.2 Transverse and longitudinal aberration.

proportional to the pupil function. Therefore, the longitudinal aberration has a quadratic dependence upon the pupil function. This is illustrated in Eq. (3.5).

$$LA \propto p^2 \tag{3.5}$$

In fact, if the radius of the pupil aperture is r_0 and the lens focal length is f, then the longitudinal and transverse aberration are related in the following way:

$$LA = \frac{f}{pr_0}TA = \frac{TA}{pNA} \tag{3.6}$$

NA is the numerical aperture of the lens.

A plot of the transverse aberration against the pupil function is referred to as a '***ray fan***'. Ray fans are widely used to provide a simple description of the fidelity of optical systems. If one views the transverse aberration at the paraxial focus, then the transverse aberration should show a purely cubic dependence upon the pupil function. This is illustrated in Figure 3.3a which shows the aberrated ray fan. If, one the other hand, the transverse aberration is plotted away from the paraxial focus, then an additional linear term is present in the plot. This is because pure defocus (i.e. without third order aberration) produces a transverse aberration that is linear with respect to pupil function. This is illustrated in Figure 3.3b which shows a ray fan where both the linear defocus and third order aberration terms are present.

The underlying amount of third order aberration is the same in both plots. However, the overall transverse aberration in Figure 3.3b (plotted on the same scale) is significantly lower than that seen in Figure 3.3a. This is because defocus can, to some extent, be used to 'balance' the original third order aberration. As a result, by moving away from the paraxial focus, the size of the blurred spot is reduced. In fact, there is a point at which the size (root mean square radius) of the spot is minimised. This optimum focal position is referred to as the ***circle of least confusion***. This is illustrated in Figure 3.4.

Most generally, the transverse aberration where third order aberration is combined with defocus can be represented as:

$$TA = TA_0(p^3 + \alpha p) \tag{3.7}$$

TA_0 is the nominal third order aberration and α represents the defocus

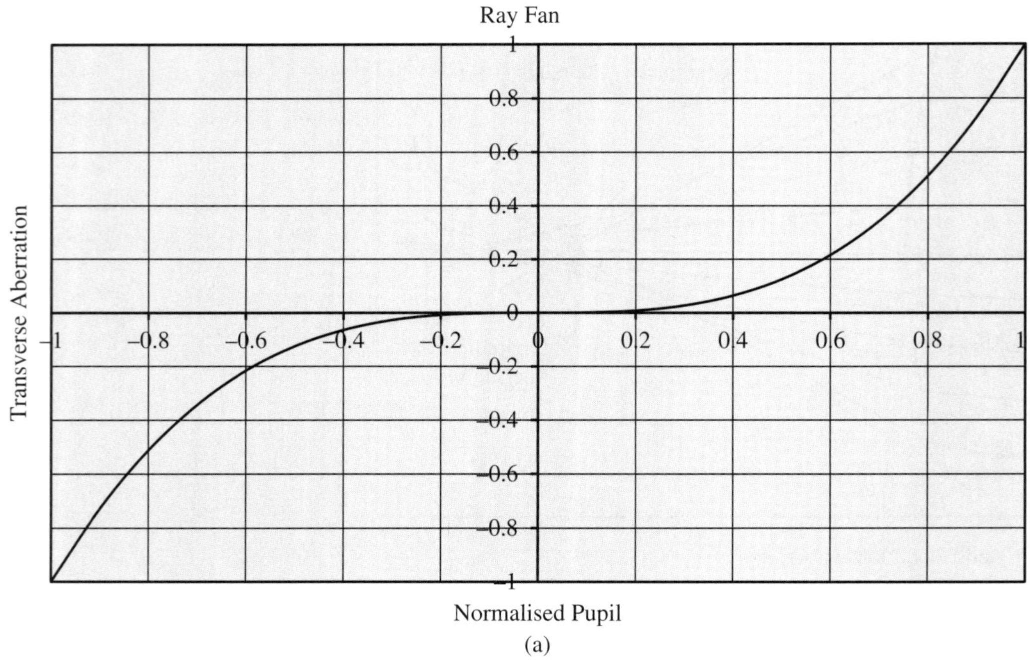

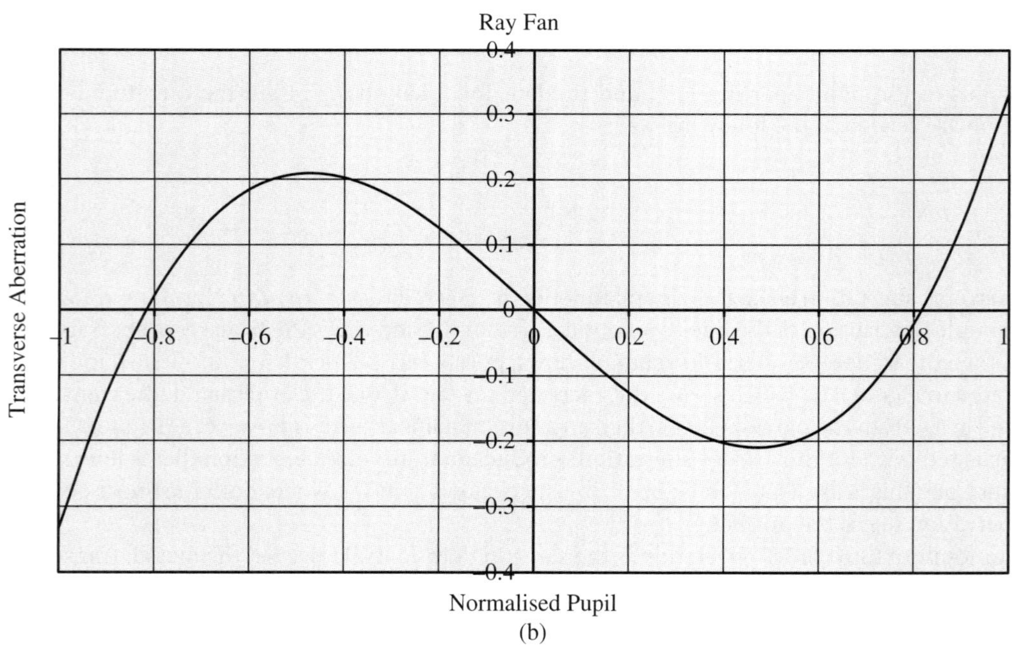

Figure 3.3 (a) Ray fan for pure third order aberration. (b) Ray fan with third order aberration and defocus.

Since the geometry is assumed to be circular, to calculate the rms (root mean square) aberration, one must introduce a weighting factor that is proportional to the pupil function, p. The mean squared transverse aberration is thus:

$$\langle TA^2 \rangle = \frac{TA_0^2}{2}\int_0^1 (p^3 + \alpha p)^2 p dp = \frac{TA_0^2}{2}\left(\frac{1}{8} + \frac{\alpha}{3} + \frac{\alpha^2}{4}\right) \tag{3.8}$$

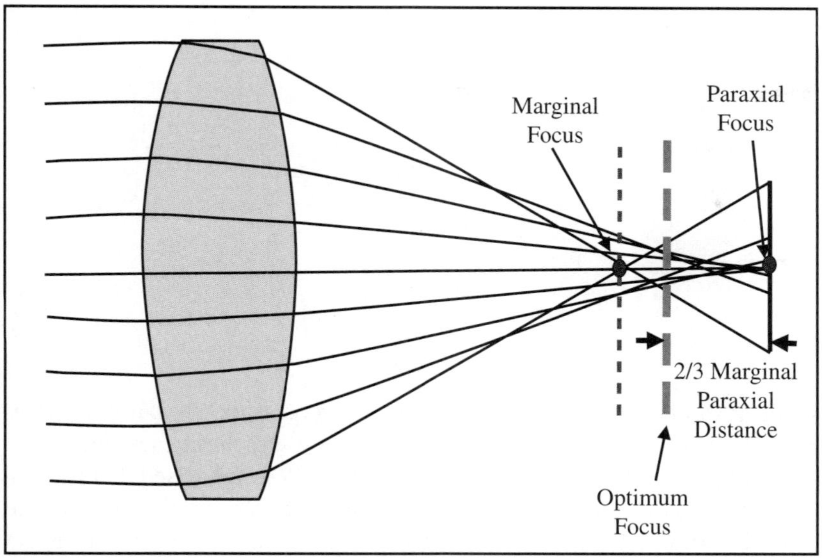

Figure 3.4 Balancing defocus against aberration – optimal focal position.

The expression is minimised where $\alpha = -2/3$. To understand the significance of this, examination of Eq. (3.6) suggests that, without defocus, the marginal ray ($p = 1$) has a longitudinal aberration of TA_0/NA. The defocus term itself produces a constant longitudinal aberration or defocus of $\alpha TA_0/NA$. Therefore, the optimum defocus is equivalent to placing the adjusted focus at 2/3 of the distance between the paraxial and marginal focus, as shown in Figure 3.4. Without this focus adjustment, with the third order aberration viewed at the paraxial focus, the rms aberration is $TA_0/4$. However, adding the optimum defocus reduces the rms aberration to $TA_0/12$, a reduction by a factor of 3.

This analysis provides a very simple introduction to the concept of third order aberrations. In the basic illustration so far considered, we have looked at the example of a simple lens focussing an on axis object located at infinity. In the more general description of monochromatic aberrations that we will come to, this simple, on-axis aberration is referred to as ***spherical aberration***. In developing a more general treatment of aberration in the next sections we will introduce the concept of optical path difference (OPD).

3.3 Aberration and Optical Path Difference

In the preceding section, we considered the impact of optical imperfections on the transverse aberration and the construction of ray fans. Unfortunately, this treatment, whilst providing a simple introduction, does not lead to a coherent, generalised description of aberration. At this point, we introduce the concept of ***optical path difference*** (OPD). For a perfect imaging system, with no aberration, if all rays converge onto the paraxial focus, then all ray paths must have the same optical path length from object to image. This is simply a statement of Fermat's principle. We now consider an aberrated system where we accurately (*not relying on the paraxial approximation*) trace all rays through the system from object to image. However, at the last surface, we (*hypothetically*) force all rays to converge onto the paraxial focus. For all rays, we compute the optical path from object to image. The OPD is the difference between the integrated optical path of a specified ray with respect to the optical path of the chief ray. Of course, if there were no aberration present, the OPD would be zero. Thus, the OPD represents a quantitative description of the violation of Fermat's principle.

The general concept is shown in Figure 3.5. Rays are accurately traced from the object through the system, emerging into image space. That is to say, ray tracing proceeds until the last optical surface, mirror or lens

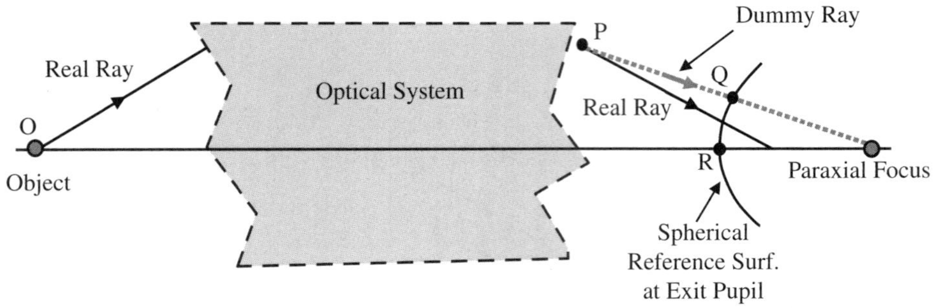

Figure 3.5 Illustration of optical path difference.

etc. Following the preceding discussion, at some point, we force all rays to converge upon the paraxial focus. However, the convention for computing OPD is that all rays are traced back to a spherical surface centred on the paraxial focus and which lies at the exit pupil of the system. Of course, it must be emphasised that the real rays do not actually follow this path. In the generic system illustrated, the real ray is traced to point P located in object space and the optical path length computed. Thereafter, instead of tracing the real ray into the object space, a dummy ray is traced, as shown by the dotted line. This dummy ray is traced from point P to point Q that lies on the reference surface – a sphere located at the exit pupil and centred on the paraxial focus. The optical path length of this segment is then added to the total.

After calculating the optical path length for the dummy ray **OPQ**, we need to calculate the OPD with respect to the chief ray. The chief ray path is calculated from the object to its intersection with the reference sphere at the pupil, represented, in this instance, by the path **OR**. In calculating the OPD, the convention is that the OPD is the chief ray optical path (OR) minus the dummy ray optical path (OPQ). Note the sign convention.

OPD = Chief Ray Optical Path – Dummy Ray Optical Path

Having established an additional way of describing aberrations in terms of the violation of Fermat's principle, the question is what is the particular significance and utility of this approach? The answer is that, when expressed in terms of the OPD, aberrations are additive through a system. As a consequence of this, this treatment provides an extremely powerful general description of aberrations and, in particular, third order aberrations. Broadly, aberrations can be computed for individual system elements, such as surfaces, mirrors, or lenses and applied additively to the system as a whole. This generality and flexibility is not provided by a consideration of transverse aberrations.

There is a correspondence between transverse aberration and OPD. This is illustrated in Figure 3.6. At this point, we introduce a concept that is related to that of OPD, namely ***wavefront error (WFE).*** We must remember that, according to the wave description, the rays we trace through the system represent normals to the relevant wavefront. The wavefront itself originates from a single object point and represents a surface of equal phase. As such, the wavefront represents a surface of equal optical path length. For an aberrated optical system, the surface normals (rays) do not converge on a single point. In Figure 3.6, this surface is shown as a solid line. A hypothetical spherical surface, shown as a dashed line, is now added to represent rays converging on the paraxial focus. This surface intersects the real surface at the chief ray position. The distance between these two surfaces is the WFE.

In terms of the sign convention, the wavefront error, WFE, is given by:

WFE = Aberrated wavefront – Reference Wavefront (in direction of propagation).

The sign convention is important, as it now concurs with the definition of OPD. As the wavefronts form surfaces of constant optical path length, there is a direct correspondence between OPD and WFE. A positive OPD indicates the optical path of the ray at the reference sphere is ***less*** than that of the chief ray. Therefore,

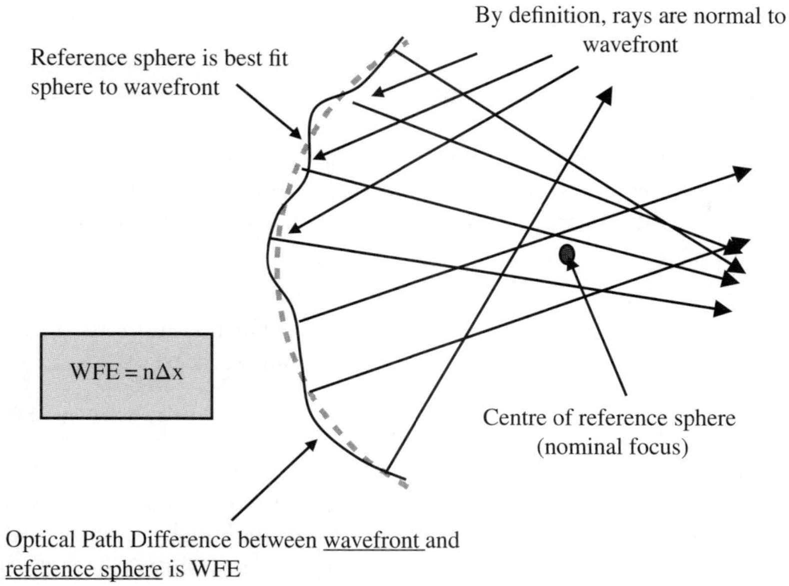

Figure 3.6 Wavefront representation of aberration.

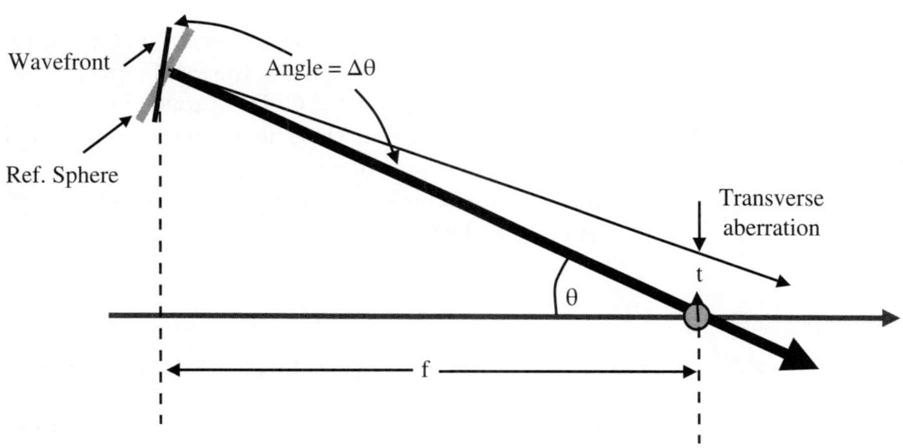

Figure 3.7 Simplified wavefront and ray geometry.

this ray has to travel a small positive distance to 'catch up' with the chief ray to maintain phase equality. Hence, the WFE is also positive.

Both OPD and WFE quantify the violation of Fermat's principle in the same way. OPD is generally used to describe the path length difference of a specific ray. WFE tends to be used when describing OPD variation across an assembly of rays, specifically across a pupil. The concept of WFE enables us to establish the relationship between OPD and transverse aberration in that it helps define the link between wave (phase and path length) geometry and ray geometry. This is shown in Figure 3.7. It is clear that the transverse aberration is related to the angular difference between the wavefront and reference sphere surfaces.

We now describe the WFE, Φ, as a function of the reference sphere (paraxial ray) angle, θ. The radius of the reference sphere (distance to the paraxial focus) is denoted by f. This allows us to calculate the difference in angle, $\Delta\theta$, between the real and paraxial rays. This is simply equal to the difference in local slope between the two surfaces.

$$\Delta\theta = \frac{1}{nf}\frac{d\Phi}{d\theta} \tag{3.9}$$

n is the medium refractive index.

In this analysis, the WFE represents the difference between the real and reference surfaces with the positive axial direction represented by the propagation direction (from object to image). In this convention, the WFE has the opposite sign to the OPD. The transverse aberration, t, can be derived from simple trigonometry.

$$t = -\frac{1}{n\cos\theta}\frac{d\Phi}{d\theta} \tag{3.10}$$

If θ describes the angle the ray makes to the chief ray, then Eq. (3.10) may be reformed in terms of the numerical aperture, NA. The numerical aperture is equal to nsinθ, and Eq. (3.11) may be recast as:

$$t = -\frac{d\Phi}{dNA} \tag{3.11}$$

So, the transverse aberration may be represented by the first differential of the WFE with respect to the numerical aperture. In terms of third order aberration theory, the numerical aperture of an individual ray is directly proportional to the normalised pupil function, p. If the overall system, or marginal ray, numerical aperture is NA_0, then the individual ray numerical aperture is simply NA_0p. The transverse aberration is then given by:

$$t = -\frac{1}{NA_0}\frac{d\Phi}{dp} \tag{3.12}$$

Equation (3.12) provides a simple direct relationship between OPD and transverse aberration. Of course, we know that, for third order aberration, the transverse aberration is proportional to the third power of the pupil function, p. If this is the case, then it is apparent, from Eq. (3.12), that the OPD is proportional to the fourth power of the pupil function. So, for third order aberration, the transverse aberration shows a third power dependence upon the pupil function whereas the OPD shows a fourth power dependence.

Applying these arguments to the analysis of the simple on-axis example illustrated earlier, with the object placed at the infinite conjugate, then the WFE can be represented by the following equation:

$$OPD = \Phi_0 p^4 \tag{3.13}$$

p is the normalised pupil function.

Figure 3.8 shows a plot of the OPD against the normalised pupil function; such a plot is referred to as an OPD fan.

Despite the fact that this simple aberration has a quartic dependence on the pupil function, it is still referred to as third order aberration after the transverse aberration dependence. As with the optimisation of transverse aberration, the OPD can be balanced by applying defocus to offset the aberration. We saw earlier that a simple defocus produces a linear term in the transverse aberration. Referring to Eq. (3.12), it is clear that defocus may be represented by a quadratic term. Equation (3.14) describes the OPD when some defocus has been added to the initial aberration.

$$OPD = \Phi_0(p^4 + \alpha p^2) \tag{3.14}$$

An OPD fan with aberration plus balancing defocus is shown in Figure 3.9.

In this instance, the plot has a characteristic 'W' shape, with the curve in the vicinity of the origin dominated by the quadratic defocus term. As with the case for transverse aberration, the defocus can be optimised to produce the minimum possible OPD value when taken as a root mean squared value over the circular pupil. Again, using a weighting factor that is proportional to the pupil function, p, (to take account of the circular geometry), the mean squared OPD is given by:

$$\langle OPD^2\rangle = \frac{\Phi_0^2}{2}\int_0^1 (p^4 + \alpha p^2)^2 p\,dp = \frac{\Phi_0^2}{2}\left(\frac{1}{10} + \frac{\alpha}{4} + \frac{\alpha^2}{6}\right) \tag{3.15}$$

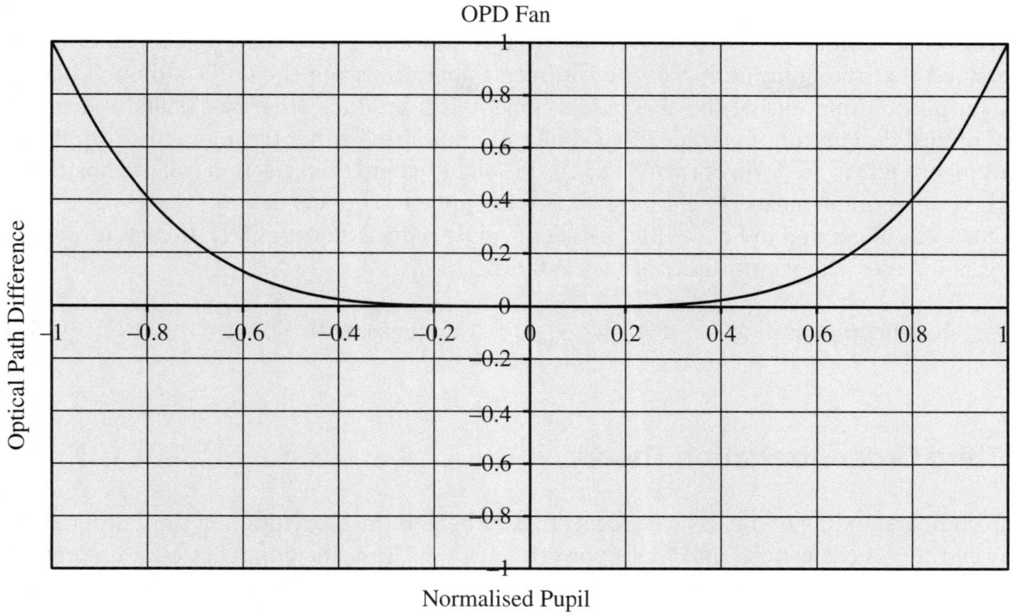

Figure 3.8 Quartic OPD fan.

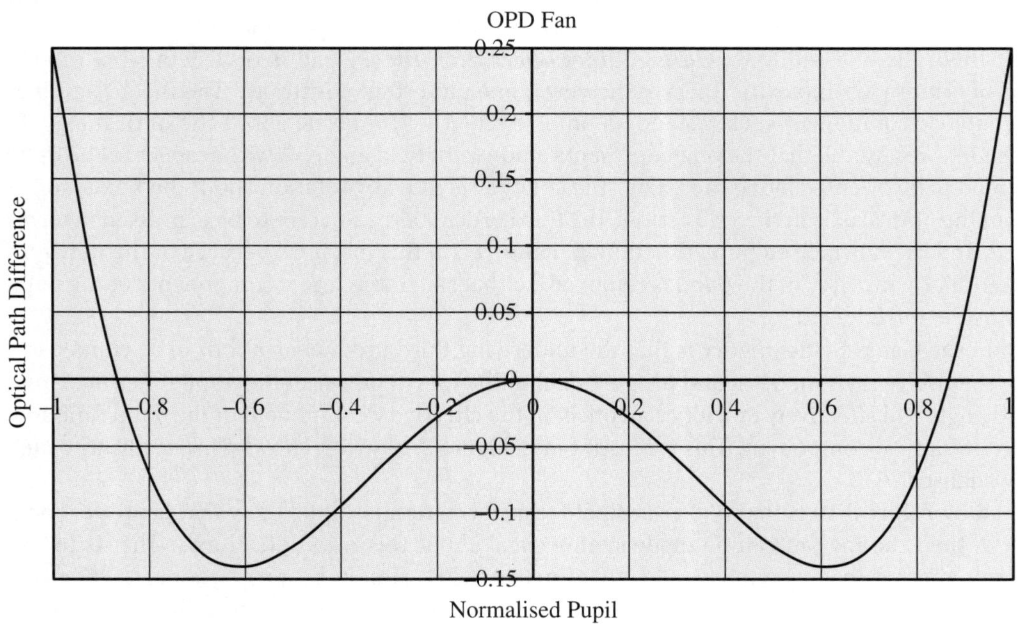

Figure 3.9 OPD fan with balancing defocus.

The above expression has a minimum where $\alpha = -3/4$. To understand the magnitude of this defocus, it is useful first to convert the new OPD expression into a transverse aberration using Eq. (3.12).

$$TA = -\frac{\Phi_0}{NA_0}\left(4p^3 - \frac{3}{2}p\right) = -\frac{4\Phi_0}{NA_0}\left(p^3 - \frac{3}{8}p\right) \tag{3.16}$$

From Eq. (3.16), it can be seen that the optimum defocus is 3/8 of the distance between the paraxial and marginal ray foci. This value is different to that derived for the optimisation of the transverse aberration itself. It should be understood that the optimisation of the transverse aberration and the OPD, although having the same ultimate purpose in minimising the aberration, nonetheless produce different results. Indeed, in the optimisation of optical designs, one is faced with a choice of minimising either the geometrical spot size (transverse aberration) or OPD in the form of rms WFE. The rationale behind this selection will be considered in later chapters when we examine measures of image quality, as applied to optical design.

The balanced defocus, as illustrated in Eq. (3.15) does significantly reduce the rms OPD. In fact, it reduces the OPD by a factor of four. Resultant rms values are set out in Eq. (3.17).

$$\Phi_{RMS} = \frac{\Phi_0}{2\sqrt{5}} \quad \text{(uncompensated)} \qquad \Phi_{RMS} = \frac{\Phi_0}{8\sqrt{5}} \quad \text{(compensated)} \tag{3.17}$$

3.4 General Third Order Aberration Theory

Armed with a simple understanding of the basic concepts that lie behind the description of third order aberration, we can proceed to a more general and more powerful analysis. This analysis relies on a theoretical treatment of OPD as a measure of aberration. As pointed out earlier, although the lowest order aberration (beyond the paraxial approximation) has a fourth order dependence upon pupil function, this theory is still referred to as third order aberration theory. In the example we have hitherto considered, we have analysed an on axis object located at the infinite conjugate. For the more general treatment, we must consider off-axis objects with the chief ray having some non-zero field angle with respect to the optical axis. In addition, the object may have an arbitrary axial location and we must also consider the axial position of the pupil.

This third order theory is referred to as ***Gauss-Seidel aberration theory*** and is of general applicability to optical systems of arbitrary complexity. There is, however, one important constraint. The theory assumes that the entire geometry, component surfaces and so on, is circularly symmetric about the optical axis. In formulating the theory, we assume that the object presents a non-zero field angle, θ, with respect to the optic axis which is assumed to be oriented along the z axis. The chief ray is tilted by rotation about the x axis, so the object is offset from the optical axis in the y direction. The third order aberrations are to be expressed in terms of the field angle, θ, and the normalised pupil function, p. However, in this instance, because of the non-zero field angle, the rotational symmetry of the pupil is removed, so that separate x and y components of the pupil function, p_x, p_y, must be introduced.

The assumption in the Gauss-Seidel theory is that the underlying third order aberrations in a symmetrical optical system are themselves symmetrical and proportional to the fourth power of the pupil function. However, the finite field angle will effectively introduce an offset in the effective y component of the pupil function, Δp_y, at some arbitrary optical component. This is illustrated in Figure 3.10 which shows generically how such an offset may be visualised.

What is suggested by Figure 3.10 is that if a co-ordinate transformation is applied in y that is proportional to the field angle, θ, then the ray fan can be made symmetrical about this new optical axis. That is to say, in Figure 3.10b, any aberration generated would, in terms of OPD, simply be proportional to p^4, ***with respect to the new axis***. In arguing that the required offset is proportional to θ, rather than some other trigonometrical function, we are making an approximation based on linearization in θ. This is justified for third order analysis, since any error produced would only be visible in higher order aberration terms (than third order). In Figure (3.10), the pupil is shown at the optical surface under consideration. However, this is not a necessary condition; wherever the pupil is located a symmetrical ray fan may be produced by simple offset of the co-ordinate system in the Y axis.

Thus, by the argument presented here, any third order aberration may be represented by a pupil dependence of p^4 augmented by a shift in the y component of the pupil function, Δp_y that is proportional to the field

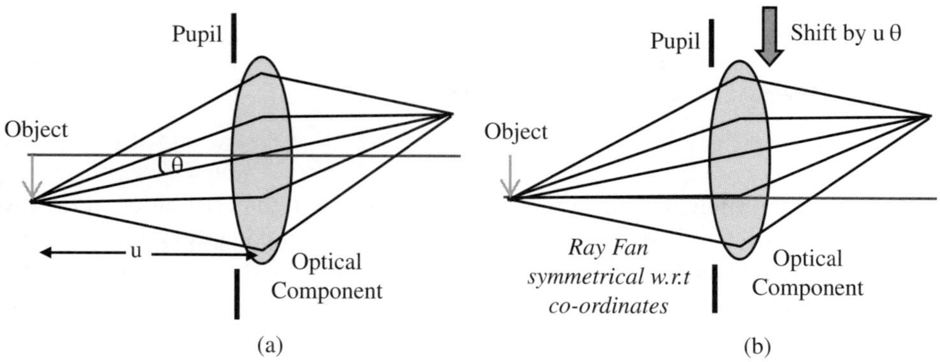

Figure 3.10 (a) Generic layout. (b) Layout with y co-ordinate transformation.

angle, θ. This is set out in Eq. (3.18), which describes the WFE, Φ, in terms of θ, p, and p_y. From this point, we use WFE, rather than OPD as the key descriptor, as we are describing OPDs across the entire pupil. The offset pupil is now denoted by p'

$$\Phi = \Phi_0 p'^4 = \Phi_0(p_x^2 + (p_y + c\theta)^2)^2 \tag{3.18}$$

c is a constant of proportionality for the pupil offset.

Equation (3.18) may be expanded as follows:

$$\Phi = \Phi_0(p_x^2 + p_y^2 + 2cp_y\theta + c^2\theta^2)^2 = \Phi_0(p^2 + 2cp_y\theta + c^2\theta^2)^2 \tag{3.19}$$

Finally expanding Eq. (3.19) gives an expression for all third order aberrations:

$$\Phi = \Phi_0(p^4 + 4cp^2p_y\theta + 4c^2p^2\theta^2 + 2c^2(p_y^2 - p_x^2)\theta^2 + 4c^3p_y\theta^3 + c^4\theta^4) \tag{3.20}$$

Equation (3.20) contains six distinct terms describing the WFE across the pupil. However, the final term, $c^4\theta^4$, for a given field position, simply describes a constant offset in the optical path or phase of the rays originating from a particular point. That is to say, for a specific ray bundle, no OPD or violation of Fermat's principle could be ascribed to this term, when the difference with respect to the chief ray is calculated. Therefore, the final term in Eq. (3.20) cannot describe an optical aberration. We are thus left with five distinct terms describing third order aberration, each with a different dependence with respect to pupil function and field angle. These are the so called five third order ***Gauss-Seidel aberrations***. Of course, in terms of the WFE dependence, all terms show a fourth order dependence with respect to a combination of pupil function and field angle. That is to say, the sum of the exponents in p and in θ must always sum to 4.

3.5 Gauss-Seidel Aberrations

3.5.1 Introduction

In this section we will describe each of the fundamental third order aberrations in turn. Re-iterating Eq. (3.20) below, it is possible to highlight each of the aberration terms:

$$\Phi = \Phi_0\left(\underbrace{p^4}_{\text{Spherical Aberration}} + \underbrace{4cp^2p_y\theta}_{\text{Coma}} + \underbrace{4c^2p^2\theta^2}_{\text{Field Curvature}} + \underbrace{2c^2(p_y^2 - p_x^2)\theta^2}_{\text{Astigmatism}} + \underbrace{4c^3p_y\theta^3}_{\text{Distortion}} + c^4\theta^4\right)$$

We will now describe each of these five terms in turn.

3.5.2 Spherical Aberration

The first term, spherical aberration, has a simple fourth order dependence upon pupil function and no dependence upon field. This is illustrated in Eq. (3.21):

$$\Phi_{SA} = \Phi_0 p^4 \tag{3.21}$$

This aberration shows no dependence upon field angle and no dependence upon the orientation of the ray fan. Since, in the current analysis and for a non-zero field angle, the object is offset along the y axis, then the pupil orientation corresponding to p_y defines the tangential ray fan and the pupil orientation corresponding to p_x defines the sagittal ray fan. This is according to the nomenclature set out in Chapter 2. So, the aberration is entirely symmetric and independent of field angle. In fact, the opening discussion in this chapter was based upon an illustration of spherical aberration.

Spherical aberration characteristically produces a circular blur spot. The transverse aberration may, of course, be derived from Eq. (3.21) using Eq. (3.12). For completeness, this is re-iterated below:

$$t_{SA} = 4\frac{\Phi_0}{NA_0}p^3 \tag{3.22}$$

A 2D geometrical plot, of ray intersection at the paraxial focal plane, as produce by an evenly illuminated entrance pupil is referred to as a ***geometrical point spread function***. Due to the symmetry of the aberration, this spot is circular. However, since the transformation in Eq. (3.22) is non-linear, the blur spot associated with spherical aberration is non uniform. For spherical aberration alone (no defocus or other aberrations), the density of the geometrical point spread function is inversely proportional to the pupil function, p. That is to say, spherical aberration manifests itself as a blur spot with a pronounced peak at the centre, with the density declining towards the periphery. This is illustrated in Figure 3.11. The spot, as shown in Figure 3.11, with a pronounced central maximum, is characteristic of spherical aberration and should be recognised as such by the optical designer.

As suggested earlier, the size of this spot can be minimised by moving away from the paraxial focus position. The ray fan and OPD fan for this aberration look like those illustrated in Figures 3.3 and 3.8. Overall, the characteristics of spherical aberration and the balancing of this aberration is very much as described in the treatment of generic third order aberration, as set out earlier.

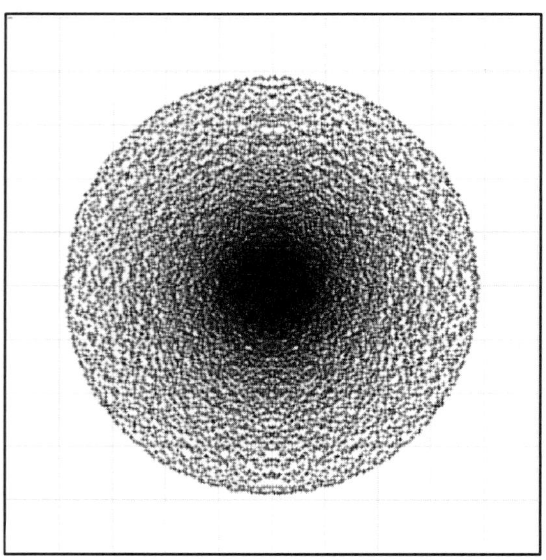

Figure 3.11 Geometrical spot associated with spherical aberration.

3.5.3 Coma

The second term, coma, has an WFE that is proportional to the field angle. Its pupil dependence is third order, but it is not symmetrical with respect to the pupil function. The WFE associated with coma is as below:

$$\Phi_{CO} = 4\Phi_0 c p^2 p_y \theta \tag{3.23}$$

In the preceding discussions, the transverse aberration has been presented as a scalar quantity. This is not strictly true, as the ray position at the paraxial focus is strictly a vector quantity that can only be described completely by an x component, t_x and a y component t_y. Equation (3.12) should strictly be rendered in the following vectorial form:

$$t_x = \frac{1}{NA_0}\frac{\partial(OPD)}{\partial p_x} \qquad t_y = \frac{1}{NA_0}\frac{\partial(OPD)}{\partial p_y} \tag{3.24}$$

The transverse aberration relating to coma may thus be written out as:

$$t_{xCO} = \frac{8\Phi_0}{NA_0} c p_x p_y \theta \qquad t_{yCO} = \frac{4\Phi_0}{NA_0} c(p_x^2 + 3p_y^2)\theta \tag{3.25}$$

From the perspective of both the OPD and ray fans the behaviour of the tangential (y) and sagittal ray fans are entirely different. As an optical designer, the reader should ultimately be familiar with the form of these fans and learn to recognise the characteristic third order aberrations. For a given field angle, the tangential OPD fan ($p_x = 0$) shows a cubic dependence upon pupil function, whereas, for the sagittal ray fan ($p_y = 0$), the OPD is zero. The OPD fan for coma is shown below in Figure 3.12.

The picture for the ray fans is a little more complicated. For both the tangential and sagittal ray fans, there is no component of transverse aberration in the x direction. On the other hand, for both ray fans, there is a quadratic dependence with respect to pupil function for the y component of the transverse aberration. The problem, in essence, it that transverse aberration is a vector quantity. However, when ray fans are computed for optical designs they are presented as scalar plots for each (tangential and sagittal) ray fan. The convention,

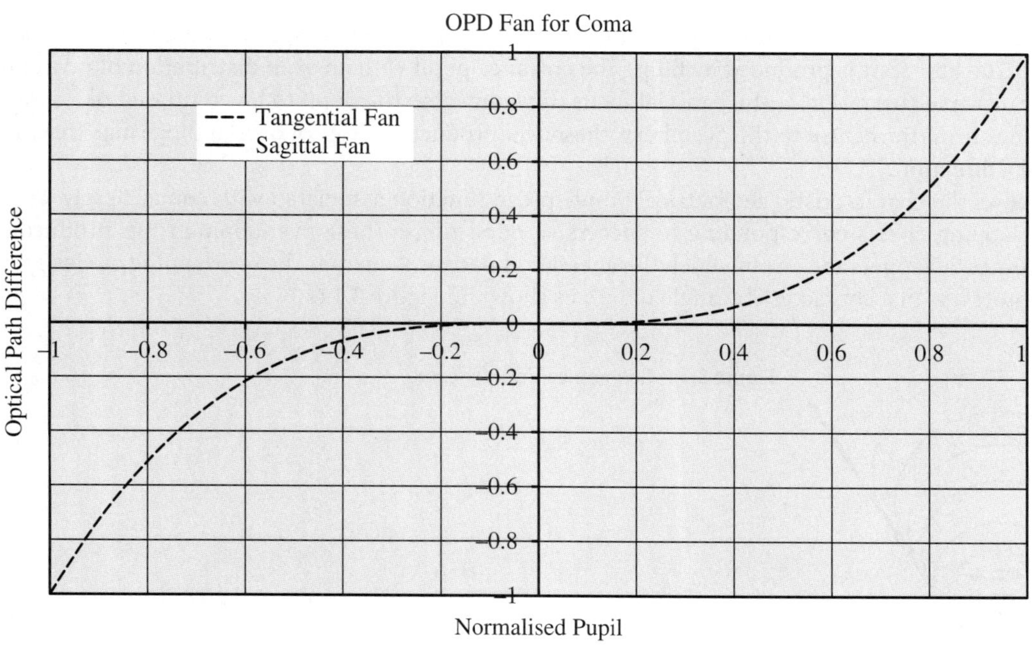

Figure 3.12 OPD fan for coma.

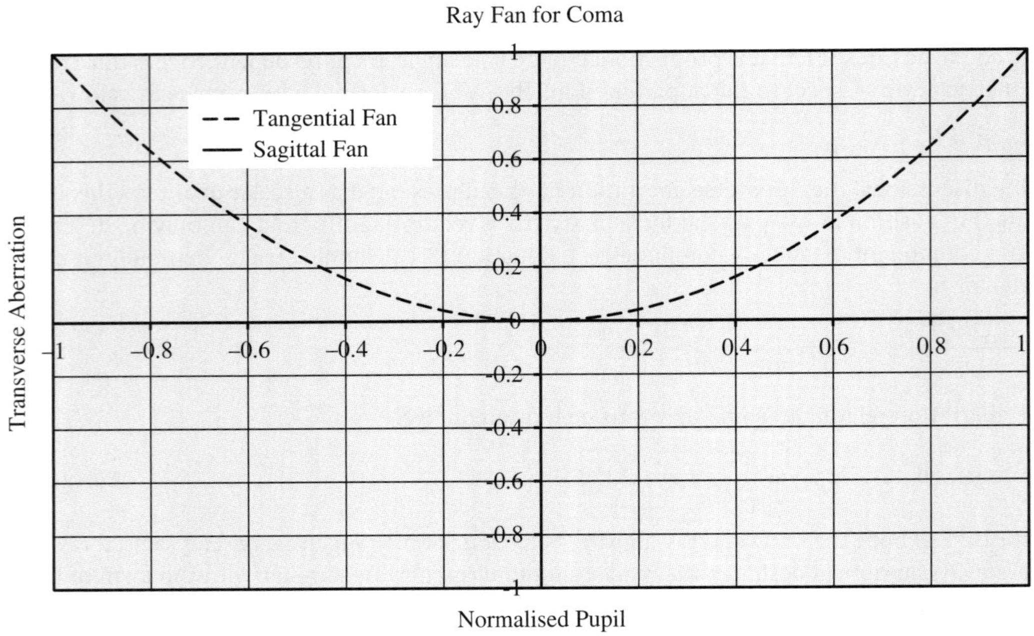

Figure 3.13 Ray fan for coma.

therefore, is to plot only the y (tangential) component of the aberration in a tangential ray fan, and only the x (sagittal) component of the aberration in a sagittal ray fan. With this convention in mind, the tangential ray fan shows a quadratic variation with respect to pupil function, whereas there is no transverse aberration for the sagittal ray fan. Tangential and sagittal ray fan behaviour is shown in Figure 3.13 which shows relevant plots for coma.

Since the (vector) transverse aberration for coma is non-symmetric, the blur spot relating to coma has a distinct pattern. The blur spot is produced by filling the entrance pupil with an even distribution of rays and plotting their transverse aberration at the paraxial focus. If we imagine the pupil to be composed of a series of concentric rings from the centre to the periphery, these will produce a series of overlapping rings that are displaced in the y direction.

Figure 3.14 shows the characteristic geometrical point spread function associated with coma, clearly illustrating the overlapping circles corresponding to successive pupil rings. These overlapping rings produce a characteristic comet tail appearance from which the aberration derives its name. The overlapping circles produce two asymptotes, with a characteristic angle of 60°, as shown in Figure 3.14.

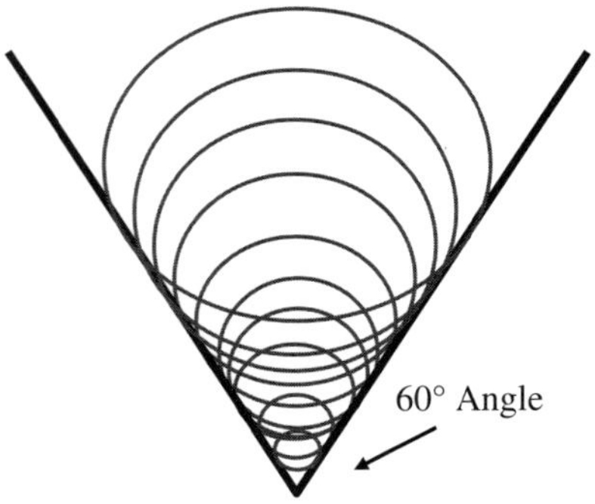

Figure 3.14 Geometrical spot for coma.

To see how these overlapping circles are formed, we introduce an additional angle, the ray fan angle, ϕ, which describes the angle that the plane of the ray fan makes with respect to the y axis. For the tangential ray fan, this angle is zero. For the sagittal ray fan, this angle is 90°. We can now describe the individual components of the pupil function, p_x and p_y in terms of the magnitude of the pupil function, p, and the ray fan angle, ϕ:

$$p_x = p \sin \phi \qquad p_y = p \cos \phi \tag{3.26}$$

From (3.25) we can express the transverse aberration components in terms of p and ϕ. This gives:

$$t_x = Ap^2 \sin 2\varphi \qquad t_y = Ap^2(2 + \cos 2\varphi) \tag{3.27}$$

A is a constant

It is clear from Eq. (3.27) that the pattern produced is a series of overlapping circles of radius $A\sqrt{2p^2}$ offset in y by $2Ap^2$. Coma is not an aberration that can be ameliorated or balanced by defocus. When analysing transverse aberration, the impact of defocus is to produce an odd (anti-symmetrical) additional contribution with respect to pupil function. The transverse aberration produced by coma, is, of course, even with respect to pupil function, as shown in Figure 3.12. Therefore, any deviation from the paraxial focus will only increase the overall aberration.

Another important consideration with coma is the location of the geometrical spot ***centroid***. This represents the mean ray position at the paraxial focus for an evenly illuminated entrance pupil taken with respect to the chief ray intersection. The centroid locations in x and y, C_x, and C_y, may be defined as follows.

$$C_x = \langle t_x \rangle \qquad C_y = \langle t_y \rangle \tag{3.28}$$

By symmetry considerations, the coma centroid is not displaced in x, but it is displaced in y. Integrating over the whole of the pupil function, p (from 0 to 1) and allowing for a weighting proportional to p (the area of each ring), the centroid location in y, Cy may be derived from Eq. (3.27):

$$C_y = 2\int_0^1 t_y p dp = 2\int_0^1 2Ap^3 dp = A \tag{3.29}$$

(the term cos2ϕ is ignored as its average is zero)

So, coma produces a spot centroid that is displaced in proportion to the field angle. The constant A is, of course, proportional to the field angle.

3.5.4 Field Curvature

The third Gauss-Seidel term produced is known as field curvature. The OPD associated with field curvature is second order in both field angle and pupil function. Furthermore, there is no dependence upon ray fan angle, as the WFE is circularly symmetric. Unlike in the case for coma, behaviour is identical for the tangential and sagittal ray fans.

$$\Phi_{FC} = 4\Phi_0 c^2 p^2 \theta^2 \tag{3.30}$$

From Eq. (3.30), in the case of a single field point, the effect of a quadratic dependence of WFE on pupil function is to produce a uniform defocus. That is to say, a uniform defocus produces a characteristic quadratic pupil dependence in the WFE. The extent of this defocus is proportional to the square of the field angle, producing a curved surface which intersects the paraxial focal plane at zero field angle – the optical axis. If this field curvature were the only aberration, then this curved surface would produce a perfectly sharp image for all these field points. That is to say, with the presence of field curvature, the ideal focal surface is a curved surface or sphere rather than a plane. This is illustrated in Figure 3.15.

Figure 3.15 shows both the tangential and sagittal focal surfaces (S and T), with the optimum focal surface lying between the two. Ideally, for field curvature, the imaging surface should be curved, following the ideal

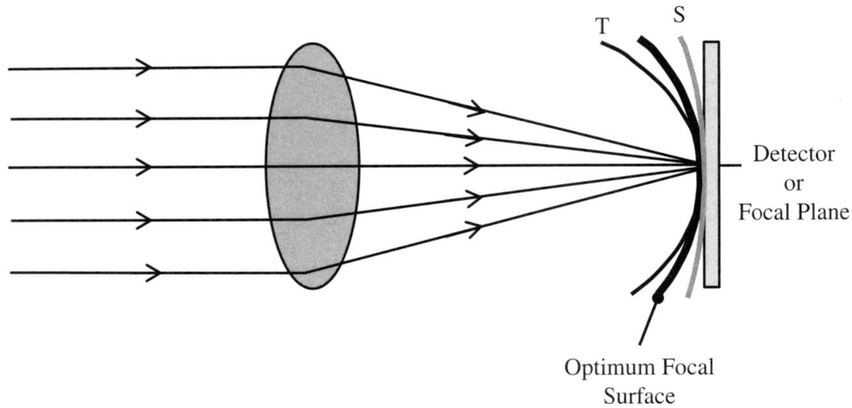

Figure 3.15 Field curvature.

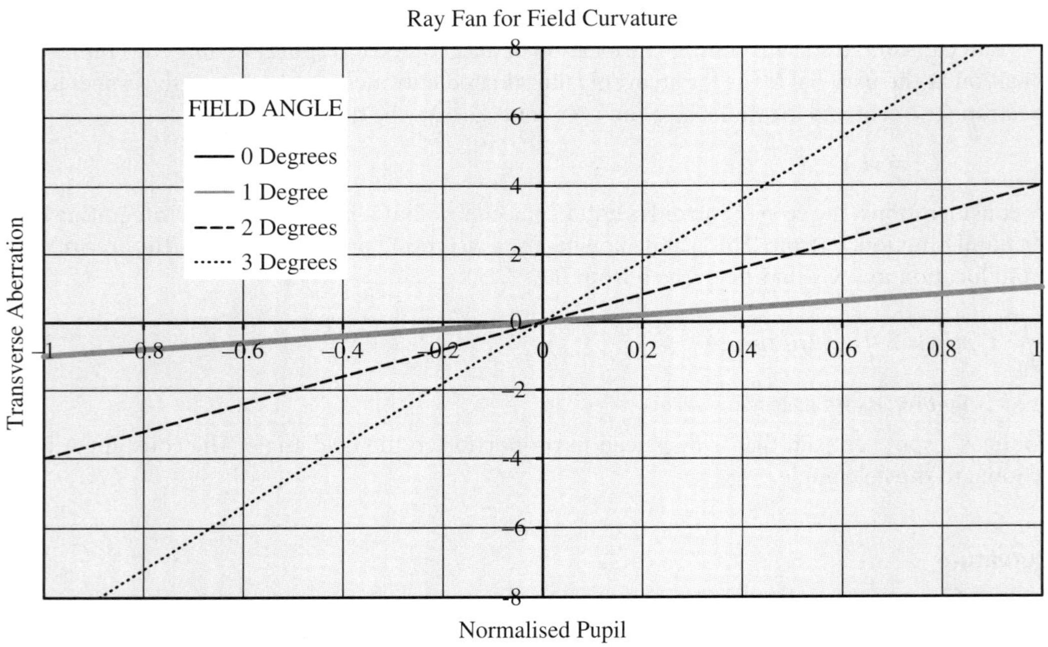

Figure 3.16 Ray fan plots illustrating field curvature.

focal surface. If, for instance, only a plane imaging surface is available, then this need not be located at the paraxial focus. This surface can, in principle, be located at an offset, such that the rms WFE is minimised across all fields. In calculating the rms WFE, this would be weighted according to area across all object space, as represented by a circle centred on the optical axis whose radius is the maximum object height.

Clearly, the OPD fan for field curvature is a series of parabolic curves whose height is proportional to the square of the field angle. There is no distinction between the sagittal and tangential fans. Similarly, the ray fans show a series of linear plots whose magnitude is also proportional to the square of the field angle. A series of ray fan plots for field curvature is shown in Figure 3.16.

In view of the symmetry associated with field curvature, the geometrical spot consists of a uniform blur spot whose size increases in proportion to the square of the field angle. In addition, this spot is centred on the chief ray; unlike in the case for coma, there is no centroid shift with respect to the chief ray.

3.5.5 Astigmatism

The fourth Gauss-Seidel term produced is known as astigmatism, literally meaning '*without a spot*'. Like field curvature, the WFE associated with astigmatism is second order in both field angle and pupil function. It differs from field curvature in that the WFE is non-symmetric and depends upon the ray fan angle as well as the magnitude of the pupil function. That is to say, the behaviour of the tangential and sagittal ray fans is markedly different.

$$\Phi_{AS} = 4\Phi_0 c^2(p_y^2 - p_x^2)\theta^2 = 4\Phi_0 c^2 p^2 \cos 2\varphi\theta^2 \tag{3.31}$$

In some respects, the OPD behaviour is similar to field curvature, in that, for a given ray fan, the quadratic dependence upon pupil function implies a uniform defocus. However, the degree of defocus is proportional to $\cos 2\phi$. Thus, the defocus for the tangential ray fan ($\cos 2\phi = 1$) and the sagittal ray fan ($\cos 2\phi = -1$) are equal and opposite. Clearly, the tangential and sagittal foci are separate and displaced and this displacement is proportional to the square of the field angle. The displacement of the ray fan focus is set out in Eq. (3.32):

$$\Delta f = A \cos 2\phi\theta^2 \tag{3.32}$$

A is a constant

As suggested previously, for a given field angle, the OPD fan would be represented by a series of quadratic curves whose magnitude varies with the ray fan angle. Similarly, the ray fan itself is represented by a series of linear plots whose magnitude is dependent upon the ray fan angle. This is shown in Figure 3.17, which shows the ray fan for a given field angle for both the tangential and sagittal ray fans.

For a general ray, it is possible to calculate the two components of the transverse aberration as a function of the pupil co-ordinates.

$$t_{xAS} = -\frac{8\Phi_0}{NA_0} c^2 p_x \theta^2 \qquad t_{yAS} = \frac{8\Phi_0}{NA_0} c^2 p_y \theta^2 \tag{3.33}$$

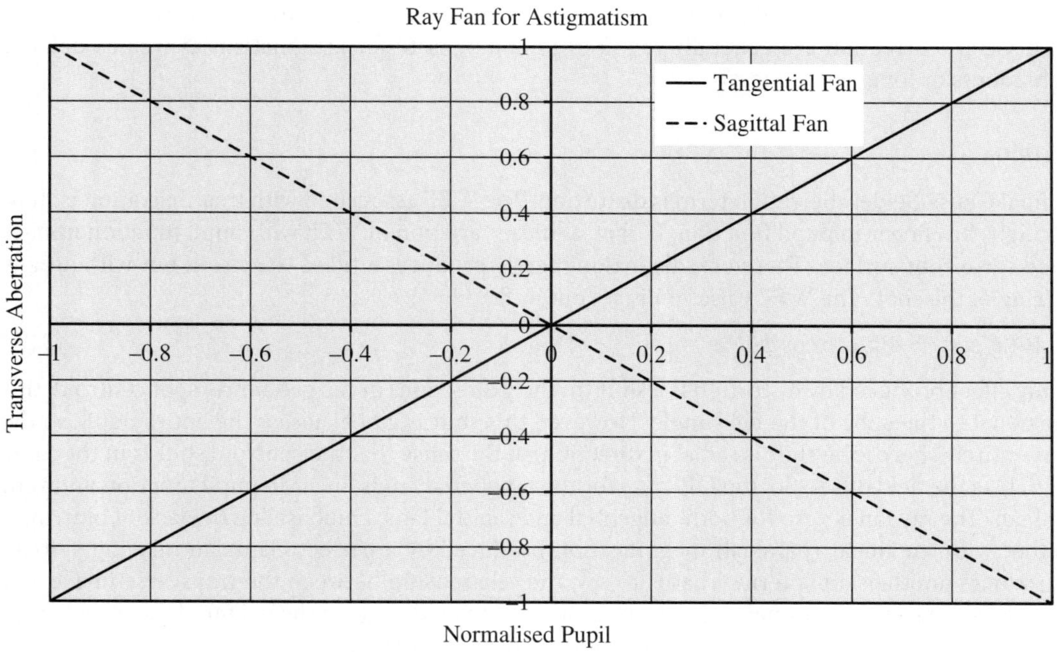

Figure 3.17 Ray fan for astigmatism showing tangential and sagittal fans.

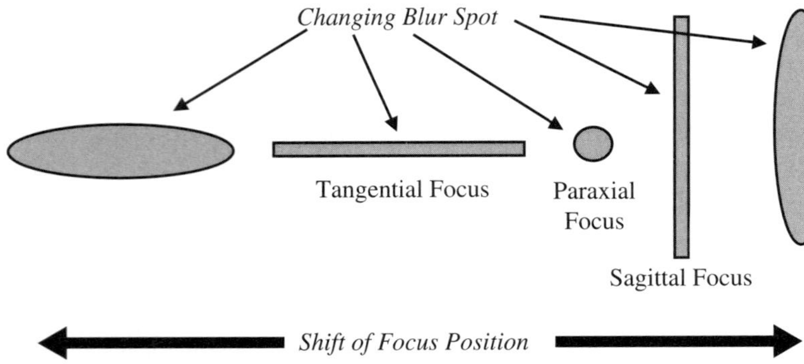

Figure 3.18 Geometric spot vs. defocus for astigmatism.

According to Eq. (3.33), the blur spot produced by astigmatism (at the paraxial focus) is simply a uniform circular disc. Each point in the uniform pupil function simply maps onto a similar point on the blur spot, but with its x value reversed. However, when a uniform defocus is added, similar linear terms (in p) will be added to both t_x and t_y, having both the same magnitude and sign. As a consequence, the relative magnitude of t_x and t_y will change producing a uniform elliptical pattern. Indeed, as mentioned earlier, there are distinct and separate tangential and sagittal foci. At these points, the blur spot is effectively transformed into a line, with the focus along one axis being perfect and the other axis in defocus. This is shown in Figure 3.18.

Due to the even (second order) dependence of OPD upon pupil function, there is no centroid shift evident for astigmatism. For Gauss-Seidel astigmatism, its magnitude is proportional to the square of the field angle. Thus, for an on-axis ray bundle (zero field angle) there can be no astigmatism. This Gauss-Seidel analysis, however, assumes all optical surfaces are circularly symmetric with respect to the optical axis. In the important case of the human eye, the validity of this assumption is broken by the fact that the shape of the human eye, and in particular the cornea, is not circularly symmetrical. The slight cylindrical asymmetry present in all real human eyes produces a small amount of astigmatism, even at zero field angle. That is to say, even for on-axis ray bundles, the tangential and sagittal foci are different for the human eye. For this reason, spectacle lenses for vision correction are generally required to compensate for astigmatism as well as defocus (i.e. short-sightedness or long-sightedness).

3.5.6 Distortion

The fifth and final Gauss-Seidel aberration term is distortion. The WFE associated with this aberration is third order in field angle, but linear in pupil function. In fact, a linear variation of WFE with pupil function implies a flat, but tilted wavefront surface. Therefore, distortion merely produces a tilted wavefront but without any apparent blurring of the spot. The WFE variation is set out in Eq. (3.34).

$$\Phi_{DI} = 4\Phi_0 c^3 p_y \theta^3 = 4\Phi_0 c^3 p \cos\phi\theta^3 \quad (3.34)$$

Thus, the only effect produced by distortion is a shift (in the y direction) in the geometric spot centroid; this shift is proportional to the cube of the field angle. However, this shift is global across the entire pupil, so the image remains entirely sharp. The shift is radial in direction, in the sense that the centroid shift is in the same plane (tangential) as the field offset. So, the OPD fan for the tangential fan is linear in pupil function and zero for the sagittal fan. The ray fan is zero for both tangential and sagittal fans, emphasising the lack of blurring.

Taken together with the linear (paraxial) magnification produced by a perfect Gaussian imaging system, distortion introduces another cubic term. That is to say, the relationship between the transverse image and object locations is no longer a linear one; magnification varies with field angle. If the height of the object is h_{ob} and that of the image is h_{im}, then the two quantities are related as follows:

$$h_{im} = M_0 h_{ob} + \zeta h_{ob}^3 \quad (3.35)$$

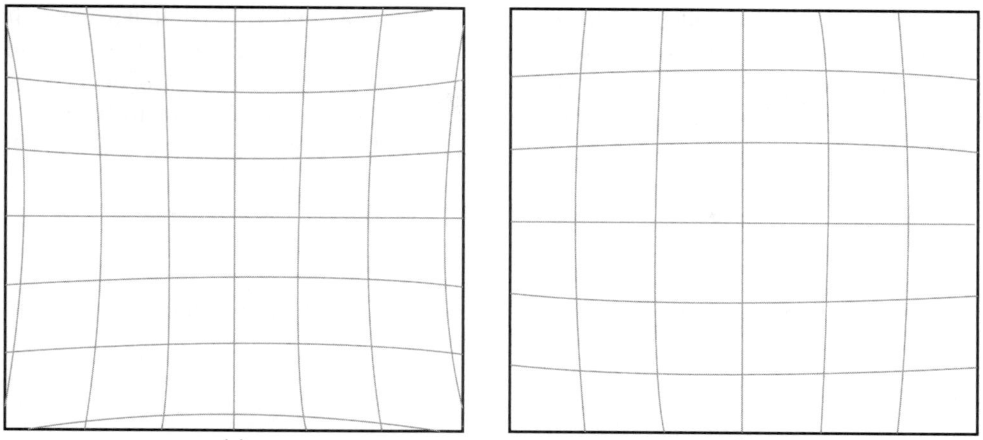

Figure 3.19 (a) Pincushion (positive) distortion. (b) Barrel (negative) distortion.

M_0 is the paraxial magnification; ζ is a constant quantifying distortion

If we denote the x and y components of the object and image location by x_{ob}, y_{ob} and x_{im}, y_{im} respectively, then we obtain:

$$x_{im} = M_0 x_{ob} + \zeta x_{ob}\left(x_{ob}^2 + y_{ob}^2\right) \qquad y_{im} = M_0 y_{ob} + \zeta y_{ob}\left(x_{ob}^2 + y_{ob}^2\right) \tag{3.36}$$

From Eq. (3.35), it is clear that an object represented by straight line that is offset from the optical axis in object space will be presented as a parabolic line in image space. As such, the image is clearly distorted. The sense and character of the distortion is governed by the sign and magnitude of ζ. This is shown in Figures 3.19a,b.

Where ζ and the distortion is positive, the distortion is referred to as pincushion distortion, as suggested by the form shown in Figure 3.19a. On the other hand, if ζ is negative, the resultant image is distended in a form suggested by Figure 3.19b; this is referred to as barrel distortion.

Worked Example 3.1 The distortion of an optical system is given as a WFE by the expression, $4\Phi_0 c^3 p\cos\phi\theta^3$, where Φ_0 is equal to 50 μm and c = 1. The radius of the pupil, r_0, is 10 mm. What is the distortion, expressed as a deviation in percent from the paraxial angle, at a field angle of 15°? From Eq. (3.12) and when expressed as an ***angle***, the transverse aberration generated is given by:

$$\Delta\theta = -\frac{1}{r_0}\frac{d\Phi}{dp} = \frac{4\Phi_0 \cos\phi\theta^3}{r_0}$$

The cosϕ term expresses the fact that the direction of the transverse aberration is in the same plane as that of the object/axis. The proportional distortion is therefore given by:

$$\frac{\Delta\theta}{\theta} = \frac{4\Phi_0\theta^2}{r_0} = \frac{4 \times 5x10^{-2} \times (0.262)^2}{10} = 0.0013$$

(dimensions in mm; angles in radians)

The proportional distortion is therefore 0.13%.

3.6 Summary of Third Order Aberrations

At this stage it will be useful to summarise the five Gauss-Seidel aberrations in terms of the pupil and field dependence of their OPD and ray fans. It should be noted that for all Gauss-Seidel aberrations, the order of the pupil dependence and the order of the field angle dependence sum to four (for the OPD). In particular, it is important for the reader to understand how the different types of aberration vary with both pupil size and field angle. For example, in many optical systems, such as telescopes and microscopes, the range of field angles tend to be significantly smaller than the larger angles subtended to the pupil. Therefore, for such instruments, those aberrations with a higher-order pupil dependence, such as spherical aberration (4) and coma (3), will predominate.

3.6.1 OPD Dependence

The list below sets out the WFE dependence of the five Gauss-Seidel aberrations on pupil function, p, and field angle, θ.

- Spherical Aberration: $\Phi_{SA} \propto p^4$
- Coma: $\Phi_{CO} \propto p^3\theta$
- Field Curvature: $\Phi_{FC} \propto p^2\theta^2$
- Astigmatism: $\Phi_{AS} \propto p^2\theta^2$
- Distortion: $\Phi_{DI} \propto p\theta^3$

To quantify each aberration, we can define a coefficient, K, which describes the magnitude (in units of length) of the aberration. In addition, as well as normalising the pupil function, we can also normalise the field angle by introducing the quantity, h, which represents the ratio, θ/θ_0, the ratio of the field angle to the maximum field angle.

$$\Phi_{SA} = K_{SA}p^4 \tag{3.37}$$

$$\Phi_{CO} = K_{CO}hp^3 \cos\phi \tag{3.38}$$

$$\Phi_{FC} = K_{FC}h^2p^2 \tag{3.39}$$

$$\Phi_{AS} = K_{AS}h^2p^2 \cos 2\phi \tag{3.40}$$

$$\Phi_{DI} = K_{DI}h^3p \cos\phi \tag{3.41}$$

The reader should take particular note of the form of Eq. (3.40). The description of astigmatism here is such that the mean defocus over all orientations of the ray fan is taken to be zero. However, other representations adopt the convention that the defocus is zero for the sagittal ray and the balance of the astigmatism is incorporated into the field curvature. That is to say, in these conventions, the astigmatism is taken to be proportional to $\cos^2\phi$, rather than $\cos2\phi$, as in Eq. (3.40). Of course, in using $\cos^2\phi$, an average defocus of the same form as field curvature is introduced, hence the reason for adopting the convention used here. If the field curvature and astigmatism were redefined according to that convention, then the following revised description would apply:

$$\Phi_{FC} = (K_{FC} - K_{AS})h^2p^2 \tag{3.42}$$

$$\Phi_{AS} = 2K_{AS}h^2p^2 \cos^2\phi \tag{3.43}$$

3.6.2 Transverse Aberration Dependence

The ray fan or transverse aberration dependence upon pupil function and field angle is such that the order of the two variables sum to three, as opposed to four for OPD. The dependence of transverse aberration is listed below:

- Spherical Aberration: $t_{SA} \propto p^3$
- Coma: $t_{CO} \propto p^2\theta$
- Field Curvature: $t_{FC} \propto p\theta^2$
- Astigmatism: $t_{AS} \propto p\theta^2$
- Distortion: $\Phi_{AS} \propto \theta^3$

3.6.3 General Representation of Aberration and Seidel Coefficients

The analysis presented in this chapter has demonstrated the power of using the OPD as a way of describing aberrations. More generally, when expressed as a WFE, it can be used to describe the deviation of a specific wavefront from an ideal wavefront that converges on a specific reference point. As such, this deviation can be used to describe defocus, which shows a quadratic dependence on pupil function and tilt, where the WFE is plane surface that is tilted about the x or y axis (the optical axis being the z axis). The standard representation for describing and quantifying generic WFE and aberration behaviour is shown in Eq. (3.44).

$$WFE = \underbrace{W_{020}p^2}_{\textbf{Defocus}} + \underbrace{W_{111}hp\cos\varphi}_{\textbf{Tilt}} + \underbrace{W_{040}p^4}_{\textbf{Spherical Aberration}} + \underbrace{W_{131}hp^3\cos\varphi}_{\textbf{Coma}} + \underbrace{W_{222}h^2p^2\cos^2\varphi}_{\textbf{Astigmatism}} + \underbrace{W_{220}h^2p^2}_{\textbf{Field Curvature}} + \underbrace{W_{311}h^3p\cos\varphi}_{\textbf{Distortion}} + \ldots\ldots\ldots \tag{3.44}$$

p is the pupil function and h is the object height (proportional to field angle θ); ϕ is the ray fan angle.

In the general term, W_{abc}, 'a' describes the order of the object height (field angle dependence), 'b' describes the order of the pupil function dependence and 'c' describes the dependence on the ray fan angle. The defocus and tilt, are of course paraxial terms. Overall, the dependence of each coefficient is given by Eq. (3.45):

$$WFE = W = W_{abc}h^a p^b \cos^n\phi \tag{3.45}$$

It should be noted that this convention incorporates powers of cosϕ, so the astigmatism term contains some average field curvature. Describing each of the aberration coefficients introduced earlier in terms of these coefficients gives the following:

$$W_{040} = K_{SA} \quad \textit{Spherical Aberration} \tag{3.46}$$

$$W_{131} = K_{CO} \quad \textit{Coma} \tag{3.47}$$

$$W_{220} = K_{FC} - K_{AS} \quad \textit{Field Curvature} \tag{3.48}$$

$$W_{222} = 2_{KAS} \quad \textit{Astigmatism} \tag{3.49}$$

$$W_{311} = K_{DI} \quad \textit{Distortion} \tag{3.50}$$

Another convention exists of which the reader should be aware. These are the so called Seidel coefficients, named after the nineteenth century mathematician, Phillip Ludwig von Seidel, who first elucidated the five monochromatic aberrations. The coefficients are usually denominated, S_I, S_{II}, S_{III}, S_{IV}, and S_V, referring to spherical aberration, coma, astigmatism, field curvature, and distortion. They nominally quantify the WFE, as the other coefficients do, but their magnitude is determined by the size of the blur spot that the aberration creates. The correspondence of these terms is as follows:

$$S_I = 8 \times K_{SA} \quad \textit{Spherical Aberration} \tag{3.51}$$

$$S_{II} = 2 \times K_{CO} \quad \textit{Coma} \tag{3.52}$$

$$S_{III} = 4 \times K_{AS} \quad \textit{Astigmatism} \tag{3.53}$$

$$S_{IV} = 4 \times K_{CO} - 8 \times K_{AS} \qquad \textit{Field Curvature (Petzval Curvature)} \tag{3.54}$$

$$S_V = 2 \times K_{DI} \qquad \textit{Distortion} \tag{3.55}$$

The form of Eq. (3.54) is interesting. When compared to the definition of W_{220} in Eq. (3.48), an additional amount of astigmatism has been compounded with the field curvature. As such, this new representation of field curvature, S_{IV} represents a fundamental and important property of an aberrated optical system and is referred to as the Petzval curvature. Its significance will be discussed more fully in the next chapter.

The treatment of aberrations, thus far, has been entirely generic. We have introduced the five Gauss-Seidel aberrations without specific reference to how they are generated at specific optical surfaces and by individual optical components. This will be discussed in detail in the next chapter. The most important feature of this treatment is that the third order aberrations are additive through a system when described in terms of OPD. That is to say, the five aberrations may be calculated independently at each optical surface and summed over the entire optical system. This analysis is an extremely powerful tool for characterisation of aberration in a complex system.

Further Reading

Born, M. and Wolf, E. (1999). *Principles of Optics*, 7e. Cambridge: Cambridge University Press. ISBN: 0-521-642221.

Hecht, E. (2017). *Optics*, 5e. Harlow: Pearson Education. ISBN: 978-0-1339-7722-6.

Kidger, M.J. (2001). *Fundamental Optical Design*. Bellingham: SPIE. ISBN: 0-81943915-0.

Kidger, M.J. (2004). *Intermediate Optical Design*. Bellingham: SPIE. ISBN: 978-0-8194-5217-7.

Longhurst, R.S. (1973). *Geometrical and Physical Optics*, 3e. London: Longmans. ISBN: 0-582-44099-8.

Mahajan, V.N. (1991). *Aberration Theory Made Simple*. Bellingham: SPIE. ISBN: 0-819-40536-1.

Mahajan, V.N. (1998). *Optical Imaging and Aberrations: Part I. Ray Geometrical Optics*. Bellingham: SPIE. ISBN: 0-8194-2515-X.

Mahajan, V.N. (2001). *Optical Imaging and Aberrations: Part II. Wave Diffraction Optics*. Bellingham: SPIE. ISBN: 0-8194-4135-X.

Slyusarev, G.G. (1984). *Aberration and Optical Design Theory*. Boca Raton: CRC Press. ISBN: 978-0852743577.

Smith, F.G. and Thompson, J.H. (1989). *Optics*, 2e. New York: Wiley. ISBN: 0-471-91538-1.

4

Aberration Theory and Chromatic Aberration

4.1 General Points

In the previous chapter, we developed a generalised description of third order aberration, introducing the five Gauss-Seidel aberrations. The motivation for this is to give the reader a fundamental understanding and a feel for the underlying principles. At the same time, it is fully appreciated that optical system design and detailed analysis of aberrations is underpinned by powerful optical software tools. Nevertheless, a grasp of the underlying principles, including an appreciation of the form of ray fans and optical path difference (OPD) fans, greatly facilitates the application of these sophisticated tools.

The treatment presented here is restricted to consideration of third order aberrations. Before the advent of powerful software analysis tools, the designer was compelled to resort to a much more elaborate and complex analysis, in particular introducing an analytical treatment of higher order aberrations. For all the labour that this would involve, the reader would gain little in terms of a useful understanding that could be applied to current design tools. As the third order aberrations are third order in transverse aberration and fourth order in OPD, so succeeding higher order aberrations are fifth, seventh etc. order in transverse aberration, but sixth, eighth order in OPD. That is to say, aberrations, whose order is expressed conventionally in terms of the transverse aberration, can only be odd. One can re-iterate the analysis of Section 3.4 to generate the form and number of terms involved in the higher order aberrations. This is left to the reader, but it is straightforward to derive the number of distinct terms N_n as a function of aberration order, n:

$$N_n = \frac{1}{8}(n+1)(n+7) \tag{4.1}$$

In concentrating on third order aberrations, we shall, in the remainder of this chapter, seek to determine the impact of refractive surfaces, mirrors, and lenses on all the Gauss-Seidel aberrations. This analysis will proceed, initially, on the assumption that the surface in question lies at the pupil position. Subsequently, the impact of changing the position of the stop will be analysed. Manipulation of the stop position is an important variable in the optimisation of an optical design. The concept of the ***aplanatic geometry*** will be introduced where specific, simple optical geometries may be devised that are wholly free from either spherical aberration (SA) or coma (CO). These aplanatic building blocks feature in many practical designs and are significant because, in many instruments, such as telescopes and microscopes, there is a tendency for spherical aberration and coma to dominate the other aberrations. The elimination of spherical aberration and coma is thus a priority. Furthermore, by the same token, astigmatism (AS) and field curvature (FC) are more difficult to control. In particular, the control of field curvature is fundamentally limited by Petzval curvature, as alluded to in the previous chapter.

Optical Engineering Science, First Edition. Stephen Rolt.

Companion website: www.wiley.com/go/Rolt/opt-eng-sci

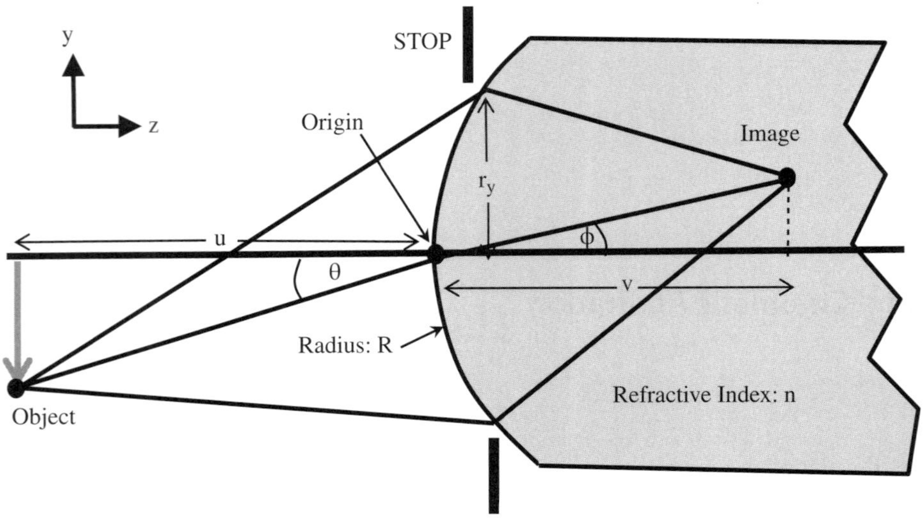

Figure 4.1 Calculation of OPD for refractive surface.

4.2 Aberration Due to a Single Refractive Surface

The analysis of the aberrations of a single refractive surface is based on the computation of the OPD of a generalised field point to the appropriate order (4th) in terms of field angle, θ and ray height, r, at the pupil. For this analysis, we will assume that the pupil is located at the lens surface. In calculating the OPD, we force all rays to go to the paraxial focus and compute the OPD with respect to the chief. Figure 4.1 shows an object with a field angle, θ, located at a distance, u from a ***spherical*** refractive surface of radius R. It must be emphasised, in this instance, that this analysis applies specifically to a spherical surface. In this geometry, it is assumed that the object is displaced from the optical axis in the y direction. The paraxial image is itself located at a distance v from the surface and the position of a ray at the surface (and stop) is described by its components in x and y – h_x and h_y.

The image in this case is the paraxial image and from the paraxial theory, the angle ϕ may be expressed in terms of θ as θ/n. To compute the optical path of a general ray as it passes from object to paraxial image, we need to define the ray co-ordinates at three points:

At image: Position vector $= -u\theta\mathbf{j} - u\mathbf{k}$

At stop: Position vector $= r_x\mathbf{i} + r_y\mathbf{j} + [(r^2/2R) + (r^4/8R^3)]\mathbf{k} \quad r = \sqrt{r_x^2 + r_y^2}$

At paraxial image: Position vector $= v\theta/n\mathbf{j} + v\mathbf{k}$

The z co-ordinate of the stop position is derived from the binomial expansion for the axial sag of a sphere including terms up to the fourth power. In making this approximation, it is assumed that h is significantly less than R. If we were to adopt the paraxial approximation we would only consider the first r^2 term in the expansion. In the case of third order aberration, we need to consider the next term. It is then very straightforward to calculate the total optical path, Φ, for a general ray in passing from object to paraxial image:

$$\Phi = \sqrt{r_x^2 + (r_y + u\theta)^2 + \left(u + \frac{r^2}{2R} + \frac{r^4}{8R^3}\right)^2} + n\sqrt{r_x^2 + \left(r_y - \frac{v\theta}{n}\right)^2 + \left(v - \frac{r^2}{2R} - \frac{r^4}{8R^3}\right)^2} \tag{4.2}$$

The two square root terms represent the optical path of two 'legs' of the journey, with the path through the glass adding a multiplicative factor of n. The next stage of the process is an extension of the paraxial theory. It is assumed that r_x, r_y, and $u\theta$ are all significantly less than u. We can now approximate Φ from Eq. (4.2) using the binomial theorem. In the meantime collecting terms we get:

$$\Phi \approx \sqrt{u^2 + \left(1 + \frac{u}{R}\right) r^2 + \left(\frac{1}{4R^2} + \frac{u}{4R^3}\right) r^4 + 2ur_y\theta + u^2\theta^2}$$
$$+ n\sqrt{v^2 + \left(1 - \frac{v}{R}\right) r^2 + \left(\frac{1}{4R^2} - \frac{v}{4R^3}\right) r^4 - \frac{2v}{n} r_y\theta + \frac{v^2}{n^2}\theta^2}$$

Before deriving the third order aberration terms, we examine the paraxial contribution which contain terms in h up to order r^2.

$$\Phi_{parax} \approx u + nv + \left(\frac{1}{2u} + \frac{n}{2v} - \frac{n-1}{2R}\right) r^2 \text{ Since } \frac{1}{u} + \frac{n}{v} = \frac{n-1}{R} \quad \Phi_{parax} \approx u + nv \tag{4.3}$$

As one would expect, in the paraxial approximation, the optical path length is identical for all rays. However, for third order aberration, terms of up to order h^4 must be considered. Expanding Eq. (4.2) to consider all relevant terms, we get:

$$\Phi \approx \underbrace{\left[\left(\frac{1}{8uR^2} + \frac{n}{8vR^2} + \frac{1-n}{8R^3}\right) + \frac{1}{8u^3}\left(1 + \frac{u}{R}\right)^2 + \frac{n}{8v^3}\left(1 - \frac{v}{R}\right)^2\right]}_{\text{Spherical Aberration}} r^4 + \underbrace{\left[\frac{1}{2u^2} + \frac{1}{2uR} - \frac{1}{2v^2} + \frac{1}{2vR}\right]}_{\text{Coma}} r^2 r_y\theta$$
$$+ \underbrace{\left[\frac{1}{2u} + \frac{1}{2nv}\right]}_{\text{Astigmatism}} r_y^2\theta^2 + \underbrace{\left[\frac{1}{4u} + \frac{1}{4R} + \frac{1}{4nv} - \frac{1}{4nR}\right]}_{\text{Field Curvature}} r^2\theta^2 \tag{4.4}$$

Four of the five Gauss-Seidel terms are present – spherical aberration, coma, astigmatism, and field curvature. However, clearly there is no distortion. In fact, as will be seen later, distortion can only occur where the stop is not at the surface as it is here. Of course, Eq. (4.4) can be simplified if one considers that u, v, and R are dependent variables, as related in Eq. (4.3). Substituting v for u, and R, we can express the OPD in terms of u and R alone. Furthermore, it is useful, at this stage to split the OPD contributions in Eq. (4.4) into Spherical Aberration (SA), Coma (CO), Astigmatism (AS), and Field Curvature (FC). With a little algebraic manipulation this gives:

$$\Phi_{SA} = -\left(\frac{n-1}{8n^2}\right)\left[\left(\frac{1}{u} + \frac{1}{R}\right)^2 \left(\frac{(n+1)}{u} + \frac{1}{R}\right)\right] r^4 \text{ (Spherical Aberration)} \tag{4.5a}$$

$$\Phi_{CO} = -\frac{(n-1)}{2n^2}\left(\left(\frac{1}{u} + \frac{1}{R}\right)\left(\frac{(n+1)}{u} + \frac{1}{R}\right)\right) r^2 r_y\theta \text{ (Coma)} \tag{4.5b}$$

$$\Phi_{AS} = -\frac{(n-1)}{2n^2}\left[\frac{(n+1)}{u} + \frac{1}{R}\right] r_y^2\theta^2 \text{ (Astigmatism)} \tag{4.5c}$$

$$\Phi_{FC} = -\frac{(n^2-1)}{4n^2}\left[\frac{1}{u} + \frac{1}{R}\right] r^2\theta^2 \text{ (Field Curvature)} \tag{4.5d}$$

4.2.1 Aplanatic Points

It is worthwhile, at this juncture, to examine the four expressions in Eqs. (4.5a)–(4.5d) in some detail and, in particular, those for spherical aberration and coma. Before examining these expressions further, it is worthwhile to cast them in the form outlined in Chapter 3:

$$K_{SA} = -\left(\frac{n-1}{8n^2}\right)\left[\left(\frac{1}{u} + \frac{1}{R}\right)^2 \left(\frac{(n+1)}{u} + \frac{1}{R}\right)\right] r_0^4 \tag{4.6a}$$

$$K_{CO} = -\frac{(n-1)}{2n^2}\left(\left(\frac{1}{u} + \frac{1}{R}\right)\left(\frac{(n+1)}{u} + \frac{1}{R}\right)\right) r_0^3\theta \tag{4.6b}$$

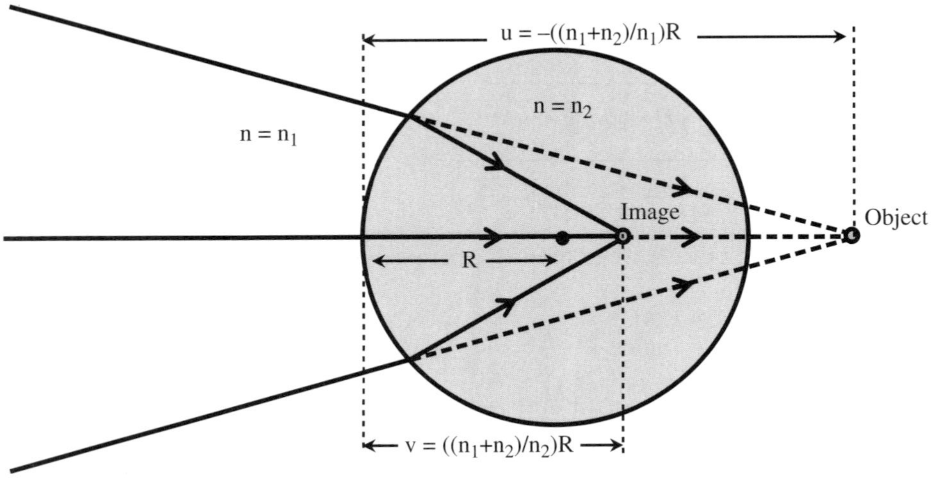

Figure 4.2 Aplanatic points for refraction at single spherical surface.

There is a clear pattern in these expressions in that both spherical aberration and coma can be reduced to zero for specific values of the object distance, u. Examining Eqs. (4.6a) and (4.6b), it is evident that this condition is met where $u = -R$. That is to say, where the object is located at the centre of the spherical surface. However, this is a somewhat trivial condition where rays are undeviated by the surface and where the surface would not provide any useful additional refractive power to the system. Most significantly, another condition does exist for $u = -(n+1)R$. Here, for this non-trivial case, both third order spherical aberration and coma are absent. This is the so-called ***aplanatic condition*** and the corresponding conjugate points are referred to as ***aplanatic points*** (Figure 4.2). From Eq. (4.3) we can derive the image distance, v, as $(n+1)R/n$. That is to say, the object is virtual and the image is real if R is positive and vice-versa if R is negative .

To be a little more rigorous, we might suppose that refractive index in object space is n_1 and that in image space is n_2. The location of the aplanatic points is then given by:

$$u_{AP} = -\left(\frac{n_1 + n_2}{n_1}\right) R \quad v_{AP} = \left(\frac{n_1 + n_2}{n_2}\right) R \tag{4.7}$$

Fulfilment of the aplanatic condition is an important building block in the design of many optical systems and so is of great practical significance. As pointed out in the introduction, for those systems where the field angles are substantially less than the marginal ray angles, such as microscopes and telescopes, the elimination of spherical aberration and coma is of primary importance. Most significantly, not only does the aplanatic condition eliminate third order spherical aberration, but it also ***provides theoretically perfect imaging for on axis rays***.

Worked Example 4.1 ***Microscope Objective***

The 'front end' of many high power microscope objectives exploits the principle of single surface aplanatic points through the use of a hyperhemisphere co-located with the object. The hyperhemisphere consists of a sphere that has been truncated at one of the aplanatic points which also coincides with the object location, as illustrated in Figure 4.3.

Using the hyperhemisphere, we wish to create a ×20 microscope objective for a standard optical tube length of 200 mm. In this example, it is assumed that two thirds of the optical power resides in the hyperhemisphere itself; other components collimate the beam. In other words:

$$\frac{1}{f_{hyper}} = \frac{2}{3} \times \frac{1}{f_{objective}}$$

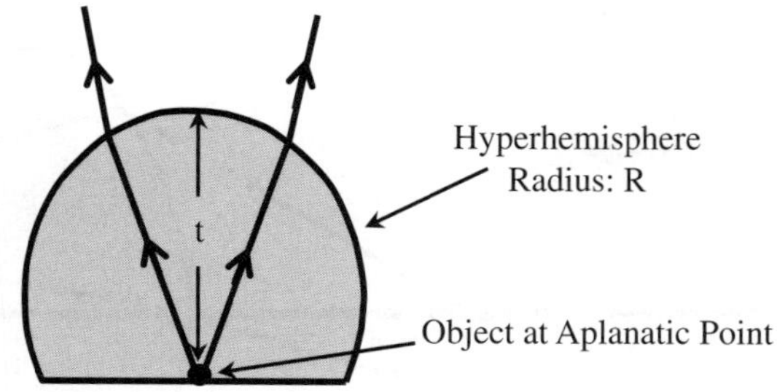

Figure 4.3 Hyperhemisphere objective.

The refractive index of the hyperhemisphere is 1.6. What is the radius, R, of the hyperhemisphere and what is its thickness?

For a tube length of 200 mm, a ×20 magnification corresponds to an objective focal length of 10 mm. As two thirds of the power resides in the hyperhemisphere, then the focal length of the hyperhemisphere must be 15 mm. Inspecting Figure 4.2, it is clear that the thickness of the hyperhemisphere is $-R\times(n+1)/n$, or $-1.625\times R$. To calculate the value of R, we set up a matrix for the system. The first matrix corresponds to refraction at the planar air/glass boundary, the second to translation to the spherical surface and the final matrix to the refraction at that surface. On this occasion, translation to the original reference is not included.

$$M = \begin{bmatrix} 1 & 0 \\ \frac{(1.6-1)}{R} & 1.6 \end{bmatrix} \begin{bmatrix} 1 & 1.625\times R \\ 0 & 1 \end{bmatrix} \begin{bmatrix} 1 & 0 \\ 0 & \frac{1}{1.6} \end{bmatrix} = \begin{bmatrix} 0.391 & 1.016\times R \\ \frac{0.6}{R} & 1 \end{bmatrix}$$

From the above matrix, the focal length is $-R/0.6$ and hence $R = -9.0$ mm. The thickness, t, we know is $-1.625\times R$ and is 14.625. In this sign convention, R is negative, as the sense of its sag is opposite to the direction of travel from object to image space.

The (virtual) image is at $(n+1)\times R$ from the sphere vertex or $2.6\times 9 = 23.4$ mm.

In summary:

$$R = -9\text{ mm};\ t = 14.625\text{ mm};\ v = -23.4\text{ mm}.$$

4.2.2 Astigmatism and Field Curvature

Unlike spherical aberration and coma, there is less scope for correction of astigmatism and field curvature. In Eqs. (4.5c) and (4.5d), astigmatism is corrected at the aplanatic point and field curvature at the radial points. However, the convention used in Eq. (4.5c) to describe astigmatic correction corresponds to zero sagittal ray defocus. On the other hand, using the alternative convention set out in Chapter 3 we have:

$$K_{AS} = -\frac{(n-1)}{4n^2}\left[\frac{(n+1)}{u} + \frac{1}{R}\right] r_0^2\theta_0^2 \tag{4.8a}$$

$$K_{FC} = -\frac{(n-1)}{4n^2}\left[\frac{2(n+1)}{u} + \frac{n+2}{R}\right] r_0^2\theta_0^2 \tag{4.8b}$$

From Eq. (4.8a), it is evident that at the aplanatic condition where $u = -(n+1)R$, the astigmatism vanishes, as does the spherical aberration and coma. It is interesting to see what might happen to the field curvature where this condition is fulfilled:

$$K_{PETZ} = -\frac{(n-1)}{4nR} r_0^2\theta_0^2 \tag{4.9}$$

This is related to the ***Petzval field curvature***, which, by definition, is the field curvature that arises when the astigmatism in the system is zero. Relating this to Eq. 4.8b, then the field curvature may be expressed as:

$$K_{FC} = K_{PETZ} + 2K_{AS} \tag{4.10}$$

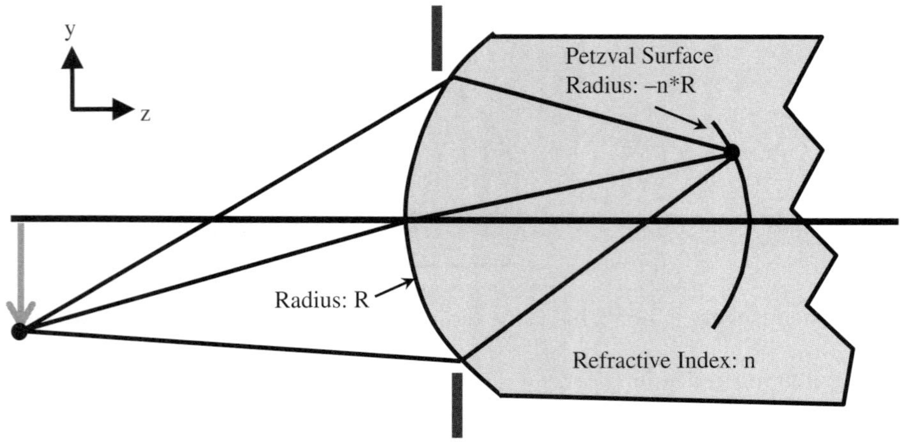

Figure 4.4 Field curvature for single refraction.

It is clear that Eq. (4.9) represents, with its quadratic dependence upon the pupil location, r, a degree of defocus, Δf, or longitudinal aberration, that is quadratic in the field angle. This defocus is given by:

$$\Delta f_{PETZ} = -\frac{(n-1)}{2nR}\theta^2 \tag{4.11}$$

The systematic field dependent defocus can be represented as a spherical surface where the each field point is in focus. The curvature of this surface, C_{PETZ} and equivalent to $1/R_{PETZ}$ where R_{PETZ} is the ***Petzval Radius***, is given by:

$$C_{PETZ} = \frac{1}{R_{PETZ}} = -\frac{(n-1)}{nR} \tag{4.12}$$

The sign is important, in that the Petzval curvature is in the opposite sense to that of the surface itself. This point is illustrated in Figure 4.4.

The most significant point about Petzval curvature is, in common with the underlying wavefront error, that it is additive through a system. To illustrate this, we might consider a system with N surfaces with radius of curvature R_i. The material that follows each surface has a refractive index of n_i. The Petzval curvature associated with the system is simply the sum of the individual curvatures and is referred to as the ***Petzval sum***. This is given by:

$$PetzvalSum = \sum_{1}^{N} -\frac{(n_i - n_{i-1})}{n_i}\frac{1}{R_i} \tag{4.13}$$

The practical implication of Eq. (4.13) is that if a system consists of elements with entirely positive or entirely negative focal power, then that system will always exhibit field curvature. To achieve a flat field, or a zero Petzval sum, then any positive optical elements must be balanced by negative elements elsewhere in the system.

It must be emphasised that the condition for perfect image formation on the Petzval surface applies specifically to the scenario where astigmatism has been removed.

4.3 Reflection from a Spherical Mirror

The third order analysis for a spherical mirror proceeds in very much the same way as the single refractive surface. That is to say, a ray is traced from the object location to the mirror and thence to the paraxial focus regardless as to whether the real ray actually terminates there. The general layout is shown in Figure 4.5. The

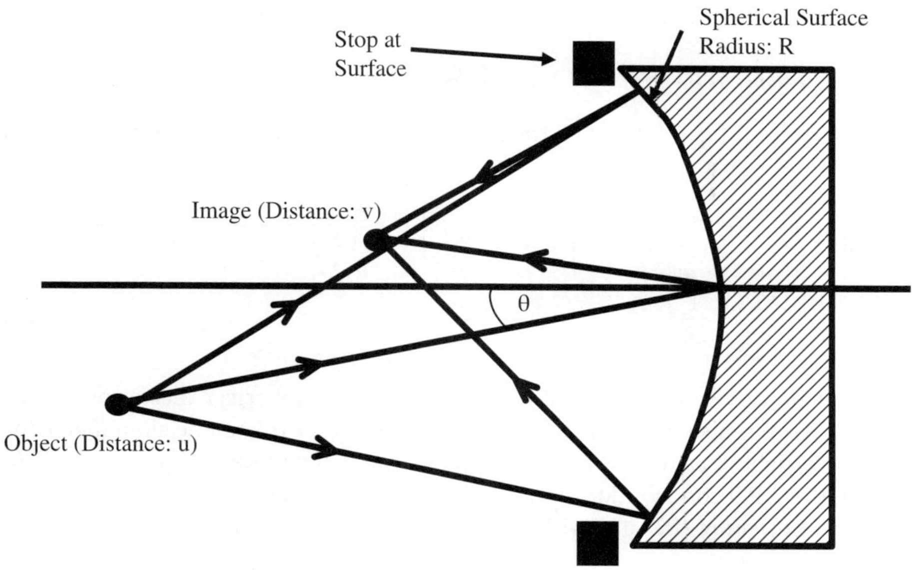

Figure 4.5 Reflection at spherical mirror.

sign convention used here is the same as applied to all previous analyses. That is to say, positive image distance is with the image to the right, and the image distance, as shown in Figure 4.5, is actually negative. However, it must be accepted that, as rays physically converge on this image point, then this image is actually real, despite v being negative. In addition, the same convention is applied to mirror curvature; the mirror depicted in Figure 4.5 has negative curvature.

The analysis proceeds as previously. Firstly, we set out the object and image positions and the ray intercept at the stop.

At image: Position vector $= -u\theta\mathbf{j} - u\mathbf{k}$

At stop: Position vector $= r_x\mathbf{i} + r_y\mathbf{j} + [(r^2/2R) + (r^4/8R^3)]\mathbf{k} \quad r = \sqrt{r_x^2 + r_y^2}$

At paraxial image: Position vector $= -v\theta\mathbf{j} + v\mathbf{k}$

The optical path is given by:

$$\Phi = \sqrt{r_x^2 + (r_y + u\theta)^2 + \left(u + \frac{r^2}{2R} + \frac{r^4}{8R^3}\right)^2} + \sqrt{r_x^2 + (r_y + v\theta)^2 + \left(v - \frac{r^2}{2R} - \frac{r^4}{8R^3}\right)^2} \tag{4.14}$$

Rearranging:

$$\Phi \approx \sqrt{u^2 + \left(1 + \frac{u}{R}\right)r^2 + \left(\frac{1}{4R^2} + \frac{u}{4R^3}\right)r^4 + 2ur_y\theta + u^2\theta^2}$$
$$+ \sqrt{v^2 + \left(1 - \frac{v}{R}\right)r^2 + \left(\frac{1}{4R^2} - \frac{v}{4R^3}\right)r^4 + 2vr_y\theta + v^2\theta^2}$$

In applying the binomial approximation, one needs to be careful with regard to the sign convention. It should be accepted that each of the square root terms in Eq. (4.14) is positive for a real object and real image. That is to say, all rays are physically traced to the appropriate location. In the case of a mirror surface, the definition of a real image corresponds to a negative image distance, u. Once again, we examine the paraxial terms

$$\Phi_{parax} \approx u - v + \left(\frac{1}{2u} - \frac{1}{2v} + \frac{1}{R}\right) r^2 \text{ since } \frac{1}{u} - \frac{1}{v} = -\frac{2}{R} \quad \Phi_{parax} \approx u - v \tag{4.15}$$

As for the refractive surface we expand Eq. (4.14) using the binomial theorem to give terms of the fourth order in OPD.

$$\Phi \approx \underbrace{\left[\left(\frac{1}{8uR^2} - \frac{1}{8vR^2} + \frac{2}{8R^3}\right) + \frac{1}{8u^3}\left(1 + \frac{u}{R}\right)^2 - \frac{1}{8v^3}\left(1 - \frac{v}{R}\right)^2\right] r^4}_{\text{Spherical Aberration}} + \underbrace{\left[\frac{1}{2u^2} + \frac{1}{2uR} - \frac{1}{2v^2} + \frac{1}{2vR}\right] r^2 r_y \theta}_{\text{Coma}}$$
$$+ \underbrace{\left[\frac{1}{2u} - \frac{1}{2v}\right] r_y^2 \theta^2}_{\text{Astigmatism}} + \underbrace{\left[\frac{1}{4u} - \frac{1}{4v} + \frac{1}{2R}\right] r^2 \theta^2}_{\text{Field Curvature}} \tag{4.16}$$

As with the refractive case, four of the five Gauss-Seidel terms are present – spherical aberration, coma, astigmatism, and field curvature. There is also no distortion. As previously, Eq. (4.16) can be simplified considering u, v, and R as dependent variables, as related in Eq. (4.15). We can, once more, express the OPD in terms of u and R alone. Splitting the OPD contributions in Eq. (4.16) into Spherical Aberration (SA), Coma (CO), Astigmatism (AS), and Field Curvature (FC) and with a little algebraic manipulation we have:

$$\Phi_{SA} = -\frac{1}{4R}\left(\frac{1}{u} + \frac{1}{R}\right)^2 r^4 \tag{4.17a}$$

$$\Phi_{CO} = -\frac{1}{R}\left(\frac{1}{u} + \frac{1}{R}\right) r^2 r_y \theta \tag{4.17b}$$

$$\Phi_{AS} = -\frac{1}{R} r_y^2 \theta^2 \tag{4.17c}$$

$$\Phi_{FC} = 0 \tag{4.17d}$$

Equations (4.17a)–(4.17c) bear some striking similarities with respect to those for the refractive surface. In fact, if one substitutes $n = -1$ in the corresponding refractive formulae, one obtains expressions similar to those listed above. Thus, in some ways, a mirror behaves as a refractive surface with a refractive index of minus one. Once again, there are aplanatic points where both spherical aberration and coma are zero. This occurs only where both object and image are co-located at the centre of the spherical surface. The apparent absence of field curvature may appear somewhat surprising. However, the Petzval curvature is non-zero, as will be revealed. We can now cast all terms in the form set out in Chapter 3 and introduce the Lagrange invariant, which is equal to the product of r_0 and θ_0 (the maximum field angle):

$$K_{SA} = -\frac{1}{4R}\left(\frac{1}{u} + \frac{1}{R}\right)^2 r_0^4 \tag{4.18a}$$

$$K_{CO} = -\frac{1}{R}\left(\frac{1}{u} + \frac{1}{R}\right) r_0^2 \left[\frac{\theta}{\theta_0}\right] H \tag{4.18b}$$

$$K_{AS} = -\frac{1}{2R}\left[\frac{\theta}{\theta_0}\right]^2 H^2 \tag{4.18c}$$

$$K_{FC} = -\frac{1}{2R} H^2 \left[\frac{\theta}{\theta_0}\right]^2 \tag{4.18d}$$

The Petzval curvature is simply given by subtracting twice the K_{AS} term in Eq. (4.18c) from the field curvature term in Eq. (4.18d). This gives:

$$K_{PETZ} = \frac{1}{2R} H^2 \left[\frac{\theta}{\theta_0}\right]^2 \tag{4.19}$$

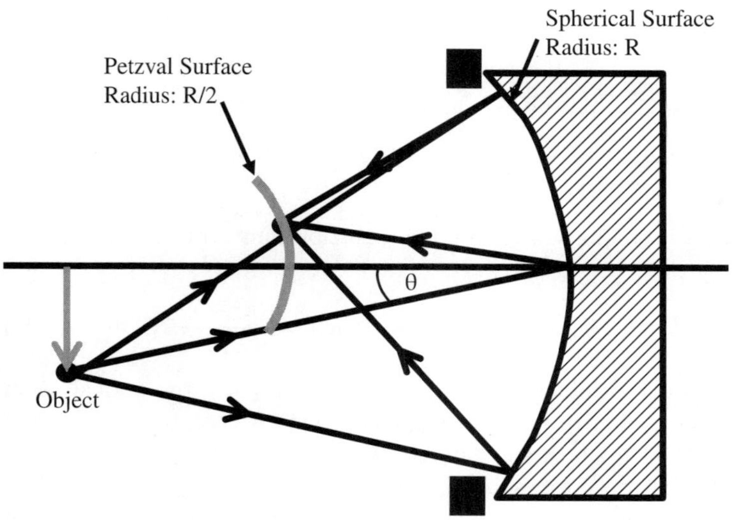

Figure 4.6 Petzval curvature for mirror.

In this instance, the Petzval surface has the same sense as that of the mirror itself. However, the radius of the Petzval surface is actually half that of the original surface. This is illustrated in Figure 4.6.

Calculation of the Petzval sum proceeds more or less as the refractive case. However, there is one important distinction in the case of a mirror system. For a system comprising N mirrors, each successive mirror surface inverts the sense of the wavefront error imparted by the previous mirrors.

$$PetzvalSum = \sum_{1}^{N} (-1)^{i+1} \frac{2}{R_i} \tag{4.20}$$

4.4 Refraction Due to Optical Components

4.4.1 Flat Plate

Equations (4.5a)–(4.5d) give the Gauss-Seidel aberration terms for a spherical reflector. However, for a flat surface, where $1/R = 0$, the aberration is non zero.

$$\Phi_{SA} = -\left(\frac{n^2-1}{8n^2u^3}\right) r^4; \Phi_{CO} = -\frac{(n^2-1)}{2n^2u^2} r^2 r_y \theta; \Phi_{AS} = -\frac{(n^2-1)}{2n^2u} r_y^2 \theta^2; \Phi_{FC} = -\frac{(n^2-1)}{4n^2u} r^2 \theta^2 \tag{4.21}$$

If we now make the approximation that $r_0/u \sim NA_0$ and express all wavefront errors in terms of the normalised pupil function, we obtain the following expressions.

$$\Phi_{SA} = -\left(\frac{n^2-1}{8n^2}\right) NA_0^4 p^4 u; \Phi_{CO} = -\frac{(n^2-1)}{2n^2} NA_0^3 p^2 p_y \theta u$$

$$\Phi_{AS} = -\frac{(n^2-1)}{2n^2} NA_0^2 p_y^2 \theta^2 u; \Phi_{FC} = -\frac{(n^2-1)}{4n^2} p^2 \theta^2 u \tag{4.22}$$

In all expressions, the wavefront error is proportional to the object distance. Equation (4.22) only considers refraction at a single surface. For a flat plate whose thickness is vanishingly small, it is clear that refraction at the second (glass-air) boundary will produce a wavefront error that is equal and opposite to that induced at the first surface. Furthermore, it is also clear that the form of wavefront error contribution will be identical to Eq. (4.22), but reversed in sign. For a glass plate of finite thickness, t, the effective object distance, expressed as the object distance in air, will be given by $u + t/n$. Therefore, the relevant wavefront error contributions at the second surface are given by:

$$\Phi'_{SA} = \left(\frac{n^2-1}{8n^2}\right) NA_0^4 p^4 \left(u + \frac{t}{n}\right); \Phi'_{CO} = \frac{(n^2-1)}{2n^2} NA_0^3 p^2 p_y \theta \left(u + \frac{t}{n}\right)$$

$$\Phi'_{AS} = \frac{(n^2-1)}{2n^2} NA_0^2 p_y^2 \theta^2 \left(u + \frac{t}{n}\right); \Phi'_{FC} = \frac{(n^2-1)}{4n^2} p^2 \theta^2 \left(u + \frac{t}{n}\right) \quad (4.23)$$

The total wavefront error is then simply given by the sum of the two contributions. This is expressed in standard format, as below:

$$K_{SA} = \left(\frac{n^2-1}{8n^3}\right) tNA_0^4; K_{CO} = \frac{(n^2-1)}{2n^3} tNA_0^3\theta_0; K_{AS} = \frac{(n^2-1)}{4n^3} tNA_0^2\theta_0^2; K_{FC} = \frac{(n^2-1)}{2n^3} tNA_0^2\theta_0^2 \quad (4.24)$$

The important conclusion here is that a flat plate will add to system aberration, unless the optical beam is collimated (object at infinite conjugate). This is of great practical significance in microscopy, as a thin flat plate, or 'cover slip' is often used to contain a specimen. A standard cover slip has a thickness, typically, of 0.17 mm. Examination of Eq. (4.24) suggests that this cover slip will add significantly to system aberration. In practice, it is the spherical aberration that is of the greatest concern, as θ_0 is generally much smaller than NA_0 in most practical applications. As a consequence, some microscope objectives are specifically designed for use with cover slips and have built in aberration that compensates for that of the cover slip. Naturally, a microscope objective designed for use with a cover slip will not produce satisfactory imaging when used without a cover slip.

Worked Example 4.2 ***Microscope Cover Slip***

A microscope cover slip 0.17 mm thick is to be used with a microscope objective with a numerical aperture of 0.8. The refractive index of the cover slip is 1.5. What is the root mean square (rms) spherical aberration produced by the cover slip? The aberration is illustrated in Figure 4.7.

From Eq. (4.24):

$$K_{SA} = \left(\frac{n^2-1}{8n^3}\right) tNA_0^4$$

$$n = 1.5; t = 0.17\text{ mm}; NA = 0.8$$

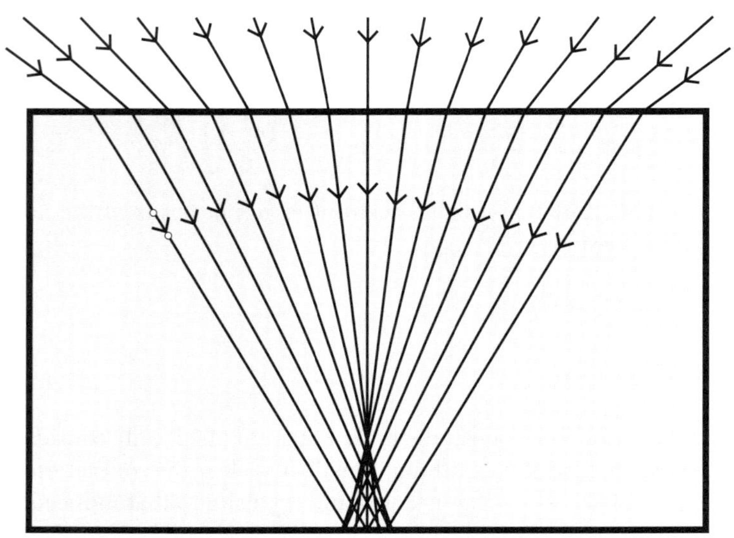

Figure 4.7 Spherical aberration in cover slip.

Substituting the above values we get: $K_{sa} = 0.003\,22$ mm or 3.2 μm.

The wavefront error (in microns) is thus given by:

$$\Phi_{SA} = 3.22 \times p^4$$

where p is the normalised pupil function.

$$\Phi_{RMS} = \frac{\Phi_0}{6\sqrt{5}} \qquad \Phi_{\mathrm{RMS}} = 0.24\ \mu m$$

For reasons that will become apparent later, in practice, wavefront errors are usually expressed as a fraction of some standard wavelength, for example 589 nm. The above wavefront error represents about $0.4 \times \lambda$ when expressed in this way. An rms wavefront error of about $\lambda/14$ is considered consistent with good image quality. This level of aberration is, therefore, significant and measures must be taken (within the objective) to correct for it.

4.4.2 Aberrations of a Thin Lens

We extend the treatment already outlined to analyse a thin lens. A thin lens can be considered as combination of two refractive surfaces, where the distance between the two surfaces is ignored. In practice, this is a reasonable assumption, provided the thickness is much less than the radii of the surfaces in question. Of course, the wavefront error produced by the two surfaces is simply the sum of the aberrations of the individual surfaces. A schematic for the analysis is shown in Figure 4.8.

The wavefront error contribution for the first surface is very easy to compute; it is simply that set out in Eqs. (4.5a)–(4.5d). To compute the contribution for the second surface, one can analyse this using the same methodology as in Section 4.2, but exploiting natural symmetry. That is to say, one can analyse the second surface by rotating the whole surface about the y axis, such that z → −z and x → −x. In this event, for the second surface, $R \to -R_2$, $u \to v$, $\theta \to -\theta$. It is then simply a case of substituting these values into the formulae in Eqs. (4.5a)–(4.5d) and adding the wavefront error contribution of the first surface. The total wavefront error

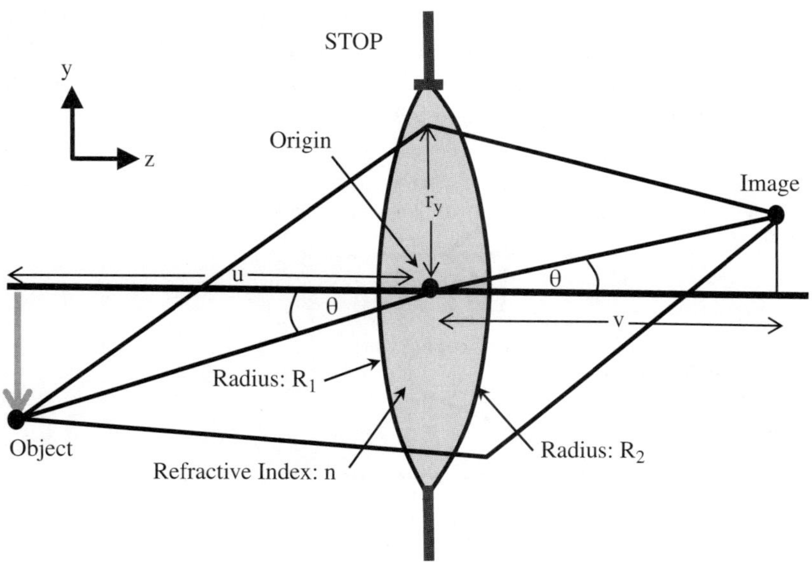

Figure 4.8 Aberration analysis for thin lens.

for the thin lens is then:

$$\Phi_{SA} = -\left(\frac{n-1}{8n^2}\right)\left[\left(\frac{1}{u}+\frac{1}{R_1}\right)^2\left(\frac{(n+1)}{u}+\frac{1}{R_1}\right)+\left(\frac{1}{v}-\frac{1}{R_2}\right)^2\left(\frac{(n+1)}{v}-\frac{1}{R_2}\right)\right]r^4 \tag{4.25a}$$

$$\Phi_{CO} = -\frac{(n-1)}{2n^2}\left(\left(\frac{1}{u}+\frac{1}{R_1}\right)\left(\frac{(n+1)}{u}+\frac{1}{R_1}\right)-\left(\frac{1}{v}-\frac{1}{R_2}\right)\left(\frac{(n+1)}{v}-\frac{1}{R_2}\right)\right)r^2 r_y\theta \tag{4.25b}$$

$$\Phi_{AS} = -\frac{(n-1)}{2n^2}\left[\left(\frac{(n+1)}{u}+\frac{1}{R_1}\right)+\left(\frac{(n+1)}{v}-\frac{1}{R_2}\right)\right]r_y^2\theta^2 \tag{4.25c}$$

$$\Phi_{FC} = -\frac{(n^2-1)}{4n^2}\left[\frac{1}{u}+\frac{1}{R_1}+\frac{1}{v}-\frac{1}{R_2}\right]r^2\theta^2 \tag{4.25d}$$

4.4.2.1 Conjugate Parameter and Lens Shape Parameter

In terms of gaining some insight into the behaviour of a thin lens, the formulae in Eqs. (4.25a)–(4.25d) are a little opaque. It would be somehow useful to express the aberrations of a thin lens directly in terms of its focusing power and some other parameters. The first of these other parameters is the so called **conjugate parameter**, t. The conjugate parameter is defined as below:

$$t = \frac{1/u - 1/v}{1/u + 1/v} = \frac{v-u}{v+u} \tag{4.26}$$

As we are dealing with a thin lens, we can use the thin lens formula to calculate the focal length, f, of the lens:

$$\frac{1}{f} = \left(\frac{1}{u}+\frac{1}{v}\right)$$

This, in turn, leads to expressions for u and v:

$$\frac{1}{u} = \left(\frac{1+t}{2}\right)\left(\frac{1}{f}\right) \quad \frac{1}{v} = \left(\frac{1-t}{2}\right)\left(\frac{1}{f}\right) \tag{4.27}$$

Figure 4.9 illustrates the conjugate parameter schematically. The infinite conjugate is represented by a conjugate parameter of ±1. If the conjugate parameter is $+1$, then the image is at infinity. Conversely, a conjugate parameter of -1 is associated with an object located at the infinite conjugate. In the symmetric scenario where object and image distances are identical, then the conjugate parameter is zero. As illustrated in Figure 4.9, where the conjugate parameter is greater than 1, then the object is real and the image is virtual. Finally, where the conjugate parameter is less than -1, then the object is virtual and the image is real.

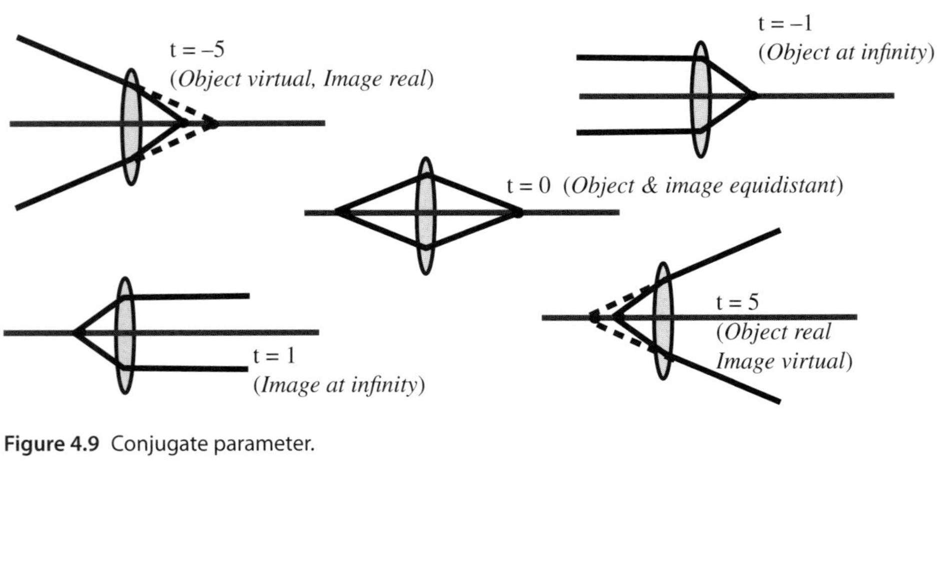

Figure 4.9 Conjugate parameter.

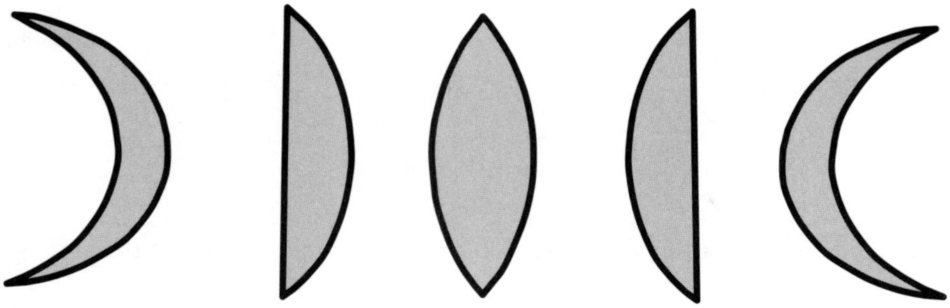

Figure 4.10 Coddington lens shape parameter.

We have thus described object and image location in terms of a single parameter. By analogy, it is also useful to describe a lens in terms of its focal power and a single parameter that describes the shape of the lens. The lens, of course, is assumed to be defined by two spherical surfaces, with radii R_1 and R_2, defining the first and second surfaces respectively. The shape of a lens is defined by the so-called ***Coddington lens shape factor***, s, which is defined as follows:

$$s = \frac{1/R_1 + 1/R_2}{1/R_1 - 1/R_2} = \frac{R_1 + R_2}{R_2 - R_1} \tag{4.28}$$

As before, the power of the lens may be expressed in terms of the lens radii:

$$\frac{1}{f} = (n-1)\left(\frac{1}{R_1} - \frac{1}{R_2}\right)$$

where n is the lens refractive index.

As with the conjugate parameter and the object and image distances, the two lens radii can be expressed in terms of the lens power and the shape factor, *s*.

$$\frac{1}{R_1} = \left(\frac{s+1}{2(n-1)}\right)\left(\frac{1}{f}\right) \quad \frac{1}{R_2} = \left(\frac{s-1}{2(n-1)}\right)\left(\frac{1}{f}\right) \tag{4.29}$$

Figure 4.10 illustrates the lens shape parameter for a series of lenses with positive focal power. For a symmetric, bi-convex lens, the shape factor is zero. In the case of a plano-convex lens, the shape factor is 1 where the plane surface faces the image and is −1 where the plane surface faces the object. A shape factor of greater than 1 or less than −1 corresponds to a ***meniscus lens***. Here, both radii have the same sense, i.e. they are either both positive or both negative. For a shape parameter of greater than 1, the surface with the greater curvature faces the object and for a shape parameter of less than −1, the surface with the greater curvature faces the image. Of course, this applies to lenses with positive power. For (diverging) lenses with negative power, then the sign of the shape factor is opposite to that described here.

4.4.2.2 General Formulae for Aberration of Thin Lenses

Having parameterised the object and image distances and the lens radii in terms of the conjugate parameter, shape parameter, and lens power, we can recast the expressions in Eqs. (4.25a)–(4.25d) in a more generic form. With a little algebraic manipulation, we obtain the following expressions for the Gauss-Seidel aberration of a lens with the stop at the lens surface:

$$\Phi_{SA} = -\left(\frac{1}{32f^3}\right)\left[\left(\frac{n}{n-1}\right)^2 - \left(\frac{n}{n+2}\right)t^2 + \frac{(n+2)}{n(n-1)^2}\left[s + 2\left[\frac{n^2-1}{n+2}\right]t\right]^2\right]r^4 \tag{4.30a}$$

$$\Phi_{CO} = -\frac{1}{4nf^2}\left((2n+1)t + \frac{(n+1)}{(n-1)}s\right) r^2 r_y \theta \tag{4.30b}$$

$$\Phi_{AS} = -\frac{r_y^2}{2f}\theta^2 \tag{4.30c}$$

$$\Phi_{FC} = -\frac{(n+1)}{4nf}r^2\theta^2 \tag{4.30d}$$

Again, casting all expressions in the form set out in Chapter 3, as for the expressions for the mirror we have

$$K_{SA} = -\left(\frac{1}{32f^3}\right)\left[\left(\frac{n}{n-1}\right)^2 - \left(\frac{n}{n+2}\right)t^2 + \frac{(n+2)}{n(n-1)^2}\left[s + 2\left[\frac{n^2-1}{n+2}\right]t\right]^2\right] r_0^4 \tag{4.31a}$$

$$K_{CO} = -\frac{1}{4nf^2}\left((2n+1)t + \frac{(n+1)}{(n-1)}s\right) r_0^2 \left[\frac{\theta}{\theta_0}\right] H \tag{4.31b}$$

$$K_{AS} = -\frac{1}{4f}\left[\frac{\theta}{\theta_0}\right]^2 H^2 \tag{4.31c}$$

$$K_{FC} = -\frac{(2n+1)}{4nf}\left[\frac{\theta}{\theta_0}\right]^2 H^2 \tag{4.31d}$$

Once again, the Petzval curvature is simply given by subtracting twice the K_{AS} term in Eq. (4.31c) from the field curvature term in Eq. (4.31d). This gives:

$$K_{PETZ} = -\frac{1}{4nf}\left[\frac{\theta}{\theta_0}\right]^2 H^2 \tag{4.32}$$

That is to say, a single lens will produce a Petzval surface whose radius of curvature is equal to the lens focal length multiplied by its refractive index. Once again, the Petzval sum may be invoked to give the Petzval curvature for a system of lenses:

$$PetzvalSum = -\sum_{1}^{N}\frac{1}{n_1 f_i} \tag{4.33}$$

It is important here to re-iterate the fact that for a system of lenses, it is impossible to eliminate Petzval curvature where all lenses have positive focal lengths. For a system with positive focal power, i.e. with a positive effective focal length, there must be some elements with negative power if one wishes to 'flatten the field'.

Before considering the aberration behaviour of simple lenses in a little more detail, it is worth reflecting on some attributes of the formulae in Eqs. (4.30a)–(4.30d). Both spherical aberration and coma are dependent upon the lens shape and conjugate parameters. In the case of spherical aberration there are second order terms present for both shape and conjugate parameters, whereas the behaviour for coma is linear. However, the important point to recognise is that the field curvature and astigmatism are independent of both lens shape and conjugate parameter and only depend upon the lens power. Once again, it must be emphasised that this analysis applies only to the situation where the stop is situated at the lens.

4.4.2.3 Aberration Behaviour of a Thin Lens at Infinite Conjugate

We will now look at a simple special case to apply to a thin lens with the stop at the lens. This is the common situation where a lens is being used to focus an object located at the infinite conjugate, such as a telescope objective or a lens focusing a parallel laser beam. From Eq. (4.26), the conjugate parameter, t, is equal to -1. Substituting $t = -1$ into Eq. (4.31a) gives the spherical aberration as:

$$K_{SA} = -\left(\frac{1}{32f^3}\right)\left[\frac{4n^2+n}{(n-1)^2(n+2)} + \frac{(n+2)}{n(n-1)^2}\left[s - 2\left[\frac{n^2-1}{n+2}\right]\right]^2\right] r_0^4 \tag{4.34}$$

The important point to note about Eq. (4.34) is that the spherical aberration can never be equal to zero and that for a positive lens, K_{SA} is always negative. This means that the longitudinal aberration for a positive lens is also negative and that, for all single lenses, more marginal rays are brought to a focus closer to the lens. Whilst Eq. (4.34) asserts that the spherical aberration in this case can never be zero, its magnitude can be minimised for a specific lens shape. Inspection of Eq. (4.34) reveals that this condition is met where:

$$s_{\min} = 2\left[\frac{n^2 - 1}{n + 2}\right] \tag{4.35}$$

This optimum shape factor corresponds to the so-called '***best form singlet***' and is generally available from optical component suppliers, particularly with regard to applications in the focusing of laser beams. For a refractive index of 1.5, the optimum shape factor is around 0.7. This is close in shape to a plano-convex lens. However, it is important to emphasise, that optimum focusing is obtained where the more steeply curved surface is facing the infinite conjugate. Generally, also, where a plano-convex lens is used to focus a collimated beam, the curved surface should face the infinite conjugate. This behaviour is shown in Figure 4.11, which emphasises the quadratic dependence of spherical aberration on lens shape factor.

Coma for the infinite conjugate also depends upon the shape factor. However, in this instance, the dependence is linear. Once more, substituting $t = -1$ into Eq. (4.31b), we get:

$$K_{CO} = -\frac{1}{4nf^2}\left(-(2n+1) + \frac{(n+1)}{(n-1)}s\right) r_0^3\theta \tag{4.36}$$

Unlike in the case for spherical aberration, there exists a shape factor for which the coma is zero. This is simply given by:

$$s_0 = \frac{(2n+1)(n-1)}{(n+1)} \tag{4.37}$$

For a refractive index of 1.5, this minimum condition is met for a shape factor of 0.8. This is similar, but not quite the same as the optimum for spherical aberration. Again, the most curved surface should face the infinite conjugate. Overall behaviour is illustrated in Figure 4.12.

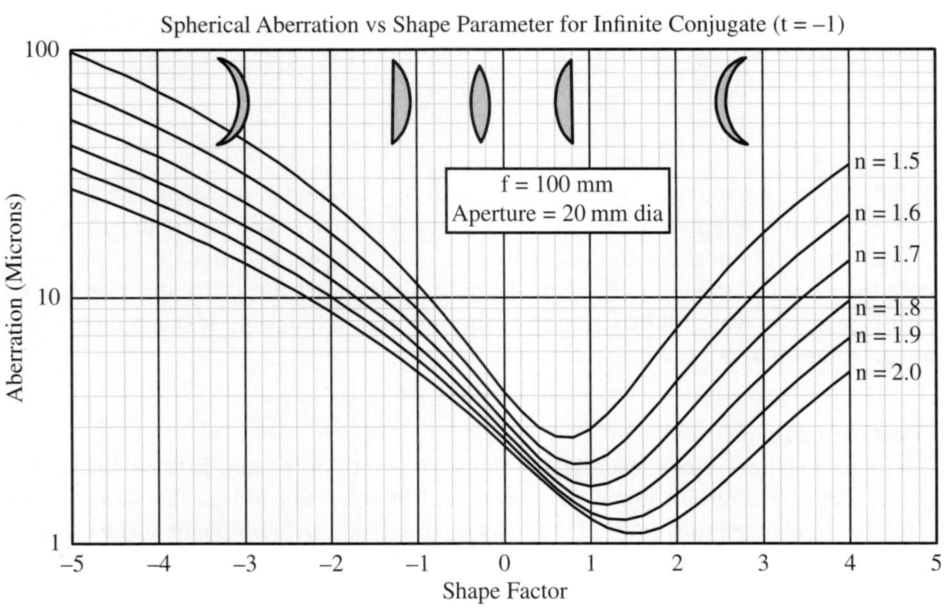

Figure 4.11 Spherical aberration vs. shape parameter for a thin lens.

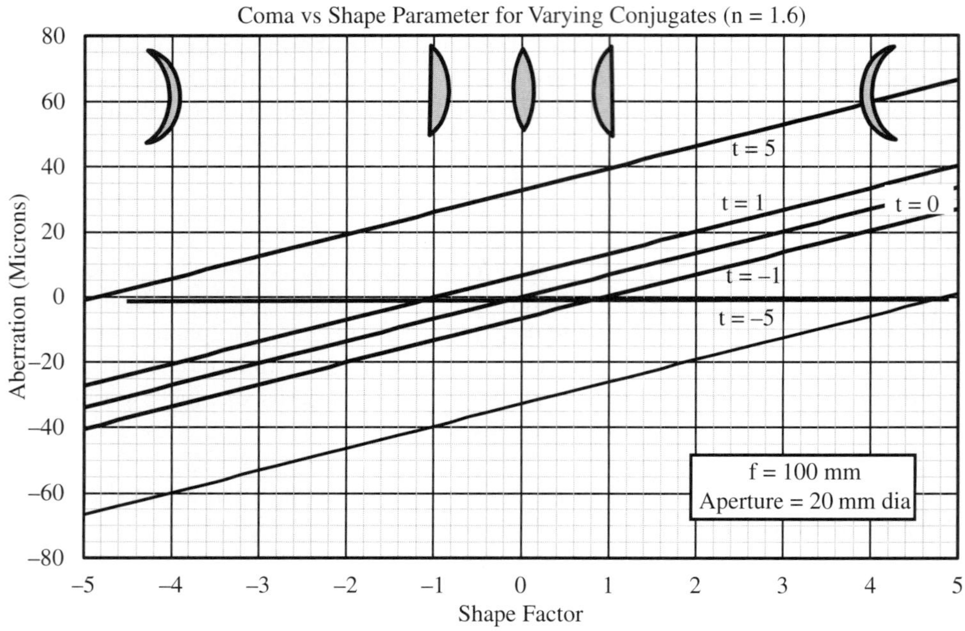

Figure 4.12 Coma vs lens shape for various conjugate parameters.

Once again, this specifically applies to the situation where the stop is at the lens surface. Of course, as stated previously, neither astigmatism nor field curvature are affected by shape or conjugate parameter.

Although it is impossible to reduce spherical aberration for a thin lens to zero at the infinite conjugate, it is possible for other conjugate values. In fact, the magnitude of the conjugate parameter must be greater than a certain specific value for this condition to be fulfilled. This magnitude is always greater than one for reasonable values of the refractive index and so either object or image must be virtual. It is easy to see from Eq. (4.31a) that this threshold value should be:

$$|t| >= \frac{\sqrt{n(n+2)}}{n-1} \tag{4.38}$$

For n = 1.5, this threshold value is 4.58. That is to say for there to be a shape factor where the spherical aberration is reduced to zero, the conjugate parameter must either be less than −4.58 or greater than 4.58. Another point to note is that since spherical aberration exhibits a quadratic dependence on shape factor, where this condition is met, there are two values of the shape factor at which the spherical aberration is zero. This behaviour is set out in Figure 4.13 which shows spherical aberration as a function of shape factor for a number of difference conjugate parameters.

Worked Example 4.3 ***Best form Singlet***

A thin lens is to be used to focus a Helium-Neon laser beam. The focal length of the lens is to be 20 mm and the lens is required to be 'best form' to minimise spherical aberration. The refractive index of the lens is 1.518 at the laser wavelength of 633 nm. Calculate the required shape factor and the radii of both lens surfaces. From Eq. (4.35) we have:

$$s_{min} = 2\left[\frac{n^2-1}{n+2}\right] = 2\times\left[\frac{1.518^2-1}{1.518+2}\right] = 2\times\frac{1.304}{3.518} = 0.742$$

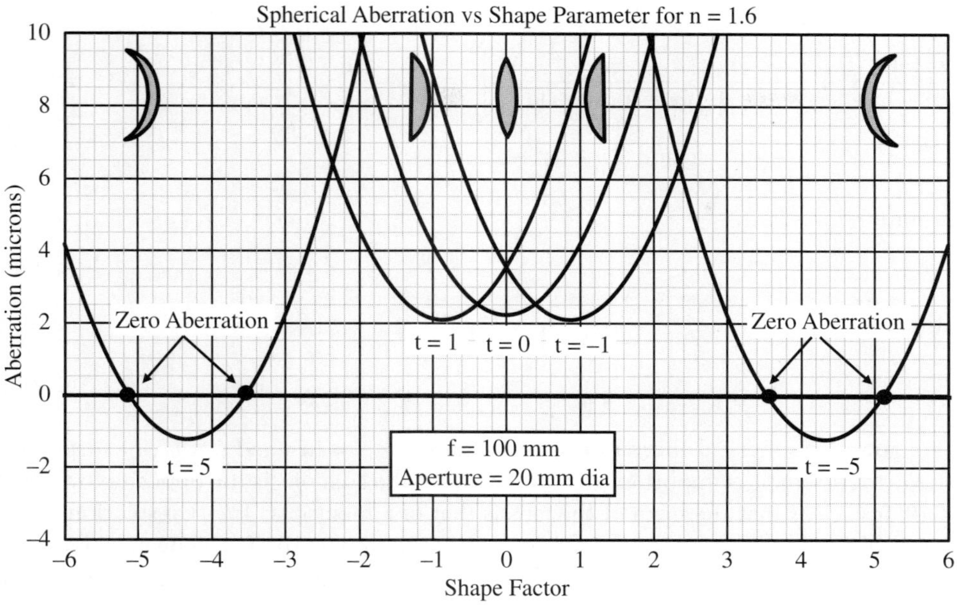

Figure 4.13 Spherical aberration vs shape factor for various conjugate parameter values.

The optimum shape factor is 0.742 and we can use this to calculate both radii given knowledge of the required focal length. Rearranging Eq. (4.29) we have:

$$R_1 = \frac{2(n-1)}{1+s}f \quad R_2 = \frac{2(n-1)}{1-s}f$$

$$R_1 = \frac{2 \times 0.518}{1.742} \times 20 \quad R_2 = \frac{2 \times 0.518}{0.258} \times 20$$

This gives:

$$R_1 = 11.9\text{ mm and } R_2 = 80.2\text{ mm}$$

It is the surface with the greatest curvature, i.e. R1, that should face the infinite conjugate (the parallel laser beam).

4.4.2.4 Aplanatic Points for a Thin Lens

Just as in the case of a single surface, it is possible to find a conjugate and lens shape pair that produce neither spherical aberration nor coma. For reasons outlined previously, it is not possible to eliminate astigmatism or field curvature for a lens of finite power. If the spherical aberration is to be zero, it must be clear that for the aplanatic condition to apply, then either the object or the image must be virtual. Equations (4.31a) and (4.31b) provide two conditions that uniquely determine the two parameters, s and t. Firstly, the requirement for coma to be zero clearly relates s and t in the following way:

$$t = -\frac{(n+1)}{(n-1)(2n+1)}s$$

Setting the spherical aberration to zero and substituting for t we have the following expression given entirely in terms of s:

$$\left(\frac{n}{n-1}\right)^2 - \left(\frac{n}{n+2}\right)\frac{(n+1)^2}{(n-1)^2(2n+1)^2}s^2 + \frac{(n+2)}{n(n-1)^2}\left[s - 2\left[\frac{(n+1)^2}{(n+2)(2n+1)}\right]s\right]^2 = 0$$

and

$$n^2 - \left(\frac{n}{n+2}\right)\frac{(n+1)^2}{(2n+1)^2}s^2 + \frac{(n+2)}{n}\left[\left(\frac{n}{(n+2)(2n+1)}\right)s\right]^2 = 0$$

$$(2n+1)^2 - \frac{(n+1)^2}{n(n+2)}s^2 + \frac{1}{n(n+2)}s^2 = 0 \text{ and } (2n+1)^2 - s^2 = 0$$

Finally this gives the solution for s as:

$$s = \pm(2n+1) \tag{4.39a}$$

Accordingly the solution for t is

$$t = \mp\frac{(n+1)}{(n-1)} \tag{4.39b}$$

Of course, since the equation for spherical aberration gives quadratic terms in s and t, it is not surprising that two solutions exist. Furthermore, it is important to recognise that the sign of t is the opposite to that of s. Referring to Figure 4.10, it is clear that the form of the lens is that of a meniscus. The two solutions for s correspond to a meniscus lens that has been inverted. Of course, the same applies to the conjugate parameter, so, in effect, the two solutions are identical, except the whole system has been inverted, swapping the object for image and vice-versa.

An aplanatic meniscus lens is an important building block in an optical design, in that it confers additional focusing power without incurring further spherical aberration or coma. This principle is illustrated in Figure 4.14 which shows a meniscus lens with positive focal power.

It is instructive, at this point to quantify the increase in system focal power provided by an aplanatic meniscus lens. Effectively, as illustrated in Figure 4.14, it increases the system numerical aperture in (minus) the ratio of the object and image distance. For the positive meniscus lens in Figure 4.14, the conjugate parameter is negative and equal to $-(n+1)/(n-1)$. From Eq. (4.27) the ratio of the object and image distances is given by:

$$\frac{u}{v} = \left(\frac{1-t}{1+t}\right) = \left(\frac{(n-1)+(n+1)}{(n-1)-(n+1)}\right) = -n$$

As previously set out, the increase in numerical aperture of an aplanatic meniscus lens is equal to minus the ratio of the object and image distances. Therefore, ***the aplanatic meniscus lens increases the system power by a factor equal to the refractive index of the lens***. This principle is of practical consequence in many system designs. Of course, if we reverse the sense of Figure 4.14 and substitute the image for the object and vice versa, then the numerical aperture is effectively ***reduced*** by a factor of n.

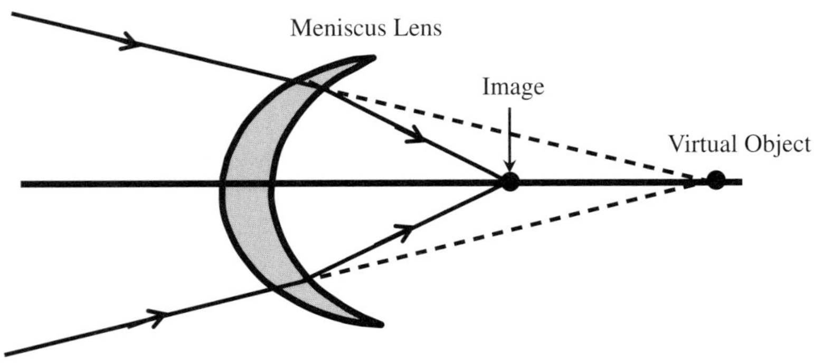

Figure 4.14 Aplanatic meniscus lens.

Worked Example 4.4 ***Microscope Objective – Hyperhemisphere Plus Meniscus Lens***
We now wish to add some power to the microscope objective hyperhemisphere set out in Worked Example 4.1. We are to do so with an extra meniscus lens situated at the vertex of the hyperhemisphere with a negligible separation. As with the hyperhemisphere, the meniscus lens is in the aplanatic arrangement. The meniscus lens is made of the same material as the hyperhemisphere, that is with a refractive index of 1.6. All properties of the hyperhemisphere are as set out in Worked Example 4.1.

What are the radii of curvature of the meniscus lens and what is the location of the (virtual) image for the combined system? The system is as illustrated below.

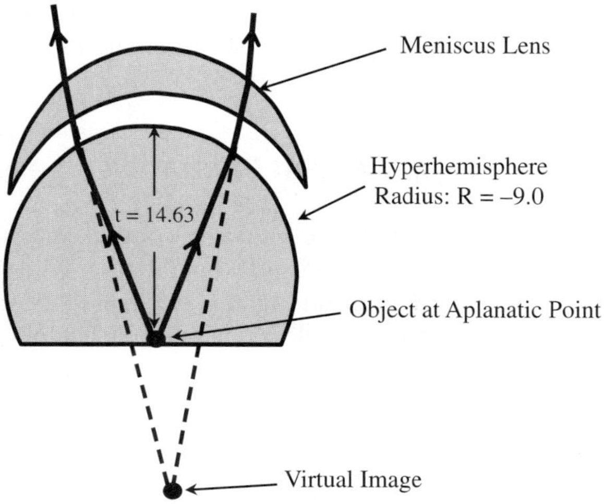

We know from Worked Example 4.1 that the original image distance produced by the hyperhemisphere is −23.4 mm. The object distance for the meniscus lens is thus 23.4 mm. From Eq. (4.39a) we have:

$$t = \pm\frac{n+1}{n-1} = \pm\frac{2.6}{0.6} = \pm 4.33$$

There remains the question of the choice of the sign for the conjugate parameter. If one refers to Figure 4.14, it is clear that the sense of the object and image location is reversed. In this case, therefore, the value of t is equal to +4.33 and the numerical aperture of the system is reduced by a factor of 1.6 (the refractive index). In that case, the image distance must be equal to minus 1.6 times the object distance. That is to say:

$$v = -1.6 \times u = -1.6 \times 23.4 = -37.44\text{ mm}$$

We can calculate the focal length of the lens from:

$$\frac{1}{f} = \frac{1}{u} + \frac{1}{v} \quad \frac{1}{f} = \frac{1}{23.4} - \frac{1}{37.44} = \frac{1}{62.4}$$

Therefore the focal length of the meniscus lens is 62.4 mm. If the conjugate parameter is +4.33, then the shape factor must be −(2n + 1), or −4.2 (note the sign). It is a simple matter to calculate the radii of the two surfaces from Eq. (4.29):

$$R_1 = \left(\frac{2(n-1)}{s+1}\right)f \quad R_2 = \left(\frac{2(n-1)}{s-1}\right)f$$

$$R_1 = \left(\frac{1.2}{-3.2}\right) * 62.4 \quad R_2 = \left(\frac{1.2}{-5.2}\right) * 62.4$$

Finally, this gives R_1 as **−23.4 mm** and R_2 as **−14.4 mm**. The signs should be noted. This follows the convention that positive displacement follows the direction from object to image space.

If the microscope objective is ultimately to provide a collimated output – i.e. with the image at the infinite conjugate, the remainder of the optics must have a focal length of 37.44 mm (i.e. 23.4 × 1.6). This exercise illustrates the utility of relatively simple building blocks in more complex optical designs. This revised system has a focal length of 9 mm. However, the 'remainder' optics have a focal length of 37.4 mm, or only a quarter of the overall system power. Spherical aberration increases as the fourth power of the numerical aperture, so the 'slower' 'remainder' will intrinsically give rise to much less aberration and, as a consequence, much easier to design. The hyperhemisphere and meniscus lens combination confer much greater optical power to the system without any penalty in terms of spherical aberration and coma. Of course, in practice, the picture is complicated by chromatic aberration caused by variations in refractive properties of optical materials with wavelength. Nevertheless, the underlying principles outlined are very useful.

4.5 The Effect of Pupil Position on Element Aberration

In all previous analysis, it is assumed that the stop is located at the optical surface in question. This is a useful starting proposition. However, in practice, this is most usually not the case. With the stop located at a spherical surface, by definition, the chief ray will pass directly through the vertex of that surface. If, however, the surface is at some distance from the stop, then the chief ray will, in general, intersect the surface at some displacement from the surface vertex. This displacement is, in the first approximation, proportional to the field angle of the object in question. The general concept is illustrated in Figure 4.15.

Instead of the stop being located at the surface in question, the stop is displaced by a distance, s, from the surface. The chief ray, passing through the centre of the stop defines the field angle, θ. In addition, the pupil co-ordinates defined at the stop are denoted by r_x and r_y. However, if the stop were located at the optical surface, then the field angle would be θ', as opposed to θ. In addition, the pupil co-ordinates would be given by r_x' and r_y'. Computing the revised third order aberrations proceeds upon the following lines. All the previous analysis, e.g. as per Eqs. (4.31a)–(4.31d), has enabled us to express all aberrations as an OPD in terms of θ', r_x', and r_y'. It is clear that to calculate the aberrations for the new stop locations, one must do so in terms of the new parameters θ, r_x, and r_y. This is done by effecting a simple linear transformation between the two sets of parameters. Referring to Figure 4.15, it is easy to see:

$$r_x' = \left(\frac{u}{u-s}\right) r_x \qquad (4.40a)$$

$$r_y' = \left(\frac{u}{u-s}\right) r_y + s\theta \qquad (4.40b)$$

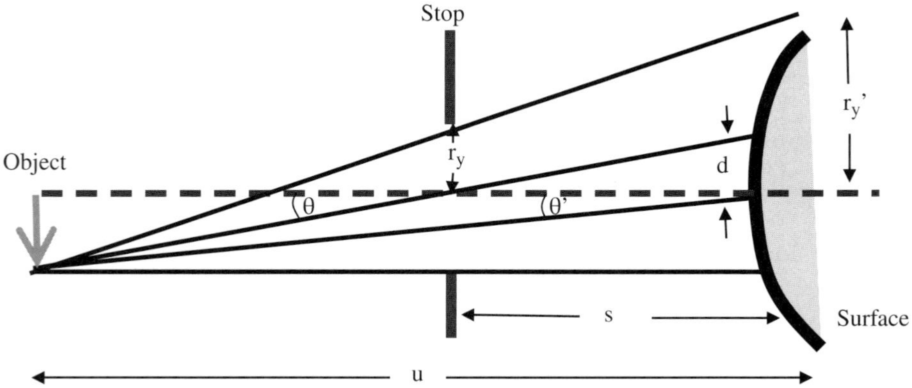

Figure 4.15 Impact of stop movement.

$$\theta' = \left(\frac{u-s}{u}\right)\theta \tag{4.40c}$$

The effective size of the pupil at the optic is magnified by a quantity M_p and the pupil offset set out in Eq. (4.40b) is directly related to the eccentricity parameter, E, described in Chapter 2. Indeed, the product of the eccentricity parameter and the Lagrange invariant, H is simply equal to the ratio of the marginal and chief ray height at the pupil. That is to say:

$$M_p = \frac{r_0'}{r_0} = \left(\frac{u}{u-s}\right) \quad EH = \frac{s}{r_0}\left(\frac{u-s}{u}\right) \tag{4.41}$$

In this case, r_0 refers to the pupil radius at the stop and r_0' to the effective pupil radius at the surface in question. As a consequence, we can re-cast all three equations in a more convenient form.

$$r_x' = M_p r_x \quad r_y' = M_p\left(r_y + EHr_0\frac{\theta}{\theta_0}\right) \quad \theta' = \frac{\theta}{M_p} \tag{4.42}$$

The angle, θ_0 is representative of the maximum system field angle and helps to define the eccentricity parameter and the Lagrange invariant. We already know the OPD when cast in terms of r_x', r_y', and θ, as this is as per the analysis for the case where the stop is at the optic itself. That is to say, the expression for the OPD is as given in Eqs. (4.17a)–(4.17d) and Eqs. (4.30a)–(4.30d) and these aberrations defined in terms of K_{SA}', K_{CO}', K_{AS}', K_{FC}', and K_{DI}'. Therefore, the total OPD attributable to the five Gauss-Seidel aberrations is given by:

$$\Phi_{Seidel} = \frac{K_{SA}'}{r_0'^4}r'^4 + \frac{K_{CO}'}{r_0'^3}r'^2\,r_y'\,\theta' + \frac{K_{AS}'}{r_0'^2}(r_y'^2 - r_x'^2)\theta'^2 + \frac{K_{FC}'}{r_0'^2}r'^2\,\theta'^2 + \frac{K_{DI}'}{r_0}r_y'\,\theta'^3 \tag{4.43}$$

To determine the aberrations as expressed by the pupil co-ordinates for the new stop location, it is a simple matter of substituting Eq. (4.42) into Eq. (4.43). This results in the so-called stop shift equations:

$$K_{SA} = K_{SA}' \tag{4.44a}$$

$$K_{CO} = \frac{4EH}{\theta_0}K_{SA}' + K_{CO}' \tag{4.44b}$$

$$K_{AS} = \frac{2E^2H^2}{\theta_0^2}K_{SA}' + \frac{EH}{\theta_0}K_{CO}' + K_{AS}' \tag{4.44c}$$

$$K_{FC} = \frac{4E^2H^2}{\theta_0^2}K_{SA}' + \frac{2EH}{\theta_0}K_{CO}' + K_{FC}' \tag{4.44d}$$

$$K_{DI} = \frac{4E^3H^3}{\theta_0^3}K_{SA}' + \frac{3E^2H^2}{\theta_0}K_{CO}' + \frac{2EH}{\theta_0}K_{AS}' + \frac{2EH}{\theta_0}K_{FC}' + K_{DI}' \tag{4.44e}$$

What this set of equations reveals is that there exists a 'hierarchy' of aberrations. Spherical aberration may be transmuted into coma, astigmatism, field curvature, and distortion by shifting the stop position. Similarly, coma may be transformed into astigmatism, field curvature, and distortion and both astigmatism and field curvature may produce distortion. However, coma can never produce spherical aberration and neither astigmatism nor field curvature is capable of generating spherical aberration or coma. Equation (4.44e) reveals, for the first time, that it is possible to generate distortion by shifting the stop. Our previous idealised analysis clearly suggested that distortion is not produced where the lens or optical surface is located at the stop.

Another important conclusion relating to Eqs. (4.44a)–(4.44e) is the impact of stop shift on the astigmatism and field curvature. Inspection of Eqs. (4.44c) and (4.44d) reveals that the change in field curvature produced by stop shift is precisely double that of the change in astigmatism in all cases. Therefore, the Petzval curvature, which is given by $K_{FC}-2K_{AS}$ remains unchanged by stop shift. This further serves to demonstrate the fact that the Petzval curvature is a fundamental system attribute and is unaffected by changes in stop location and, indeed component location. Petzval curvature only depends upon the system power. Thus, it is important to recognise that the quantity $K_{FC}-2K_{AS}$ is preserved in any manipulation of existing components within a system. If we express the Petzval curvature in terms of the tangential and sagittal curvature we find:

$$K_{Petz} = K_{FC} - 2K_{AS} \sim (\Phi_{\tan} + \Phi_{sag}) - 2(\Phi_{\tan} - \Phi_{sag}) \quad K_{Petz} \sim 3\Phi_{sag} - \Phi_{\tan} \tag{4.45}$$

Since K_{Petz} is not changed by any manipulation of component or stop positions, Eq. (4.45) implies that ***any change in the sagittal curvature*** is accompanied by ***a change three times as large in the tangential curvature***. This is an important conclusion.

For small shifts in the position of the stop, the eccentricity parameter is proportional to that shift. Based on this and examining Eqs. (4.44a)–(4.44e), one can come to some general conclusions. For a system with pre-existing spherical aberration, additional coma will be produced in linear proportion to the stop shift. Similarly, the same spherical aberration will produce astigmatism and field curvature proportional to the square of the stop shift. The amount of distortion produced by pre-existing spherical aberration is proportional to the cube of the displacement. Naturally, for pre-existing coma, the additional astigmatism and field curvature produced is in proportion to the shift in the stop position. Additional distortion is produced according to the square of the stop shift. Finally, with pre-existing astigmatism and field curvature, only additional distortion may be produced in direct proportion to the stop shift.

As an example, a simple scenario is illustrated in Figure 4.16. This shows a symmetric system with a biconvex lens used to image an object in the $2f - 2f$ configuration. That is to say, the conjugate parameter is zero. In this situation, the coma may be expected, by virtue of symmetry, to be zero. For a simple lens, the distortion is also zero. The spherical aberration is, of course, non-zero, as are both the astigmatism and field curvature.

Using basic modelling software, it is possible to analyse the impact of small stop shifts on system aberration. The results are shown in Figure 4.17.

Clearly, according to Figure 4.17, the spherical aberration remains unchanged as predicted by Eq. (4.44a). For small shifts, the amount of coma produced is in proportion to the shift. Since there is no coma initially, the only aberration that can influence the astigmatism and field curvature is the pre-existing spherical aberration. As indicated in Eqs. (4.44c) and (4.44d), there should be a quadratic dependence of the astigmatism and field curvature on stop position. This is indeed borne out by the analysis in Figure 4.17. Similarly, the distortion shows a linear trend with stop position, mainly influenced by the initial astigmatism and field curvature that is present.

Although, in practice, these stop shift equations may not find direct use currently in optimising real designs, the underlying principles embodied are, nonetheless, important. Manipulation of the stop position is a key part in the optimisation of complex optical systems and, in particular, multi-element camera lenses. In these complex systems, the pupil is often situated between groups of lenses. In this case, the designer needs to be aware also of the potential for vignetting, should individual lens elements be incorrectly sized.

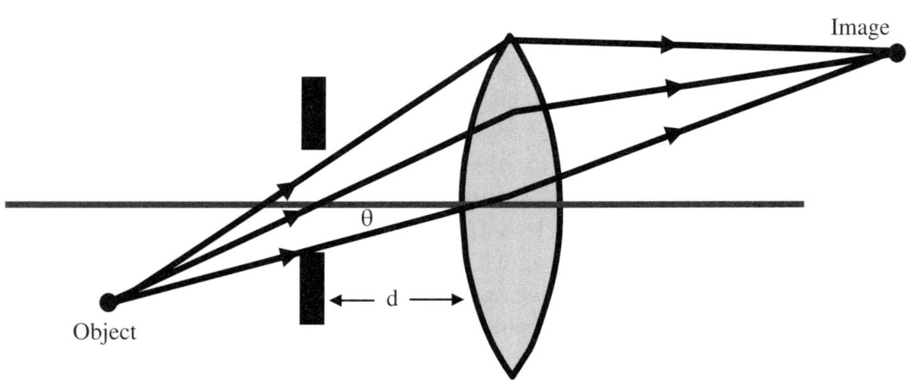

Figure 4.16 Simple symmetric lens system with stop shift.

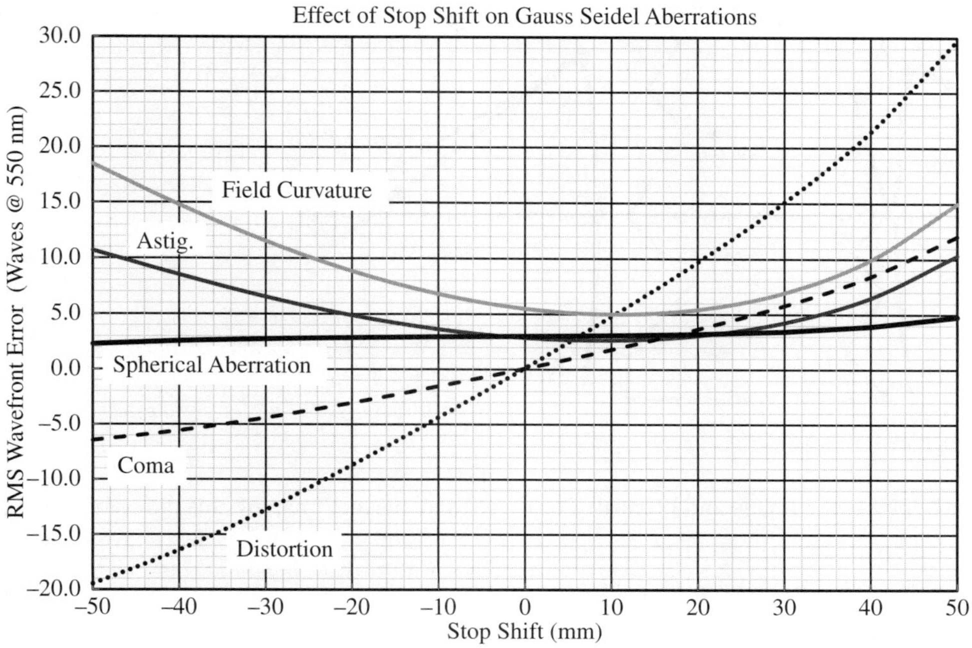

Figure 4.17 Impact of stop shift for simple symmetric lens system.

The stop shift equations provide a general insight into the impact of stop position on aberration. Most significant is the hierarchy of aberrations. For example, no fundamental manipulation of spherical aberration may be accomplished by the manipulation of stop position. Otherwise, there some special circumstances it would be useful for the reader to be aware of. For example, in the case of a spherical mirror, with the object or image lying at the infinite conjugate, the placement of the stop at the mirror's centre of curvature altogether removes its contribution to coma and astigmatism; the reader may care to verify this.

4.6 Abbe Sine Condition

Long before the advent of powerful computer ray tracing models, there was a powerful incentive to develop simple rules of thumb to guide the optical design process. This was particularly true for the complex task of ameliorating system aberrations. Working in the nineteenth century, Ernst Abbe set out the ***Abbe sine condition***, which directly relates the object and image space numerical apertures for a 'perfect', unaberrated system. Essentially, the Abbe sine condition articulates a specific requirement for a system to be free of spherical aberration and coma, i.e. aplanatic. The Abbe sine condition is expressed for an infinitesimal object and image height and its justification is illustrated in Figure 4.18.

In the representation in Figure 4.18 we trace a ray from the object to a point, P, located on a reference sphere whose centre lies on axis at the axial position of the object and whose vertex lies at the entrance pupil. At the same time, we also trace a marginal ray from the object location to the entrance pupil. The conjugate point to P, designated, P′, is located nominally at the exit pupil and on a sphere whose centre lies at the paraxial image location. For there to be perfect imaging, then the OPD associated with the passage of the marginal ray must be zero. Furthermore, the OPD of the ray from object to image must also be zero. It is also further assumed that the relative OPD of the object to image ray when compared to the marginal ray is zero on passage from points P to P′. This assumption is justified for an infinitesimal object height. Therefore, it is possible to compute the total object to image OPD by simply summing the path differences relative to the marginal ray between the

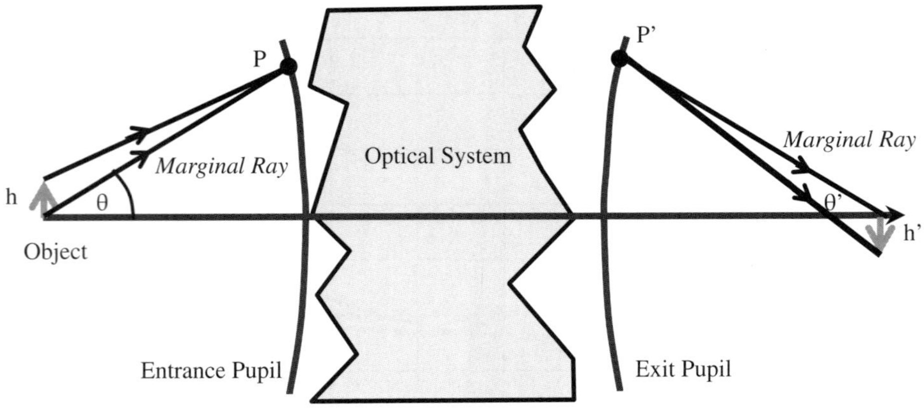

Figure 4.18 Abbe sine condition.

object and point P and between the image and point P′. For there to be perfect imaging this difference must, of course be zero.

$$nh\sin\theta - n'h'\sin\theta' = 0 \text{ or } nh\sin\theta = n'h'\sin\theta' \quad (4.46)$$

n is the refractive index in object space and n′ is the refractive index in image space.

Equation (4.46) is one formulation of the Abbe sine condition which, nominally, applies for all values of θ and θ', including paraxial angles. If we represent the relevant paraxial angles in object and image space as θ_p and θ_p' then the Abbe sine condition may be rewritten as:

$$\frac{\sin\theta}{\theta_p} = \frac{\sin\theta'}{\theta'_p} \quad (4.47)$$

One specific scenario occurs where the object or image lies at the infinite conjugate. For example, one might imagine an object located on axis at the first focal point. In this case, the height of any ray within the collimated beam in image space is directly proportional to the numerical aperture associated with the input ray.

Figure 4.19 illustrates the application of the Abbe sine condition for a specific example. As highlighted previously, the sine condition effectively seeks out the aplanatic condition in an optical system. In this example, a meniscus lens is to be designed to fulfil the aplanatic condition. However, its conjugate parameter is adjusted around the ideal value and the spherical aberration and coma plotted as a function of the conjugate parameter. In addition, the departure from the Abbe sine condition is also plotted in the same way. All data is derived from detailed ray tracing and values thus derived are presented as relative values to fit reasonably into the graphical presentation. It is clear that elimination of spherical aberration and coma corresponds closely to the fulfilment of the Abbe sine condition.

The form of the Abbe sine condition set out in Eq. (4.46) is interesting. It may be compared directly to the Helmholtz equation which has a similar form. However, instead of a relationship based on the sine of the angle, the Helmholtz equation is defined by a relationship based on the tangent of the angle:

$$nh\sin\theta = n'h'\sin\theta'(Abbe) \quad nh\tan\theta = n'h'\tan\theta'(Helmholtz)$$

It is quite apparent that the two equations present something of a contradiction. The Helmholtz equation sets the condition for perfect imaging in an ideal system for ***all pairs of conjugates***. However, the Abbe sine condition relates to aberration free imaging for a ***specific conjugate pair***. This presents us with an important conclusion. It is clear that aberration free imaging for a specific conjugate (Abbe) fundamentally denies the possibility for perfect imaging across all conjugates (Helmholtz). Therefore, ***an optical system can only be designed to deliver aberration free imaging for one specific conjugate pair***.

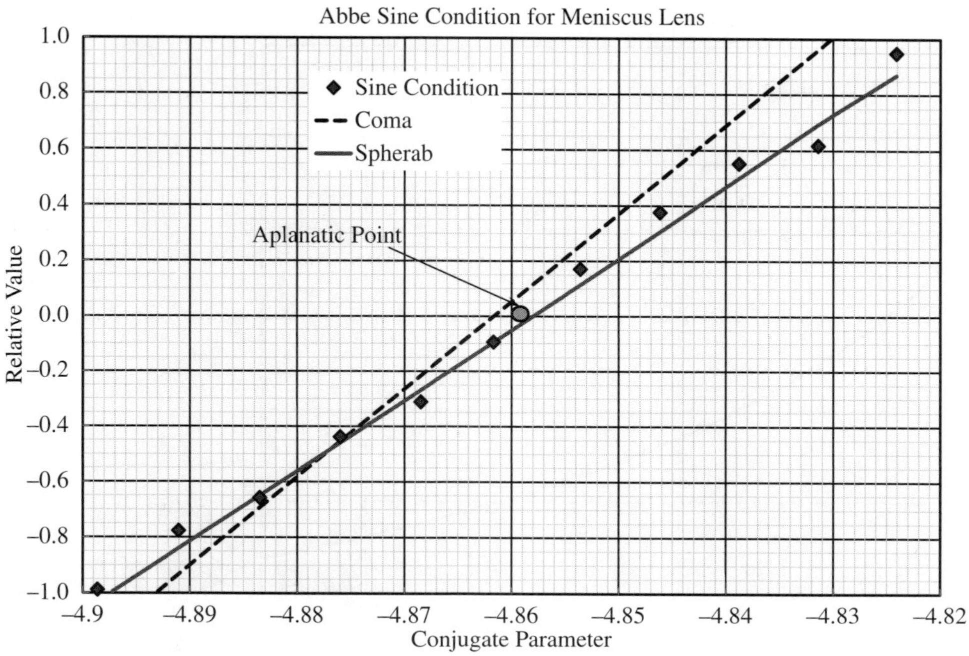

Figure 4.19 Fulfilment of Abbe sine condition for aplanatic meniscus lens.

4.7 Chromatic Aberration

4.7.1 Chromatic Aberration and Optical Materials

Hitherto, we have only considered the classical monochromatic aberrations. At this point, we must introduce the phenomenon of ***chromatic aberration*** where imperfections in the imaging of an optical system are produced by significant variation in optical properties with wavelength. All optical materials are ***dispersive*** to some degree. That is to say, their refractive indices vary with wavelength. As a consequence, all first order properties of an optical system, such as the location of the cardinal points, vary with wavelength. Most particularly, the paraxial focal position of an optical system with dispersive components will vary with wavelength, as will its effective focal length. Therefore, for a given axial position in image space, only one wavelength can be in focus at any one time.

Dispersion is a property of transmissive optical materials, i.e. glasses. On the other hand, mirrors show no chromatic variation and their incorporation is favoured in systems where chromatic variation is particularly unwelcome. Such a system, where the optical properties do not vary with wavelength, is said to be ***achromatic***. As argued previously, a mirror behaves as an optical material with a refractive index of minus one, a value that is, of course, independent of wavelength. In general, the tendency in most optical materials is for the refractive index to decrease with increasing wavelength. This behaviour is known as ***normal dispersion***. In certain very specific situations, for certain materials at particular wavelengths, the refractive index actually decreases with wavelength; this phenomenon is known as ***anomalous dispersion***.

Although dispersion is an issue of concern covering all wavelengths of interest from the ultraviolet to the infrared, for obvious reasons, historically, there has been particular focus on this issue within the visible portion of the spectrum. Across the visible spectrum, for typical glass materials, the refractive index variation might amount to 0.7–2.5%. This variation in the dispersive properties of different materials is significant, as it affords a means to reduce the impact of chromatic aberration as will be seen shortly. Figure 4.20 shows a typical dispersive plot, for the glass material, SCHOTT BK7®.

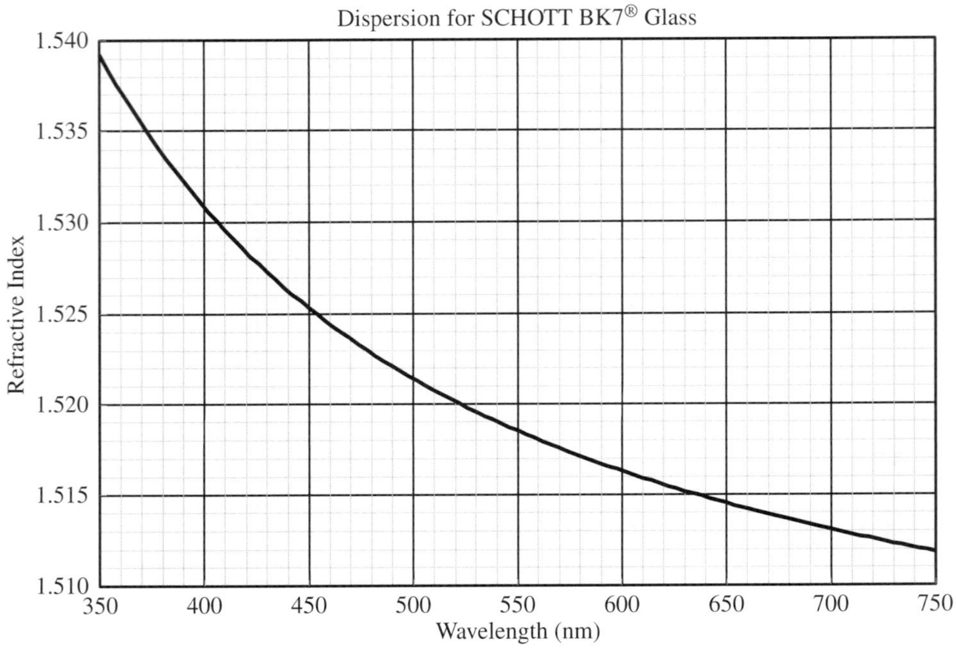

Figure 4.20 Refractive index variation with wavelength for SCHOTT BK7 glass material.

Because of the historical importance of the visible spectrum, glass materials are typically characterised by their refractive properties across this portion of the spectrum. More specifically, glasses are catalogued in terms of their refractive indices at three wavelengths, nominally 'blue', 'yellow', and 'red'. In practice, there are a number of different conventions for choosing these reference wavelengths, but the most commonly applied uses two hydrogen spectral lines – the 'Balmer-beta' line at 486.1 nm and the 'Balmer-alpha' line at 656.3, plus the sodium 'D' line at 589.3 nm. The refractive indices at these three standard wavelengths are symbolised as n_F, n_C, and n_D respectively. At this point, we introduce the ***Abbe number***, V_D, which expresses a glass's dispersion by the ratio of its optical power to its dispersion:

$$V_D = \frac{n_D - 1}{n_F - n_C} \tag{4.48}$$

The numerator in Eq. (4.48) represents the effective optical or focusing power at the 'yellow' wavelength, whereas the denominator describes the dispersion of the glass as the difference between the 'blue' and the 'red' indices. It is important to recognise that the higher the Abbe number, then the less dispersive the glass, and vice versa. Abbe numbers vary, typically between about 20 and 80. Broadly speaking, these numbers express the ratio of the glass's focusing power to its dispersion. Hence, for a material with an Abbe number of 20, the focal length of a lens made from this material will differ by approximately 5% (1/20) between 486.1 and 656.3 nm.

4.7.2 Impact of Chromatic Aberration

The most obvious effect of chromatic aberration is that light is broad to a different focus for different wavelengths. This effect is known ***as longitudinal chromatic aberration*** and is illustrated in Figure 4.21.

As can be seen from Figure 4.21, light at the shorter, 'blue' wavelengths are focused closer to the lens, leading to an axial (longitudinal) shift in the paraxial focus for the different wavelengths. In summary, longitudinal chromatic aberration is associated with a shift in the paraxial focal position as a function of wavelength. Thus

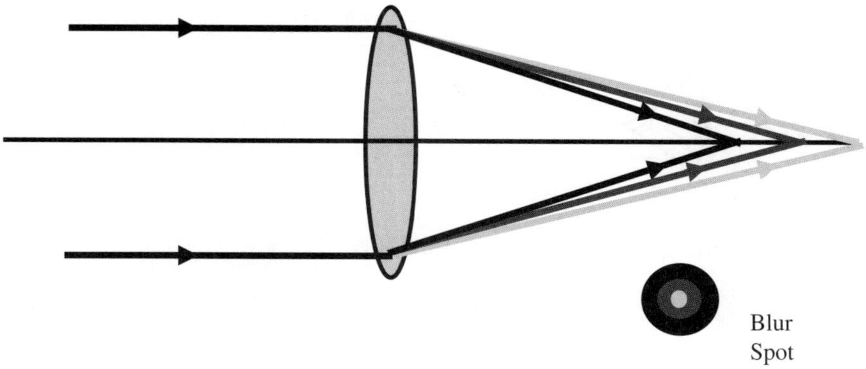

Figure 4.21 Longitudinal chromatic aberration.

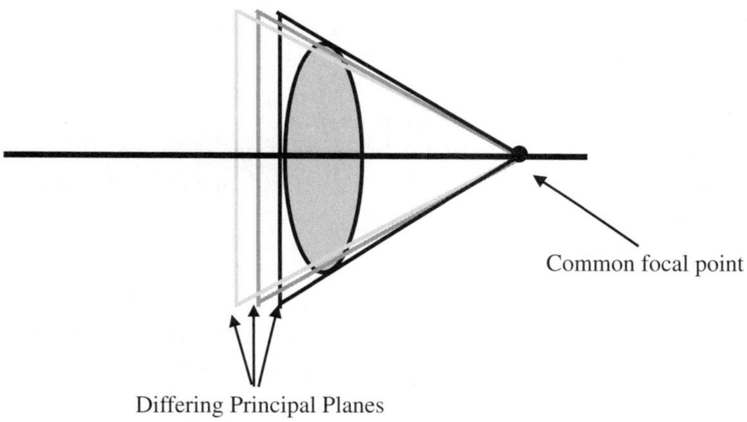

Figure 4.22 Transverse chromatic aberration.

the effect of longitudinal chromatic aberration is to produce a blur spot or transverse aberration whose magnitude is directly proportional to the aperture size, but is independent of field angle. However, there are situations where, to all intents and purposes, all wavelengths share the same paraxial focal position, but the principal points are not co-located. That is to say, whilst all wavelengths are focused at a common point, the effective focal length corresponding to each wavelength is not identical. This scenario is illustrated in Figure 4.22.

The effect illustrated is known as ***transverse chromatic aberration*** or ***lateral colour***. Whilst no distinct blurring is produced by this effect, the fact that different wavelengths have different focal lengths inevitably means that system magnification varies with wavelength. As a result, the final image size or height of a common object depends upon the wavelength. This produces distinct coloured fringing around an object and the size of the effect is proportional to the field angle, but independent of aperture size.

Hitherto, we have cast the effects of chromatic aberration in terms of transverse aberration. However, to understand the effect on the same basis as the Gauss-Seidel aberrations, it is useful to express chromatic aberration in terms of the OPD. When applied to a single lens, longitudinal chromatic aberration simply produces defocus that is equal to the focal length divided by the Abbe number. Therefore, the longitudinal chromatic aberration is given by:

$$\Phi_{LC} = \frac{r^2}{V_D f} \tag{4.49a}$$

f is the focal length of the lens and r the pupil position.

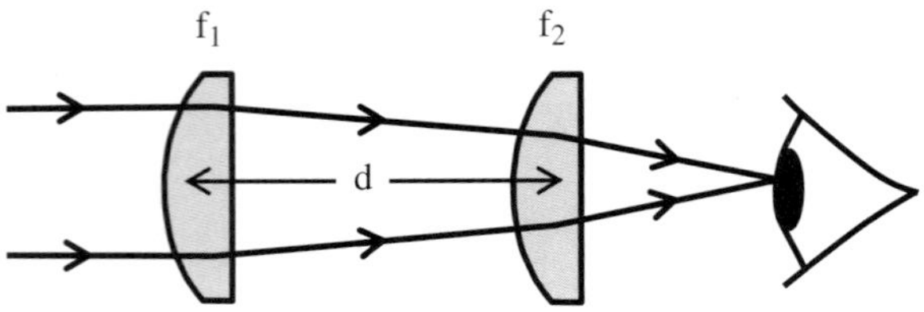

Figure 4.23 Huygens eyepiece.

Similarly, the transverse chromatic aberration can be expressed as an OPD:

$$\Phi_{TC} = \frac{r_y}{V_D}\theta \tag{4.49b}$$

Examining Eqs. (4.49a) and (4.49b) reveals that the ratio of transverse to longitudinal aberration is given by the ratio of the field angle to the numerical aperture. In practice, for optical elements, such as microscope and telescope objectives, the field angle is very much smaller than the numerical aperture and thus longitudinal chromatic aberration may be expected to predominate. For eyepieces, the opposite is often the case, so the imperative here is to correct lateral chromatic aberration.

Worked Example 4.5 ***Lateral Chromatic Aberration and the Huygens Eyepiece***
A practical example of the correction of lateral chromatic aberration is in the Huygens eyepiece. This very simple, early, eyepiece uses two plano-convex lenses separated by a distance equivalent to half the sum of their focal lengths. This is illustrated in Figure 4.23.

$$d = \frac{f_1 + f_2}{2}$$

Since we are determining the impact of lateral chromatic aberration, we are only interested in the effective focal length of the system comprising the two lenses. Using simple matrix analysis as described in Chapter 1, the system focal length is given by:

$$\frac{1}{f_{sys}} = \frac{1}{f_1} + \frac{1}{f_2} - \frac{d}{f_1 f_2}$$

If we assume that both lenses are made of the same material, then their focal power will change as a function of wavelength by a common proportion, α. In that case, the system focal power at the new wavelength would be given by:

$$\frac{1}{f_{sys}} = \frac{1+\alpha}{f_1} + \frac{1+\alpha}{f_2} - \frac{(1+\alpha)^2 d}{f_1 f_2}$$

For small values of α, we can ignore terms of second order in α, so the change in system power may be approximated by:

$$\frac{\Delta}{f_{sys}} \approx \left[\frac{1}{f_1} + \frac{1}{f_2}\right]\alpha - \frac{2d}{f_1 f_2}\alpha = 0$$

The change in system power should be zero and this condition unambiguously sets the lens separation, d, for no lateral chromatic aberration:

$$d = \frac{f_1 + f_2}{2} \tag{4.50}$$

If this condition is fulfilled, then the Huygens eyepiece will have no transverse chromatic aberration. However, it must be emphasised that this condition does not provide immunity from longitudinal chromatic aberration.

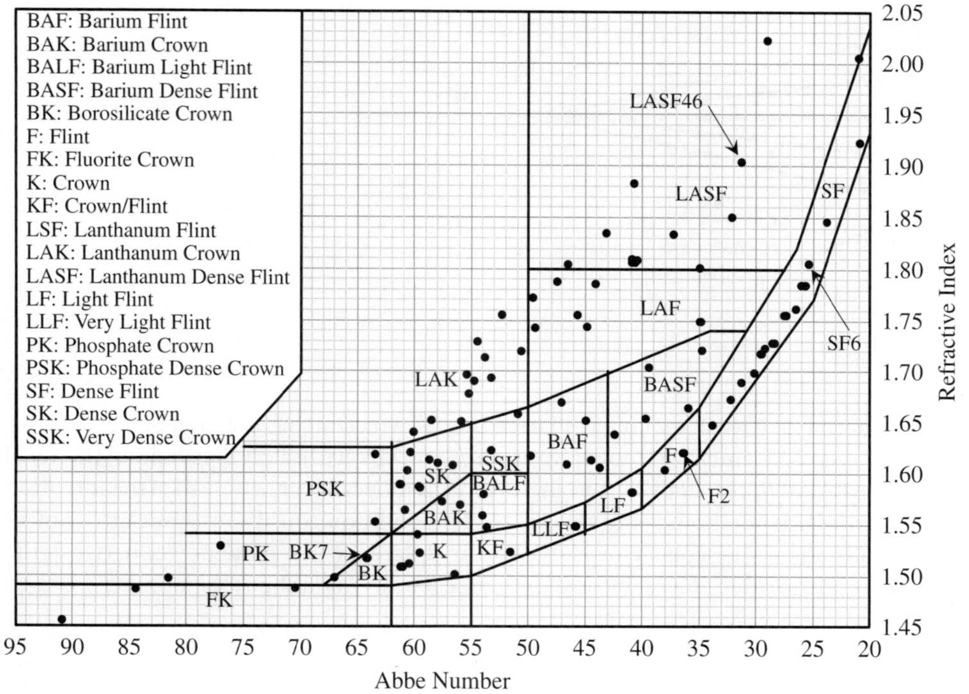

Figure 4.24 Abbe diagram.

4.7.3 The Abbe Diagram for Glass Materials

For visible applications, the Abbe number for a glass is of equal practical importance as the refractive index itself. The Abbe diagram is a simple graphic tool that captures the basic refractive properties of a wide range of optical glasses. It comprises a simple 2D map with the horizontal axis corresponding to the Abbe number and the vertical axis to the glass index. A representative diagram is shown in Figure 4.24.

By referring to this diagram, the optical designer can make appropriate choices for specific applications in the visible. In particular, it helps select ***combinations*** of glasses leading to a substantially achromatic design. One special and key application is the ***achromatic doublet***. This lens is composed of two elements, one positive and one negative. The positive lens is a high power (short focal length) element with low dispersion and the negative lens is a low power element with high dispersion. Materials are chosen in such a way that the net dispersion of the two elements cancel, but the powers do not. This will be considered in more detail in the next section.

The different zones highlighted in the Abbe diagram replicated in Figure 4.24 refer to the elemental composition of the glass. For example, 'Ba' refers to the presence of barium and 'La' to the presence of lanthanum. Originally, many of the dense, high index glasses used to contain lead, but these are being phased out due to environmental concerns. The Abbe diagram reveals a distinct geometrical profile with a tendency for high dispersion to correlate strongly with refractive index. In fact, it is the presence of absorption features within the glass (at very much shorter wavelengths) that give rise to the phenomenon of refraction and these features also contribute to dispersion.

4.7.4 The Achromatic Doublet

As introduced previously, the achromatic doublet is an extremely important building block in a transmissive (non-mirror) optical design. The function of an achromatic doublet is illustrated in Figure 4.25.

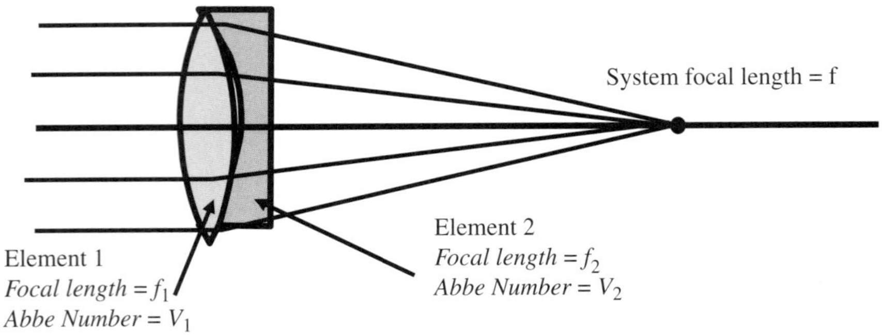

Figure 4.25 The achromatic doublet.

The first element, often (on account of its shape) referred to as the 'crown element', is a high power positive lens with low dispersion. The second element is a low power negative lens with high dispersion. The focal lengths of the two elements are f_1 and f_2 respectively and their Abbe numbers V_1 and V_2. Since the intention is that the dispersions of the two elements should entirely cancel, this condition constrains the relative power of the two elements. Individually, the dispersion as measured by the difference in optical power between the red and blue wavelengths is proportional to the reciprocal of the focal power and the Abbe number for each element. Therefore:

$$Dispersion \propto \frac{1}{f_1V_1} + \frac{1}{f_2V_2} = 0 \quad \frac{f_1}{f_2} = -\frac{V_2}{V_1} \tag{4.51}$$

From Eq. (4.51), it is clear that the ratio of the two focal lengths should be minus the inverse of the ratio of their respective Abbe numbers. In other words, the ratio of their powers should be minus the ratio of their Abbe numbers. The power of the system comprising the two lenses is, in the thin lens approximation, simply equal to the sum of their individual powers. Therefore, it is possible to calculate these individual focal lengths, f_1 and f_2, in terms of the desired system focal length of f:

$$\frac{1}{f_1} + \frac{1}{f_2} = \frac{1}{f} \text{ and (from Eq. [4.51])} \quad \frac{1}{f_1} - \frac{V_2}{V_1}\frac{1}{f_1} = \frac{1}{f} \text{ and } f_1 = \left(\frac{V_1 - V_2}{V_1}\right)f$$

Thus, the two focal lengths are simply given by:

$$f_1 = \left(\frac{V_1 - V_2}{V_1}\right)f \quad f_2 = \left(\frac{V_2 - V_1}{V_2}\right)f \tag{4.52}$$

In the thin lens approximation, therefore, light will be focused at the same point for the red and blue wavelengths. Consequentially, in this approximation, this system will be free from both longitudinal and transverse chromatic aberration. The simplicity of this approach may be illustrated in a straightforward worked example.

Worked Example 4.6 ***Simple Achromatic Doublet***
We wish to construct and achromatic doublet with a focal length of 200 mm. The two glasses to be used are: SCHOTT N-BK7 for the positive crown lens and SCHOTT SF2 for the negative lens. Both these glasses feature on the Abbe diagram in Figure 4.24 and the Abbe number for these glasses are 64.17 and 33.85 respectively. The individual focal lengths may be calculated using Eq. (4.52):

$$f_1 = \left(\frac{V_1 - V_2}{V_1}\right)f = \frac{64.17 - 33.85}{64.17} \times 200 = 94.5$$

$$f_1 = \left(\frac{V_2 - V_1}{V_2}\right)f = \frac{33.85 - 64.17}{33.85} \times 200 = -179$$

Therefore, the focal length of the first 'crown lens' should be 94.5 mm and the focal length of the second diverging lens should be −179 mm.

Thus far, the analysis design of an achromatic doublet has been fairly elementary. In the previous worked example, we have constrained the focal lengths of the two lens elements to specific values. However, we are still free to choose the shape of each lens. That is to say, there are two further independent variables that can be adjusted. Achromatic doublets can either be cemented or air spaced. In the case of the cemented doublet, as presented in Figure 4.25, the second surface of the first lens must have the same radius as the first surface of the second lens. This provides an additional constraint; thus, for the cemented doublet, there is only one additional free variable to adjust. However, introduction of an air space between the two lenses removes this constraint and gives the designer an extra degree of freedom to play with. That said, the cemented doublet does offer greater robustness and reliability with respect to changes in alignment and finds very wide application as a standard optical component.

As a 'stock component' achromatic doublets are designed, generally, for the infinite conjugate. For cemented doublets, with the single additional degree of design freedom, these components are optimised to have zero spherical aberration at the central wavelength. This is an extremely important consideration, for not only are these doublets free of chromatic aberration, but they are also well optimised for other aberrations. Commercial doublets are thus extremely powerful optical components.

4.7.5 Optimisation of an Achromatic Doublet (Infinite Conjugate)

An air spaced achromatic doublet may be optimised to eliminate both spherical aberration and coma. The fundamental power of the wavefront approach in describing third order aberration is reflected in the ability to calculate the total system aberration as the sum of the aberration of the two lenses. In the thin lens approximation, we may simply use Eqs. (4.30a) and (4.30b) to express the spherical aberration and coma contribution for each lens element. We simply ascribe a ***variable*** shape parameter, s_1 and s_2 to each of the two lenses. The two conjugate parameters are fixed. In the particular case of a doublet designed for the infinite conjugate, the conjugate parameter for the first lens, t_1, is −1. In the case of the second lens, the conjugate parameter, t_2, is determined by the relative focal lengths of the two lenses and thus fixed by the ratio of the two Abbe numbers and, from Eq. (4.52), we get:

$$t_2 = \frac{v_2 - u_2}{v_2 + u_2} = \frac{f + f_1}{f - f_1} = -\frac{2V_1 - V_2}{V_2} \qquad t_2 = \frac{2V_1}{V_2} - 1 \tag{4.53}$$

Without going through the algebra in detail, it is clear that having determined both t_1 and t_2, Eqs. (4.30a) and (4.30b) give us two expressions solely in terms of s_1 and s_2. These expressions for the spherical aberration and coma must be set to zero and can be solved for both s_1 and s_2. The important point to note about this procedure is that because Eq. 4.30a contains terms that are quadratic in shape factor, this is also reflected in the final solution. Therefore, in general, we might expect to find two solutions to the equation and this, in general, is true.

Worked Example 4.7 *Detailed Design of 200 mm Focal Length Achromatic Doublet*

At this point we illustrate the design of an air spaced achromat by looking more closely at the previous example where we analysed a 200 mm achromat design. We are to design an achromat with a focal length of 200 mm working at the infinite conjugate, using SCHOTT N-BK7 and SCHOTT SF2 as the two glasses, with the less dispersive N-BK7 used as the positive 'crown' element. Again, the Abbe numbers for these glasses are 64.17 and 33.85 respectively and the n_d values (refractive index at 589.6 nm) 1.5168 and 1.647 69. From the previous example, we know that focal lengths of the two lenses are:

$$f_1 = 94.5\text{ mm}; f_2 = -179\text{ mm}$$

The two conjugate parameters are straightforward to determine. The first conjugate parameter, t_1, is naturally −1. Eq. (4.53) can be used to determine the second conjugate parameter, t_2. This gives:

$$t_1 = -1; t_2 = -2.79$$

We now substitute the conjugate parameter values together with the refractive index values (ND) into Eq. (4.30a). We sum the contributions of the two lenses giving the total spherical aberration which we set to zero. Calculating all coefficients we get a quadratic equation in terms of the two shape factors, s_1 and s_2.

$$1.212s_1^2 - 0.108s_2^2 - 1.793s_1 - 0.568s_2 + 1 = 0 \tag{4.54}$$

We now repeat the same process for Eq. (4.30b), setting the total system coma to zero. This time we get a linear equation involving s_1 and s_2.

$$-5.061s_1 - 1.088s_2 + 1 = 0 \text{ or } s_2 = -4.651s_1 + 0.919 \tag{4.55}$$

Substituting Eq. (4.55) into Eq. (4.54) gives the desired quadratic equation:

$$-1.127s_1^2 + 1.771s_1 + 0.387 = 0 \tag{4.56}$$

There are, of course, two sets of solutions to Eq. (4.56), with the following values:

Solution 1: $s_1 = -0.194; s_2 = 1.823$
Solution 2: $s_1 = 3.198; s_2 = 2.929$

There now remains the question as to which of these two solutions to select. Using Eq. (4.29) to calculate the individual radii of curvature from the lens shapes and focal length we get:

Solution 1: $R_1 = 121.25$ mm; $R_2 = -81.78$ mm; $R_{3\_}81.29$ mm; $R_4 = -281.88$ mm
Solution 2: $R_1 = 23.26$ mm; $R_2 = 44.43$ mm; $R_{3\_}58.91$ mm; $R_4 = -119.68$ mm

The radii R_1 and R_2 refer to the first and second surfaces of lens 1 and R_3 and R_4 to the first and second surfaces of lens 2. It is clear that the first solution contains less steeply curved surfaces and is likely to be the better solution, particularly for relatively large apertures. In the case of the second solution, whilst the solution to the third order equations eliminates third order spherical aberration and coma, higher order aberrations are likely to be enhanced.

The first solution to this problem comes under the generic label of the ***Fraunhofer doublet***, whereas the second is referred to as a ***Gauss doublet***. It should be noted that for the Fraunhofer solution, R_2 and R_3 are almost identical. This means that should we constrain the two surfaces to have the same curvature (in the case of a cemented doublet) and just optimise for spherical aberration, then the solution will be close to that of the ideal aplanatic lens. To do this, we would simply use Eq. (4.29), forcing R_2 and R_3 to be equal and to replace Eq. (4.55) constraining the total coma, providing an alternative relation between s_1 and s_2. However, the fact that the cemented doublet is close to fulfilling the zero spherical aberration and coma condition further illustrates the usefulness of this simple component.

The analysis presented applies only strictly in the thin lens approximation. In practice, optimisation of a doublet such as presented in the previous example would be accomplished with the aid of ray tracing software. However, the insights gained by this exercise are particularly important. For instance, in carrying out a computer-based optimisation, it is critically important to understand that two solutions exist. Furthermore, in setting up a computer-based optimisation, an exercise, such as this, provides a useful 'starting point'.

4.7.6 Secondary Colour

The previous analysis of the achromatic doublet provides a means of ameliorating the impact of glass dispersion and to provide correction at two wavelengths. In the case of the standard visible achromat, correction is provided at the F and C wavelengths, the two hydrogen lines at 486.1 and 656.3 nm. Unfortunately, however, this does not guarantee correction at other, intermediate wavelengths. If one views dispersion of

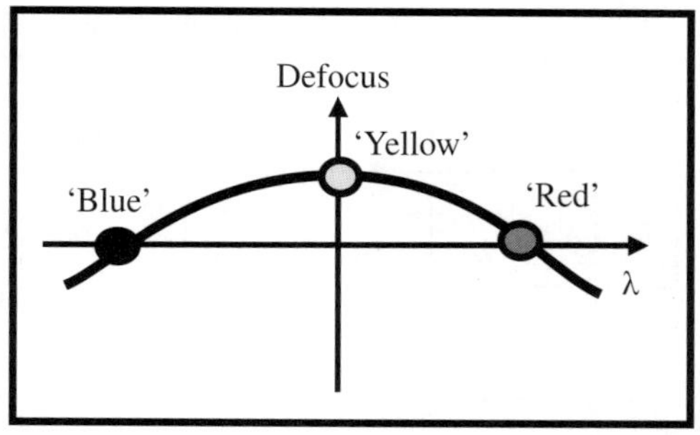

Figure 4.26 Secondary colour.

optical materials as a 'small signal' problem, and that any difference in refractive index is small across the region of interest, then correction of the chromatic focal shift with a doublet may be regarded as a 'linear process'. That is to say we might approximate the dispersion of an optical material by some pseudo-linear function of wavelength, ignoring higher order terms. However, by ignoring these higher order terms, some residual chromatic aberration remains. This effect is referred to as ***secondary colour.*** The effect is illustrated schematically in Figure 4.26 which shows the shift in focus as a function of wavelength.

Figure 4.26 clearly shows the effect as a quadratic dependence in focal shift with wavelength, with the 'red' and 'blue' wavelengths in focus, but the central wavelength with significant defocus. In line with the notion that we are seeking to quantify a quadratic effect, we can define the ***partial dispersion coefficient***, *P*, as:

$$P_{F,D} = \frac{n_F - n_D}{n_F - n_C} \tag{4.57}$$

If we measure the impact of secondary colour as the difference in focal length, Δf, between the 'blue' and 'red' and the 'yellow' focal lengths for an achromatic doublet corrected in the conventional way we get:

$$\Delta f = \frac{P_2 - P_1}{V_1 - V_2} f \tag{4.58}$$

where f is the lens focal length.

The secondary colour is thus proportional to the difference between the two partial dispersions. For simplicity, we have chosen to represent the partial dispersion in terms of the same set of wavelengths as used in the Abbe number. However, whilst the same central (n_d) wavelength might be used, some wavelength other than the n_F, hydrogen line might be chosen for the partial dispersion. Nevertheless, this does not alter the logic presented in Eq. (4.58). Correcting secondary colour is thus less straightforward when compared to the correction of primary colour. Unfortunately, in practice, there is a tendency for the partial dispersion to follow a linear relationship with the Abbe number, as illustrated in the partial dispersion diagram shown in Figure 4.27, illustrating the performance of a range of glasses.

Thus, in the case of the achromatic doublet, judicious choice of glass pairs can minimise secondary colour, but without eliminating it. In principle, secondary colour can be entirely corrected in a triplet system employing lenses of different materials. More formally, if we describe the three lenses as having focal powers of P_1, P_2, and P_3, with the Abbe numbers represented as V_1, V_2, and V_3 and the partial dispersions as, α_1, α_2, α_3, then the lens powers may be uniquely determined from the following set of equations:

$$P_1 + P_2 + P_3 = P_0 \tag{4.59a}$$

$$\frac{P_1}{V_1} + \frac{P_2}{V_2} + \frac{P_3}{V_3} = 0 \tag{4.59b}$$

$$\frac{\alpha_1 P_1}{V_1} + \frac{\alpha_2 P_2}{V_2} + \frac{\alpha_3 P_3}{V_3} = 0 \tag{4.59c}$$

As indicated previously, Figure 4.27 exemplifies the close link between primary and secondary dispersion, with a linear trend observed linking the partial dispersion and the Abbe number for most glasses. It is easy

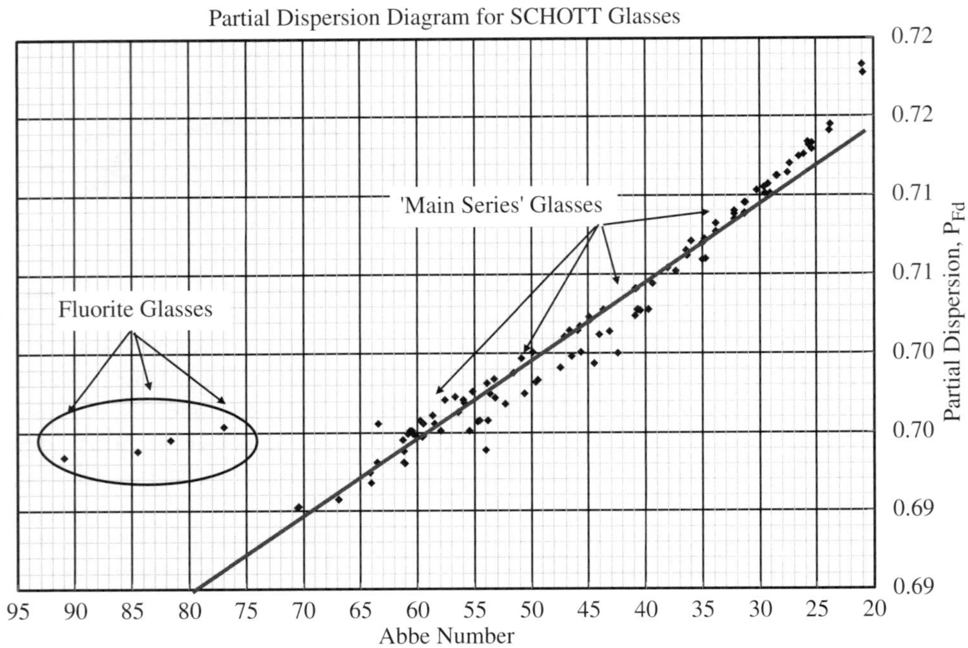

Figure 4.27 Plot of partial dispersion against Abbe number.

to demonstrate by presenting Eqs. (4.59a)–(4.59c) in matrix form that, if a wholly linear relationship exists between partial dispersion and Abbe number, then the matrix determinant will be zero. In this instance, a triplet solution is therefore impossible. Furthermore, the same analysis suggests that for a set of glasses lying close to a straight line on the partial dispersion plot will necessitate the deployment of lenses with very high countervailing powers. It is clear, therefore, that an optimum triplet design is afforded by selection of glasses that depart as far as possible from a straight-line plot on the partial dispersion diagram. In this context, the isolated group of glasses that appear in Figure 4.27, the fluorite glasses, are especially useful in correcting for secondary colour. These glasses lie particularly far from the general trend line for the 'main series' of glasses. Lenses which are corrected for both primary and secondary colour are referred to as ***apochromatic*** lenses. These lenses invariably incorporate fluorite glasses.

4.7.7 Spherochromatism

In the previous analysis we learned that the basic design of simple doublet lenses allowed for the correction of both chromatic aberration and spherical aberration. Furthermore, this flexibility for correction could be extended to coma for an air spaced lens. However, since the refractive index of the two glasses in a doublet lens varies with wavelength, then inevitably, so does the spherical aberration. As such, spherical aberration can only be corrected at one wavelength, e.g. at the 'D' wavelength. This means that there will be some uncorrected spherical aberration at the extremes of the spectrum. This effect is known as ***spherochromatism***. It is generally less significant in magnitude when compared with secondary colour.

4.8 Hierarchy of Aberrations

For some specific applications, such as telescope and microscope objective lenses, the field angles tend to be very much smaller than the angles associated with the system numerical aperture. In these instances, the off-axis aberrations, such as coma, are much less significant than the on-axis aberrations. Therefore, as far as the Gauss-Seidel aberrations are concerned, there exists a hierarchy of aberrations that can be placed in order of their significance or importance:

i) Spherical Aberration
ii) Coma
iii) Astigmatism and Field Curvature
iv) Distortion

That is to say, it is of the greatest importance to correct spherical aberration and then coma, followed by astigmatism, field curvature, and distortion. This emphasises the significance and use of aplanatic elements in optical design.

Of course, for certain optical systems, this logic is not applicable. For instance, in both camera lenses and in eyepieces, the field angles are very substantial and comparable to the angles associated with the numerical aperture. Indeed, in systems of this type, greater emphasis is placed upon the correction of astigmatism, field curvature, and distortion than in other systems.

With these comments in mind, it would be useful to summarise all the aberrations covered in this chapter and to classify them by virtue of their pupil and field angle dependence. Table 4.1 sets out the wavefront error dependence upon pupil and field angle for each of the aberrations.

It would be instructive, at this point, to take the example of the 200 mm doublet and to plot the wavefront aberrations attributable to some of the aberrations listed in Table 4.1 against numerical aperture. Spherochromatism is expressed as the difference in spherical aberration wavefront error between the n_F and n_C wavelengths (486.1 and 656.3 nm). Secondary colour is expressed as the wavefront error attributable to the difference in defocus between the n_F and n_D wavelengths (486.1 and 589.3 nm). A plot is shown in Figure 4.28.

It is clear that for the simple achromat under consideration, at least for modest lens apertures, the impact of secondary colour predominates. If a wavefront error of about 50 nm is consistent with 'high quality' imaging, then secondary colour has a significant impact for numerical apertures in excess of 0.05 or f#10. With numerical apertures in excess of 0.2 (f#2.5), higher order spherical aberration starts to make a significant contribution. On the other hand the effect of spherochromatism is more modest throughout. In this context, the impact of spherochromatism would only be a significant issue if secondary colour were first corrected.

Table 4.1 Pupil and field dependence of principal aberrations.

Aberration	Pupil exponent	Field angle exponent
Defocus	2	0
Spherical aberration	4	0
Coma	3	1
Astigmatism	2	2
Field curvature	2	2
Distortion	1	3
Lateral colour	1	1
Longitudinal colour	2	0
Secondary colour	2	0
Spherochromatism	4	0
5th order spherical aberration	6	0

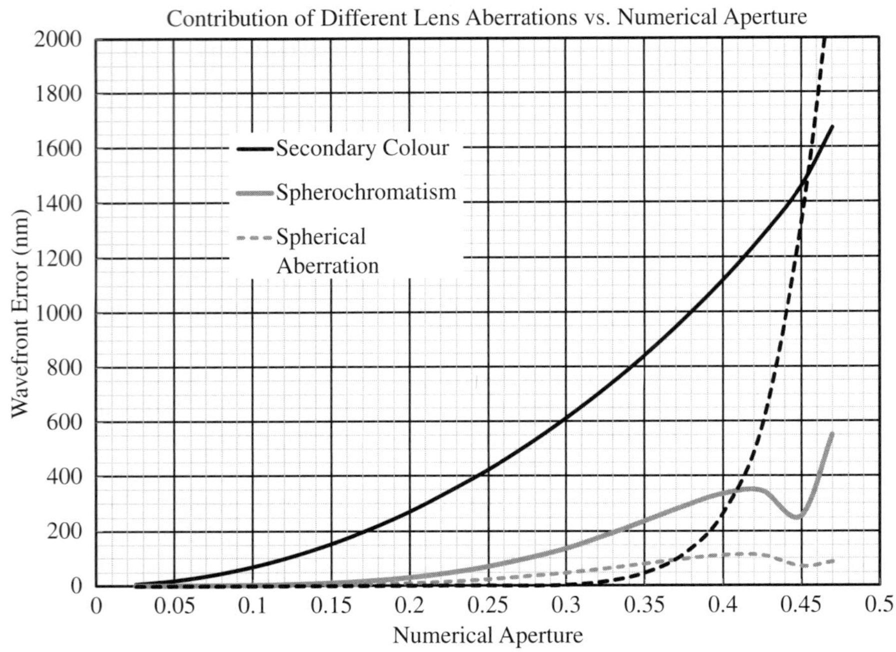

Figure 4.28 Contribution of different aberrations vs. numerical aperture for 200 mm achromat.

Of course, in practice, the design of such lens systems will be accomplished by means of ray tracing software or similar. Nonetheless, an understanding of the basic underlying principles involved in such a design would be useful in the initiation of any design process.

Further Reading

Born, M. and Wolf, E. (1999). *Principles of Optics*, 7e. Cambridge: Cambridge University Press. ISBN: 0-521-642221.

Hecht, E. (2017). *Optics*, 5e. Harlow: Pearson Education. ISBN: 978-0-1339-7722-6.

Kidger, M.J. (2001). *Fundamental Optical Design*. Bellingham: SPIE. ISBN: 0-81943915-0.

Kidger, M.J. (2004). *Intermediate Optical Design*. Bellingham: SPIE. ISBN: 978-0-8194-5217-7.

Longhurst, R.S. (1973). *Geometrical and Physical Optics*, 3e. London: Longmans. ISBN: 0-582-44099-8.

Mahajan, V.N. (1991). *Aberration Theory Made Simple*. Bellingham: SPIE. ISBN: 0-819-40536-1.

Mahajan, V.N. (1998). *Optical Imaging and Aberrations: Part I. Ray Geometrical Optics*. Bellingham: SPIE. ISBN: 0-8194-2515-X.

Mahajan, V.N. (2001). *Optical Imaging and Aberrations: Part II. Wave Diffraction Optics*. Bellingham: SPIE. ISBN: 0-8194-4135-X.

Slyusarev, G.G. (1984). *Aberration and Optical Design Theory*. Boca Raton: CRC Press. ISBN: 978-0852743577.

Smith, F.G. and Thompson, J.H. (1989). *Optics*, 2e. New York: Wiley. ISBN: 0-471-91538-1.

Welford, W.T. (1986). *Aberrations of Optical Systems*. Bristol: Adam Hilger. ISBN: 0-85274-564-8.

5

Aspheric Surfaces and Zernike Polynomials

5.1 Introduction

The previous chapters have provided a substantial grounding in geometrical optics and aberration theory that will provide the understanding required to tackle many design problems. However, there are two significant omissions.

Firstly all previous analysis, particularly with regard to aberration theory, has assumed the use of spherical surfaces. This, in part, forms part of a historical perspective, in that spherical surfaces are exceptionally easy to manufacture when compared to other forms and enjoy the most widespread use in practical applications. Modern design and manufacturing techniques have permitted the use of more exotic shapes. In particular, conic surfaces are used in a wide variety of modern designs.

The second significant omission is the use of ***Zernike circle polynomials*** in describing the mathematical form of wavefront error across a pupil. Zernike polynomials are an orthonormal set of polynomials that are bounded by a circular aperture and, as such, are closely matched to the geometry of a circular pupil. There are, of course, many different sets of orthonormal functions, the most well known being the ***Fourier series***, which, in two dimensions, might be applied to a rectangular aperture. As the wavefront pattern associated with defocus forms one specific Zernike polynomial, the orthonormal property of the series means that all other terms are effectively optimised with respect to defocus. This topic was touched on in Chapter 3 when seeking to minimise the wavefront error associated with spherical aberration by providing balancing defocus. The optimised form that was derived effectively represents a Zernike polynomial.

5.2 Aspheric Surfaces

5.2.1 General Form of Aspheric Surfaces

In this discussion, we will restrict ourselves to surfaces that are symmetric about a central axis. Although more exotic surfaces are used, such symmetric surfaces predominate in practical applications. The most general embodiment of this type of surface is the so-called ***even asphere***. Its general form is specified by its ***surface sag***, z, which represents the axial displacement of the surface with respect to the axial position of the vertex, located at the axis of symmetry. The surface sag of an even asphere is given by the following formula:

$$z = \frac{cr^2}{1 + \sqrt{1 - (1+k)c^2r^2}} + \alpha_1 r^2 + \alpha_2 r^4 + \alpha_3 r^6 + \alpha_4 r^8 + \alpha_5 r^{10} + \alpha_6 r^{12} \quad (5.1)$$

c = 1/R is the surface curvature (R is the radius); k is the conic constant; α_n *is the even polynomial coefficient.*

The curvature parameter, c, essentially describes the spherical radius of the surface. The ***conic constant***, k, is a parameter that describes the shape of a conic surface. For k = 0, the surface is a sphere. More generally, the conic shapes are as set out in Table 5.1.

Optical Engineering Science, First Edition. Stephen Rolt.

Companion website: www.wiley.com/go/Rolt/opt-eng-sci

Table 5.1 Form of conic surfaces.

Conic constant	Surface description
$k > 0$	Oblate ellipsoid
$k = 0$	Sphere
$-1 < k < 0$	Prolate ellipsoid
$k = -1$	Paraboloid
$k < -1$	Hyperboloid

Without the further addition of the even polynomial coefficients, α_n, the surfaces are pure conics. Historically, the paraboloid, as a parabolic mirror shape, has found application as an objective in reflective telescopes. As will be seen subsequently, use of a parabolic mirror shape entirely eliminates spherical aberration for the infinite conjugate. The introduction of the even aspheric terms add further useful variables in optimisation of a design. However, this flexibility comes at the cost of an increase in manufacturing complexity and cost. Strictly speaking, at the first approximation, the terms, α_1 and α_2 are redundant for a general conic shape. Adding the conic term, k, to the surface prescription and optimising effectively allows local correction of the wavefront to the fourth order in r. In this context, the first two even polynomial terms are, to a significant degree, redundant.

5.2.2 Attributes of Conic Mirrors

There is one important attribute of conic surfaces that lies in their mathematical definition. To illustrate this, a section of an ellipsoid, i.e. an ellipse, is shown in Figure 5.1. An ellipse is defined by its two foci and has the property that a line drawn from one focus to any point on the ellipse and thence to the other focus has the same total length regardless of which point on the ellipse was included.

The ellipsoid is defined by its two foci, F_1 and F_2. In this instance, the shape of the ellipsoid is defined by its semi-major distance, a, and its semi-minor distance, b. As suggested, the key point about the ellipsoid shape sketched in Figure 5.1 is that the aggregate distance $F_1P + PF_2$ is always constant. By virtue of Fermat's principle, this inevitably implies that, since the optical path is the same in all cases, F_1 and F_2, from an optical perspective, represent perfect focal points with no aberration whatsoever generated by reflection from the

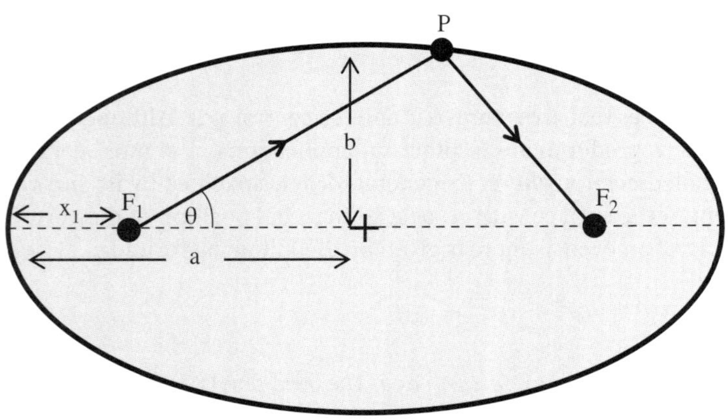

Figure 5.1 Ellipsoid of revolution.

ellipsoidal surface. In describing the ellipsoid above, it is useful to express it in terms of polar coordinates defined with respect to the focal points. If we label the distance F_1P as d, then this distance may be expressed in the following way in terms of the polar angle, θ:

$$d = \frac{d_0}{1 + \varepsilon \cos \theta} \tag{5.2}$$

The parameter, ε, is the so-called eccentricity of the ellipse and is related to the conic parameter, k. In addition, the parameter, d_0 is related to the base radius, R, as defined in the conic section formula in Eq. (5.1). The connection between the parameters is as set out in Eq. (5.3):

$$k = -\varepsilon^2 \tag{5.3}$$

From the perspective of image formation, the two focal points, F_1 and F_2 represent the ideal object and image locations for this conic section. If x_1 in Figure 5.1 represents the object distance u, i.e. the distance from the object to the nearest surface vertex, then it is also possible to calculate the distance, v, to the other focal point. These distances are presented below in the form of Eq. (5.2):

$$u = \frac{d_0}{1 + \varepsilon} \qquad v = -\frac{2d_0}{1 - \varepsilon} \tag{5.4}$$

From the above, it is easy to calculate the conjugate parameter for this conjugate pair:

$$t = \frac{u - v}{u + v} = \frac{1/(1 + \varepsilon) + 1/(1 - \varepsilon)}{1/(1 + \varepsilon) - 1/(1 - \varepsilon)} = -\frac{1}{\varepsilon} = -\frac{1}{\sqrt{-k}}$$

In fact, object and image conjugates are reversible, so the full solution for the conic constant is as in Eq. (5.5):

$$t = \pm\frac{1}{\sqrt{-k}} \tag{5.5}$$

Thus, it is straightforward to demonstrate that for a conic section, there exists one pair of conjugates for which perfect image formation is possible. Of course, the most well known of these is where k = −1, which defines the paraboloidal shape. From Eq. (5.5), the corresponding conjugate parameter is −1 and relates to the infinite conjugate. This forms the basis of the paraboloidal mirror used widely (at the infinite conjugate) in reflecting telescopes and other imaging systems.

As for the spherical mirror, the effective focal length of the mirror remains the same as for the paraxial relationship:

$$\frac{1}{u} - \frac{1}{v} = -2c = -\frac{2}{R} \tag{5.6}$$

More generally, the spherical aberration produced by a conic mirror is of a similar form as for the spherical mirror but with an offset:

$$\Phi_{SA} = -\left(\frac{1}{4R^3}\left(k + \frac{1}{t^2}\right)\right)(x^2 + y^2)^2 \tag{5.7}$$

Worked Example 5.1 ***Simple Mirror-Based Magnifier***

We wish to construct a simple magnification system with a simple conic mirror. The system magnification is to be two and the object distance 100 mm. There is to be no on axis aberration. What is the prescription of the mirror, i.e. base radius and conic constant?

It is assumed that object and image are located the same side of the mirror, so that, in this context, the image distance is −200 mm. The overall set up is illustrated in the diagram:

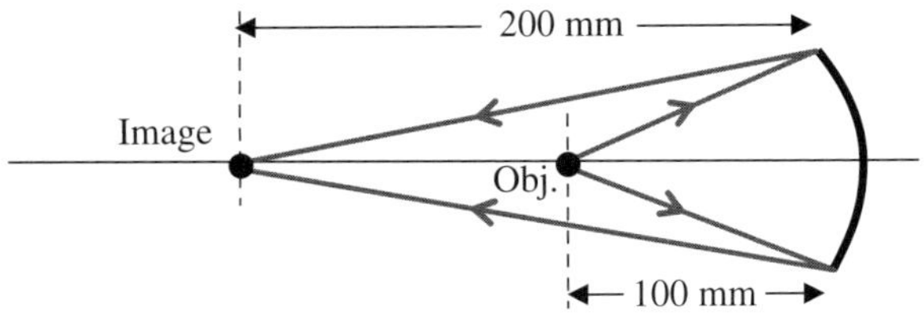

The base radius of the conic mirror is very simple to calculate as it follows the simple paraxial formula, as replicated in Eq. (5.6):

$$-\frac{2}{R} = \frac{1}{u} - \frac{1}{v} \text{ and } -\frac{2}{R} = \frac{1}{100} - \frac{1}{-200}$$

This gives $R = -133$ mm.

We now need to calculate the conjugate parameter, t:

$$t = \frac{v-u}{v+u} = \frac{-200-100}{-200+100} = 3$$

From Eq. (5.5) it is straightforward to see that $k = -(1/t)^2$ and thus $k = -0.1111$. The shape is that of a slightly prolate ellipsoid.

The practical significance of a perfect on axis set up described in this example, is that it forms the basis of an ideal manufacturing test for such a conic surface. This will be described in more detail later in this text.

5.2.3 Conic Refracting Surfaces

There is no generic rule for conic refracting surfaces that generate perfect image formation for an arbitrary conjugate. However, there is a special condition for the infinite conjugate where perfect image formation results, as illustrated in Figure 5.2.

If the refractive index of the surface is n, assuming that the object is in air/vacuum, then the conic constant of the ideal surface is $-n^2$. In fact, the shape is that of a hyperboloid. The abscissa of the hyperboloid effectively produce grazing incidence for rays originating from the object. By definition, therefore, the angle that the surface normal makes with the optical axis at the abscissa is equal to the critical angle. This restricts the maximum numerical aperture that can be collected by the system. With this constraint, it is clear that the maximum numerical aperture is equal to $1/n$. In summary therefore:

$$k_\infty = -n^2 \qquad NA_{MAX} = 1/n \tag{5.8}$$

Unfortunately, no other general condition for perfect image formation results for a conic surface. However, for perfect image correction, all orders of (on axis) aberration are corrected. Thus, although no condition for perfect image formation is possible, it is still possible, nevertheless, to correct for third order spherical aberration with a single refractive surface.

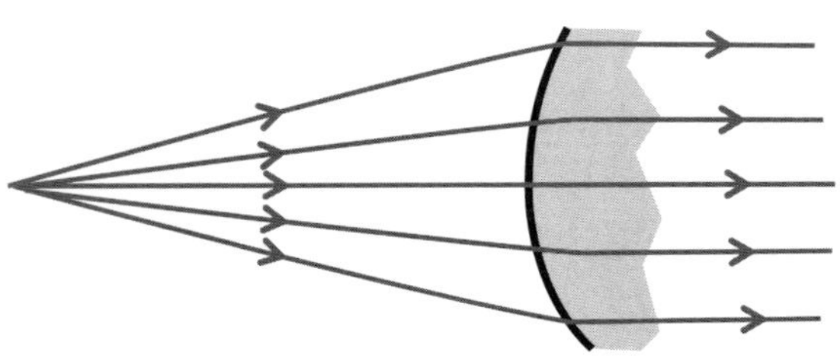

Figure 5.2 Single refractive surface at infinite conjugate.

5.2.4 Optical Design Using Aspheric Surfaces

The preceding discussion largely focused on perfect imaging in specific and restricted circumstances. However, even where perfect imaging is not theoretically possible, aspheric surfaces are extremely useful in the correction of system aberrations with a minimum number of surfaces. For more general design problems, therefore, even asphere terms may be added to the surface prescription. With the stop located at a specific surface, adding aspheric terms to the form of that surface can only control the spherical aberration at that surface. One perspective on the form of a surface is that second order terms only add to the power of that surface, whereas fourth order terms control the third order (in transverse aberration) aberrations. The reasoning behind this assertion may be viewed a little more clearly by expanding the sag of a conic surface in terms of even polynomial terms:

$$z \approx \frac{1}{2}cr^2 + \frac{1}{8}(1+k)c^3r^4 + \frac{1}{16}(1+k)^2c^5r^6 + \frac{5}{128}(1+k)^3c^7r^8 + \ldots\ldots \tag{5.9}$$

Adding a conic term to the surface, in addition to defining the curvature of the surface by its base radius, effectively adds an independent term to Eq. (5.9), effectively controlling two polynomial orders in Eq. (5.9). To this extent, adding separate additional second order and fourth order terms to the even asphere expansion in Eq. (5.1) is redundant. From the perspective of controlling third order aberrations, Eq. (5.9) confirms the utility of a conic surface in adding a controlled amount of fourth order optical path difference (OPD) to the system. In fact, the amount of OPD added to the system, to fourth order, is simply given by the change in sag produced by the conic surface multiplied by the difference in refractive indices. If the refractive index of the first medium is n_0, and that of the second medium, n_1, then the change in OPD produced by introducing a conic parameter of k is given by:

$$\Delta OPD = \frac{1}{8}(n_0 - n_1)kc^3r^4 = \frac{1}{8}(n_0 - n_1)k\frac{r^4}{R^3} \tag{5.10}$$

Equation (5.10) allows estimation of the spherical aberration produced by a conic surface introduced at the stop position. However, by virtue of the stop shift equations introduced in the previous chapter, providing fourth order sag terms at a surface remote from the stop not only influences spherical aberration, but also the other third order aberrations as well. In principle, therefore, by using aspheric surfaces, it is possible to eliminate all third order aberrations with fewer surfaces that would be possible with using just spherical surfaces alone. In fact, assuming that a system has been designed with zero Petzval curvature, it is only necessary to eliminate spherical aberration, coma, and astigmatism. Therefore, only three surfaces are strictly necessary. This represents a considerable improvement over a system employing only spherical surfaces. Notwithstanding the difficulties in manufacturing aspheric surfaces, some commercial camera systems are designed with this principal in mind.

Having introduced the underlying principles, it must be stated that design using aspheric surfaces is not especially amenable to analytical solution. In principle, of course, Eq. (5.10) could be used together with the relevant stop shift equations to compute analytically all third order aberrations. However, in practice, this is a rather cumbersome procedure and design of such systems proceeds largely by computer optimisation. Nevertheless, a clear understanding of the underlying principles is of invaluable help in the design process. An example, a simple two lens system, employing aspheric surfaces is sketched in Figure 5.3. This lens system replicates the performance of a three lens Cooke triplet with an aperture of f#5 and a field of view of 40°. Figure 5.3 is not intended to present a realistic and competitive design, but it merely illustrates the flexibility introduced by the incorporation of aspheric surfaces. In particular, it offers the potential to achieve the same performance with fewer surfaces.

Whilst aspheric components represent a significant enhancement to the toolkit of an optical designer, they represent something of a headache to the component manufacturer. As will be revealed later, in general, aspheric components are more difficult to manufacture and test and hence more costly. As such, their use is restricted to those situations where the advantage provided is especially salient. At the same time,

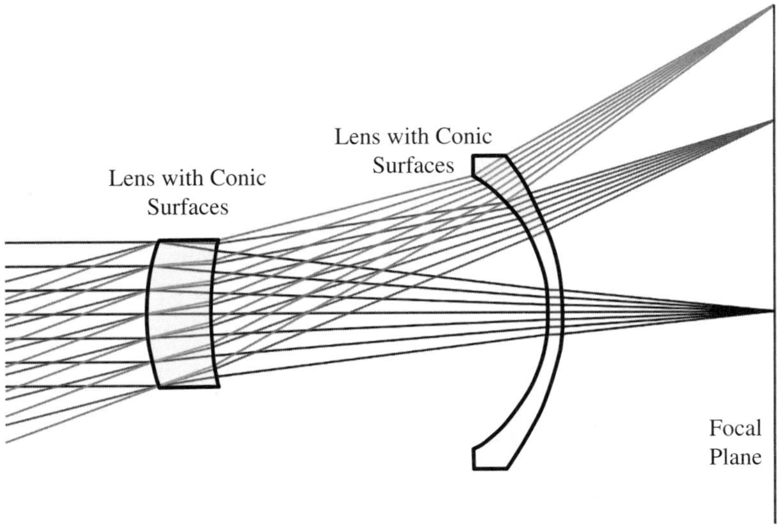

Figure 5.3 Simple two lens system employing aspheric components.

advanced manufacturing techniques have facilitated the production of aspheric surfaces and their application in relatively commonplace designs, such as digital cameras, is becoming a little more widespread. Of course, the presence of conic and aspheric surfaces in large reflecting telescope designs is, by comparison, relatively well established.

5.3 Zernike Polynomials

5.3.1 Introduction

In describing wavefront aberrations at any surface in a system, it is convenient to do so by expressing their value in terms of the two components of normalised pupil functions P_x and P_y. Where the magnitude of the pupil function is equal to unity, this describes the position of a ray at the edge of the pupil. With this description in mind, we now proceed to describe the normalised pupil position in terms of the polar co-ordinates, ρ and θ. This is illustrated in Figure 5.4.

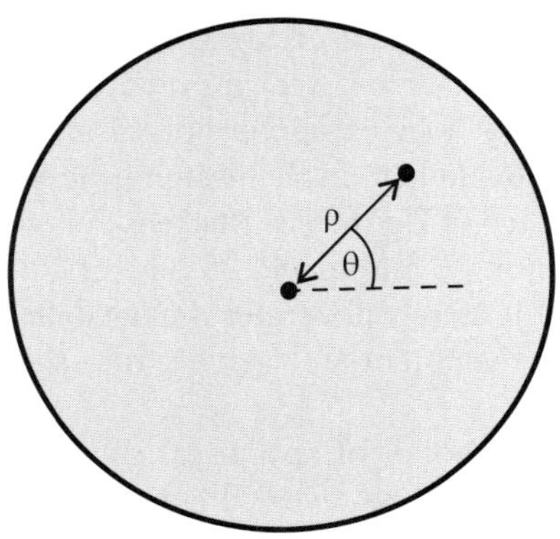

$$P_x = \rho\cos\theta; \; P_y = \rho\sin\theta$$

Figure 5.4 Polar pupil coordinates.

The wavefront error across the pupil can now be expressed in terms of ρ and θ. What we are seeking is a set of polynomials that is orthonormal across the circular pupil described. Any continuous function may be represented in terms of this set of polynomials as follows:

$$F(\rho, \theta) = A_1 f_1(\rho, \theta) + A_2 f_2(\rho, \theta) + A_3 f_3(\rho, \theta) + \ldots \tag{5.11}$$

The individual polynomials are described by the term $f_i(\rho,\theta)$, and their magnitude by the coefficient, A_i. The property of ***orthonormality*** is significant and may be represented in the following way:

$$\iint f_i(\rho, \theta) f_j(\rho, \theta) d\rho d\rho = \delta_{ij} \tag{5.12}$$

The symbol, δ_{ij} is the ***Kronecker delta***. That is to say, when i and j are identical, i.e. the two polynomials in the integral are identical, then the integral is exactly one. Otherwise, if the two polynomials in the integral are different, then the integral is zero. The first property is that of normality, i.e. the polynomials have been normalised to one and the second is that of ***orthogonality***, hence their designation as an ***orthonormal*** polynomial set.

Equations (5.11) and (5.12) give rise to a number of important properties of these polynomials. Initially we might be presented with a problem as to how to represent a known but arbitrary wavefront error, $\Phi(\rho,\theta)$ in terms of the orthonormal series presented in Eq. (5.11). For example, this arbitrary wavefront error may have been computed as part of the design and analysis of a complex optical system. The question that remains is how to calculate the individual polynomial coefficients A_i. To calculate an individual term, one simply takes the cross integral of the function, $\Phi(\rho,\theta)$, with respect to an individual polynomial, $f_i(\rho, \theta)$:

$$\iint \Phi(\rho, \theta) f_i(\rho, \theta) d\rho d\theta = \delta_{i1} A_1 + \delta_{i2} A_2 + \delta_{i3} A_3 + \ldots\ldots$$

By definition we have:

$$\iint \Phi(\rho, \theta) f_i(\rho, \theta) d\rho d\theta = A_i \tag{5.13}$$

So, any coefficient may be determined from the integral presented in Eq. (5.13). The coefficients, A_i, clearly express, in some way, the magnitude of the contribution of each polynomial term to the general wavefront error. In fact, the magnitude of each component, A_i, represents the root mean square (rms) contribution of that component. More specifically, the total rms wavefront error is given by the square root of the sum of the squares of the individual coefficients. That this is so is clearly evident from the orthonormal property of the series:

$$\langle \Phi^2(\rho, \theta) \rangle = \iint \Phi(\rho, \theta) * \Phi(\rho, \theta) d\rho d\theta = A_1^2 + A_2^2 + A_3^2 + \ldots\ldots \tag{5.14}$$

5.3.2 Form of Zernike Polynomials

Following this general discussion about the useful properties of orthonormal functions, we can move on to a description of the Zernike circle polynomials themselves. They were initially investigated and described by Fritz Zernike in 1934 and are admirably suited to a solution space defined by a circular pupil. We will suppose initially, that the polynomial may be described by a component, $R(\rho)$, that is dependent exclusively upon the normalised pupil radius and a component $G(\phi)$ that is dependent upon the polar angle, ϕ. That is to say:

$$f_i(\rho, \varphi) = R(\rho) \times G(\varphi) \tag{5.15}$$

We can make the further assumption that $R(\rho)$ may be represented by a polynomial series in ρ. The form of $G(\phi)$ is easy to deduce. For physically realistic solutions, $G(\phi)$ must repeat identically every 2π radians. Therefore $G(\phi)$ must be represented by a periodic function of the form:

$$G(\phi) = e^{im\phi} \tag{5.16}$$

where m is an integer

This part of the Zernike polynomial clearly conforms to the desired form, since not only does it have the desired periodicity, but it also possesses the desired orthogonality. The parameter, m, represents the angular frequency of the polar dependence.

Having dealt with the polar part of the Zernike polynomial, we turn to the radial portion, $R(\rho)$. The radial part of the Zernike polynomial, $R(\rho)$, comprises of a series of polynomials in ρ. The form of these polynomials, $R(\rho)$, depends upon the angular parameter, m, and the maximum radial order of the polynomial, n. Furthermore, considerations of symmetry dictate that the Zernike polynomials must either be wholly symmetric or anti-symmetric about the centre. That is to say, the operation $r \to -r$ is equivalent to $\phi \to \phi + \pi$. For the Zernike polynomial to be equivalent for both (identical) transformations, for even values of m, only even polynomials terms can be accepted for $R(\rho)$. Similarly, exclusively odd polynomial terms are associated with odd values of m.

Overall, the entirety of the set of Zernike polynomials are continuous and may be represented in powers of P_x and P_y or $\rho\cos(\phi)$ and $\rho\sin(\phi)$. It is not possible to construct trigonometric expressions of order, m, i.e. $\cos(m\phi)$ and $\rho\sin(m\phi)$ where the order of the corresponding polynomial is less than m. Therefore, the polynomial, $R(\rho)$, cannot contain terms in ρ that are of lower order than the angular parameter, m.

To describe each polynomial, $R(\rho)$, it is customary to define it in terms of the maximum order of the polynomial, n, and the angular parameter, m. For all values of m (and n), the polynomial, $R(\rho)$, may be expressed as per Eq. (5.17).

$$R_{n,m}(\rho) = N_{n,m} \sum_{i=0}^{i=(n-m)/2} C_{n,m,i}\rho^{(n-2i)} \tag{5.17}$$

$C_{n,m,i}$ represents the value of a specific coefficient

The parameter, $N_{n,m}$, is a normalisation factor. Of course, any arbitrary scaling factor may be applied to the coefficients, $C_{n,m,i}$, provided it is compensated by the normalisation factor. By convention, the base polynomial has a value of unity for $\rho = 1$. Of course, with this in mind, the purpose of the normalisation factor is to ensure that, in all cases, the rms value of the polynomial is normalised to one. It now remains only to calculate the values of the coefficients, $C_{n,m,i}$. These are determined from the condition of orthogonality which applies separately for $R_{n,m}(\rho)$ and may be set out as follows:

$$\int_0^1 R_{n,m}(\rho)R_{n',m}(\rho)\rho d\rho = \delta_{nn'} \tag{5.18}$$

The general formula for the coefficients $C_{n,m,i}$ is set out in Eq. (5.18).

$$C_{n,m,i} = (-1)^i \left[\frac{(n-i)!}{i!\left(\frac{(n+m)}{2} - i\right)!\left(\frac{(n-m)}{2} - i\right)!} \right] \tag{5.19}$$

For $i = n = 0$, the value of the coefficient, $C_{n,m,i}$, as prescribed for the piston term, is unity. The value of the normalisation factor, $N_{n,m}$, is given in Eq. (5.20).

$$N_{n,m} = \sqrt{(n+1)} \qquad \text{For } m = 0; \qquad N_{n,m} = \sqrt{2(n+1)} \qquad \text{For } m <> 0 \tag{5.20}$$

More completely we can express the entire polynomial:

$$R(\rho)_{n,m} = \sqrt{(n+1)} \sum_{i=0}^{i=(n-m)/2} (-1)^i \left[\frac{(n-i)!}{i!\left(\frac{(n+m)}{2} - i\right)!\left(\frac{(n-m)}{2} - i\right)!} \right] \rho^{(n-2i)} \qquad \mathrm{m} = 0 \tag{5.21a}$$

$$R(\rho)_{n,m} = \sqrt{(2n+1)} \sum_{i=0}^{i=(n-m)/2} (-1)^i \left[\frac{(n-i)!}{i!\left(\frac{(n+m)}{2} - i\right)!\left(\frac{(n-m)}{2} - i\right)!} \right] \rho^{(n-2i)} \qquad m <> 0 \tag{5.21b}$$

The parameter, m, can take on positive or negative values as can be seen from Eq. (5.16). Of course, Eq. (5.16) gives the complex trigonometric form. However, by convention, negative values for the parameter m are ascribed to terms involving sin(mϕ), whilst positive values are ascribed to terms involving cos(mϕ).

Zernike polynomials are widely used in the analysis of optical system aberrations. Because of the fundamental nature of these polynomials, all the Gauss-Seidel wavefront aberrations clearly map onto specific Zernike polynomials. For example, spherical aberration has no polar angle dependence, but does have a fourth order dependence upon pupil function. This suggests that this aberration has a radial order, n, of 4 and a polar dependence, m, of zero. Similarly, coma has a radial order of 3 and a polar dependence of one. Table 5.2 provides a list of the first 28 Zernike polynomials.

In Table 5.2, each Zernike polynomial has been assigned a unique number. This is the 'Standard' numbering convention adopted by the American National Standards Institute, (***ANSI***). It has the benefit of following the Born and Wolf notation logically, starting from the piston term which is denominated the zeroth term. If the ANSI number is represented as j, and the Born and Wolf indices as n, m, then the ANSI number may be derived as follows:

$$j = \frac{n(n+1)+m}{2} \tag{5.22}$$

Unfortunately, a variety of different numbering conventions prevail, leading to significant confusion. This will be explored a little later in this chapter. As a consequence of this, the reader is advised to be cautious in applying any single digit numbering convention to Zernike polynomials. By contrast, the n, m numbering convention used by Born and Wolf is unambiguous and should be used where there is any possibility of confusion.

5.3.3 Zernike Polynomials and Aberration

As outlined previously, there is a strong connection between Zernike polynomials and primary aberrations when expressed in terms of wavefront error. Table 5.2 clearly shows the correspondence between the polynomials and the Gauss Seidel aberrations, with the 3rd order Gauss-Seidel aberrations, such as spherical aberration and coma clearly visible.

The power of the Zernike polynomials, as an orthonormal set, lies in their ability to represent any arbitrary wavefront aberration. Using the approach set out in Eq. (5.13), it is possible to compute the magnitude of any Zernike term by the cross integral of the relevant polynomial and the wavefront disturbance. Furthermore, the total root mean square (rms) wavefront error, as per Eq. (5.14), may be calculated from the RSS (root sum square) of the individual Zernike magnitudes. That is to say, the Zernike magnitude of each term represents its contribution to the rms wavefront error, as averaged over the whole pupil.

The use of defocus to compensate spherical aberration was explored in Chapters 3 and 4. In this instance, for a given amount of fourth order wavefront error, we sought to minimise the rms wavefront error by applying a small amount of defocus.

$$\Phi(\rho) = A\rho^6 = \underbrace{\frac{A}{6\sqrt{5}}[\sqrt{5}(6\rho^4 - 6\rho^2 + 1)]}_{\text{Spherical Aberration}} + \underbrace{\frac{A}{2\sqrt{3}}[\sqrt{3}(2\rho^2 - 1)]}_{\text{Defocus}} + \underbrace{\frac{A}{3}}_{\text{Piston}}$$

Hence, without defocus, adjustment, the raw spherical aberration produced in a system may be expressed as the sum of three Zernike terms, one spherical aberration, one defocus and one piston term. The total aberration for an uncompensated system is simply given by the RSS of the individual terms. However, for

Table 5.2 First 28 Zernike polynomials.

ANSI#	N	m	$N_{n,m}$	$R(\rho)$	$G(\varphi)$	Name
0	0	0	1	1	1	Piston
1	1	−1	$\sqrt{2}$	ρ	$\sin\phi$	Tilt X
2	1	1	$\sqrt{2}$	ρ	$\cos\phi$	Tilt Y
3	2	−2	$\sqrt{6}$	ρ^2	$\sin 2\phi$	45° Astigmatism
4	2	0	$\sqrt{3}$	$2\rho^2-1$	1	Defocus
5	2	2	$\sqrt{6}$	ρ^2	$\cos 2\phi$	90° Astigmatism
6	3	−3	$\sqrt{8}$	ρ^3	$\sin 3\phi$	Trefoil
7	3	−1	$\sqrt{8}$	$3\rho^3-2\rho$	$\sin\phi$	Coma Y
8	3	1	$\sqrt{8}$	$3\rho^3-2\rho$	$\cos\phi$	Coma X
9	3	3	$\sqrt{8}$	ρ^3	$\cos 3\phi$	Trefoil
10	4	−4	$\sqrt{10}$	ρ^4	$\sin 4\phi$	Quadrafoil
11	4	−2	$\sqrt{10}$	$4\rho^4-3\rho^2$	$\sin 2\phi$	5th Order astigmatism 45°
12	4	0	$\sqrt{5}$	$6\rho^4-6\rho^2+1$	1	Spherical aberration
13	4	2	$\sqrt{10}$	$4\rho^4-3\rho^2$	$\cos 2\phi$	5th Order astigmatism 90°
14	4	4	$\sqrt{10}$	ρ^4	$\cos 4\phi$	Quadrafoil
15	5	−5	$\sqrt{12}$	ρ^5	$\sin 5\phi$	Pentafoil
16	5	−3	$\sqrt{12}$	$5\rho^5-4\rho^3$	$\sin 3\phi$	High order trefoil
17	5	−1	$\sqrt{12}$	$5\rho^5-4\rho^3$	$\sin\phi$	5th Order coma Y
18	5	3	$\sqrt{12}$	$10\rho^5-12\rho^3+3\rho$	$\cos\phi$	5th Order coma X
19	5	−5	$\sqrt{12}$	ρ^5	$\cos 3\phi$	High order trefoil
20	5	−5	$\sqrt{12}$	ρ^5	$\cos 5\phi$	Pentafoil
21	6	−6	$\sqrt{14}$	ρ^6	$\sin 6\phi$	Hexafoil
22	6	−4	$\sqrt{14}$	$6\rho^6-5\rho^4$	$\sin 4\phi$	High order quadrafoil
23	6	−2	$\sqrt{14}$	$15\rho^6-20\rho^4+6\rho^2$	$\sin 2\phi$	7th Order astigmatism 45°
24	6	0	$\sqrt{7}$	$20\rho^6-30\rho^4+12\rho^2-1$	1	5th Order spherical aberration
25	6	2	$\sqrt{14}$	$15\rho^6-20\rho^4+6\rho^2$	$\cos 2\phi$	7th Order astigmatism 90°
26	6	4	$\sqrt{14}$	$6\rho^6-5\rho^4$	$\cos 4\phi$	High order quadrafoil
27	6	6	$\sqrt{14}$	ρ^6	$\cos 6\phi$	Hexafoil

a compensated system only the Zernike $n = 4$, $m = 0$ term needs be considered. This then gives the following fundamental relationship:

$$\Phi(\rho) = A\rho^4 \qquad \Phi_{RMS}(Uncompensated) = \frac{A}{\sqrt{5}} \qquad \Phi_{RMS}(Compensated) = \frac{A}{\sqrt{180}} \tag{5.23}$$

The rms wavefront error has thus been reduced by a factor of six by the focus compensation process. Furthermore, this analysis feeds in to the discussion in Chapter 3 on the use of balancing aberrations to minimise wavefront error. For example, if we have succeeded in eliminating third order spherical aberration and are presented with residual fifth order spherical aberration, we can minimise the rms wavefront error by balancing this aberration with a small amount of third order aberration in addition to defocus. Analysis using Zernike polynomials is extremely useful in resolving this problem:

$$\Phi(\rho) = A\rho^6 = \underbrace{\frac{A}{20\sqrt{7}}[\sqrt{7}(20\rho^6 - 30\rho^4 + 12\rho^2 - 1)]}_{\text{5th order Spherical Aberration}} + \underbrace{\frac{A}{4\sqrt{5}}[\sqrt{5}(6\rho^4 - 6\rho^2 + 1)]}_{\text{Spherical Aberration}} + \underbrace{\frac{9A}{20\sqrt{3}}[\sqrt{3}(2\rho^2 - 1)]}_{\text{Defocus}} + \underbrace{\frac{A}{4}}_{\text{Piston}}$$

As previously outlined, the uncompensated rms wavefront error may be calculated from the RSS sum of all the four Zernike terms. Naturally, for the compensated system, we need only consider the first term.

$$\Phi(\rho) = A\rho^6 \qquad \Phi_{RMS}(Uncompensated) = \frac{A}{\sqrt{7}} \qquad \Phi_{RMS}(Compensated) = \frac{A}{\sqrt{2800}} \tag{5.24}$$

For the fifth order spherical aberration, the rms wavefront error has been reduced by a factor of 20 through the process of aberration balancing. In terms of the practical application of this process, one might wish to optimise an optical design by minimising the rms wavefront error. Although, in practice, the process of optimisation will be carried out using software tools, nonetheless, it is useful to recognise some key features of an optimised design. By virtue of the previous example, optimisation of spherical aberration should lead to an OPD profile that is close to the 5th order Zernike term. This is shown in Figure 5.5 which illustrates the profile of an optimised OPD based entirely on the relevant fifth order Zernike term. The graph plots the nominal OPD again the normalised pupil function with the form given by the Zernike polynomial, $n = 6$, $m = 0$.

In the optimisation of an optical design it is important to understand the form of the OPD fan displayed in Figure 5.5 in order recognise the desired endpoint of the optimisation process. It displays three minima and two maxima (or vice versa), whereas the unoptimised OPD fan has one fewer maximum and minimum. Thus, although the design optimisation process itself might be computer based, nevertheless, understanding and recognising the how the process works and its end goal will be of great practical use. That is to say, as the computer-based optimisation proceeds, on might expect the OPD fan to acquire a greater number of maxima and minima.

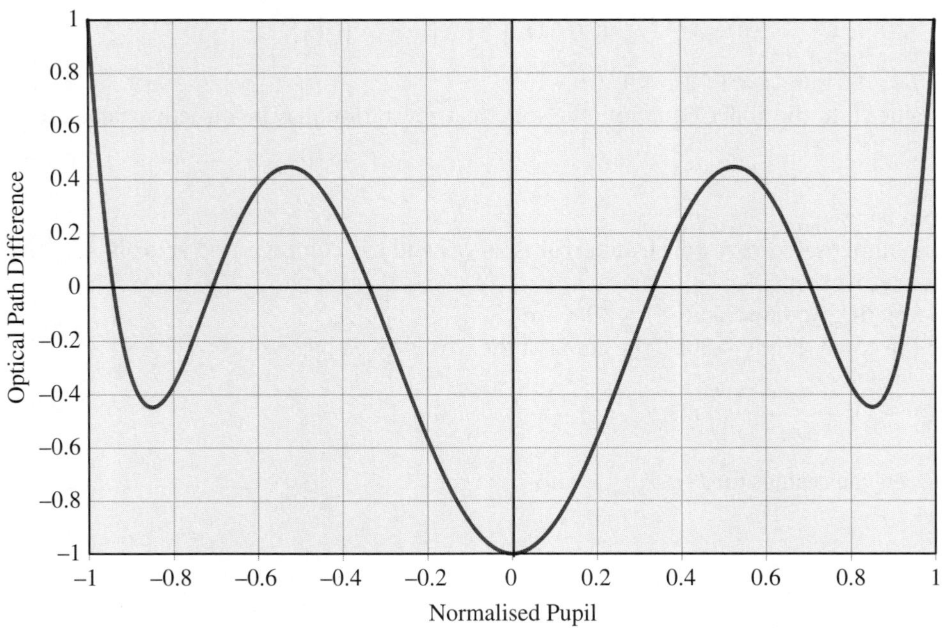

Figure 5.5 Fifth order Zernike polynomial and aberration balancing.

One can apply the same analysis to all the Gauss-Seidel aberrations and calculate its associated rms wavefront error.

$$\text{Spherical Aberration}: \quad \Phi(\rho) = A\rho^4 \quad \Phi_{RMS} = \frac{A}{\sqrt{180}} \tag{5.25a}$$

$$\text{Coma}: \quad \Phi(\rho,\varphi,\theta) = A\theta\rho^3 \sin\varphi \quad \Phi_{RMS} = \frac{A\theta}{\sqrt{72}} \tag{5.25b}$$

$$\text{Astigmatism}: \quad \Phi(\rho,\varphi,\theta) = A\theta^2\rho^2 \sin 2\varphi \quad \Phi_{RMS} = \frac{A\theta^2}{\sqrt{6}} \tag{5.25c}$$

$$\text{Field Curvature}: \quad \Phi(\rho,\varphi,\theta) = A\theta^2\rho^2 \quad \Phi_{RMS} = \frac{A\theta^2}{\sqrt{12}} \tag{5.25d}$$

θ represents the field angle

Equations (5.25a)–(5.25d) are of great significance in the analysis of image quality, as the rms wavefront error is a key parameter in the description of the optical quality of a system. This will be discussed in more detail in the next chapter.

Worked Example 5.2 A plano-convex lens, with a focal length of 100 mm is used to focus a collimated beam; the refractive index of the lens material is 1.52. It is assumed that the curved surface faces the infinite conjugate. The pupil diameter is 12.5 mm and the aperture is situated at the lens. What is the rms spherical aberration produced by this lens – (i) at the paraxial focus; (ii) at the compensated focus? What is the rms coma for a similar collimated beam with a field angle of one degree?

Firstly, we calculate the spherical aberration of the single lens. With the object at infinity and the image at the first focal point, the conjugate parameter, t, is equal to -1. The shape parameter, s, for the plano convex lens is equal to 1 since the curved surface is facing the object. From Eq. (4.30a) the spherical aberration of the lens is given by:

$$\Phi_{SA} = -\left(\frac{1}{32f^3}\right)\left[\left(\frac{n}{n-1}\right)^2 - \left(\frac{n}{n+2}\right)t^2 + \frac{(n+2)}{n(n-1)^2}\left[s + 2\left[\frac{n^2-1}{n+2}\right]t\right]^2\right] r^4$$

$r_{\max} = 6.25$ mm (12.5/2); $f = 100$ mm; $n = 1.52$; $s = 1$; $t = -1$

By substituting these values into the above equation, the spherical aberration may be directly calculated:

$$\Phi_{SA} = A\rho^4$$

where $A = 4.13 \times 10^{-4}$ mm $\rho = r/r_{\max}$

From Eq. (5.23), the uncompensated rms wavefront error is $A/\sqrt{5}$ and the compensated error is $A/\sqrt{180}$. Therefore the rms values are given by:

Φ_{rms}(paraxial) = 185 nm; Φ_{rms}(compensated) = 30.8 nm

Secondly, we calculate the coma. From (4.30b), the coma of the lens is given by:

$$\Phi_{CO} = -\frac{1}{4nf^2}\left((2n+1)t + \frac{(n+1)}{(n-1)}s\right) r^2 r_y \theta$$

Again, substituting the relevant values for f, n, r_{max}, s, and t, we get:

$$\Phi_{CO} = A\theta\rho^3 \sin\varphi$$

where $A = 3.24 \times 10^{-3}$ mm $\rho = r/r_{\max}$ $\quad r_y = r\sin\varphi$

From (5.25b) $\Phi_{RMS} = \frac{A\theta}{\sqrt{72}} = 3.81 \times 10^{-4}$ mm × θ (in radians)

We are told that θ = 1° or 0.0174 rad. Therefore, **$\Phi_{rms} = 6.66 \times 10^{-6}$ or 6.66 nm**

5.3.4 General Representation of Wavefront Error

We have emphasised the synergy between Zernike polynomials and the classical treatment of aberrations in an axially symmetric optical system, i.e. the Gauss-Seidel aberrations. However, in practice, in real optical systems, these axial symmetries are often compromised, either by accident or by design. Some systems are deliberately designed whereby not all optical surfaces are aligned to a common axis. These will inevitably introduce non-standard wavefront aberrations into the system. Most significantly, even with a symmetrical design, component manufacturing errors and system alignment may introduce more complex wavefront errors into the system. Naturally, alignment errors create an off-axis optical system 'by accident'. Manufacturing or polishing errors might produce an optical surface whose shape departs from that of an ideal sphere or conic in a somewhat complex fashion. For example, the effects of these errors may be to introduce a trefoil term ($n = 3$, $m = 3$) into the wavefront error; this is not a standard Gauss-Seidel term.

As argued, Zernike polynomials are widely used in the analysis of wavefront error both in the design and testing of optical systems. From a strictly analytical and theoretical point of view the description of wavefront error in terms of its rms value is the most meaningful. However, for largely historical reasons, wavefront error is often presented as a '***peak to valley***' error. That is to say, the value presented is the difference between the maximum and minimum OPD across the pupil. Historically, the wavefront error for a system might have been derived from a visual inspection of a fringe pattern in an interferogram. The maximum deviation of fringes is relatively straightforward to estimate visually from a fringe pattern which might have been produced photographically. However, the rms wavefront error is more directly related to system performance. Calculation of the rms wavefront error across a pupil is a mathematical process that requires computational data acquisition and analysis and has only been universally available in more recent times. Therefore, the use of the peak to valley description still persists.

One particular disadvantage of the peak to valley description is that it is unusually responsive to large, but highly localised excursions in the wavefront error. More generally, as a rule of thumb, the peak to valley is considered to be 3.5 times the rms value. Of course, this does depend upon the form of the wavefront error. Table 5.3 sets out this relationship for the first 11 Zernike terms (apart from piston). For comparison, a standard statistical measure is also presented – namely for a normally distributed wavefront error profile, the limits containing 95% of the wavefront error distribution (± 1.96 standard deviations).

The values presented in Table 5.3 are simply the ratio of the peak to valley (p-to-v) error for that particular distribution. To overcome the principal objection to the p-to-v measure, namely its heightened sensitivity to local variation a new peak to valley measure has been proposed by the Zygo Corporation. This measure is known as ***P to Vr*** or peak to valley robust. In this measure, the wavefront error is fitted to a set of 36 Zernike polynomials. Although this process is carried out by computational analysis, the procedure is very simple. Essentially the calculation process exploits the orthonormal properties of the polynomial set and calculates the contribution of each Zernike term using the relation set out in Eq. (5.12). Following this process, the

Table 5.3 Peak to valley: Root mean square (rms) ratios for different wavefront error forms.

Noll#	n	m	Description	P_to_V multiplier
2 and 3	1	±1	Tilt	2.83
4	2	0	Defocus	3.46
5 and 6	2	±2	Astigmatism	4.90
7 and 8	3	±1	Coma	5.66
9 and 10	3	±3	Trefoil	5.66
11	4	0	Spherical aberration	3.35
—	—	—	95% Gaussian	3.92

Table 5.4 Comparison of Zernike numbering systems.

n	m	ANSI	Noll	Fringe	n	m	ANSI	Noll	Fringe	n	m	ANSI	Noll	Fringe
0	0	0	1	0	6	−4	22	25	28	8	8	44	44	64
1	−1	1	3	2	6	−2	23	23	21	9	−9	45	55	82
1	1	2	2	1	6	0	24	22	15	9	−7	46	53	67
2	−2	3	5	5	6	2	25	24	20	9	−5	47	51	54
2	0	4	4	3	6	4	26	26	27	9	−3	48	49	43
2	2	5	6	4	6	6	27	28	36	9	−1	49	47	34
3	−3	6	9	10	7	−7	28	35	50	9	1	50	46	33
3	−1	7	7	7	7	−5	29	33	39	9	3	51	48	42
3	1	8	8	6	7	−3	30	31	30	9	5	52	50	53
3	3	9	10	9	7	−1	31	29	23	9	7	53	52	66
4	−4	10	15	17	7	1	32	30	22	9	9	54	54	81
4	−2	11	13	12	7	3	33	32	29	10	−10	55	66	101
4	0	12	11	8	7	5	34	34	38	10	−8	56	64	84
4	2	13	12	11	7	7	35	36	49	10	−6	57	62	69
4	4	14	14	16	8	−8	36	45	65	10	−4	58	60	56
5	−5	15	21	26	8	−6	37	43	52	10	−2	59	58	45
5	−3	16	19	19	8	−4	38	41	41	10	0	60	56	35
5	−1	17	17	14	8	−2	39	39	32	10	2	61	57	44
5	1	18	16	13	8	0	40	37	24	10	4	62	59	55
5	3	19	18	18	8	2	41	38	31	10	6	63	61	68
5	5	20	20	25	8	4	42	40	40	10	8	64	63	83
6	−6	21	27	37	8	6	43	42	51	10	10	65	65	100

maximum and minimum of the fitted surface is calculated and the revised peak to valley figure calculated. Of course, the reduced set of 36 polynomials cannot possibly replicate localised asperities with a high spatial frequency content. Therefore, the fitted surface is effectively a smoothed version of the original and the peak to valley value derived is more representative of the underlying physics.

It must be stated, at this point, that the 36 polynomials used, in this instance, are not those that would be ordered as in Table 5.1. That is to say, they are not the first 36 ANSI standard polynomials. As mentioned earlier, there are, unfortunately, a number of competing conventions for the numbering of Zernike polynomials. The convention used in determining the P to Vr figure is the so called ***Zernike Fringe*** polynomial convention. The logic of ordering the polynomials in a different way is that this better reflects, in the case of the fringe polynomial set, the spatial frequency content of the polynomial and its practical significance in real optical systems.

5.3.5 Other Zernike Numbering Conventions

The ordering convention adopted by the Fringe polynomials expresses, to a significant degree, the spatial frequency content of the polynomial. As a consequence, the polynomials are ordered by the sum of their radial and polar orders, rather than primarily by the radial order. That is to say, the polynomials are ordered by the sum $n + m$, as opposed to n alone. For polynomials of equal 'fringe order' they are then ordered by descending values of the modulus of m, i.e. $|m|$, with the positive or cosine term presented first.

Another convention that is very widely used is the ***Noll convention***. The Noll convention proceeds in a broadly similar way to the ANSI convention, in that it uses the radial order, n, as the primary parameter for sorting. However, there are a number of key differences. Firstly, the sequence starts with the number one, as opposed to zero, as is the case for the other conventions. Secondly, the ordering convention for the polar order, m, as in the case of the fringe polynomials, follows the modulus of m rather its absolute value. However, the ordering is in ascending sequence of $|m|$, unlike the fringe polynomials. The ordering of the sine and cosine terms is presented in such a way that all positive m (cosine terms) are allocated an even number. In consequence, sometimes the sine term occurs before the cosine term in the sequence and sometimes after. Table 5.4 shows a comparison of the different numbering systems up to ANSI number 65.

Further Reading

American National Standards Institute (2017). *Methods for Reporting Optical Aberrations of Eyes*, ANSI Z80.28:2017. Washington DC: ANSI.

Born, M. and Wolf, E. (1999). *Principles of Optics*, 7e. Cambridge: Cambridge University Press. ISBN: 0-521-642221.

Fischer, R.E., Tadic-Galeb, B., and Yoder, P.R. (2008). *Optical System Design*, 2e. Bellingham: SPIE. ISBN: 978-0-8194-6785-0.

Hecht, E. (2017). *Optics*, 5e. Harlow: Pearson Education. ISBN: 978-0-1339-7722-6.

Noll, R. (1976). Zernike polynomials and atmospheric turbulence. *J. Opt. Soc. Am.* 66 (3): 207.

Zernike, F. (1934). Beugungstheorie des Schneidenverfahrens und Seiner Verbesserten Form, der Phasenkontrastmethode. *Physica* 1 (8): 689.

6

Diffraction, Physical Optics, and Image Quality

6.1 Introduction

Hitherto, we have presented optics purely in terms of the geometrical interpretation provided by the propagation and tracing of rays. Notwithstanding this rather simplistic foundation, this conveniently simple picture is ultimately derived from an understanding of the wave nature of light. More specifically, Fermat's principle, which underpins geometrical optics is itself ultimately derived from Maxwell's famous wave equations, as introduced in Chapter 1. However, in this chapter, we shall focus on the circumstances where the assumptions underlying geometrical optics breakdown and this convenient formulation is no longer tractable. Under these circumstances, we must look to another approach, more explicitly tied to the wave nature of light, the study of ***physical optics***. To look at this a little more closely, we must further examine Maxwell's equations. The ubiquitous vector form in which Maxwell's equations are now cast is actually due to Oliver Heaviside and these are set out below:

$$\nabla.\mathbf{D} = \rho\,(Gauss's\ law) \tag{6.1a}$$

$$\nabla.\mathbf{B} = 0\,(Gauss's\ law\ for\ magnetism) \tag{6.1b}$$

$$\nabla\times\mathbf{E} = -\frac{\partial\mathbf{B}}{\partial t}\,(Faraday's\ law\ of\ electromagnetic\ induction) \tag{6.1c}$$

$$\nabla\times\mathbf{H} = \mathbf{J} + \frac{\partial\mathbf{D}}{\partial t}\,(Ampère's\ law\ with\ addition\ of\ displacement\ current) \tag{6.1d}$$

D, **B**, **E**, **H**, and **J** are all vector quantities, where **D** is the *electric displacement*, **B** the *magnetic field*, **E** the *electric field*, **H** the *magnetic field strength* and **J** the *current density*.

The quantities D and E and B and H are themselves interrelated:

$$\mathbf{D} = \varepsilon\varepsilon_0\mathbf{E} \quad \mathbf{B} = \mu\mu_0\mathbf{H} \tag{6.2}$$

The quantities, ε_0 and μ_0, are the permittivity and magnetic permeability of free space respectively. These quantities are associated specifically with free-space (vacuum). The quantities ε and μ are the relative permittivity and relative permeability of a specific medium or substance.

These equations may be greatly simplified if we assume that the local current and charge density is zero and we are ultimately presented with the classical wave equation.

$$\nabla^2\mathbf{E} = \mu\mu_0\varepsilon\varepsilon_0\frac{\partial^2\mathbf{E}}{\partial t^2} = \frac{1}{c^2}\frac{\partial^2\mathbf{E}}{\partial t^2}\ \text{where c is the speed of light and}\ c = \frac{1}{\sqrt{\mu\mu_0\varepsilon\varepsilon_0}} \tag{6.3}$$

The next stage in this critique of geometrical optics is to use Maxwell's equation to derive the Eikonal equation, that was briefly introduced in Chapter 1.

Optical Engineering Science, First Edition. Stephen Rolt.

Companion website: www.wiley.com/go/Rolt/opt-eng-sci

6.2 The Eikonal Equation

In Eq. (6.3), we have presented that wave equation in its true vector format. That is to say, the equation describes the electric field, **E**, as a vector quantity. However, much of what we will present in this chapter is a simplification of the wave equation, known as ***scalar theory***. In this case, it is assumed that the electric field may be represented as a pseudo-scalar quantity. That is to say, the electric field, although varying in magnitude, is confined to one specific orientation and may be treated as if it were a scalar quantity. In fact, this approximation is reasonable where light is closely confined to some axis of propagation, i.e. consistent with the paraxial approximation. Thus, we are to understand that there are some limitations to this treatment.

In presenting the Eikonal equation according the scalar view, we assume that solutions to the wave equation are of the form:

$$E = E_0(x, y, z)e^{i(kS(x,y,z)-\omega t)} \tag{6.4}$$

$E_0(x, y, z)$ is a slowly varying envelope function and $S(x, y, z)$ is the spatially varying phase of the wave. In fact $S(x, y, z)$ has dimensions of length and when it is equal to the wavelength the phase term it describes is equal to 2π. The angular frequency is denoted by ω and the spatial frequency by k.

The scalar form of the wave equation may be written as

$$\frac{\partial^2 E}{\partial x^2} + \frac{\partial^2 E}{\partial y^2} + \frac{\partial^2 E}{\partial z^2} = \frac{n^2}{c^2}\frac{\partial^2 E}{\partial t^2}$$

From the above, we can derive the Eikonal equation, but we must assume the $E_0(x, y, z)$ and the first differential of $S(x, y, z)$ vary slowly with respect to position. The classical Eikonal equation is set out in Eq. (6.5).

$$\left(\frac{\partial S}{\partial x} + \frac{\partial S}{\partial y} + \frac{\partial S}{\partial z}\right)^2 = n^2 \tag{6.5}$$

It is clear that by differentiating Eq. (6.4) twice with respect to x, y, and z, that in deriving Eq. (6.5), we are neglecting terms containing the second differential with respect to S. We are also ignoring changes in the envelope function. Thus it is clear that in deriving Eq. (6.5), we are making the following assumptions:

$$\frac{\frac{\partial E_0}{\partial x} + \frac{\partial E_0}{\partial y} + \frac{\partial E_0}{\partial z}}{E_0} << k\left(\frac{\partial S}{\partial x} + \frac{\partial S}{\partial y} + \frac{\partial S}{\partial z}\right) \tag{6.6a}$$

and

$$\frac{\partial^2 S}{\partial x^2} + \frac{\partial^2 S}{\partial y^2} + \frac{\partial^2 S}{\partial z^2} << k\left(\frac{\partial S}{\partial x} + \frac{\partial S}{\partial y} + \frac{\partial S}{\partial z}\right)^2 \tag{6.6b}$$

What Eq. (6.6a) suggests is that the envelope function must vary slowly compared to the wavelength. In addition, Eq. (6.6b) suggests that the curvature of the wavefront must be small when compared to the spatial frequency, k. In other words, the assumptions underlying the Eikonal equation are only justified where the radius of any wavefront is much greater than the wavelength. As the Eikonal equation underpins geometrical optics, this sets the limits on the applicability of this methodology, and we must then seek other, more general, means to describe the behaviour of light. These methods are, of course, based on a more rigorous application of Maxwell's equations and are generally categorised under the heading of ***physical optics***.

6.3 Huygens Wavelets and the Diffraction Formulae

Although Maxwell's equations form the rigorous description of electromagnetic wave propagation, we will first proceed from the rather more intuitive description by ***Huygens' principle***. Huygens' principle states

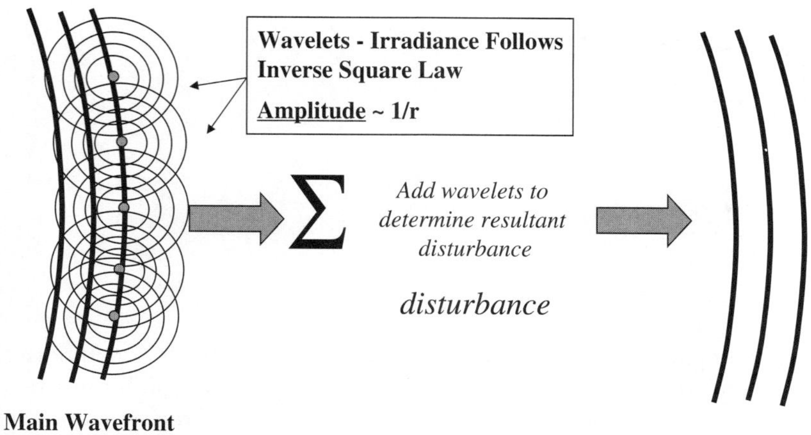

Figure 6.1 Conceptual illustration of Huygens' principle.

that, given a known wave disturbance described by a continuous surface of equal phase – the wavefront, then the amplitude of the wave at any point in space may be determined as the sum of the amplitude of ***forward propagating*** wavelets from that surface. This is illustrated in Figure 6.1.

The amplitude of the wave represents the strength of the local electric or magnetic field. In this case, in our scalar representation, we consider the amplitude as the magnitude of the vector electric field. The flux or power per unit area transmitted by the wave is determined by the Poynting vector, which is the cross product of the electric and magnetic fields. In the context of this scalar treatment, the flux density is proportional to the square of the electric field. In the Huygens' representation, as illustrated in Figure 6.1, the amplitude of the secondary waves emerging from some point on the original wavefront is inversely proportional to the distance from that point. It follows, therefore, that the flux density associated with that secondary wave follows an inverse square dependence with distance. This is further illustrated in Figure 6.2 which summarises the geometry.

$$Amplitude = f(\chi)A(x, y, z)\frac{e^{iks}}{s}$$

Figure 6.2 describes the contribution to the wave amplitude at point P′ made by a single point, P, on the original wavefront. The original wavefront has an amplitude, A(x, y, z) which may be complex. The angle, χ, is the angle the line from P to P′ makes to the normal to the wavefront. As indicated in Figure 6.2, there is

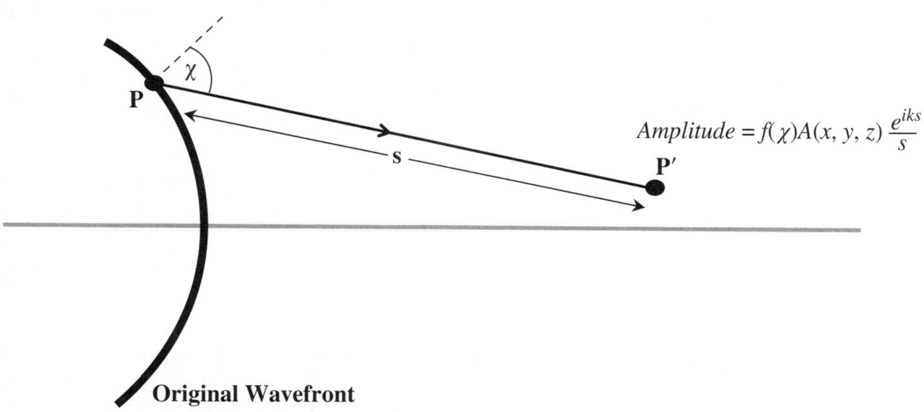

Figure 6.2 Huygens secondary wave geometry.

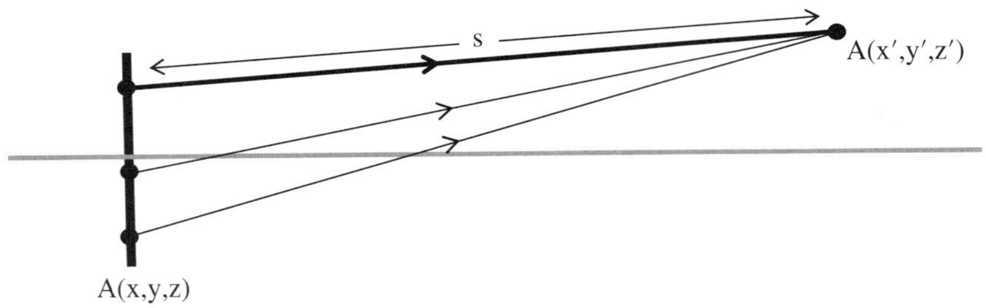

Figure 6.3 Geometry for Rayleigh diffraction equation of the first kind.

some dependence of the secondary wave amplitude upon this angle, in the form of f(χ). There is no intuitive process that can shed further light on the precise form of this function. Elucidation of this can only be provided by a proper application of Maxwell's equation. Re-iterating the description of the Huygens' representation in Figure 6.2, it can be described more formally, as in Eq. (6.7).

$$WaveletAmplitude = f(\chi)A(x,y,z)\frac{e^{iks}}{s} \tag{6.7}$$

Proper application of Maxwell's equations gives rise to a series of equations that are similar in form to the Huygens' representation shown in Eq. (6.7). These include the so-called ***Rayleigh diffraction formulae*** of the first and second kinds. In the first case, it is assumed that the amplitude of the wave disturbance A(x, y, z) is known across some semi-infinite plane. We now seek to determine the amplitude, A(x′, y′, z′) at some other point in space. The geometry of this is illustrated in Figure 6.3.

Equation (6.8) shows the Rayleigh diffraction formula of the first kind.

$$A(x',y',z') = \frac{1}{2\pi}\iint_{z=0} A(x,y,z)\frac{\partial}{\partial z}\left(\frac{e^{iks}}{s}\right)dx'dy' \tag{6.8}$$

Equation (6.8) is referred to as the ***Rayleigh diffraction formula of the first kind***. In form, Eq. (6.8) is very similar to what one might expect from the summation of an expression of the form shown in the Huygens' representation in Eq. (6.7). We have formally expressed the summation of the Huygens wavelets as a surface integral over the plane, as shown in Figure 6.3. Note, however, instead of the decay of the wavelet amplitude with distance being expressed as in Eq. (6.7), a differential with respect to the axial distance is added. This is crucial, since it gives an insight into the formulation of the inclination term f(χ) which will be explored further a little later.

The other condition covered by the Rayleigh formulae occurs where the axial gradient of the amplitude is known rather than the amplitude itself. In this instance, we have the ***Rayleigh diffraction formula of the second kind***.

$$A(x',y',z') = -\frac{1}{2\pi}\iint_{z=0}\left[\frac{\partial A(x,y,z)}{\partial z}\right]\frac{e^{iks}}{s}dxdy \tag{6.9}$$

If we combine these two solutions and make the qualifying assumption that k ≫ 1/s, then we obtain the so-called ***Kirchoff diffraction formula***, which is replicated in Eq. (6.10).

$$A(x',y',z') = \frac{1}{4\pi}\iint_{z=0}(1+\cos\chi)A(x,y,z)\frac{e^{iks}}{s}dxdy \tag{6.10}$$

The Kirchoff diffraction formula lacks the generality of the Rayleigh formulae, as it only applies where the secondary wave propagation distance is much greater than the wavelength. However, it provides a useful reference point for comparison with the Huygens approach. The factor, $1+\cos\chi$ is the inclination factor that was alluded to previously. A further approximation may be made where the system is paraxial, i.e. where $\cos\chi \sim 1$. In this case, there is no inclination factor to speak of. Furthermore, if the axial displacement s is very much larger than the lateral extent of the illuminated area defined by A(x, y, z), then for all intents and purposes, s is constant and the inverse term may be taken outside the integral. This is the so-called ***Fraunhofer approximation*** and may be written as:

$$A(x',y',z') = \frac{1}{2\pi s}\iint\limits_{z=0} A(x,y,z)e^{iks}dxdy \tag{6.11}$$

6.4 Diffraction in the Fraunhofer Approximation

The assumptions underlying the Fraunhofer approximation are relevant to a wealth of problems in optical engineering. In particular, the approximation relates the behaviour and distribution of electromagnetic radiation in two distinct zones, the so-called ***near field*** and ***far field***. Separation of these two zones must be such that the preceding approximations apply, i.e. that the axial displacement is much larger than the lateral extent of the radiation field and, of course, is much greater than the wavelength. We now wish to calculate the amplitude on a sphere whose vertex is located at $z' = z_0$ and whose centre is located at $z = 0$, where the near field is located. Figure 6.4 shows the general scheme.

Choice of the reference sphere centred on the near field location places the following constraint upon the variables, x, y, x′, y′, and z′:

$$x'^2 + y'^2 + z'^2 = z_0^2$$

Expanding the propagation distance, s, in terms of the variables, x, y, and x′, y′ we can re-write Eq. (6.11):

$$A(x',y') = \frac{1}{2\pi z_0}\iint\limits_{z=0} A(x,y)e^{ik\sqrt{z'^2+(x-x')^2+(y-y')^2}}dxdy = \frac{1}{2\pi z_0}\iint\limits_{z=0} A(x,y)e^{ik\sqrt{z_0^2+x^2+y^2-2(xx'+yy')}}dxdy$$

In the Fraunhofer approximation, we are seeking to calculate the amplitude at the limit where z_0 tends to infinity. We wish to know the far field distribution at some angle, θ, where z_0 tends to infinity. Therefore, we can assume that $x' \gg x$ and $y' \gg y$. Hence, the diffraction formula may be recast in the following form:

$$A(x',y') = \frac{1}{2\pi z_0}e^{ikz_0}\iint\limits_{z=0} A(x,y)e^{-ik\frac{(xx'+yy')}{z_0}}dxdy \tag{6.12}$$

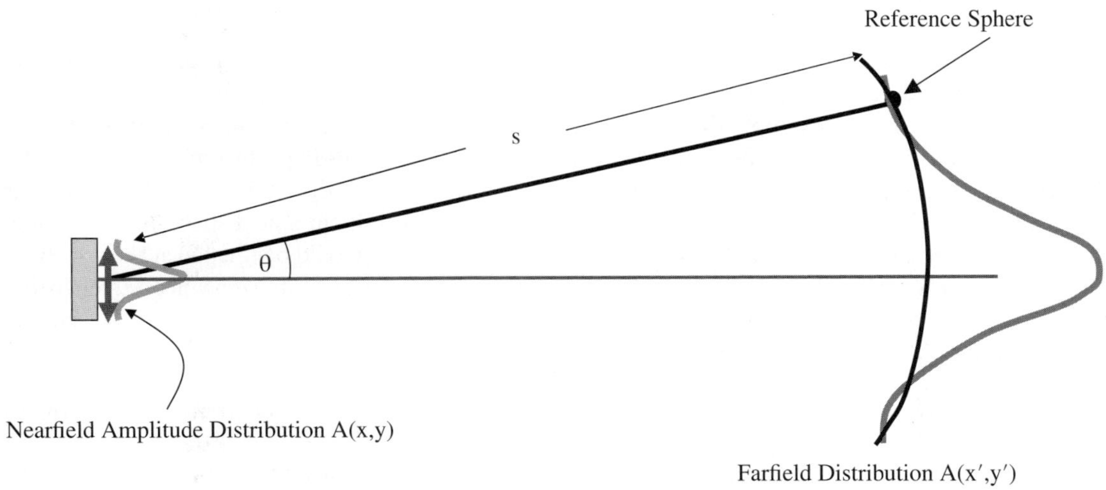

Figure 6.4 Far field diffraction.

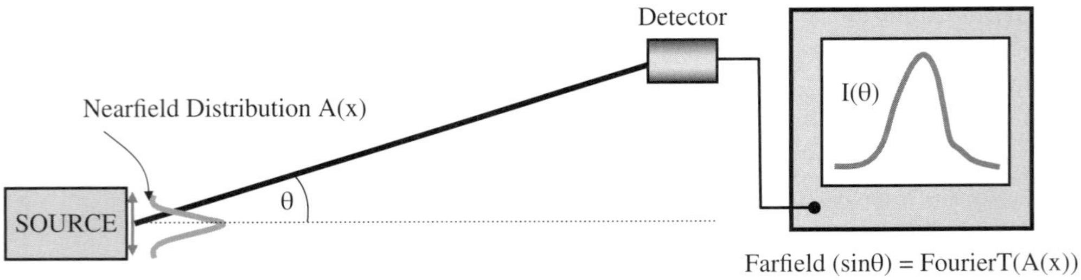

Figure 6.5 Far field diffraction of laser beam emerging from fibre.

Equation (6.12) has the form of a Fourier transform. So, the far field diffraction pattern of a near field amplitude distribution is simply given by the Fourier transform of that near field distribution. Of course, we must understand all the caveats that apply to this treatment, namely that the far field distribution must imply that the distance of the 'far field' location from the near field location must be sufficiently great. Finally, we might like to cast Eq. (6.12) more conveniently in terms of the angles involved:

$$A(x', y') = \frac{1}{2\pi z_0} e^{ikz_0} \iint_{z=0} A(x, y) e^{-ik(xNA_x + yNA_y)} dxdy \tag{6.13}$$

NA_x and NA_y are the numerical apertures (sine of the angles) in the x and y directions respectively.

A typical example of the application of Fraunhofer diffraction might be the emergence of a laser beam from a very small, single mode optical fibre a few microns across. As the beam emerges from the fibre, it will have some near field distribution. In fact, the spatial variation of this amplitude may be ***approximated*** by a Gaussian distribution. In the light of the previous analysis, the angular distribution of the emitted radiation far from the fibre will be the Fourier transform of the near field distribution. This is shown in Figure 6.5.

We will be returning to the subject of diffraction and laser beam propagation later in this chapter. A more traditional concern is the impact of diffraction upon image formation in real systems. As far as the design of optical systems is concerned, hitherto we have only been concerned with the impact of aberrations in limiting optical performance. In the next section, we will examine the application of Fraunhofer diffraction to the study of image formation in optical system and the way in which the presence of diffraction limits optical resolution.

6.5 Diffraction in an Optical System – the Airy Disc

In the Fraunhofer approximation, we considered the effect of diffraction by considering a near field amplitude distribution and a far field, nominally located at infinity. However, it is not necessary for the far field to be physically located at infinity. For example the (second) focal point of an optical system is conjugated to an object plane located at infinity. In this instance, the relation of the two planes is perfectly described by the Fraunhofer approximation. This is illustrated schematically in Figure 6.6 which shows the realisation of the far field of a laser source. The focus of the lens in Figure 6.6 is conjugate to infinity in object space and, assuming the lens aberration is not significant, then the Fraunhofer diffraction pattern would be imaged at this location.

In the case of the near field distribution associated with the laser, the far field distribution will be given by the Fourier transform of the near field distribution, but mediated by the focal length of the lens. In other words, the spatial distribution, $A(x', y')$ of the far field at the lens focus is given by:

$$Farfield\left(\frac{x'}{f}\right) = FourierT(A(kx)) \tag{6.14}$$

In practice, the quantity, x'/f, can be regarded as equivalent to the numerical aperture (NA) associated with the far field distribution.

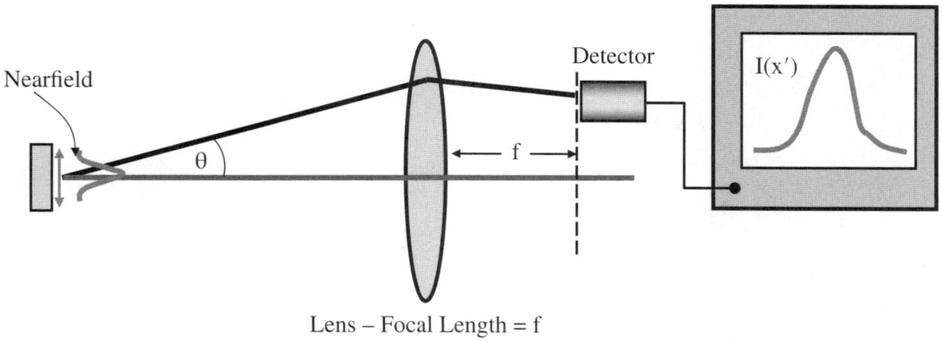

Figure 6.6 Imaging of a Fraunhofer diffraction pattern by a simple lens.

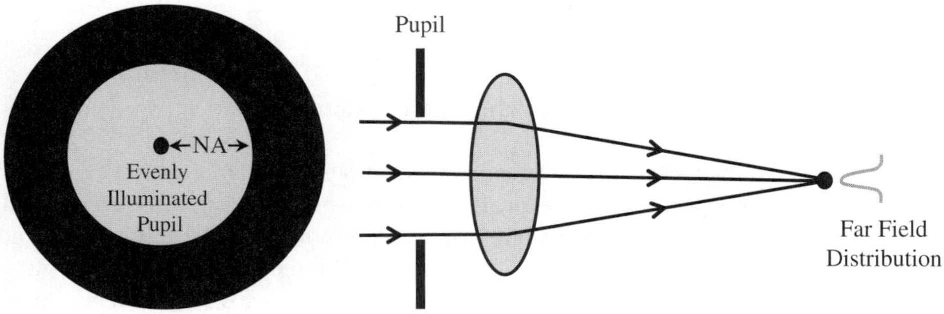

Figure 6.7 Diffraction of evenly illuminated pupil.

In terms of a real optical system, the greatest practical interest is invested in the diffraction produced by the pupil. For an object located at infinity and the physical stop located in object space, the far field diffraction pattern of the pupil will be formed at the focal point of the system. Of course, the pupil, or its image, the exit pupil, is of great significance in the analysis of an optical system as the system optical path difference (OPD) is, by convention, referenced to a sphere whose vertex lies at the exit pupil. As such, the diffraction pattern produced by a uniformly illuminated circular disc is of prime importance in the analysis of optical systems.

We will now assume that an optical system is seeking to image a point object, and the exit pupil size can be expressed as an even cone of light with a numerical aperture, NA. It will produce a diffraction pattern at the focus of the system, whose extent and form we wish to elucidate. A schematic of this scenario is shown in Figure 6.7.

We are now simply required to determine the Fourier transform of a circular disc. In fact, the Fourier transform of a circular disc is described in terms of $J_1(x)$, a Bessel function of the first kind. Proceeding along the lines set out in Eq. (6.14), we find that the far field distribution at the system focus is given by:

$$A(r') = \left[\frac{2J_1(r'/r_0)}{(r'/r_0)}\right] \quad r' = \sqrt{x'^2 + y'^2} \quad r_0 = \frac{1}{kNA} = \frac{\lambda}{2\pi NA} \tag{6.15}$$

It is natural, of course, that the far field distribution retains the circular symmetry of the near field. We have to remember that, in this analysis, we have calculated the amplitude (electric field) of the far field distribution. The flux density, $I(r')$, is proportional to the square of the electric field and this is given by:

$$I(r') = \left[\frac{2J_1(r'/r_0)}{(r'/r_0)}\right]^2 \tag{6.16}$$

The pattern produced at the far field location, as defined by Eq. (6.16) is known as the Airy disc. For $r' \to 0$, Eq. (6.16) tends to one. Thus, all values computed by Eq. (6.16) represent the local flux taken in ratio to the

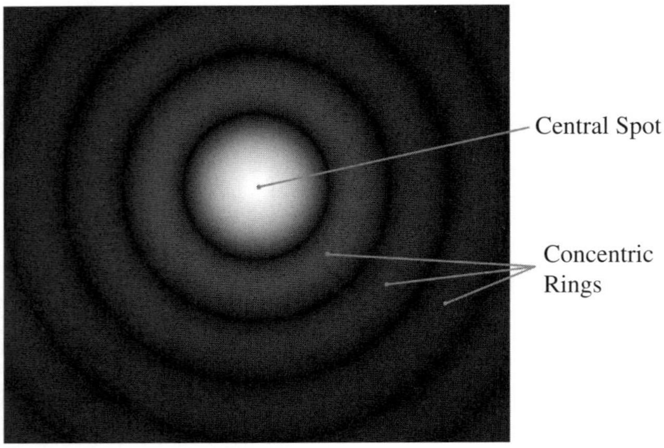

Figure 6.8 Airy disc.

central maximum. The form of the Airy disc consists of a bright central region surrounded by a number of weaker rings. This is shown in Figure 6.8.

The importance of the Airy disc lies in the fact that it represents the ideal replication of a point source in a totally unaberrated system. Hitherto, in the idealised geometrical optics representation, a point source would be replicated as a point image. The presence of diffraction, therefore, critically compromises resolution. That is to say, even in a perfect optical system, the lateral resolution of the system is limited by the extent of the Airy disc. At this point it is useful to examine the form of the Airy disc in more detail. Figure 6.9 shows a graphical trace of the Airy disc, expressed in terms of the ratio r'/r_0.

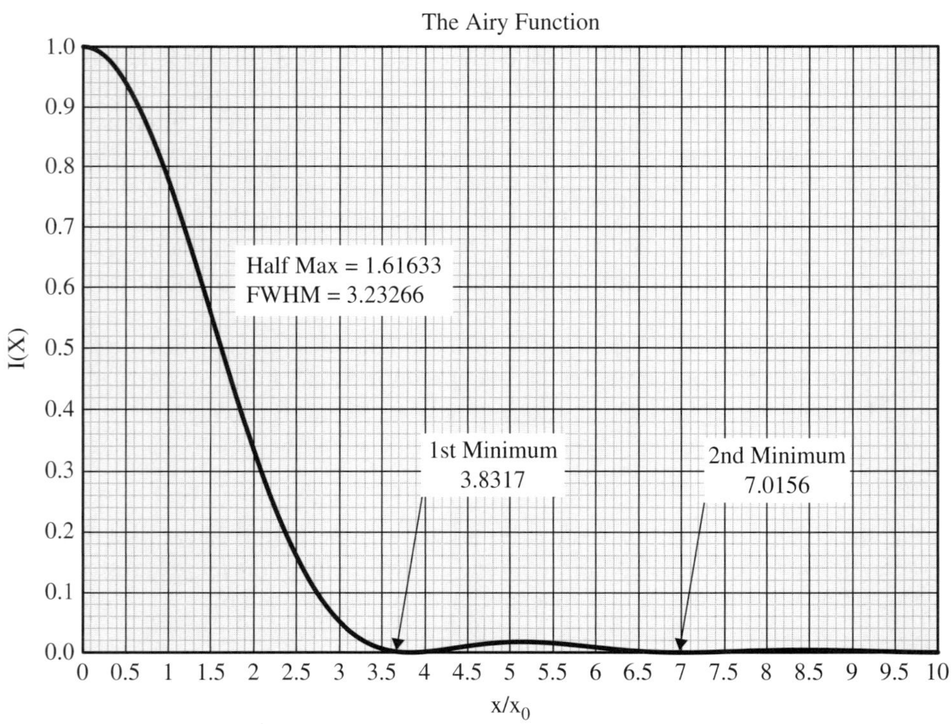

Figure 6.9 Graphical trace of Airy disc.

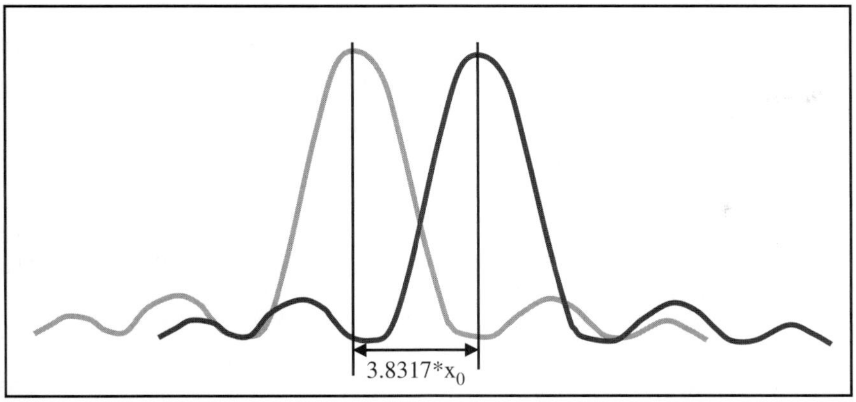

Figure 6.10 The Rayleigh criterion and ideal diffraction limited resolution.

As illustrated in Figure 6.9, the full width half maximum (FWHM) is equal to $3.233r_0$. Equally significant is the presence of local minima at $3.832r_0$ and $7.016r_0$. It is more informative to express these values in terms of the wavelength and numerical aperture. This gives the FWHM as 0.514λ/NA and the locations of the minima as 0.610λ/NA and 1.117λ/NA. At first sight, the FWHM may seem a useful indication of the ideal optical system resolution. In practice, it is the location of the first minimum that forms the basis for the conventional definition of ideal resolution. The rationale for this is shown in Figure 6.10.

Considering two adjacent point sources, these are said to be resolved when the maximum of one Airy disc lies at the minimum of the other. Therefore the separation of the two images must be equal to 0.610λ/NA. This is the so-called ***Rayleigh criterion*** for ***diffraction limited imaging***. Under the Rayleigh criterion, two separated and resolved peaks are seen with a local minimum between them at 73.5% of the maximum. This is illustrated in Figure 6.11

At this point, we will re-iterate the formula describing diffraction limited resolution under the Rayleigh criterion, as it is fundamental to the enumeration of resolution in a perfect optical system. This is set out in Eq. (6.17).

$$Resolution = \frac{0.61\lambda}{NA} \tag{6.17}$$

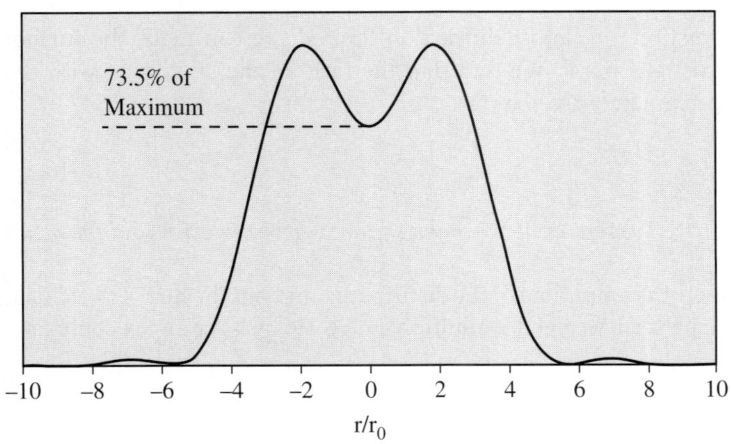

Figure 6.11 Profile of two point sources just resolved under Rayleigh criterion.

Worked Example 6.1 ***Microscope Objective***

A microscope objective has a numerical aperture of 0.8. What is the diffraction limited resolution at the D wavelength of 589.3 nm?

Calculation is very straightforward, we simply need to substitute the relevant values into Eq. (6.17).

$$Resolution = \frac{0.61 \times 0.5893 \mu m}{0.8} = 0.45 \mu \mathrm{m}$$

The resolution is 0.45 μm.

This figure only applies to 'perfect' or diffraction limited in an aberration-free system. The presence of aberrations will affect the resolution, as will be considered in the next section.

6.6 The Impact of Aberration on System Resolution

6.6.1 The Strehl Ratio

In the preceding analysis we examined the diffraction pattern produced by a circular disc – namely the pupil. This produced the Airy diffraction pattern. In this analysis, we ignored the impact of phase, i.e. the possibility that the amplitude across the pupil might have a complex component. In fact, for a point source, the phase across the pupil is, by definition, directly related to the OPD. That is to say, if we assume that the modulus of the near field amplitude, $A(x, y)$ is unity, the complex amplitude is given by:

$$A(x, y) = \cos(k\Phi(x, y)) + i \sin(k\Phi(x, y))$$

where $\Phi(x, y)$ is the wavefront error across the pupil.

The final diffraction pattern is given by the Fourier transform of the above which, from Eq. (6.13), is given by:

$$A(x', y') = \frac{1}{2\pi z_0} e^{ikz_0} \iint_{z=0} (\cos(k\Phi(x, y)) + i \sin(k\Phi(x, y)))e^{-ik(xNA_x + yNA_y)} dxdy \tag{6.18}$$

We now wish to compute the amplitude at the central location of the far field pattern, i.e. where NA_x and $Na_y = 0$. In this case the Fourier transform can be further simplified:

$$A(0, 0) = \frac{1}{2\pi z_0} e^{ikz_0} \iint_{z=0} (\cos(k\Phi(x, y)) + i \sin(k\Phi(x, y)))dxdy \tag{6.19}$$

For an optical system that is close to perfection, or almost diffraction limited, we can make the further assumption that $k\Phi \ll 1$ at all locations across the pupil. We find that the ratio of the amplitude with the presence of aberration to that without is approximately given by:

$$\frac{A(0, 0)}{A_0(0, 0)} \approx 1 - k^2 \frac{\langle \Phi^2(x, y) \rangle}{2} + ik\langle \Phi(x, y) \rangle \tag{6.20}$$

The expressions in the pointed brackets in Eq. (6.19) *represent the mean square wavefront error and the mean wavefront error respectively.*

However, the expression in Eq. (6.20) is merely the amplitude of the disturbance and not the flux. To calculate the flux density at the centre of the diffraction pattern, we need to multiply Eq. (6.19) by its complex conjugate. This gives:

$$\frac{I(0, 0)}{I_0(0, 0)} \approx 1 - k^2(\langle \Phi^2(x, y) \rangle - \langle \Phi(x, y) \rangle^2) \tag{6.21}$$

The expression contained within the brackets is merely the variance of the wavefront error taken across the pupil. If we define the root mean square (rms) wavefront error, Φ_{rms}, as the rms value computed under the assumption that the average wavefront error has been normalised to zero, we get the following fundamental relationship:

$$\frac{I(0,0)}{I_0(0,0)} \approx 1 - k^2\Phi_{rms}^2 \tag{6.22}$$

Equation (6.22) is of great significance. The ratio expressed in Eq. (6.22), the ratio of the aberrated and unaberrated flux density, is referred to as the ***Strehl ratio***. The Strehl ratio is a measure of the degradation produced by the introduction of system imperfections. Of course, Eq. (6.22) only applies where $k\Phi^2_{rms} \ll 1$. The fact that the peak flux of a diffraction pattern is reduced by the introduction of aberration necessarily implies that the distribution is in someway broadened, i.e. the resolution is reduced. For example, if the Strehl ratio is 0.8, then the area associated with the diffraction pattern is likely to have increased by about 20% and the linear dimension by about 10%.

6.6.2 The Maréchal Criterion

The Strehl ratio is of great practical significance. It affords a useful, practical, but somewhat arbitrary definition of 'diffraction limited' imaging. Where the Strehl ratio is 0.8 or greater, the image is said to be ***diffraction limited***. This is the so-called ***Maréchal criterion***. It can be expressed as follows:

$$1 - 4\pi^2\frac{\Phi_{rms}^2}{\lambda^2} > 0.8 \text{ or } \Phi_{rms} < \frac{\lambda}{2\pi\sqrt{5}} \text{ or } \Phi_{rms} < \frac{\lambda}{14.05} \tag{6.23}$$

This condition is widely accepted as the basis for the definition of diffraction limited imaging. As a measure of system wavefront error, the peak to valley wavefront error, $\Phi_{pv} \ll \lambda/4$ is often preferred, in practice. This is very much a traditional description of system wavefront error, preserved for historical reasons, primarily on account of the ease of reckoning peak to valley fringe displacements on visual images of interferograms. This consideration has, of course, been displaced by the ubiquitous presence of computer-based interferometry which has rendered the calculation of rms wavefront errors a trivial process. Nonetheless, the peak to valley description remains prevalent. Although dependent upon the precise distribution of wavefront error, as a rule of thumb, the peak to valley wavefront error is about 3.5 times the standard deviation or rms value. Therefore, we can set out another condition for diffraction limited imaging.

$$\Phi_{pv} < \frac{\lambda}{4} \tag{6.24}$$

Worked Example 6.2 A simple ×10 microscope objective is to be made from a BK7 plano-convex lens and is to operate at the single wavelength of 589.3 nm. The refractive index of BK7 at 589.3 nm is 1.518 and the assumed microscope tube length is 160 mm. We also assume that only the microscope objective contributes to system aberration. What is the maximum objective numerical aperture consistent with diffraction limited performance, assuming on-axis spherical aberration as the dominant aberration?

Firstly, for a wavelength of 589.3 nm, from Eq. (6.22):

$$\Phi_{rms} < \frac{\lambda}{14.05} = \frac{589.3}{14.05} = 41.94nm \text{ RMS WFE should be less than 41.94 nm}$$

For a ×10 objective, the focal length should be 160/10 = 16 mm for a 160 mm tube length. The tube length is much longer than the objective focal length, so, for all intents and purposes, the image is at the infinite conjugate, as illustrated.

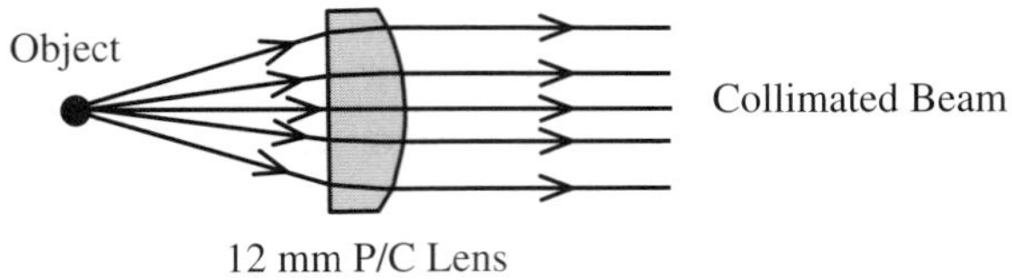

Note that the curved surface faces the infinite conjugate in order to minimise spherical aberration. Thus, in this instance, the conjugate parameter is 1 and the lens shape factor is −1. From Eq. (4.30a)

$$\Phi_{SA} = -\left(\frac{f}{32}\right)\left[\left(\frac{n}{n-1}\right)^2 - \left(\frac{n}{n+2}\right)t^2 + \frac{(n+2)}{n(n-1)^2}\left[s + 2\left[\frac{n^2-1}{n+2}\right]t\right]^2\right] NA^4$$

Substituting the values of s and t:

$$\Phi_{SA} = -\left[\frac{n^4 - 4n^2 - 2n + 4}{n(n+2)(n-1)^2}\right]\left(\frac{f}{8}\right) NA^4$$

Substituting the values of n (1.518) and f (16 mm) into the equation, we get:

$$\Phi_{SA} = -4.37 * NA^4 mm$$

Further assuming that we are using defocus to 'balance' and minimise aberrations, we can relate the rms wavefront error to the maximum OPD:

$$\Phi_{RMS} = \frac{\Phi_0}{6\sqrt{5}}$$

$$\text{Finally}: \Phi_{RMS} = 0.325 \times NA^4 mm$$

For the 'diffraction limited' condition to be fulfilled the rms wavefront error must be less than 41.94 nm.

$$0.325 \times NA^4 < 4.194E - 5$$

The minimum numerical aperture is thus 0.107. Substituting this value into Eq. (5.17), then it is clear that the resolution of the objective is 3.37 μm. The design of complex objectives does, of course, generally proceed by virtue of ray tracing. However, this illustration does provide some insight into the limitations of simple optics and the value added by more complex designs.

6.6.3 The Huygens Point Spread Function

The ***Huygens point spread function*** is the diffraction pattern produced in the image plane by a point source located at the object plane. Determination of the point spread function proceeds as per Eq. (6.18) and so is fundamentally influenced by the system wavefront error across the pupil. Of course, as previously argued, any wavefront error reduces the flux at the axial location in proportion to the Strehl ratio. In concert with this, the width of the diffraction pattern increases with increasing wavefront error.

As will be discussed in more detail later, where the wavefront error is small, purely geometrical ray tracing produces a very different spatial distribution when compared to the point spread function. Naturally, in the limit where the wavefront error becomes large, the Huygens point spread function tends to the geometrical spot distribution. This is quite an important consideration, as it impacts how optical systems are optimised with regard to optical performance. If the intention is to design a diffraction limited system, the most efficient optimisation proceeds by minimising the wavefront error. However, where the intended wavefront error is large when compared to the diffraction limit, it is best to optimise the system by minimising the geometrical spot size.

6.7 Laser Beam Propagation

6.7.1 Far Field Diffraction of a Gaussian Laser Beam

The far field divergence of a laser beam may be accounted for by the Fraunhofer approximation according to Eq. (6.13). In the treatment of a laser beam, we consider the beam to have a 'near field' location, or ***beam waist***, where the phase is uniform across a plane perpendicular to the propagation direction. That is to say, the wavefronts are planar at this beam waist location. In addition, it must be assumed that the wavefront is 'coherent', i.e. that an unambiguous phase relation exists across the wavefront at all times. We now wish to know the disturbance produced by this near field distribution in the far field. To make the problem tractable, we assume that the near field distribution of the laser beam waist may be described by a Gaussian function. This is a useful approximation, however, it must be emphasised that this is an approximation. In practice, the profile of a laser beam is not quite Gaussian, with more flux being in the wings, far from the centre, than would be expected from a Gaussian distribution.

In the Gaussian approximation, we may, for example, describe the laser beam emerging from the end of a single mode fibre or from the output of a semiconductor laser facet or from a helium neon laser. The size of the beam waist is described by the parameter, w_0, namely the radial distance at which the amplitude, A(r), falls to 1/e, or 37% of the peak value. When expressed in terms of flux, the size of the beam waist, r_0 defines the radius at which the flux, $I(r)$ falls to $(1/\mathrm{e})^2$ or 13% of the peak value. For example, in the case of a single mode telecoms fibre, a laser beam emerging from the end of the fibre might have a beam waist, w_0 of 5.5 μm. Equation (6.25) expresses the form of the laser beam profile:

$$A(r) = A_0 e^{-\left[\frac{r}{w_0}\right]^2} \text{ (Amplitude) } I(r) = I_0 e^{-2\left[\frac{r}{w_0}\right]^2} \text{ (Flux)} \tag{6.25}$$

Having characterised the beam waist in this way, it is useful to relative it to both the FWHM, d_{FWHM}, and the rms radius, r_{rms}.

$$d_{FWHM} = w_0\sqrt{2\ln(2)} = 1.177w_0 \quad r_{rms} = \frac{w_0}{\sqrt{2}} \tag{6.26}$$

In the Fraunhofer approximation, the far field may be derived from the Fourier transform of the near field. At this point, the significance of the Gaussian representation becomes clear, as the Fourier transform of a Gaussian is another Gaussian. The far field representation is thus given by another Gaussian, with the divergence expressed by a characteristic numerical aperture, NA_0.

$$A(NA) = A_0 e^{-\left[\frac{NA}{NA_0}\right]^2} \text{ (Amplitude) } I(NA) = I_0 e^{-2\left[\frac{NA}{NA_0}\right]^2} \text{ (Flux)} \tag{6.27}$$

From the Fourier transform, it is possible to derive a clear relationship between the near field beam waist, w_0, and the far field divergence, NA_0. This is given by Eq. (6.28).

$$NA_0 = \frac{\lambda}{\pi w_0} \tag{6.28}$$

As one might expect, the far field divergence is inversely proportional to the size of the beam waist, a smaller beam waist effecting a larger divergence.

Worked Example 6.3 – ***Beam Divergence of a Fibre Laser***
A laser beam with a wavelength of 1.55 μm emerges from a single mode fibre. The laser beam, at this point can be characterised as a beam with a waist size of 5.25 μm. What is the characteristic numerical aperture, NA_0, associated with the far field divergence?

Substituting the relevant values into Eq. (6.28), we get:

$$NA_0 = \frac{1.55}{\pi \times 5.25} = 0.094$$

The numerical aperture is thus 0.094 and this corresponds to a divergence angle of 5.39°.

Expressed as a FWHM, the near field beam width is 6.18 μm, and the rms radius is 4.37 μm. Similarly in the far field the FWHM divergence angle is 6.35° and the rms divergence angle is 3.81°.

6.7.2 Gaussian Beam Propagation

Thus far, we have analysed the relationship between the near field and far field dispositions of a Gaussian beam. We now turn to the more general case of the propagation of a Gaussian beam. As in the previous analysis, the laser beam is defined by a characteristic Gaussian radius. In this case, the characteristic radius, $w(z)$, is a function of the axial propagation distance, z. To describe the laser beam, we introduce an envelope function, $A_0(x, y, z)$ that describes the radial profile of a propagating laser beam.

$$A(x, y, z) = A_0(x, y, z)e^{j(\omega t - kz)} \tag{6.29}$$

In practice, in most cases, the size, $w(z)$, of the laser beam is much larger than the wavelength, and we can use the ***slowly varying envelope approximation***. This is, in effect, a paraxial approximation, where far field divergence angles are necessarily small and can be formally expressed as:

$$\frac{\partial A(x, y, z)}{\partial x} << kA(x, y, z) \quad \frac{\partial A(x, y, z)}{\partial y} << kA(x, yz) \tag{6.30}$$

Applying Eq. (6.29) to the scalar version of Maxwell's equation and taking into account the above approximation, we obtain the so-called paraxial Helmholtz equation:

$$\frac{\partial^2 A}{\partial x^2} + \frac{\partial^2 A}{\partial y^2} - i2k\frac{\partial A}{\partial z} = 0 \text{ or } \frac{1}{r}\frac{\partial}{\partial r}\left(r\frac{\partial A}{\partial r}\right) - i2k\frac{\partial A}{\partial z} = 0 \quad r^2 = x^2 + y^2 \tag{6.31}$$

As suggested, in the case of a Gaussian beam, the amplitude may be expressed a characteristic width, $w(z)$ which varies slowly with respect to z. Most significantly, $w(z)$ can be complex, giving rise to wavefront curvature. This may be easily seen if the complex part of the envelope is subsumed within the sinusoidal propagation term, leading to a quadratic phase variation across the beam. In the Fresnel and paraxial approximation, these quadratic wavefronts may be viewed as approximately spherical. This is illustrated in Figure 6.12.

Mathematically, the Gaussian beam envelope is expressed as follows:

$$A(x, y, z) = A_0(z)e^{-\left[\frac{1}{w^2(z)} + i\frac{k}{2R(z)}\right](x^2+y^2)} \tag{6.32}$$

The component $A_0(z)$ represents the peak on axis amplitude and this would be expected to diminish as the beam expands. To make the analysis more tractable, we subsume this axial variation into the exponential function as a complex function of z, $\beta(z)$. Furthermore, both real and imaginary parts of the Gaussian are combined into a single complex function, $\alpha(z)$. Hence, to solve the paraxial Helmholtz equation in this instance we re-cast Eq. (6.32) in the following form:

$$A(x, y, z) = A_0 e^{-\left(\beta(z) + \frac{r^2}{\alpha(z)}\right)}$$

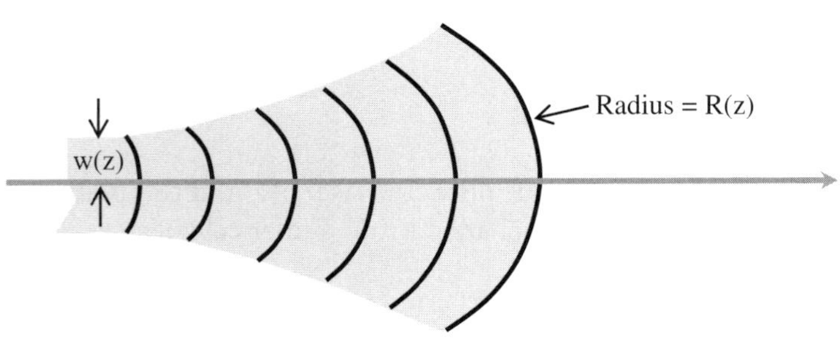

Figure 6.12 Gaussian beam.

And substituting this into the paraxial Helmholtz equation:

$$\frac{4r^2}{\alpha^2(z)} - \frac{4}{\alpha(z)} + 2ik\beta'(z) - 2ik\frac{\alpha'(z)}{\alpha^2(z)}r^2 = 0$$

This equation must hold for all values of r. Both the quadratic terms and those with no dependence in r must sum to zero. Taking the quadratic element only, we may derive a very simple relationship for $\alpha(z)$.

$$\alpha'(z) = -\frac{2}{k}i \quad \alpha(z) = C - \frac{2z}{k}i \quad \text{Where C is a constant} \tag{6.33}$$

By viewing Eq. (6.32) it is straightforward to relate $\beta(z)$ to the beam size, $w(z)$ and the radius $R(z)$:

$$\frac{1}{\alpha(z)} = \frac{1}{w^2(z)} + \frac{k}{2R(z)}i \quad w(z) = \sqrt{C + \frac{4z^2}{Ck^2}} \quad R(z) = z + \frac{C^2k^2}{4z} \tag{6.34}$$

To interpret the constant C, we assume that there exists a minimum beam size, the beam waist, where the wavefront curvature is zero. We denote this beam waist by the symbol, w_0. It is clear, then, that the constant C is equal to the square of the beam waist. Eliminating the constant, C, gives:

$$w(z) = w_0\sqrt{1 + \frac{\lambda^2}{\pi^2 w_0^4}z^2} \quad R(z) = z + \frac{\pi^2 w_0^4}{\lambda^2 z} \tag{6.35}$$

We note that there some characteristic distance, R_z, over which the beam expands around the beam waist. This distance is known as the ***Rayleigh distance*** and the Eq. (6.35) may be finally re-cast to give the following:

$$w(z) = w_0\sqrt{1 + \left(\frac{z}{Z_R}\right)^2} \quad R(z) = z + \frac{Z_R^2}{z} \quad Z_R = \frac{\pi w_0^2}{\lambda} \tag{6.36}$$

The Rayleigh distance is of particular significance. In effect, it sets the demarcation between the near field and the far field. In the case of the far field, $z \gg Z_R$ the expressions in Eq. (6.36) revert to the Fraunhofer diffraction pattern of a Gaussian beam:

$$w(z) \approx \frac{w_0 z}{Z_R} \approx \frac{\lambda}{\pi w_0}z \quad R(z) \approx z \quad z \gg Z_R \tag{6.37}$$

This may be compared with Eq. (6.28) and is very similar in form. Thus Eq. (6.37) represents the far field diffraction pattern of a Gaussian beam. In the near field where $z \ll Z_R$, the beam is parallel and the size constant at w_0 and, of course, the radius tends to infinity. At the Rayleigh distance, the beam size is increased by a factor corresponding to the square root of two and the wavefront radius is equal to twice the Rayleigh distance. This is illustrated more formally in Figure 6.13.

For values of z that are of a similar magnitude to the Rayleigh distance, then the beam is in an intermediate zone between the near and far fields.

Worked Example 6.4 – ***Rayleigh Distance of Fibre Laser***

In the previous worked example, we introduced a fibre laser with a wavelength of 1.55 μm, with a Gaussian beam size of 5.25 μm. If we assume that the beam waist is located at the exit from the fibre, what is the Rayleigh distance of the laser beam? In addition, what is the beam size, $w(z)$, 50 μm from the exit from the fibre and what is the wavefront radius at that location?

From Eq. (6.36):

$$Z_R = \frac{\pi w_0^2}{\lambda} = \frac{\pi 5.25^2}{1.55} = 55.9$$

The Rayleigh distance is 55.9 μm

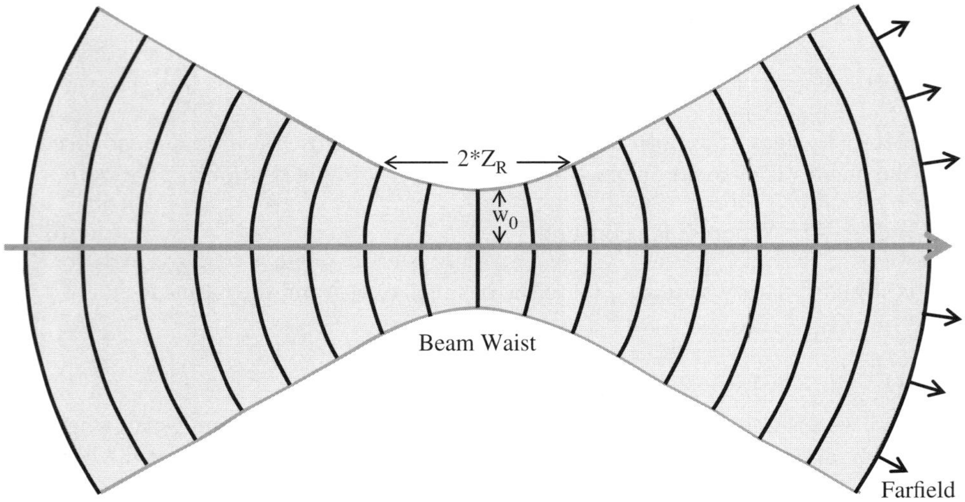

Figure 6.13 Form of expanding Gaussian beam and beam waist.

Calculation of the beam size and the wavefront radius also proceeds from Eq. (6.36).

$$w(z) = w_0\sqrt{1+\left(\frac{z}{Z_R}\right)^2} = 5.25\sqrt{1+\left(\frac{50}{55.9}\right)^2} = 7.05$$

The beam size, w(z), is 7.05 μm

$$R(z) = z + \frac{Z_R^2}{z} = 50 + \frac{55.9^2}{50} = 112.4$$

The wavefront radius is 112.4 μm

6.7.3 Manipulation of a Gaussian Beam

The preceding analysis has presented an analysis of the propagation of a Gaussian beam through free space. However, in most practical instances, the beam will be manipulated by optical components, lenses, and mirrors and so on. Therefore it would be useful to be able to understand the impact of an individual optical component or system on the propagation of a Gaussian beam. In the analysis presented here, the component or system is simply represented in its paraxial form by a ray tracing matrix, as presented in Chapter 1. The matrix is populated by four elements, A, B, C, and D and these transform the wavefront radius according to the following equation:

$$R_2 = \frac{AR_1 + B}{CR_1 + D} \tag{6.38}$$

(R_1 and R_2 are the input and output radii respectively)

For a Gaussian beam, the wavefront curvature can be represented by the complex quantity, q, where $q = z + iZ_R$. The distance from the beam waist is represented by z and Z_R is the Rayleigh distance. Expression (6.38) may be re-cast in the following form:

$$q_2 = \frac{Aq_1 + B}{Cq_1 + D} \tag{6.39}$$

where $q = z + iZ_R$

This is the so-called ***ABCD law*** for the manipulation of Gaussian beams.

Worked Example 6.5 – ***Gaussian Beam Manipulation***
A helium neon laser beam at 633 nm has a beam waist of 0.6 mm, located 80 mm from a positive lens of focal length 60 mm. Calculate the size of the beam waist following the lens and its location with respect to the lens.

If we take the origin of the co-ordinate system to be at the original beam waist (z = 0), then the system matrix consists of a translation by 80 mm followed by a 60 mm focal length lens.

$$\begin{bmatrix} A & B \\ C & D \end{bmatrix} = \begin{bmatrix} 1 & 0 \\ -1/60 & 1 \end{bmatrix} \begin{bmatrix} 1 & 80 \\ 0 & 1 \end{bmatrix} = \begin{bmatrix} 1 & 80 \\ -1/60 & -0.333 \end{bmatrix}$$

$A = 1; B = 80; C = -0.0167; D = -0.333$

The Rayleigh distance of the original beam is given by:

$$R_Z = \frac{\pi w_0^2}{\lambda} = \frac{\pi 0.6^2}{0.000633} = 1787mm \quad q_1 = z + iR_z = i1787$$

Using the ABCD law:

$$q_2 = \frac{Aq_1 + B}{Cq_1 + D} = \frac{i1787 + 80}{-0.0167 \times i1787 - 0.333} = -59.9 + i2.2012$$

The lens lies at −59.9 mm from the beam waist and hence the beam waist is 59.9 mm in advance of the lens. The Rayleigh distance is 2.201 mm. This corresponds to a beam waist size (from Eq. (6.36)) of 0.02 mm or 20 μm.

The new beam waist lies approximately at the focus of the lens. Since the lens lies much closer to the original beam waist than the corresponding Rayleigh distance, then the beam is almost parallel and the new beam waist should lie very close to the focal position.

6.7.4 Diffraction and Beam Quality

All the analysis presented thus far has assumed a Gaussian beam that possesses perfect ***spatial coherence***. Perfect spatial coherence implies that an unambiguous phase relationship exists between all points across the wavefront. A less than perfect wave disturbance is composed of a number of different components whose phase relationship is entirely random. As such, spatial coherence is defined more formally as the correlation between the wave disturbance at two points. The complex amplitude of a wave at a point, $A(t)$, may be expressed in Fourier space in terms of a frequency distribution, $S(f)$. The coherence between two points, x, y is simply given by the cross correlation of the two disturbances:

$$C(x, y) = \frac{\left| \int S_x(f) \times S_y(f) \right|^2}{\left| \int S_x(f) \times S_x(f) \right| \left| \int S_y(f) \times S_y(f) \right|} \tag{6.40}$$

Perfect coherence is represented by a correlation of one; complete incoherence is represented by a correlation of zero. In many practical problems in Gaussian beam propagation, it may be assumed that the coherence of the laser beam is one. However, this is dependent upon the number of independent 'modes' that characterise the laser beam. A single mode is effectively one unique solution to the wave equation and laser devices are often engineered in such a way that only one of these modes is allowed to propagate. The extent to which this is true is a measure of the laser's ***beam quality***. The beam quality of a laser is generally expressed by the parameter ***M^2 or M squared*** and is indicative of the number of modes supported by the beam. If this value is one or close to one, then the beam quality is high and any propagation analysis will proceed as previously described. For a beam with a beam quality defined by the M^2 parameter, then the spatial coherence is given by:

$$C(x, y) = \frac{1}{M^2} \tag{6.41}$$

Returning to the practical question of Gaussian beam propagation, the beam propagation may be expressed entirely in the original form given by Eq. (6.36), except with a revised Rayleigh distance, Z'_R.

$$w(z) = w_0\sqrt{1+\left(\frac{z}{Z_R}\right)^2} \quad Z'_R = \frac{\pi w_0^2}{M^2\lambda} \tag{6.42}$$

It is clear from Eq. (6.42), that, where M^2 is significantly greater than one, the divergence of the laser beam in the far field is greater than would be expected from a perfect beam. The revised equivalent of 5.68, giving the beam divergence is set out in Eq. (6.43).

$$NA_0 = \frac{M^2\lambda}{\pi w_0} \tag{6.43}$$

In practice, the M^2 value for a laser beam is measured and then analysed using the relationships set out in Eqs. (6.42) and (6.43). The parameter is generally specified for many commercial laser systems.

6.7.5 Hermite Gaussian Beams

Further exploring the theme of multiple modes as individual solutions to the wave equation, we must recognise that the simple Gaussian beam previously defined is not the only solution to the paraxial Helmholtz equation. In fact, a complete set of orthonormal solutions exist of which the simple Gaussian solution is the first member. Orthonormal, in this context, means that the cross-integral of two different solutions is always zero and that involving the same solutions is always unity. This is an important property, as will be seen later. This set of solutions are defined by the Hermite-Gaussian polynomials, where the original Gaussian amplitude envelope is multiplied by a unique Hermite polynomial in x and y. Each Hermite polynomial solution is defined by its maximum order in x and y, which we will refer to as l and m respectively. Overall, the solution may be represented as:

$$A[w(z)] \times H_l\left[\frac{x}{\sqrt{2}w(z)}\right] \times H_m\left[\frac{y}{\sqrt{2}w(z)}\right] \times e^{ik\left[\left(\frac{i}{kw^2(z)}+\frac{1}{2R(z)}\right)(x^2+y^2)+z+(1+l+m)\tan^{-1}\left(\frac{z}{Z_R}\right)\right]} \tag{6.44}$$

The expression $w(z)$ is simply the Gaussian beam radius for any specific value of z, as given in Eq. (6.36) and $A[w(z)]$ is a normalising factor. The first few polynomials are set out Table 6.1.

Figure 6.14 shows graphically the form of some low order Hermite polynomials.

The orthogonal nature of the polynomials provides a suggestion as to their utility. As with a Fourier series, any arbitrary beam profile may be represented as a summation of a series of Hermite polynomials. If we represent the full expression contained in Eq. (6.44) as $G_{l,m}(x, y, z)$, then the series may be represented as:

$$A(x,y,z) = \sum_{l,m} C_{l,m}G_{l,m}(x,y,z) \tag{6.45}$$

$C_{l,m}$ are coefficients describing the amplitude of each term

Table 6.1 Low order Hermite polynomials.

Order	Polynomial	Order	Polynomial
0	1	3	$8x^3 - 12x$
1	$2x$	4	$16x^4 - 48x^2$
2	$4x^2-2$	5	$32x^5 - 160x^3 + 120x$

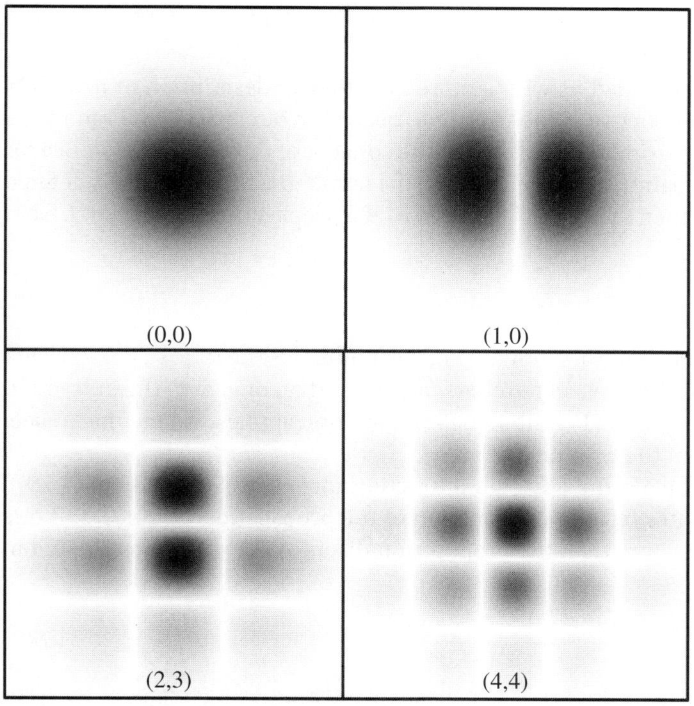

Figure 6.14 A selection of low order Hermite polynomials.

Assuming the beam profile is known as some plane, z_0, then each coefficient may be calculated by exploiting the orthonormal property of the series:

$$C_{l,m} = \int A(x, y, z_0) G_{l,m}(x, y, z_0) dxdy \tag{6.46}$$

Thus, Gaussian-Hermite polynomials represent a powerful tool for physical optics propagation. Assuming a beam profile is known at some point, the relevant coefficients may be calculated according to Eq. (6.46) and summed according to Eq. (6.45) and then propagated in free space according to Eq. (6.44).

6.7.6 Bessel Beams

An interesting solution to the paraxial Helmholtz equation, Eq. (6.31), is the so-called Bessel beam. Generally, Bessel functions of the first kind form a series of solutions to the equation. Most specifically for an axially symmetric form, the solution is represented by a Bessel function of the first kind and zeroth order. The unique feature of this solution is that the wavefronts are planar and the amplitude envelope of the beam does not change as it propagates. That is to say, the beam appears to be diffractionless and does not diverge. The solution is of the form given by:

$$A(x, y, z) = A_0 J_0(k_r r) e^{-i\beta z} \tag{6.47}$$

J_0 is a Bessel function of the first kind and zeroth order; $r = \sqrt{x^2 + y^2}$; k_r is the effective transverse wavevector; β is the propagation wavevector given by: $\beta^2 = k^2 - k_r^2$

Another interesting type of beam is the ***Talbot beam***. Rather than retaining a constant profile as it propagates, the Talbot beam replicates itself at specific propagation distances. For further details, the reader is advised to consult the Further Reading section at the end of the chapter.

6.8 Fresnel Diffraction

The study of Gaussian beam propagation has provided us with a more quantitative description of near field and far field propagation and where the boundary between the two zones occurs. In our analysis of Fraunhofer diffraction, we considered only the far field approximation. Related to the concept of the Rayleigh distance for Gaussian beam propagation is a dimensionless parameter called the Fresnel number. If the near field is defined by some aperture with a radial dimension of a, then the Fresnel number, F, at a propagation distance of L from the aperture is given by:

$$F = \frac{a^2}{L\lambda} \quad \text{where } \lambda \text{ is the wavelength} \tag{6.48}$$

Referring to Eq. (6.36), then the Gaussian beam equivalent of the Fresnel number is the ratio of the Rayleigh distance, Z_R, to the beam propagation distance. For Fresnel numbers much less than one, then the diffraction pattern may be considered as a far field pattern and the Fraunhofer approximation applies. Where the Fresnel number is much greater than one, then one is in the near field.

The analysis of Fresnel diffraction is derived from the Rayleigh diffraction formulae (Eqs. (6.8) and (6.9)). The key assumption in the Fresnel analysis relates to an approximation of the propagation distance, s. If one assumes that the near field object is located at $z = 0$, then the propagation distance may be approximated in the following manner:

$$s = \sqrt{((x' - x)^2 + (y' - y)^2 + z'^2)} \text{ and } s \approx z' + \frac{(x' - x)^2}{2z'} + \frac{(y' - y)^2}{2z'} \tag{6.49}$$

In making the above approximation, based on a Taylor series expansion, we are choosing to ignore terms of fourth order in x and y. These terms cannot be permitted to make a significant contribution to the phase when the approximation is applied to the Rayleigh formulae. Setting out the fourth order terms more explicitly, it is straightforward to delineate the approximation more clearly:

$$k\Delta s << 2\pi \quad k\left(\frac{(x' - x)^4}{8z^3} + \frac{(y' - y)^4}{8z^3}\right) << 2\pi$$

If we re-cast z as the propagation distance, L, and represent the ratio x/z as θ, the angular size of the near field and also denominate the near field radius as a, then the Fresnel condition is given by:

$$\frac{\theta^2 a^2}{4L\lambda} << 1 \text{ and } \frac{\theta^2 F}{4} << 1 \tag{6.50}$$

The value of the Fresnel approximation is that it now permits us to treat the axial propagation distance, z, as a constant and to remove it from the integral in the diffraction equation, producing a more simple expression involving integration with respect to x and y. This is the so-called Fresnel integral and it is set out below.

$$A(x', y', z') = \frac{e^{ikz'}}{i\lambda z} \iint A(x, y, 0) e^{\frac{ik}{2z'}((x-x')^2+(y-y')^2)} dxdy \tag{6.51}$$

It would be useful, at this point, to illustrate the assumptions underlying Fresnel diffraction with a practical example. An optical system populated with components with a standard diameter of 25 mm would have an effective radius of 12.5 mm. For a wavelength of 500 nm, the Fresnel approximation applies to distances much greater than 250 mm. At that distance, the Fresnel number is about 1000, so we are clearly in the near field zone.

To illustrate the application of Fresnel diffraction, we might now apply it to a uniformly illuminated slit of width w. Without loss of generality, this provides a simple illustration of the application of Fresnel diffraction in one dimension. For a given source point, e.g. $x = 0$, then the phase of the sinusoidal component of Eq. (6.51) is of critical interest. In particular, we are concerned with points where the phase expressed in Eq. (6.51) is a half period number of waves. That is to say:

$$\frac{kx^2}{2z'} = \pi n \text{ or } x = \sqrt{n\lambda z} \tag{6.52}$$

The effect of diffraction at an edge or an aperture is to produce an alternating series of light and dark rings. The disposition of these rings is affected by the relative phases of contributions from the source. As such, Eq. (6.52) provides some indication of the location of these rings. The locations of these points, as set out in Eq. (6.52) are referred to as the ***Fresnel zones***. Based on application of Eq. (6.51), for the one dimension, the diffracted amplitude from the slit is proportional to:

$$A(x',y',z') = \int_{-w/2}^{w/2} \cos\left(\frac{k(x-x')^2}{2z}\right)dx + i\int_{-w/2}^{w/2} \sin\left(\frac{k(x-x')^2}{2z}\right)dx \tag{6.53}$$

We make the substitution $s = x - x'$ and make the further assumption that the diffraction pattern is symmetrical about the centre of the slit. In doing this, we may be permitted, without loss of generality in assuming that $x > 0$. The integral now becomes:

$$A(x',y',z') = \int_{-w/2-x'}^{w/2-x'} \cos\left(\frac{ks^2}{2z}\right)ds + i\int_{-w/2-x'}^{w/2-x'} \sin\left(\frac{ks^2}{2z}\right)ds \tag{6.54}$$

We will now refer to the quantity $w/2 - x'$ as Δ. The quantity Δ now represents the distance in x from the positive edge of the slit.

$$A = \mp\int_0^{\Delta} \cos\left(\frac{ks^2}{2z}\right)ds + \int_0^{w-\Delta} \cos\left(\frac{ks^2}{2z}\right)ds \mp i\int_0^{\Delta} \sin\left(\frac{ks^2}{2z}\right)ds + i\int_0^{w-\Delta} \sin\left(\frac{ks^2}{2z}\right)ds \tag{6.55}$$

The sign of the first and third terms in Eq. (6.55) is dependent upon the sign of Δ. If Δ is greater than 0, then the sign is negative and vice versa. The structure of the integral above is of great importance, as it can be decomposed into two relatively simple integrals of the form:

$$\int_0^{\Delta} \cos\left(\frac{ks^2}{2z}\right)ds + i\int_0^{\Delta} \sin\left(\frac{ks^2}{2z}\right)ds \tag{6.56}$$

The above integral is of great importance and is known as the ***Fresnel integral***. Plotting both components of amplitude in Figure 6.15, we produce the familiar form of the ***Cornu spiral***.

Progression around the Cornu spiral in Figure 6.15 is marked by increasing values of Δ, the distance from the slit boundaries. Each successive Fresnel zone is marked in Figure 6.15 and the numbering of the zones is as per Eq. (5.52). Most importantly, it is clear from Figure 6.15 that an asymptote is reached for large values of Δ. At large values of Δ, the integral tends to 0.25 + 0.25i. If, in Eq. (6.55), one assumes that w-Δ is large, then this asymptotic value must be added to the integral. In this case, we can now reasonably approximate the integral expression in Eq. (6.55) in the following manner:

$$A \approx 0.25 \mp \int_0^{\sqrt{k/2z}*\Delta} \cos(s^2)ds + 0.25i \mp i\int_0^{\sqrt{k/2z}*\Delta} \sin(s^2)ds \tag{6.57}$$

When Δ is large and positive, the integral part of Eq. (6.57) cancels out the constant asymptotic values, so the amplitude is zero. Of course, that the amplitude is zero away from the illuminated portion of the slit is to be expected. In the opposite scenario, where a position within the illuminated area is viewed, then the flux levels tend to a uniform value. Around the edge position, and towards the illuminated area, a series of light and dark bands emerge. One can see from the disposition of the Cornu spiral, that the contrast of these bands diminishes as the effective Fresnel zone number increases and, from Eq. (6.52), they also become more tightly packed.

We can now illustrate this process by considering a slit with a width of 2 mm which is illuminated by a 500 nm source. By reference to the example of Gaussian beam propagation, we assume in this analysis that the illumination is significantly spatially coherent. This does not necessarily imply the use of a laser beam; in practice it means that the slit is illuminated by a parallel beam with very small angular divergence. We now view the slit at a distance of 100 mm. The Fresnel number is 20 and, as the effective angle, θ, is 0.01 rad, the

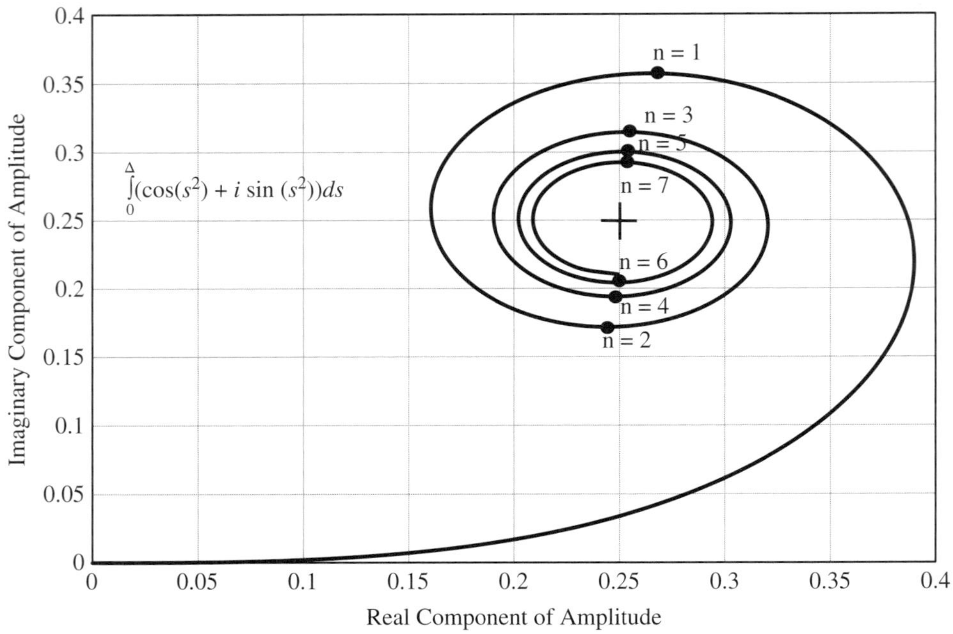

Figure 6.15 Fresnel integral and Cornu spiral.

Fresnel approximation is clearly justified by applying Eq. (6.50). Applying the Fresnel integral to this specific problem, we obtain the flux distribution described by Figure 6.16.

As previously indicated, the illuminated portion is described a series of fringes characterised by the spacing of the Fresnel zones. In the obscured region, the flux tails off to zero. At the slit boundary, the flux is one half of the nominal. Of course, if the set up were reversed, and an obscuration substituted for the slit, then the pattern in Figure 6.16 would be reversed.

Generally, in problems associated with Fresnel diffraction, the diffraction pattern produced by sharp edges broadly follows that illustrated in Figure 6.16. The characteristic diffraction pattern away from the sharp edge feature consists of a series of ripples denominated by the relevant Fresnel zone.

6.9 Diffraction and Image Quality

6.9.1 Introduction

The analysis of image quality is central to any analysis of an imaging system. Where the wavefront error of a system is rather larger than the operating wavelengths of the system, the performance of the system may be adequately described by geometrical optics. Metrics such as geometrical spot size, as derived directly from ray tracing, prevail in this instance. However, where the wavefront errors are very much less than this, then diffraction effects prevail. Indeed, where effects other than diffraction may be legitimately ignored, the image is said to be ***diffraction limited***. Overall, there are a number of metrics that quantitatively describe image quality and these are summarised below:

- **Geometric spot size** (rms spot size, 90% encircled energy etc.) – Geometric optics
- **Point spread function** (rms spot size, 90% encircled energy etc.) – Wave optics
- **Strehl Ratio** – Wave optics
- **Modulation Transfer Function** (MTF)

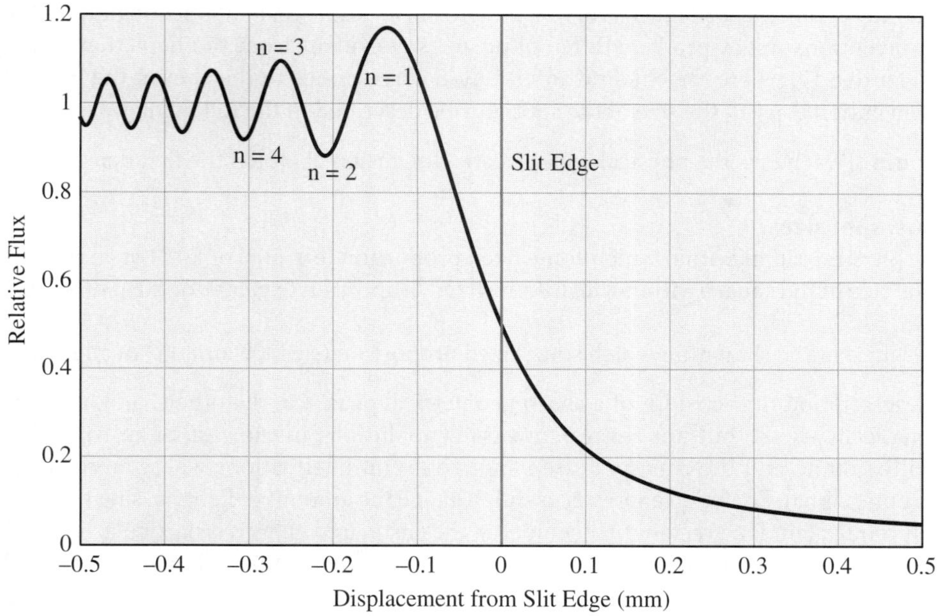

Figure 6.16 Fresnel diffraction at 100 mm from 2 mm Slit – $\lambda = 500$ nm.

6.9.2 Geometric Spot Size

This is perhaps the most straightforward of the image quality metrics to visualise. By virtue of ray tracing, for example, using ray tracing software, a number of representative rays that uniformly illuminate the entrance pupil are traced to the image plane. Of course, in an ideal image formation system, all rays would be traced to a common image point. However, deviation from this ideal behaviour is a measure of the image quality. Furthermore, this process would be attempted for a number of different field positions and, inevitably, for a dispersive system, for a number of different wavelengths. An example geometric spot is shown in Figure 6.17, illustrating the impact of spherical aberration and coma.

In order to quantify the data depicted in Figure 6.17, a number of different measures may be adopted. Measurements are characterised typically with respect to some central location. This central location may ***either*** be

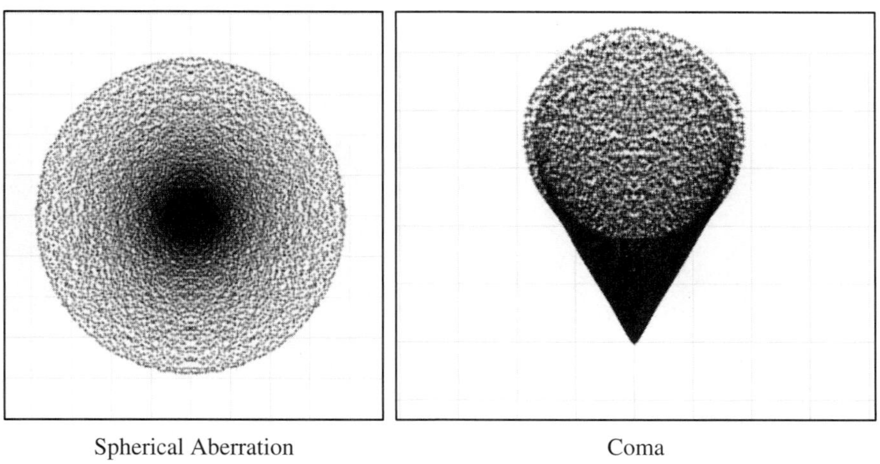

Figure 6.17 Geometric spots for spherical aberration and coma.

the intersection of the chief ray at the image plane ***or*** the weighted mean location of all intersecting rays – the ***centroid***. That the two conventions might produce different answers is evident from the depiction of the comatic spot diagram in Figure 6.17, where the chief ray intersection corresponds to the apex at the bottom of the spot. Whichever convention is used, the size of the spot may be described in the following ways:

- **Full width half maximum** (FWHM) – the physical width in one dimension at which the flux density falls to half of the maximum
- **Root mean square (rms) spot size**
- **Encircled energy** – the physical radius within which some fixed proportion (e.g. 50% or 80%) of all rays lie.
- **Ensquared energy** – the size of the square within which some fixed proportion (e.g. 50% or 80%) of all rays lie
- **Enslitted energy** – the width of the slit within which some fixed proportion (e.g. 50% or 80%) of all rays lie

The FWHM is a useful description of the width of a sharp geometrical peak. On the other hand, the rms spot size is more mathematically useful, but not always universally applicable. In the case of an Airy disc, the rms spot size is actually infinite and the rms spot size is not so useful in situations where an image is attended by a large background signal. Encircled energy is useful to gauge the amount of light passing through a small circular aperture. Its equivalent for a rectangular geometry, the ensquared energy, is particularly useful for pixelated detectors whose sensor elements are naturally either square or rectangular. Similarly, for slitted instruments, such as spectrometers, enslitted energy is a useful metric.

Where the overall wavefront error is significantly larger than the wavelength, this geometric description of image quality is perfectly adequate. However, where this is not the case, we must look to a new approach.

6.9.3 Diffraction and Image Quality

In Section 6.5 we examined briefly the diffraction pattern produced by a uniformly illuminated pupil. This is the so-called Airy disc. The Airy disc is the diffraction pattern that one would obtain in the absence of any system aberration. In Section 6.6, we included the effect of system aberration, introducing the Huygens point spread function. The Huygens point spread function is the flux distribution pattern produced at the image plane produced by a point source at the object plane. Figure 6.18 shows an example of an aberrated pupil, where the OPD is mapped in two dimensions across a circular pupil.

The Huygens point spread function (PSF) of the same system is shown in Figure 6.19.

The PSF shown in Figure 6.19 shows much deeper rippling when compared to the Airy distribution and, unlike the geometrical analysis, represents an accurate solution for the local flux distribution at the image. In

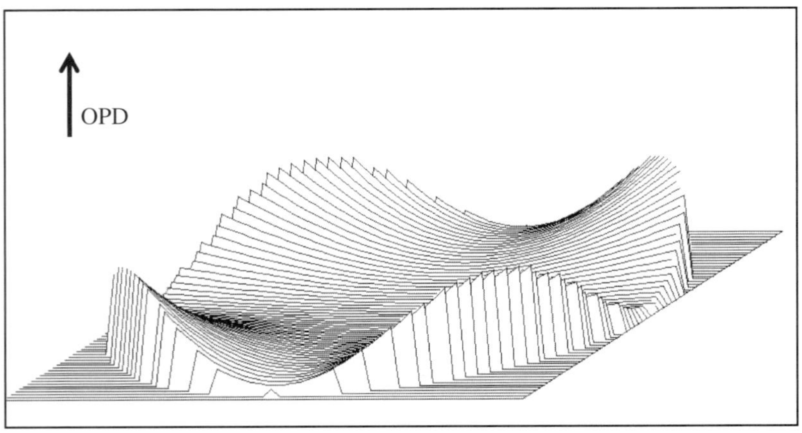

Figure 6.18 OPD map across pupil.

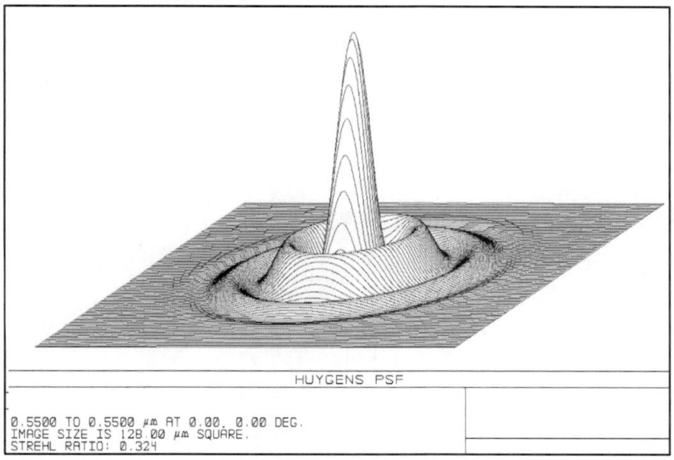

Figure 6.19 Huygens point spread function.

analysing the PSF, one can use similar metrics as for the geometric spot size, with the addition of the Strehl ratio:

- **Strehl Ratio**
- **Full width half maximum**
- **Root mean square (rms) spot size**
- **Encircled energy**
- **Ensquared energy**
- **Enslitted energy**

As mentioned previously, the Strehl ratio describes the ratio of the aberrated peak flux to the unaberrated peak flux. A ratio of 0.8 or greater, by virtue of the Maréchal criterion, is considered to be 'diffraction limited'. This is consistent with an rms wavefront error of lambda/14 or a peak to valley wavefront error of lambda/4. This measure was introduced earlier in Section 6.6, and is an exceptionally important metric to keep in mind when designing a system that is diffraction limited or near diffraction limited.

6.9.4 Modulation Transfer Function

The **MTF** expresses the ability of an imaging system to replicate the contrast of a specific object pattern. In the case of the MTF, the object is represented by a sinusoidally varying pattern of light and dark, described by some kind of spatial frequency, k_o. That is to say, the spatial variation of the object illumination is represented by:

$$I_{Object} = I_0(1 + \sin k_o x) \tag{6.58}$$

The contrast ratio of the illumination is defined as the ratio of the difference of the maximum and minimum fluxes to the sum of those fluxes. The illumination pattern represented in Eq. (6.58) has a contrast ratio of unity, with the minimum flux being zero. However, because of imaging imperfections, this is not fully represented at the image plane and the contrast is somewhat reduced. Assuming the system magnification is *M*, then

$$I_{Image} = I_1\left(1 + a \sin \frac{k_o}{M} x\right) \tag{6.59}$$

For an input contrast ratio of unity, the MTF is defined as the contrast ratio of the image:

$$MTF = \frac{I_{max} - I_{min}}{I_{max} + I_{min}} = a \tag{6.60}$$

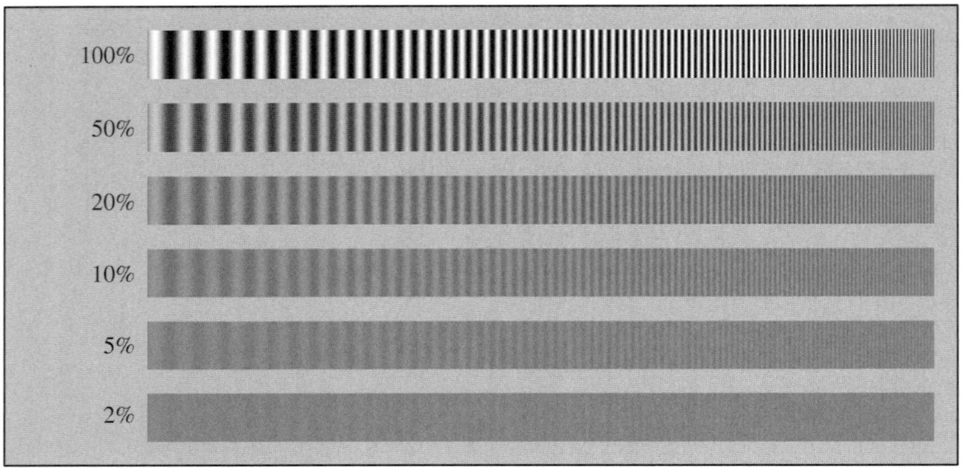

Figure 6.20 MTF pattern.

A typical example of an MTF pattern is shown in Figure 6.20.

The MTF response shown in Figure 6.20 illustrates different final contrast levels, varying from 2% to 100%. In addition, a range of input spatial frequencies is shown. In practice, there is a tendency for the contrast ratio to reduce at higher spatial frequencies; a typical imaging system has a reduced capacity for replicating fine details. A typical MTF plot against spatial frequency is shown in Figure 6.21.

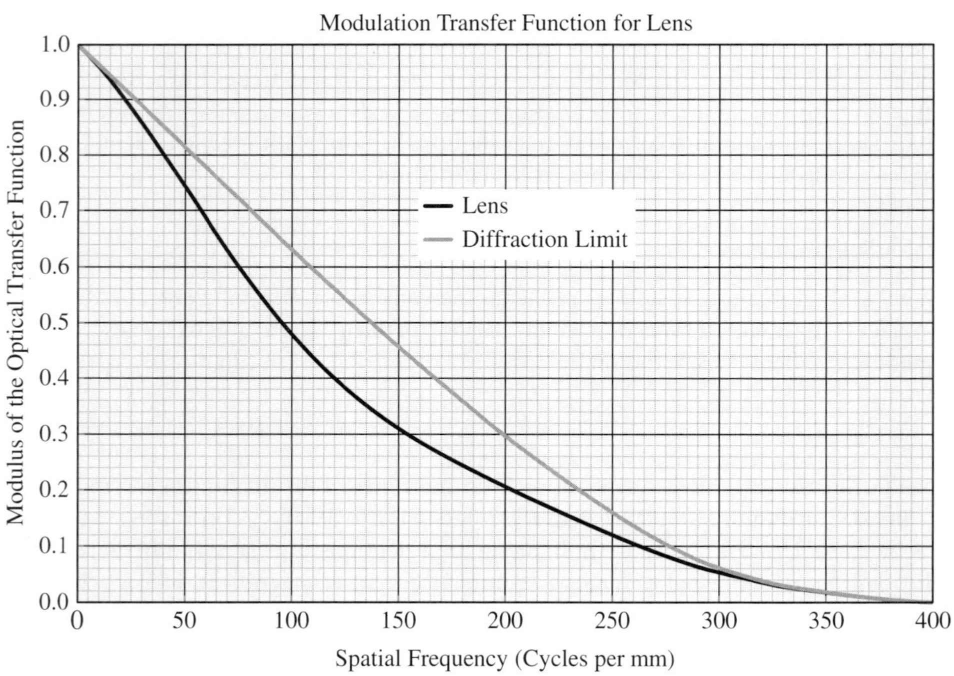

Figure 6.21 Typical MTF plot.

It is evident, from Figure 6.21, that the MTF declines with spatial frequency. Also included in the plot is the MTF of the diffraction limited system. In fact, the MTF is the absolute value of the complex ***optical transfer function*** (OTF). The OTF of a system is related to the Fourier transform of the point spread function. In fact, for the diffraction limited system, the MTF follows a fairly simple mathematical prescription. There is some maximum spatial frequency, υ_{max}, above which the MTF is zero and this is defined by the system numerical aperture, *NA*. In this case, the diffraction limited MTF is simply given by:

$$MTF = \frac{2}{\pi}\left(\cos^{-1}\left(\frac{\upsilon}{\upsilon_{max}}\right) - \left(\frac{\upsilon}{\upsilon_{max}}\right)\sqrt{1-\left(\frac{\upsilon}{\upsilon_{max}}\right)^2}\right) \quad \upsilon_{max} = \frac{2NA}{\lambda} \tag{6.61}$$

The MTF is widely used in the testing and analysis of camera systems. One particular attribute of the MTF is especially useful. For a system composed of a number of subsystems, the MTF of the system is simply given by the product of the individual MTFs:

$$MTF_{TOTAL} = MTF_1 \times MTF_2 \times MTF_3 \times \ldots. \tag{6.62}$$

Analysis of the MTF is also useful in incorporating the behaviour of the detector. In a traditional context, where photographic film had been used, the contrast provided by the film media would be defined by the spatial frequency at which its effective MTF fell to 50%. For high contrast black and white film, this spatial frequency might have been of the order of 100 cycles per mm, although this would vary with film type and sensitivity. On the whole, colour film had poorer contrast with the equivalent spatial frequency being less than 50 cycles per mm. Of course, modern cameras base their detection upon pixelated sensors. In this instance, the characteristic spatial frequency is defined by Nyquist sampling where the equivalent spatial frequency covers two whole pixels. That is to say, for a pixel spacing of 5 μm, the equivalent spatial frequency is 100 cycles per mm.

Worked Example 6.6 We are designing a camera system to give an MTF of 0.5 at 100 cycles per mm. The camera has a pixelated detector with a pixel spacing of 5 μm. It may be assumed that the effective MTF of this detector is 0.75. The remainder of the system may be assumed to be diffraction limited. For a working wavelength of 500 nm, what is the minimum numerical aperture that the system needs to have to fulfil its requirement?

From Eq. (6.62) – the MTF of the remainder of the system is equal to 0.5/0.75 = 0.67.

Using this MTF figure we can calculate υ/υ_{max} from Eq. (6.61) and this amounts to 0.265, giving υ_{max} as 377 cycles per mm. Again from Eq. (6.61), given a wavelength of 500 nm, we can calculate the minimum numerical aperture of the system as 0.095 or about f#5.

6.9.5 Other Imaging Tests

The MTF provides a clearly mathematically defined test pattern for testing and subsequent analysis. However, there are other image resolution tests based upon the replication of reticulated patterns, often consisting of sharply delineated features, such as lines or line pairs. One example of this is the ***1951 USAF resolution test chart*** which is a standard reticle placed at the object location. Broadly speaking, this consists of a set of line features whose characteristic size reduces by the sixth root of two when progressing from feature to feature. Visual inspection of the final image enables determination of the minimum line spacing resolution. The standard USAF pattern is illustrated in Figure 6.22.

Although, these types of test are inherently simpler and less capital intensive, the reliance on human visual inspection is, in itself, a weakness. Where, at one time, analytical complexity precluded the widespread use of MTF and other more abstruse processes, the availability of high performance computational power overcomes this obstacle nowadays.

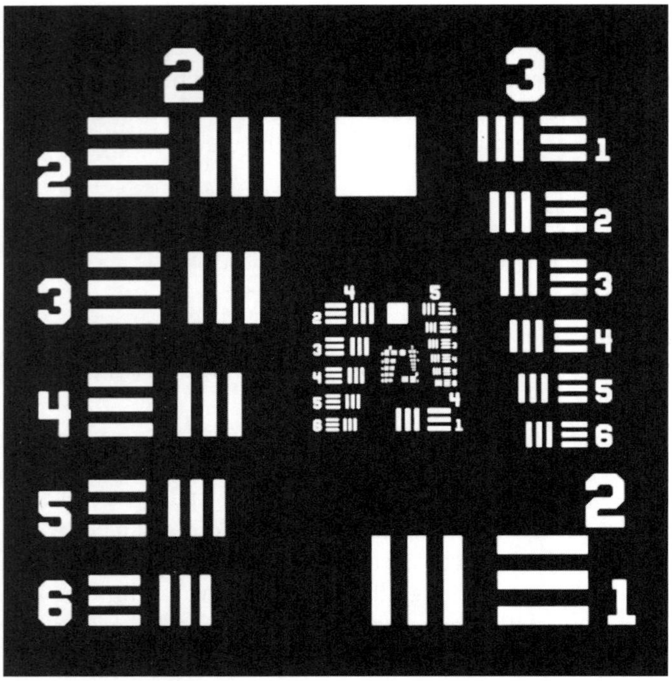

Figure 6.22 1951 USAF resolution test chart.Source: Image Provided by Thorlabs Inc.

Further Reading

Bleaney, B.I. and Bleaney, B. (1976). *Electricity and Magnetism*, 3e. Oxford: Oxford University Press. ISBN: 978-0-198-51141-0.

Born, M. and Wolf, E. (1999). *Principles of Optics*, 7e. Cambridge: Cambridge University Press. ISBN: 0-521-642221.

Lipson, A., Lipson, S.G., and Lipson, H. (2011). *Optical Physics*. Cambridge: Cambridge University Press. ISBN: 978-0-521-49345-1.

Saleh, B.E.A. and Teich, M.C. (2007). *Fundamentals of Photonics*, 2e. New York: Wiley. ISBN: 978-0-471-35832-9.

Wolf, E. (2007). *Introduction to the Theory of Coherence and Polarisation of Light*. Cambridge: Cambridge University Press. ISBN: 978-0-521-82211-4.

Yariv, A. (1989). *Quantum Electronics*, 3e. New York: Wiley. ISBN: 978-0-471-60997-1.

7

Radiometry and Photometry

7.1 Introduction

In the preceding chapters, we have been concerned with the general behaviour of light in an optical system, as described by ray and wave propagation. Hitherto, there has been no interest in the ***absolute*** magnitude of the wave disturbance. On the other hand, radiometry and photometry is intimately concerned with the absolute flux of light within a system, its analysis and, above all, its measurement.

At this point we will make a distinction between the two terms, radiometry and photometry. ***Radiometry*** relates to analysis of the absolute magnitude of optical flux, as defined by the relevant SI unit, e.g. Watts or Watts per square metre. In contrast, ***photometry*** is concerned with the measurement of flux ***as mediated by the sensitivity of some detector***. Most notably, although not exclusively, the detector in question might be the human eye. So, from a radiometric perspective 1 W of ultraviolet or infrared emission is worth 1 W of visible emission. However, from a photometric view (as referenced to the human eye) the ultraviolet and infrared emissions are worthless.

In the study of radiometry, we are interested in the emission of light from a physical source that might have some area dS and subtend some solid angle, $d\Omega$. The light may either be directly emitted from a luminous source, such as a lamp filament, or scattered indirectly. The generic geometry for this is illustrated in Figure 7.1.

The geometry above may be applied both to the emission of light from a surface or to the absorption/scattering of light at a surface. The distinction between these two scenarios simply implies a reversal of the direction of travel of the rays.

7.2 Radiometry

7.2.1 Radiometric Units

For the purposes of this introduction, we will confine the initial discussion to radiometry, as opposed to photometry, where we are able to quantify the optical power of a source simply in terms of its output in watts. Fundamental to the analysis of radiometry are the radiometric quantities and their associated radiometric units. The most basic measure of an optical source is its ***radiant flux***, Φ, measured in watts. Associated with the radiant flux is the ***radiant flux density***, E. This refers to the total flux per unit area that is incident upon or is leaving a surface element and is measured in watts per square metre. If the radiant flux is incident upon a surface, then the radiant flux density is more usually referred to as the ***irradiance***. If, on the other hand, the flux is emitted from the surface, it is referred to as ***exitance***. It is of the utmost importance to apprehend that according to the strict definitions of radiometry, ***flux per unit area is never described as intensity***. There is often a 'colloquial' tendency to describe flux per unit as intensity. However, this term is reserved rather for ***flux per unit solid angle***. As such, the ***radiant intensity*** of a (point) source, I, is defined as its flux per unit solid angle and is measured in watts per steradian.

Optical Engineering Science, First Edition. Stephen Rolt.

Companion website: www.wiley.com/go/Rolt/opt-eng-sci

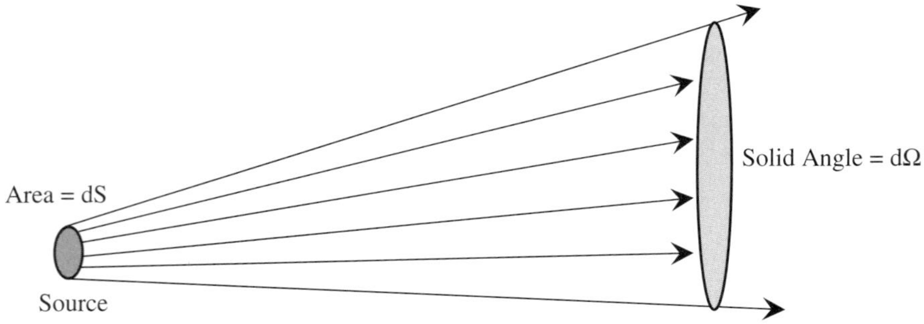

Figure 7.1 Emission from a generic source.

Table 7.1 Radiometric units.

Quantity	Also called	Description	Symbol	Unit
Flux		Total radiated power	Φ	W
Radiant flux density	Irradiance, exitance	Power per unit area	E	W m^{-2}
Radiant intensity		Power of source per unit solid angle	I	W sr^{-1}
Radiance		Power per unit area per unit solid angle	L	W m^{-2} sr^{-1}

Radiance, L, is the flux arriving at or leaving a surface associated with a pencil of rays, per unit solid angle per unit surface area projected onto a plane normal to the direction of travel of those rays. It is measured in watts per square metre per steradian. Radiance is intimately related to 'how bright' an extended object appears and is not affected by distance from the object. For example, in the case of the sun, as one moves away from the sun, the irradiance of the solar illumination inevitably diminishes. However, the angle subtended by the solar disc reduces proportionally and the smaller solar disc would appear just as bright if one were so ill advised as to view it.

All the radiometric quantities and associated units are summarised in Table 7.1.

7.2.2 Significance of Radiometric Units

The radiant flux density can be taken as the differential of the flux with respect to area. Expressing this mathematically:

$$E = \frac{d\Phi}{dS} \tag{7.1}$$

The important point to recognise about Eq. (7.1), is that the area, dS, is not only described by a scalar area, but also by a vector that defines the surface normal. Thus, orientation of the surface is of importance, and this will be described in more detail presently. Radiant intensity, may also be expressed mathematically, in this case as the differential of the flux with respect to the solid angle.

$$I = \frac{d\Phi}{d\Omega} \tag{7.2}$$

Most usually, intensity is used to describe the output of a point source. Simple geometry may be used to establish the relationship between the intensity of a point source and the irradiance it produces at a surface

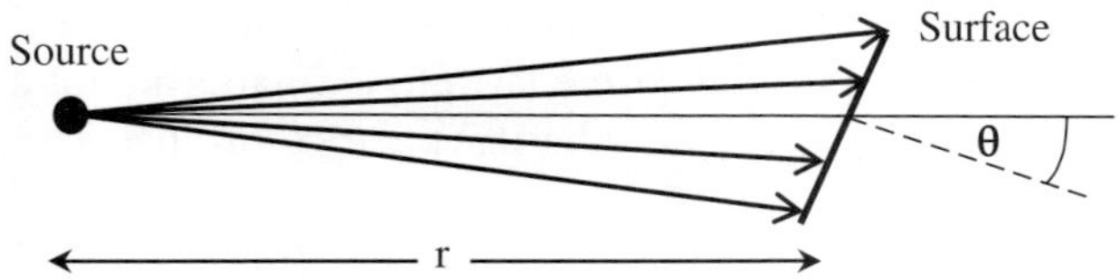

Figure 7.2 Operation of the inverse square law.

Figure 7.3 Radiance and exitance from a surface.

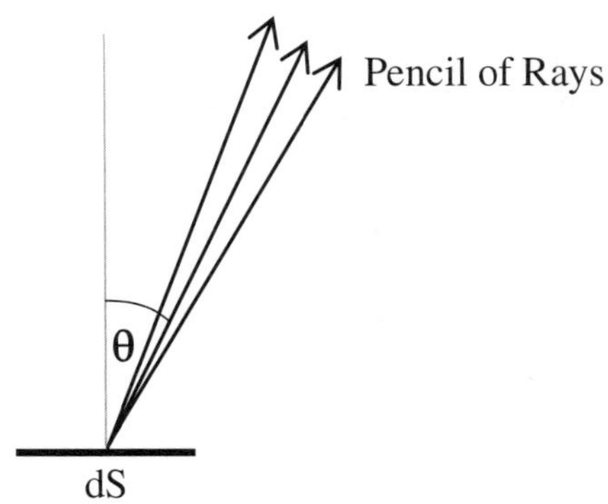

located at some distance from the source. This gives rise to the so-called ***inverse square law***. The inverse square law states that the irradiance delivered by a point source to a distant object is inversely proportional to the square of the separation. Operation of the inverse square law is illustrated in Figure 7.2.

In the geometry illustrated in Figure 7.2, the irradiated surface is situated at a distance r from the source and its normal is at an angle θ with respect to the line joining the source at the surface. As alluded to earlier, the orientation of the surface is of some relevance. Assuming that the radiant intensity of the source is I, then the irradiance at the surface is given by:

$$E = \frac{I\cos\theta}{r^2} \tag{7.3}$$

Radiance is the flux arriving at a surface or leaving a surface per unit area, per unit solid angle. The area, in this case, is the projected area whose normal is aligned with the ray pencil, rather than the surface normal. This is illustrated schematically in Figure 7.3.

Expressing the intensity in terms of the area of an element of surface, dS, we obtain the following:

$$L = \frac{dI}{dS\cos\theta} = \frac{d^2\Phi}{d\Omega dS\cos\theta} \tag{7.4}$$

L is the radiance and I the radiant intensity.

7.2.3 Ideal or Lambertian Scattering

An ideal or ***Lambertian scatterer*** scatters light from a surface with uniform radiance irrespective of the scattering angle, θ. An imperfect approximation to a Lambertian surface might be a blank sheet of paper or a matt surface, such as a painted wall. In practice, for most surfaces, the radiance has a tendency to decline with θ. However, for a Lambertian surface, the radiant intensity of the scattered light is (from Eq. (7.4)) given by:

$$I = L\cos\theta dS \tag{7.5}$$

Hence for a Lambertian scatterer, the radiant intensity emitted from a surface element is proportional to the cosine of the angle with respect to the surface normal. In many instances, we are interested in the total amount of light scattered from a surface. This is the ***total hemispherical scatter*** or ***total hemispherical***

exitance from a surface. At first sight, this would seem to amount to the product of the solid angle, 2π, and the radiance, L. However, the radiant intensity declines according to the cosine law Eq. (7.5) and the total hemispherical exitance may be derived from the following integral, based on spherical polar co-ordinates:

$$E = 2\pi \int_0^{\pi/2} L \cos\theta \sin\theta \, d\theta = L\pi \tag{7.6}$$

Hence, the total hemispherical exitance is half what would be expected if the radiant intensity were constant as a function of polar angle.

7.2.4 Spectral Radiometric Units

In most instances, the output of an optical source varies very significantly with wavelength. As such, we are generally interested in the radiometric flux within a very narrow band of wavelengths and how this quantity varies with wavelength. In this case, flux becomes spectral flux and radiance becomes spectral radiance, ad so on. If λ is the wavelength of interest, the corresponding spectral quantities may be defined as follows:

$$\text{Spectral Flux} \quad \Phi_\lambda = \frac{d\Phi}{d\lambda} \tag{7.7a}$$

$$\text{Spectral irradiance/exitance} \quad E_\lambda = \frac{dE}{d\lambda} \tag{7.7b}$$

$$\text{Spectral radiant intensity} \quad I_\lambda = \frac{dI}{d\lambda} \tag{7.7c}$$

$$\text{Spectral radiance} \quad L_\lambda = \frac{dL}{d\lambda} \tag{7.7d}$$

By way of illustration, we will examine the spectral intensity produced by a commonly used illumination source. The xenon arc lamp is extensively used in commercial and laboratory applications as a 'point source', with a spectrum similar to that of the sun. Such (nominally) point sources are generally described by their radiant intensity, which gives a useful measure of the overall output of the source. In the case of the spectral measure, spectral radiant intensity is measured in Watts per steradian per nm. Figure 7.4 shows a plot of spectral radiant intensity versus wavelength for a 1000 W xenon lamp.

Similarly, the solar flux arriving at the earth's surface may be denominated in terms of the spectral irradiance – that is the solar flux per unit area of the earth's surface per unit bandwidth. In the case of Figure 7.5, the data presented represents the spectral irradiance of the sun above the earth's atmosphere, as signified by the parameter 'AM0' or air mass zero.

Of course, Figure 7.5 does not present the solar irradiance as it would be at the sun's surface; this would be very much greater and would fall off according to the inverse square law, Eq. (7.3). When calculating the spectral radiance associated with the data in Eq. (7.5), one would have to divide the irradiance by the solid angle subtended by the 0.5° solar disc, i.e. 6.8×10^{-5} sr. Peak solar spectral radiance (at ~500 nm) would be about 30 000 W m^{-2} sr^{-1} nm^{-1}.

7.2.5 Blackbody Radiation

Thermal radiation is associated with the thermal emission of electromagnetic radiation from an incandescent source. In particular, ***blackbody emission*** occurs when a solid surface is in thermal equilibrium with the surrounding electromagnetic radiation. The exitance associated with a black body emitter at an absolute temperature of T, is proportional to the fourth power of the temperature and given by the well known ***Stefan's law***:

$$E = \varepsilon\sigma T^4 \tag{7.8}$$

σ is Stefan's constant (5.67×10^{-8} W m^{-2} K^{-4}); ε is the surface emissivity (1 for a perfect black body).

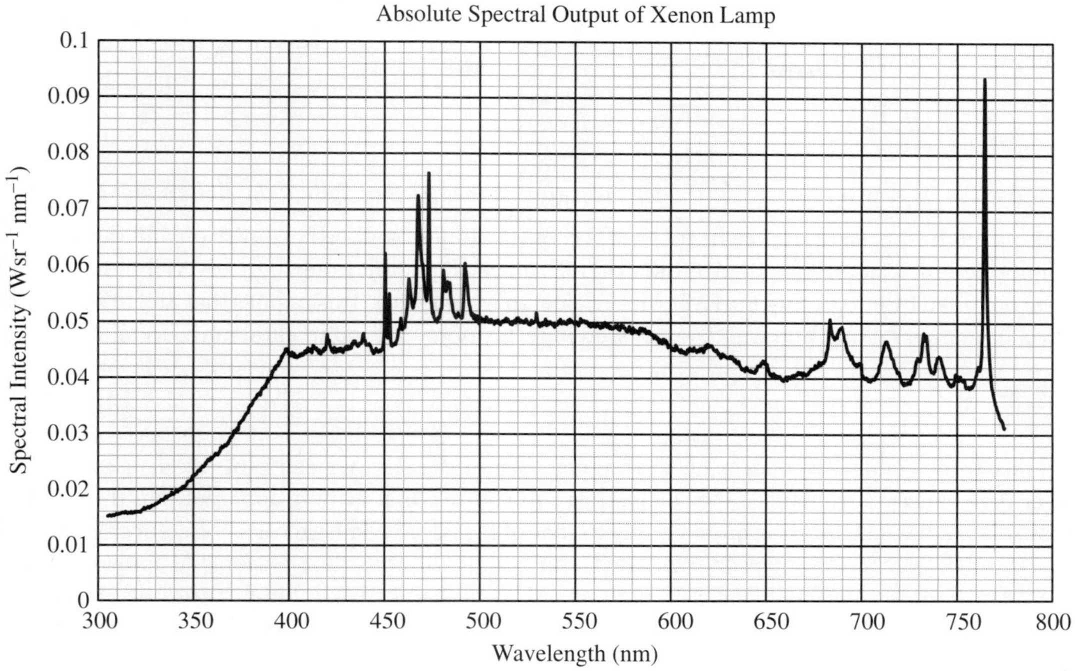

Figure 7.4 Xenon arc lamp spectral intensity.

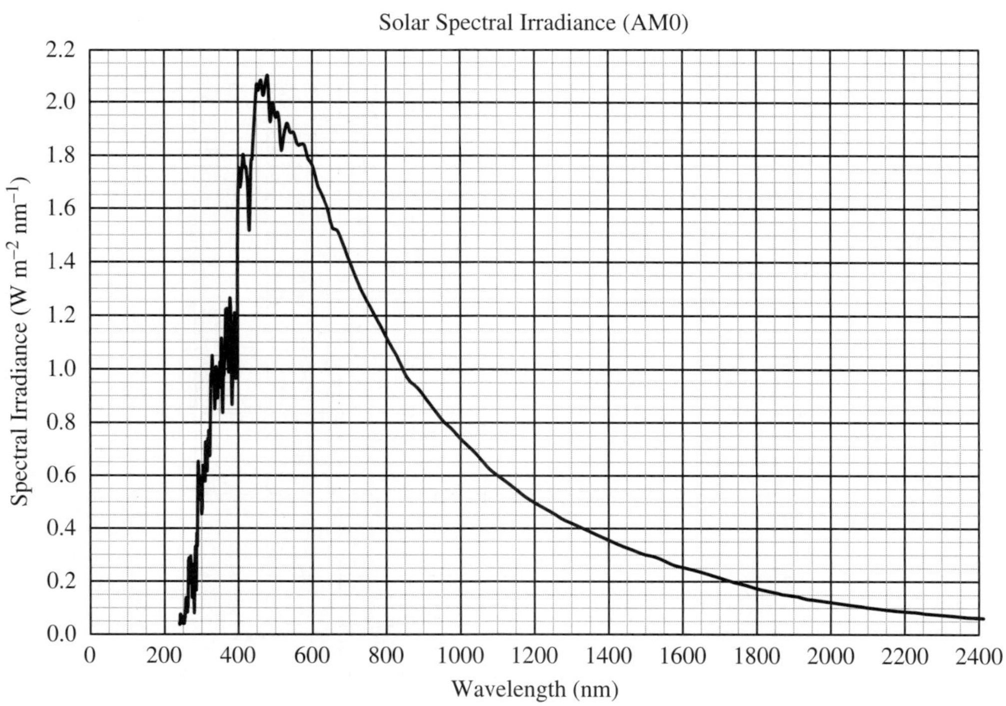

Figure 7.5 Solar spectral irradiance. Source: NASA SORCE Satellite Data – Courtesy University of Colorado.

Most importantly, blackbody emission has a characteristic spectral distribution, quantified by its spectral radiance which depends only upon the wavelength and the surface temperature. The spectral radiance of blackbody emission is defined by Planck's law:

$$L_\lambda(\lambda, T) = \frac{2hc^2}{\lambda^5}\frac{1}{e^{hc/\lambda kT} - 1} \tag{7.9}$$

L_λ is expressed in SI units – $W\,m^{-3}\,sr^{-1}$; h is Planck's constant; c the speed of light; k the Boltzmann constant.

To convert Eq. (7.9) to spectral exitance from a surface, one assumes Lambertian emission and the spectral radiance is multiplied by a factor of π to give the exitance, as per Eq. (7.6). Indeed, the overall radiance and exitance can be obtained by integrating Eq. (7.9) with respect to wavelength. This implies that Stefan's constant is not actually a fundamental unit and can be expressed in terms of more fundamental units, as follows:

$$\sigma = \frac{2\pi^5 k^4}{15c^2h^3} \tag{7.10}$$

Taking the data from Figure 7.5, and using the angular size of the sun, we can plot the data as spectral radiance rather than spectral irradiance. This is illustrated in Figure 7.6. It is quite apparent that the spectral distribution of solar radiance conforms quite closely to that of blackbody emission. For reference, Figure 7.6 shows a plot of 5800 K blackbody emission generated using Eq. (7.9). Thus, to a reasonable approximation, solar radiation can be described as blackbody emission with a characteristic temperature of 5800 K. As stated previously, radiance describes the effective brightness of a surface and, for blackbody emission is purely related to the physical characteristics of the source, temperature, and so on and not to geometry. So, as stated earlier, although the spectral irradiance of solar emission is reduced as one moves away from the sun, the corresponding reduction in the angular size of the sun maintains the spectral radiance at a constant level.

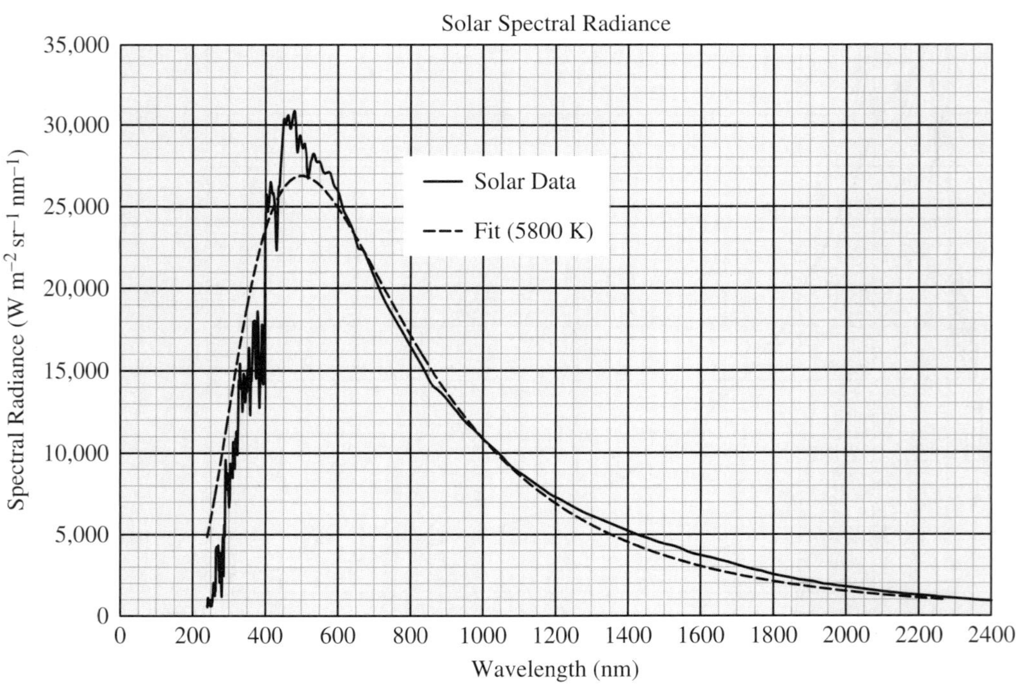

Figure 7.6 Solar spectral radiance and 5800 K blackbody radiance.

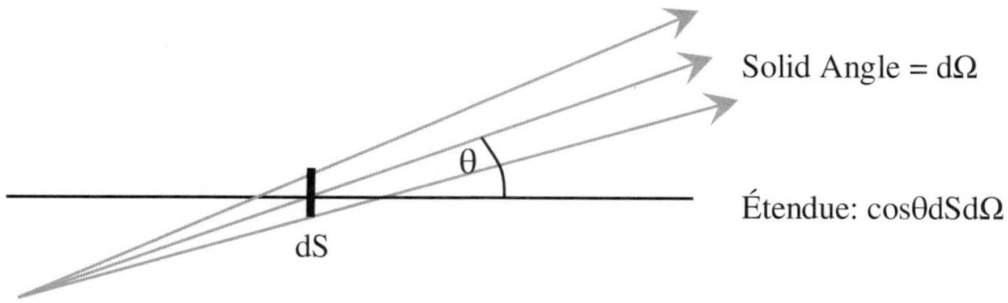

Figure 7.7 Étendue of a pencil of rays.

7.2.6 Étendue

Étendue is the product of the area and solid angle of a pencil of rays in an optical system. The concept of étendue is central to the understanding of the radiometry of an optical system together with many other aspects of optical system performance. As applied to an optical system, its étendue may be represented as the product of the entrance pupil area and the solid angle of the input field. A critical aspect of the behaviour of étendue in an optical system is the operation of the Lagrange invariant. Effectively, the Lagrange invariant and the inverse relationship between linear angular magnification implies that ***étendue must be preserved in an ideal optical system***. That is to say, for a perfect paraxial system, as the imaged (exit) pupil size is increased, the corresponding field angle will be reduced proportionately. Of course, this only applies to a perfect optical system and any image degradation due to aberration has a tendency to increase the étendue. The concept of étendue is illustrated in Figure 7.7.

More formally, as illustrated in Figure 7.7, the étendue of a pencil of rays is given by:

$$dG = \cos\theta dSd\Omega \tag{7.11}$$

G is the étendue and θ is the tilt of the surface normal with respect to the ray pencil.

As outlined earlier, for a generalised and perfect optical system, its étendue is a system invariant. Describing the pupil size of a generalised optical system by its numerical aperture, *NA* and the field by its total area, *S*, then the system étendue is given by:

$$G = \pi NA^2 S \tag{7.12}$$

The similarity of Eqs. (7.11) and (7.4) which denominates the connection between radiance and flux, brings us to the fundamental utility of étendue in radiometric calculations. It is easy to appreciate that, for an optical system, the radiance associated with a pencil of rays is the derivative of the flux with respect to the étendue.

$$L = \frac{d\Phi}{dG} \tag{7.13}$$

If the étendue of a pencil of rays is invariant through an ideal system, then the implication of Eq. (7.13) is that the radiance associated with the object and image must be identical. This is very important, as it conveys a fundamental thermodynamic truth. If one considers a blackbody object, any reduction in étendue through the system would imply that the radiance of the image is higher than that of the object. In the context of blackbody radiation, the associated temperature of the image would be higher than that of the object. Therefore, the effect of this would be to take energy from the lower temperature source (the object) and convey it to a higher temperature body (the image) without doing work. This is in violation of the second law of thermodynamics. Any imperfections in the optical system (aberrations) tend to increase the étendue and so reduce the radiance at the image.

The practical utility of étendue lies in its assistance in expediting radiometric calculations in complex optical systems. If one has a source with some known spectral radiance, L_{source}, a system with étendue, G_{system}, and a system throughput of ξ, then the flux, Φ_{image}, arriving at the image is simply given by:

$$\Phi_{image} = \xi G_{system} L_{source} \tag{7.14}$$

The throughput, ξ, is simply a measure of how much light is transmitted through an optical system as mediated by any scattering, absorption, or reflection that occurs. If, as in the case of an ideal system, none of the optical surfaces were to absorb, scatter or reflect any light, then the throughput would be 100%.

Worked Example 7.1 ***Flux Calculation***
To illustrate the power of the foregoing analysis, we will now examine a practical example. An optical system is designed to view the filament of a tungsten halogen lamp. A camera with an aperture of f#2 images the filament onto the square pixels of a detector; the size of the pixels is 10 μm. For a single pixel of interest, only a small part of the incandescent filament is imaged which fills the entire pixel. The filament itself may be regarded as a blackbody emitter with a temperature of 3000 K. A narrowband filter is included in the optical train which only admits light in a 5 nm wide band around 500 nm. With the exception of the filter, the system throughput, ξ, is 80%. What is the flux arriving at a single pixel?

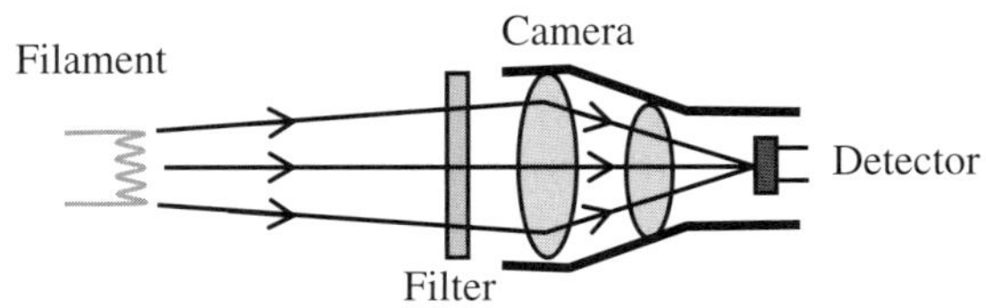

We are told that the source is a 3000 K blackbody emitter; therefore we should be able to calculate the spectral radiance from Eq. (7.9). In fact, we are interested in the spectral radiance at 500 nm. From Eq. (7.9), the spectral radiance is 2.6×10^{11} W m^{-3} sr^{-1} or 260 W m^{-2} sr^{-1} nm^{-1} at 500 nm. The radiance transmitted by the 5 nm bandwidth filter is 5×260 or 1300 W m^{-2} sr^{-1} at. We now need to calculate the system étendue from Eq. (7.9). The numerical aperture of the system in image space is 0.25 (for f#2) and the area, S, of a single pixel is $10^{-5} \times 10^{-5} = 10^{-10}$ m^2.

$$G = \pi NA^2 S = 1.96 \times 10^{-11}\ \text{m}^2\ \text{sr}^{-1}.$$

The solution is now almost complete, we only need to apply Eq. (7.14), making an allowance for the throughput, ξ, of 80%:

$$\Phi_{image} = \xi G_{system} L_{source} = 0.8 \times 1.96 \times 10^{-11}\ \text{m}^2\ \text{sr} \times 1300 = 2.04 \times 10^{-8}\ \text{W}$$

Thus, the power arriving at a single pixel is 2.04×10^{-8} W.

The essential point of the previous analysis is that the same fundamental logic and analysis applies, ***irrespective of the complexity of the optical system under investigation***. In this example, we are not given any details of the optical design, only the pupil and field size. Nevertheless, we are able to estimate the flux arriving at the detector pixel. Of course, we are assuming that system aberrations do not play a significant role.

7.3 Scattering of Light from Rough Surfaces

Much of the preceding analysis has focused on self-luminous sources. These sources, such as blackbody emitters, more or less emit light in a random fashion. In the study of radiometry, we are also interested in surfaces that scatter light in a more or less random fashion. This is distinct from ***specular*** surfaces, i.e. mirrors, which reflect light in a deterministic, ordered fashion. The topic was touched on very briefly when we touched on perfect or Lambertian scattering where the radiance of the scattered light is independent of the scattered angle. Unfortunately, this condition is not realised in real materials, so an alternative approach is needed. Here we shall now describe a more generalised treatment, which describes the scattering from a surface using the so-called ***bi-directional reflection distribution function*** (BRDF).

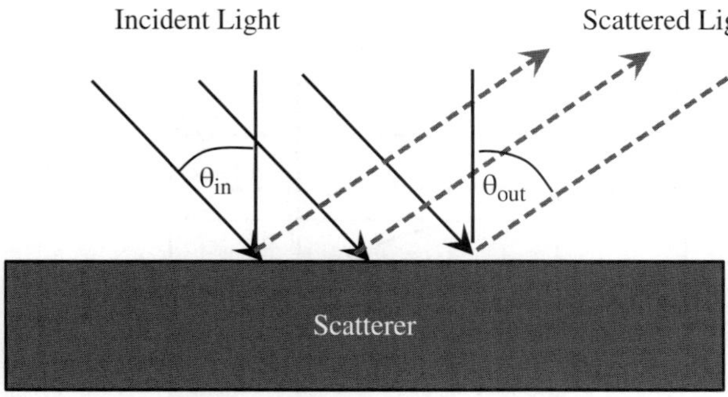

Figure 7.8 Illustration of BRDF.

Light that is incident upon a surface is described by its irradiance and its incident angle, θ_{in}, as depicted in Figure 7.8. The scattered light is described by its radiance and its output polar angle, θ_{out}. Significantly, since the incident light breaks the symmetry of the scattering surface about the normal, the azimuthal angle, ϕ, of the scattered light also needs to be described. The BRDF is simply the derivative of the output radiance with respect to the input irradiance.

Naturally, the BRDF is a function of wavelength, so the input irradiance might be defined as E(θ_{in}, λ) and the output radiance as E(θ_{out}, ϕ, λ). In this case, the BRDF is given by:

$$BRDF(\theta_{in}, \theta_{out}, \varphi, \lambda) = \frac{dL_{out}(\theta_{out}, \varphi, \lambda)}{dE_{in}(\theta_{in}, \lambda)} \tag{7.15}$$

Units for BRDF are sr^{-1} and a perfect Lambertian scatterer, with a total hemispherical reflectance of unity, would have a uniform BRDF of $1/\pi$. Interest in the radiometry of scattering arises from two principal practical considerations. Firstly, in many applications in optical imaging, there is a requirement to provide uniform illumination over a specific input field. Secondly, there is a diverging motivation in that the optical designer is keen to avoid the deleterious impact of scattered light on image contrast. Therefore, it is important to understand not only the impact of the optical components themselves in manipulating light, but also the effect of the optical mounts and surrounding enclosures and other non-optical surfaces in scattering light.

The preceding chapters have given a clear understanding as to the underlying principles of optical design in so far as the optical components and surfaces are concerned. Ultimately, as will be discussed in detail later, in contemporary design, this proceeds by the use of optical modelling software. For the optical components themselves, the process is referred to as ***sequential modelling*** where rays progress in a deterministic fashion and in a clear sequence from one optical surface to the next. In contrast, scattering is an inherently stochastic process, with the scattered distribution described by the BRDF which is essentially a probability distribution for scattering. In the light of these random processes, there is no inherent, ordered sequence of surfaces through which the light progresses. As such, any modelling in this scenario must account for the ***non-sequential*** nature of light propagation. Such modelling, of course, must account for the geometrical distribution of any scattering and the study of BRDF distributions is of considerable practical utility.

An example of the BRDF of a real material, Spectralon®, is shown in Figure 7.9. The data is for a wavelength of 900 nm and normal incidence, i.e. $\theta_{in} = 0$. Spectralon, is based upon sintered polytetrafluoroethylene (PTFE) and represents the closest approximation to an ideal scatterer of any material. Even so, there is a tendency for the BRDF to decline with increasing polar angle.

7.4 Scattering of Light from Smooth Surfaces

The foregoing analysis is entirely appropriate for the scattering of light from matt or rough surfaces. However, polished surfaces, such as those in lenses and mirrors, can contribute to the unintended scattering of light, even though their roughness is very low. Analysis of this type of scattering is of exceptional importance where

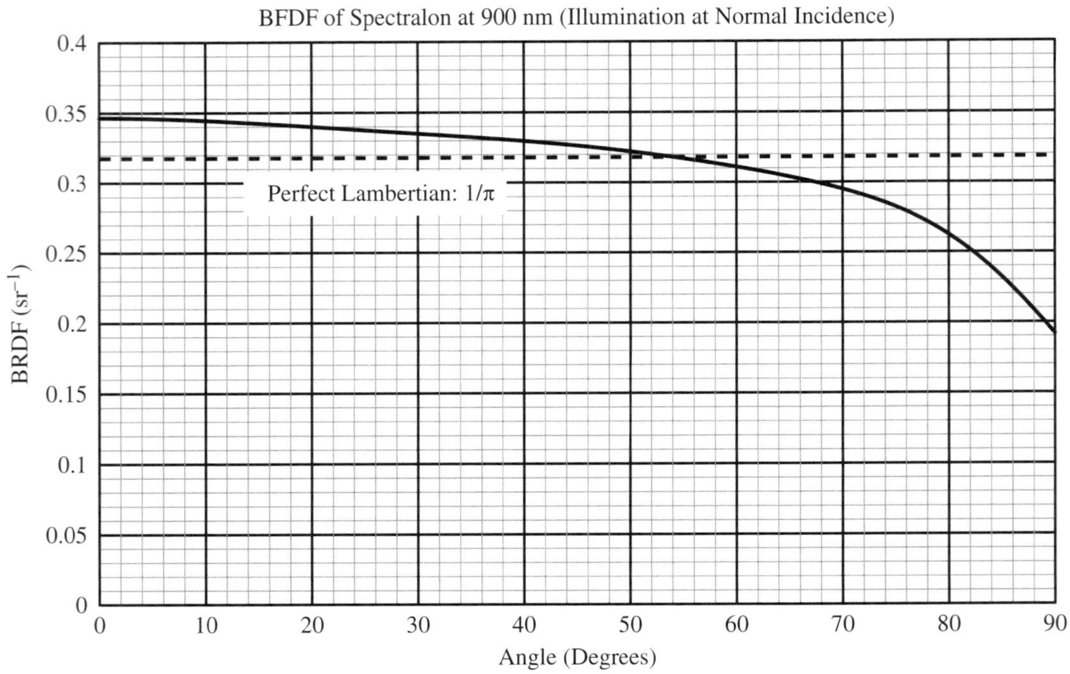

Figure 7.9 BRDF of Spectralon at 900 nm for normal illumination.

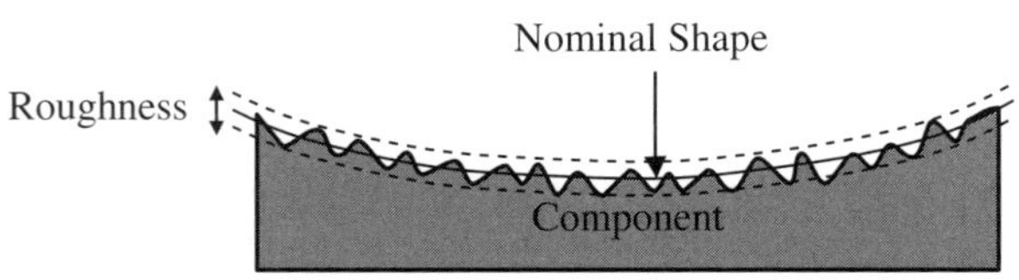

Figure 7.10 Surface roughness.

low levels of stray light might degrade faint images. For such surfaces, it is useful to quantify the roughness of the surface in terms of the root mean square roughness, σ_{rms}, which expresses the rms departure of the surface from the ideal surface, whether that be a plane, spherical, or aspherical surface. This is illustrated in Figure 7.10, showing the high spatial frequency departure from the nominal shape.

For polished optical surfaces, such as mirrors and lenses, σ_{rms} is very low, typically a fraction of a nanometre. The surface roughness is thus a very small fraction of the wavelength of light and, in this case, surface scattering may be presented as a diffraction problem. That is to say, a perfect surface would produce the reflection of a perfect wavefront and the surface roughness imposes a wavefront error equal to twice the surface roughness (due to the reflective double pass). In classical diffraction analysis, we would analyse the additional wavefront error induced in terms of the image quality degradation. That is to say the scattered light caused by the departure from nominal surface shape would cause some kind of change in the clarity of the image itself. However, in the case of surface roughness, the scattered light is considered entirely separately from image degradation. In terms of the departure of the surface from the nominal shape, only high spatial frequency variations are considered to contribute to scattering and are included in the definition of surface roughness. If Fraunhofer diffraction is considered, then the high spatial frequency components of surface roughness scatter the light far away from the nominal image. As such, this produces an irradiance distribution that is clearly separated from the imaged spot at the image focal plane. In practice, spatial wavelengths of less than 0.1–1.0 mm are considered as surface roughness; longer wavelength departures are analysed as ***'form error'*** and contribute to image degradation. The analysis of scattering proceeds in a similar way to the calculation of the Strehl ratio

(Chapter 6) for small system wavefront errors and gives a total hemispherical reflection of:

$$R = \left[\frac{4\pi\sigma_{rms}}{\lambda}\right]^2 \tag{7.16}$$

It is tempting to proceed with an analysis of scattering on the assumption that this 'small signal' scattering is Lambertian in character. However, this is very far from the truth. The angle of scattering, from simple Fraunhofer diffraction analysis is proportional to the spatial frequency of the surface roughness component. Of course, Fourier analysis may be used to express the roughness deviation of any surface in terms of the sum or integral of a series of sinusoidal terms of varying frequency. The random surface roughness of the type depicted in Figure 7.10 may be thus analysed and its power spectrum (i.e. square of the amplitude) may be expressed as a ***power spectral density*** (PSD) as a function of spatial frequency. As such, the PSD represents surface deviation power per unit spatial frequency bandwidth. The 'power' of a surface deviation is proportional to the square of the amplitude and might be measured in mm^2 and since the surface is represented by Fourier components in two dimensions (x and y), spatial frequency bandwidth might be measured in mm^{-2}. Therefore, for an ***area*** based description, as opposed to a linear one, PSD has dimensions of $length^4$, e.g. mm^4. The relevance of this discussion is for all polished surfaces, the PSD falls off very rapidly with spatial frequency and, as a consequence, the scattering amplitude or BRDF diminishes rapidly with angle (with respect to the main beam).

To a reasonable approximation, the PSD follows an inverse power law dependence upon spatial frequency. For a two dimensional Fourier description, for typical polished surfaces, this power law exponent is around −3. In the corresponding linear Fourier description, which is sometimes used, this exponent is around −2 and the PSD dimensions are mm^3, rather than mm^4. However, in this text we will retain the two dimensional description. Figure 7.11 shows an idealised PSD spectrum for a polished surface with nominal frequency exponent of −3. The total integrated surface roughness for the plot in Figure 7.11 is 5 nm rms. Apart from the simple exponent in Figure 7.11, we have introduced a '***corner frequency***', f_0, where the PSD reaches a maximum value. Without the introduction of a corner frequency, the integrated roughness would tend to infinity when

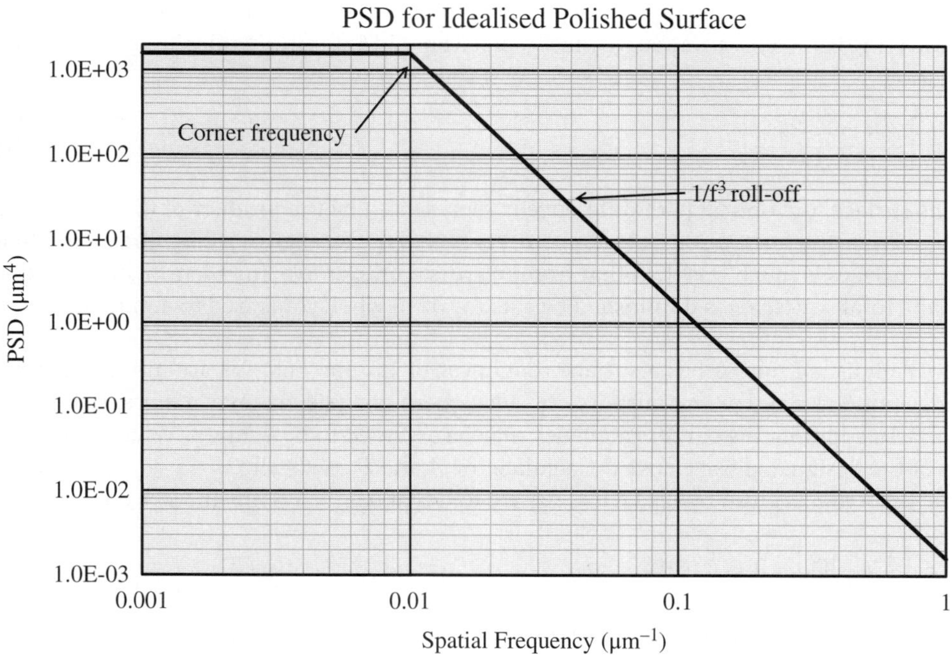

Figure 7.11 PSD for idealised polished surface (note units are in microns).

the integral proceeds to zero spatial frequency. In the context of our discussion on scattering, this corner frequency relates more to the somewhat arbitrary demarcation between scattering and image degradation, as previously outlined. This boundary may typically be between spatial frequencies of 1 and 10 mm^{-1} or spatial wavelengths between 0.1 and 1 mm.

With the introduction of the corner frequency, f_0, surface roughness power dependence upon spatial frequency may be modelled in a very specific way, as set out in Eq. (7.17):

$$PSD = \frac{PSD_0}{(f_0^2 + f^2)^{3/2}} \tag{7.17}$$

A more generalised formulation of Eq. (7.17) is the so-called ***k correlation model*** which introduces the ABC parameters:

$$PSD = \frac{A}{(1 + (Bf)^2)^{C/2}} \tag{7.18}$$

In our specific model, as outlined in Eq. (7.17), the C parameter in Eq. (7.18) is three. The parameter, B, is effectively the inverse of the corner frequency. In terms of the utility of this model with regard to scattering, the spatial frequencies may be directly translated into scattering angles or, more strictly, the sine of the scattering angles. As a consequence, the ABC model may be re-cast to given an explicit solution for the BRDF in terms of the scattering angle, θ:

$$\lambda \textit{ is the scattering wavelength} \quad BRDF = \frac{A}{(1 + (B\sin\theta)^2)^{C/2}} \qquad B = \frac{1}{\lambda f_0} \tag{7.19}$$

Of course, the ABC coefficients in Eq. (7.19) are not the same as those in Eq. (7.18). Equation (7.19) may be integrated across all polar angles to give the total hemispherical reflection. This gives:

$$R = \frac{2\pi A}{B^2(C-2)} = \left[\frac{4\pi\sigma_{rms}}{\lambda}\right]^2 \text{ and } A = \frac{8\pi\sigma_{rms}^2}{\lambda^4 f_0^2}(C-2) \tag{7.20}$$

Equations (7.19) and (7.20) gives us the ability to model scattering from mirror surfaces. However, when modelling the direct scattering from lens surfaces, we must replace Eq. (7.16) for the hemispherical scattering with the following equation:

$$R = \frac{2\pi A}{B^2(C-2)} = \left[\frac{2(n-1)\pi\sigma_{rms}}{\lambda}\right]^2 \tag{7.21}$$

In Eq. (7.21) for a lens surface, the optical path difference is represented by the product of $(n - 1)$ and the form error, as opposed to twice the form error, as in a mirror. As such, Eq. (7.21) gives a clear indication that the scattering from lens surfaces is much less than that from mirrors. For example, for a lens material with a refractive index of 1.5, the total scattering is diminished by a factor of 16 when compared to a mirror.

Worked Example 7.2 A polished mirror has a surface roughness of 1.5 nm rms. We are interested in its scattering at a wavelength of 633 nm. For the purposes of subsequent analysis, we may assume that the C exponent has a value of 3. In addition, the corner frequency may be assumed to be 4 mm^{-1}. What is the total hemispheric reflection at the designated wavelength? Calculate the A and B coefficients.

The total hemispheric reflection is given by Eq. (7.16). We are told that $\sigma_{\text{rms}} = 1.5$ nm and $\lambda = 633$ nm.

$$R = \left[\frac{4\pi\sigma_{rms}}{\lambda}\right]^2 = \left[\frac{4\pi \times 1.5}{633}\right]^2 \quad \mathbf{R = 0.089\%}.$$

The corner frequency, f_0 is 4 mm^{-1} and the B coefficient is given by the simple Eq. (7.19):

$$B = \frac{1}{\lambda f_0} = \frac{1}{6.33x10^{-4} mm \times 4 mm^{-1}} \quad \mathbf{B = 395}$$

Finally, from (7.20) we have:

$$A = \frac{B^2(C-2)}{2\pi}R = \frac{395^2(3-2)}{2\times 3.142}8.9\times 10^{-4} \quad \mathbf{A} = 22.0$$

Thus, in the full representation, $A = 22.0$, $B = 395$, and $C = 3$.

In terms of practical application, models such as the *ABC* model are extremely useful in the validation of designs, such as cameras and telescopes where restriction of scattered light is of paramount importance. This topic will be considered further when we look in more detail at the optical design process in later chapters.

7.5 Radiometry and Object Field Illumination

7.5.1 Köhler Illumination

Hitherto, in all discussions of image formation, no attention has been paid to the illumination of the object. It is assumed, quite arbitrarily, that the object spontaneously emits rays. This may be perfectly proper for a self-luminous object. However, in many cases, the object is not luminous and needs to be illuminated ***evenly*** across the entire field. The earliest investigation of this problem is due to ***August Köhler***, resulting in the development of the ***Köhler illumination*** system, still in use today. Most light sources, such as filaments lamps or arc sources have a highly spatially non-uniform irradiance. Traditionally, Köhler illumination was developed with a filament lamp source in mind. In this scheme, the light from the filament is collected by two lenses, the collector lens and the condenser lens and presented to the object. However, instead of imaging the filament at the object, which would produce uneven illumination, the filament is imaged at the nominal pupil location which it overfills. The Köhler illumination scheme is shown in Figure 7.12.

The field stop is located close to the collector lens which images the filament onto the aperture stop location. The condenser lens is separated from the aperture stop by its focal length and thus images the filament at the infinite conjugate. In this way, the object plane is uniformly illuminated. Of course, the pupil itself is not uniformly illuminated. However, this is not an impediment to image formation, provided the pupil is well filled. Uniform illumination of both image and pupil conjugates from an uneven source, such as a filament can only be achieved through division of amplitude, e.g. by scattering. This will be dealt with in the next section.

7.5.2 Use of Diffusers

The problem with using an imaging system for illumination, as in Köhler illumination, is that the uneven illumination source must be imaged at some conjugate in the system. This problem may be circumvented

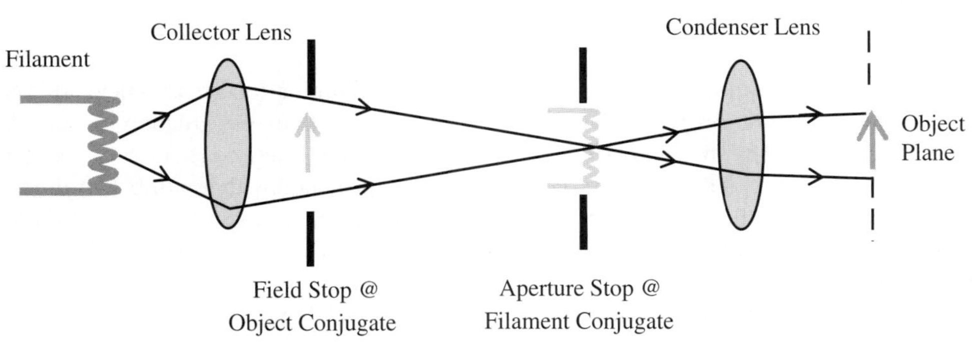

Figure 7.12 Köhler illumination.

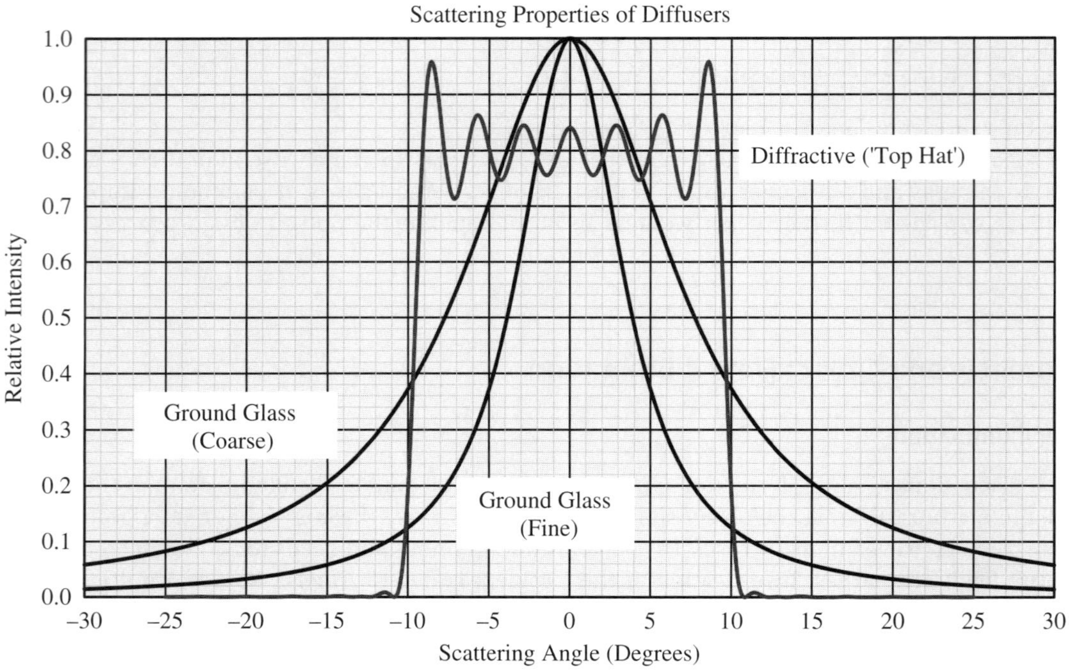

Figure 7.13 Diffuser scattering profile.

by use of a diffusing component within an optical system. Diffusers scatter light in a random but controlled fashion and take the form of transmissive components, such as ground glass screens and opal diffusers, and reflective screens, such as Spectralon diffusers. Reflective materials can approach Lambertian behaviour, but transmissive materials such as ground glass scatter light into relatively narrow angles. Ground glass screens produce a broadly Gaussian BRDF distribution with a full-width half-maximum scattering angle of between 5° and 20° depending upon the coarseness of the ground surface. 'Engineered diffusers' based on diffractive surfaces, can be used to create tailor made scattering profiles, such as a ***top hat profile***, where the scattered flux is constant up to a specific scattering angle whence is falls to zero. Figure 7.13 shows the scattering profile of some diffusers.

Overall, diffusers are very useful in re-arranging light by division of amplitude to promote even illumination. However, it must be understood, in a radiometric context, that diffusers inevitably increase system étendue and their use is inevitably accompanied by a significant reduction in radiance at the final image plane.

7.5.3 The Integrating Sphere

7.5.3.1 Uniform Illumination

Some exacting technical applications require the creation of highly uniform illumination across a field. This is particularly the case in instrument calibration, where even illumination to better than ±1% might be required. Such even illumination may be provided by an ***integrating sphere***. An integrating sphere consists of a spherical cavity coated with some high reflectivity, diffusing material. The sphere is provided with a number of ***ports***, which are apertures in the spherical shell and significantly smaller than the sphere diameter. One of these ports is designated as the ***input port*** and one as the ***output port***. The design of the integrating sphere is such that input and output ports are not intervisible and light can only reach the output port by scattering off the internal walls of the integrating sphere. This is shown in Figure 7.14.

The internal coating of the integrating sphere is made of some nominally white coating that scatters efficiently. Traditionally, classic white paint pigments, such as titania (TiO_2) and barium sulphate ($BaSO_4$),

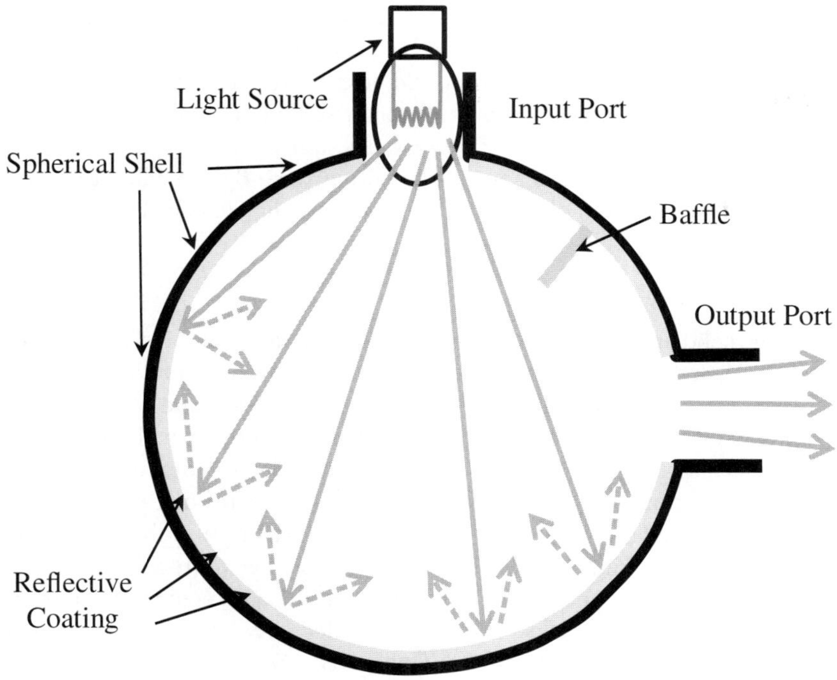

Figure 7.14 Integrating sphere.

were used. More recently, this has been replaced by Spectralon (sintered PTFE) for ultraviolet and visible applications and gold coating for infrared applications. These materials have a hemispherical reflectance of over 99% over wide regions of the spectrum. The integrating sphere is designed with a combined port area much smaller than the internal surface area of the sphere. In this way, before exiting the output port, the light must undergo a large number of scattering events. For Lambertian scattering at some point on the internal surface of the sphere, it can be demonstrated, for a spherical geometry, that the irradiance produced at other points of the sphere is entirely uniform.

However, in practice, no real material is perfectly Lambertian. Nevertheless, in theory, for an infinite number of scattering events, the radiance distribution of the light exiting the output port tends to the Lambertian distribution, even if the internal coating is non-Lambertian. Therefore, as the area of the ports is reduced, as a fraction of the sphere area, then the emission from the output port becomes more Lambertian. As a rule of thumb, the port area should make up no more than 5% of the total sphere area. Thus, for a reasonably small ***port fraction***, the integrating sphere has the property of considerably enhancing the Lambertian quality of the emission from the output port, over and above that of the reflective coating of the sphere itself.

If a light source injects a specific flux into the integrating sphere, as indicated in Figure 7.13, then the irradiance seen at a point on the sphere's surface is not merely the flux divided by the internal area of the sphere. By making the assumption that the integrating sphere is effective in promoting uniform internal illumination, the internal irradiance may be calculated by assuming the flux input is balanced by flux loss from the ports and absorption in the sphere coating. If the internal area of the sphere, including port area, is A, the fractional area occupied by the ports, f, the reflectivity of the sphere coating, R, and the flux input, Φ, then the internal irradiance, E, is given by:

$$\Phi = AfE + (1-f)(1-R)AE \quad E = \left[\frac{1}{1-R(1-f)}\right]\left(\frac{\Phi}{A}\right) = M\left(\frac{\Phi}{A}\right) \tag{7.22}$$

The quantity M is the so-called 'multiplier'. In practice, for many integrating sphere applications, $R > 0.99$ and thus M is approximately the inverse of the port fraction. Thus, with a port fraction of 5%, the multiplier is 20. That is to say, the internal irradiance is 20 times greater than would be expected from dividing the flux by the sphere area. The 5% port area restriction means that the port diameter should be smaller than 45% of the sphere diameter and less than a third of the sphere diameter for two ports, and so on.

It is clear that the integrating sphere delivers uniform radiance at the output port. By providing the integrating sphere with a calibrated source, or by calibration of its output radiance and irradiance, it can provide a standard calibrated (spectral) radiance.

7.5.3.2 Integrating Sphere Measurements

By integrating flux uniformly over a large solid angle, integrating spheres can provide an unbiased measurement of flux for diverging sources. That is to say, the integrating sphere integrates emission from these sources across all angles. Examples of such sources might include light emitting diodes (LEDs) and incandescent lamps, and so on. In such measurements, radiation from a lamp is directed into the input port with a photodetector situated at an exit port. This setup is illustrated in Figure 7.15a. In this example, the source is placed at the input port, although for lamps, the source is often placed ***inside*** the sphere. A detector placed at the output port is used to monitor integrating sphere radiance. By calibrating the detector using a source (e.g. laser) of known flux, the absolute flux may be calculated. Figure 7.15b illustrates the principle of (total) reflectance measurement. A source irradiates a sample situated opposite the input port. Again, a detector at the output port monitors the integrating sphere radiance. Reference reflectors of known reflectivity are available and such reflectors may be substituted for the sample for calibration purposes. Comparison of the two measurements will give the reflectivity of the sample.

7.5.4 Natural Vignetting

In many respects, the Lambertian illumination of an entrance pupil as would be provided by an integrating sphere represents an ideal situation. However, the irradiance produced at an image plane is actually non-uniform, assuming perfect imaging. In this context, 'perfect imaging' means perfect replication of the entrance pupil at the exit pupil. The effect described is known as natural ***natural vignetting***. In the perfect

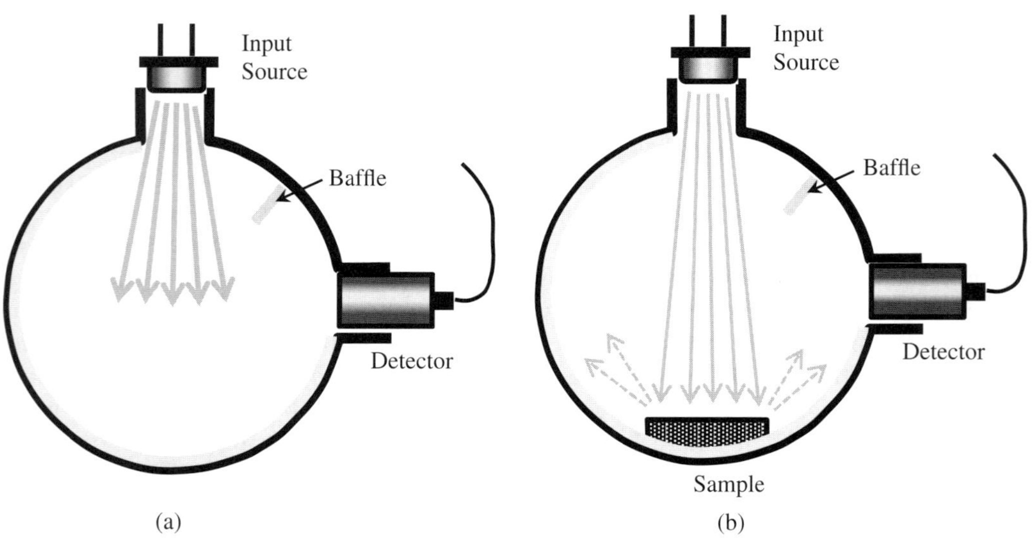

Figure 7.15 (a) Flux measurement. (b) Reflectance measurement.

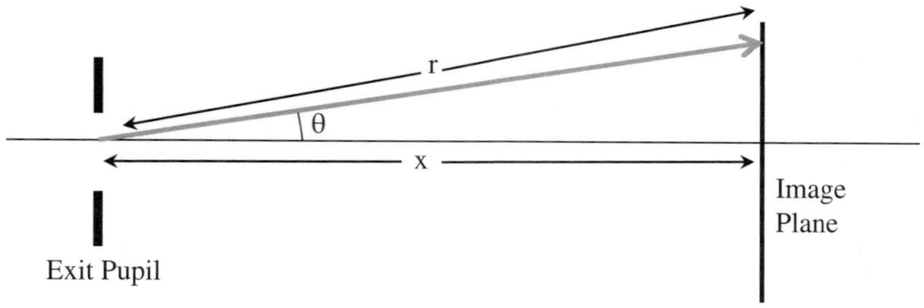

Figure 7.16 Natural vignetting.

realisation of this phenomenon, the irradiance produced at the image plane is proportional to the fourth power of the cosine of the field angle. The logic of this is illustrated in Figure 7.16.

If the (Lambertian) radiance at the exit pupil is L, from Eq. (7.5), the radiant intensity emerging at a normal angle of θ from an area element, dS, of the pupil is given by:

$$I = L\cos\theta dS$$

However, from the inverse square law Eq. (7.3) we know that the irradiance produced at the image plane is equal to:

$$E = \frac{I\cos\theta}{r^2}$$

θ is the angle of the ray to the image plane normal (same as angle to pupil normal)

Since $r = x/\cos\theta$, we finally arrive at the following relationship for natural vignetting:

$$E = \frac{L\cos^4\theta}{x^2} \tag{7.23}$$

Equation (7.23) summarises the phenomenon of natural vignetting. The reason for the term 'natural vignetting' is this effect replicates artificial vignetting, i.e. the darkening of an image towards the edges of a wide field caused by obstruction of light by physical apertures, other than the main stop. In the case of natural vignetting, however, there is no physical obstruction of the light path.

7.6 Radiometric Measurements

7.6.1 Introduction

In any real application, we are interested in the measurement of radiometric quantities, such as irradiance and radiant intensity. However, ***absolute*** measurements of these quantities are, in practice, extremely challenging. As an example, absolute measurement of flux or irradiance and so on to ±1% represents a high precision measurement. Although calibration plays an important role in any measurement, this is especially true for radiometric measurements. Absolute radiometric measurement generally proceeds by the use of ***calibrated detectors***. These detectors convert the optical flux into an electrical or thermal signal which can be directly monitored. Critically, the sensitivity of these detectors has been carefully calibrated using a reference source providing a known spectral output. Hence, the signal can be directly converted into flux or radiance, and so on. The reference sources are generally maintained by, or derived from, National Measurements Institutions (NMIs), such as the National Physical Laboratory (NPL) or the National Institute of Standards and Technology (NIST).

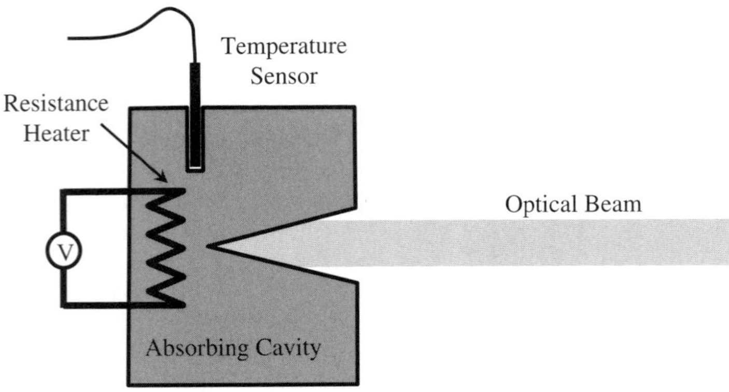

Figure 7.17 Substitution radiometer.

7.6.2 Radiometric Calibration

7.6.2.1 Substitution Radiometry

Calibrated measurements of optical flux are ultimately derived from the principle of ***substitution radiometry***. In this measurement, optical radiation is wholly absorbed in a specially designed black cavity and the temperature increase measured by a thermal transducer. Thereafter, the optical power is substituted by electrical input derived from a resistance heater. The original optical flux is given by the electrical input required to produce the same temperature change. The principle is illustrated in Figure 7.17.

The optical beam in Figure 7.16 may, for example, be derived from a stabilised laser beam. This laser beam, thus characterised, may then be used to calibrate the sensitivity of a detector. Ultimately, the temperature rise with respect to the surroundings provides the signal for this measurement. As a consequence, any drift in the ambient temperature interferes with the fidelity of the measurements. For this reason, the highest precision measurements are obtained with a ***cryogenic radiometer***, where the cavity, sensor, and heater are enclosed within a vacuum and cooled to a few degrees Kelvin.

7.6.2.2 Reference Sources

The primary reference source for flux, or rather for spectral irradiance, is a carefully maintained blackbody source. For the ultraviolet, visible, and near infrared spectral regions, the blackbody source is based upon a pyrolytic graphite cavity. Such sources can operate up to a temperature of 3500 K. In order to capture the spectral irradiance, the output from the blackbody is characterised by a number of filtered detectors previously calibrated by a substitution radiometer. A filtered detector is comprised of a sensor with a bandpass filter which only admits radiation within a narrow range of wavelengths. The general setup is shown in Figure 7.18.

The pyrolytic graphite discs that comprise the cavity are heated electronically, as indicated. Fully calibrated, this is a precision, broadband radiometric source. However, it is not practical for use in a standard laboratory setting. Therefore, practical calibration is generally carried out using ***transfer standards***. These are simpler light sources whose spectral irradiance has been calibrated (ultimately) against a primary source at an NMI. One very commonly used example of a transfer standard is a filament emission lamp or ***FEL***. This lamp is simply a well characterised and calibrated quartz halogen lamp. Generally the FEL is a 1000 W lamp whose irradiance at a nominal distance of 500 mm has been measured and calibrated at an NMI. These emission lamps approximate to a 3200 K blackbody source. Table 7.2 shows calibrated spectral radiance levels for a typical lamp.

The process of transferring the standard from the primary to the transfer standard does increase the uncertainty of the calibration of the FEL lamp and the calibration uncertainty for this type of lamp is of the order of 1–2%, depending upon the wavelength. Any subsequent use of the FEL in laboratory calibration of

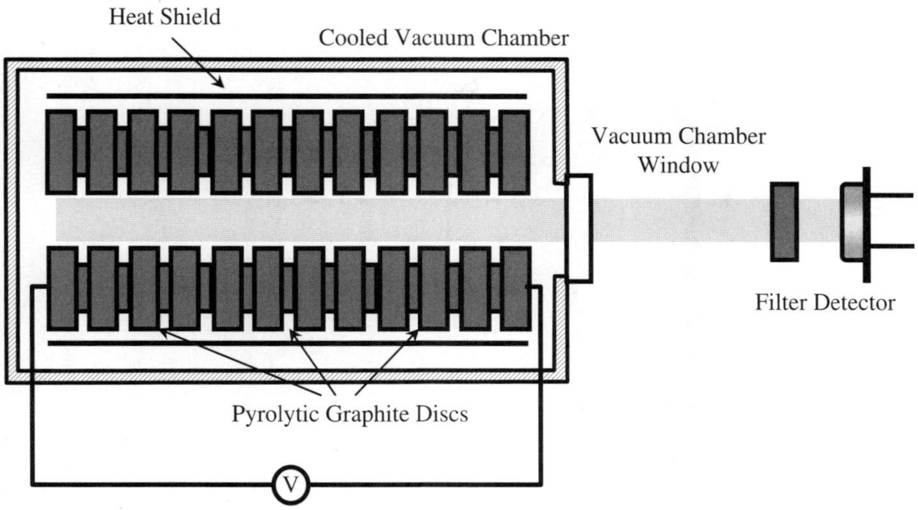

Figure 7.18 Blackbody radiometric source.

Table 7.2 Spectral radiance for typical calibrated FEL lamp.

Wavelength (nm)	Spectral irradiance (m Wm^{-2} nm^{-1})	Wavelength (nm)	Spectral irradiance (m Wm^{-2} nm^{-1})
250	0.15	700	185.6
300	1.56	800	218.8
350	7.02	900	231.1
400	19.6	1000	227.8
450	40.5	1200	196.5
500	68.2	1400	156.1
600	131.6	1600	119.8

photodetector sensitivity must faithfully replicate the NMI calibration set up. The laboratory set up might look like that shown in Figure 7.19.

Great care must be taken to minimise the contribution from scattered light from the surroundings. The original calibration is based entirely on direct radiation from the lamp; any contribution from scattered light would compromise this.

7.6.2.3 Other Calibration Standards

Measurement, characterisation, and modelling of reflection represent an important part of radiometry as the preceding discussions illustrate. Measurement of reflectivity, might be on either polished (specular) or diffuse surfaces. In the former case, the reflectivity is a simple function of the incident angle as, for specular reflection, the reflected angle is pre-determined. For laboratory measurements of specular reflectance, reference standards may be obtained that have been calibrated at NMIs. These might be aluminised mirrors or polished glass blanks with low, but measurable reflectivity. For diffuse reflection, the interest is not only in total reflection (total hemispherical reflectivity) but also in its distribution with angle. In this case, full characterisation of BRDF is of interest. Again, routine laboratory measurements are facilitated by the provision of calibrated

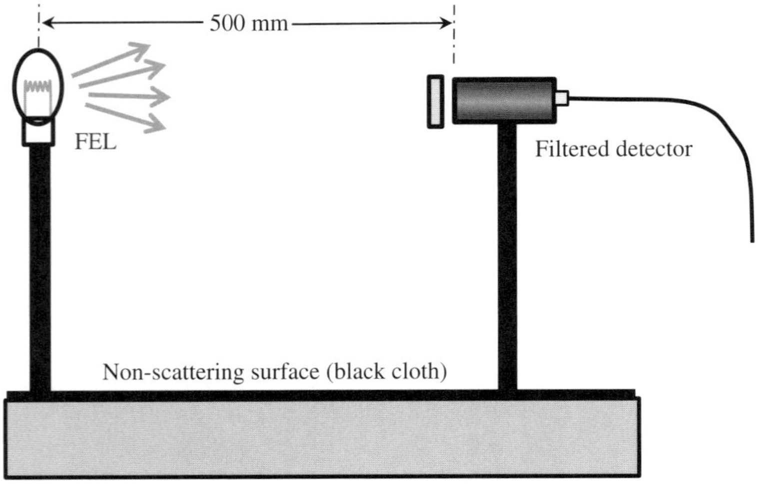

Figure 7.19 FEL lamp calibration.

artefacts. These might include ~100% reflectance standards in addition to matt black standards to provide a nominal zero reference.

7.7 Photometry

7.7.1 Introduction

Radiometry is concerned with the measurement of ***absolute*** flux levels of optical radiation. However, in many practical instances, we are rather concerned with the ***effect*** of these flux levels on detection systems, most notably the human eye. For instance, real, tangible radiometric fluxes in the infrared are of no relevance to human vision. Therefore, ***photometry*** is concerned with optical fluxes as mediated by some detection sensitivity, most particularly of the human eye. Naturally most of the discussion here relates to ***visual photometry***, although there are other areas of photometry, such as astronomical photometry.

7.7.2 Photometric Units

For visual photometry, each radiometric unit has its corresponding photometric unit. The photometric equivalent of radiant flux is the luminous flux expressed in lumens and the equivalent of radiant intensity is luminous intensity whose base unit is the ***candela***. Similarly, the radiometric quantities, irradiance and radiance correspond to illuminance and luminance respectively in photometry. The base unit for illuminance is lux and that for luminance is candela per square metre. Units for luminance are occasionally referred to as ***nits***. Comparison of the radiometric and photometric quantities is set out in Table 7.3.

Each photometric quantity is derived from the respective radiometric quantity by integration across the visible spectrum using a spectrally dependent weighting function V(λ). This weighting function is a standardised representation of the sensitivity of the human eye. Normally, this standard weighting function is taken to represent photopic (daytime) vision as opposed to scotopic (dark adapted) vision. This standard weighting function, V(λ), or luminous efficiency function, was originally established by the Commission Internationale de l'Éclairage (CIE) in 1924. By definition, V(λ) has a maximum value of unity and, for the photopic function, this occurs at a wavelength of 555 nm, corresponding to the peak sensitivity of the human eye. The function has since been revised slightly on a number of occasions most notably in 1978 and 2005. Figure 7.20 shows the plot for both photopic and scotopic sensitivity.

Table 7.3 Photometric quantities.

Photometric quantity	Radiometric equivalent	Photometric unit
Luminous flux	Flux	Lumen (lm)
Luminous intensity	Radiant intensity	Candela (cd) – (lm sr^{-1})
Illuminance	Irradiance/exitance	Lux (lm m^{-2})
Luminance	Radiance	Candela (m^{-2} or nit)

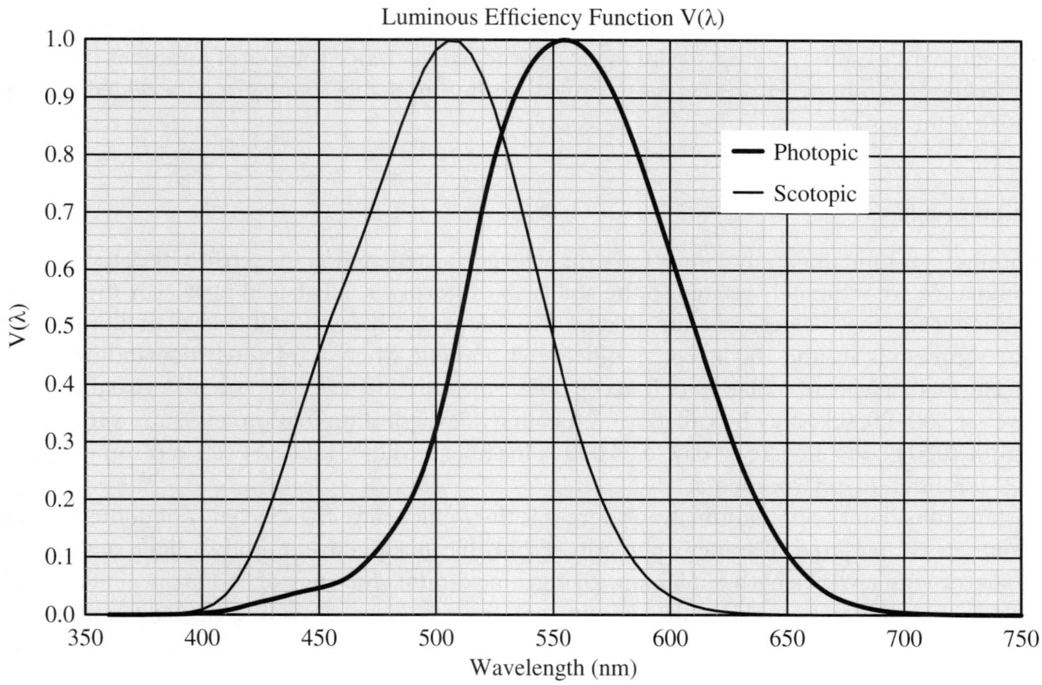

Figure 7.20 Luminous efficiency function.

However, V(λ) is only a relative measurement of luminous efficiency. To link photometric units to their corresponding radiometric quantities a constant of proportionality, K_M, must be added to relate the two. That is to say, if the radiometric spectral flux is $\Phi_r(\lambda)$ and the corresponding luminous flux is $\Phi_p(\lambda)$ then the two may be linked by the following equation:

$$\Phi_p(\lambda) = K_M \int_0^{\infty} \Phi_r(\lambda) V(\lambda) d\lambda \tag{7.24}$$

The value of K_M is defined as 683.002 lm W^{-1}. That is to say, an optical beam with a wavelength of 555 nm (actually 5.4×10^{14} Hz or 555.17 nm) and having a luminous flux of 1 lm, actually has a radiant flux of 1/683.002 W. At first sight, this might seem a rather curious definition. The reason for this is essentially historical. It is the candela, rather than the lumen that forms the base SI photometric unit. All other photometric units are derived from the candela. As such, the candela is defined as the luminous intensity of a source of monochromatic radiation of frequency 5.4×10^{14} Hz having a radiant intensity of 1/683.002 W sr^{-1}. However, originally the definition of luminous intensity was related directly to the output of a standard hydrocarbon burning lamp. In fact, the candela was historically related to an earlier unit of luminous flux,

Table 7.4 Typical illuminance levels for difference environments.

Environment	Illuminance (Lux)	Environment	Illuminance (Lux)
Starlit night	0.002	Cinema (darkened)	1
Moonlit night	0.5	General domestic	100
Overcast day	1000	General office	400
Sunny day	20 000	Visual inspection	1000

candlepower. So, for historical consistency, a radiometric intensity 1/683 W sr^{-1} at 555 nm is broadly related to the output of a 'standard candle'. Attempts were made to produce reference sources of luminous intensity using standard blackbody emitters. However, these proved to be unreliable and were superseded by the current radiometric definition.

7.7.3 Illumination Levels

Since optical photometry is fundamentally connected to light levels as mediated by the sensitivity of the human eye, levels of illuminance are intimately related to the ability to perform visually based tasks. For the indoor environment, lighting levels may be designed for specific areas. A generally comfortable level of illuminance for a domestic environment is around 100 lx. For an office environment, where moderately demanding visual tasks are to be performed, a level of 300–500 lx is acceptable. For more critical tasks, such as visual inspection, a higher level of 500–2000 lx may be called for. Of course, daylight illumination levels are very much higher, ranging from 1000 lx on an overcast day to 25 000 lx for full sunshine. Table 7.4 sets out some typical illumination levels for different environments:

Another important consideration in illumination sources is their efficiency. The efficiency of domestic and industrial light sources is measured in lumens per watt. From that perspective, the ideal light source is a monochromatic source with a wavelength of 555 nm, giving a maximum efficiency of 683 lm W^{-1} of ***optical*** output. A reasonable approximation to this is the sodium vapour street lamp, providing virtually monochromatic light at 589 nm with an ***electrical*** efficiency of 200 lm W^{-1}. However, such a highly coloured source is not acceptable for domestic and industrial applications where broadband or nominally 'white' sources are preferred. The least efficient sources are incandescent tungsten sources which are being replaced in domestic and industrial applications due to their poor energy efficiency. Their immediate successors, fluorescent mercury lamps create broadband emission from fluorescent phosphor coatings irradiated by ultraviolet emission from mercury spectral lines. More latterly, these are being replaced by white light LEDs which rely on ultraviolet emission from gallium nitride diodes to create broad band fluorescence from phosphors. Efficiencies of these sources are set out in Table 7.5.

Table 7.5 Luminous efficiencies of different sources.

Source	Efficiency (lm W^{-1})	Source	Efficiency (lm W^{-1})
Tungsten lamp	14	White light LED	50–100
Tungsten halogen lamp	24	Xenon lamp	40
Fluorescent lamp	70–100	5800 K blackbody (sun)	93

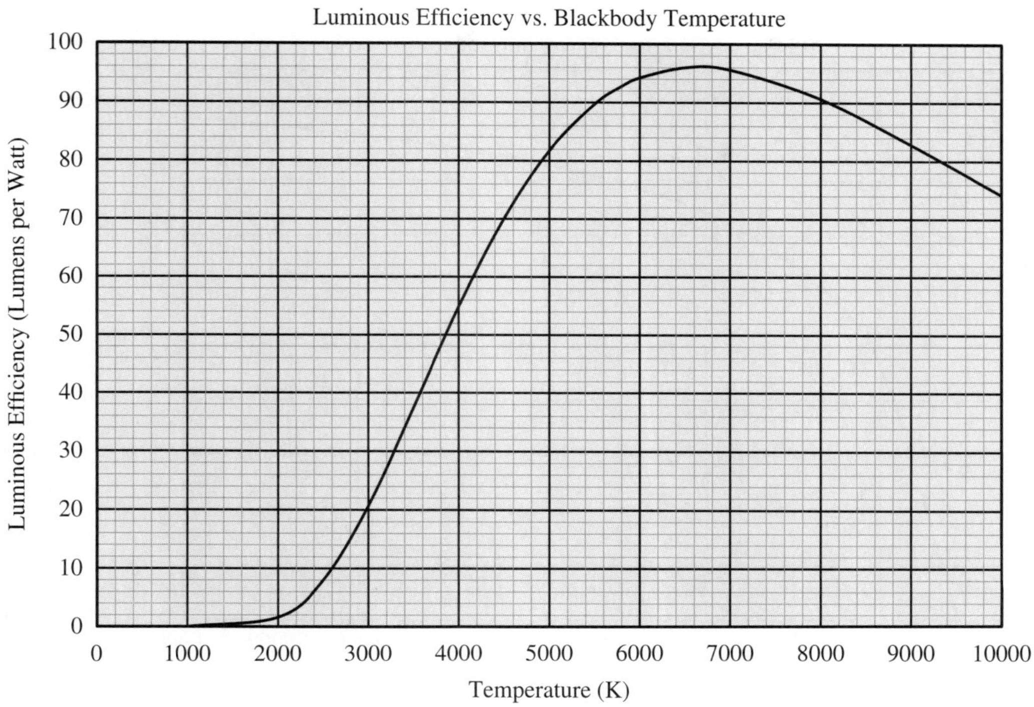

Figure 7.21 Luminous efficiency vs. blackbody temperature.

In fact, the luminous efficiency of a blackbody source may be calculated directly from the Planck distribution set out in Eq. (7.9) and the luminous efficiency function, $V(\lambda)$. The plot is shown in Figure 7.21. The peak efficiency occurs around 6000 K and it is, of course, no coincidence that this is close to the solar blackbody temperature of 5800 K. Clearly, the human eye has been 'designed' to efficiently harvest light from its primary illumination source.

The brightness of different sources (to the human eye), or luminance, is expressed in candelas per square metre or nits. Representative values range from 80 cdm^{-2} for a typical cinema screen to 7×10^6 cdm^{-2} for the filament of an incandescent lamp and 1.6×10^9 cdm^{-2} for the solar disk. As for the luminous efficiency plot, the luminance of a blackbody source may be derived directly from the Planck distribution and the luminous efficiency curve, $V(\lambda)$. This plot is shown in Figure 7.22.

7.7.4 Colour

7.7.4.1 Tristimulus Values

The preceding discussion has been wholly concerned with the level of illumination rather than (human) perception of the spectral distribution. This spectral distribution is described by the notion of colour, as perceived by humans. From the perspective of human vision, colour is discerned by the relative stimulation of three types of colour receptors (the cones). To model this process, the CIE, in 1931, proposed a set of colour matching functions, effectively mimicking the relative sensitivity of each type of sensor. The colour matching functions are represented as three separate curves, $x(\lambda)$, $y(\lambda)$, and $z(\lambda)$, and operate, in principle, in a similar manner to the $V(\lambda)$ curve for photopic efficiency. However, each curve is shifted with respect to the others. The form of these curves is illustrated in Figure 7.23.

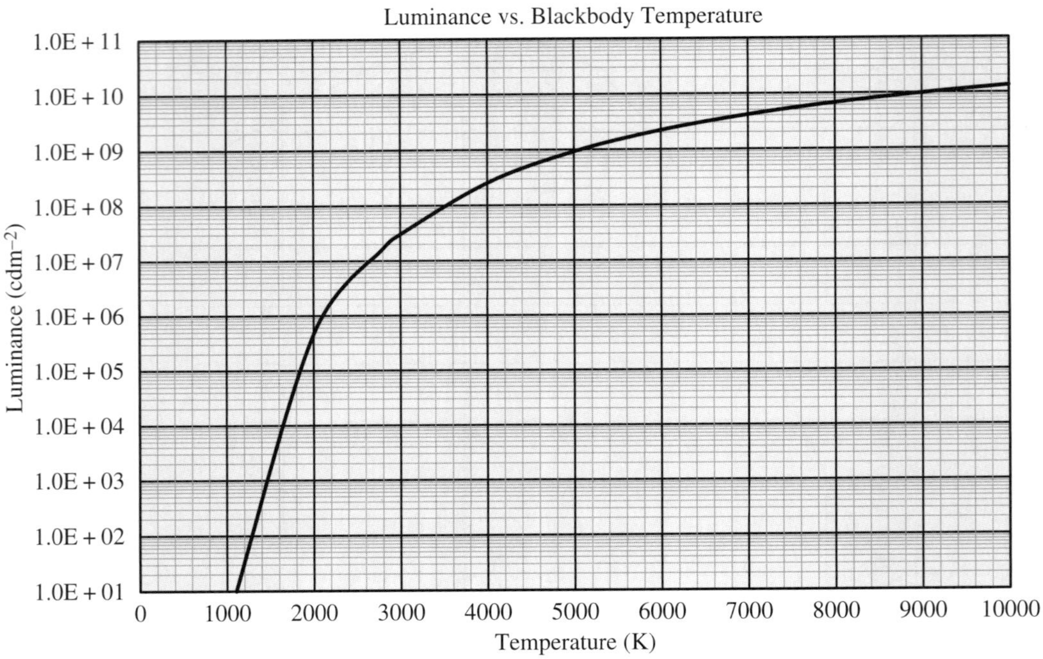

Figure 7.22 Luminance vs. blackbody temperature.

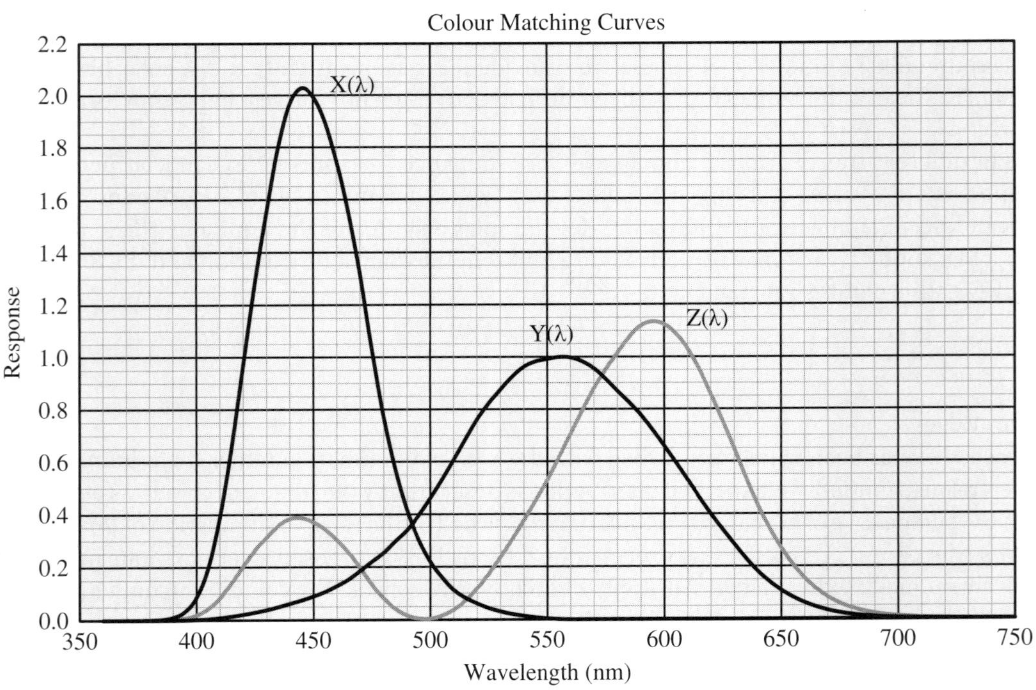

Figure 7.23 Colour matching curves.

It must be emphasised that the colour matching curves and the luminous efficiency curve are merely ***representative*** of human visual perception. These curves represent the fruits of sustained efforts to find a representative average of human perception. However, not surprisingly, there are considerable variations in spectral sensitivity between individuals.

Quite significantly, the $y(\lambda)$ curve follows that of the standard $V(\lambda)$ curve. As for the basic photometric quantities, an input spectral radiance is transformed by integrating across the spectral range using the colour matching functions. However, instead of producing a single luminous flux value, three separate ***tristimulus values***, X, Y, and Z are derived, as below.

$$X = \int_0^\infty \Phi_r(\lambda)x(\lambda)d\lambda \quad Y = \int_0^\infty \Phi_r(\lambda)y(\lambda)d\lambda \quad Z = \int_0^\infty \Phi_r(\lambda)z(\lambda)d\lambda \tag{7.25}$$

From the preceding arguments, the Y tristimulus value is a measure of the luminance of the source. Normalisation of the tristimulus values provides a two dimensional description of colour.

$$x = \frac{X}{X+Y+Z} \quad y = \frac{Y}{X+Y+Z} \quad z = \frac{Z}{X+Y+Z} \tag{7.26}$$

Only the x and y ordinates are used in the standard CIE ***chromaticity diagram*** which provides a standardised quantification of the human perception of colour. The chromaticity diagram provides a plot in these two dimensions, with the third degree of freedom effectively corresponding to the luminous flux or intensity. Although it is perhaps obvious from the preceding discussion, this tripartite description of colour is purely an artefact of human vision and in no sense related to any property of light. Indeed, in recording any manifestly complex or subtle spectral distribution, the human eye can only, in effect, describe these by three independent parameters. It is clear that it is very possible for different spectral distributions to produce the same X, Y, Z stimulus values. This effect is known as ***metamerism***. This highlights the limited spectral information that is provided by the three different sensor types. Indeed, two surface coatings (e.g. painted) can appear to be the same colour under one illumination (e.g. fluorescent) but different under another illumination source (e.g. tungsten) because of this effect.

7.7.4.2 RGB Colour

In many instances in describing colour we are interested in the effect of adding or blending colours. As a result of the three sensor types it is clear, in principle, that a linear combination of three different colours may be used to create a wide range colour sensations. The three colours themselves are described by a linear combination of the tristimulus values, X, Y, and Z and are known as ***primary colours***. Definition of the suite of primary colours is arbitrary and established by virtue of convention. The guiding principle is that these three colours must be capable of being admixed to create as wide as possible a range of colours without recourse to negative coefficients in the linear combination. This range is referred to as a ***gamut***. The original standard (1931 CIE) colour representation is the so-called ***RGB*** system (Red – Green – Blue) of primary colours using three standard monochromatic stimuli at 700 nm (R), 546.1 nm (G), and 435.8 nm (B). In this scheme, definition of the RGB primary colours from the tristimulus values is given by the following (X, Y, Z) vectorial representation:

$$R = \begin{bmatrix} 2.769 \\ 1 \\ 0 \end{bmatrix} \quad G = \begin{bmatrix} 0.633 \\ 1.658 \\ 0 \end{bmatrix} \quad B = \begin{bmatrix} 0.645 \\ 0.0343 \\ 3.194 \end{bmatrix} \tag{7.27}$$

The inverse transformation between the two representations is effected by the following matrix:

$$\begin{bmatrix} R \\ G \\ B \end{bmatrix} = \begin{bmatrix} 0.41847 & -0.15866 & 0.082835 \\ -0.09117 & 0.25243 & 0.01571 \\ 0.00092 & -0.00255 & 0.1786 \end{bmatrix} \begin{bmatrix} X \\ Y \\ Z \end{bmatrix} \tag{7.28}$$

Presentation of this RGB colour convention is intended for illustrative purposes only. This simple scheme has been largely superseded. In reality, there are a plethora of different primary colour conventions designed with specific applications, such computer screen rendition and so on, in mind. Some conventions take account of the non-linearity of the eye's response. That is to say, we abandon the linear convention hitherto prescribed. Other conventions ensure that uniform movements across colour space correspond to uniform changes in human perception of colour. These are called ***perceptually uniform colour spaces***.

In principle, an equal admixture of primary colour components leads some form of standard white colouration or '***white point***'. The concept of whiteness as a chromatic descriptor is purely associated with the human perception of colour, rather than a fundamental property of a source. However, definition of whiteness, is convention dependent. Rather than defining a colour sensation by virtue of the admixture of RGB, it may also be defined by another three parameter set, ***HSL*** or ***hue***, ***saturation***, and ***luminosity***. Hue is a measure of the undiluted colour, loosely corresponding to the equivalent monochromatic wavelength of stimulation. Saturation describes the purity of the colour, or the extent to which white must be admixed with a pure monochromatic colour to achieve the desired colour. The final degree of freedom is provided by luminosity which correlates to the brightness of the sensation, effectively the sum of the RGB components.

Colour difference, ΔE, is a measure of the absolute difference in colour between two different colours. It is generally expressed as the root sum of squares of the difference in each of the three colour ordinates and dependent upon the convention adopted. With this in mind, the concept of ***colour temperature*** describes the temperature of the blackbody radiator that most closely matches the colour of interest, i.e. with the smallest colour difference. This is particularly associated with the characterisation of light sources. Somewhat ironically, the term 'cool' describes a source with a bluer spectral distribution whereas a 'warm' light source refers to illumination with a larger red contribution. This is rather based on human perception and psychology; a 'cool', bluer blackbody source is, of course, hotter than a redder 'warm' source.

Much of the preceding discussion introduces the topic of colour with treatment of just one, antecedent, colour convention. As such, this provides a useful description of the basic underlying principles. However, the topic in itself is much too broad to provide any comprehensive treatment here and the reader is referred to specialist texts for further study. Some guidance is provided in the short bibliography and the end of the chapter.

7.7.5 Astronomical Photometry

Astronomical photometry is concerned with the measurement of the magnitude of electromagnetic flux from stellar objects, stars galaxies, and so on. For modern observations, these measurements are almost exclusively dependent upon semi-conductor detectors. For a given stellar object, the ideal measurement might involve the high resolution capture of its spectral irradiance at the earth across a wide range of wavelengths. That is to say, a detailed spectrum of each object should be obtained that can be related to absolute spectral irradiance. However, as stellar objects of interest are almost invariably faint, the amount of flux that is captured in any given wavelength band is necessarily very small. For the majority of measurements, therefore, this approach is impractical. A more practical solution is to use a number of spectrally filtered detectors to monitor flux from the star (via a telescope). Each filter has a relatively broad passband, e.g. 100 nm and centred on some specific wavelength, e.g. 555 nm. Using a small number of these filtered detectors across the ultraviolet, visible, and infrared spectral ranges, provides what amounts to low resolution spectral information for the source. However, since interpretation of the spectral quality of a stellar object is based on a limited number (e.g. 3) of spectrally filtered measurements, there is a clear correlation with visual photometry.

As with visual photometry, conventions for stellar photometry are varied. By way of illustration, we invoke here one set of standard filter response curves which follow the same broad purpose as the colour matching functions in visible photometry. Standard filters include U (ultraviolet), B (blue), V (visible), R (red), I (infrared) as well as Z, Y, J, H, and *K* bands further into the infrared. The visible filter function follows the standard visible

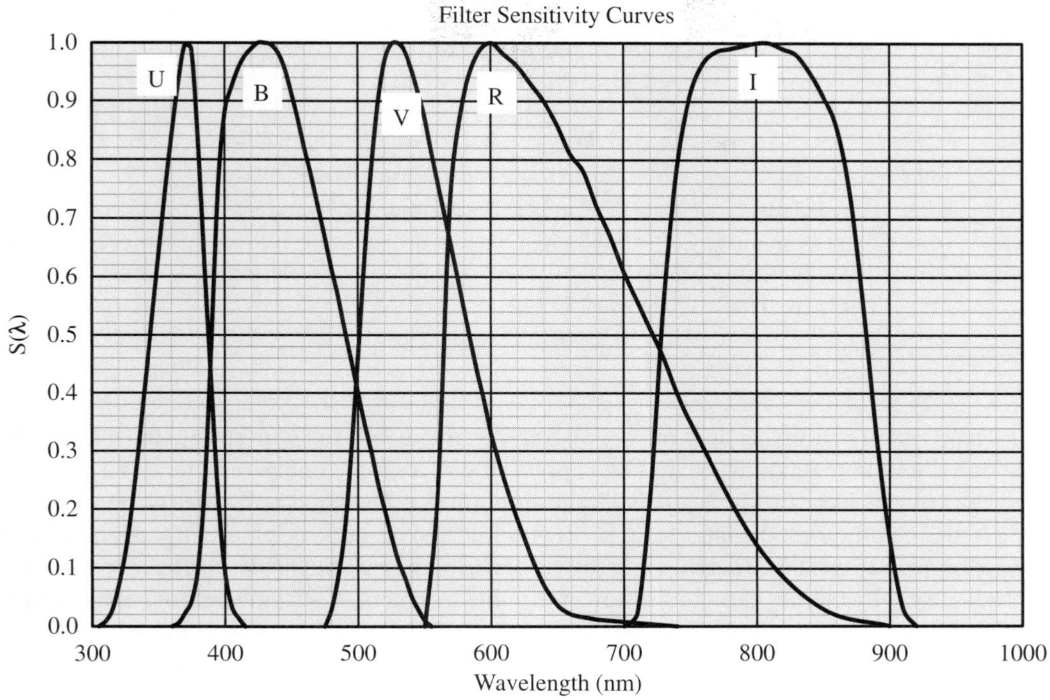

Figure 7.24 Standard astronomical filter response curves.

photopic curve with a maximum response at about 555 nm. Standard filter response curves are illustrated in Figure 7.24.

As for visible photometry, the spectral irradiance of the source, $E(\lambda)$ is mediated by the filter transmission function, $S(\lambda)$, to give the effective illuminance, E_p.

$$E_p = \int E(\lambda)S(\lambda)d\lambda \tag{7.29}$$

Hitherto, we have not introduced any absolute scale into this discussion. In fact, it will be appreciated, as part of a strand running through this chapter, that ***absolute*** radiometric and photometric measurements are extremely challenging. That is to say, it is difficult (not impossible) to relate the brightness of stellar objects to fundamental units of flux such as watts. Therefore, in most practical applications, stellar photometry relates the brightness of specific stellar objects to a small number of reference stellar objects. Pre-eminent amongst these reference objects is the star Vega or α Lyrae. The brightness of other stellar objects is expressed as a ratio of their effective illuminance, E_p to that of Vega, E_{p0}. However, a simple ratio does not provide a useful impression of the brightness of a source. This is because the human eye is effectively a logarithmic detector, with geometric ratios in flux appearing as a uniform progression in brightness. By the same token, the sensitivity of the human eye covers many orders of magnitude of flux. For this reason, the brightness of a star is described by its ***apparent magnitude, M,*** which is based on the logarithm of the illuminance ratio (to the standard). For historical reasons, the base of the logarithm is the fifth root of one hundred.

$$M = 2.5\log_{10}\left(\frac{E_{p0}}{E_p}\right) \tag{7.30}$$

Note that the reference star illuminance forms the numerator and the measured illuminance the denominator. As a consequence, in this convention, brighter stars have a lower magnitude. Nominally Vega (a bright star) has a magnitude of zero. However, attempts to reconcile these measurements with absolute and other

Table 7.6 Visual magnitudes of several astronomical objects.

Source	M_V	Source	M_V	Source	M_V
Sun	−26.74	Sirius	−1.46	8 m telescope limit	28
Moon (Full)	−12.6	Vega	0.03	40 m telescope limit	36
Venus (Brightest)	−4.4	Polaris	1.97	Dark sky/sq. degree	5–7
Jupiter (Brightest)	−2.9	Andromeda Galaxy	3.44		
Neptune	7.8	Naked Eye Limit	~6–6.5		

photometric scales have required a small adjustment. For example, in the Johnson convention, the magnitude of Vega is 0.03. So far, the term magnitude, M, has been presented as a simple one parameter description of stellar brightness. However, as indicated in Figure 7.23, this brightness is mediated by a range of standard filters. In presenting stellar magnitudes, the filter band is always indicated by a subscript, e.g. M_U, M_B, M_V, M_R, and so on. The most frequently used is M_V, reflecting the visible brightness of the object. Difference between two of the magnitudes, e.g. M_V and M_B, is an indication of the object's colour. Visual magnitudes of a number of astronomical objects are sketched out in Table 7.6.

Apparent magnitude describes the relative brightness or illuminance of a stellar object ***as it appears at the Earth's surface***. By virtue of the inverse square law, differences in apparent magnitude might arise from differences in distance ***or*** the absolute luminous intensity of the source. The ***absolute magnitude*** of a source is equivalent to the relative magnitude of that source if it were placed at a standard reference distance from the Earth (10 pc or 32.6 light years). In this scheme, our own sun is a relatively non-descript source with an absolute magnitude of 4.83.

The relative photometry of stellar sources may, with some difficulty be related to absolute radiometric units. In stellar photometry a standard unit of spectral irradiance is used, the ***Jansky***. However, instead of expressing the spectral irradiance in irradiance per unit wavelength it is expressed as irradiance per unit frequency (Hertz). The Jansky unit is 10^{-26} W m^{-2} Hz^{-1}. In the so-called ***AB Magnitude convention*** the zero magnitude brightness, ***for the visible passband***, is defined for an imaginary source whose spectral irradiance is flat in the frequency domain and amounts to 3631 Jansky units. Converting from spectral radiance in the frequency domain, $E_f(f)$ to spectral radiance in the wavelength domain, $E_\lambda(\lambda)$ is straightforward:

$$E_\lambda(\lambda) = \frac{c}{\lambda^2} E_f(f) \tag{7.31}$$

To gain an appreciation of how stellar magnitudes relate to absolute irradiance it would be useful to compute and effective irradiance of a zero magnitude star in the AB system. We may compute an effective irradiance using Eq. (7.29) and the visible transmission function $S_V(\lambda)$ illustrated in Figure 7.23. It is also assumed that over the passband of the filter, the spectral irradiance is flat at 3631 Jansky units. This gives the effective irradiance of a zero magnitude star as about 3.2×10^{-9} W m^{-2}.

Further Reading

Hengtsberger, F. (1989). *Absolute Radiometry: Electrically Calibrated Detectors of Thermal Radiation*. Orlando: Academic Press. ISBN: 978-0-323-15786-5.

ISO 23539:2005(E)/CIE S 010/E:2004 (2004). *Photometry – The CIE System of Physical Photometry*. Vienna: Commission Internationale d'Eclairage.

McCluney, W.R. (2014). *Introduction to Radiometry and Photometry, 2e*. Washington, DC: Artech House. ISBN: 978-1-608-07833-2.
Palmer, J.M. and Grant, B.G. (2009). *The Art of Radiometry*. Bellingham: SPIE. ISBN: 978-0-819-47245-8.
Parr, A., Datla, R., and Gardner, J. (2005). *Optical Radiometry*. Cambridge, MA: Academic Press. ISBN: 978-0-124-75988-6.
Wolfe, W.J. (1998). *Introduction to Radiometry*. Bellingham: SPIE. ISBN: 978-0-819-42758-8.

8

Polarisation and Birefringence

8.1 Introduction

In our treatment of electromagnetic wave propagation we have maintained the convenient fiction that the amplitude of the wave disturbance is a scalar quantity. However, the physical quantities that underlie the amplitude are the electric and magnetic fields. These are unambiguously vector quantities. Indeed, the set of equations (Maxwell's equations) defining electromagnetic propagation are a set of differential equations that establish the relationship between vector quantities. However, in the scalar theory that we have presented to this point, analytical convenience dictates that we imagine an electromagnetic wave to be described by a scalar quantity. This has the benefit of making the analysis somewhat more tractable. We applied this to the analysis of optical diffraction. It is inherently a useful approximation that is applicable under certain constraints. Most particularly, if light is constrained to some optical axis, then the scalar approximation is largely valid provided that all propagation angles with respect to this axis are relatively small. In effect, this is a 'paraxial approximation'.

Notwithstanding this, we must ultimately accept that the quantities describing the amplitude of electromagnetic radiation are in fact vector quantities. It is the direction of the electric field, **E**, associated with this radiation that, by convention defines the direction of ***polarisation*** of light. Of course, there is also the magnetic field, **H**, and the two are inextricably linked via Maxwell's equations. As will be seen later, in plane polarised light, the magnetic field vector is orthogonal to the electric field vector and both are orthogonal to the direction of propagation. In reality ***all*** light is polarised and what is described as ***unpolarised*** light is, in fact, ***randomly polarised***. That is to say, there is a complete lack of correlation or coherence between different vector components of polarisation producing random shifts in the direction of polarisation over an optical cycle.

In this chapter we will also look in a little more detail at the underlying structure of optical materials that contribute to refractive properties. Previously we had understood the refractive property of a material to be associated only with its modification of the local speed of light. The refractive index of a material is, of course, defined as the ratio of the speed of light in vacuo to that in the medium. Refractive effects in almost all materials are produced by the interaction of the ***electric field*** of the radiation with the internal atomic structure of the material. This local electric field causes charge separation at the atomic level, leading to the production of ***electric dipoles***. In effect, these dipoles produce an additional electric field which interacts with the imposed electric field. It is this effect that leads ultimately to refraction. In the simplest scenario, in an ***isotropic material***, where 'all directions are equal', we might imagine that these dipoles will simply be aligned to the imposed electric field. However, in anisotropic materials, such as crystals, not all directions are equivalent, and these internal dipoles are more readily created where the imposed electric field is in certain specific orientations. The effect of this is that the refractive index of the material varies with the direction of the imposed electric field, or polarisation. This effect is known as ***birefringence***.

Optical Engineering Science, First Edition. Stephen Rolt.

Companion website: www.wiley.com/go/Rolt/opt-eng-sci

8.2 Polarisation

8.2.1 Plane Polarised Waves

To understand polarisation, we need to return briefly to Maxwell's equations which we encountered earlier. As a partial differential equation, it describes both the electric field, **E**, and the magnetic field, **H**.

$$\nabla^2\mathbf{E} = \frac{1}{n^2c^2}\frac{\partial^2\mathbf{E}}{\partial t^2} \text{ and } \nabla^2\mathbf{H} = \frac{1}{n^2c^2}\frac{\partial^2\mathbf{H}}{\partial t^2} \quad \text{where } c = \frac{1}{\sqrt{\varepsilon_0\mu_0}} \text{ and } n = \frac{1}{\sqrt{\varepsilon\mu}} \tag{8.1}$$

The two vector quantities are themselves linked by the following equations:

$$\nabla \times \mathbf{H} = \frac{\partial\mathbf{D}}{\partial t} \text{ (assuming zero electric current density)} \quad \mathbf{D} = \varepsilon\varepsilon_0\mathbf{E} \tag{8.2}$$

One possible set of solutions to Eq. (8.1) is the ***plane polarised wave***. The plane polarised wave consists of a wave disturbance whose amplitude (electric field) is uniform across an infinite plane, e.g. the X, Y plane. Spatial variation of the amplitude only occurs in the perpendicular direction of propagation, in this case, the z direction. As a wave disturbance, this spatial variation is sinusoidal and is defined by the wavelength, λ.

$$\mathbf{E} = \mathbf{E_0}\sin(\mathbf{k}z - \omega t) \quad k = 2\pi/\lambda \quad \omega/k = c/n \tag{8.3}$$

The direction of propagation is itself described by another vector, **k**, the wave-vector. In the example illustrated, the propagation direction is aligned to the z axis. For a given propagation direction, e.g. z, we can define two possible independent solutions by virtue of the direction of the electric field. In this case, the electric field may be oriented in either the x or y directions and these form ***two independent solutions***. From a linear combination of these two independent solutions we can produce any plane wave solution whose propagation direction is in the z direction.

Another important point to note is that from Eq. (8.2), the magnetic field should be perpendicular to the electric field. So, if the electric field is oriented along the x axis, then the magnetic field must be oriented along the y axis. Similarly, for an electric field orientation along the y axis, the magnetic field must be aligned parallel to the x axis. So, the propagation vector, **k**, the electric field **E**, and the magnetic field, **H**, form a set of three mutually perpendicular vectors. There is one caveat to this however. For this conclusion to be drawn, we are making the assumption that the relationship between **D** and **E** (in Eq. [(8.2)]) is entirely isotropic. That is to say, **D** and **E** are always parallel. This is absolutely true for a vacuum and isotropic materials, such as air, water, and glass, but is not true for crystal materials. Most significantly, in the event that **D** is not parallel to **E**, Eq. (8.2), suggests that it is the electric displacement, rather than the electric field that is perpendicular to the magnetic field and propagation direction. We will discover the significance of this when we consider birefringent materials.

In the meantime, however, we can assume that a plane polarised wave propagating in the z direction may be described by two independent solutions, as previously asserted. This is illustrated in Figure 8.1.

As advised previously, then any wave propagating in a given direction, e.g. z, may be represented as a linear combination of two independent solutions, one, for example, with the electric field along the x axis and the other with the electric field oriented along the y axis. Choice of which axes to adopt for these independent solutions is arbitrary. This arises as a feature of Maxwell's equations themselves which, being linear, conform to the principle of linear superposition. That is to say, any linear combination of known solutions to the equation is also itself a solution. Not only may we apply this principle to waves propagating in a specific direction, this means that any electromagnetic disturbance may be represented by a superposition of plane waves with complex coefficients.

8.2.2 Circularly and Elliptically Polarised Light

We introduced the topic of polarisation through the concept of linear polarisation. For a given direction of propagation, any electromagnetic wave disturbance may be represented by two independent linear

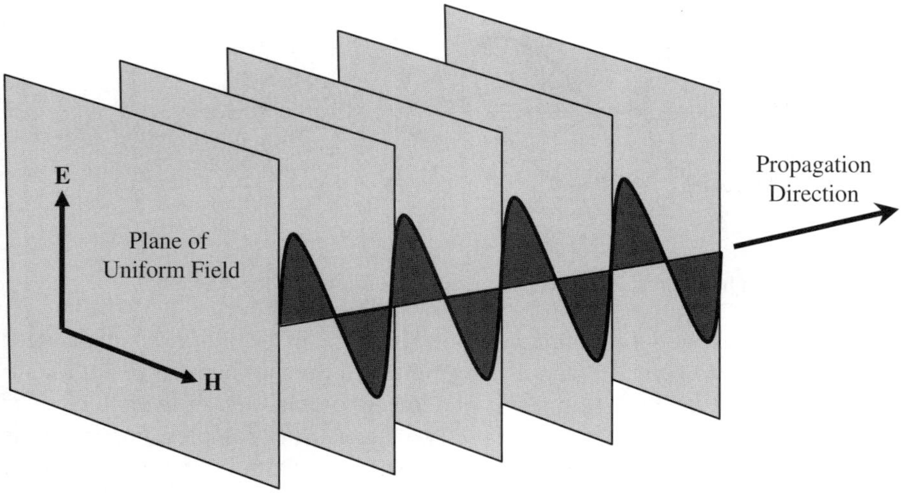

Figure 8.1 Plane polarised waves.

polarisation states whose electric field vectors are orthogonal. The contribution of each polarisation state is described by an amplitude coefficient which, for a wave oscillation, is necessarily complex. That is to say, each coefficient is described by a phase as well as a magnitude. If a wave disturbance is to be considered linearly polarised, either one of the two linear polarisation states must be zero, or both contributions must be in phase. On the other hand, if the two linear states are equal in magnitude and 90° out of phase with respect to the other, then we have a state that is described as ***circular polarisation***. In this case, the electric field, instead of oscillating at the optical frequency in a constant direction, rotates at the optical frequency. This rotation can be either clockwise or anticlockwise depending upon the sign of the phase difference. By convention, these two cases are respectively referred to as ***right handed circularly polarised*** and ***left handed circularly polarised*** light. In adopting such a convention one must be clear in which direction one is viewing the rotating field. In the convention adopted in this text, we are viewing the electric field rotation ***from the point of view of the source***, looking in the direction of propagation.

If now we consider and electromagnetic wave with a frequency, ω, propagating in the z direction, we may describe the oscillating electric field, **E**, in terms of the two unit vectors, **i**, **j**, in the x and y directions:

$$\text{Linearly polarised: } \mathbf{E} = A\cos\omega t\mathbf{i} + B\cos\omega t\mathbf{j} \tag{8.4a}$$

$$\text{Right handed circular: } \mathbf{E} = A\cos\omega t\mathbf{i} + A\sin\omega t\mathbf{j} \tag{8.4b}$$

$$\text{Left handed circular: } \mathbf{E} = A\cos\omega t\mathbf{i} - A\sin\omega t\mathbf{j} \tag{8.4c}$$

Note that in Eqs. (8.4b) and (8.4c), the magnitudes of both components are identical. If on the other hand, this is not the case, we have ***elliptical polarisation***. The electric field, in this case, sweeps out an elliptical path, rather than a circular path, as it rotates at the optical frequency. The major and minor axes of the ellipse are oriented along the primary x and y axes. Elliptical polarisation will also result if the phase difference is anything other than 90°. However, in this more general case, the major and minor axes will not be oriented along the x and y axes. Figure 8.2 illustrates the different polarisation states.

In Figure 8.2 is also included ***random polarisation***. Random polarisation is a more accurate description of so called unpolarised light. This is the general state of polarisation applicable to general illumination. In this case, there is simply no defined phase relationship between the two polarisation states. Both linearly and circularly polarisation have a measure of coherence between the two polarisation states.

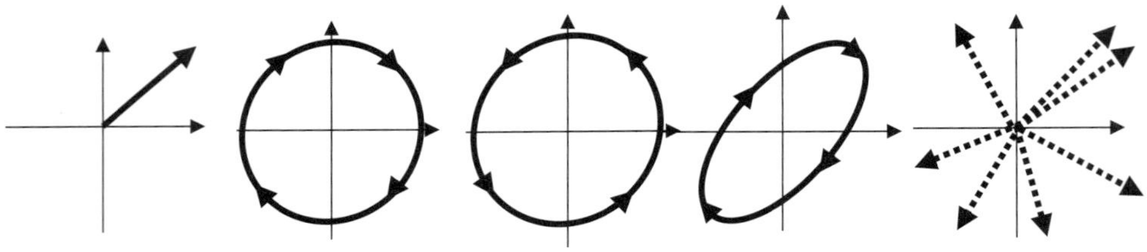

Figure 8.2 Polarisation states (a) Linear, (b) Right hand circular, (c) Left hand circular, (d) Elliptical, and (e) Random.

During the course of the foregoing analysis, we defined circularly polarised light in terms of two independent and orthogonal linear polarisation states. However, in fact, it is legitimate to pursue the reverse argument. That is to say, linear polarisation states may be defined in terms of a linear combination of the two circular polarisation states. Indeed, by making a linear combination of the two states using complex amplitudes as coefficients, we may define any arbitrary polarisation state. This description of polarisation is more in tune with the quantum description of light and matter. Circularly polarised light has a quantised angular momentum, with right hand and left hand polarisation having an angular momentum (in the direction of propagation) of $\pm h$ for each photon. However, from an engineering perspective, the convention is to describe the polarisation state of light in terms of the two linear components.

8.2.3 Jones Vector Representation of Polarisation

We now seek a more formal description of the polarisation state of optical radiation. It is intuitively sensible that this representation might take a vector format, as polarisation essentially describes the vector representation of the electric field. In the ***Jones vector*** representation, the light is assumed to be propagating along the z direction and that the field may be described by its components in the x and y directions. Most importantly, this field is represented as a complex amplitude and takes account of any relative phase delay between the x and the y components. So the generalised representation of the Jones vector may be written as:

$$\begin{bmatrix} \mathbf{E}_x \\ \mathbf{E}_y \end{bmatrix} = \begin{bmatrix} E_{0x}e^{i\theta_x} \\ E_{0y}e^{i\theta_y} \end{bmatrix} \tag{8.5}$$

The convention is that the Jones vector is normalised to a magnitude of one. A number of illustrative polarisation states is shown below:

$$\underset{\text{Linear X}}{\begin{bmatrix} 1 \\ 0 \end{bmatrix}} \quad \underset{\text{Linear Y}}{\begin{bmatrix} 0 \\ 1 \end{bmatrix}} \quad \underset{\text{Linear 45°}}{\frac{1}{\sqrt{2}}\begin{bmatrix} 1 \\ 1 \end{bmatrix}} \quad \underset{\text{Linearq}}{\begin{bmatrix} \cos\theta \\ \sin\theta \end{bmatrix}} \quad \underset{\text{Right Hand Circ.}}{\frac{1}{\sqrt{2}}\begin{bmatrix} 1 \\ i \end{bmatrix}} \quad \underset{\text{Left Hand Circ.}}{\frac{1}{\sqrt{2}}\begin{bmatrix} 1 \\ -i \end{bmatrix}}$$

Absent from this list, of course, is random polarisation which cannot be represented in the Jones vector notation. That is to say, Jones vectors can only be applied for fully polarised light and assumes that the phase relationship between the two components is entirely coherent. It is, therefore, useful in specific instances for the treatment of fully polarised light but not for partially polarised or randomly polarised light. This is a weakness of the system.

This deficiency may be addressed by another vector convention, the ***Stokes vector*** representation which also quantifies the degree of coherence between the two polarisation directions.

8.2.4 Stokes Vector Representation of Polarisation

The Stokes vector representation is based on expressing polarisation in the most general terms as elliptical polarisation. At first sight, this may appear a little contradictory, as linear polarisation seems to be quite distinct

Figure 8.3 Polarisation ellipse.

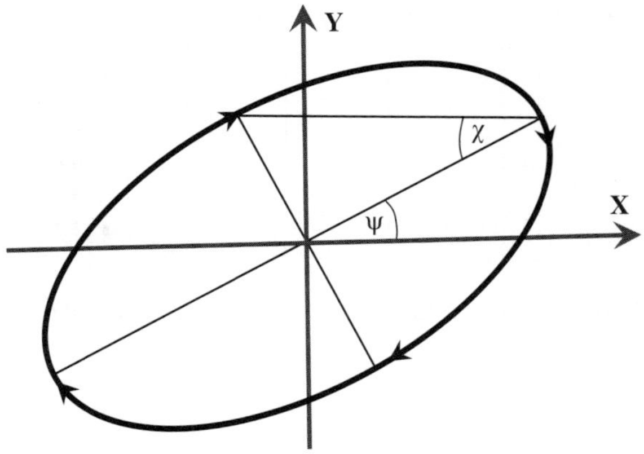

from circular and elliptical polarisation. However, linear polarisation may itself be considered as a special case of elliptical polarisation where the minor axis of the ellipse is actually zero. This ellipse is described by the angle, ψ, that its lobe, or major axis, makes with the x axis and by the angle, χ, defining the degree of ellipticity.

A full description of generalised elliptical polarisation is provided by the ***Poincaré sphere***. The Poincaré sphere is based on a Cartesian coordinate system with axes x, y, z, but represented in spherical polar form. The spherical polar system is described by two angles, the polar angle ('latitude' with respect to the x, y, plane) and the second the azimuthal angle. Because the ellipse has two lobes with twofold rotational symmetry, the actual polar and azimuthal angles are defined as 2χ and 2ψ. The polar angle, 2χ, for the Poincare sphere, describes the degree of ellipticity or the ratio of minor to major axes. At the 'equator' the ellipticity is zero, and we have linear polarisation. At the 'poles', the ellipticity is one and the polarisation is circular. The azimuthal angle, 2ψ, describes the location of the lobe or major axis of the ellipse. To understand this, it is useful to view a generalised diagram depicting elliptical polarisation, as shown in Figure 8.3.

To complete the Stokes representation of polarisation, we now introduce the four components of the Stokes vector, S_0, S_1, S_2, S_3. It must be emphasised that the Stokes vector description is based on the irradiance or flux of the radiation and ***not*** the amplitude. S_0 simply represents the irradiance, E, of the radiation. S_2 and S_3 represent the X and Y components that define the orientation of the major axis or lobe of the ellipse. Finally, S_3 represents the ellipticity of the polarisation. Most significantly, at this point, we introduce a parameter, p, to describe the '***degree of polarisation***', ranging from 0 for randomly polarised light, to 1 for fully polarised light. The components of the Stokes vector are given as follows:

$$S = \begin{bmatrix} S_0 \\ S_1 \\ S_2 \\ S_3 \end{bmatrix} \quad \begin{matrix} S_0 = E \\ S_1 = Epcos2\chi cos2\psi \\ S_2 = Epcos2\chi sin2\psi \\ S_3 = Epsin2\chi \end{matrix} \tag{8.6}$$

More formally, we may regard p as the cross correlation between the two independent components of polarisation, E_x and E_y:

$$p = \frac{\langle E_x E_y^* \rangle}{\sqrt{\langle E_x E_x^* \rangle \langle E_y E_y^* \rangle}} \tag{8.7a}$$

We can also see that from Eq. (8.6), the polarisation, p may be expressed in terms of S_1, S_2, and S_3:

$$p = \sqrt{\frac{S_1^2 + S_2^2 + S_3^2}{S_0^2}} \tag{8.7b}$$

It is the usual convention to normalise E to one. Some typical Stokes vectors are shown below:

$$\underset{\text{Linear X}}{\begin{bmatrix}1\\1\\0\\0\end{bmatrix}} \quad \underset{\text{Linear Y}}{\begin{bmatrix}1\\-1\\0\\0\end{bmatrix}} \quad \underset{\text{Linear 45°}}{\begin{bmatrix}1\\0\\1\\0\end{bmatrix}} \quad \underset{\text{Linear }\theta}{\begin{bmatrix}1\\\cos 2\theta\\\sin 2\theta\\0\end{bmatrix}} \quad \underset{\text{RH Circ.}}{\begin{bmatrix}1\\0\\0\\1\end{bmatrix}} \quad \underset{\text{LH Circ}}{\begin{bmatrix}1\\0\\0\\-1\end{bmatrix}} \quad \underset{\text{Random}}{\begin{bmatrix}1\\0\\0\\0\end{bmatrix}}$$

Worked Example 8.1 An elliptically polarised beam of light has a degree of polarisation equal to 0.4.

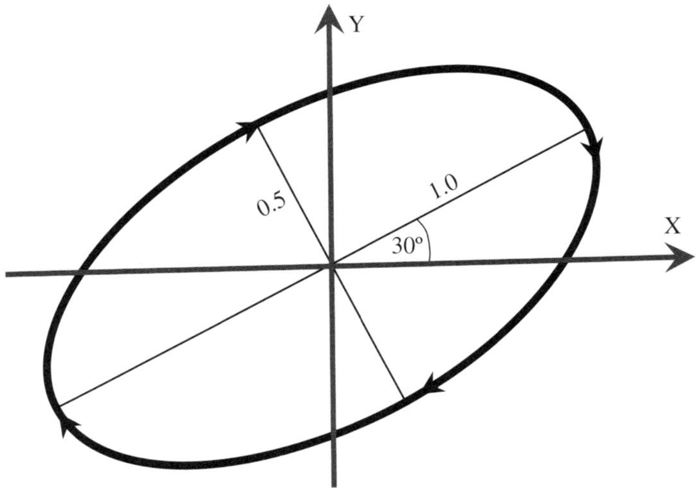

The major axis is tilted at 30° to the x axis, as shown and the ratio of the minor axes to the major axis of the ellipse is 0.5–1. What is the Stokes vector for this polarisation state, assuming the vector is normalised (i.e. $E = 1$)?

Firstly, $\psi = 30°$. We can see clearly that $\tan(\chi) = 0.5$, from the ratio of the axes and hence $\chi = 26.57°$. From this, $2\psi = 60°$ and $\chi = 53.13°$. We are told that p = 0.4 and so it is quite straightforward to derive the Stokes vector from Eq. (8.6).

$$S_0 = E = 1$$

$$S_1 = Ep\cos 2\chi \cos 2\psi = 1 \times 0.4 \times 0.6 \times 0.5 = 0.12$$

$$S_2 = Ep\cos 2\chi \sin 2\psi = 1 \times 0.4 \times 0.6 \times 0.8667 = 0.208$$

$$S_3 = Ep\sin 2\chi = 1 \times 0.4 \times 0.8 = 0.32$$

In summary:

$$\mathbf{S} = \begin{bmatrix}1\\0.12\\0.208\\0.32\end{bmatrix}$$

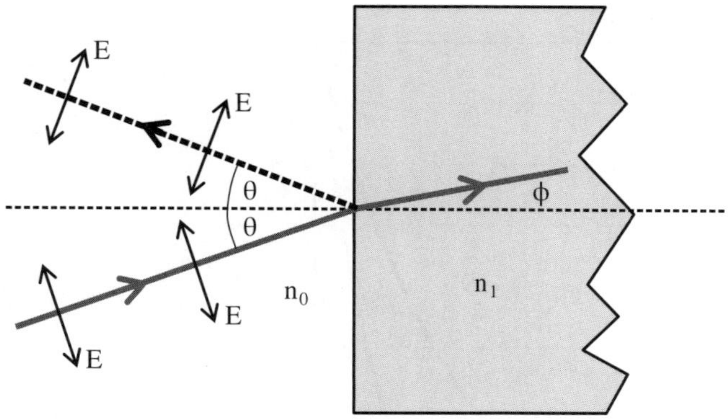

Figure 8.4 Reflection at an interface.

8.2.5 Polarisation and Reflection

From our treatment of simple paraxial optics, we are very familiar with the concept of reflection at an interface between two different materials. The key to this process of reflection is that there is a contrast in refractive index between the media; it is this that drives the reflection process. For the most part, we have considered the reflection process to be total. In one case, we consider 'a mirror' which rather simplistically is considered to reflect 100% of the incident light. More realistically, ***total*** internal reflection does indeed occur where light is incident at an optical interface at an angle exceeding the critical angle. However, ***partial*** reflection always occurs at the interface of two materials of contrasting index. This general process is referred to as ***Fresnel reflection***. For example, reflection at a glass air interface is of the order of 4%, with the remainder being transmitted. Crucially, for non-zero angles of reflection, this reflection is dependent upon the polarisation state of the incident light. This is sketched in Figure 8.4.

The electric field direction or polarisation is also sketched out in Figure 8.4. In this case, the direction of polarisation is in the plane of incidence. It is clear that the direction of polarisation could also be orthogonal to this, or out of the plane of the paper. Thus, it is the underlying geometry of reflection that determines the two independent directions of polarisation. This provides us with another convention for describing polarisation. Where the electric field is in the plane of incidence, as in Figure 8.4, it is described as ***p polarisation*** or parallel polarisation. Conversely, where the electric field is orthogonal to the plane of incidence it is described as ***s polarisation***, from the German, senkrecht, for perpendicular. In describing polarisation in this context, there is yet another convention. Polarisation may be described as ***transverse electric (TE)*** or ***transverse magnetic (TM)***. In the case of TE polarisation the electric field is perpendicular to the plane of incidence and this is equivalent to s polarisation. Conversely, in the case of TM polarisation it is the magnetic field which is perpendicular to the plane of incidence and hence the electric field is in the plane of incidence. This is equivalent to p polarisation.

The ***reflection coefficient***, R, for reflection at an interface is the ratio of the exitance of the reflected beam to the irradiance of the incident beam. It may be derived by considering Maxwell's equations together with the relevant boundary conditions – the electric field, E, in the plane of the surface is continuous across the interface; the electric displacement, D, normal to the plane of the surface is continuous across the interface. However, full derivation of the formulae quoted here is beyond the scope of this text. Of course, for non-zero angles of incidence, the reflection coefficients for the two polarisations are different. If we consider the first medium having a refractive index of n_0 and the second a refractive index of n_1, we may set out the so-called ***Fresnel equations***.

$$R_s = \left[\frac{n_0 \cos\theta - n_1 \cos\phi}{n_0 \cos\theta + n_1 \cos\phi}\right]^2 \quad R_p = \left[\frac{n_0 \cos\phi - n_1 \cos\theta}{n_0 \cos\phi + n_1 \cos\theta}\right]^2 \tag{8.8a}$$

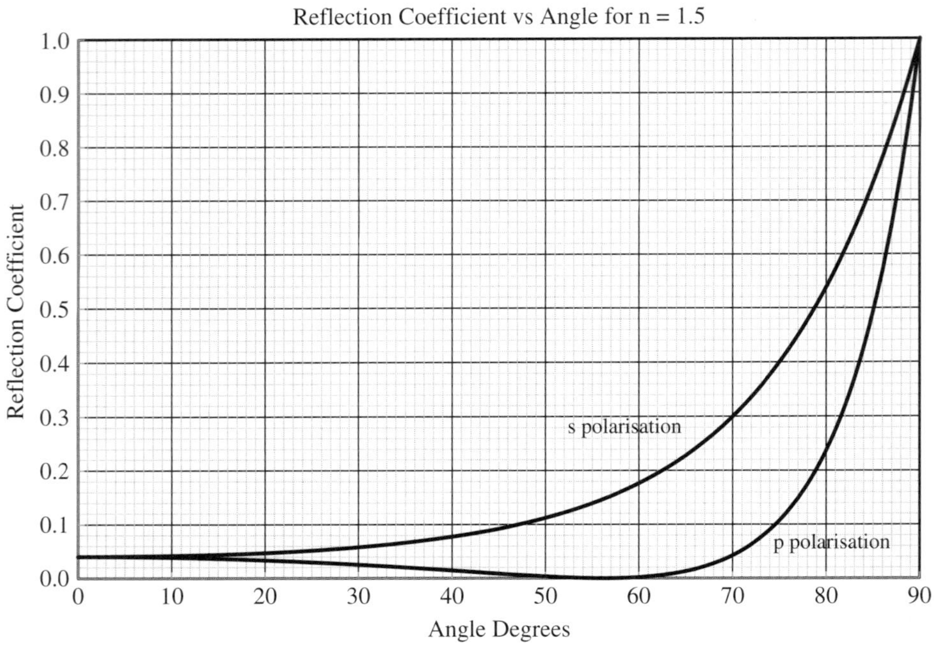

Figure 8.5 Reflection coefficient vs angle for n = 1.5.

Rs and Rp are the reflection coefficient for s and p polarisation respectively, θ is the angle of incidence and ϕ is the angle of refraction.

Where the angle of incidence is zero, then the reflection coefficients for both polarisations are identical. This is quite intuitive for reasons of symmetry. In this case, if we assume that the first medium is air/vacuum (i.e. $n_0 = 1$), and $n_1 = n$, then the equations simplify to:

$$R = \left[\frac{n-1}{n+1}\right]^2 \tag{8.8b}$$

If we substitute n = 1.5 (for a typical glass) into Eq. 8.8b, we find that R = 0.04 or 4%, in line with what we had argued previously. Figure 8.5 shows a plot of the reflection coefficient vs angle for light incident upon glass with a refractive index of 1.5

First, it is clear that, whilst the reflection coefficient for both coefficients is identical for zero angle of incidence, thereafter they diverge. Initially, the reflection coefficient for the p polarisation diminishes until reaching a minimum value, before increasing sharply. By contrast, the reflection coefficient for s polarisation increases monotonically with angle. For both polarisations, the reflection coefficient tends to 100% at grazing incidence ($\theta = 90°$). If one looks carefully at Eq. (8.8a), it is evident that the form of the equation is reversible. That is to say, if ϕ were substituted for θ and n_1 substituted for n_0, then the reflection coefficients are identical. From the perspective of internal reflection, i.e. progressing from the high index material to the low index material, the grazing incidence reflection is equivalent to an incident angle equal to the critical angle. Therefore, at the critical angle, the internal reflection coefficient is 100%, i.e. total internal reflection. Of course, for angles greater than the critical angle, where no explicit refraction solution exists, the reflection coefficient is also 100%. This is the total internal reflection that we have encountered before.

What is noticeable about Figure 8.5 is the minimum in the reflection coefficient that exists for p polarisation. In fact that minimum is zero, and this results from the following equality derived from Eq. (8.8a).

$$\cos\phi - n\cos\theta = 0 \tag{8.9}$$

Manipulating Eq. (8.9), it is easy to show that the reflection coefficient equals zero where the tangent of the incident angle is equal to the refractive index:

$$\tan\theta = n \tag{8.10}$$

This angle is known as the ***Brewster angle*** and for a refractive index of 1.5 is equal to 56.3°. So for the Brewster angle, the reflection coefficient for p polarisation is zero. This has some practical significance in optics. ***Brewster windows*** are optical windows that are, by design, inclined to an incoming beam at the Brewster angle. Their application features where exceptionally low loss transmission is required. Examples include windows on certain types of laser, e.g. helium neon lasers.

Another point to note about Fresnel reflection is that for non-zero incident angles, random polarised light is converted to partially polarised light. Quite obviously, at the Brewster angle, any reflected light is wholly polarised and so the degree of polarisation, p, becomes equal to one. More generally the degree of polarisation is equal to:

$$p = \frac{R_s - R_p}{R_s + R_p} \tag{8.11}$$

It is clear from Figure 8.5 that reflection of polarised light at high angles of incidence produces light that is substantially polarised. This circumstance affords one very specific practical application. Where light is reflected, for instance, from a horizontal surface, such as a road, it is likely to be significantly polarised. This is particularly true where sunlight is reflected from small quantities of water on the surface of the road. That is to say, glare from a road surface is preponderantly s polarised, or horizontally polarised. Therefore, by employing a device (polarising sunglasses) that preferentially blocks this horizontal polarisation, then the amount of glare will be substantially reduced.

Worked Example 8.2 The glass, BK7, has a refractive index of 1.5168 at 589.3 nm. What is the Brewster angle at this wavelength? What is the degree of polarisation produced by reflection from randomly polarised light that is incident at angle of 45°.

The Brewster angle is simply given by:

$$\tan\theta = n = 1.5168.$$

This gives: $\theta_{Brewster} = 56.6°$

The Brewster angle is 56.6°

To calculate the degree of polarisation at 45°, we first need to calculate R_s and R_p from Eq. (8.8a):

$$R_s = \left[\frac{n_0\cos\theta - n_1\cos\phi}{n_0\cos\theta + n_1\cos\phi}\right]^2 \quad R_p = \left[\frac{n_0\cos\phi - n_1\cos\theta}{n_0\cos\phi + n_1\cos\theta}\right]^2$$

$n_0 = 1; n_1 = 1.5168; \theta = 45°; \phi = 27.79°$ (from Snell's Law)

$$R_s = \left[\frac{n_0\cos\theta - n_1\cos\varphi}{n_0\cos\theta + n_1\cos\varphi}\right]^2 = \left[\frac{0.707 - 1.5168\times 0.8847}{0.707 + 1.5168\times 0.8847}\right]^2 = 0.096$$

$$R_p = \left[\frac{n_0\cos\varphi - n_1\cos\theta}{n_0\cos\varphi + n_1\cos\theta}\right]^2 = \left[\frac{0.8847 - 1.5168\times 0.707}{0.8847 + 1.5168\times 0.707}\right]^2 = 0.0092$$

Thus the R_s coefficient is 9.6% and the R_p coefficient 0.92%.

Finally, the degree of polarisation, p, is given by (8.11):

$$p = \frac{R_s - R_p}{R_s + R_p} = \frac{0.096 - 0.0092}{0.096 + 0.0092} = 0.825$$

The degree of polarisation is 0.825.

8.2.6 Directional Flux – Poynting Vector

Just as the amplitude of an electromagnetic wave is a vector quantity and defined by the electric and magnetic fields, **E** and **H**, then the energy flux produced by the radiation field is also a vector quantity. The ***Poynting vector***, **S**, is a vector that describes the local irradiance at a point and also the direction associated with that energy flux. The Poynting vector is given by the cross product of the electric and magnetic fields.

$$\mathbf{S} = \mathbf{E} \times \mathbf{H} \tag{8.12}$$

We would expect, under normal circumstances, that the Poynting vector, **S**, is aligned to the wavevector, **k**. Both **E** and **H** are generally perpendicular to **k**, so it is reasonable to suppose that **k** and **S** are parallel. That is to say, for a plane wave, the direction of energy propagation is normal to the planes of constant phase. However, the significance of the Poynting vector, as far is this chapter is concerned, is that for a birefringent material, **S** and **k** are generally not aligned. As will be revealed in the next section, this leads to the phenomenon of '***walk-off***' where the energy propagation and phase vectors are no longer co-parallel.

As well as directional energy flux in a wave, there is also momentum flux. This produces 'radiation pressure'. Radiation pressure is simply given by the Poynting vector divided by the speed of light.

$$\mathbf{P}_{rad} = \frac{\mathbf{S}}{c} \tag{8.13}$$

Of course, the radiation pressure in most practical cases is vanishingly small. Taking one specific example, the solar irradiance above the Earth's atmosphere averages about 1370 $\mathrm{Wm^{-2}}$. The radiation pressure associated with this amounts to 4.6 μPa, or over 10 orders of magnitude less than atmospheric pressure at the Earth's surface. For the most part, this is wholly insignificant. However, for Earth orbiting satellites with a very low mass to surface area ratio, radiation pressure can, over long periods of time, have a significant perturbing effect on orbital dynamics.

In addition, as we have observed previously, light also conveys angular momentum. This is directly related to polarisation, with right handed and left-handed circular polarisation associated with different directions of polarisation.

8.3 Birefringence

8.3.1 Introduction

Birefringence occurs in non-isotropic materials where the refractive properties of a material are dependent upon the direction of polarisation. Before we can understand this phenomenon, we first need to grasp the underlying atomic processes that give rise to refractive properties in a material. One can image that a material is composed of a balance of positively and negatively charge centres – atomic nuclei and surrounding electron clouds. The effect of an externally imposed electric field is to pull these charges apart. This creates a very large number of small ***electric dipoles*** within the body of a material. An electric dipole consists of equal and opposite charges that are separated by a small distance. It is quantified by the product of the charge and the separation and this quantity is referred to as the ***electric dipole moment***, **p**. The electric dipole moment is, of course a vector quantity and is given more formally by:

$$\mathbf{p} = q^{*}\mathbf{r} \tag{8.14}$$

*where q is the charge and **r** is the vector separation.*

We might imagine a refractive material consisting of a large number of these nascent dipoles, formed of charged pairs bound by 'elastic springs'. An external electric field tends to pull the charged pairs apart, which is resisted by the spring, so that the separation and induced dipole moment increases linearly with the external electric field. This process is shown in Figure 8.6.

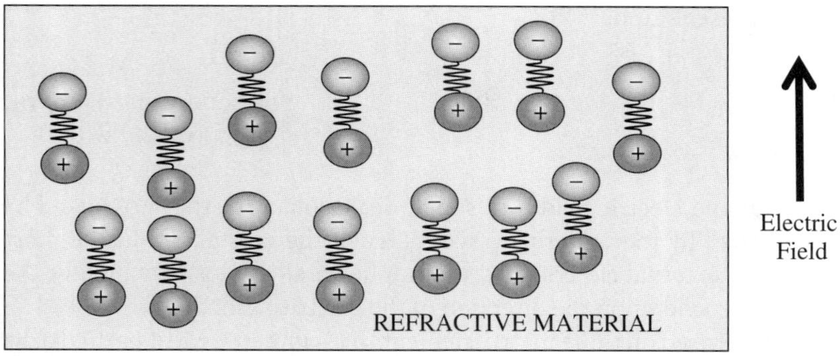

Figure 8.6 Induced dipole formation in refractive material.

In the model in Figure 8.6, the electric dipole moment is proportional to the electric field and the constant of proportionality, α, is known as the polarizability.

$$\mathbf{p} = \alpha \mathbf{E} \tag{8.15}$$

In the case of a real material, the number of electric dipoles created, as illustrated in Figure 8.6, is extremely large. Thus, averaged over the material's volume, the effect of an applied field is to create a dipole moment per unit volume. This is referred to as the ***polarisation density*** or simply the ***polarisation, P***. As for the individual dipoles, the polarisation is proportional to the electric field and the constant of proportionality, χ, is referred to as the ***electric susceptibility***.

$$\mathbf{P} = \varepsilon_0 \chi \mathbf{E} \tag{8.16}$$

ε_0 is the permittivity of free space

The effect of these internal dipoles is to create an electric field of their own. Although the electric field associated with an individual charge follows an inverse square law, the electric field associated with a dipole is proportional to the inverse of the ***cube*** of the distance from the dipole. Strictly, this only applies where the distance concerned is larger than the dipole separation; however, this is true in this specific case. While the derivation is beyond the scope of this text, the effect is to create an internal electric field that is proportional to the polarisation. This internal electric field is quantified by the electric displacement, **D**:

$$\mathbf{D} = \varepsilon_0 \mathbf{E} + \mathbf{P} \text{ and } \mathbf{D} = \varepsilon_0 (1 + \chi) \mathbf{E} = \varepsilon_0 \varepsilon \mathbf{E} \tag{8.17}$$

The quantity, ε, is referred to as the relative permittivity and is equal to the square of the refractive index. Thus, the refractive index, n, may now be written:

$$n = \sqrt{1 + \chi} = \sqrt{\varepsilon} \tag{8.18}$$

Thus far, in this discussion of internal polarisation we have implicitly assumed an isotropic material. That is to say, the vector polarisation produced by an electric field is always parallel to that imposed electric field. However, in the case of a birefringent material, that assumption breaks down. Most specifically, the assumption breaks down in crystal materials where the geometry of electric polarisation is driven by the internal structure of the material. The vector, **D**, is an important parameter in the description of a plane wave. From Eq. (8.2), it is specifically the vector **D** that is perpendicular to the magnetic field and the direction of propagation direction as defined by the wavevector, **k**, rather than the electric field, **E**. So, therefore, in the most general case, for a birefringent material, one cannot assume that the electric field is perpendicular to the propagation direction. Therefore, to take account of this, Eq. (8.16), which relates two vector quantities, should be defined in terms of the following matrix representation:

$$\begin{bmatrix} \mathbf{P}_x \\ \mathbf{P}_x \\ \mathbf{P}_z \end{bmatrix} = \varepsilon_0 \begin{bmatrix} \chi_{xx} & \chi_{xy} & \chi_{xz} \\ \chi_{yx} & \chi_{yy} & \chi_{yz} \\ \chi_{zx} & \chi_{zy} & \chi_{zz} \end{bmatrix} \begin{bmatrix} \mathbf{E}_x \\ \mathbf{E}_y \\ \mathbf{E}_z \end{bmatrix} \tag{8.19}$$

Hence instead of relating the polarisation and electric field by a simple scalar quantity, they are linked by a tensor of rank two, whose form is governed by the underlying symmetry of the crystal. So, in the most general case, the electric polarisation and the external electric field will not be parallel. Most pertinently, the susceptibility and thus refractive index will depend upon the direction of the electric field.

There are some fundamental constraints imposed on these matrix elements by symmetry. Most particularly:

$$\chi_{xy} = \chi_{yx} \quad \chi_{xz} = \chi_{zx} \quad \chi_{yz} = \chi_{zy} \tag{8.20}$$

However, of fundamental importance is the behaviour of Eq. (8.19) with respect to rotation of the co-ordinate frame. Equation (8.19) is cast in a generalised co-ordinate frame. However, we are ***always*** able to find a specific Cartesian co-ordinate frame whereby the matrix in Eq. (8.19) is diagonalised:

$$\begin{bmatrix} \mathbf{P}_x \\ \mathbf{P}_x \\ \mathbf{P}_z \end{bmatrix} = \varepsilon_0 \begin{bmatrix} \chi_{xx} & 0 & 0 \\ 0 & \chi_{yy} & 0 \\ 0 & 0 & \chi_{zz} \end{bmatrix} \begin{bmatrix} \mathbf{E}_x \\ \mathbf{E}_y \\ \mathbf{E}_z \end{bmatrix} \tag{8.21}$$

In this representation, the x, y, and z axes are referred to as the ***principal axes*** of the crystal. We may choose to represent the electric displacement, **D**, in a similar way:

$$\begin{bmatrix} \mathbf{D}_x \\ \mathbf{D}_x \\ \mathbf{D}_z \end{bmatrix} = \varepsilon_0 \begin{bmatrix} \varepsilon_{xx} & 0 & 0 \\ 0 & \varepsilon_{yy} & 0 \\ 0 & 0 & \varepsilon_{zz} \end{bmatrix} \begin{bmatrix} \mathbf{E}_x \\ \mathbf{E}_y \\ \mathbf{E}_z \end{bmatrix} \tag{8.22}$$

As a consequence of Eq. (8.22), the refractive index associated with a specific electric field orientation is described by the so-called ***index ellipsoid***. That is to say, the refractive index of a birefringent material may be mathematically defined by an ellipsoidal surface with principal axes, n_x, n_y, n_z.

$$\frac{x^2}{n_x^2} + \frac{y^2}{n_y^2} + \frac{z^2}{n_z^2} = 1 \tag{8.23}$$

x, y, and z are refractive index components along these axes, as defined by the electric field vector such that:

$$x^2 + y^2 + z^2 = n^2 \tag{8.24}$$

where n is the refractive index for that specific polarisation direction.

8.3.2 The Index Ellipsoid

Figure 8.7 shows a graphical representation of the index ellipsoid, with the refractive indices along the principal axes being n_x, n_y, and n_z. To re-iterate, the principal axes are determined entirely by the internal symmetry of the optical material. Of course, for an isotropic material, all indices are equal and $n_x = n_y = n_z = n$. For many crystal materials, however, their underling symmetry dictates that two of these indices are equal. In this case, $n_x = n_y \neq n_z$. Such crystals are referred to as ***uniaxial crystals*** and represent a significantly common and important group of optical materials in terms of practical applications. However, in the most general case, all principal indices are different; such crystals are referred to as ***biaxial crystals***.

From Figure 8.7, it is quite straightforward to see what happens when light propagates in a medium with its wavevector aligned with one of the principal axes, e.g. the z axis. In this case, as with the general treatment of

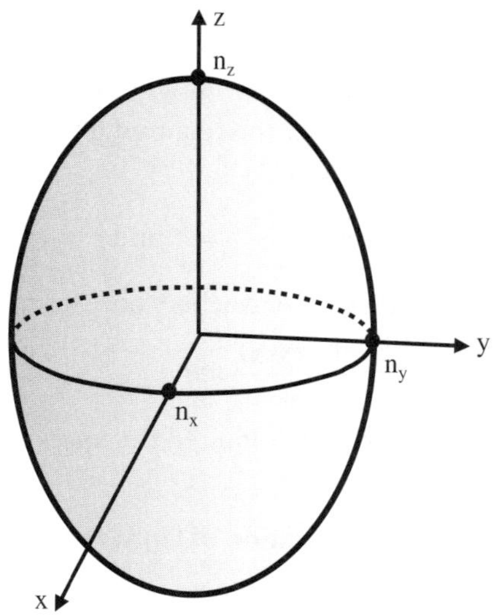

Figure 8.7 Index ellipsoid.

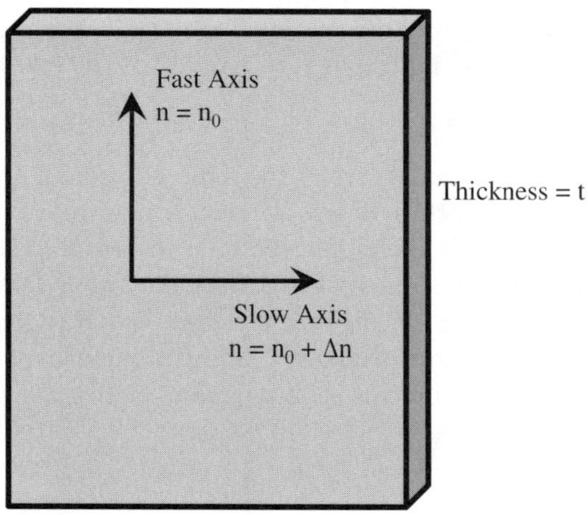

Figure 8.8 Phase delay and propagation through a birefringent crystal.

polarisation, there are two independent plane wave solutions. However, in the case of propagation in a crystal material, these independent solutions are constrained to the crystal x and y axes. On account of their different refractive indices, the two polarisations propagate at different velocities through the medium. The axis with the higher index is referred to as the ***'slow axis'*** as that polarisation travels more slowly when compared to the other lower index axis which is referred to as the ***'fast axis'***.

The important aspect of the different phase velocities for the two polarisations is that the two polarisations experience a relative phase delay as they propagate through a material. This is illustrated in Figure 8.8. Since the referactive index of the two polarisations is different, then their respective optical path lengths will be different as they propagate through a common thickness of material. If one considers a birefringent material of thickness, t, and a refractive index difference of Δn between the two polarisations, then the phase difference produced (in radians) is equal to:

$$\Delta\phi = 2\pi\frac{\Delta n}{\lambda}t \tag{8.25}$$

λ is the wavelength

Worked Example 8.3 At 590 nm, a thin plate of quartz has a refractive index of 1.553 along its slow axis and a refractive index of 1.544 along its fast axis. How thick should the plate be made such that the phase difference produced between the two polarisations is equal to one half of a wavelength at 590 nm?

It is clear that the required phase difference produced between the two polarisations is equal to π radians. From Eq. (8.25), we have:

$$\Delta\phi = 2\pi\frac{\Delta n}{\lambda}t = \pi \text{ and } t = \frac{\lambda}{2\Delta n}$$

The difference, Δn, between the two indices is equal to 1.533–1.544 or 0.009. Therefore:

$$t = \frac{0.00059}{2 \times 0.009}mm$$

This gives the required thickness as 0.033 mm.

8.3.3 Propagation of Light in a Uniaxial Crystal – Double Refraction

Thus far, we have only considered the specific case where light propagates along one of the principal axes of the crystal. In considering a more general treatment, we restrict ourselves to the special, but important, case of uniaxial crystals. For uniaxial crystals, two of the indices are equal. For reasons that will become apparent later, these indices are equal to what is referred to as ***the ordinary refractive index,*** n_0. The refractive index of the remaining axis is referred to as ***the extraordinary refractive index,*** n_e. That is to say:

$$n_x = n_y = n_o \quad n_z = n_e$$

For a uniaxial crystal, the index ellipsoid is symmetric about the z axis. Hence we might describe the propagation direction of any plane wave in terms of the angle, ϕ, that it makes with respect to this axis. In a uniaxial crystal, this axis is referred to as the ***optic axis***. For the case of propagation along a principal axis, we identified two fundamental polarisation modes related to the crystal symmetry. To do this in this more general case, we generate a plane that is perpendicular to the propagation direction. The intersection of this plane with the index ellipsoid produces an ellipse. The two modes are then described by the semi-major and semi-minor axes of this ellipse. This is illustrated in Figure 8.9.

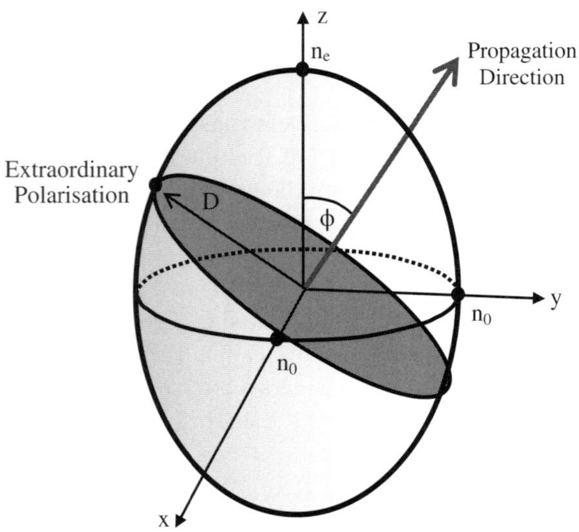

Figure 8.9 Propagation of light in a uniaxial crystal.

The two modes are referred to as the ***ordinary ray*** and the ***extraordinary ray***. Irrespective of the angle, ϕ, with respect to the optic axis, one of the modes has a refractive index equal to the ordinary index, n_0. This is the ordinary ray. The reason this ray is so labelled is that it can be demonstrated that, in this specific instance, the electric field and electric displacement are parallel. However, for the other polarisation mode, polarised perpendicularly to the ordinary ray, the electric field and electric displacement are not parallel. This explains the choice of terminology, as it is this ray that is described as the extraordinary ray. It is this ray that is illustrated in Figure 8.9, showing the direction of polarisation in terms of the **D** vector.

As previously discussed, the refractive index of the ordinary ray is always n_0. For the extraordinary ray, the refractive index, n_x depends upon the propagation angle, ϕ, with respect to the optical axis. It is straightforward to calculate this by using the index ellipsoid. For reasons of symmetry, we only need to consider one of the two 'ordinary' axes, for example, the x axis. Applying Eq. (8.23), we get:

$$\frac{x^2}{n_0^2} + \frac{z^2}{n_e^2} = 1 \text{ and } \frac{n_x^2\cos^2\phi}{n_0^2} + \frac{n_x^2\sin^2\phi}{n_e^2} = 1$$

Manipulating the above expression we get:

$$n_x^2 = n_0^2\left[\frac{1}{1 + ((n_o/n_e)^2 - 1)sin^2\varphi}\right] \tag{8.26}$$

Of course, when the angle, ϕ, is zero, where the propagation direction is parallel to the optic axis, then the refractive index of the 'extraordinary ray' is the same as that of the ordinary ray, i.e. n_0. When the angle is 90°, then n_x is equal to the extraordinary index n_e. At other angles, the refractive index lies between the two extremes. The most important consequence of this is that the two fundamental polarisation modes have different refractive indices. Therefore where light is incident at an angle upon a birefringent material, the two polarisations will, in general, be refracted differently. So for randomly polarised light, the light will be split into the ordinary and extraordinary ray which are refracted at different angles. This process of ***double refraction*** is the defining quality of a birefringent material. A common birefringent material is calcite and the phenomenon of double refraction is illustrated figuratively for this material in Figure 8.10.

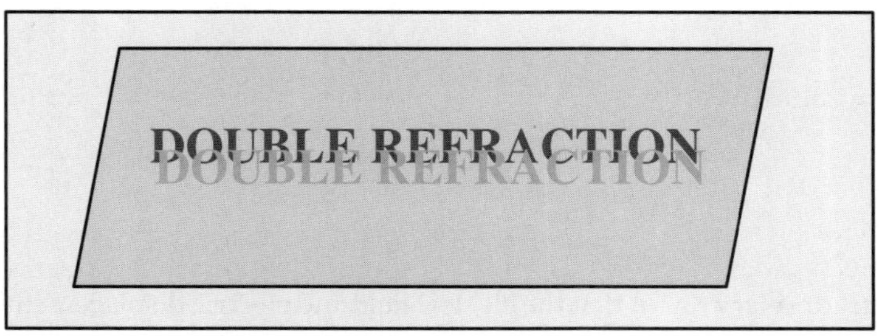

Figure 8.10 Double refraction in calcite.

Worked Example 8.4 ***Double Refraction in Calcite***

Randomly polarised light at a wavelength of 590 nm is incident upon a slab of calcite at an angle of incidence of 45°. The optic axis of the crystal is normal to the surface. At this wavelength, n_0 is equal to 1.658 and n_e is equal to 1.468. Calculate the refracted angles of both the ordinary and the extraordinary ray.

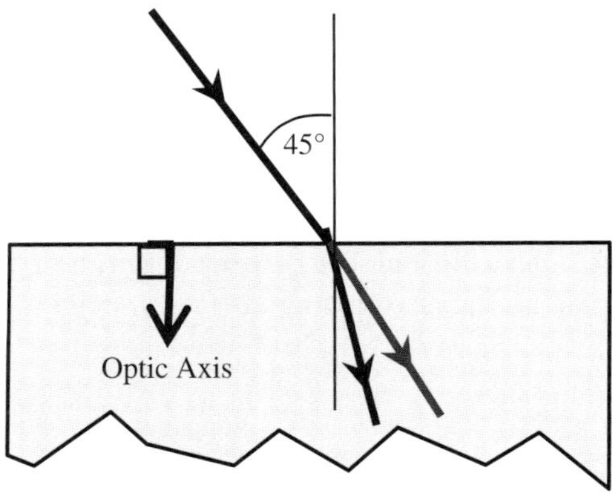

The picture for the ordinary ray is very straightforward, as we know the refractive index at the outset is equal to n_0 or 1.658. By virtue of Snell's law we have:

$$\sin\phi = \frac{\sin\theta}{n_0} = \frac{0.7071}{1.658} = 0.4265 \quad \text{i.e. } \sin(\Theta) = 0.4265 \text{ and } \Theta_{ord} = 25.24°$$

The picture is more complicated for the extraordinary ray where we have:

$$\sin\phi = \frac{\sin\theta}{n_x} \text{ and } n_x^2 = n_0^2\left[\frac{1}{1 + ((n_e/n_0)^2 - 1)\sin^2\phi}\right]$$

Manipulating the above, we have:

$$\sin^2\phi = \left[\frac{1 + ((n_e/n_0)^2 - 1)\sin^2\phi}{n_0^2}\right]\sin^2\theta \text{ and } [n_0^2 + (1 - (n_e/n_0)^2)\sin^2\theta]\sin^2\phi = \sin^2\theta$$

We are told ne = 1.468 and hence:

$$\sin\phi = \frac{\sin\theta}{\sqrt{n_0^2 + (1 - (n_e/n_0)^2)\sin^2\theta}} \text{ and } \sin\varphi = \frac{0.7071}{\sqrt{2.155 + (1 - 0.7894) \times 0.5}}$$

Hence $\sin(\Theta) = 0.47$ and $\Theta_{extra} = 28.04°$

In summary: $\boldsymbol{\theta_{ord} = 25.24°}$ ***and*** $\boldsymbol{\theta_{extra} = 28.04°}$

8.3.4 'Walk-off' in Birefringent Crystals

We have alluded to a defining feature of the extraordinary ray, in that the electric field and electric displacement are not parallel. By contrast, for the ordinary ray, this strange behaviour does not occur. This gives a clue as to the extraordinary properties of the extraordinary ray. As we have previously stated, it is the electric displacement, **D**, rather than the electric field, **E**, that is perpendicular to the direction of wave propagation, as defined by the wavevector, **k**. The consequence of this is that the electric field is **not** necessarily perpendicular

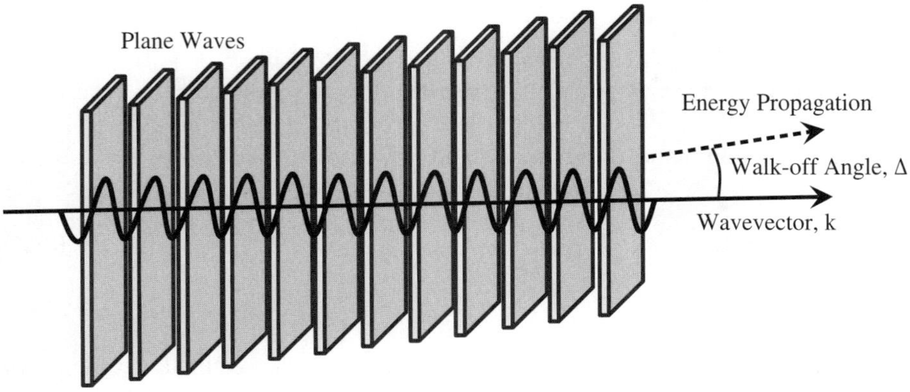

Figure 8.11 Phenomenon of walk-off in birefringent crystals.

to the wavevector. With this in mind, the fundamental property of extraordinary rays has a significant impact on the direction of energy propagation, as defined by the Poynting vector, **S** Eq. (8.12).

$$\mathbf{S} = \mathbf{E}x\mathbf{H}$$

The implication of this is that since **S** is perpendicular to **E** and **k** is perpendicular to **D**, then **S** and **k** are not co-parallel. So we now have the extraordinary proposition that the direction of wave propagation and the direction of energy propagation are not identical. This phenomenon is known as ***walk-off***. This is illustrated in Figure 8.11, which shows a depiction of a plane wave effectively 'walking off' at some angle, Δ, with respect to the normal to equiphase planes.

It is clear that the ***walk-off angle*** is equivalent to the angle between the electric and electric displacement vectors. In a uniaxial crystal, without loss of generality, we are only interested in the relationship between **D** and **E** along the x and z axes (x and y are equivalent). From Eq. (8.22) we get:

$$\mathbf{E}_x = \frac{\mathbf{D}_x}{\varepsilon_{ord}} \quad \mathbf{E}_z = \frac{\mathbf{D}_z}{\varepsilon_{extra}}$$

If the propagation direction is at an angle of ϕ with respect to the optical (z) axis, then the perpendicular **D** vector lies at an angle of ϕ with respect to the **x** axis. So we might re-cast the expression for E, in terms of the following vector notation:

$$\mathbf{E} = D\left[\frac{\cos\phi}{\varepsilon_{ord}}\mathbf{i} + \frac{\sin\phi}{\varepsilon_{extra}}\mathbf{k}\right]$$

If the walk off angle is Δ, then the electric field vector would be at an angle of $\phi + \Delta$ to the original electric displacement. Therefore it is clear to see:

$$\tan(\phi + \Delta) = \frac{\varepsilon_{ord}}{\varepsilon_{extra}}\tan\phi \text{ and (from Eq. 8.18)} \quad \tan(\phi + \Delta) = \frac{n_o^2}{n_e^2}\tan\phi \tag{8.27}$$

From Eq. (8.26) it is straightforward to calculate the walk-off angle, Δ:

$$\frac{\tan\phi + \tan\Delta}{1 - \tan\phi\tan\Delta} = \frac{n_o^2}{n_e^2}\tan\phi \Rightarrow \tan\Delta = \left[\frac{(n_o/n_e)^2 - 1}{1 + (n_o/n_e)^2\tan^2\phi}\right]\tan\phi \tag{8.28}$$

Examination of Eq. (8.28) demonstrates that the walk-off angle for $\phi = 0$ is zero, as is for the case where $\phi = 90°$, as tanϕ tends to infinity. In between these two extremes, the walk-off angle is clearly non-zero. Therefore, a question that might be posed by Eq. (8.28) is what is the maximum walk-off angle? If we now substitute the variable, x, for $(n_o/n_e)\tan(\phi)$ we get the following expression:

$$\tan\Delta = \left[\frac{(n_o/n_e)^2 - 1}{1 + x^2}\right]\left(\frac{n_e}{n_0}\right)x = \left[\frac{(n_o/n_e) - (n_e/n_o)}{1/x + x}\right]$$

The expression is maximised where denominator is minimised, and this clearly occurs where $x = 1$. Therefore:

$$(\tan\Delta)_{max} = \frac{n_o}{2n_e} - \frac{n_e}{2n_o} \text{ this occurs where } \tan\phi_{max} = \frac{n_e}{n_o} \tag{8.29}$$

Worked Example 8.5 – ***Walk off Angle***
In the previous worked example, we looked at a calcite crystal with randomly polarised light at an angle of incidence of 45°. We calculated that the refracted angle of the extraordinary ray was 28.04° If the ordinary index is 1.658 and the extraordinary index is 1.468 at the wavelength of interest, what is the walk-off angle? For this material and wavelength, what is the maximum walk-off angle and at what angle does the ray make with the optic axis for this scenario?

For an angle of 28.04°, $\tan\phi = 0.5326$ and from Eq. (8.27):

$$\tan\Delta = \left[\frac{(n_o/n_e)^2 - 1}{1 + (n_o/n_e)^2\tan^2\phi}\right]\tan\phi = \left[\frac{(1.1294)^2 - 1}{1 + (1.1294)^2 \times (0.5326)^2}\right] \times 0.5326$$

And $\tan\Delta = 0.1078$ ***Therefore*** $\boldsymbol{\Delta = 6.15^\circ}$

The maximum walk-off angle is given by:

$$(\tan\Delta)_{max} = \frac{1.1294}{2} - \frac{1}{2 \times 1.1294} \text{ and } \tan\Delta_{max} = 0.122 \quad \textit{Therefore } \boldsymbol{\Delta_{max} = 6.96^\circ}$$

This condition occurs where $\tan\phi = n_e/n_o = 1/1.1294$ ***Therefore*** $\boldsymbol{\phi_{max} = 41.52^\circ}$

8.3.5 Uniaxial Materials

In the common example we have considered, calcite was described as a uniaxial material with an ordinary index of 1.658 and an extraordinary index of 1.468. That is to say, the extraordinary index is less than the ordinary index. This is an example of a ***negative uniaxial crystal***. By contrast, there are uniaxial crystals where the extraordinary index is ***greater*** than the ordinary index. These materials are referred to as ***positive uniaxial crystals***. A list of some common uniaxial crystals is shows in Table 8.1.

Table 8.1 Some common uniaxial crystals.

Positive/Negative	Material	n_o	n_e
Positive	Quartz	1.544 296	1.553 398
	Tellurite	2.274 935	2.430 453
	Rutile	2.614 13	2.910 31
Negative	Calcite	1.658 643	1.486 498
	Barium borate	1.670 737	1.552 518
	Lithium niobate	2.300 158	2.214 539

8.3.6 Biaxial Crystals

Biaxial crystals lack the rotational axis symmetry of uniaxial crystals and, as such, all three principal refractive indices are different. That is to say, $n_x \neq n_y \neq n_z$. The detailed treatment of biaxial crystals is beyond the scope of this text. However, as for uniaxial crystals, for each propagation direction, there are two unique polarisation modes, as determined by the crystal symmetry. However, in contrast to uniaxial crystals, both modes are extraordinary. That is to say, for neither mode are the electric field and electric displacement aligned and both modes experience walk-off. The two modes are defined, as before, by the direction of the electric displacement vector, **D**. To calculate these modes, we refer to the index ellipsoid defined by Eq. (8.23). We first draw a line through the centre of the ellipsoid that defines the propagation direction and then a plane perpendicular to this line that itself also intersects the ellipsoid centre. This plane will intersect the ellipsoid, generating an ellipse. The semi-major and semi-minor axes then define the two modes of the biaxial crystal.

8.4 Polarisation Devices

8.4.1 Waveplates

Earlier, we discussed the propagation of a beam along the optic axis of the crystal. In this case, the two polarisation modes had a refractive index equal to the ordinary index, n_o and the extraordinary index, n_e. As previously outlined, the polarisation direction with the higher refractive index is referred to as the slow axis and the other axis is referred to as the fast axis. A ***waveplate*** is an optical component where the phase delay is deliberately engineered to be some specific value. Common waveplates are ***quarter waveplates*** or ***half waveplates***. For a quarter waveplate, the phase difference between the two axes is chosen to be a quarter wavelength or an integer number of wavelengths plus a quarter wavelength. Similarly, the half waveplate has a phase difference equal to one half a wavelength (plus integral number of wavelengths). So if Δn represents the difference in refractive index between the fast and slow axes and t is the thickness of the waveplate, then the following relationships apply:

$$\Delta n \times t = (i + 1/2)\lambda \; (Half\ Waveplate) \qquad \Delta n \times t = (i + 1/4)\lambda \; (Quarter\ Waveplate) \tag{8.30}$$

λ is the wavelength and i is a positive integer.

The most interesting application of a waveplate occurs when it is illuminated by linearly polarised light with the direction of polarisation bisecting the two axes, i.e. at 45°. This is shown in Figure 8.12 for a half waveplate.

The incoming polarisation at 45° may be thought of as a linear combination of the two 'fast' and 'slow' modes. On propagation through the half-wave plate, the slow axis experiences a phase delay of 180° with respect to the fast axis. That is to say, the phase of this component is reversed. Therefore, as indicated in Figure 8.12, the transmitted polarisation direction is at 90° to the original polarisation direction. This analysis assumes that

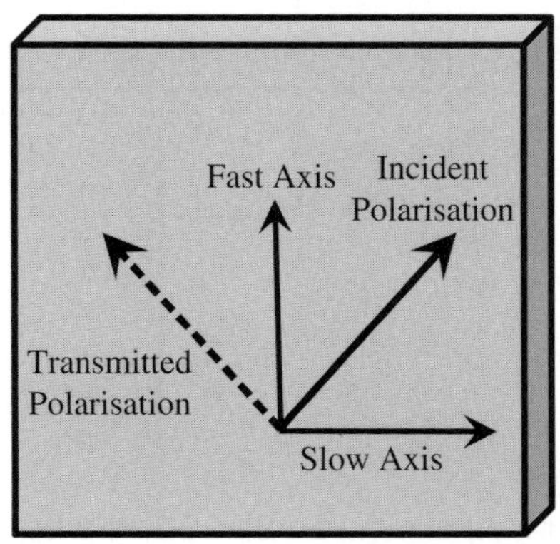

Figure 8.12 Operation of half-waveplate.

the phase relationship between the two modes is coherent; that is to say, the input polarisation is linear. For random polarisation, the waveplate will have no effect. In the case of a quarter waveplate, incident light that is linearly polarised at 45° to the main axes will be converted to circularly polarised light. Clearly, the effect of the quarter waveplate would be to produce fast and slow axis components which are out of phase by 90°. Therefore, one would expect the light to be circularly polarised.

The waveplate is a very useful component for manipulating polarised light. It can rotate the axis of polarisation and convert linear polarised light into circularly polarised light and vice versa. Most generally, for a waveplate of arbitrary thickness and for an arbitrary (coherent) input polarisation state, elliptically polarised light will be produced.

It is possible to formulate a more general expression for the polarisation state of light upon passage through a waveplate. If the complex amplitude of the slow axis component is denoted by A and that for the fast axis component by B, then the input polarisation state is described by the vector $A\mathbf{x} + B\mathbf{y}$. The vectors $\mathbf{x}$ and $\mathbf{y}$ are unit vectors in the direction of the slow and fast axes respectively. For a phase delay of ϕ between fast and slow axes, then the polarisation state of the transmitted beam is described by the following vector:

$$\mathbf{P} = A\mathbf{x} + Be^{i\phi}\mathbf{y} \tag{8.31}$$

Waveplates find widespread application, particularly in the laboratory environment. The transformation of linear polarisation into circular polarisation (and vice versa) is, in itself an extremely useful attribute of a quarter waveplate. Perhaps the most useful feature of a waveplate is its ability to rotate the axis of linear polarisation through 90°. This forms the basis of an optical set up where one can divert a normally reflected optical beam into a different path, thus apparently defeating the principle of reversibility. A common such scenario is illustrated in Figure 8.13.

In Figure 8.13, a linearly polarised beam is incident upon a polarising beamsplitter. The function of this beamsplitter is to separate the two orthogonal linear polarisations. That is to say, as shown, horizontally polarised light is transmitted straight through the beamsplitter, whereas vertically polarised light is diverted by 90°. Imposition of a quarter waveplate prior to the retroreflector transmutes the beam into vertically polarised light upon reflection. The effect of two passes through the quarter waveplate is equivalent to a single passage through a half waveplate. As a consequence, when this reflected beam meets the beamsplitter again, instead of being transmitted, it is reflected at 90°. There are many other configurations that exploit this useful property in waveplates, both for the combination and separation of optical beams.

8.4.2 Polarising Crystals

The separation of the two fundamental linear polarisation modes is a very important function performed by a variety of optical components that exploit the phenomenon of birefringence. Such devices can either use

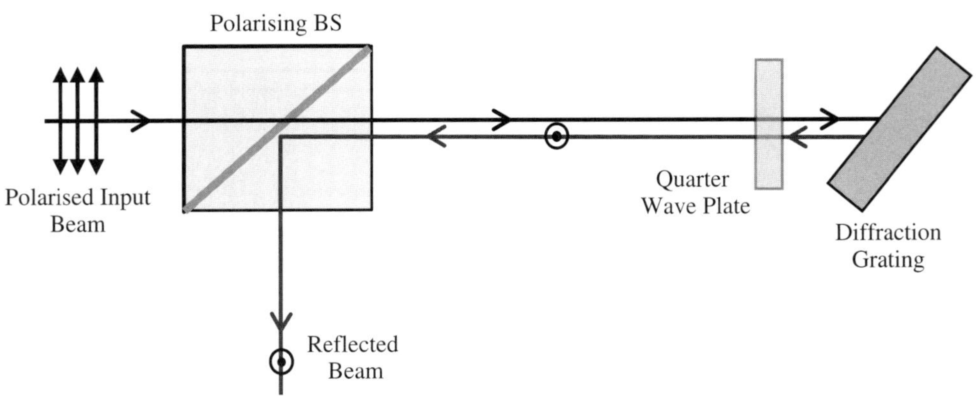

Figure 8.13 Common experimental configuration for quarter waveplate.

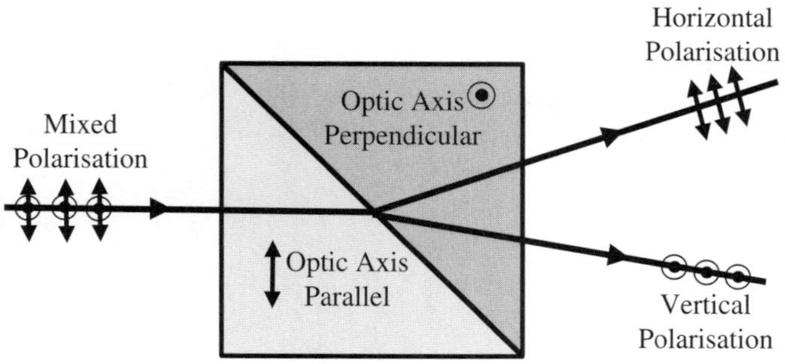

Figure 8.14 Woolaston prism.

the index difference to separate the components by differential refraction, or, alternatively separation can be performed by virtue of total internal reflection. In the latter case, one of the modes is totally internally reflected whilst the other, by virtue of the refractive index difference, is partially transmitted. One such arrangement is the so-called ***Wollaston prism***.

The Wollaston prism consists of a pair of right angle prisms made from a uniaxial crystal material and cemented together at their hypotenuse. Most particularly, the orientation of their optic axes is orthogonal. That is to say, the ordinary polarisation direction for one prism corresponds to the extraordinary in the other. As can be seen in Figure 8.14, on passing from the first prism to the second, for the horizontal polarisation, the relevant change is from the extraordinary to the ordinary index. For the vertical polarisation, the index contrast is reversed. Therefore, for the two polarisation directions, the refraction is equal and opposite, as shown, and the two polarisation components are unambiguously separated.

Another type of polarising prism is the ***Glan-Taylor polariser***. As for the Wollaston prism, this component consists of two separate pieces uniaxial crystal. However, in the case of the Glan Taylor polariser, a small air gap is maintained between the two crystals, and the orientation of the optic axis is the same for both pieces. Most specifically, the angle of incidence at the air interface is designed to be greater than the critical angle for one polarisation and less than the critical angle for the other. In this way, one polarisation orientation is entirely reflected whereas the other is partially transmitted. Of course, the reflected beam, whilst predominantly of one polarisation, by virtue of Fresnel reflection contains elements of the other polarisation. On the other hand, the transmitted beam is composed exclusively of one polarisation. This is illustrated in Figure 8.15.

In the example shown, the ordinary ray is vertically polarised and is totally internally reflected. Therefore, the ordinary ray must possess higher refractive index and the extraordinary ray the lower index. Therefore, the material, in this instance, is a negative birefringent material.

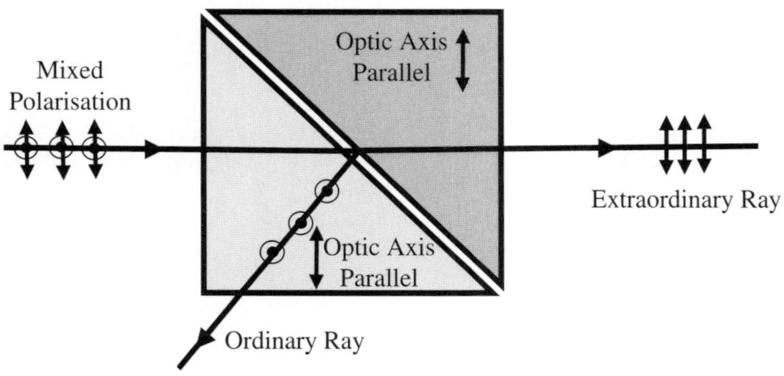

Figure 8.15 Glan taylor polariser.

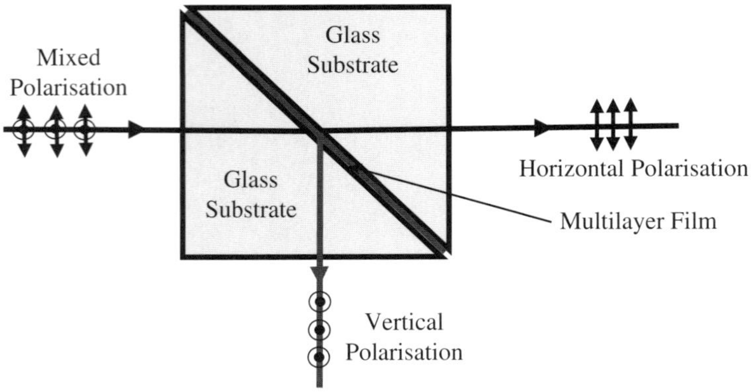

Figure 8.16 Polarising beamsplitter cube.

8.4.3 Polarising Beamsplitter

A polarising device not dependent upon birefringence is the multilayer ***polarising beamsplitter***. Such a device uses multiple layers on a glassy substrate to accentuate the tendency for angled reflection to be highly polarisation sensitive. As shown in Figure 8.5, Fresnel reflection at non-zero angles of incidence is highly dependent upon polarisation. It is possible, by careful design, involving the deposition of thin layers of material with contrasting refractive indices, to reflect almost 100% of one polarisation whilst exclusively transmitting the other polarisation. In most practical applications, it is convenient to divert one polarisation by 90°, so the most common configuration involves a 45° beamsplitter. A particularly common embodiment of the polarising beamsplitter is the ***beamsplitter cube***, where the reflecting layer is effectively sandwiched between two 45° prisms. This is shown in Figure 8.16.

8.4.4 Wire Grid Polariser

A commonly used polarising device, particularly for longer wavelengths is the ***Wire Grid Polariser***. This device consists of a series of very closely spaces parallel conductors, often deposited or patterned onto a glass substrate. Because of the directional geometry of this grid, the reflective properties of the metal coating are polarisation dependent. In effect, the refractive/reflective properties of a metal conductor depend upon the movement of charge carriers within the material. In the case of the wire-grid polariser movement of charge carriers is only possible along the lines of the conductors. Therefore, light that is polarised with the electric vector parallel to the wires is efficiently reflected whereas the orthogonal polarisation is transmitted. The principle of this is illustrated in Figure 8.17.

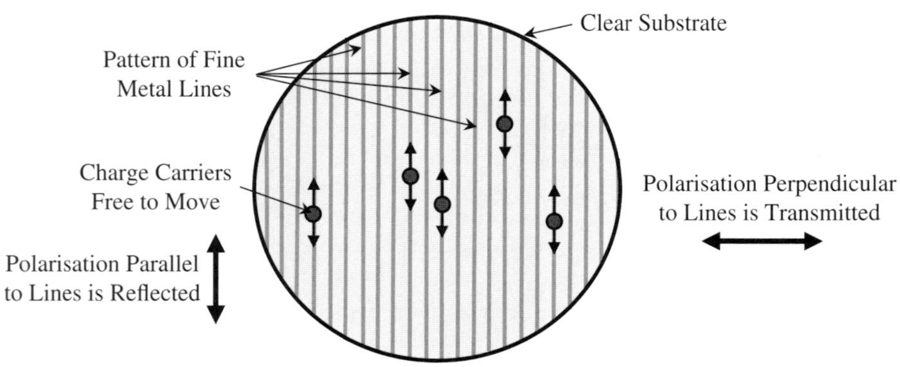

Figure 8.17 Wire grid polariser.

8.4.5 Dichroitic Materials

Some materials work by selective absorption of specific polarisation orientations, rather than by reflection. The well known ***Polaroid H type sheet*** is an example of a ***dichroitic*** material. It was first commercially developed by Edwin Land in 1929. The sheet consists of a substrate of stretched PVA (polyvinyl alcohol) polymer that has been doped with iodine molecules. Stretching the sheet helps to orient the polymer chains and further acts to align the iodine dopant. Thus aligned, the iodine molecules, in some respects, act like the oriented wires in the wire grid polariser and absorb light whose electric field is oriented parallel to this direction.

8.4.6 The Faraday Effect and Polarisation Rotation

Hitherto, all material properties impacting upon polarisation that we have considered have been related to their internal ***electric*** polarisability. As such, we have largely ignored the impact of the internal magnetic properties these materials. For the most part, this is justified. However, there is one specific optical phenomenon that is directly related to the exploitation of underlying magnetic properties. This is the so called ***Faraday Effect*** which is produced when a suitable material is subjected to an external magnetic field. The impact of the Faraday Effect is to produce a phase difference between the two circular components of electrical polarisation that is proportional to the magnetic field and the distance propagated. As a consequence of this, for linear polarised light, a rotation in the axis of linear polarisation is produced. The rotation angle, ϕ (in radians) is related to the applied magnetic field, **B** and the propagation length, l, by the ***Verdet constant***, V.

$$\phi = V\mathbf{B}l \tag{8.32}$$

Typical materials are Yttrium-Iron-Garnet (YIG), Terbium-Gallium-Garnet (TGG) and Terbium-Aluminium-Garnet (TbAlG). The Verdet constant of TGG is -40 $\text{radT}^{-1}\ \text{m}^{-1}$ (at 1064 nm).

One specific application of the Faraday Effect is in the ***optical isolator***. The purpose of an optical isolator is to shield sensitive optical devices, such as lasers, from the deleterious effects of unwanted back-reflections. Light incident upon the isolator is linearly polarised by a polarisation device and the axis of polarisation is subsequently rotated by 45° upon transmission through the isolator. Another linear polariser arranged at 45° transmits this light, as shown in Figures 8.18a and 8.18b. However, any back reflected light passing through this polariser has its axis of polarisation rotated through a further 45°, i.e. at 90° to the original polarisation direction. As a consequence, the back-reflected light is not transmitted by the original polariser.

8.5 Analysis of Polarisation Components

8.5.1 Jones Matrices

In Section 8.2.3 we introduced Jones vectors to describe the polarisation state of light in terms of the two independent components of polarisation. The impact of polarisation components described in the previous section is to transform the polarisation from one state to another. In the majority of practical applications this transformation is a linear process. Therefore, we are justified in using a set of linear equations to describe this transformation. In particular, we may describe such a linear transformation in a matrix representation. Such matrices are referred to as ***Jones Matrices***, as illustrated in Figure 8.19.

The transformation between the two vectors is simply given by the Jones matrix:

$$\begin{bmatrix} x_2 \\ y_2 \end{bmatrix} = \begin{bmatrix} M_{11} & M_{12} \\ M_{12} & M_{22} \end{bmatrix} \begin{bmatrix} x_1 \\ y_1 \end{bmatrix} \tag{8.33}$$

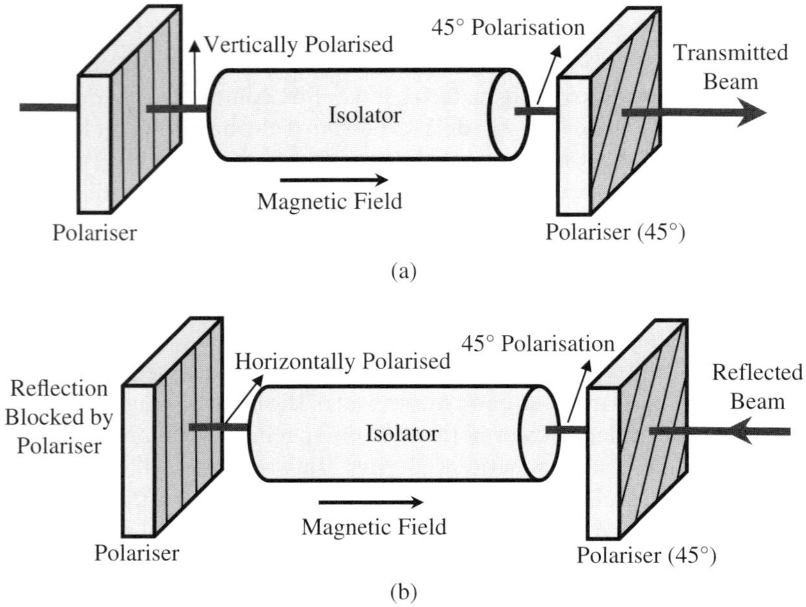

Figure 8.18 (a) Optical isolator in transmission. (b) Blocking by optical isolator in reflection.

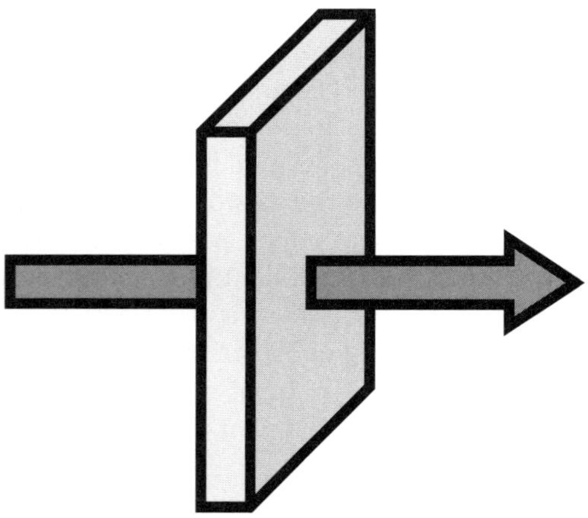

Figure 8.19 Application of jones matrices.

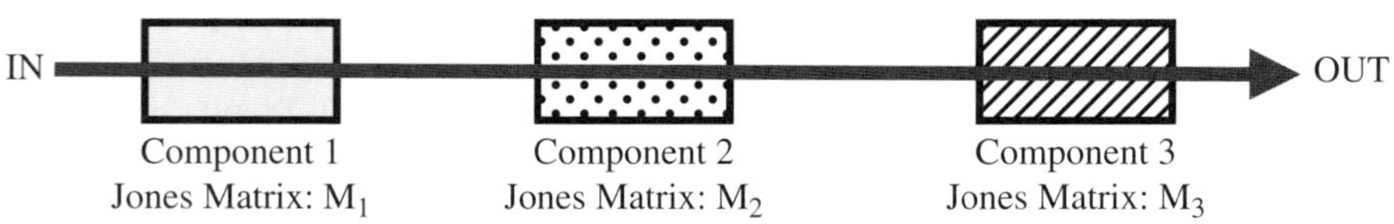

Figure 8.20 Jones matrix multiplication.

Jones Matrices are applied to a system in the same manner as ray tracing matrices are applied. That is to say, the system matrix may be represented by the sequential multiplication of the individual component matrices. This is illustrated in Figure 8.20.

In the example given in Figure 8.20, it is quite clear that the system matrix is given by:

$$M_{system} = M_3 \times M_2 \times M_1 \tag{8.34}$$

At this point, it is useful to set out the Jones Matrices for some common polarisation components. For example, a simple linear polariser in the x and y directions would be represented by the following matrices:

$$\begin{bmatrix} 1 & 0 \\ 0 & 0 \end{bmatrix}_{(linear\ polariser\ x)} \qquad \begin{bmatrix} 0 & 0 \\ 0 & 1 \end{bmatrix}_{(linear\ polariser\ y)} \tag{8.35}$$

More generally, for a polariser set at an angle of θ with respect to the x axis, the matrix is given by:

$$\begin{bmatrix} cos^2\theta & cos\theta\sin\theta \\ cos\theta\sin\theta & sin^2\theta \end{bmatrix} \tag{8.36}$$

It should be noted that Eq. (8.36) operates on the ***amplitude*** of the polarisation components. As such, the ***flux*** is insensitive to rotation through 180°. That is to say, if linear polarised light were incident on a polarising sheet whose axis of polarisation is θ with respect to the original polarisation direction, then the transmitted flux would be given by:

$$\Phi = \Phi_0 \cos^2\theta \tag{8.37}$$

Matrices for the quarter and half waveplates are given by:

$$\begin{bmatrix} 1 & 0 \\ 0 & i \end{bmatrix}_{(quarter\ wave\ plate)} \qquad \begin{bmatrix} 1 & 0 \\ 0 & -1 \end{bmatrix}_{(half\ wave\ plate)} \tag{8.38}$$

A more general expression for an arbitrary waveplate, whose phase delay is ϕ is given by:

$$\begin{bmatrix} 1 & 0 \\ 0 & e^{i\phi} \end{bmatrix} \tag{8.39}$$

Finally, rotation of co-ordinate frame or rotation of the direction of polarisation by θ is given by the following matrix:

$$\begin{bmatrix} \cos\theta & -\sin\theta \\ \sin\theta & \cos\theta \end{bmatrix} \tag{8.40}$$

In the classical deployment of the half-wave plate, the input linear polarisation is at 45° to the fast and slow axes of the waveplate. We may represent this scenario by an initial 45° rotation before invoking the half-wave plate, followed by a reverse rotation to the original co-ordinate frame. Thus we have three matrices describing this scenario:

$$\begin{bmatrix} 1/\sqrt{2} & 1/\sqrt{2} \\ -1/\sqrt{2} & 1/\sqrt{2} \end{bmatrix} \begin{bmatrix} 1 & 0 \\ 0 & -1 \end{bmatrix} \begin{bmatrix} 1/\sqrt{2} & -1/\sqrt{2} \\ 1/\sqrt{2} & 1/\sqrt{2} \end{bmatrix} = \begin{bmatrix} 0 & -1 \\ -1 & 0 \end{bmatrix} \tag{8.41}$$

It is clear from Eq. (8.41), if light is polarised along either the x or y (fast or slow) axes, then the direction of polarisation will be rotated by 90°.

Worked Example 8.6 ***Twisted Nematic Liquid Crystal***

A twisted nematic liquid crystal consists of a solution of a polar organic molecule sandwiched between two polished glass plates. The organic molecule may be regarded as behaving like a uniaxial crystal, with the pole of the molecule acting as the optic axis. This is illustrated in Figure 8.21.

The solution is sandwiched between two polished glass plates. Most importantly, the polishing process creates linear striations in the substrate which have a tendency to align the polar molecules to the direction of those striations. Furthermore, the polishing direction of the two glass plates is orthogonal. As a result, under normal circumstances, in progressing from one plate to the other, the polar molecules twist through 90°, as

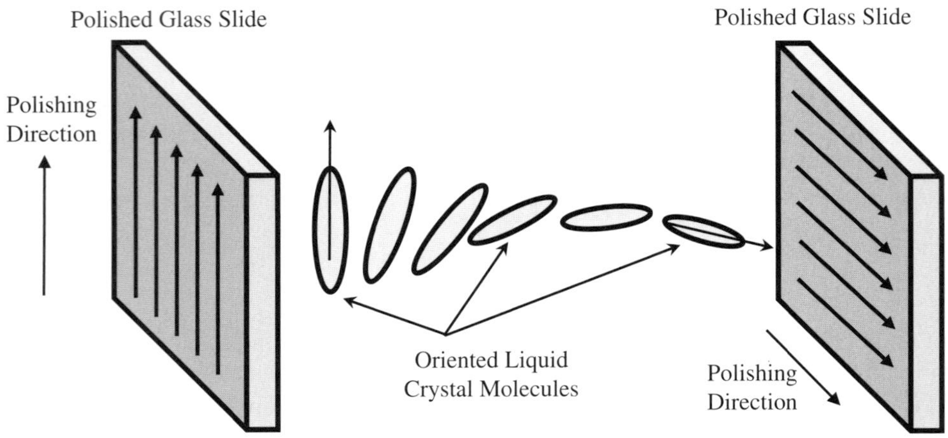

Figure 8.21 Twisted nematic liquid crystal.

shown in Figure 8.21. We can use Jones matrices to analyse this system, modelling each molecule as a very thin layer of birefringent crystal.

The effective liquid crystal indices are n_e and n_0. The uniaxial axis rotates uniformly on propagation through the crystal according to

$$\theta = \alpha z$$

The Jones Matrix for a single increment of depth, Δz, may be presented as:

$$M = \begin{bmatrix} \exp\left(\frac{-j\beta\Delta z}{2}\right) & 0 \\ 0 & \exp\left(\frac{j\beta\Delta z}{2}\right) \end{bmatrix} \begin{bmatrix} \cos(\alpha\Delta z) & \sin(\alpha\Delta z) \\ -\sin(\alpha\Delta z) & \cos(\alpha\Delta z) \end{bmatrix} \qquad \beta = 2\pi(n_e - n_0)/\lambda$$

We may assume that the twisting takes place on a scale that is much longer than that of the wavelength, λ, and therefore, $\alpha \ll \beta$. Therefore, for a series of N elements of thickness Δz, the matrix may be represented as:

$$M = \begin{bmatrix} \cos(N\alpha\Delta z) & -\sin(N\alpha\Delta z) \\ \sin(N\alpha\Delta z) & \cos(N\alpha\Delta z) \end{bmatrix} \begin{bmatrix} \exp\left(\frac{-j\beta\Delta z}{2}\right) & 0 \\ 0 & \exp\left(\frac{j\beta\Delta z}{2}\right) \end{bmatrix}^N$$

For $\Delta z \rightarrow 0$ this becomes

$$M = \begin{bmatrix} \cos(\alpha d) & -\sin(\alpha d) \\ \sin(\alpha d) & \cos(\alpha d) \end{bmatrix} \begin{bmatrix} \exp\left(\frac{-j\beta d}{2}\right) & 0 \\ 0 & \exp\left(\frac{j\beta d}{2}\right) \end{bmatrix} \tag{8.42}$$

Thus, from Eq. (8.42), the liquid crystal acts as a combination of polarisation rotator (by αd) and phase retarder. If the input polarisation is aligned to the x or y axis, the only impact is as a polarisation rotator. In practice, the liquid crystal device, as analysed, is sandwiched between two crossed polarisers. As the impact of the nematic liquid crystal is to rotate the polarisation through αd (in this case 90°), then the light will be transmitted. However, when an external electric field is applied in the z direction, all molecules orient themselves along the z axis and no rotation in the polarisation direction is produced. Therefore the second, crossed polariser will absorb the light; this describes the operation of a liquid crystal device.

8.5.2 Müller Matrices

In the previous subsection we looked at Jones Matrices which model optical components in the transformation of Jones vectors. By analogy, when we consider Stokes vectors, then these are transformed by 4 × 4 matrices, referred to as ***Müller Matrices.***

It has to be remembered, that unlike Jones Matrices, Müller Matrices operate on flux or irradiance, rather than amplitude. A straightforward example of a Müller Matrix is a linear polariser in the x direction. This is given by:

$$M = \frac{1}{2}\begin{bmatrix} 1 & 1 & 0 & 0 \\ 1 & 1 & 0 & 0 \\ 0 & 0 & 0 & 0 \\ 0 & 0 & 0 & 0 \end{bmatrix} \text{ Müller Matrix for polariser aligned to x} - \text{axis} \tag{8.43}$$

If the input light is linearly polarised in x, then the new value of S_0 (flux) will be unchanged at unity, as expected. If, however, the input light is randomly polarised, the flux is reduced to 0.5, by virtue of absorption of the other polarisation component. A more general expression for a polariser inclined at an angle of θ to the x axis is given below:

$$M = \frac{1}{2}\begin{bmatrix} 1 & \cos 2\theta & \sin 2\theta & 0 \\ \cos 2\theta & (1+\cos 4\theta)/2 & (\sin 4\theta)/2 & 0 \\ \sin 2\theta & (\sin 4\theta)/2 & (1-\cos 4\theta)/2 & 0 \\ 0 & 0 & 0 & 0 \end{bmatrix} \tag{8.44}$$

For example, a polariser that is polarised at 45° to the x axis is given by:

$$M = \frac{1}{2}\begin{bmatrix} 1 & 0 & 1 & 0 \\ 0 & 0 & 0 & 0 \\ 1 & 0 & 1 & 0 \\ 0 & 0 & 0 & 0 \end{bmatrix} \tag{8.45}$$

For simple rotation of the axis of linear polarisation or rotation of the co-ordinate system by θ, then the matrix is given by:

$$M = \begin{bmatrix} 1 & 0 & 0 & 0 \\ 0 & \cos 2\theta & -\sin 2\theta & 0 \\ 0 & \sin 2\theta & \cos 2\theta & 0 \\ 0 & 0 & 0 & 1 \end{bmatrix} \tag{8.46}$$

A quarter waveplate will turn linearly polarised light into circularly polarised light and the matrix for a quarter waveplate with the fast axis aligned to the x-axis is given by:

$$M = \begin{bmatrix} 1 & 0 & 0 & 0 \\ 0 & 1 & 0 & 0 \\ 0 & 0 & 0 & 1 \\ 0 & 0 & -1 & 0 \end{bmatrix} \tag{8.47}$$

A more general expression for a retarder, where Δ is the phase difference between the fast and slow axes, and θ is the angle of the fast axis with respect to the x-axis, is given by:

$$M = \begin{bmatrix} 1 & 0 & 0 & 0 \\ 0 & (a+b\cos 4\theta)/2 & b\sin 4\theta/2 & \sin\Delta\sin 2\theta \\ 0 & b\sin 4\theta/2 & (a-b\cos 4\theta)/2 & -\sin\Delta\cos 2\theta \\ 0 & -\sin\Delta\sin 2\theta & \sin\Delta\cos 2\theta & \cos\Delta \end{bmatrix} \quad (8.48)$$

Where $a = 1 + \cos\Delta$; $b = 1 - \cos\Delta$

As for the Jones Matrices, a system may be built by multiplying the Müller Matrices for individual components, as in Eq. (8.34). As such, the simple matrices listed here provide the building blocks for more complex systems and should be possible to represent any system using a combination of these matrices.

8.6 Stress-induced Birefringence

In our examination of birefringent materials, we restricted our attention to crystalline materials. It is clear that these materials are highly anisotropic with preferred axial alignment. However, even in nominally isotropic materials, such as glasses and plastics, isotropic symmetry may be destroyed by the imposition of external stresses. By analogy with uniaxial materials, a small uniaxial stress condition may be imposed on a glassy material which, in turn, produces a uniaxial strain in the material. As a consequence, these materials display a very small degree of birefringence, with the effective optic axis aligned to the external stress. The degree of birefringence produced is, of course, very small, being of a similar order of magnitude to the strains produced, e.g. tens or hundreds of parts per million. This phenomenon is referred to as ***stress induced birefringence***.

The degree of birefringence produced is described by a linear parameter referred to as the stress optic coefficient. This is a fundamental material property and directly relates the index change to the applied stress. In a manner similar to the description of refractive index, stress can always be described in terms of three principal axes. If we describe the two relevant principal stresses as σ_1 and σ_2, then the refractive index difference, Δn, produced is equal to:

$$\Delta n = C(\sigma_1 - \sigma_2) \quad (8.49)$$

C is the stress optic coefficient.

If the stress is entirely uniaxial, then σ_2 is necessarily zero. Of course, Δn represents the difference in refractive index between the two polarisations, 'ordinary' and 'extraordinary'. For a sample of thickness, t, and for a wavelength of λ, then the phase difference, $\Delta\phi$, (in radians) produced between the two polarisations is:

$$\Delta\phi = \frac{2\pi C(\sigma_1 - \sigma_2)t}{\lambda} \quad (8.50)$$

The effect of stress induced birefringence is also described as the ***photoelastic*** or ***photoacoustic effect*** and the stress optic coefficient is also described as the photoelastic coefficient. Typical values for glass are 2.5×10^{-12} m^2 N^{-1}. Values for polymers are typically higher, of the order of 10^{-11} m^2 N^{-1}. Broadly, these figures are of the order of the inverse of the Young's modulus of the material in question.

Stress induced birefringence may be viewed by placing a material under stress between crossed polarisers. Plastic materials, in particular, display large stress birefringence. The impact of the stress induced birefringence is to produce phase retardance (as in wave plates) in the material. This is the consequence of Eq. (8.50) and the magnitude of the phase difference depends on the stress and its orientation. Of course, under certain conditions, a partial or a 90° rotation in the polarisation axis is effected which in turn produces a variation in the transmission which is dependent upon the local material stress. This is illustrated in Figure 8.22.

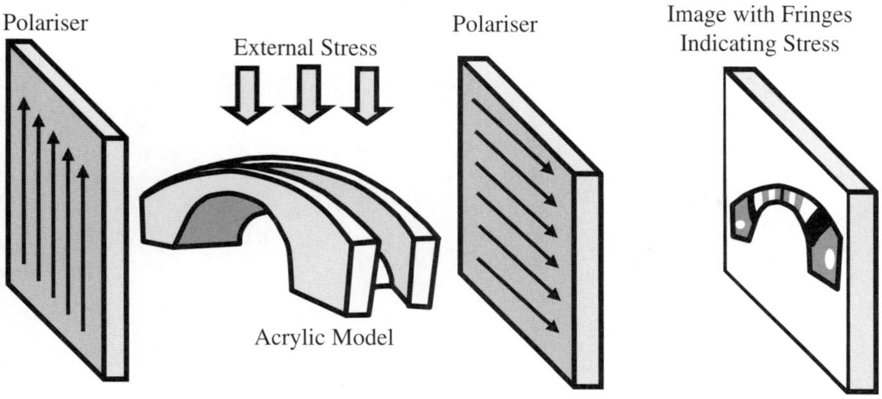

Figure 8.22 Use of stress induced birefringence to analyse patterns of stress.

Before the advent of sophisticated computer modelling, arrangements, such as depicted in Figure 8.22, were used to analyse mechanical stresses in complex structures such as bridges and buildings. Plastic models would have been made of the structure in question and subjected to a representative load and then viewed between crossed polarisers as in Figure 8.22. Certain stress values and orientations would produce sufficient phase delay to rotate the axis of polarisation by 90°. As a result, a series of light and dark fringes are produced which map the local stress, clearly highlighting areas of stress concentration.

Small amounts of internal stress are 'frozen in' to any glassy material as it cools and solidifies during the manufacturing process. As a result, all amorphous materials, such as glasses, have a certain amount of natural birefringence. In the vast majority of cases, this is an undesirable property, especially in precision applications, such as interferometry or ***polarimetry*** (the measurement of polarisation). Therefore, in the manufacture of glasses, great care is taken to reduce internal stresses to an absolute minimum, by very slow and controlled cooling of molten glass and subsequent annealing processes. Birefringence in optical glasses is specified by the maximum optical path difference per unit thickness produced by relative phase retardance. Figures are usually presented in nanometres per centimetre. A stress induced birefringence of less than two nanometres per centimetre represents exceptionally high quality material.

Further Reading

Born, M. and Wolf, E. (1999). *Principles of Optics*, 7e. Cambridge: Cambridge University Press. ISBN: 0-521-642221.

Huard, S. (1997). *Polarisation of Light*. New York: Wiley. ISBN: 0-471-96536-7.

Saleh, B.E.A. and Teich, M.C. (2007). *Fundamentals of Photonics*, 2e. New York: Wiley. ISBN: 978-0-471-35832-9.

Wolf, E. (2007). *Introduction to the Theory of Coherence and Polarisation of Light*. Cambridge: Cambridge University Press. ISBN: 978-0-521-82211-4.

9

Optical Materials

9.1 Introduction

In this chapter, we will be looking in more detail at the wide range of materials used in optical applications. Whilst, hitherto, we have focused almost exclusively on the optical properties of glasses and other materials, an optical designer must also be concerned with a host of other relevant properties. Many of these are mechanical properties, such as fracture toughness, stiffness, and density. In addition, we are also interested in thermal properties, such as the expansion coefficient, thermal conductivity, and temperature dependence of the refractive index. Other important properties are the relative ease of polishing (important for an optical glass!) and resistance to chemical attack.

Whilst the determining optical property of a material is its refractive index, we are also interested in other properties that impact its performance. For example, a glassy material may be non-uniform, containing ***bubbles***, ***inclusions***, and streaks of index non-uniformity referred to as ***striae***. As discussed in the previous chapter, glass materials may also exhibit stress induced birefringence. Although the bulk properties of an optical material are of exceptional importance, the presence of surface defects is also significant. Polishing and grinding processes inevitably create some damage to optical surfaces, leaving behind ***scratches***, and ***pits*** or ***digs***.

Optical materials may be broadly divided into three material categories, namely glassy or amorphous materials, crystalline materials, and metals. Although glasses, in general, are brittle, they are also hard and easy to work. Most significantly, glasses are exceptionally dimensionally stable over time and during processing, especially if the glass has most of its internal stresses removed by annealing. This is of critical importance in many optical applications where precise geometrical replication is essential. By contrast, internal stresses in relatively ductile metals may lead to deformation by creep over time, leading to dimensional change. The same difficulty pertains to the application of plastics and resins in optics. Plastic materials, in general, are glassy and amorphous and the transparency and ease of replication of optical polymers favours their use in a range of low cost, consumer applications. However, they cannot be used in precision applications where dimensional stability is critical.

Conventional glass formulation is based on amorphous silica (SiO_2) to which a range of alkali oxides (e.g. Ca, Na) or other oxides (B, Pb) have been added. Pure silica has a very high softening temperature and is difficult to work; the addition of these other oxides substantially reduces the softening temperature, making processing easier. Addition of specific metal oxides, such as lanthanum, barium, and lead, in a variety of admixtures, enables the formulation of a wide range of glass types. However, the use of lead in glass formulations is being increasingly phased out for environmental reasons.

Crystalline materials are generally more fragile and more difficult to work in comparison to glasses. They are, however, particularly useful because they transmit in regions of the spectrum (mid infrared and deep ultraviolet) where glasses are opaque. Metals are largely used on account of their high reflectivity. They are, of course, tough, but they are not as easy to polish as glasses or as mechanically stable. Therefore, most generally, metals

Optical Engineering Science, First Edition. Stephen Rolt.

Companion website: www.wiley.com/go/Rolt/opt-eng-sci

tend to be applied as thin (500 nm) coatings onto glass substrates. As such, the highly desirable reflective properties of the metal may be combined with the dimensional stability of the glass substrate.

In the case of glasses and crystalline materials, we are chiefly concerned about their transmissive properties. Conversely, for metals, it is their reflectivity that concerns us. For most glasses, their useful transmission range extends from the ultraviolet, from 300–350 to 2000–3000 nm in the mid infrared. Therefore, for wavelengths outside this region, more exotic, often crystalline, materials must be used. These include fused silica and calcium fluoride for deep or vacuum ultraviolet applications and silicon and germanium for the deep infrared. Metals, on account of their high electrical conductivity, are generally very good reflectors in the infrared and beyond. However, absorption bands in the visible and ultraviolet tend to degrade their reflectivity at shorter wavelengths. Gold exhibits progressively stronger absorption below 600 nm, hence its distinctive colour. On the other hand, aluminium has good reflectivity that extends significantly into the ultraviolet.

As outlined, it is the absorptive properties of optical materials that principally limits their useful spectral range. Before considering this in a little more detail, we will examine their refractive properties.

9.2 Refractive Properties of Optical Materials

9.2.1 Transmissive Materials

9.2.1.1 Modelling Dispersion

In Chapter 4, we considered the phenomenon of chromatic aberration in glassy materials, as caused by their wavelength-dependent refractive index. We now wish to examine this behaviour in a little more detail. In Chapter 8, we sought to explain the polarisation effects in crystalline materials in terms of locally induced electric dipoles. That is to say, the electric field associated with incident light creates, at the atomic level, small electric dipoles in the material which contributes to the local electric field. In a rather simplified, classical model, we might consider these nascent dipoles as discrete charges of opposite polarity, which are somehow attached to each other by a 'spring'. Any external electric field will produce a charge separation, and hence dipole moment, that is proportional to the applied field. The applied field is, of course, an oscillating field, with the frequency, ω, being the optical frequency. In this model, we might understand the behaviour of the 'dipole springs' in terms of forced oscillation. In other words, the frequency behaviour of the induced dipoles is understood in terms of a resonant system with a specific resonant frequency. To complete the model we might also consider the introduction of a 'damping term' which adds a damping force proportional to the instantaneous velocity. The scenario is illustrated in Figure 9.1.

In this simple forced oscillation model, the induced dipole moment, p, as a function of the optical frequency, ω, is simply given by:

$$p = \frac{AE_0}{\sqrt{(\omega_0^2 - \omega^2)^2 + \omega^2\omega_d^2}} \tag{9.1}$$

E_0 is the electric field amplitude; ω_0 is the resonance frequency; ω_d is damping frequency; A is a constant.

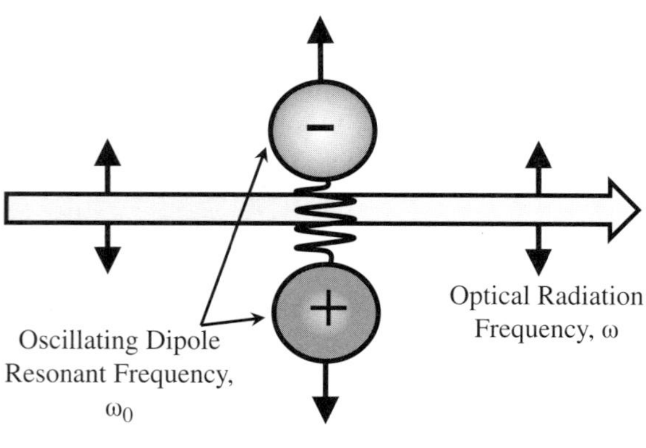

Figure 9.1 Resonant dipole.

In Chapter 8, we learned that the dipole moment per unit volume, or polarisation P, is directly related to the dielectric permittivity of the material and thence to the refractive index. Equation (9.1) describes the behaviour of an individual dipole. However, we would expect the frequency dependence of the electric susceptibility, χ, to follow the same pattern. From Eq. (8.18), we can express the electric susceptibility in terms of the refractive index:

$$\chi = n^2 - 1 = \frac{A'}{\sqrt{(\omega_0^2 - \omega^2)^2 + \omega^2\omega_d^2}} \tag{9.2}$$

Equation (9.2) provides a basic indication of how the refractive index of an optical material varies with frequency. As such Eq. (9.2) represents a starting point for the modelling of refractive index dispersion. In practice, an optical material is a little more complex then this basic description. By analogy to mechanical systems, one might envisage an optical material having multiple resonances. Therefore, a more general description of dispersive behaviour is given by:

$$n^2 - 1 = \frac{A_1}{\sqrt{(\omega_1^2 - \omega^2)^2 + \omega^2\omega_{d1}^2}} + \frac{A_2}{\sqrt{(\omega_2^2 - \omega^2)^2 + \omega^2\omega_{d2}^2}} + \frac{A_3}{\sqrt{(\omega_3^2 - \omega^2)^2 + \omega^2\omega_{d3}^2}} + \ldots\ldots \tag{9.3}$$

In practice, the damping frequency is considerably less than the relevant resonance frequencies and also optical transmission occurs only at frequencies some distance from the resonance frequency. Therefore, the following approximation may be made, ignoring the damping term:

$$n^2 - 1 = \frac{A_1}{\omega_1^2 - \omega^2} + \frac{A_2}{\omega_2^2 - \omega^2} + \frac{A_3}{\omega_3^2 - \omega^2} + \ldots\ldots \tag{9.4}$$

It is more usual to cast Eq. (9.4) in terms of wavelength rather than frequency. By basic manipulation of Eq. (9.4), we get:

$$n^2 - 1 = \frac{A_1\lambda^2}{\lambda^2 - \lambda_1^2} + \frac{A_2\lambda^2}{\lambda^2 - \lambda_2^2} + \frac{A_3\lambda^2}{\lambda^2 - \lambda_3^2} + \ldots\ldots \text{ or } n^2 = 1 + \sum_{i=1}^{i=N} \frac{A_i\lambda^2}{\lambda^2 - \lambda_i^2} \tag{9.5}$$

Equation (9.5) is referred to as the Sellmeier equation and is the most widely used expression for the modelling of dispersion in glasses. By way of example, Table 9.1 sets out the coefficients for the three Sellmeier terms used for modelling the common SCHOTT BK7® glass:

It is clear from Table 9.1 that the first two terms are associated with resonances in the vacuum ultraviolet and the third term is in the mid infrared, at about 10 μm. These are, of course, regions in which the material is strongly absorbing. Resonance features are naturally associated with extremely efficient interaction between the material and any incident electromagnetic field. When taken together with a finite damping coefficient, it is inevitable that the material is strongly absorbing in those regions.

The Sellmeier model is generally very accurate, with a precision of a few parts in 10^6. Figure 9.2 illustrates the application of the Sellmeier formula over an artificially wide spectral range from 10 to 20 000 nm. The point of this illustration is to show the operation of the resonance features. Moving from short wavelength to long wavelength, the impact of each resonance is to add an incremental amount to the refractive index. For

Table 9.1 Sellmeier coefficients for SCHOTT BK7®.

Coefficient	Value	Coefficient	Value (nm)
A_1	1.039 612 12	λ_1	77.464 177
A_2	0.231 792 344	λ_2	141.484 68
A_3	1.010 469 45	λ_3	10 176.475

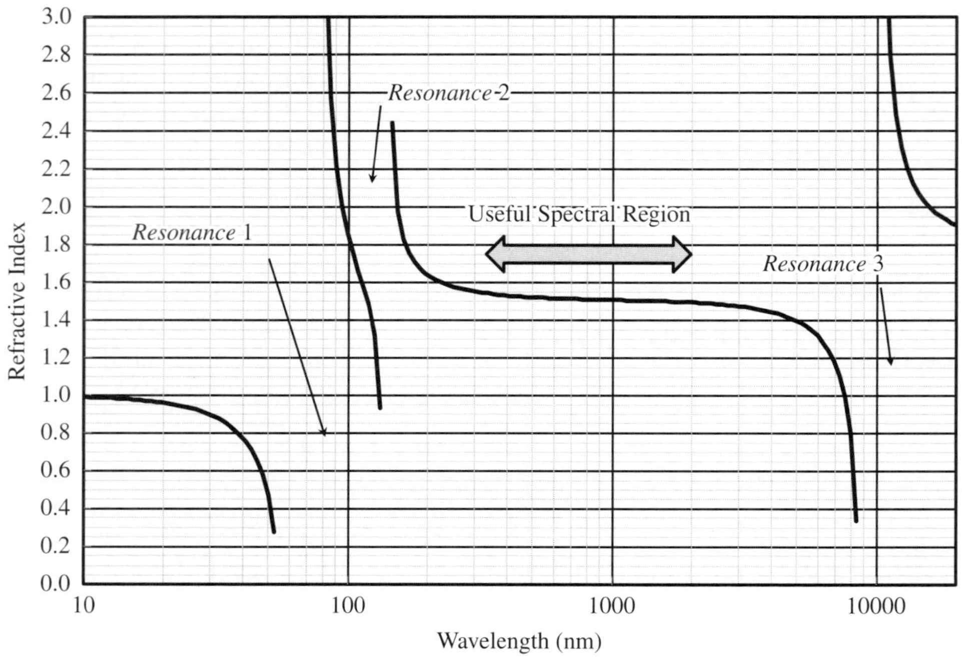

Figure 9.2 Modelled index of SCHOTT BK7®.

each resonance, in the limit of long wavelength, a quantity, A_i is added to the square of the refractive index. Thus we have the following limiting behaviours:

$$n \rightarrow 1 \text{ as } \lambda \rightarrow 0; n \rightarrow \sqrt{1 + \sum_{i=1}^{i=N} A_i} \text{ as } \lambda \rightarrow \infty \tag{9.6}$$

Of course, in the limit of long wavelength, the square of the refractive index should tend to the DC relative permittivity of the material. One example of this is water, where its DC relative permittivity is 81, yet its refractive index at visible wavelengths is only 1.33. That is to say, the effect of numerous intervening resonances is sufficient to lift the effective dielectric coefficient of water from about 1.7 in the visible spectrum to its DC value of 81.

In our example, illustrated in Figure 9.2, the useful spectral range occupies a small portion. This useful range is situated between two resonances and the refractive index is actually in decline across this entire region. This reduction in refractive index with wavelength is referred to as ***normal dispersion*** and is characteristic of most materials in practical applications. As can be seen from Figure 9.2, there are regions close to resonance features where the refractive index has a tendency to increase with wavelength. This behaviour is referred to as anomalous dispersion.

Worked Example 9.1 *Abbe Number of SCHOTT BK7*

We can now apply this knowledge to a practical example – calculating the Abbe number of SCHOTT BK7. Of course most optical design tools, software programs, etc. use formulae, such as the Sellmeier equation in their computation of relevant material indices.

From Eq. (4.48) $V_D = \frac{n_D - 1}{n_F - n_C}$ *(D is 589.3 nm; F is 486.1 nm; C is 656.3 nm)*

From Eq. (9.5) we have:

$n_D = 1.516\,728$; $n_F = 1.522\,379$; $n_C = 1.514\,321$

Hence

$$V_D = \frac{1.516728 - 1}{1.522379 - 1.514321} = 64.13$$

Hence, the Abbe Number for BK7 is 64.13.

The Sellmeier formula is the most prevalent expression for the modelling of dispersion. However, a number of other formulae do exist which tend to trade simplicity of form for accuracy. These are:

$$\textbf{Cauchy Formula}: n(\lambda) = A + \frac{10^4 B}{\lambda^2} + \frac{10^9 C}{\lambda^4} \quad (\lambda\ in\ nm) \tag{9.7}$$

$$\textbf{Briot Formula}: n^2(\lambda) = A_0 + 10^{-2}A_1\lambda^2 + \frac{10^{-2}A_2}{\lambda^2} + \frac{10^{-4}A_3}{\lambda^4} + \frac{10^{-6}A_4}{\lambda^6} + \frac{10^{-7}A_5}{\lambda^8} \quad (\lambda\ in\ nm) \tag{9.8}$$

$$\textbf{Hartman Formula}: n(\lambda) = A + \frac{C}{\lambda - B} \quad (\lambda\ in\ nm) \tag{9.9}$$

$$\textbf{Conrady Formula}: n(\lambda) = A + \frac{10^2 B}{\lambda} + \frac{10^9 C}{\lambda^{3.5}} \quad (\lambda\ in\ nm) \tag{9.10}$$

9.2.1.2 Temperature Dependence of Refractive Index

Since it is well known that most materials expand with increasing temperature, it is reasonable to expect that, as a consequence, the volume polarizability of such materials will also decline. As such, one might conclude that, in general, the refractive index of most materials would tend to decline with temperature. However, the picture is more complex than this. Of course, it is true that thermal expansion, by itself, has a tendency to reduce the refractive index. However, the effect of temperature changes not only impacts material density, but also alters the position and strength of the resonance features that underlie the modelling of dispersion. As a consequence, some materials exhibit a positive dependence of refractive index with temperature, whereas as others exhibit a negative dependence.

As with modelling the refractive index dispersion, modelling of temperature dependence of refractive index is based on characterising the variation of the electric susceptibility with temperature. A formula for modelling glass has been developed by SCHOTT and is reproduced here:

$$\Delta n = \frac{n^2 - 1}{2n}\left[D_0\Delta T + D_1\Delta T^2 + D_2\Delta T^3 + \frac{E_0\Delta T + E_1\Delta T^2}{\lambda^2 - \lambda^2_{thermal}}\right] \tag{9.11}$$

Δn is the change in index; ΔT is the change in temperature; D_0, D_1, D_2, E_0, E_1 are constants.

The formula is largely empirical, but the bracketed expression represents the proportional change in electric susceptibility, χ.

$$\frac{\Delta\chi}{\chi} = \left[D_0\Delta T + D_1\Delta T^2 + D_2\Delta T^3 + \frac{E_0\Delta T + E_1\Delta T^2}{\lambda^2 - \lambda^2_{thermal}}\right]$$

Often, we are interested in the differential of the refractive index with respect to temperature. From Eq. (9.11), we get:

$$\beta = \frac{dn}{dT} = \frac{n^2 - 1}{2n}\left[D_0 + 2D_1\Delta T + 3D_2\Delta T^2 + \frac{E_0 + 2E_1\Delta T}{\lambda^2 - \lambda^2_{thermal}}\right] \tag{9.12}$$

The above expression gives the ***temperature coefficient of refractive index,*** β. In this case, ΔT represents the difference in temperature with respect to some nominal temperature at which the glass has been characterised. It might represent the temperature at which the Sellmeier formula applied. The need to incorporate non-linear terms in Eq. (9.11) suggests that there is a significant temperature dependence in the differential of index with respect to temperature. Nevertheless, is it useful to set out here the temperature coefficient

Table 9.2 Temperature coefficient of refractive index for some common glasses.

Glass	Index	Tempco (ppm)	Glass	Index	Tempco (ppm)
N-PK51	1.53019	−6.70	N-LAK12	1.68083	−0.40
N-FK58	1.45600	−6.20	N-BK7	1.51872	3.00
N-PK52A	1.49845	−6.40	N-LAF2	1.74791	1.00
N-FK51A	1.48794	−5.70	F2	1.62408	4.40
N-FK5	1.48914	−1.00	SF57	1.85504	12.50
N-PSK53A	1.62033	−2.40			

for some common glasses for the temperature range 20–40 °C (546.1 nm). It is important to note that the temperature coefficient listed in Table 9.2 is the ***relative coefficient,*** that is to say the coefficient of the glass index relative to that of air (i.e. not vacuum).

The most significant impact of the change in refractive index in temperature is the shift in the paraxial focal positions, leading to defocus as the temperature changes. Where a high performance optical system is to be designed to fulfil exacting requirements over a wide temperature range, an ***athermal*** design is preferred. An athermal design is achieved where there is negligible first order change in key paraxial parameters. At first sight, an athermal design might be achieved by employing a glass with a very low coefficient, such as N-LAK12 from Table 9.2. However, this is not the whole picture. Most glasses expand to some significant degree with increasing temperature. As a consequence of this, critical mechanical dimensions, such as lens radii etc. increase with temperature, leading to a reduction in focusing power. The thermal expansion coefficient, α, expresses the proportional length increase with temperature and hence the reduction in focal power. To describe the overall impact of temperature change on focusing power, we introduce a parameter, Γ, the ***optical power coefficient*** of the material. If we express the change in focusing power as $-\Delta f/f$, then we have:

$$\Gamma = -\frac{\Delta f}{f} = \frac{\beta}{n-1} - \alpha \tag{9.13}$$

For N-LAK12, α is 7.6 ppm K^{-1} and the overall optical power coefficient is −8.2 ppm K^{-1}. However, in the case of SCHOTT BK7, the expansion coefficient is 7.1 ppm and the optical power coefficient is −2.7 ppm K^{-1}. Thus, in the case of BK7, the effect of the thermal expansion coefficient is to balance out some of the refractive index change. Although this analysis gives a clear picture of the behaviour of the material, as far as the system is concerned it omits the behaviour of the substrate or optical bench upon which the lens is mounted. Just as thermal expansion of the lens material shifts the focal plane of a system, the physical location of that focal plane moves with the thermal expansion of the 'optical bench'. This is illustrated in Figure 9.3.

The ***effective optical power coefficient***, Γ', is modified by the thermal expansion coefficient of the substrate, α_{sub}. This is now given by:

$$\Gamma' = -\frac{\Delta f}{f} = \frac{\beta}{n-1} - \alpha_{lens} + \alpha_{sub} \tag{9.14}$$

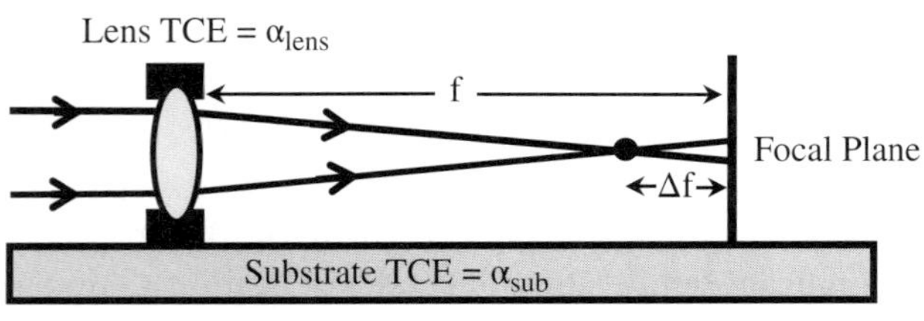

Figure 9.3 Thermal sensitivity and effect of substrate.

If the optical bench were to comprise aluminium with a coefficient of thermal expansion (CTE) of about 23 ppm at ambient temperatures, then, from Eq. (9.14), the change in effective focal power for a SCHOTT BK7 lens amounts about +20 ppm K^{-1}. In other words, in this instance, it is the thermal property of the substrate material that is dominant in determining thermal behaviour. So, if we now re-examine the glass N-LAK12, with an optical power coefficient of −8.2 ppm, we can create an athermal design by employing a substrate with a CTE of 8.2 ppm In practice, ferritic stainless steel has a CTE of about 10 ppm, giving an effective optical power coefficient of 1.8 ppm Thus, the design is substantially athermal, especially when compared with a combination of BK7 and aluminium.

9.2.1.3 Temperature Coefficient of Refraction for Air

Hitherto, we have largely ignored the impact of air as a refractive medium. In practice, for the majority of real applications, optical components are immersed in air and its presence does have some effect on refraction at surface interfaces. There are, of course, exceptions in scientific applications, where optical systems or components are placed in a vacuum environment. Under normal circumstances, the impact of atmospheric environment is negligible. The refractive index of air is very close to one, the departure being of the order of 270 ppm If we wish to take account of the refractive index of air, by subsuming it into an effective index, n_{eff}, for a material, then this effective index is simply the ratio of the material index and that of air:

$$n_{eff} = \frac{n}{n_{air}} \tag{9.15}$$

n is the material index

Temperature variations in the index of air are relatively straightforward to measure. The relative permittivity and hence, (n−1), is proportional to the density of the air. In turn, the density of air, assuming it behaves as an ideal gas, is proportional to the pressure and inversely proportional to the absolute temperature:

$$(n - 1) = (n_0 - 1)\frac{P}{T}\frac{T_0}{P_0} \tag{9.16}$$

P and T are the pressure and absolute temperature. P_0 and T_0 are the reference temperature and pressure, 101 325 Pa and 388.15 K respectively. n_0 is the index of air under the reference conditions.

It is straightforward to calculate the temperature coefficient for refractive index of the air, β_{air}:

$$\beta_{air} = -(n_0 - 1)\frac{P}{T^2}\frac{T_0}{P_0} \tag{9.17}$$

Under ambient conditions, β_{air} is approximately −1 ppm Returning to the problem of the relative temperature coefficient of a material, we revisit Eq. (9.15). If we denote the absolute temperature coefficient of the material as β and the relative value as β' we can see that:

$$\beta' = \beta - n\beta_{air} \tag{9.18}$$

n is the material index

Hence, under ambient conditions, the presence of air adds about 1.5 ppm to the relative temperature coefficient.

The refractive index of air follows a Sellmeier type dependence versus wavelength. Equation (9.19) sets out a simple formula for n_0 for dry air.

$$n_0 - 1 = A_0 + \frac{A_1\lambda^2}{\lambda^2 - \lambda_1^2} + \frac{A_2\lambda^2}{\lambda^2 - \lambda_2^2} \tag{9.19}$$

$A_0 = 83.4078\,ppm$, $A_1 = 184.9723\,ppm$, $A_2 = 4.1116\,ppm$; $\lambda_1 = 87.71\,nm$, $\lambda_2 = 160.33\,nm$

It must be emphasised that Eq. (9.19) strictly applies to dry air and without CO_2. The presence of water vapour and carbon dioxide does modify the index slightly. This is, of course, complicated by the fact that water vapour is a significant and highly variable constituent of the Earth's atmosphere. The refractive index of air is sketched out in Figure 9.4.

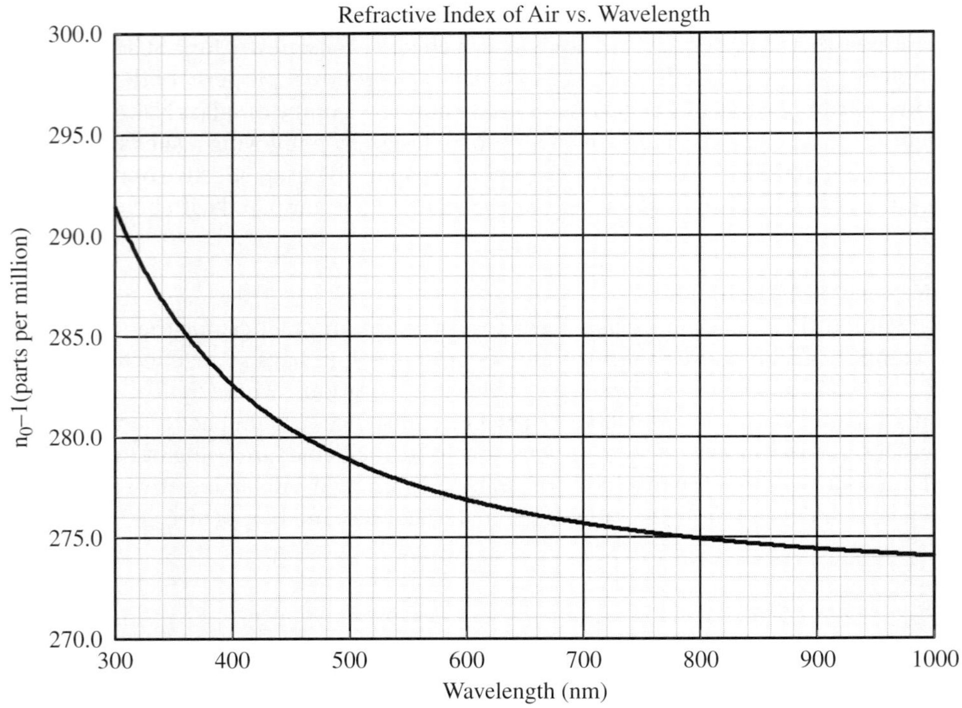

Figure 9.4 Refractive index of air.

9.2.2 Behaviour of Reflective Materials

Our treatment of specular surfaces, such as plane or spherical mirrors, has largely taken for granted that there are materials that can efficiently reflect light. We are, of course, excluding total internal reflection from this discussion, which is essentially a refractive phenomenon. We are also excluding low level reflection arising from Fresnel reflection in purely dielectric materials. Aside from these considerations, efficient reflection of light is a property generally associated with metals. For example, aluminium reflects electromagnetic radiation efficiently over a range that extends from the ultraviolet to the infrared. However, for reasons alluded to previously, in general, bulk metal pieces are not used to fabricate optical components. In most cases, practical use of metallic materials involves the application of a thin, e.g. 0.5 μm, coating onto a polishable substrate, such as glass.

The question might then be asked – why do metals in general possess good reflective properties? In fact, this is associated with one specific property common to all metals – their high electrical conductivity. The electrical conductivity of metals is linked to the presence of a 'free electron gas'. That is to say, the electrons in a metal are, within reason, entirely free to move without any constraint other than their inertia. This is significantly different from the model we built for a refractive material, where opposing charge carriers were constrained and linked with 'springs'. In effectively discarding these 'springs', we change the model dramatically. The most basic model consists of an electron gas with a number density of N electrons per unit volume. Each electron is then subject to a force equal to the product of the oscillating electric field, $E_0e^{i\omega t}$ at angular frequency, ω, and the electric charge, e. Solving the relevant differential equation gives the effective electric susceptibility:

$$\chi = n^2 - 1 = -\frac{Ne^2}{m\varepsilon_0\omega^2} \tag{9.20}$$

m is the electron mass, ε_0 *the permittivity of free space*

It is important to grasp the significance of the sign on the right hand side of Eq. (9.20). The negative susceptibility is a direct consequence of the motion of the electron being in anti-phase to the field. As a consequence, should the RHS of Eq. (9.20) become smaller than −1, then n^2 must be negative and hence n must be imaginary. This is easier to grasp if we re-cast Eq. (9.20):

$$n = \sqrt{\frac{\omega_0^2}{\omega^2} - 1} \quad \text{where } \omega_0 = \sqrt{\frac{Ne^2}{\varepsilon_0 m}} \tag{9.21}$$

The angular frequency, ω_0 is referred to as the ***plasma frequency***. For frequencies less than the plasma frequency, the refractive index is entirely imaginary. To appreciate the significance of this, we might express the refractive index of a metal as $n = i\kappa$. Based on Eq. (8.9), the Fresnel reflection coefficient for normal incidence is given by:

$$R = \left[\frac{1-n}{1+n}\right]^2 = \left[\frac{1-i\kappa}{1+i\kappa}\right]^2 = \left[\frac{1+\kappa^2}{1+\kappa^2}\right]^2 = 1 \tag{9.22}$$

In other words, for frequencies less than the plasma frequency, a metal should be a perfect reflector. To explore the practical significance of this, the conduction electron density of aluminium is 6×10^{28} m^{-3}. This corresponds to a plasma frequency of about 2×10^{15} Hz or a wavelength of about 140 nm. This suggests that aluminium should behave as a perfect reflector for wavelengths in excess of this. Naturally, this ideal model is somewhat unrealistic, as the phenomenon of electrical resistance suggests that the movement of electrons is not entirely free. A basic extension to the simplistic model described here models the effect of finite electrical conductance by introducing a damping term. That is to say, an electron experiences an effective viscous damping force proportional to its velocity that inhibits its motion. This is the so-called ***Drude model***.

$$n^2 = 1 - \frac{\omega_0^2}{\omega^2 - i\omega_D\omega} \tag{9.23}$$

ω_D is a damping frequency or coefficient.

For convenience, we can express Eq. (9.23) in terms of wavelength as follows:

$$n^2 = 1 - \frac{\lambda^2}{\lambda_0^2 - i\lambda_D\lambda} \tag{9.24}$$

The consequence of Eqs. (9.23) and (9.24) is that the refractive index of a metal must, in general, be described by a complex number. This is the so-called ***complex refractive index*** and is generally expressed as:

$$\underline{n} = n + i\kappa \tag{9.25}$$

It is now a simple matter to calculate the Fresnel reflection coefficient for normal incidence for a material with a complex index:

$$R = \left[\frac{1-n-i\kappa}{1+n+i\kappa}\right]^2 = \left[\frac{1-n^2-\kappa^2-i2k}{(1+n)^2+\kappa^2}\right]^2 = \frac{(1-n)^2+\kappa^2}{(1+n)^2+\kappa^2} \tag{9.26}$$

Worked Example 9.2 ***Reflectivity of Aluminium***

The complex refractive index of aluminium at 800 nm is 2.7075 + i8.2713. What is the reflectivity (for normal incidence) at this wavelength? Substituting the relevant values into Eq. (9.26), we have:

$$R = \frac{(1-n)^2+\kappa^2}{(1+n)^2+\kappa^2} = \frac{(1-2.7075)^2+8.2713^2}{(1+2.7075)^2+8.2713^2} = 0.868$$

The reflectivity of aluminium at 800 nm is 86.8%.

Figure 9.5 shows a plot of the complex index of aluminium between 200 and 5000 nm. In addition to the data, the best fit Drude approximation is also added. Whilst the fit is, for the most part, tolerable, there is

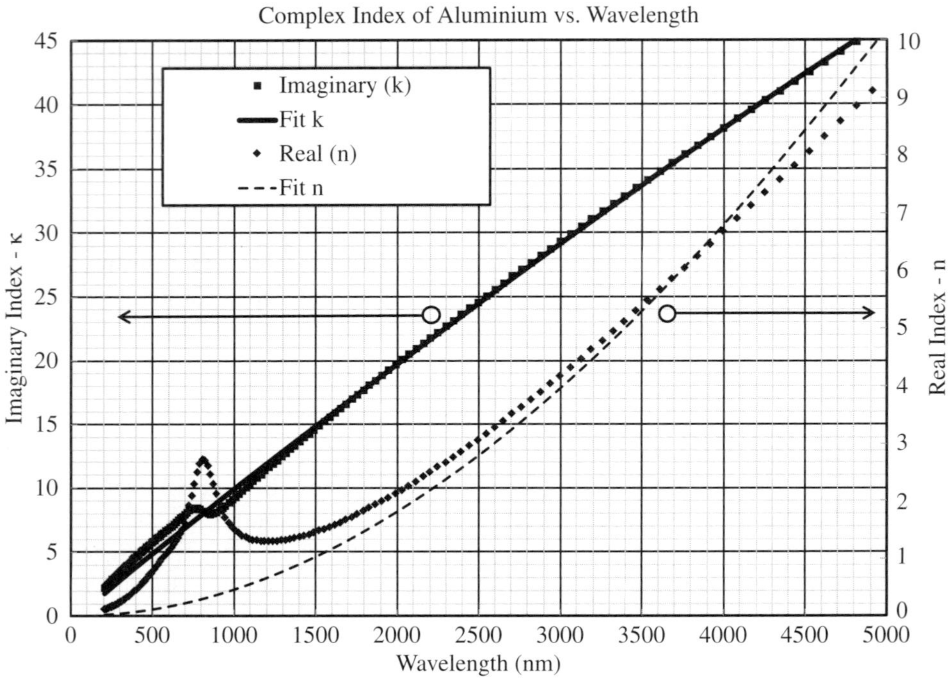

Figure 9.5 Complex index of aluminium.

some departure from the Drude relationship, particularly at the shorter wavelengths. Indeed, there is the appearance of a resonance feature around 800 nm which the simple Drude model is not able to account for. This suggests that there is some more complex band structure that the simple free electron model cannot account for. That this is so, is clearly illustrated by the example of other metals, such as gold and copper. These metals exhibit strong colouration which would not be predicted in the Drude model and is suggestive of more complex resonance/absorption features.

In terms of practical applications, aluminium has good coverage from the ultraviolet to the infrared, albeit with reduced reflectivity in the visible and near infrared regions. Better visible and infrared reflectivity is provided by silver, at the expense of reduced reflectivity in the ultraviolet. Unfortunately, silver is extremely vulnerable to atmospheric degradation (tarnishing) and must be 'protected' by an overcoat, usually of silica or magnesium fluoride. For specifically infrared applications, gold is generally the material of choice with superior reflectivity in this region. Figure 9.6 shows a plot of reflectivity versus wavelength for these three materials.

Another important aspect of reflectivity in metals is the propensity to preferentially reflect one polarisation. Based on Eq. (8.8), which describes the Fresnel reflection for a beam at oblique incidence, we may present the formula for the more general case of a complex refractive index. In this case, the reflection coefficients for s and p polarisations differ.

$$R_s = \left[\frac{\cos\theta - (n + i\kappa)\cos\varphi}{\cos\theta + (n + i\kappa)\cos\varphi}\right]^2 \quad R_p = \left[\frac{\cos\varphi - (n + i\kappa)\cos\theta}{\cos\varphi + (n + i\kappa)\cos\theta}\right]^2 \tag{9.27}$$

The angle θ is the angle of incidence and the angle of 'refraction' is denoted as ϕ. In applying Snell's law to a material of complex refractive index, we apply the real part of the inverse of the complex index. Thus, the 'index' that is applied to determine ϕ is equal to $(n^2 + \kappa^2)/n$. For metals, polarisation sensitivity is greatest at wavelengths where there are moderately strong extinction features, such as that for aluminium at 800 nm. Figure 9.7 shows a plot of the reflectivity of aluminium vs angle of incidence for both polarisations. The data

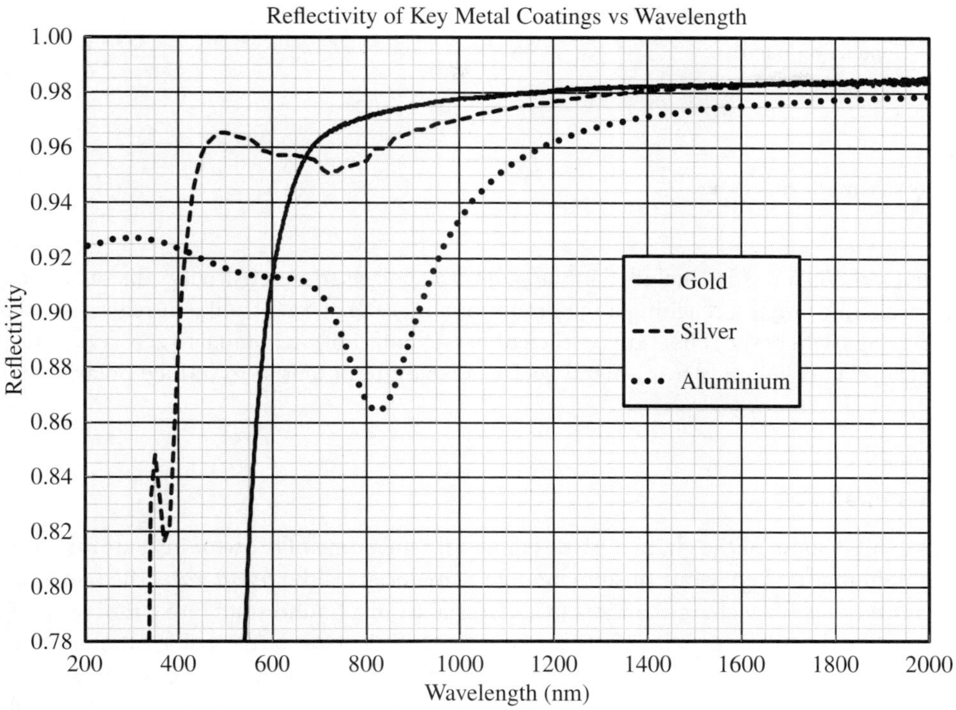

Figure 9.6 Reflectivity of principal metal coatings.

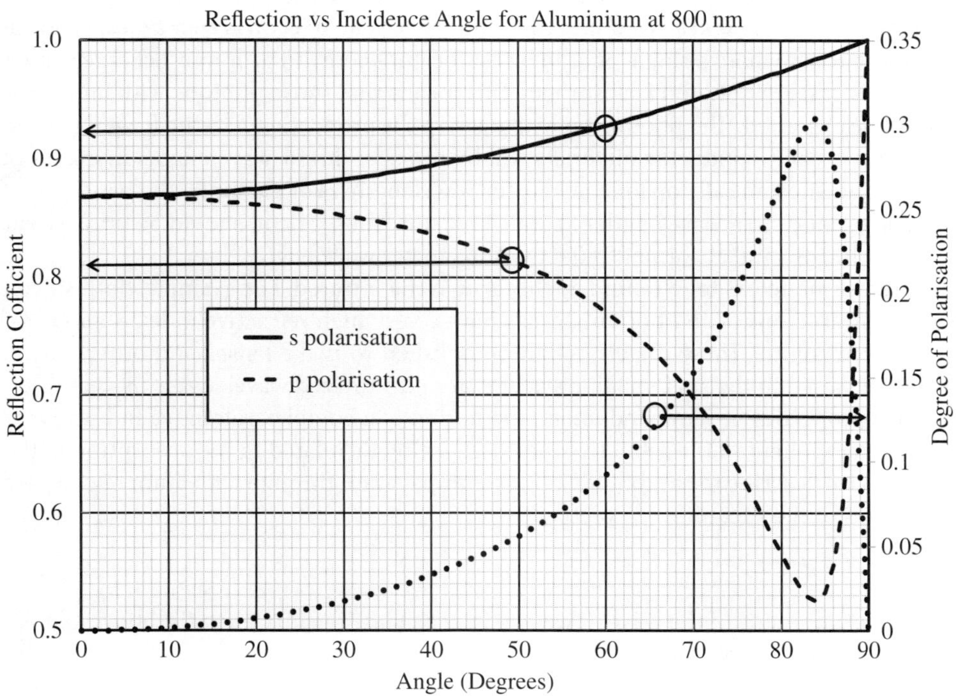

Figure 9.7 Reflection coefficient vs angle for aluminium at 800 nm.

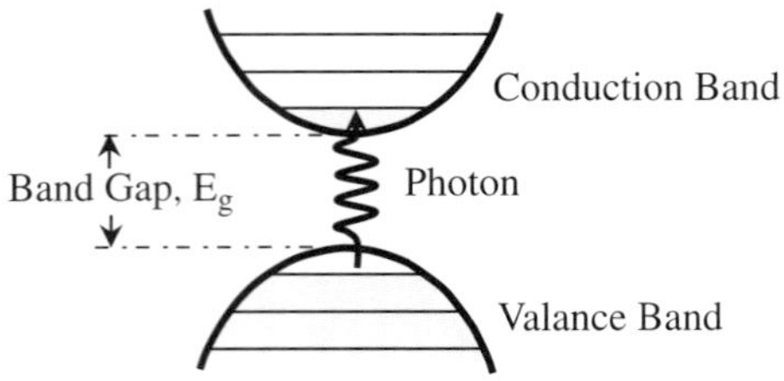

Figure 9.8 Band gap in semiconductors.

presented is for a wavelength of 800 nm. Also included in the plot is the polarisation sensitivity. The sensitivity is particularly high at very shallow angles, reaching a maximum of about 0.3 at 84°. Unlike purely refractive media, there is no Brewster angle where the reflection coefficient of one polarisation is equal to zero. One must therefore be aware, that glancing reflections from metal surfaces have the potential to induce polarisation effects.

9.2.3 Semiconductor Materials

Semiconductor materials exhibit some of the characteristics of metals and some of those of insulator materials. Of critical interest in defining the behaviour of semiconductor materials is the ***band gap***. The band gap represents the energy required to elevate charge carriers in a material from a bound state into a mobile, conductive state. Of course, for metals, this energy is effectively zero, so that electrons are capable of interacting with external electric fields under normal conditions. Conversely, for insulator materials, the band gap is so large that there are no 'ordinary circumstances' under which charge carriers may be induced to interact with external fields. Semi-conductors represent an intermediate scenario where relatively modest input of energy, thermal, or optical, is sufficient to generate charge carriers. That is to say, semi-conductors are characterised by their intermediate band gap energy. The principal is illustrated in Figure 9.8.

The band gap energy of a material is usually denominated in electron volts, the energy required to promote a single charge carrier from the bound or ***valence band*** into the conductive state or ***conduction band***. From an optical standpoint, the interest arises where a single photon has sufficient energy to promote a charge carrier to the conduction band. In this case, the photon is absorbed. This means that those photons with an energy in excess of the bandgap will be absorbed. The photon energy, of course, is proportional to the frequency and the inverse of the wavelength. So semi-conductor materials will be transparent for wavelengths in excess of some critical wavelength, λ_g and substantially opaque at shorter wavelengths. For example, the band gap of silicon is 1.14 eV and this corresponds to a wavelength of about 1100 nm. Thus, silicon should be transparent for wavelengths greater than 1100 nm and absorbing/reflective for shorter wavelengths.

In practical terms, most semiconductors tend to find application in the infrared. For example, both silicon and germanium are used in infrared optics and both materials have a very high refractive index – about 3.4 for silicon and about 4 for germanium. These materials can be polished to make lenses and other optical components and are extensively used in thermal imaging systems at mid-infra red wavelengths (3–7 μm).

In describing the refractive properties of semiconductors, the notion of a complex refractive index is very useful. In fact, the imaginary component of the refractive index, κ, is directly related to the absorption of the material. It is easy to see this if we describe the amplitude of wave propagation through a medium of complex refractive index, $n + i\kappa$, in terms of the vacuum wavevector, k:

$$A = A_0 e^{i(n+i\kappa)kx} = A_0 e^{-\kappa kx} \times e^{inkx} \tag{9.28}$$

Equation (9.28) describes the amplitude of the wave. If we need to calculate the flux, then the above expression must be multiplied by its complex conjugate:

$$\Phi = \Phi_0 e^{-2\kappa kx} = \Phi_0 e^{-\alpha x} = \Phi_0 e^{-x/x_0} \quad \alpha = \frac{4\pi\kappa}{\lambda} \quad x_0 = \frac{\lambda}{4\pi\kappa} \tag{9.29}$$

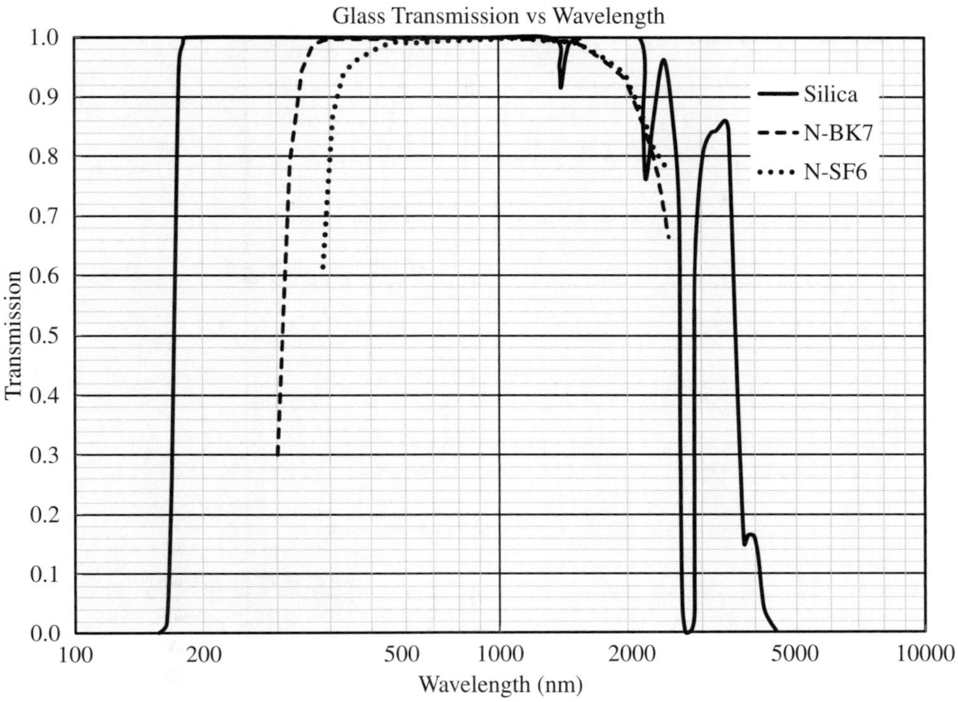

Figure 9.9 Glass transmission vs wavelength (internal transmission for 10 mm thickness).

The exponential attenuation of light in a medium is referred to as ***Beer's Law***. The coefficient, α, is known as the absorption coefficient and the distance, x_0, as the absorption depth. By convention, these quantities are usually denominated in cm^{-1} and cm respectively. Naturally, for a semiconductor material, one would expect κ to increase in the region of the band gap and at shorter wavelengths. Figure 9.9 plots the data for κ vs. wavelength for silicon, together with the absorption depth. It is only for wavelengths in excess of the 1100 nm bandgap that the absorption depth increases beyond a centimetre, opening up applications for transmissive optical components, such as lenses. Figure 9.10 shows a plot of the real component of the index, n. The high refractive index of silicon is an asset in optical design. Although the Fresnel reflection losses are high, the large deviations produced by comparatively gentle lens curvatures means that third order aberrations are substantially reduced. In any case, Fresnel losses can be ameliorated by special measures, such as surface coating.

Table 9.3 lists a number of semiconductors together with their bandgaps. The bandgaps of the majority of true semiconductors tend to lie in the infrared region of the spectrum. However, there are a number of 'wide bandgap' materials which, from an electronic perspective, are insulators, but have interesting and useful optical properties.

Many of the semiconductors listed are not elemental semiconductors, such as silicon and germanium, but are compound semiconductors, such as gallium arsenide or indium phosphide. An interesting example of a compound semiconductor is mercury cadmium telluride. Its chemical formula is actually $Hg_xCd_{1-x}Te$. The relative proportions of cadmium and mercury can be adjusted to 'engineer' the bandgap. This is an example of a ***ternary compound*** and, from Table 9.3, its bandgap can be adjusted to anywhere between 0 and 1.5 eV. As well as being used in passive optical components, such as lenses, these semiconductor materials are also useful as detectors of optical radiation. Absorption of a photon with an energy greater than that of the band gap produces mobile charge carriers which may then be detected. For some, though not all, semiconductor

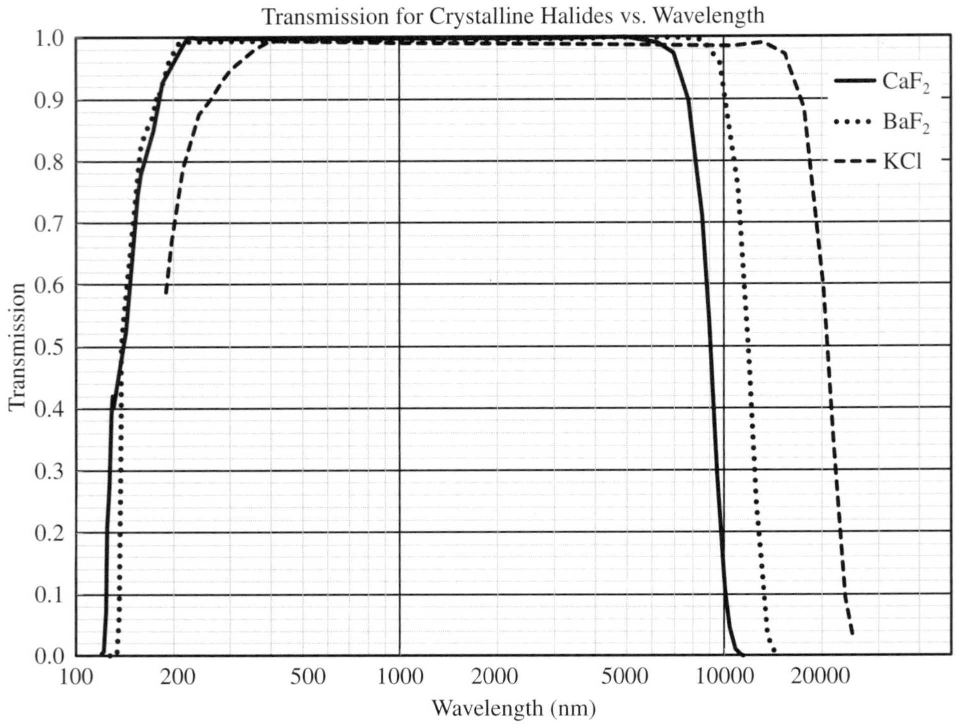

Figure 9.10 Internal transmission for crystalline halides (10 mm thickness).

Table 9.3 Semiconductor materials and bandgaps.

Material	Bandgap (eV)	λ_g (nm)	Material	Bandgap (eV)	λ_g (nm)
ZnS	3.6	344	GaAs	1.43	867
GaN	3.5	354	InP	1.27	976
SiC	3.26	380	Si	1.11	1120
ZnO	3.2	387	GaSb	0.68	1820
ZnSe	2.7	459	Ge	0.66	1880
GaP	2.25	551	InAs	0.36	3440
CdSe	1.74	713	InSb	0.17	7290
CdTe	1.44	861	HgCdTe	0–1.5	>827

materials, the reverse is possible, the injection of electrical charge producing optical emission. This is the basis of ***light emitting diodes*** or ***LEDs*** and also ***semiconductor lasers***.

9.3 Transmission Characteristics of Materials

9.3.1 General

In our treatment of semiconductor materials, we have considered their absorption characteristics as well as their refractive properties. We now turn our attention to the more general question of absorption, or transmission, in all materials. For the semiconductors it was shown that their transmission properties are strongly determined by their bandgap. In the case of glassy materials one might consider these to be, in effect, wide band gap materials. For the majority of glasses, the 'cut off' in transmission at short wavelengths falls into the region of 300–350 nm; this significantly limits the application of conventional glasses in the deep ultraviolet. Fused silica, which is an amorphous variant of crystalline quartz will transmit down to 170 nm in the vacuum ultraviolet, depending upon material purity. For wavelengths shorter than this, crystalline fluorides, such as calcium fluoride and barium fluoride, are useful materials, extending transmission down to 130 nm.

In characterising transmission, we have focused on short wavelength absorption produced by interaction with electronic states of the material. At longer (infrared) wavelengths, absorption results from excitation of lattice vibrations. For glasses and silica, transmission ceases above about 2500 nm. In the case of the fluorides (barium and calcium), transmission extends further into the infrared to 10 000 nm or so, giving a wide range of transparency from the vacuum ultraviolet to the mid-infrared.

9.3.2 Glasses

Figure 9.9 shows the transmission vs wavelength for two glasses (SCHOTT N-BK7® and N-SF6) and for fused silica. All plots are for a sample thickness of 10 mm. The graph depicts the ***internal transmission*** which only includes losses due to absorption. By contrast, ***external transmission*** includes the Fresnel losses caused by reflection at the two interfaces. As outlined earlier, fused silica has an extended transmission range, particularly in the ultraviolet. It is economical and relatively easy to work and is the preferred material for wavelengths outside the transmission envelope for glasses. This is particularly true for ultraviolet wavelengths below about 320 nm. In fact, there are a variety of different 'grades' of fused silica. In particular, there are grades of silica specifically tailored to ultraviolet applications; these tend to contain very low concentrations of metallic contaminants. The plot shown in Figure 9.9 is for a representative ultraviolet grade of fused silica. It is clear that there are significant band absorption structures in the 2000–3000 nm region. These absorption bands are associated with trace contamination by water. By preferentially focusing on the reduction of water contamination, these features may be substantially eliminated. This process produces infrared grade fused silica.

9.3.3 Crystalline Materials

There are a number of halide crystals that have excellent transmission from the deep ultraviolet to the mid infrared. These include calcium fluoride and barium fluoride which are reasonably robust materials and can be shaped and polished to produce standard optical components, such as prisms and lenses. Other halides, such as sodium chloride and potassium chloride, have an even greater transmission range. However, these materials are soft, difficult to work, and hygroscopic. The latter difficulty is a substantial one. On the other hand, such hygroscopic materials are cheap and are often used in 'throwaway' applications, such as cell windows in infrared spectroscopy, where the window can be discarded after a few uses. Internal transmission curves for CaF_2, BaF_2 and KCl are shown in Figure 9.10. As with the previous plot, internal transmission for a thickness of 10 mm is shown. For calcium fluoride and barium fluoride, transmission extends from about 130 nm in the ultraviolet to 10 000 nm in the mid infrared. Infrared transmission for potassium chloride extends to beyond 20 000 nm.

Sapphire (aluminium oxide) is another material with a wide transmission range – from about 160 to 4500 nm that is used relatively extensively in the fabrication of bulk components. It also has a high thermal conductivity, which is a useful characteristic in severe thermal environments. In addition, there are a number of other crystalline materials, such as calcite (calcium carbonate) barium borate, lithium niobate, potassium dihydrogen phosphate (KDP) which may be considered as wide bandgap materials with extensive ranges of transmission. For the most part, these materials are not used as bulk optical components, but rather in specialist applications, such as in polarising and non-linear crystals.

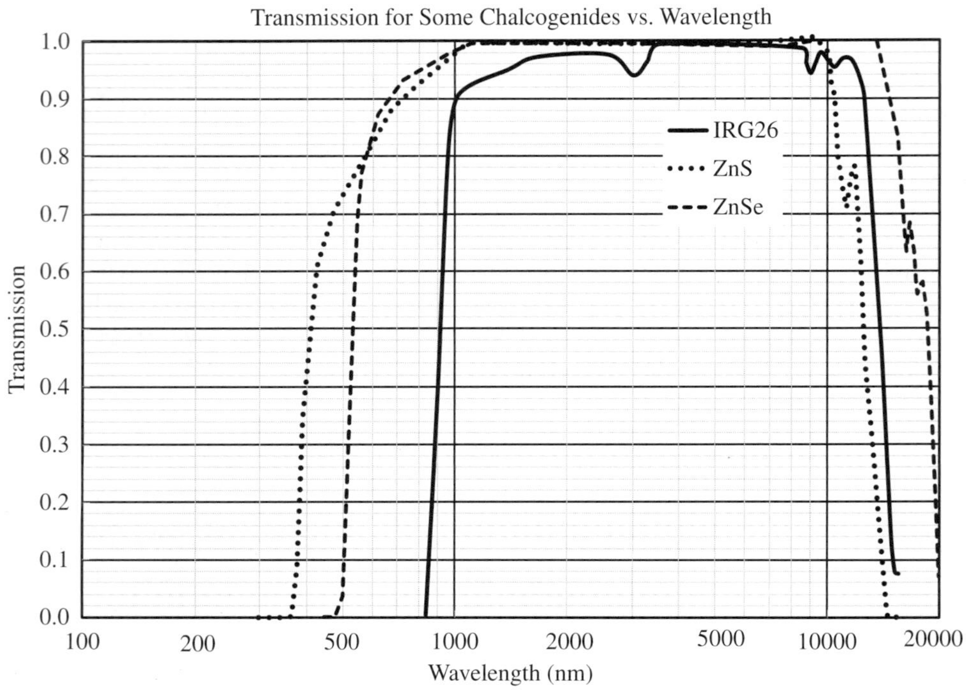

Figure 9.11 Internal transmission for some chalcogenides (10 mm thickness).

9.3.4 Chalcogenide Glasses

Chalcogenide glasses have some chemical similarity to standard glasses, except the oxygen content in these glasses is replaced by the other group six elements, sulfur, selenium, and tellurium. Simple examples of the chalcogenides, include zinc sulfide and zinc selenide. In terms of their applications, they are specifically characterised by their excellent infrared transmission. They are also comparatively easy to work. Proprietary glasses can be formulated from an admixture of various elements, including germanium, arsenic, selenium, and tellurium. Figure 9.11 shows the internal transmission for some chalcogenide glasses, including IR26, a proprietary glass from SCHOTT, plus zinc sulfide and zinc selenide. Zinc selenide, in particular, is a robust material widely used as an optical material for focusing the output of high power infrared lasers, including the CO_2 laser at 10 600 nm.

9.3.5 Semiconductor Materials

As previously outlined, the transmission of semiconductor materials is largely dictated by their bandgap. Transmission of the most prominent optical semiconductors, silicon, germanium, and gallium arsenide is shown in Figure 9.12. The data is for a thickness of 10 mm.

9.3.6 Polymer Materials

A variety of polymer materials are used in the fabrication of optical components, although the range is not as wide as for glasses. Many of these applications lie in the domain of consumer optics. For example, spectacle lenses are largely made from resin materials, such as CR 39 and tough thermoplastics, such as polycarbonates. Low cost components may be moulded from acrylates (e.g. perspex). Alternatively, precision components can be produced from higher specification materials, such as cyclo-olefins (Zeonex) and polysulfones (Ultem).

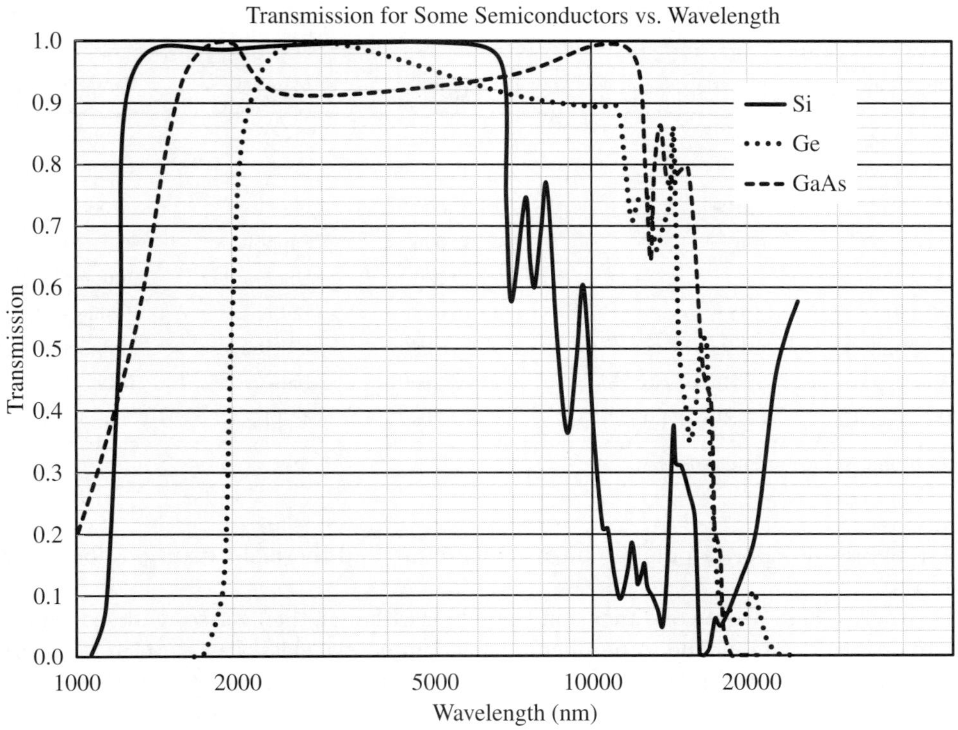

Figure 9.12 Internal transmission for some semiconductors (10 mm thickness).

Polymer materials have refractive indices in the ranges from 1.5 to 1.6 and Abbe numbers ranging from about 30 to 60. There is less scope for chromatic correction in polymers than there is for glasses.

The transmission range of optical polymers is largely restricted to the visible and the near infrared. Organic polymers are generally opaque in the ultraviolet region due to the presence of strong electronic transition. In addition, because of the restricted elemental composition of polymers, carbon–carbon and carbon–hydrogen bonds give rise to distinctive absorption features in the near infrared. Substitution of fluorine for hydrogen in polymer formulations tends to shift these features further into the infrared, producing extended transmission in the near infrared. Overall, in terms of their transmission, polymers behave largely like glasses.

9.3.7 Overall Transmission Windows for Common Optical Materials

At this point it will be useful to tabulate the useful window of transmission for common optical materials. These are set out in Table 9.4.

9.4 Thermomechanical Properties

9.4.1 Thermal Expansion

In discussing the refractive index variation of glasses, we touched on the significance of thermal expansion, in addition to the variation of refractive index with temperature. For the most part, the thermal expansion of glasses falls within a narrow range. Many commercial glasses tend to have a thermal expansion coefficient of between 5 and 10 ppm per degree centigrade. On the other hand fused silica has an expansion coefficient

Table 9.4 Transmission range for common optical materials.

Material	Range (nm)	Material	Range (nm)	Material	Range (nm)
BK7® Glass	320–2 500	KCl	210–20 000	GaAs	1 200–15 000
SF6 Glass	370–2 500	KBr	230–25 000	Si	1 200–7 000
Silica (UV Gr.)	170–2 500	CsBr	280–35 000	Ge	20 000–14 000
Silica (IR Gr.)	200–3 500	CsI	380–45 000	Sapphire	250–5 000
LiF	100–7 000	ZnS	400–12 000	Diamond	300–100 000[a)]
MgF_2	120–7 000	ZnSe	600–16 000	Rutile	2 200–4 000
CaF_2	130–7 000	CdTe	1 000–25 000	Calcite	300–2 300
BaF_2	150–12 500	SCHOTT IRG26	1 000–14 000	$BaB0_3$	190–3 500
NaCl	200–20 000	SCHOTT IRG22	1 200–14 000	YVO_4	500–3 400

a) Diamond has a narrow absorption feature around 5 000 nm.

of around 0.5 ppm These figures do apply around ambient temperature and thermal expansion does vary somewhat with temperature.

For some applications, where a high degree of dimension stability is essential, then glasses with extremely low thermal expansion are in particular demand. A number of such commercial glasses exist. SCHOTT produce ZERODUR®, Ohara produce CLEARCERAM® and Corning produce ULE® (Ultralow expansion glass). For these glasses, the expansion coefficient is typically less than 0.1 ppm per °C. However, the expansion coefficient does vary around room temperature and these glasses are generally optimised to have a low expansion coefficient in a particular temperature regime (usually ambient).

Polymers, in general, are distinguished by their much higher thermal expansion coefficients, usually an order of magnitude higher than glasses. Generally, the thermal expansion coefficient of optical polymers lies in the region of 50–100 ppm

9.4.2 Dimensional Stability Under Thermal Loading

For some critical optical applications, we wish to know the dimensional stability of an optical system with respect to external thermal loading. This is of particular interest for the deployment of optical instruments in uncontrolled outside environments, or in space applications. Having briefly examined the impact of thermal expansion, it is clear that dimensional stability under thermal loading is directly related. However, the temperature excursion produced by a given thermal loading is impacted by the thermal conductivity of the material. The higher the thermal conductivity, then the lower the temperature excursion for a given heat loading. Therefore, dimension stability is afforded by a combination of low thermal expansion and high thermal conductivity. In fact, it is useful to define a 'figure of merit' for thermal stability which is the ratio of the thermal conductivity to the thermal expansion. It is important to recognise that dimensional stability of an optical system is affected not only by the optical materials themselves, but also mechanical materials used in support structures etc. Figures of merit are listed for a range of useful materials in Table 9.5.

At the top of this least we see silicon carbide. It is not surprising, therefore, that silicon carbide finds use in space applications, both as mirror substrates and as material for stable optical benches.

9.4.3 Annealing

Glasses are amorphous material and are characterised by their high degree of dimensional stability. In many respects, they are not true solids; they do not, unlike metals and other solid materials, undergo liquid-solid

Table 9.5 Material thermal stability – Figures of merit.

Material	CTE (ppm K^{-1})	Thermal conductivity ($Wm^{-1}\ K^{-1}$)	Figure of merit
Silicon carbide	2.4	170	1.000
Silicon	2.33	150	0.909
Copper	8.9	398	0.631
ULE	0.03	1.3	0.612
Graphite epoxy	0.3	6	0.282
Super invar	0.65	10.5	0.228
Zerodur®	0.1	1.6	0.226
Beryllium	11.4	160	0.198
Nickel	8.9	90	0.143
Aluminium	24	167	0.099
Titanium	4.43	21	0.067
Silica	0.55	1.3	0.033
BK7	8.2	1.1	0.002
Acrylic	45	0.2	8×10^{-5}
Epoxy	50	0.2	8×10^{-5}

phase change. Instead, they exhibit a progressive and monotonic increase in viscosity on cooling. As cooling in all materials is accompanied by dimensional change, usually contraction, there is the potential for this to be translated into local stress, if this cannot be accommodated by internal movement. For glasses, this can occur if a glass is cooled quickly and a rapid increase in viscosity inhibits the relaxation of internal stress. The presence of internal stresses in the material and their accommodation through creep, or irreversible, time dependent strain, may lead to unpredictable changes in component geometry. This is especially the case for metals. In the case of glasses, particularly optical glasses, very careful, slow, and controlled cooling of molten glass minimises the internal stresses locked into the material. In addition, following initial manufacture, glass blanks may be carefully heated in a controlled manner with the specific objective of minimising these internal stresses. This process is referred to as ***annealing***. The annealing temperature is dictated by the viscosity of the glass. A number of critical temperatures, expressed in terms of the material viscosity, η, are used to define the thermal processing characteristics of each glass:

Strain point. Temperature at which internal stresses are not relieved – $\eta = 3.1 \times 10^{13}$ Pa s
Annealing point. Temperature at which internal stresses are just relieved – $\eta = 10^{12}$ Pa s
Softening point. Glass starts to slump under its own weight – $\eta = 3.1 \times 10^{6}$ Pa s

At a temperature around the annealing point lies the ***glass transition temperature, T_g***. This marks the transition from a hard and brittle state to a more deformable or rubbery state. It is not a first order phase change, but is sometimes described as a second order phase change, where the specific heat capacity, rather than the total heat (latent heat) is discontinuous with respect to temperature. Not surprisingly T_g for most glasses is high, in the region of 500–600 °C. For fused silica T_g is even higher – 1200 °C. For hard thermoplastics it is around 100 °C, whereas for softer plastics, T_g is less than 50 °C or even sub-zero.

9.4.4 Material Strength and Fracture Mechanics

For optical systems deployed in extreme environments, such as those that pertain to aerospace applications, the strength of the underlying materials becomes an important issue. Direct mechanical loading of optical components is not the only issue that must be considered. Sudden thermal loading of a component produces ***thermal shock*** leading to large internal stresses that can only be resolved by component failure. Another example where the thermal environment produces severe loading is where elements with differing thermal expansions are maintained in close contact. One example of this is the ubiquitous achromatic doublet. The two glasses in contact may have considerably different thermal expansion coefficients. For extreme applications, for example where the optical assembly is to be cooled to cryogenic temperature, the resulting material strain may lead to failure with one or both of the two components shattering.

In this discussion, we are largely concerned with the behaviour of (optical) glasses. We may assume that in all practical applications, optical materials are deployed at well below the glass transition temperature. We can therefore assume that such materials are hard and brittle. Whereas the strength of ductile materials, such as most metals, is easy to define in terms of a maximum tensile load, or the ***tensile strength***, the strength of brittle materials is less easy to define. Failure in brittle materials is largely dependent upon the catastrophic propagation of flaws or cracks. The presence of small cracks in a perfectly elastic material causes the local amplification of stress and, in the presence of relatively small tensile loads, these cracks enlarge rapidly, leading to failure. By contrast, in ductile materials, high stresses at the tip of a crack are accommodated by plastic flow and there is no catastrophic failure. Ductile materials are said to be ***tougher.*** The propensity for crack propagation leading to catastrophic failure is described by a parameter referred to as the ***fracture toughness, κ***. Units for fracture toughness are in MPam$^{1/2}$. One example of interest is fused silica. Fused silica has a respectable tensile strength of about 50 MPa; the tensile strength of aluminium is about 90 MPa. By contrast, the fracture toughness of fused silica is only 0.6 MPam$^{1/2}$. whereas that for aluminium is around 30 MPam$^{1/2}$. Clearly, fused silica is much more brittle than aluminium. The natural brittleness of glass materials is a serious practical concern in the mechanical mounting of optical components. For example, great care must be taken in the mounting of lens components in a camera barrel. Usually it is customary to use mounting designs that are, to some degree, compliant and can accommodate small movements without applying large stresses to the lens components.

As indicated earlier, glasses in general are vulnerable to thermal shock. Famously, borosilicate glasses were developed by Otto Schott in the nineteenth century, specifically for their resistance to thermal shock. These were sold under the trade name of DURAN®; PYREX® is the Corning equivalent. The salient feature of borosilicate glasses is their comparatively low thermal expansion coefficient – around 3 ppm K^{-1}. However, resistance to thermal shock is not only incurred by low thermal expansion. Thermal conductivity and fracture toughness, of course, play a key role. Glasses are naturally susceptible to thermal shock on account of their brittleness. Furthermore, a high thermal conductivity will help to dissipate heat, ameliorating any local temperature gradients and resultant thermal stresses. It must also be remembered that resolution of any unrelaxed strain into stress, should be tempered by the material's elastic or Young's modulus. Thus it can be seen that the ratio of the product of the thermal conductivity, k, and the facture toughness, κ, to the product of the Young's modulus, Y and thermal expansion, α, provides a figure of merit, F_{shock}, that describes the resistance to thermal shock.

$$F_{Shock} = \frac{k\kappa}{Y\alpha} \tag{9.30}$$

Table 9.6 sets out figures for a number of key optical materials of interest. The shock resistance of aluminium is provided for comparison.

It is clear that proprietary optical glasses are particularly susceptible to shock. Returning to a problem previously outlined, we might wish to understand the thermal reliability of joining two dissimilar materials together, as per the cementing of achromatic doublets. Any stress at the interface of two materials is driven by the differential expansion of the two materials, $\Delta\alpha$ and the material Young's modulus. Therefore a useful figure of merit for this scenario, F_{Diff}, would be given by the ratio of the fracture toughness to the product of the Young's modulus and the differential expansion coefficient:

$$F_{Diff} = \frac{\kappa}{Y(\alpha_1 - \alpha_2)} \tag{9.31}$$

Table 9.6 Resistance of optical materials to thermal shock.

Material	Fracture toughness ($MPam^{1/2}$)	Expansion coefficient ($ppm K^{-1}$)	Thermal conductivity ($Wm^{-1} K^{-1}$)	Young's modulus (GPa)	Figure of merit (Al = 1)
Aluminium	29	24	167	69	1
Silicon	0.95	2.6	150	150	0.125
Sapphire	3	6.6	25	245	0.011
Fused silica	0.61	0.55	1.38	73	0.007 2
Zinc sulfide	0.5	6.8	16.7	74	0.005 6
Zinc selenide	0.3	7	18	67	0.003 9
CaF_2	0.3	19	9.7	76	0.000 69
BK7	0.7	7.1	1.11	82	0.000 45
F2	0.45	8.2	0.78	58	0.000 25
SF5	0.47	7.9	1	87	0.000 23

Worked Example 9.3 ***Achromatic Doublet***

We wish to design a thermally robust achromatic doublet. We are faced with the choice of two possible combinations: (i) BK7 + F2; (ii) BK7 + SF5. From the point of view of the thermo-mechanical design, which is the best combination?

For the first combination, BK7 + F2, we have the choice of two possible figures of merit to work with (one for each material); we need to select the lower of the two. The differential expansion is 1.1 ppm and the two figures of merit are:

$$F_1 = \frac{0.7}{82 \times 1.1} = 0.0078 \text{ and } F_2 = \frac{0.45}{58 \times 1.1} = 0.0071$$

The lowest figure of merit is 0.0071.

For the BK7 + SF5 combination, the differential expansion is 0.8 ppm The two figures of merit are:

$$F_1 = \frac{0.7}{82 \times 0.8} = 0.011 \quad F_2 = \frac{0.47}{87 \times 0.8} = 0.0068$$

The lowest figure of merit is 0.0068

Of the two combinations, the highest figure of merit is the BK7 + F2 combination. However, the difference between the two figures is small and would probably make little difference in practice. Nonetheless, this does serve to illustrate the principle involved.

9.5 Material Quality

9.5.1 General

For optical glasses, material uniformity (refractive index) and freedom from internal stresses is of paramount importance. On the one hand, variations in the refractive index of a lens or window will directly add to the wavefront error of a system. In addition, internal strain within the glass may lead to an unacceptable degree of stress induced birefringence. Furthermore, the presence of small, gross non-uniformities, such as small bubbles or inclusions will lead to increased light scattering. Unfortunately, in practice, all these defects are present, to some degree, in any manufactured batch of glass. It can be controlled, to an extent, by careful mixing of constituents and slow and controlled cooling of the glass melt. However, to a considerable degree, the production of high quality material is based upon selection and inspection. That is to say, it is accepted that some parts of a melt will be of intrinsically lower quality and that by 'grading' portions of the melt, limited quantities of very high quality material can be produced. Naturally, this does influence the economics of the process and production of higher quality material inevitably leads to increased costs.

There are a number of key metrics that define glass quality and these are described in sequence.

9.5.2 Refractive Index Homogeneity

Refractive index homogeneity refers to the maximum acceptable refractive index variation across a part. The index homogeneity generally specified tends to refer to relatively low spatial frequency components of index variation across a part. For general commercial applications, homogeneity might be of the order of ±100 ppm For precision applications, this figure might be ±5 ppm or as low as ±1 ppm for demanding high-precision applications.

Related to the index homogeneity are variations in the absolute value of the index from batch to batch. For general applications, a departure of ±0.1% from the nominal index value is acceptable. For precision applications, this figure might be reduced to 0.05% and to 0.02% for high-precision applications. Similarly, as far as dispersion properties are concerned, a variation in the absolute value of the Abbe number of 0.8% is reasonable for general applications. This value falls to 0.5% and 0.2% for precision and high-precision applications.

9.5.3 Striae

Glass manufacture involves the admixture of controlled quantities of different constituents followed by a high temperature mixing process. The process of imperfect mixing and internal convection within the melt inevitably results in the production of streaks or striae of material inhomogeneity. In practice, the term ***striae*** refers to the presence of high spatial frequency refractive index inhomogeneity whose geometry is 'streak like'. Striae are characterised by the number of streaks present within a component capable of causing an optical path difference of 30 nm or greater. For general applications, the number of streaks can be as high as 10% of the total component cross section. For precision applications, this might be as low as 2%. High-precision applications may demand the absence of detectable striae.

9.5.4 Bubbles and Inclusions

The presence of bubbles and inclusions is captured by the total area of bubbles and inclusions greater than a certain size, within a given volume of material. For example, the density may be described by the total area of defects greater than 50 μm diameter in a volume of 100 cm^3. An acceptable value is 0.5 mm^2, whereas precision and high-precision applications may demand areas as low as 0.1 mm^2 and 0.03 mm^2 respectively.

9.5.5 Stress Induced Birefringence

As outlined in the previous chapter, the presence of internal stresses within a material leads to the creation of local birefringence. This stress-induced birefringence is characterised by the phase difference that is produced between the fast and slow axes on passage through one centimetre of material. Stress-induced birefringence is normally expressed in nm per centimetre. For general applications, 40 nm cm^{-1} is acceptable. Precision and high-precision applications might demand stress induced birefringence as low as 10 and 4 nm cm^{-1} respectively. Low stress-induced birefringence is particularly critical in instruments devoted to the measurement and characterisation of polarisation.

9.6 Exposure to Environmental Attack

Optical systems must sometimes function in unfavourable environments and are potentially subject to attack by agents such as acids and alkalis or exposed to leaching by water. Because of the complex oxide mixtures used in glasses, the response of different glasses to these agents can vary. Therefore manufacturers are keen to understand how different formulations respond to such exposure.

9.6.1 Climatic Resistance

Glasses are tested in a classic accelerated test procedure at elevated temperature and humidity, e.g. 50 °C and 100% RH. Degradation of the glass surface is characterised by measuring scattered light or haze produced by attack of the glass surface. Glasses are categorised as Class CR1 (highest) to Class CR4, according to their resistance.

9.6.2 Stain Resistance

This test assesses a glasses resistance to chemical attack. In this context, a 'stain' is a thin layer of chemically altered glass that produces colour contrast by thin film interference. As such, it represents the creation of a certain thickness, e.g. 0.1 μm of chemically altered material. Using a standard solution, e.g. an acid buffered with sodium acetate, the time taken to produce a certain discolouration (i.e. film thickness) is measured. This time is an indication of the resistance to staining. A very short time, of course, indicates a very low resistance. Glasses are categorised from Class FR0 (highest) to Class FR5 for stain resistance.

9.6.3 Resistance to Acid and Alkali Attack

For resistance to acid attack, a strong acid (pH 0.3 nitric acid) is used to test more resistant materials. The time to etch 0.1 μm of material from the glass surface is an indication of the resistance to acid attack. For more vulnerable materials, a weaker acid solution, buffered with sodium acetate is prepared. Again, resistance to attack is marked by the time to remove a specific thickness of material (0.1 μm). For resistant materials, acid resistance is marked on a scale from SR1 (highest) to SR 5 and for vulnerable materials the scale proceeds from SR 51 (highest) to SR 53.

Resistance to alkali attack is characterised in a similar way, except a sodium hydroxide solution is used. Time to remove 0.1 μm of material is again the measure, and the resistance scale ranges from AR1 (highest) to AR4. In addition, there is also a test that measure resistance to alkaline phosphate attack. In this case, the pentasodium triphosphate ($Na_5P_3O_{10}$) is substituted for sodium hydroxide. As for the alkali resistance test, phosphate resistance classes range from PR1 (highest) to PR4.

9.7 Material Processing

There is a correlation between the hardness of a material and the ability to produce a highly polished surface. Generally, it is easier to polish harder materials to produce a highly smooth finish. This statement applies to most optical glasses. On the other hand, softer materials, such as calcium fluoride, calcite, and many plastics are more difficult to polish. However, before a material can be polished to produce and optically smooth finish, it must first be shaped. Generally, this is done by grinding for glasses and moulding in the case of optical polymers. Glasses can be graded according to their grindability. This is categorised be measuring the volume of material removed by a standard grinding process in a set time and comparing this to a standard material. Materials are graded from HG 1 to HG 6, with HG 1 representing the slowest rate of material removal.

Further Reading

Nikogosyan, D.N. (1998). *Properties of Optical and Laser Related Materials: A Handbook*. New York: Wiley. ISBN: 0-471-97384-X.

Owens, J.C. (1967). *Optical refractive index of air: dependence on pressure, temperature and composition*. *Appl. Opt.* 6 (1): 51.

Palik, E.D. (1998). *Handbook of Optical Constants of Solids*. London: Academic Press. ISBN: 0-12-544420-6.

10

Coatings and Filters

10.1 Introduction

The application of a thin coating to an optical surface usually serves one of two purposes. First, it may be used to alter the underlying optical properties of the surface, particularly its transmission and reflection. For example, coating a glass surface with a thin layer of aluminium produces a specular surface. Second, mechanically hard coatings are used to protect surfaces that are vulnerable and sensitive to damage. One example of this is the use of hard coatings, such as silicon monoxide or silicon dioxide (silica), to protect soft metallic coatings, such as aluminium.

The majority of coatings used in optics are '***thin film coatings***'. That is to say, they are of the order of a fraction of a wavelength or a few wavelengths thick, ranging from tens of nanometres to a few microns in thickness. These films are deposited onto a variety of substrate materials either by ***vacuum evaporation*** or by ***sputtering***. Vacuum evaporation is carried out under vacuum by the intense heating of the source material using resistance heating or electron beam heating. Evaporated material from the source then redeposits on the substrate (and elsewhere). Sputtering is also a low pressure process and is used only for the deposition of metals. It uses material removal from the (metal) cathode of a low pressure electrical discharge.

Much of this chapter is concerned with optical filters. These filters seek to modify the spectral flux of light according to properties such as wavelength or polarisation. Particularly common are filters that admit light according to some form of spectral characteristic. For example, 'long pass' filters only transmit light with a wavelength greater than a certain critical wavelength. Bandpass filters admit light in a range of wavelengths, or band, between two wavelength values. By contrast, neutral density filters control the flux of transmitted light by attenuating the flux by a factor that is largely independent of wavelengths. Most, but not all, of these filters rely on thin film technology, as previously described. Thin film filters, for the most part, manipulate light by transmitting some portion and reflecting the other portion. However, there are a number of 'glass' filters that work by virtue of volume absorption. That is to say, the glass is impregnated with some material that only partially transmits the light.

10.2 Properties of Thin Films

10.2.1 Analysis of Thin Film Reflection

The majority of optical components used in systems, such as cameras and microscopes are coated in some way. They may be coated to promote reflection; most commonly, in optical systems, components are coated to minimise stray reflections from surfaces. Thin films play a very important role in the design of optical components and, as such, it is important to understand how they work. The simplest and perhaps most ubiquitous example of a thin film coating is the so-called single layer ***antireflection coating***. Such antireflection

Optical Engineering Science, First Edition. Stephen Rolt.

Companion website: www.wiley.com/go/Rolt/opt-eng-sci

(AR) coatings are present on the majority of consumer and commercial optics and suppress unwanted Fresnel reflections. These reflections serve to reduce throughput and image contrast through the generation of unwanted stray light. Key to understanding the operation of such thin film filters is an appreciation of the critical role played by the unique thin film building block, the ***quarter wave layer***. Such a layer represents a quarter wave thickness in the material (i.e. not air), at some nominal design wavelength. So, for example, if we assume the thin film material has a refractive index of 1.5, then a quarter wavelength layer at 600 nm is represented by a film thickness of 100 nm.

Before we can understand the specific properties of a quarter wavelength layer, we need to establish a generalised model for reflection at an interface provided with a thin film coating. The simplest example is a substrate, with a refractive index of n_0, coated with a single layer of thickness, t, whose refractive index is n_1. In concerning ourselves with the reflection (and transmission) of a beam incident on this surface, we now need to consider the propagation and ***amplitude*** of five separate beams:

1. A_1 – Incident beam (in air/vacuum)
2. A_2 – Reflected beam (in air/vacuum)
3. A_3 – Transmitted beam (forward propagation) in thin film
4. A_4 – Reflected beam (reverse propagation) in thin film
5. A_5 – Transmitted beam into substrate.

This model is illustrated in Figure 10.1.

The amplitude, of course, represents the electric field of the incident and reflected wave(s), as opposed to the flux, which is proportional to the square of the amplitude. We are concerned with the ratio of all amplitudes with respect to the incident amplitude, A_1, so there are four unknowns. These four unknowns are related by four boundary conditions that express the continuity of the electric and magnetic fields at each of the two interfaces. If the amplitudes refer to the electric fields, then these boundary conditions may be expressed as:

$$A_1 + A_2 = A_3 + A_4 \tag{10.1a}$$

$$A_1 - A_2 = n_1 A_3 - n_1 A_4 \tag{10.1b}$$

$$A_3 e^{in_1 kt} + A_4 e^{-in_1 kt} = A_5 \tag{10.1c}$$

$$A_3 n_1 e^{in_1 kt} - A_4 n_1 e^{-in_1 kt} = n_0 A_5 \tag{10.1d}$$

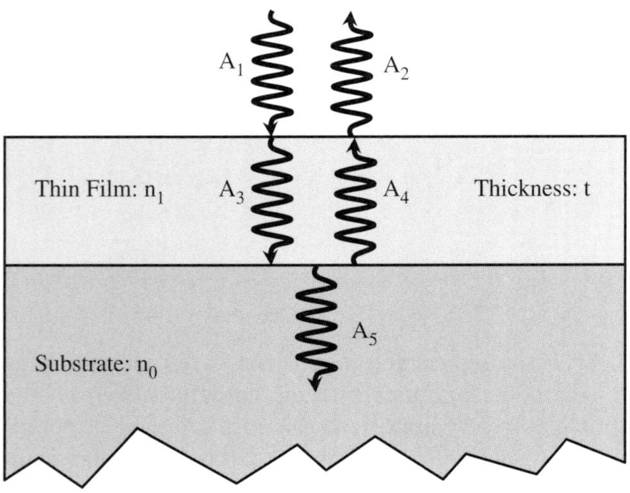

Figure 10.1 Thin film reflectance at an interface.

Solving the above simultaneous equations, we obtain a generalised expression for the reflected amplitude, A_2, in terms of the incident amplitude, A_1:

$$\frac{A_2}{A_1} = \frac{n_1(1-n_0)\cos n_1kt + i(n_1^2 - n_0)\sin n_1kt}{n_1(1+n_0)\cos n_1kt - i(n_1^2 + n_0)\sin n_1kt} \tag{10.2}$$

We might now wish to explore two scenarios. First we may consider the case where $\cos(n_1kt) = \pm 1$ and $\sin(n_1kt) = 0$, i.e. an integer multiple of half wavelengths. The second, and perhaps most critical, scenario occurs where $\cos(n_1kt) = 0$ and $\sin(n_1kt) = \pm 1$, i.e. where the layer thickness is a whole number of wavelengths plus a quarter or three quarters of a wavelength. In the first scenario, the amplitude ratio simplifies to:

$$\frac{A_2}{A_1} = \frac{(1-n_0)}{(1+n_0)} \quad \textit{Half wavelength layer} \tag{10.3}$$

Equation (10.3) is essentially the analogue of the Fresnel equations for reflection that we encountered in Chapter 8. Expressing Eq. (10.3) in terms of the classical Fresnel equation describing reflection at the interface of a medium with a refractive index n', we obtain:

$$\frac{A_2}{A_1} = \frac{(1-n')}{(1+n')} = \frac{(1-n_0)}{(1+n_0)} \text{ and } n' = n_0 \quad \textit{Half wavelength layer} \tag{10.4}$$

This is perhaps a rather trivial example, in that the effective refractive index for Fresnel reflection, of the combined system, is equal to the substrate index, where the film thickness is equal to zero. However, the example for the quarter wave layer is perhaps rather more interesting. In this case, the amplitude ratio becomes:

$$\frac{A_2}{A_1} = \frac{\left(1 - \dfrac{n_0}{n_1^2}\right)}{\left(1 + \dfrac{n_0}{n_1^2}\right)} \text{ and } n' = \frac{n_1^2}{n_0} \quad \textit{Quarter wavelength layer} \tag{10.5}$$

Equation (10.5) demonstrates that, for quarter wavelength films, it is possible to fundamentally transform the refractive properties of a substrate (from the perspective of reflection, not refraction). As we will see in the next section, an important case occurs where the film index is approximately equal to the square root of the substrate index. In this scenario, the effective index, n', is approximately one and Fresnel reflection will be substantially eliminated.

10.2.2 Single Layer Antireflection Coatings

As we pointed out, in the following scenario the reflection coefficient may be reduced to zero.

$$n_1 \approx \sqrt{n_0} \rightarrow n' \approx 1 \rightarrow A_1/A_2 \approx 0$$

This is the basis for the most simple ***antireflection (AR) coating***. A single quarter wavelength layer of low index material deposited on a glass substrate will substantially reduce the Fresnel reflection. This is a simple and highly economical process and is very widely used in a very wide range of low-cost, consumer applications. Many common glasses have a refractive index in the range from 1.5 to 1.6. This suggests that the optimum refractive index for an antireflection coating would be about 1.25. However, unfortunately, no such material exists. Therefore, it is the common material with the lowest refractive index, magnesium fluoride, which is ubiquitous in these applications. Magnesium fluoride has a refractive index of 1.38 through most of the visible and so substantially reduces the Fresnel reflection with a single quarter wave layer.

Worked Example 10.1 ***Single Layer Antireflection Coating***

We will illustrate the operation of a single layer anti-reflection coating with a worked example. A SCHOTT BK7® lens is to be provided with a single layer antireflection coating. The design wavelength is 540 nm and the

BK7 refractive index at this wavelength is 1.519 and that of magnesium fluoride is 1.379. What is the required thickness of the MgF_2 coating for this wavelength? What is the reflectivity of the coated lens at the design wavelength, and how does this compare to the uncoated lens?

First, regarding the thickness of the coating, then the optical path through the coating must be a quarter of a wavelength at the design wavelength of 540 nm:

$$nt = \lambda/4 \qquad t = \lambda/(4n) \qquad t = 540/(4 \times 1.379) = 97.9$$

The thickness of the antireflection coating is 97.9 nm.

From Eq. (10.5), the antireflection coating transforms the effective refractive index in the following way:

$$n' = \frac{n_1^2}{n_0} = \frac{1.379^2}{1.519} = 1.252$$

Thus the effective index of the coated lens is 1.252 and we must now substitute this in the expression for the Fresnel reflection coefficient, R:

$$R = \left[\frac{n'-1}{n'+1}\right]^2 = \left[\frac{1.252-1}{1.252+1}\right]^2 = 0.0125$$

Thus the reflection coefficient for the coated lens is 1.25%.

If the lens had been uncoated, then the working index would be that of the BK7 substrate, 1.519. Therefore the uncoated reflectivity is given by:

$$R = \left[\frac{n-1}{n+1}\right]^2 = \left[\frac{1.519-1}{1.519+1}\right]^2 = 0.0425$$

Thus the reflectivity of the uncoated lens is 4.25% and we have succeeded in reducing the reflectivity of the lens by almost a factor of four by virtue of a simple single layer coating.

We have illustrated the operation of a basic antireflection coating with a simple example. For many visible applications, its performance is perfectly adequate. However, away from the design wavelength, its performance does deteriorate. It is worthwhile, at this stage to illustrate this point by applying Eq. (10.2) to the previous example, across the entire visible spectrum. A plot of the performance of this single layer coating is shown in Figure 10.2.

It is clear that the reflectivity is at a minimum at the design wavelength of 540 nm. Away from this wavelength, the reflectivity steadily increases. Although the performance illustrated in Figure 10.2 is clearly superior to that of the uncoated optic, there are applications which demand substantially lower reflection losses. Indeed, in many cases, reflection losses of a fraction of a percent over a broad wavelength range are required. In this situation, a simple single layer antireflection coating is clearly inadequate. Therefore, we must study more complex coatings, with multiple layers of different materials.

10.2.3 Multilayer Coatings

The most basic extension to our consideration of a single quarter wavelength film is a stack of quarter wavelength films of contrasting refractive index. We now consider a substrate with a refractive index of n_0 and we build upon that substrate a stack of N double layer of quarter wavelength layers of index n_1 and n_2. This is shown in Figure 10.3

In Figure 10.3, only three double layers are shown. In practice, complex multilayer coatings may comprise dozens of these double layers. Because of the manner which the amplitude of reflections at successive interfaces interfere with each other, such multilayer stacks are often referred to as ***interference filters***. In addition, to distinguish them from metallic coatings or solid substrates, they are also referred to as ***dielectric filters***. To illustrate the significance of the multilayer stack, we turn our attention to the property of quarter wavelength layers defined in Eq. (10.5) and apply the first and second layers:

$$n' = \frac{n_1^2}{n_0} \quad \textit{after the first layer.}$$

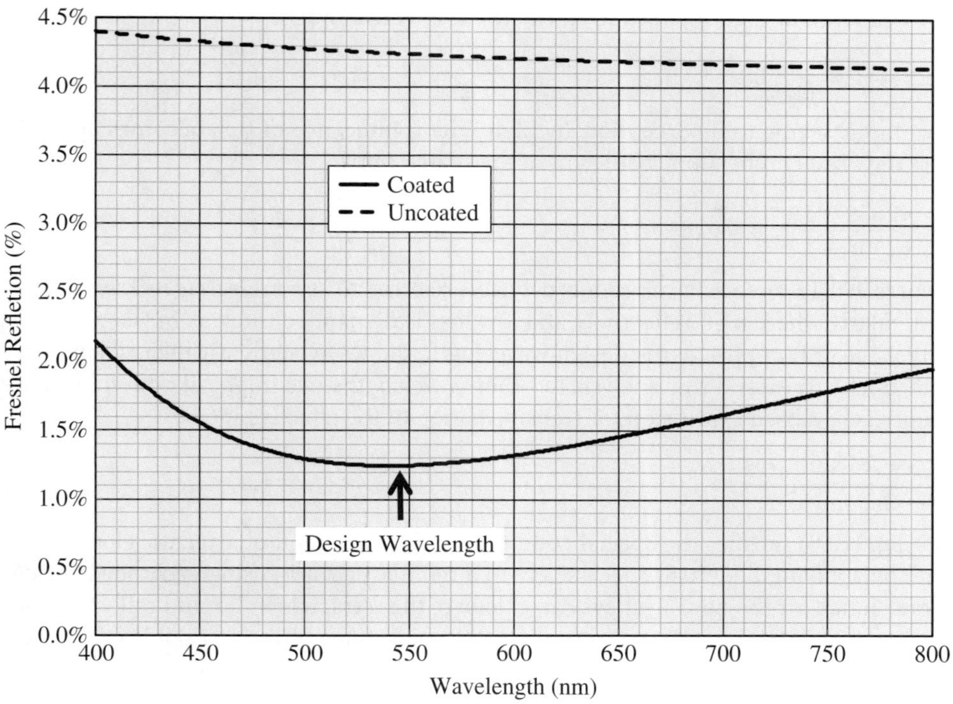

Figure 10.2 Performance of antireflection coating.

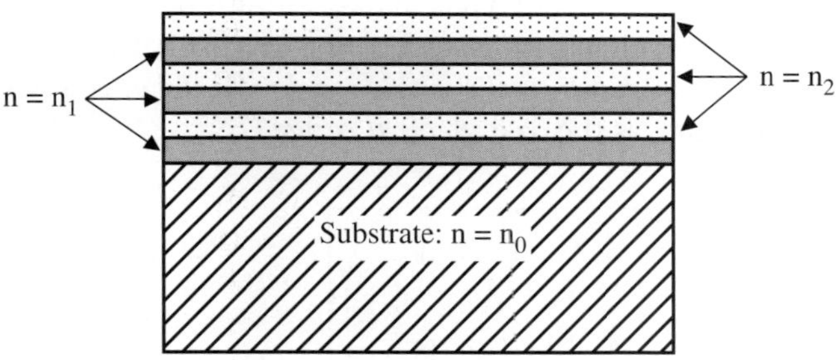

Figure 10.3 Multilayer quarter wavelength stack.

The important insight is that the above refractive index, n' may be used as the substrate index for application of the second layer. Thus:

$$n' = \left(\frac{n_2}{n_1}\right)^2 n_0 \quad \textit{after the second layer} \tag{10.6}$$

Of course, we can now add the remaining N-1 double layers in the stack to derive the following expression for the effective index of the whole stack:

$$n' = \left(\frac{n_2}{n_1}\right)^{2N} n_0 \tag{10.7}$$

To understand the significance of this expression, we now take an example of a multilayer stack with 10 double quarter wavelength layers, comprising a low index material (magnesium fluoride) with an index of 1.38 and a high index material (zinc oxide) with an index of 2.0. The substrate material in this case is BK7, with a refractive index of 1.519. It is straightforward to calculate the effective index:

$$n' = \left(\frac{2.0}{1.38}\right)^{20} \times 1.519 \quad n' = 2538$$

This effective refractive index seems to be exceptionally high. The significance of this becomes apparent when we seek to calculate the Fresnel reflection coefficient of the coated optic:

$$R = \left(\frac{n' - 1}{n' + 1}\right)^2 = \left(\frac{2538 - 1}{2538 + 1}\right)^2 = 0.9984$$

The reflection coefficient is 99.84% and we have produced what is referred to as a ***dielectric mirror***. Dielectric mirrors are used particularly where high reflectivity and low loss is demanded, such as in laser mirrors. Of course, this reflectivity value only applies to the design wavelength, the wavelength at which all layers are a quarter of a wavelength thick. To extend this analysis to other wavelengths, we might like to manipulate Eq. (10.2), to provide a more generalised expression for transformation of the effective index:

$$n' = \frac{n_0 n_1 \cos(n_1 kt) - i n_1^2 \sin(n_1 kt)}{n_1 \cos(n_1 kt) - i n_0 \sin(n_1 kt)} \tag{10.8}$$

We can see clearly that Eq. (10.8) satisfies the special conditions for the half and quarter wavelength film thicknesses. Furthermore, by applying Eq. (10.8) successively to a multilayer stack, using each newly calculated index as the 'substrate' index for the next layer, we can calculate the refractive index and reflectivity for the multilayer stack for all wavelengths. Continuing with our previous example of the 10 double layer stack, Figure 10.4 shows a plot of the reflectivity against wavelength.

Whilst the reflectivity of the filter is high, the spectral width over which this reflectance is maintained is fairly narrow. This tends to be the characteristic of simple multilayer coatings, such as this, with successive layers increasing reflectivity, but also reducing the spectral bandwidth of the reflectance. The periodicity of ripples seen in the filter response in Figure 10.4 is related to the number of layers in the coating. The greater the number of layers, then the closer the spacing between the individual ripples. In fact, the spacing between the ripples is consistent with an integral multiple of half wavelengths ***across the whole stack***. We will encounter this phenomenon a little later and this periodicity is referred to as the ***free spectral range*** *(FSR)*.

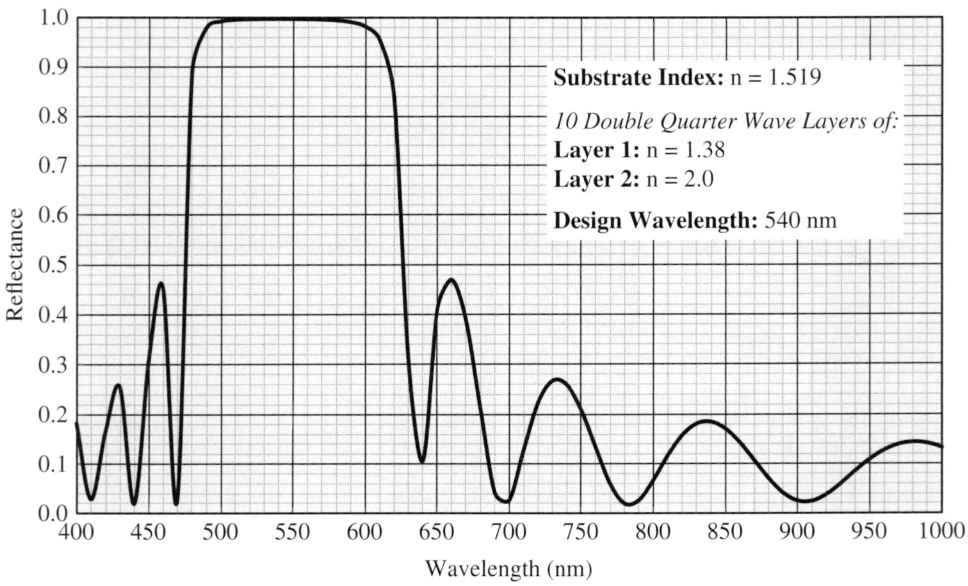

Figure 10.4 Multilayer stack – reflectivity vs wavelength (design wavelength 540 nm).

In this particular example, the coating provides high ***reflectance*** over a relatively narrow spectral range. However, it is quite possible to design a coating that will ***transmit*** over a narrow spectral range. This forms the basis of the ***bandpass filter***, which is a very widely used component in many laboratory and scientific applications. We will shortly return to the design of multilayer coatings. However, we will now further explore the properties of single layer coatings and, in particular, we will examine thin film metal coatings with complex indices.

10.2.4 Thin Metal Films

To understand the behaviour of thin metal films, we must incorporate the notion of a complex refractive index into any analysis. To that extent, definition of the problem is fairly straightforward, as the analysis proceeds along the lines of Eq. (10.2), except that a complex index is substituted for n_1.

Thin metal films are used extensively in a range of applications, particularly in ***neutral density filters***, where we are seeking to attenuate transmitted light by a factor that is broadly independent of wavelength. In this case, we are interested in understanding the amplitude of transmitted light as well as that of the reflected light. In the previous analysis, we made the implicit assumption that the total flux (i.e. the square of the amplitude) is conserved and that which is not reflected is transmitted. However, we can no longer make that assumption in this case, as the metallic layer does have some propensity towards absorption. As such, we need to calculate the transmitted flux directly from Eqs. (10.1a)–(10.1d). However, in calculating the transmitting ***amplitude***, A_5, we must be somewhat careful in establishing its connection to the transmitted ***flux***. Calculating the amplitude is straightforward and is given by:

$$\frac{A_5}{A_1} = \frac{4n_1}{(n_1 - n_0)(1 - n_1)e^{in_1kt} + (n_0 + n_1)(1 + n_1)e^{-in_1kt}} \tag{10.9}$$

Since, in this instance, the transmitting medium has a different index to that of the incident medium the transmittance, T, is not simply the square of Eq. (10.9). Rather, it is determined by the product of the square of the amplitude and the index of the medium:

$$T = n_0\left[\frac{A_5}{A_1}\right]^2 = \frac{16n_0n_1^2}{[(n_0 - n_1)(1 - n_1)e^{in_1kt} + (n_0 + n_1)(1 + n_1)e^{-in_1kt}]^2} \tag{10.10}$$

For a metal film, we must recognise, of course, that the index n_1 is complex. The metal film has a thickness, t, and a ***complex index***, $n + i\kappa$. In dealing with this scenario, it has proved convenient to cast Eqs. (10.9) and (10.10) in terms of complex exponentials, rather than trigonometric functions. Strictly speaking, of course, Eq. (10.10) rather than being the square of Eq. (10.9) represents the product of the complex conjugates. It will further help our understanding if we re-cast Eq. (10.2) in terms of complex exponentials:

$$\frac{A_2}{A_1} = \frac{(n_1 - n_0)(1 + n_1)e^{in_1kt} + (n_1 + n_0)(1 - n_1)e^{-in_1kt}}{(n_1 - n_0)(1 - n_1)e^{in_1kt} + (n_1 + n_0)(1 + n_1)e^{-in_1kt}} \tag{10.11}$$

We assume that the film is deposited on a glass substrate of index n_0, then the reflected amplitude is given by Eq. (10.2), but with the complex value, $n + i\kappa$ substituted for n_1. We will now take a specific example, that of chromium. Chromium has a complex impedance of 2.74 + i4.2 at 540 nm. Using Eqs. (10.9)–(10.11), it is possible to calculate the dependence of reflection and transmission on film thickness at 540 nm. This is shown in Figure 10.5.

What is striking about Figure 10.5 is that to produce any meaningful transmission, the film must be very thin, of the order of a few nanometres. For example, an attenuation of about 50% is produced by a film as thin as 4 nm. It is possible, at this stage to plot the attenuation for a specific thickness (e.g. 4 nm) as a function of wavelength, based on complex index data. This is shown in Figure 10.6.

As previously outlined, thin metal films form the basis of attenuating neutral density filters. An ideal neutral density filter should provide attenuation that is independent of wavelength. One can see, from Figure 10.6

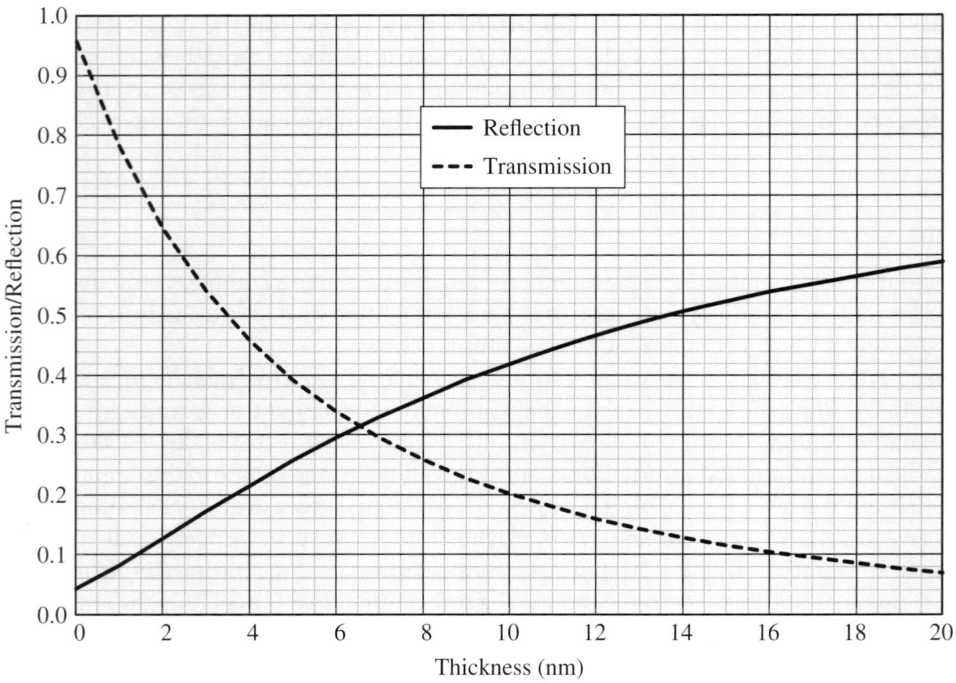

Figure 10.5 Transmission and reflection of a thin chromium film at 540 nm.

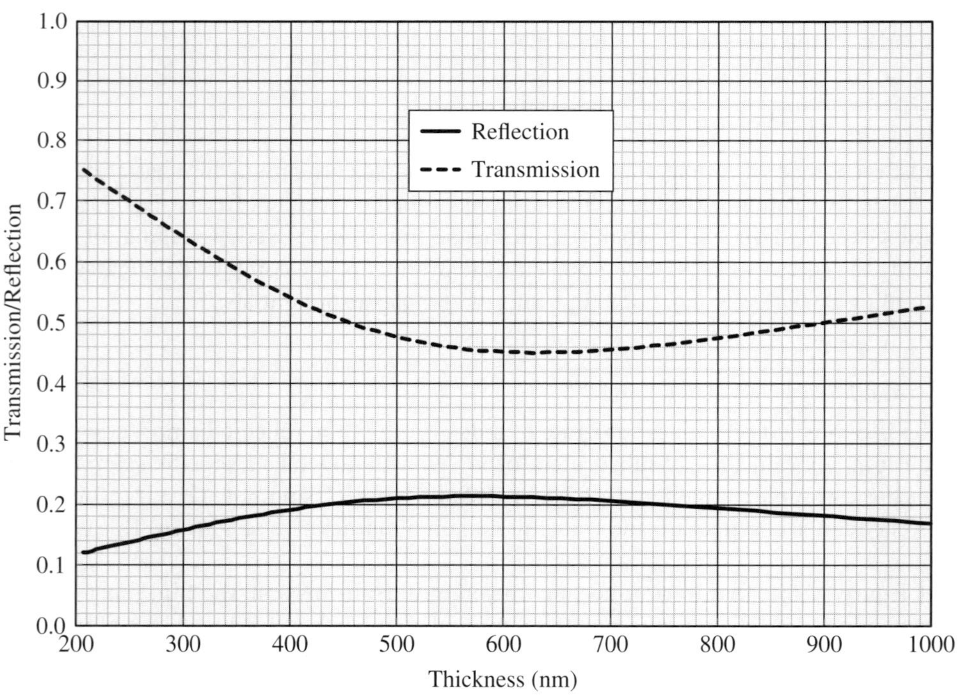

Figure 10.6 Transmission of 4 nm chromium film vs. wavelength.

that the transmission of the chromium film is modestly flat between about 400 and 1000 nm, but there is some variation. In practice, thin film neutral density filters are fabricated from a proprietary combination of chromium and nickel and other metals, such as iron and are designed specifically to produce a flat response.

10.2.5 Protected and Enhanced Metal Films

Reflective metallic coatings are formed by the evaporation of a thin layer of metal onto a polished glass or similar substrate. As demonstrated by the previous analysis, for the design of a reflective coating, the layer must be sufficiently thick to avoid reduction of efficiency through parasitic transmission. In addition, in practice, the thickness must be sufficient to mask pinholes and other defects that are inevitably associated with the deposition process. A typical thickness is in the region of 250–500 nm.

Most useful coatings, such as aluminium, silver, and gold, are relatively soft. If the optic is to be reasonably robust with respect to mechanical handling, particularly if it is to be cleaned, then the soft coating should be protected with a thin overcoating of durable material. From the perspective of simple protection, coatings such as silicon monoxide, silicon dioxide (silica) or magnesium fluoride are often used. However, addition of a thin film of material inevitably impacts the optical properties of the reflective film. Such additional coatings are often described as ***'protected'*** or ***'enhanced'***. For protected coatings, the emphasis is on ***protection*** of the underlying metal, often at the expense of optical performance. Conversely, for enhanced coatings, some effort is expended to improve the reflectivity performance of the underlying metal.

The structure of protected coatings is generally simple, generally comprising a single layer of additional material. One can regard this layer as a quarter wavelength layer at some design wavelength. To understand the impact of adding this layer to the reflectivity, we need to express the reflectivity of the original metal film in terms of its complex index, $n + \mathrm{i}\kappa$:

$$R = \frac{(1-n)^2 + \kappa^2}{(1+n)^2 + \kappa^2} \tag{10.12}$$

Equation (10.12), of course, applies to normal incidence. For most metal films, we may make a working approximation, wherein $n << \kappa$. In this case, the reflectivity is approximately given by:

$$R \approx 1 - \frac{4n}{\kappa^2} \tag{10.13}$$

If we add a single quarter wavelength layer of material with a refractive index of n_1, the complex index is transformed according to:

$$\mathbf{n} = \frac{n_1^2}{n + i\kappa}$$

The reflectivity is then given by:

$$R = \frac{(n_1^2 - n)^2 + \kappa^2}{(n_1^2 + n)^2 + \kappa^2} \text{ and } R \approx 1 - \frac{4n_1^2 n}{\kappa^2} \tag{10.14}$$

Inspection of Eq. (10.14) suggests that if $n_1 > 1$, as it inevitably will be, then the reflectivity will be reduced. Therefore, the impact of a simple single layer coating will always be to reduce the reflectivity of the metallic coating at the effective design wavelength. To improve this one might consider adding an extra quarter wave layer with an index of n_2. By virtue of the approximation previously outlined the reflectivity of this double layer may be approximated as:

$$R \approx 1 - \frac{4n_1^2 n}{n_2^2 \kappa^2} \tag{10.15}$$

From Eq. (10.15), If $n_2 > n_1$, then the reflectivity will be significantly enhanced. For example, if an aluminium film were to be coated by a single (quarter wave) layer of magnesium fluoride (n ~ 1.38) followed by a single

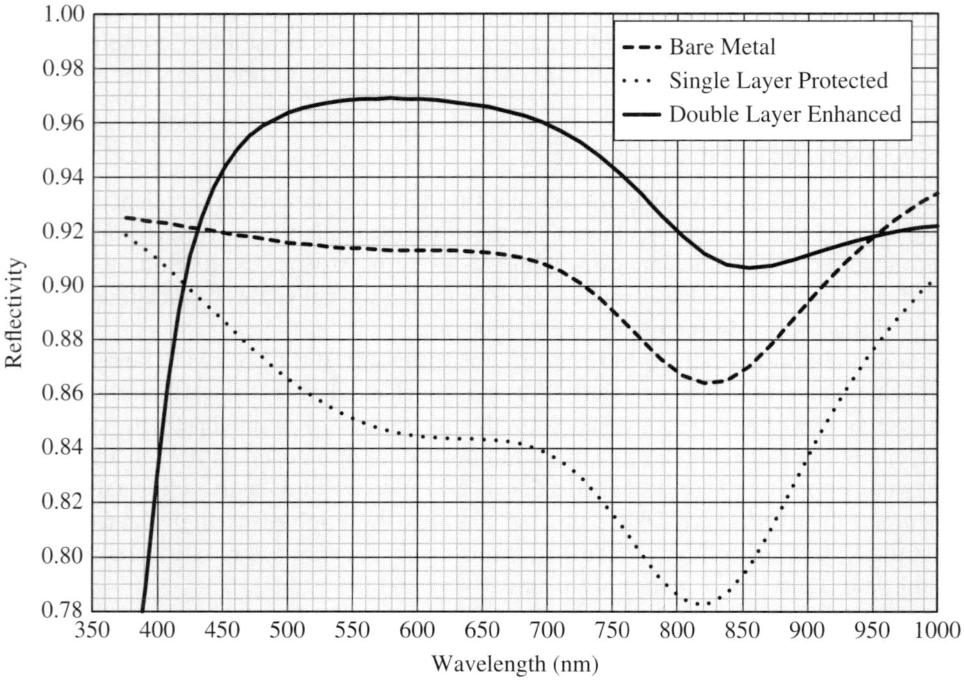

Figure 10.7 Reflectivity of aluminium coatings.

layer of titanium oxide (n ∼ 2.2), then this would represent an enhanced coating. This analysis has proceeded on the understanding that a single layer represents a quarter of a wave (or some odd multiple) at the design wavelength. On the other hand, for a half wavelength layer, or integer multiple, then addition of the layer does not amend the refractive properties of the substrate and the reflection remains unchanged. More generally, it is possible to calculate the reflectivity of these films for all wavelengths, by using Eq. (10.10) to transform the complex index of aluminium. By way of illustration, we chose three examples:

1. Bare aluminium
2. *Protected aluminium.* Quarter wavelength layer of MgF_2 – design wavelength 550 nm
3. *Enhanced aluminium.* Quarter wavelength layers of $MgF_2 + TiO_2$ – design λ 550 nm

Figure 10.7 shows the reflectivity for all three cases.

It is clear that the performance of 'protected' aluminium is worse than that of the bare metal across the whole visible spectrum and into the near infrared. By contrast, the performance of the enhanced coating is superior in the visible spectrum in the near infrared up to 900 nm. Of course, enhanced coatings may be tailored by adjusting the design wavelength.

10.3 Filters

10.3.1 General

A basic understanding of the properties of thin film coatings underpins many aspects of the technology of optical filters. At this point, we will embark on a more general discussion of optical filters and their application. It would be useful to set out the different filters into their broad application categories; these are set out below:

1. Antireflection Coatings
2. Edge filters – long pass and short pass filters

3. Bandpass filters and notch filters
4. Neutral density filters
5. Polarising filters
6. Beamsplitters
7. Dichroic filters
8. Etalon Filters

10.3.2 Antireflection Coatings

In our introduction to the analysis of thin films we examined the addition of a single quarter wavelength coating of low index material to a glass substrate. At the design wavelength, this produces a reduction in the reflection coefficient from about 4% to about 1% at the design wavelength. As shown in Figure 10.2, the performance tends to diminish away from the design wavelength. To obtain improved performance, more complex multilayer coatings are used. These coatings are described by the generic term, ***broadband antireflection coatings.*** Reflectivities are reduced to a fraction of a percent over an extended range. Of course, this improvement comes at the expense of added cost and complexity. Generally, the incorporation of more thin film layers in a design boosts performance. Performance of a typical broadband antireflection coating is shown in Figure 10.8. For a wide range of wavelengths, the reflection is less than 0.5%, averaging about 0.2% between 350 and 700 nm. Design of these coatings is, as suggested previously, based on the construction of a stack of contrasting quarter wavelength layers. We will examine the design of multilayer coatings in a little more detail later in this chapter.

10.3.3 Edge Filters

Edge filters can either be described as short pass filters or long pass filters. In the former case, the filter is designed to transmit wavelengths that are shorter than some critical value. Alternatively, a long pass filter

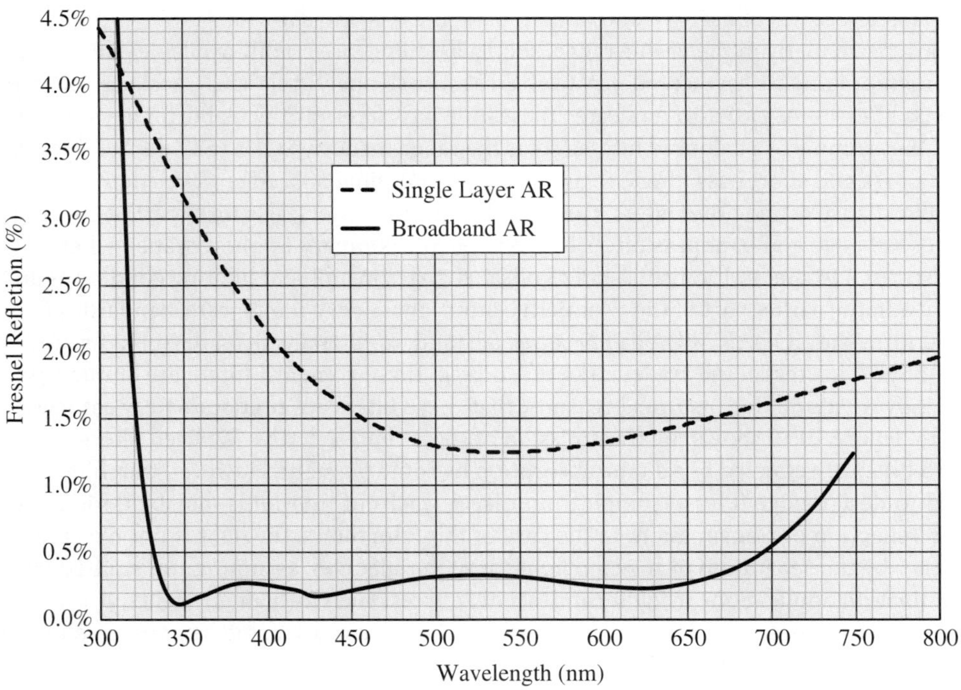

Figure 10.8 Performance of typical broadband antireflection coating.

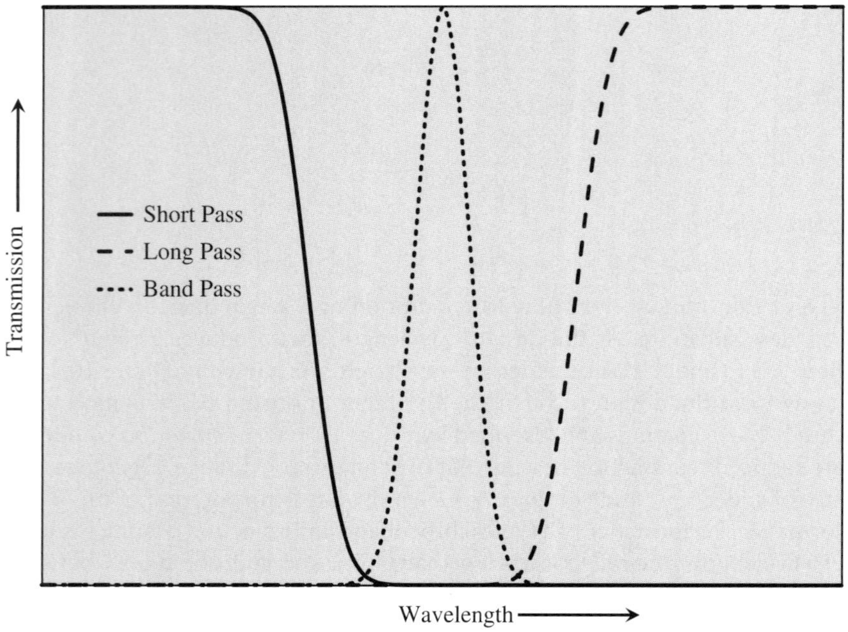

Figure 10.9 General characteristics of edge filters.

will transmit wavelength greater than some specific value. These filters fall into two categories, based on their underlying design. Firstly, there are coloured edge filters, which employ some inorganic or organic dye incorporated into the volume of a transparent, glassy material. Secondly, thin film technology may be used to design edge filters with deliberately engineered cut off values.

The general spectral characteristics of the different filter types is illustrated in Figure 10.9. In addition to edge filters, the generic characteristics of narrowband or bandpass filters is also shown.

For general applications, particularly in the visible range, solid glass filters are readily available. Since many pigments, particularly organic pigments have a tendency to absorb more strongly at shorter wavelengths, there is a tendency for long pass filters to be easier to replicate. These long pass filters are available for a wide range of cut-off wavelengths. One example of a coloured filtered set is the range made by SCHOTT Glass. Table 10.1 sets out the range of long pass and short pass filters, together will the nominal 50% edge wavelength.

The list presented is not exhaustive and other filters, particularly bandpass filters, also form part of the range of standard filters. The use of edge and bandpass filters has long been customary in photography where it is important to make adjustments to spectral transmission to offset changes in ambient lighting conditions and film or detector sensitivity. In the earlier part of the twentieth century these edge filters were standardised for the visible spectrum and form the KODAK WRATTEN series of standard filters, which are still in use today (KODAK™ and WRATTEN™ are registered trademarks of Kodak). These standard filters are summarised very briefly in Table 10.2. It is natural, for historical reasons, that these filters are predominantly associated with the visible spectrum. Figure 10.10 shows the transmission of some representative Wratten filters.

Most of these standard filters are effected by volume absorption by some pigment. However, thin film technology can also be used to design edge filters. Complex multilayer stacks of thin films may be designed to produce sharp cut-off in reflection or transmission. In the case of thin film filters, the light is not rejected by virtue of absorption, rather by the separation of rejected light through reflection, assuming transmitted light is desired. In some illumination applications using incandescent sources, such as filament lights, it is desirable to reject the large burden of infrared radiation which would otherwise cause cameras and detectors to overheat. In this case, a specialised edge filter, called a ***hot mirror***, is used to reject near infrared radiation by reflection, whilst transmitting the visible light. As such, the hot mirror acts as a short pass filter. Conversely

Table 10.1 List of short pass and long pass filters (SCHOTT glass). Reproduced with the permission for SCHOTT N-BK7®

Code	Type	Wavelength (nm)	Code	Type	Wavelength (nm)	Code	Type	Wavelength (nm)
WG225	Long	225	GG435	Long	435	RG665	Long	665
WG280	Long	280	GG445	Long	445	RG695	Long	695
WG295	Long	295	GG455	Long	455	RG715	Long	715
WG305	Long	305	GG475	Long	475	RG780	Long	780
WG320	Long	320	GG495	Long	495	RG830	Long	830
WG335	Long	335	OG515	Long	515	RG850	Long	850
WG345	Long	345	OG530	Long	530	RG1000	Long	1000
WG360	Long	360	OG550	Long	550	KG1	Short	750
GG375	Long	375	OG570	Long	570	KG2	Short	850
GG385	Long	385	OG590	Long	590	KG3	Short	700
GG395	Long	395	RG610	Long	610	KG5	Short	680
GG400	Long	400	RG630	Long	630			
GG420	Long	420	RG645	Long	645			

Table 10.2 Wratten series of filters. Reproduced with permission of Kodak and Wratten.

No.	Description	No.	Description	No.	Description	No.	Description
0	Colourless	33	Magenta	58	Green	86B	Yellowish
1	UV absorb	34	Violet	58A	Green	86C	Yellowish
1A	Blue absorb	34A	Magenta	59	Green	80A	Daylight
2B	Yellow	35	Deep violet	59A	V. Light green	81	Yellowish
3	Yellow	36	Blue-Green	60	Green	81A	Yellowish
3N5	Yellow	38	Blue-Green	61	Green	81B	Yellowish
4	Yellow	38A	Blue-Green	64	Blue-Green	81C	Yellowish
6	Yellow	38	Blue-Green	65	Blue-Green	81D	Yellowish
8	Yellow	40	Blue-Green	65A	Blue-Green	81EF	Yellowish
8N5	Yellow	44A	Blue-Green	66	Blue-Green	82	Bluish
9	Yellow	45	Blue-Green	67A	Blue-Green	82A	Bluish
11	Yellow	45A	Blue-Green	70	Dark red	82B	Bluish
12	Yellow	46	Blue	72B	Orange-Yellow	82C	Bluish
13	Yellow	47	Blue	73	Yellow-Green	83	Yellowish
15	Yellow	47B	Blue	74	Green	85	Orange
16	Yellow	48	Blue	75	Blue-Green	85B	Orange
18A	Yellow	48A	Blue	76	Violet	79	Colour balance
21	Orange	49	Dark blue	77	Mercury green	87	Infrared
22	Orange	49B	V. Dark blue	77A	Mercury green	87C	Infrared
23A	Light red	50	V. Dark blue	78	Bluish	88A	Infrared
24	Red	52	Light green	78AA	Bluish	89B	Infrared
25	Red	53	Green	78A	Bluish	90	Narrowband
26	Red	55	Green	78B	Bluish	96	Neutral
29	Red	56	Light green	78C	Bluish	97	Dichroic
31	Magenta	57	Green	86	Yellowish	102	Correction
32	Magenta	57A	Light green	86A	Yellowish	106	Correction

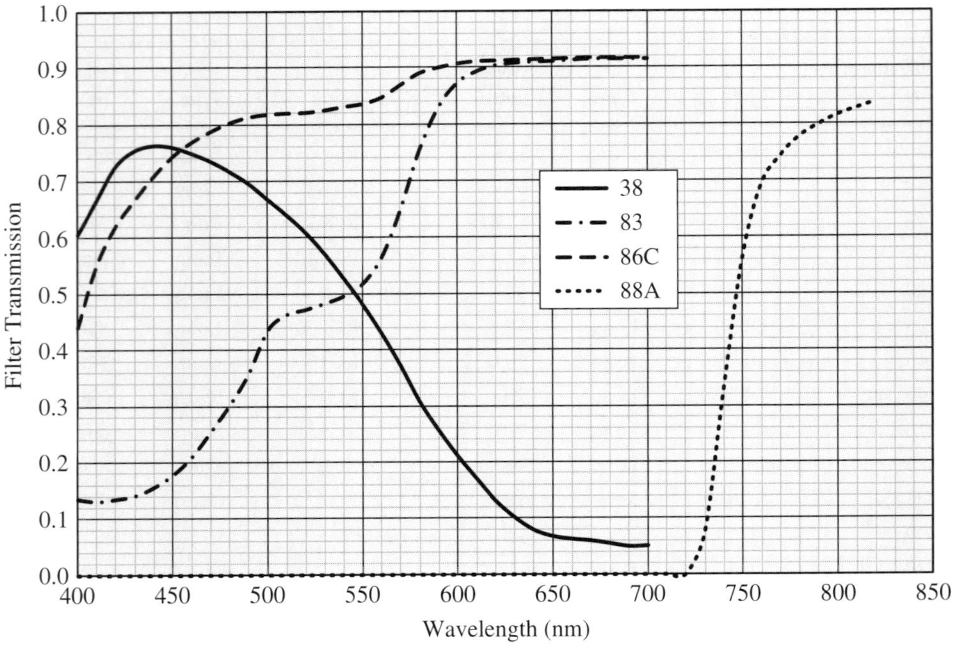

Figure 10.10 Transmission of some WRATTEN™ filters.

a ***cold mirror*** is used where visible light must be rejected by reflection, allowing for transmission of infrared radiation; this is a long pass filter.

10.3.4 Bandpass Filters

Bandpass filters are designed to transmit a range of wavelengths between two 'cut off' wavelengths. They are characterised by their central wavelength and their full-width half-maximum (FWHM) transmission and by their maximum transmittance. Where the full width half-maximum is small, they are generally referred to as ***notch filters***. Figure 10.11 shows the typical characteristics of a bandpass filter.

Bandpass filters are generally formed from multilayer dielectric coatings. As well as specifying the centre wavelength and FWHM, equally critical is the filter's ability to block wavelengths outside the passband. This is described by the ***blocking range***, the range of wavelengths over which the attenuation provided by the filter is greater than some nominal value. Attenuation or ***blocking*** is described by the ***optical density***, OD, which is equal to the logarithm (base 10) of the ratio of the input and output flux.

$$OD = \log_{10}\left[\frac{\Phi_{in}}{\Phi_{out}}\right] \tag{10.16}$$

For example, for a 500 nm bandpass filter, the blocking range might extend from 300 to 450 nm and 550 to 1000 nm and, in that range, the blocking is required to be greater than OD4, i.e. a factor of 10 000. In addition, in some applications, the steepness in the transition from transmission to blocking may be important. This is described by the ***filter slope***, which represents the wavelength difference between the 50% transmission wavelength and reaching the blocking condition (e.g. OD = 4). It is usually expressed as a percentage of the central wavelength.

Bandpass and other multilayer filters are designed for some specific angle of incidence. The quarter wavelength condition is met only for some design angle of incidence, usually 0° or 45°. In the case of filters designed for normal incidence, a deviation of ±15° is considered acceptable. Tilting a multilayer filter has

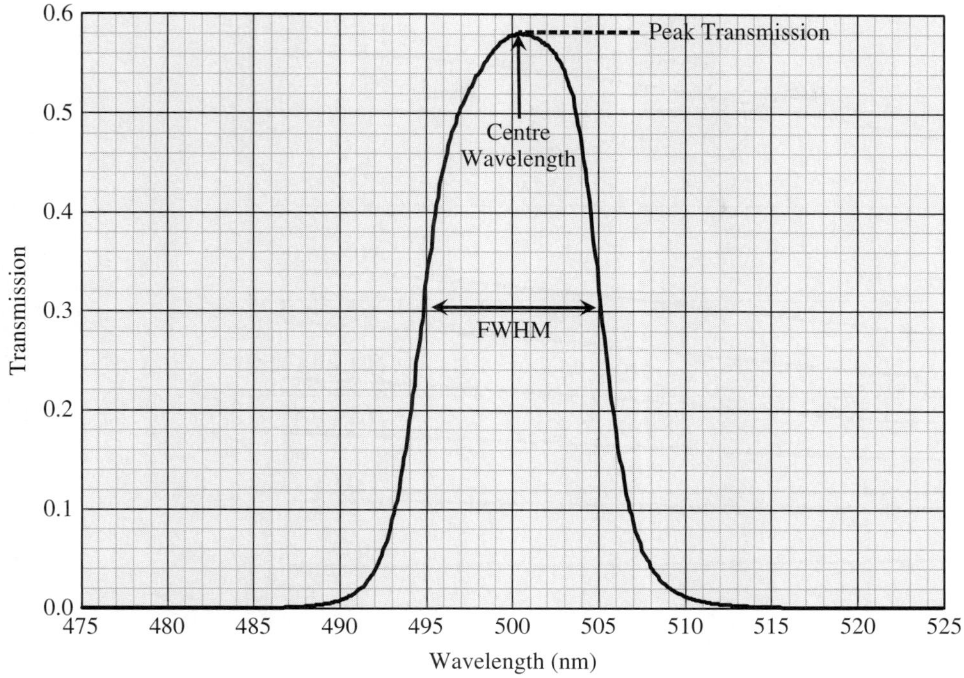

Figure 10.11 Bandpass filter characteristics.

the effect of shifting the design wavelength. Change in the angle of refraction through a given layer causes a change in the effective optical path through the medium. If the angle of refraction in a layer of refractive index n_1 is ϕ, then the path length and effective centre wavelength will change as follows:

$$\Phi_{normal} = n_1 t \qquad \Phi_{tilt} = n_1 t \cos\varphi \qquad \lambda' = \lambda_0 \cos\varphi \tag{10.17}$$

Equation (10.17) suggests that the effect of tilting is to shift the centre wavelength towards shorter wavelengths. For example, if a 500 nm bandpass filter is tilted by 15° and we assume that the aggregate refractive index of the multilayer stack is 1.7, then ϕ is equal to 8.76° and (from Eq. (10.17)), the centre wavelength shifts to 494 nm. More subtle changes in filter behaviour are produced by changes in temperature. The process is complicated as the thermal expansion of each layer is constrained by its attachment to the substrate and other layers. The refractive index of the layer will also change as we have previously seen. Overall, the effect of increasing temperature is to shift the filter response to longer wavelengths. A representative shift is around 20 parts per million per degree centigrade.

10.3.5 Neutral Density Filters

The purpose of ***neutral density filters*** is to provide controlled attenuation of light independent of wavelength. That is to say, the degree of attenuation should be spectrally flat, at least nominally. Most generally, the purpose of neutral density filters is to limit the flux seen by optical detectors or sensors to reasonable bounds.

The most common implementation of neutral density filters is the provision of a very thin metal film on either a glass or fused silica substrate. Provision of a fused silica substrate allows the filter to operate in the ultraviolet. Thin film filters are composed of a mixture of nickel and chromium and other similar metals, such as iron. Control of elemental composition helps to optimise spectral neutrality. The attenuation of neutral density filters is specified in terms of their optical density, as per Eq. (10.16). The range embraced by these filters extends from an optical density of 0.1 to an optical density as high as 4, representing an attenuation of four

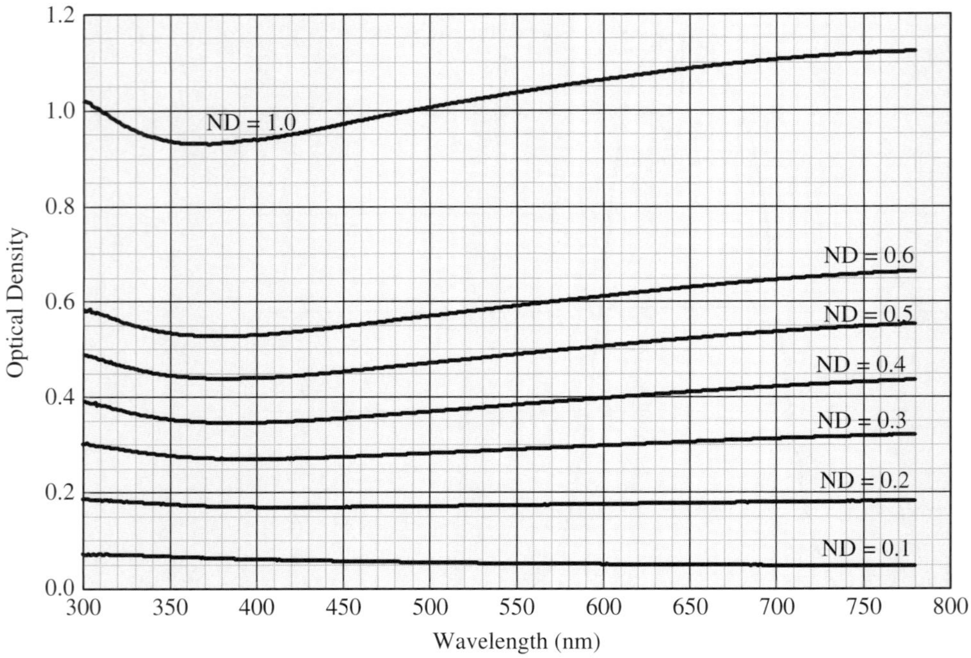

Figure 10.12 Typical characteristics of neutral density filters.

orders of magnitude. In the case of thin film filters, the attenuation is simply determined by the thickness of the film. Figure 10.10 shows the transmission versus wavelength of a range of neutral density filters. Although the filters are nominally 'neutral', in practice, there is some dependence of attenuation on wavelength.

The filters illustrated in Figure 10.12 used fused silica substrates and their usefulness extends to the ultraviolet. In addition to thin film neutral density filters, there are also volume absorbing filters. These are effectively fabricated from 'grey glass' and absorb light within the volume of the substrate; their performance does not extend to the ultraviolet. Figure 10.12 shows a range of different neutral density filters each with a different optical density, reflecting a different film thickness. It is possible, by changing film thickness across a substrate geometry to produce a ***variable neutral density filter***. These filters are usually in the form of a disc, where the film thickness varies tangentially. By rotating this disc, a variable attenuation may be produced.

10.3.6 Polarisation Filters

Thus far, we have primarily been interested in the dependence of thin film filter behaviour with respect to wavelength. Our treatment of thin film behaviour is ultimately based on an analysis of Fresnel reflection. Of course, for angles other than normal incidence, as we saw in Chapter 8, reflectance is also dependent upon polarisation. For reflection from an uncoated substrate, this dependence is relatively weak. However, multilayer coatings can be used to greatly accentuate this difference to the point where one polarisation direction is transmitted whereas the other is completely reflected. The most common format for the polarising filter is the so-called '***polarising beamsplitter***'. Here a multilayer coating is deposited onto a 45° surface. This is illustrated in Figure 10.13. In this case one polarisation is reflected by the 45° surface and deviated by 90°, whereas the other is transmitted.

To understand the polarisation dependence of thin film reflectivity, we need to recast Eq. (10.2) to cover *s* and *p* polarisations separately. To this end, we will designate the angle of incidence as θ, the angle of refraction in the film as ϕ, and the angle of refraction in the substrate as Δ. The amplitude of the reflected light for the

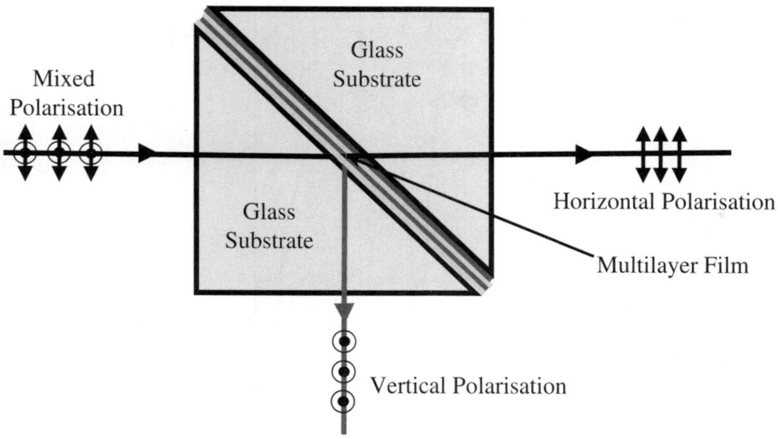

Figure 10.13 Polarising beamsplitter.

two polarisations is given by:

$$\left[\frac{A_2}{A_1}\right]_S = \frac{(n_1 \cos\theta \cos\phi - n_1 n_0 \cos\phi \cos\Delta)\cos\delta - i(n_0 \cos\theta \cos\Delta - n_1^2 \cos^2\phi)\sin\delta}{(n_1 \cos\theta \cos\phi + n_1 n_0 \cos\phi \cos\Delta)\cos\delta - i(n_0 \cos\theta \cos\Delta + n_1^2 \cos^2\phi)\sin\delta} \tag{10.18a}$$

$$\left[\frac{A_2}{A_1}\right]_P = \frac{(n_1 \cos\Delta \cos\phi - n_1 n_0 \cos\theta \cos\phi)\cos\delta - i(n_0 \cos^2\phi - n_1^2 \cos\theta \cos\Delta)\sin\delta}{(n_1 \cos\Delta \cos\phi + n_1 n_0 \cos\theta \cos\phi)\cos\delta - i(n_0 \cos^2\phi + n_1^2 \cos\theta \cos\Delta)\sin\delta} \tag{10.18b}$$

As previously, we can express the addition of one extra thin film layer as a transformation in the effective index:

$$n'_S = \frac{n_1 n_0 \cos\Delta \cos\delta - i n_1^2 \cos\phi \sin\delta}{n_1 \cos\phi \cos\delta - i n_0 \cos\Delta \sin\delta} \tag{10.19a}$$

$$n'_P = \frac{n_1 n_0 \cos\phi \cos\delta - i n_1^2 \cos\Delta \sin\delta}{n_1 \cos\Delta \cos\delta - i n_0 \cos\phi \sin\delta} \tag{10.19b}$$

In applying Eqs. (10.19a) and (10.19b) to the build up of successive layers, we need to be somewhat careful of the definition of the angle, Δ. This should be taken to be the angle of refraction in the preceding layer or the substrate. That said, our interest in designing a thin film stack lies in its behaviour at a quarter wavelength thickness. In this instance the two index transformations simplify to:

$$n'_S = \frac{n_1^2 \cos\phi}{n_0 \cos\Delta} \tag{10.20a}$$

$$n'_P = \frac{n_1^2 \cos\Delta}{n_0 \cos\phi} \tag{10.20b}$$

A successful design is based upon maximising the contrast between the two indices in Eqs. (10.20a) and (10.20b). This amounts to maximising the ratio between the cosines of the refracted angle in the two thin film media that make up the dielectric stack. This is the logic and the justification for the beamsplitter cube geometry shown in Figure 10.13. That is to say, the angle of incidence within the substrate material of refractive index, say 1.52, is 45°, rather than its being the incident angle in air. This strategy maximises the ratio of the effective indices for the two polarisations.

Worked Example 10.2 ***Polarising Beamsplitter***

We illustrate the application of Eqs. (10.19a) and (10.19b) with a representative example. A polarising beamsplitter is to be constructed with a design wavelength of 600 nm. A series of quarter wavelength layers is

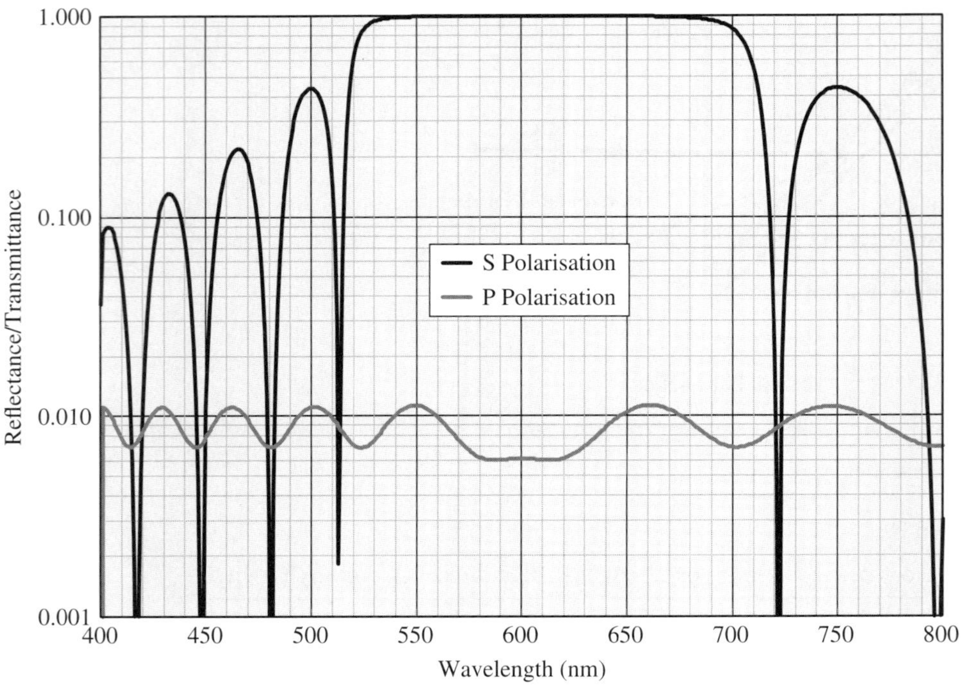

Figure 10.14 Polarising beam splitter (design wavelength 600 nm).

built up on a substrate of refractive index 1.52 (nominally SCHOTT BK7). The geometry is as per the cube beamplitter illustrated in Figure 10.13. That is to say, the layers are built up on the 45° face of a truncated cube, with the remainder of the cube placed on top of the completed dielectric stack. The first layer has a refractive index of 1.7 and the second layer a refractive index of 1.38 (nominally alumina and magnesium fluoride). A total of 10 double layers is assembled and to the top layer is attached the other portion of the cube beamsplitter. Equation (10.19a, 10.19b) may be applied sequentially, for example, using a spreadsheet tool, or other computational assistance. The reflectance at the beamsplitter for the *s* and *p* polarisations is shown in Figure 10.14. Clearly, the *s* polarisation is reflected very efficiently around the design wavelength of 600 nm, whereas the *p* polarisation is substantially transmitted. This illustrates the design of a polarising beamsplitter.

In the context of Figure 10.13, the *p* or parallel polarisation is represented by horizontally polarised light and is predominantly transmitted rather than reflected. Conversely, the vertical, or *s* polarisation, is reflected at 90°.

10.3.7 Beamsplitters

In the previous section, we examined the design of polarising beamsplitters. There are, however, a number of applications where we might wish to divide a beam into two portions without regard to polarisation. The use of amplitude division, in this way, predominates in interferometry, where we wish to characterise the phase difference between two comparable beams by interference. Historically, in early experiments, this was accomplished by a 'half-silvered mirror'. In effect this is a reflective neutral density filter, made from a very thin metal film and designed to reflect approximately 50% of the light and transmit the remainder. However, this function may be more efficiently accomplished by means of multilayer dielectric film. Such a ***beamsplitter*** is often in the form of a beamsplitter cube, as for the polarising beamsplitter. However, care is taken to minimise the polarisation sensitivity of the design. Dielectric beamsplitters are most commonly designed as 50 : 50 beamplitters, where 50% of the light is transmitted and the remainder reflected. However, other values are possible, such as 70 : 30 beamsplitters.

Interferometry is a common application area for beamsplitter devices; we will examine this topic in more detail later in the book. Other applications are found in photometry or radiometry where we wish to sample a representative portion of an optical beam by splitting off a portion and monitoring its flux. In some cases, where sampling a small percentage of the beam is adequate, a single Fresnel reflection may suffice. That is to say, a simple uncoated glass plate may suffice in this instance. An alternative is the so-called '***Pellicle Beamsplitter***'. The pellicle beamsplitter consists of a glass or other transmissive substrate coated with a fine pattern of reflective (e.g. aluminium) dots. For a 50% beamsplitter, the reflective portion is designed to cover half the substrate.

10.3.8 Dichroic Filters

In reality, a dichroic filter is, in design, identical to a thin film edge filter. That is to say, a dichroic filter will transmit (or reflect) light above a certain 'edge wavelength' and reflect (or transmit) light below this wavelength. However, in this instance, the distinction is afforded by the application, rather than the device itself. In the case of the dichroic filter, rather than wishing to reject one portion of the optical beam entirely, we wish to split the beam into two (spectrally) differing components. One example might be afforded by a specialist type of colour camera, where light is split into three psychometrically representative visible colour bands and sampled by separate detectors. A rather more straightforward application is that of fluorescence microscopy. Here a short wavelength, energetic optical beam is used to excite longer wavelength fluorescence, usually in an organic material. In many contemporary designs, the short wavelength fluorescence might be delivered by a laser beam. However, the microscope must be able to separate the short wavelength excitation from the long wavelength emission. Such an application is illustrated in Figure 10.15

10.3.9 Etalon Filters

Etalon filters are, in many respects, similar in operational principle to multilayer interference filters. The distinction, in the case of etalon filters, is that two distinct reflective layers are provided that are separated by an appreciable thickness of many wavelengths. A typical embodiment of an etalon filter is a glass slab, of the order of a millimetre or more thick, coated on both sides by a highly reflective layer. These reflective layers are usually formed by thin film dielectric coatings. The geometry of an etalon filter is shown in Figure 10.16.

At first sight, the presence of the highly reflective layers on the glass block would substantially impede transmission. However, the multiple internal reflections depicted in Figure 10.16 will, under favourable circumstances, constructively interfere, thus creating tangible output at specific wavelengths. To understand how this might work, we consider an etalon, as depicted in Figure 10.16, equipped with reflective coatings each with a reflectance of R and with a thickness of d. However, we must understand that, whilst R is high, it is actually less than unity. Therefore, a small portion incident on each coating, $1 - R$, is transmitted. The total

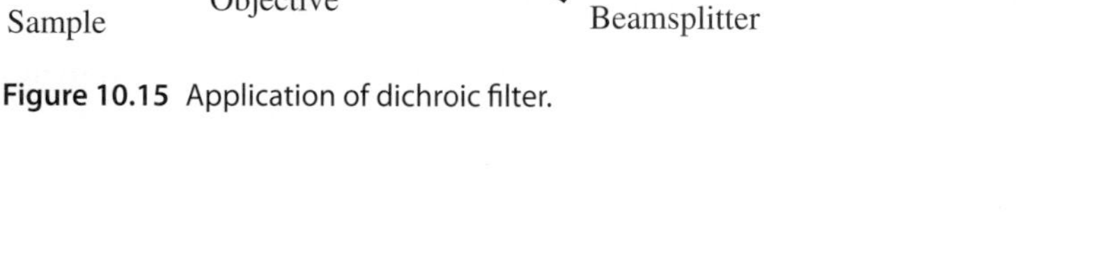

Figure 10.15 Application of dichroic filter.

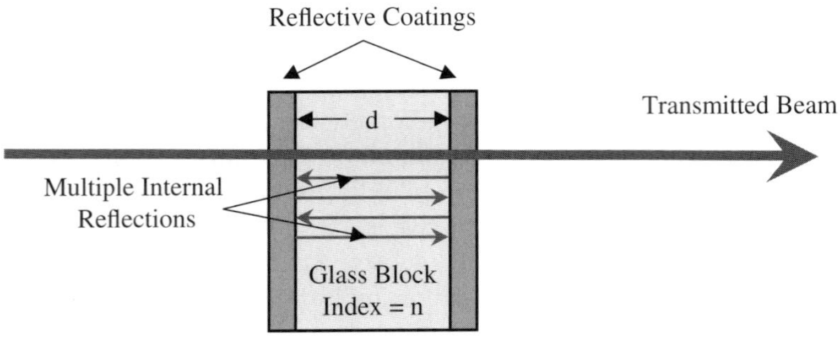

Figure 10.16 Geometry of etalon filter.

transmitted ***amplitude*** may be calculated as an infinite sum of the amplitude of successive internal reflections. However, in performing this analysis, we must remember that the amplitude reflectance is equal to the square root of the power reflectance.

In analysing the behaviour of the etalon, we assume that light is either transmitted or reflected; there is no absorption. Initially, light incident on the etalon must pass through two 'reflective' layers and the glass block. Subsequently, for each successive multiple reflection, the light must be reflected twice and pass twice through the thickness of the block in order to contribute to the transmitted light. It is straightforward to sum the components of the transmission as an (infinite) geometric sum.

$$A = (1 - R)\sum_{m=0}^{m=\infty} R^n e^{i2mnkd} = \frac{(1 - R)}{1 - \mathrm{Re}^{i2nkd}} \tag{10.21}$$

The flux is simply given by the product of the amplitude and its complex conjugate.

$$\Phi = AA^* = \frac{(1 - R)^2}{(1 - \mathrm{Re}^{i2nkd})(1 - \mathrm{Re}^{-i2nkd})} = \frac{(1 - R)^2}{(1 + R^2 - 2R\cos 2nkd)} \tag{10.22}$$

We can simplify the above expression to give:

$$\Phi = \frac{(1 - R)^2}{(1 + R^2 - 2R\cos 2nkd)} = \frac{1}{\left(1 - \frac{4R}{(1-R)^2}\sin^2 nkd\right)} \tag{10.23}$$

Equation (10.23) reaches a maximum where $\sin^2 nkd = 0$. In this case, the transmission is approximately unity. This represents a resonance condition and occurs for a whole series of wavelengths where $nkd = m\pi$. In practice for an etalon with a thickness of a millimetre or more, these resonances are closely spaced. The spacing between these resonances is uniform with respect to the inverse of the wavelength and, thus described, the spacing is referred to as the ***free spectral range***. The inverse of the wavelength, υ, is referred to as the wavenumber and the FSR or $\Delta\upsilon$ is given by:

$$\Delta\upsilon = \frac{\Delta\lambda}{\lambda^2} = \frac{1}{2nd} \tag{10.24}$$

The sharpness of the individual resonances is described by the finesse, F. We may recast Eq. (10.23) slightly, introducing the etalon finesse.

$$\Phi = \frac{1}{(1 - (2F/\pi)^2\sin^2 nkd)} \quad \text{where } F = \frac{\pi\sqrt{R}}{1 - R} \tag{10.25}$$

For large values of the finesse, the finesse is equal to the ratio of the FSR to the FWHM of the resonance. That is to say, for an etalon finesse of 100, the FWHM of the resonance is 100th of the FSR.

$$FWHM = \frac{\Delta\upsilon}{F} \tag{10.26}$$

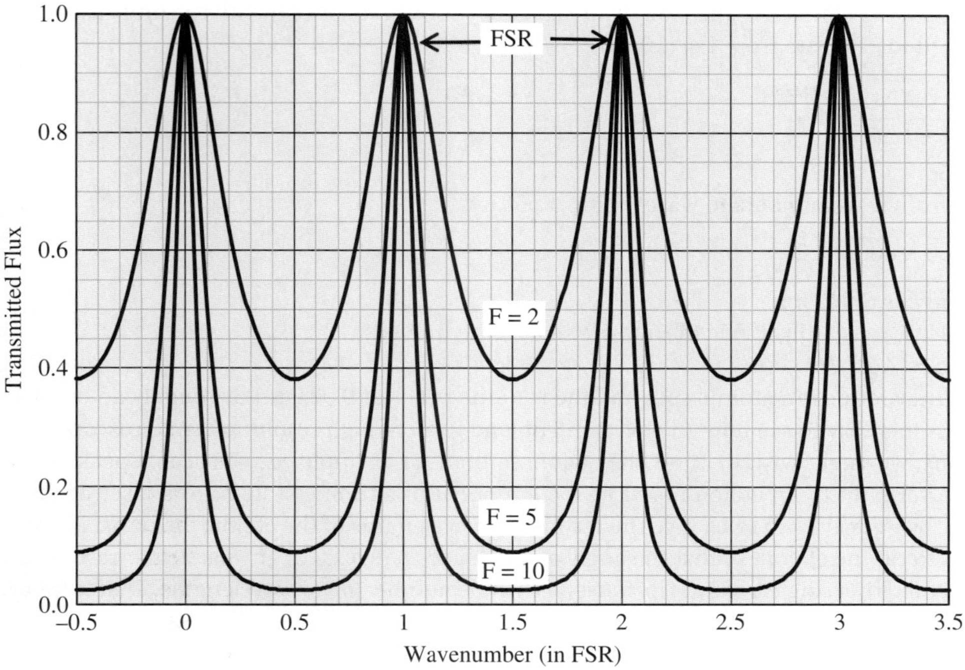

Figure 10.17 Etalon response function.

The generic etalon response function for a range of finesses is shown in Figure 10.17.

If we assume that the etalon coatings are dielectric mirrors with a reflectivity of 99.5%, then the etalon finesse (from Eq. (10.26)) is 627. In practice, however, the sharpness of the resonances is not only driven by the mirror reflectivity. The flatness of the mirrors also influences the width of the resonances. One may imagine any variation in mirror flatness producing a statistical variation in the separation, d, of the mirrors, resulting in a contribution to the spectral width. Broadly speaking for a finesse of 100, then the flatness of the mirror surfaces must be of the order of $\lambda/100$. Thus, to be effective, the optical surfaces of an etalon must be very flat.

We have described an etalon fabricated from solid glass with two mirrors coated on each surface. In addition, there are ***air spaced etalons***. Here, rather than a single solid piece of glass, there are two dielectric mirrors separated by an air gap. For each mirror, only one surface is provided with a reflective coating. Indeed, the other surface might have an anti-reflection coating.

The primary characteristic of an etalon is its very sharp resonance. The linewidth of each peak is, in general, exceptionally narrow, which generates a range of applications in high resolution spectroscopy and the precise control of spectral characteristics in lasers and other light sources. At this stage, it would be useful to consider the application of such a device through a practical example.

Worked Example 10.3 *Etalon Filter*

An etalon filter is constructed from a glass slab 10 mm thick, coated at each end with a reflective layer with a reflectance of 98%. The refractive index of the glass slab is 1.52. What is the FSR of the etalon in wavenumbers (cm^{-1})? What is the finesse of the etalon? For a central wavelength of 600 nm calculate the FSR in nm. What is the FWHM of each peak at this wavelength denominated in nm?

From Eq. (10.24), we may calculate the FSR:

$$\Delta \upsilon = \frac{1}{2nd} = \frac{1}{2 \times 1.52 \times 1} = 0.329 cm^{-1}$$

The free spectral range is 0.329 cm^{-1}. *Or 3.29×10^{-8} nm^{-1}.*

Note that we substituted the etalon thickness of 10 mm as 1 cm which is consistent with the units used. The finesse is straightforward to calculate from Eq. (10.25)

$$F = \frac{\pi\sqrt{R}}{1-R} = \pi\frac{\sqrt{0.98}}{1-0.98} = 156$$

The finesse is thus 156.

Converting the FSR from wavenumbers to wavelength at 600 nm:

$$\Delta\lambda = \lambda^2\Delta\upsilon = 600 \times 600 \times 3.29 \times 10^{-8} = 0.012\ \text{nm}.$$

The free spectral range is 0.012 nm

To calculate the FWHM, we simply divide the above value by the finesse.

FWHM = 0.012/156 = 7.6 × 10⁻⁵ nm.

By comparison, the linewidth of an atomic spectral line is of the order of 0.001 nm or roughly an order of magnitude greater. Thus it is easy to see how an etalon might lend itself to high resolution spectroscopy.

In certain applications, we might wish to 'tune' the etalon, so that, for example, we can scan a peak across an atomic transition. Alternatively, an etalon might be a component of a tuning system for a laser or other optical device. Either way, we may wish to change the resonance wavelengths of the etalon. The most obvious way of achieving this is to tilt the etalon, so that the angle of incidence changes. For any particular resonance, the optical path represented by a single 'round trip' must be a whole number, m, of wavelengths. That is to say:

$$2nd\cos\phi = m\lambda$$

We should note here the cosine term, representing the cosine of the refracted angle. If the central, untilted wavelength is λ_0, then, for a tilt angle of θ, the modified wavelength, λ, is given by:

$$\frac{(\lambda_0 - \lambda)}{\lambda_0} \approx \frac{\theta^2}{2n^2} \tag{10.27}$$

The above expression applies for small tilts only. In the above example of an etalon tuned to 600 nm, then a tilt of one degree would produce a shift (to shorter wavelengths) of 0.04 nm. This shift is rather greater than the FSR of 0.012 nm. A tilt of about 0.5° would produce a shift commensurate with the FSR. Another mechanism for tuning an etalon is by means of temperature. Adjusting the temperature of the etalon will change the refractive index of the etalon medium. Obviously, the degree of the variation is dependent upon the material, but typically the refractive index change might be in the region of 10 ppm per degree centigrade. In this instance, a 10 °C change might produce a shift of about 0.06 nm.

In the case of air spaced etalons, another mechanism for tuning the etalon is pressure tuning. By enclosing an etalon in an airtight vessel and injecting air or some other gas into the vessel at a constant flow rate, the etalon may be tuned. The refractive index of the air or gas within the etalon cavity is proportional to the pressure of the gas. The principle is illustrated in Figure 10.18.

10.4 Design of Thin Film Filters

As we have seen, the design of thin film filters is largely based on the assembly of successive quarter wavelength layers of high and low index materials. Broadly speaking, the principal filter elements consist of edge filter elements and bandpass elements. For a transmission bandpass filter, the basic building block consists of a multilayer etalon cavity. The etalon cavity consists of a half wavelength 'spacer' surrounded by two mirrors each consisting of an alternating multilayer stack of high and low index material. A simple example of such a design is shown in Figure 10.19.

In describing a thin film design, there is a simple convention for setting out the composition of a stack. The basic unit is the quarter wavelength film of some high or low index material. Each successive quarter

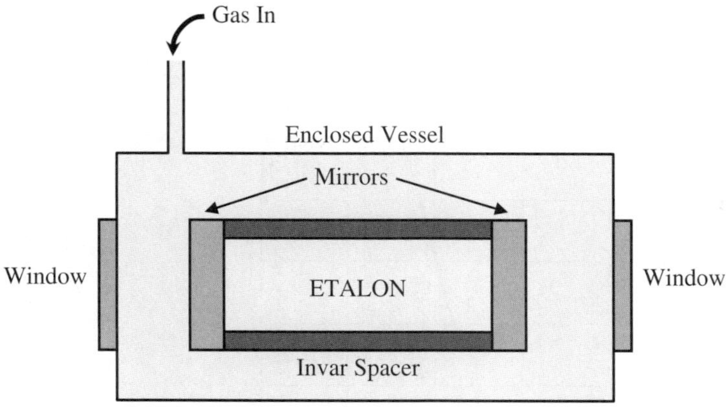

Figure 10.18 Pressure tuned etalon.

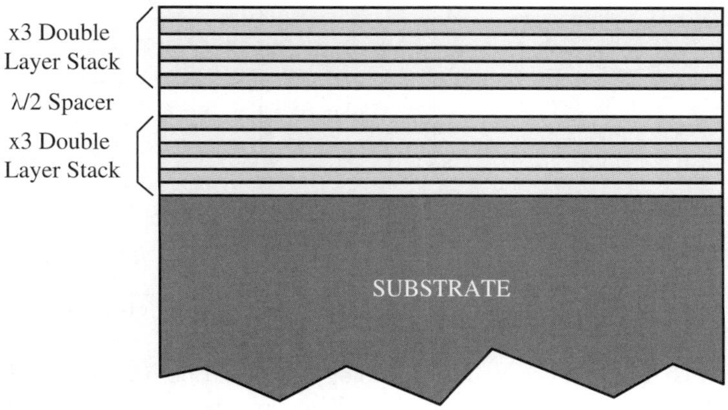

Figure 10.19 Basic bandpass filter design.

wavelength layer is denoted by the letter 'L' for a low index layer and the latter 'H' for a high index layer. The substrate is denoted by the letter 'S'. Starting with the substrate, the design in Figure 10.19 is described by the following sequence:

SLHLHLHLLHLHLHL

That is to say, the substrate is followed by three alternating double layers of low index and high index films and then two layers of low index material (the $\lambda/2$ spacer), finally followed by another three double layers of alternating index. Figure 10.20 shows the basic response of such a filter, assuming a design wavelength of 600 nm. The central peak of the bandpass filter is clearly apparent but is accompanied by side bands. These sidebands must be removed and are generally done by adding extra multilayer components to the filter. These extra layers are referred to as '***blocking layers***' and usually take the form of additional edge filter elements. Provision of these additional layers plays a key role in the control of out of band transmission.

The design shown here is purely illustrative. The spectral response afforded by the central feature is somewhat sharp. Addition of extra multilayer etalon cavities produces a somewhat flatter response at the peak. It is clear from this discussion that, in practice, the design of such filters is necessarily complex requiring the provision of very many layers. As such, meeting more stringent requirements, such as higher blocking levels and the achievement of greater filter slopes inevitably requires the provision of more layers in the design. Naturally, this increased complexity is accompanied by increased cost.

Of course, practical filter design proceeds by a process of computer optimisation. Nevertheless, it is useful to grasp the underlying principles in order to more fully understand the compromises involved in the

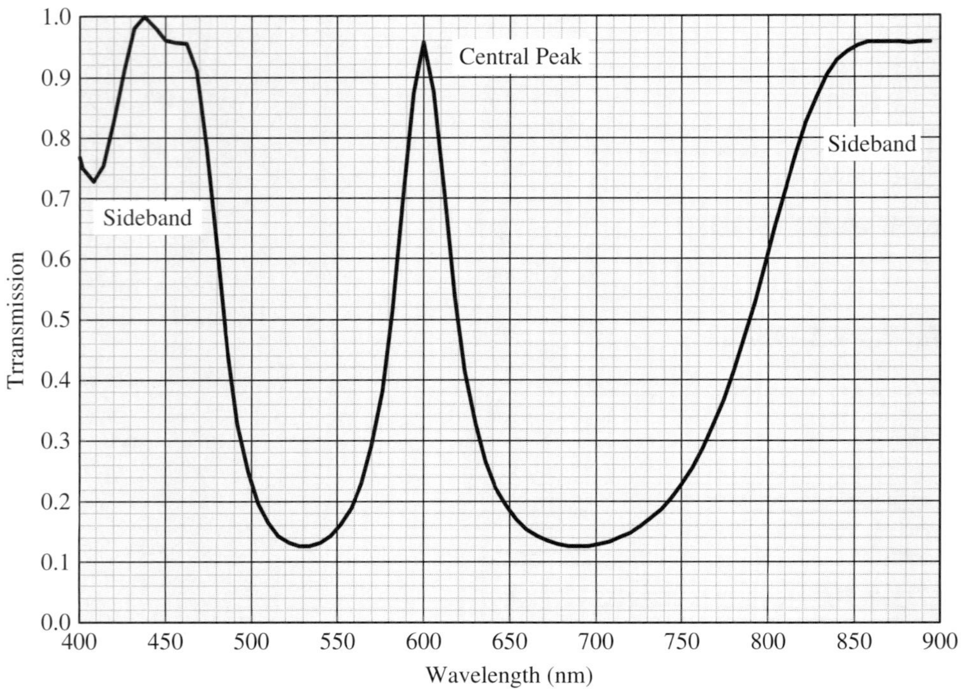

Figure 10.20 Transmission for basic bandpass filter design.

design process. It is especially important to understand such performance limitations where one is focused on requirements imposed by the end application.

Before moving on to the practical topic of coating materials and technology, it would be useful now to further illustrate the design process by undertaking the optimisation of a broadband antireflection coating. In this example we examine the performance of a five layer coating on a glass substrate with a refractive index of 1.52 and with a design wavelength of 550 nm. The coating layers are as follows:

$$\mathbf{1:}\ MgF_2\ (n = 1.38); \mathbf{2:}\ Al_2O_3\ (n = 1.77); \mathbf{3:}\ ZnO\ :\ (n = 1.98); \mathbf{4:}\ HfO_2\ (n = 1.91); \mathbf{5:}\ MgF_2\ (n = 1.38)$$

In performing the optimisation, we are free to adjust the thickness of each layer and calculate the transformed index according to Eq. (10.8). We can then choose the prescription that gives the minimum average reflection across the spectral range of interest. Of course, this is not a very sophisticated process, in this instance, but does serve to illustrate the power of computer based optimisation techniques. A very basic program was written here to minimise the average reflectivity between 400 and 750 nm. The optimum solution, in this instance, occurs for the following film thicknesses (in design wavelengths):

$$\mathbf{1:}\ 0.47; \mathbf{2:}\ 0.22; \mathbf{3:}\ 0.10; \mathbf{4:}\ 0.18; \mathbf{5:}\ 0.23$$

Figure 10.21 illustrates the performance of this design, as optimised for a design wavelength of 550 nm. The performance is clearly a great improvement on that of a single layer coating whose performance is shown for comparison.

10.5 Thin Film Materials

The preceding analysis has demonstrated the usefulness of thin film materials with contrasting refractive indices. It would be useful, at this stage, to tabulate a range of materials commonly employed in thin film coatings. As well as their refractive index, we are interested in their hardness, durability, and resistance to environmental agents, such as heat, humidity, and chemical attack. They must, of course, be amenable to

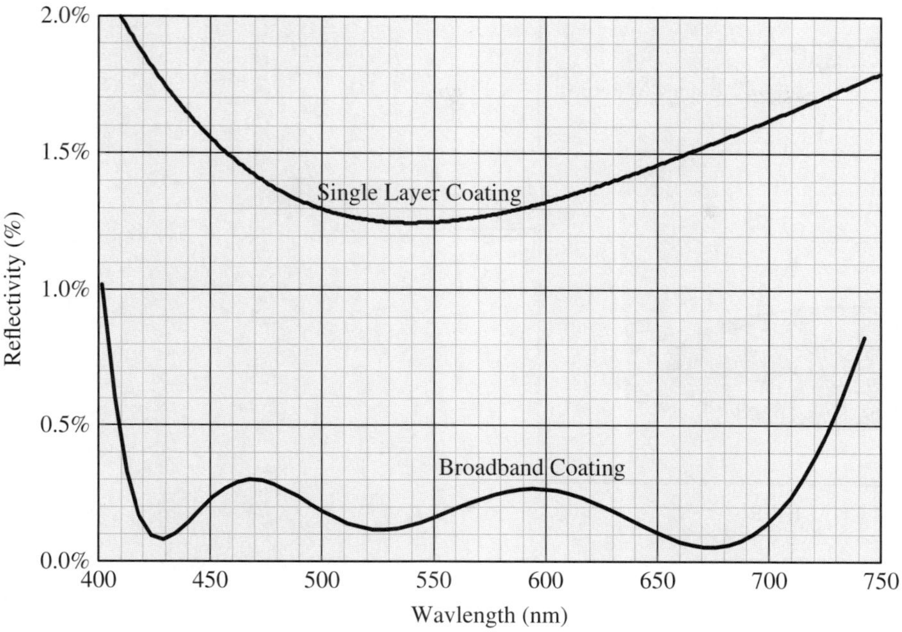

Figure 10.21 'Computer Optimised' broadband antireflection coating performance.

Table 10.3 Some common thin film materials.

Material	Formula	Index	Material	Formula	Index
Cryolite	Na_3AlF_6	1.35	Yttria	Y_2O_3	1.93
Magnesium fluoride	MgF_3	1.38	Silicon monoxide	SiO	1.97
Silica	SiO_2	1.46	Zinc oxide	ZnO	2.00
Barium fluoride	BaF_2	1.47	Silicon nitride	Si_3N_4	2.01
Yttrium fluoride	YF_3	1.50	Hafnia	HfO_2	2.11
Praseodymium fluoride	PrF_3	1.51	Tantalum pentoxide	Ta_2O_5	2.13
Thorium fluoride	ThF_4	1.52	Zirconia	ZrO_2	2.16
Lanthanum fluoride	LaF_3	1.6	Aluminium nitride	AlN	2.16
Cerium fluoride	CeF_3	1.63	Niobium pentoxide	Nb_2O_5	2.33
Lead fluoride	PbF_2	1.76	Zinc Sulfide	ZnS	2.35
Alumina	Al_2O_3	1.77	Ceria	CeO_2	2.35
Indium tin oxide		1.89	Titania	TiO_2	2.60

deposition by evaporation or sputtering. Table 10.3 gives a list of materials used in optical coatings, together with their refractive index at 600 nm.

10.6 Thin Film Deposition Processes

10.6.1 General

Many of the thin film materials in use are stable, refractory materials. This presents a number of challenges in the controlled deposition onto substrate materials. All thin film deposition processes take place under vacuum or reduced pressure. The substrate to be coated is placed into the vacuum chamber and exposed to

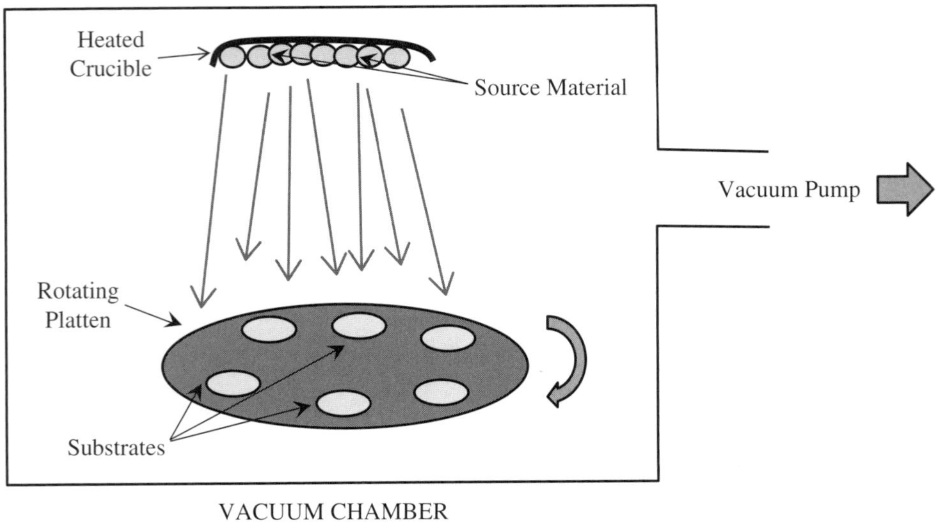

Figure 10.22 Evaporation process.

material deposition by subjecting a material source to high energy. There are two principal deposition techniques. Firstly, there is ***vacuum evaporation***, where the source material, located under vacuum is heated to a temperature sufficient to generate significant evaporation. Secondly, there is ***sputtering***, where the target material is exposed to ion bombardment in an electrical discharge. The target material forms the cathode and ion bombardment releases or sputters atoms from the cathode.

10.6.2 Evaporation

Figure 10.22 shows the general set up for vacuum evaporation.

The substrates to be coated are placed in a vacuum chamber in proximity to a heated crucible of the material to be coated. Heating of the source material is accomplished either by resistive heating of the crucible or, in the case of very refractory materials, by direct electron beam heating of the material itself. It is of prime importance that the coating be distributed evenly over the whole of the substrate. Inevitably, in practice, geometrical effects will lead to variability in the coating accumulation rates across the vacuum chamber. To ameliorate this effect, all substrates are placed upon a rotating platen, causing each substrate to sample different areas of the chamber during the coating run. This substantially reduces the thickness variation.

An important characteristic of the vacuum evaporation process is that it is 'line of sight'. The mean free path of the evaporated atoms is large and each atom proceeds directly from the source to the substrate in a straight line. As a result, it is possible to sharply delineate areas for deposition by placing a patterned mask in front of the substrate. Exposed areas, unobscured by the mask, will experience deposition, whereas areas obscured by the mask will be free from deposition. In principle, quite intricate patterned deposition may be produced in this way.

10.6.3 Sputtering

A schematic for the sputtering process is shown in Figure 10.23. In the sputtering process, the source material is, in effect, the cathode of a low pressure discharge. The process of ion bombardment of the cathode causes material to be ejected or sputtered from the cathode. Figure 10.23 shows a simple DC electrical discharge. Most sputtering systems are, in practice somewhat more complicated than shown. Most usually, the simple

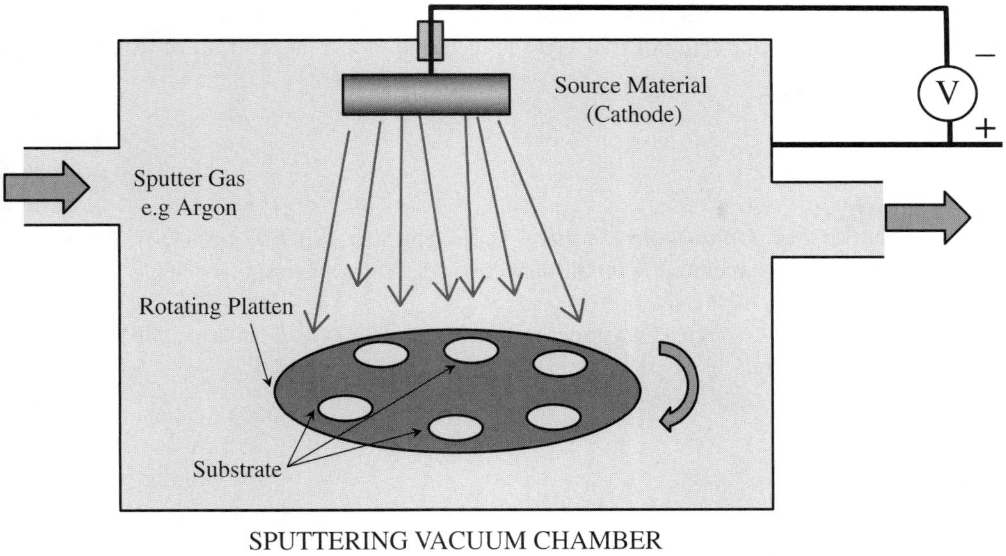

Figure 10.23 Sputtering process.

DC discharge might be replaced with a radiofrequency discharge. In any case, the overall operating principle is the same.

There is a significant distinction between the evaporation and sputtering process. Operation of the electrical discharge requires a low (a few millibar) but significant pressure of sputter gas to promote the electrical discharge. Typically, this gas might be a noble gas, such as argon. As a result of the significant gas pressure, the mean free path of the sputtered atoms is very low. Therefore, the source material does not proceed from cathode to substrate in a straight line. By contrast, it diffuses from the source to the target; its path is mediated by a large number of collisions with the sputter gas. As a consequence, it is no longer possible to use a masking process to sharply delineate patterned coatings. As for the evaporation process, placing the samples on a rotating platen enhances coating uniformity.

The range of materials that may be deposited with sputtering is more limited than for evaporation. At first sight, only metals may be directly sputtered. However, incorporation of reactive gases, such as oxygen and ammonia, into the sputter gas permits the deposition of some metal oxides and nitrides by a reactive process. Although sputtering is a more restrictive process than evaporation, this is compensated, in some instances, by the ability of sputtering to produce more durable and dense coatings.

10.6.4 Thickness Monitoring

Monitoring of the coating thickness during the coating process is of prime importance. The foundation of thin film technology is the ability to accurately deposit specific thicknesses (e.g. a quarter wavelength) of different materials. Achievement of specific film thicknesses is achieved mainly by one of three different techniques. Firstly, deposition thicknesses may be continuously monitored by a calibrated quartz crystal oscillator. Accumulation of deposited mass on the surface of the oscillator reduces its resonance frequency and this may be used to deduce the film thickness. Alternatively, a 'dead reckoning' technique may be used to establish film thickness whereby accumulated process knowledge has established a clear relationship between deposition time and film thickness. Finally, coating thicknesses may be monitored optically, using a laser or other narrow band source. In this instance, the reflectivity or transmission of a 'witness sample' is monitored continuously. Most usually, achievement of the quarter wavelength condition corresponds to a turning point (maximum or minimum) in the film transmission or reflection. In any case, analysis should always be able to establish the film transmission/reflectivity at the desired layer thickness. As soon as the condition has been achieved for a specific layer, then evaporation/sputtering of that layer is terminated and deposition of the next layer commences.

Further Reading

Kaiser, N. and Pulker, H.K. (2003). *Optical Interference Coatings*. Berlin: Springer. ISBN: 978-3-642-05570-6.

Rizea, A. and Popescu, I.M. (2012). Design techniques for all-dielectric polarizing beam splitter cubes, under constrained situations. *Rom. Rep. Phys.* 64 (2): 482.

Wolfe, W. (2003). *Optical Engineer's Desk Reference*. Washington, DC: Optical Society of America. ISBN: 1-55752-757-1.

11

Prisms and Dispersion Devices

11.1 Introduction

In studying the interaction of light and matter, we frequently wish to understand how this behaviour changes as a function of wavelength. We might wish, for instance, to view the spectrum emitted by a lamp or the solar spectrum with atomic absorption lines superimposed thereon. For a variety of reasons, we might wish to decode the spectral information in incident light and present this (usually) as a spatially dispersed spectrum. In this scenario, a beam of light is spatially dispersed in such a way as to present a specific wavelength of light at a unique spatial location. Of course, historically, the 'splitting of light into its constituent colours' was accomplished, as in the case of Newton, by means of a prism. The refractive index variation of glasses with wavelength, or dispersion, produces a variation in the refracted angle with wavelength and results in spatial dispersion of the light with respect to wavelength. However, the degree of angular dispersion produced by prisms is disappointingly small. As a result, prisms feature little in practical spectroscopic instruments. They have more or less wholly been displaced by devices that rely on diffraction, such as diffraction gratings.

Notwithstanding these considerations, we will commence this narrative with a consideration of dispersive prisms. Understanding the limitations of these components will enable their diffractive counterparts to be placed into proper context in modern applications.

11.2 Prisms

11.2.1 Dispersive Prisms

Before examining in detail the behaviour of modern diffractive components, we will consider very briefly the dispersive properties of a simple prism. In this elementary discussion we will consider the operation of a prism that is established under the so-called ***minimum deviation*** condition. In this scenario, the incidence angle at the first interface is equal to the refracted angle at the second prism interface. That is to say, the arrangement is entirely symmetrical, with an angle of incidence of θ and an apex angle α. This arrangement is shown in Figure 11.1.

It is quite apparent, under these conditions, that the incidence angle, θ, is given (from Snell's Law) by:

$$\sin\theta = n\sin(\alpha/2) \tag{11.1}$$

Of course, the maximum value for $n\sin(\alpha/2)$ should be one, giving maximum viable prism angle of $2\sin^{-1}(1/n)$. Under these conditions, both the input and output angles are grazing. For BK7 with an n_D of 1.518, this condition corresponds to a prism angle of about 82.5°. However, what we are really interested in is the deviation angle, Δ, that describes the angle from which the beam has been deviated from its original path. This is given by:

$$\Delta = 2\theta - \alpha \text{ and } \Delta/2 + \alpha/2 = \theta$$

Optical Engineering Science, First Edition. Stephen Rolt.

Companion website: www.wiley.com/go/Rolt/opt-eng-sci

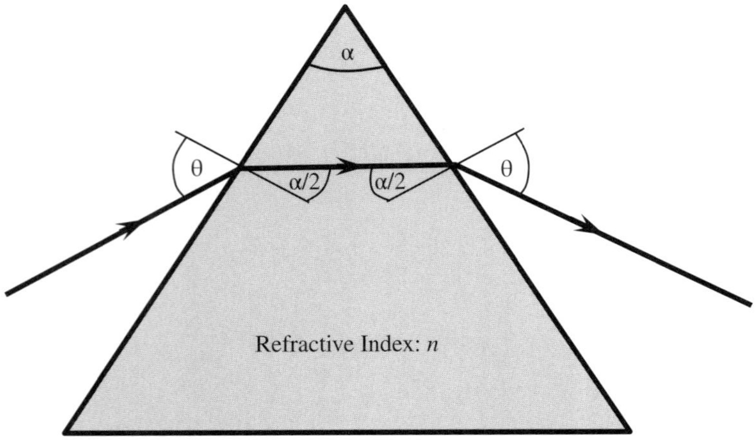

Figure 11.1 Minimum deviation refraction produced by a prism.

Finally, it is possible to express this deviation angle solely in terms of the prism apex angle and the refractive index:

$$\sin\left(\frac{\Delta+\alpha}{2}\right) = n\sin\left(\frac{\alpha}{2}\right) \tag{11.2}$$

The most pertinent measure of the utility of a dispersion device, is the angular dispersion with respect to wavelength. In this case, the angular dispersion, β, is given by:

$$\beta = \frac{d\Delta}{d\lambda} = \frac{2\tan(\alpha/2)}{\sqrt{1+(1-n^2)\tan(\alpha/2)}}\frac{dn}{d\lambda} \tag{11.3}$$

If we imagine that the input to the prism is a parallel beam of width, w, and inclined at an angle, θ, to the prism normal and assuming that beam is focused by a perfect imaging system, the angular resolution, $\Delta\theta$, by the Rayleigh criterion is given by:

$$\Delta\theta = \frac{1.22\lambda}{w} \tag{11.4}$$

On this basis and from Eq. (11.3), we can now calculate the minimum wavelength interval that can be distinguished, $\Delta\lambda$:

$$\Delta\lambda = \left(\frac{\sqrt{1+(1-n^2)\tan(\alpha/2)}}{2\tan(\alpha/2)}\left(\frac{1.22\lambda}{w}\right)\right)\bigg/\frac{dn}{d\lambda} \tag{11.5}$$

The important aspect of Eqs. (11.3) and (11.5) is that they relate the angular dispersion of the prism directly to the dispersion of the material. Ultimately, we might relate this dispersion, in some way, to the relevant Abbe number of the material. If we make the approximation that the refractive index variation across the visible spectrum is more or less linear, the following expression results:

$$\Delta\lambda = \frac{\sqrt{1+(1-n^2)\tan(\alpha/2)}}{2\tan(\alpha/2)}\left(\frac{1.22V_D\lambda_D(\lambda_C-\lambda_F)}{w(n_D-1)}\right) \tag{11.6}$$

At this point, we introduce an important parameter that defines the utility of a dispersion system. This is the so-called ***resolving power*** which is defined as the wavelength divided by the resolution:

$$R = \frac{\lambda}{\Delta\lambda} = \frac{5.67\times\tan(\alpha/2)}{\sqrt{1+(1-n_D^2)\tan(\alpha/2)}}\left(\frac{w}{V_D\lambda_D}\right) \tag{11.7}$$

The most salient feature of Eq. (11.7) is that the resolving power is driven by the ratio of the width of the beam to that of the wavelength. Looking ahead, this same feature is present in the mathematical description of diffractive components, such as diffraction gratings. However, for the prism, the resolving power is modified by its inverse relationship with the Abbe number. For practical materials, Abbe numbers range between about 20 and 80. As such, the Abbe number represents the degradation in performance of a prism dispersion system when compared to a diffraction grating.

Another aspect of the prism refraction process is the beam magnification produced by the prism. In the case of the symmetrical arrangement previously discussed, then the magnification is one. That is to say, the width of the beam is the same at the output as it is at the input. However, this condition is obviated at other incident angles. This consideration is of importance in the design of instruments which rely on dispersion, e.g. spectrometers, and also comes into play in diffraction gratings. It is important to understand, however, that this beam magnification is ***anamorphic***. It only occurs along one axis, i.e. x ***or*** y. This feature is exploited, for example, in the reshaping of laser beams from semiconductor lasers. In general, laser beams from these devices are elliptical and anamorphic magnification from prism magnifiers may be used to correct this. In proceeding from an angle of incidence, θ to an angle of refraction, ϕ, at a single surface, it is clear that the magnification M for this process is simply equal to the ratio of the cosines of these angles:

$$M = \frac{\cos\phi}{\cos\theta} \tag{11.8}$$

In the case of a prism in a symmetrical minimum deviation arrangement, then the magnification at the second refraction is the inverse of that at the first, so the overall magnification is unity. If our object is to maximise anamorphic magnification, a reasonable way to proceed is to ensure that the angle of incidence at the second surface is zero, so that all the deviation occurs at the first surface. This is illustrated in Figure 11.2.

The magnification produced by a prism with normal incidence at the second interface, as per Figure 11.2 may be expressed in terms of the prism angle, α, and the refractive index, n alone:

$$M = \frac{1}{\sqrt{1-(n^2-1)\tan^2\alpha}} \tag{11.9}$$

In many designs, it is desirable that the original direction of the beam remains un-deviated. Therefore, it is common to combine two anamorphic prisms into a matched pair. Whilst the orientation of the second prism is inverted to reverse the original deviation, the second prism further multiplies the magnification. This is the arrangement that is commonly used to provide anamorphic magnification to shape an elliptical laser beam from a semiconductor laser into a circular beam. The arrangement is shown in Figure 11.3.

The anamorphic magnification for an identical pair of prisms is simply given by the square of Eq. (11.9):

$$M_{pair} = \frac{1}{1-(n^2-1)\tan^2\alpha} \tag{11.10}$$

Worked Example 11.1 We wish to design a dual prism anamorphic magnifier to expand a 650 nm laser beam in one dimension by a factor of 2. Both prisms are made from BK7 with a refractive index of 1.5145 at that wavelength. The apex angle of both prisms is identical. What prism angle should be chosen if we assume that the angle of incidence at the second surface is zero, as in Figure 11.3?

Figure 11.2 Anamorphic magnification by prism.

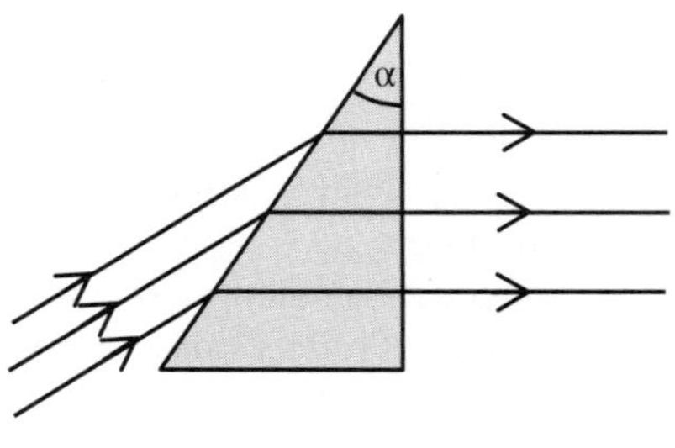

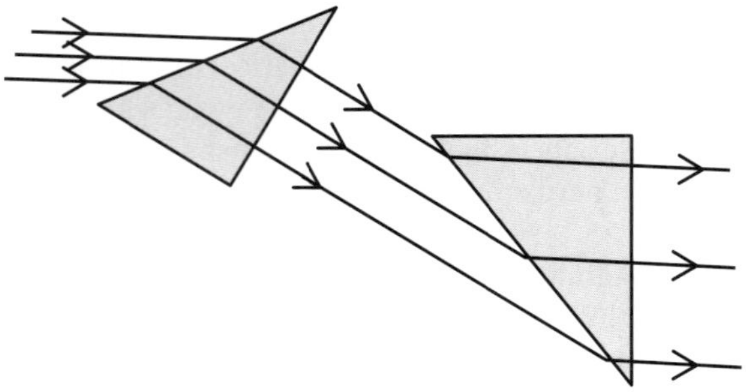

Figure 11.3 Dual prism anamorphic magnifier.

Setting the magnification equal to 2, using Eq. (11.10)

$$\frac{1}{1-(n^2-1)\tan^2\alpha} = 2 \text{ and } \tan^2\alpha = \frac{1}{2(n^2-1)}$$

$$\tan^2\alpha = \frac{1}{2(1.5145^2-1)} \text{ and } \tan\alpha = 0.622\ \alpha = 31.87°.$$

The apex angle of the prism pair should be 31.87°.

The initial discussion illustrated the simplest form of dispersive prism, namely the ***triangular prism***. Other special dispersive prism geometries were devised to produce specific angular deviations for some central wavelengths and were employed in particular instrument geometries. Examples include the ***Abbe prism*** (60°), the ***Pellin-Broca*** prism (90°) and the ***Amici prism*** (0°). The latter is a particularly convenient arrangement that permits the central wavelength to be un-deviated. It consists of a pair of cemented prisms with each prism having a different Abbe number. The deviation of the first prism element is entirely counteracted by the second prism element. However, since the Abbe numbers are different, the dispersions do not cancel out. Although dispersive prisms do not feature heavily in current applications, the last example is interesting. The Abbe prism foreshadows the modern ***grism***, a grating prism combination, where central deviation produced by the diffraction grating is countered by the deviation produced by the prism. The bulk of the dispersion is provided by the grating.

11.2.2 Reflective Prisms

Although the utility of dispersive prisms is very much diminished currently, reflective prisms that exploit total internal reflection in their design still retain their prominence in modern applications. In particular, they are useful for 'folding' a long optical path into a more convenient, if restricted spatial geometries. In some instances, prisms replicate the function of plane mirrors. However, in these cases, the efficiency of total internal reflection (~100%) is substantially superior to that of metallised mirrors. So, for a design that might include a number of these elements, incorporation of prisms might substantially boost throughput.

There is another crucial aspect of (plane) reflective elements in an optical design. Geometrically, a single reflection is a combination of a 180° rotation about the normal, combined with a pure inversion. This means that combinations of reflections may be used to effect a specific geometrical transformation on the image space. Deploying an even number of reflections will always produce a pure rotation about some axis. Where an odd number of reflections is used, then the rotation is accompanied by a pure inversion. These simple geometrical transformations are extremely useful in specific imaging applications, for instance in correcting inverted images.

As previously advised, an odd number of reflections can never be described by a simple rotation, whereas an even number of reflections is always so described. The simplest example of this is the 45° prism which diverts a beam through 90°. This is illustrated in Figure 11.4.

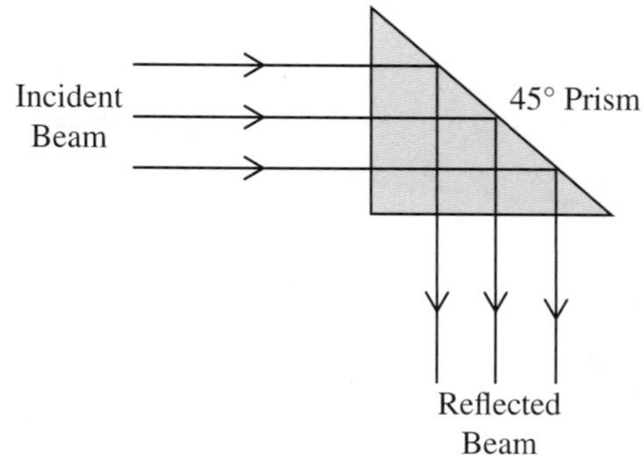

Figure 11.4 45° prism.

For the 45° prism to be viable, the critical angle needs to be less than this. This condition is true for all materials with a refractive index greater than 1.41. In practice, this means all (solid) optical materials with the exception of the fluorides may be used.

The 45° prism is perhaps a rather trivial, but nonetheless useful, example of a reflective prism. Of more interest are those examples which include at least two reflective surfaces. For each reflection, the co-ordinate transformation may be defined in terms of the product of an inversion matrix and a rotation matrix, $\mathbf{R_n}$. Each rotation matrix consists of a 180° rotation about the respective surface normal. For two reflections, the resultant matrix transformation, **M**, may be derived from the product of the two individual matrices:

$$\mathbf{M} = \begin{bmatrix} -1 & 0 & 0 \\ 0 & -1 & 0 \\ 0 & 0 & -1 \end{bmatrix} \mathbf{R_2} \begin{bmatrix} -1 & 0 & 0 \\ 0 & -1 & 0 \\ 0 & 0 & -1 \end{bmatrix} \mathbf{R_1} = \mathbf{R_2 R_1} \tag{11.11}$$

The upshot of Eq. (11.11) is that the co-ordinate transformation produced by the reflection at two plane surfaces whose surface normals are at an angle, θ, is ***equivalent to a pure reflection of 2θ about an axis produced by the intersection of the two surfaces***. So, for example, two reflective surfaces at right angles to each other will produce a transformation equivalent to a rotation of 180°. Similarly, two surfaces at 45° will produce a rotation of 90°. An example of the former is the ***Porro prism***, which is similar in geometry to the 45° prism, except that the two faces inclined at 90° are used.

In the case of the Porro prism shown in Figure 11.5, with the axes as described, the transformation is such that X′ = X; Y′ = −Y and Z′ = −Z. An interesting extension of the Porro prism is the ***double Porro prism.*** Here, two Porro prisms are combined, except the second prism is oriented with its facet intersection oriented along the Y axis, as opposed to the X axis. In this case, the overall transformation of the two prisms acting together may be described by a 180° rotation about the X axis followed by a 180° rotation about the Y axis. The combination is equivalent to a 180° rotation about the Z axis. That is to say, the double prism acts as an image

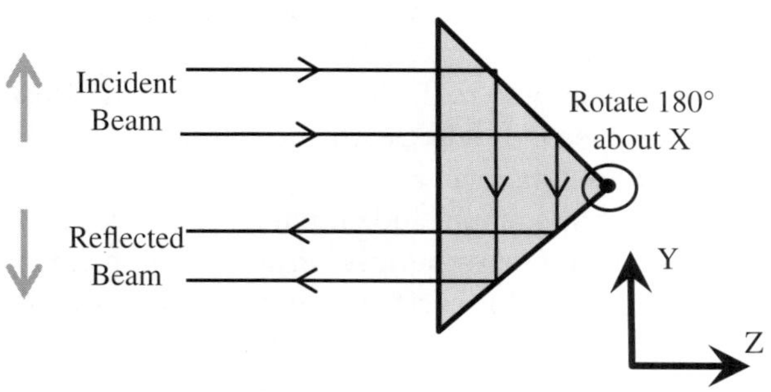

Figure 11.5 Porro prism.

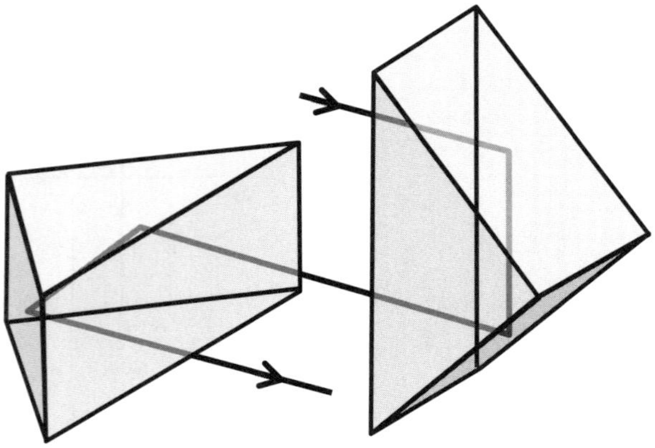

Figure 11.6 Double Porro prism.

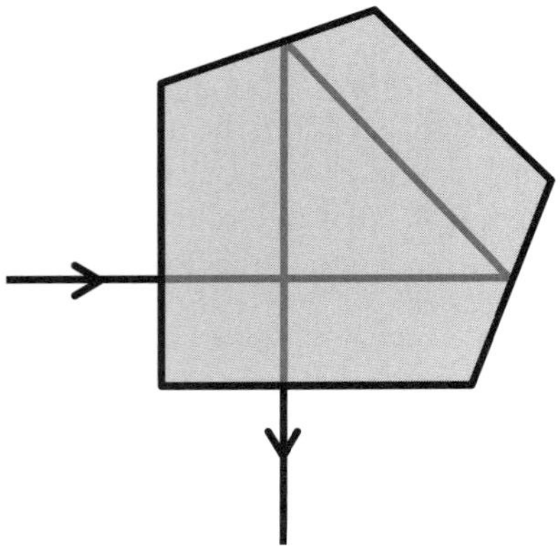

Figure 11.7 Pentaprism.

rotator. It is used in simple instruments, such as binoculars, to provide path folding and image orientation correction. This is illustrated in Figure 11.6.

Reflection at two surfaces oriented at 45° produces a 90° deviation and this is exploited in the ***pentaprism***. This 90° deviation is produced irrespective of the angle of incidence of the input beam. The geometry of the pentaprism is illustrated in Figure 11.7.

In the case of the pentaprism design, each of the two reflections occurs at an angle of incidence of 22.5°. Therefore, each of these surfaces must be coated with a reflective layer.

The first reflective facet of the pentaprism may be split into two facets inclined at 90° with respect to each other and resembling the eaves of a roof. As before, the beam is diverted by 90°, but the image is inverted. This is the so-called ***roof pentaprism***. A similar effect may be produced with the simple 45° prism, as illustrated in Figure 11.4. Again, the hypotenuse is split into two 'eaves' inclined at 90° with respect to each other. As with the roof pentaprism, the image is inverted with respect to the original design. This prism is known as the ***Amici roof prism***.

Another useful function performed by reflective prisms is the ability to rotate an image at will. The ***Dove prism*** consists of a tilted refractive surface that diverts light to a second reflecting plane at a shallow angle. After reflecting from this surface, it is refracted at the (inclined) output facet and is returned to its original course. A variant of this design is the ***Abbe-König prism***. Here, the tilted input and output facets are replaced by perpendicularly inclined surfaces. To divert light onto the second reflective surface, an additional shallowly inclined reflecting surface is interposed. Finally, after the second reflecting surface a third reflecting surface diverts the light through the output facet. These prisms are illustrated in Figures 11.8a, and 11.8b.

For both prism types, the overall geometrical transformation effected is that of a reflection about the reflective or principal reflective surface. The useful feature of this is that if the component can be rotated about

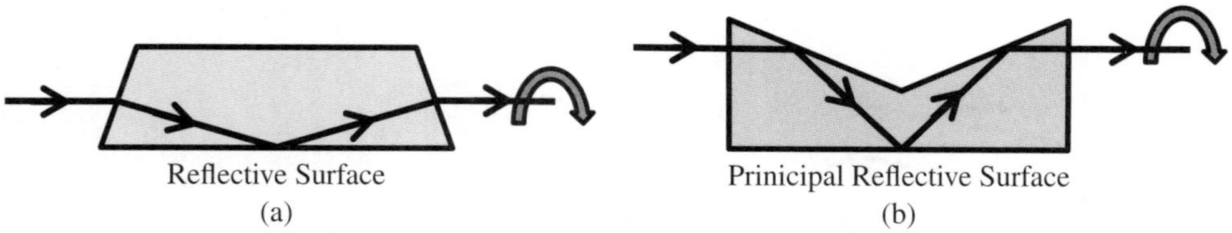

Figure 11.8 (a) Dove prism. (b) Abbe König prism.

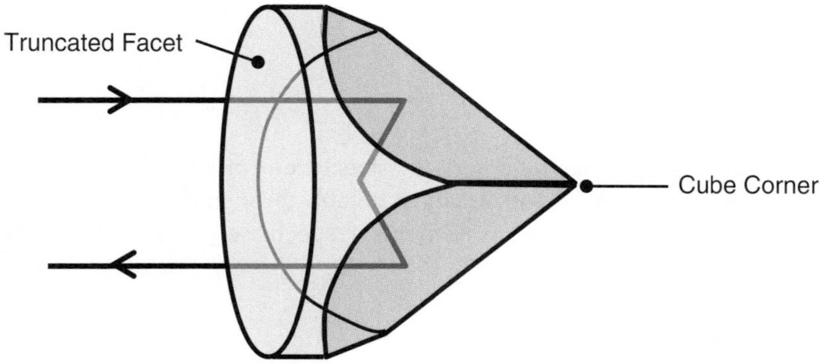

Figure 11.9 Corner cube retroreflector.

the axis of propagation, as indicated in Figure 11.6, then the plane of reflection is itself rotated. In terms of its impact upon the orientation of final image, for a component rotation angle of θ, the image is rotated by 2θ. Of course, the image is inverted as, in both cases, there are an odd number of reflections. An interesting analogue of the Abbe König prism is the so-called ***K-Mirror assembly***. If we now assume that the three reflecting surfaces are replaced by three plane mirrors, which are thus fixed with respect to each other, then the functionality of this assembly will be identical to that of the Abbe König prism. It is so called, as the shape of the mirror assembly mimics the outline of the letter K.

Finally, there is one additional reflective prism component that is perhaps the most ubiquitous. This is the so-called ***corner cube reflector*** or ***retro-reflecting prism*** or otherwise known as the ***cat's eye reflector***. As the original title suggests, the prism consists of a single corner of a glass cube that has been sliced off. Light entering the prism via the truncated facet with undergo three reflections, all of which are orthogonal. In terms of the methodology set out in Eq. (11.11), this process may be regarded as a single co-ordinate inversion and three rotations of 180° about mutually perpendicular axes. It is fairly obvious that the latter rotations simply yield the identity matrix. Therefore, the effect of a corner cube reflector is to produce a pure co-ordinate inversion about the intersection of the three surfaces. As a consequence, the direction of any ray entering the prism will be reversed irrespective of its initial direction. This is shown in Figure 11.9.

The corner cube retro-reflector finds many applications in industrial and laboratory alignment applications. Famously, the Apollo 11, 14, and 15 missions left corner cube retro-reflectors on the lunar surface as part of a laser ranging programme. Retroflection from these reflectors could be detected by using a laser beam launched from a terrestrial telescope.

11.3 Analysis of Diffraction Gratings

11.3.1 Introduction

Diffraction gratings are simple periodic structures that spatially disperse light by diffraction. This diffraction process relies on the imposition of a periodic variation in either the phase or amplitude of a plane wave. The

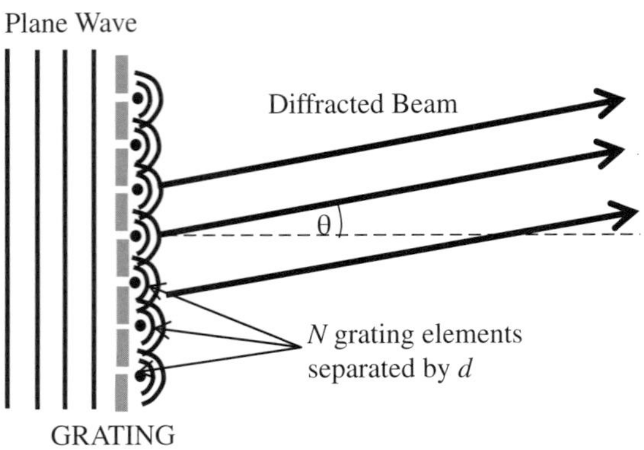

Figure 11.10 Operation of diffraction grating.

majority of these components are reflective gratings whereby a periodic structure is provided on a mirror surface. However, diffraction gratings can also work in transmission, most notably through the periodic variation of transmitted phase. Diffraction gratings are widely used as elements in spectroscopic instruments or in tunable lasers.

11.3.2 Principle of Operation

In understanding the operation of a diffraction grating, we are interested in the far field diffraction pattern of a periodic array of slits. Since the diffraction pattern is specifically to be observed in the far field, then the Fraunhofer approximation applies. As shown in Figure 11.10, we illustrate the problem by an array of N slits, each separated by a distance d and illuminated by a plane wave with a wavelength of λ.

In the Fraunhofer approximation, the far field pattern is given by the Fourier transform of the near field pattern. If, for the present, we consider each slit having negligible width, the Fourier transform may be determined by summing across the N individual slits to give the far field ***amplitude***:

$$Amplitude = \sum_{n=1}^{n=N} e^{indk\sin\theta} = \frac{1 - e^{iNdk\sin\theta}}{1 - e^{idk\sin\theta}} \tag{11.12}$$

where $k = 2\pi/\lambda$

If we wish to calculate the intensity, then we must multiply Eq. (11.12) by its complex conjugate giving:

$$I = \frac{1 - \cos(Ndk\sin\theta)}{1 - \cos(dk\sin\theta)} \tag{11.13}$$

When the denominator of Eq. (11.13) is equal to zero, then the flux reaches a maximum with respect to angle. This maximum is proportional to the square of the number of slits. At first, this may seem counter-intuitive. However, as will be seen later, the angular width of this maximum is inversely proportional to the number of slits and the integrated flux is actually proportional to the number of slits, as expected. It is clear from Eq. (11.13) that there are a number of angles for which the denominator is zero and the intensity is at a maximum. For this to occur, then the following condition must apply:

$$\mathrm{d}k\sin\theta = 2m\pi \text{ or } d\sin\theta = m\lambda \tag{11.14}$$

where m is an integer

Equation (11.14) is the so-called ***grating equation***. It must be emphasised, at this stage, that this analysis assumes that the direction of propagation of the input plane wave is normal to the grating surface. It is clear that a number of different diffraction maxima arise, as defined by the value of the integer, m. This integer is known as the ***diffraction order***. The order can take on either negative of positive integer values, however the modulus of the sine of the diffracted angle can never be greater than one. Hence the modulus of the order must always be less than the ratio of d and λ. In addition, m can be zero, or the zeroth order, which corresponds to undisturbed transmission of the plane wave.

$$|m| < d/\lambda \tag{11.15}$$

Of course, it must be understood, that the presence of multiple orders creates ambiguities when a grating is deployed in an instrument. That is to say, it is impossible to distinguish second order ($m = 2$) diffraction at 300 nm from first order diffraction at 600 nm; both wavelengths will be diffracted at the same angle. To remove this ambiguity, long-pass or short-pass filters, or ***order sorting filters*** are deployed to remove the undesired wavelength. This will be discussed in more detail in a later chapter when we cover spectroscopic instruments.

Worked Example 11.2 ***Diffraction Grating***
A transmission grating has a slit spacing of 2 μm or '500 lines per mm'. Plane waves derived from a helium neon laser at 632.8 nm are incident upon the grating. At what angle does the ***second order*** diffracted beam make with respect to the original propagation direction?

Referring to Eq. (11.14) we are told that $m = 2$, $\lambda = 0.6328$ μm and $d = 2$ μm. It is straightforward to calculate the diffraction angle:

$$d\sin\theta = m\lambda \text{ and } 2\sin\theta = 2\times 0.6328 \text{ Hence } \sin\theta = 0.6328 \text{ and } \theta = 39.26^\circ$$

The diffracted angle is 39.26°.

11.3.3 Dispersion and Resolving Power

The resolving power of a spectroscopic instrument is the minimum spectral interval that can be unambiguously distinguished. It is equivalent to the resolving power of an imaging system, except the discrimination is with respect to diffracted angle or wavelength. To illustrate this point, it is useful to graphically display the form of Eq. (11.13), showing the form of the far field diffraction pattern with respect to angle. This is shown in Figure 11.11.

Of course, a diffraction grating with 10 slits is not representative, but it does enable us to see the angular width of each diffraction order. We can see that at either side of each major maximum, there is a local minimum. As with the Rayleigh criterion for instrument resolution, we assume that adjacent patterns from different narrow spectral feature are just resolved when the peak of one is aligned to the adjacent local minimum of the next. From Eq. (11.13), the condition for this is:

$$Ndk\Delta(\sin\theta) = 2\pi \text{ and } \Delta\theta = \frac{\lambda}{Nd\cos\theta} \tag{11.16}$$

Equation (11.16) gives the ***angular*** resolution. However, we are particularly interested in the ***wavelength*** resolution. To determine this, we need to know the dispersion of the grating, β. This may be derived from Eq. (11.14):

$$\beta = \frac{d\theta}{d\lambda} = \frac{m}{d\cos\theta} \text{ or } \beta = \frac{\tan\theta}{\lambda} \tag{11.17}$$

The ***resolving power, R***, of a spectroscopic instrument is defined as the ratio of the wavelength to the smallest incremental wavelength, $\Delta\lambda$, that can be resolved. From Eqs. (11.16) and (11.17), the resolving power is given by:

$$R = \frac{\lambda}{\Delta\lambda} = \frac{w\sin\theta}{\lambda} \tag{11.18}$$

w is the width of the grating – the product of the number of slits, N, and their separation, d.

Thus, broadly, the ultimate resolving power of a diffraction grating is given by the ratio of the grating width to the wavelength. It is useful now to compare the form of Eq. (11.7), giving the resolving power of a prism

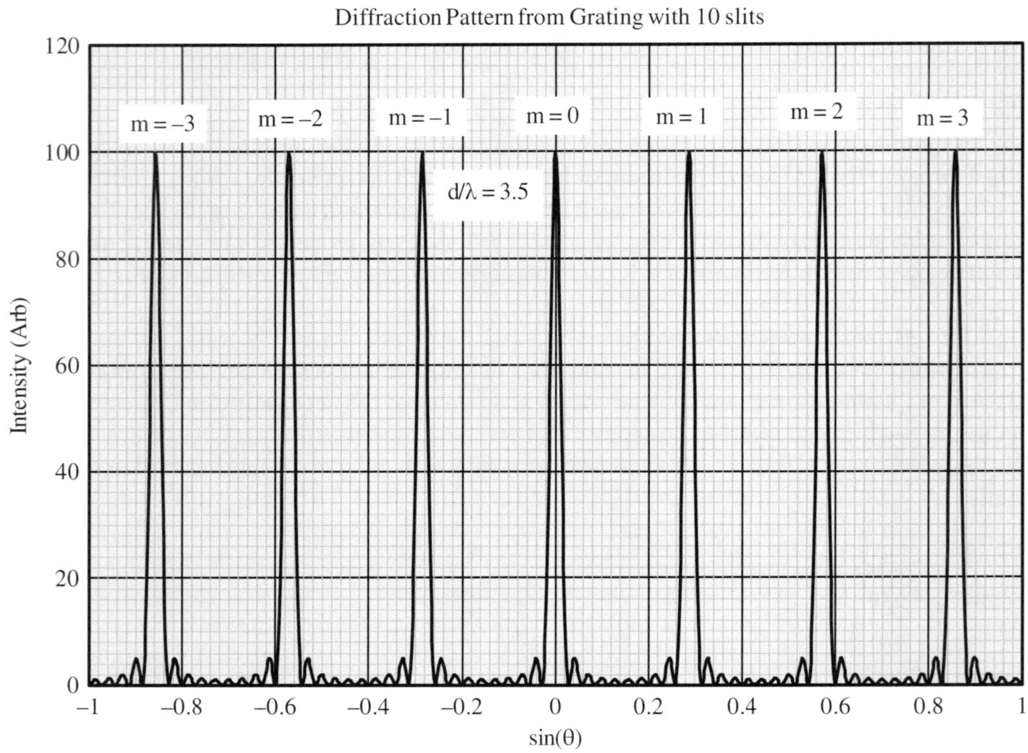

Figure 11.11 Diffraction pattern from grating with 10 slits.

and Eq. (11.18). Overall the form of the two expressions is strikingly similar, except the resolving power of the prism is mediated by its Abbe number. Thus, as previously asserted, the Abbe number of a prism's optical medium represents the degradation in resolving power when compared to a diffraction grating of equivalent width. For a typical diffraction grating with a width of tens of millimetres, then the resolving power is of the order of 10 000 or more.

Worked Example 11.3 ***Diffraction Grating Resolving Power***

A transmission diffraction grating is working at the sodium D line wavelength of 589 nm. The grating is 25 mm wide and has 800 lines per mm. It is working in first order. What is the resolving power and dispersion of the grating under these conditions?

The grating spacing d, for 800 lines per mm is equal to 1.25 μm. First, we must calculate the diffraction angle, θ, from Eq. (11.14):

$$d\sin\theta = m\lambda \text{ Hence}:\ \sin\theta = \lambda/d\ (\text{for m} = 1)\ \text{and}\ \ \sin\theta = 589/1250 = 0.471$$

This gives θ as 28.1°.

We know that the resolving power, R, is given by Eq. (11.18):

$$R = \frac{w\sin\theta}{\lambda} = \frac{25000 \times 0.471}{0.589} = 20000$$

Thus, the resolving power of the grating is 20 000.

The dispersion of the grating is given by Eq. (11.7):

$$\beta = \frac{\tan\theta}{\lambda}\ \text{and}\ \beta = \frac{0.534}{589} = 0.000907$$

The dispersion of the grating is 0.000 907 rad per nm or 0.052° per nm.

11.3.4 Efficiency of a Transmission Grating

The previous transmission grating analysis assumed that the width of the individual slits is vanishingly small. Clearly, this is not a practical proposition. Each slit must have some finite width, δ. Hence, under these conditions, the far field distribution is given by the Fourier Transform of the convolution of the point source series, as previously outlined, taken together with the finite width of the slit. The Fourier Transform of such a convolution is equal to the product of the individual Fourier Transforms. Hence to calculate the final far field pattern, we must multiply Eq. (11.13) by the far field distribution of a single slit of width, δ. This is given by:

$$I(convolve) = \frac{2(1 - \cos(\delta k \sin\theta))}{(k \sin\theta)^2} \tag{11.19}$$

Equation (11.19) is the square of the so-called ***sinc function***, which is the characteristic diffraction pattern of a single slit. It is clear from Eq. (11.19), that for $\delta = d$, then the convolved intensity is zero for all orders except the zeroth. In this instance, it is evident that where the different orders have their maxima, then the numerator in Eq. (11.19) is also zero. This is perhaps, rather trivial and not unexpected for the condition represents an uninterrupted plane wave. However, we might wish to maximise the efficiency into one specific order, e.g. the first. Perhaps, not surprisingly, the most efficient arrangement is to ensure δ is one half of the separation d. It is a feature of amplitude (as opposed to phase) gratings that the delivery into orders other than the zeroth order is rather inefficient. Under the optimum condition ($\delta = d/2$), the efficiency of diffraction into each of the first orders ($m = 1$ and $m = -1$) is $2/\pi^2$ or 20.2%. Half of the flux is retained in the zeroth order. For this condition, diffraction into the even orders ($m = 2, 4, 6$, etc.) is zero. This is illustrated in Figure 11.12 which shows the diffraction efficiency versus order for $\delta/d = 0.5$.

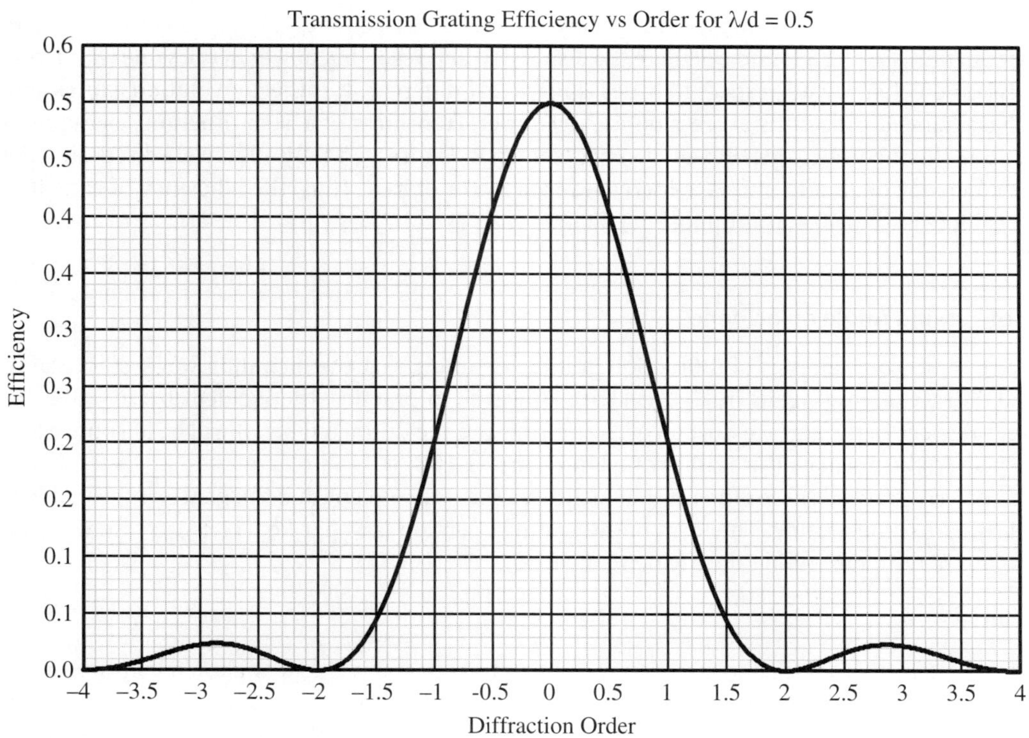

Figure 11.12 Diffraction efficiency vs order.

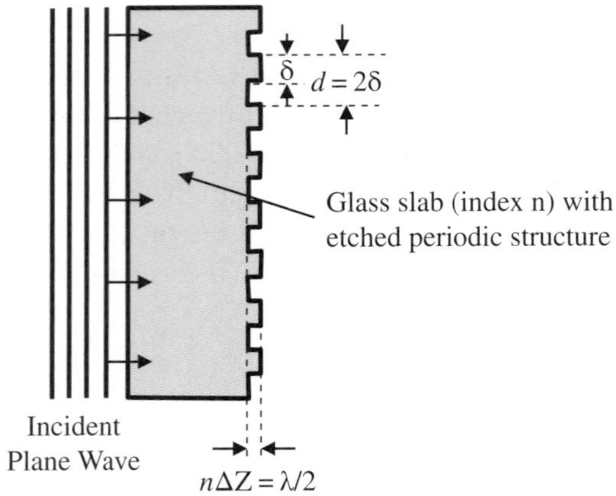

Figure 11.13 Transmission grating.

11.3.5 Phase Gratings

Thus far, we have dealt entirely with amplitude gratings. That is to say, the periodic modulation operates only on the modulus of the electric field along the grating. However, instead of modulating the amplitude, we can modulate the phase instead. The result of this is the so-called phase grating, which is generally much more efficient than the amplitude grating. For example, we might modulate the phase across the grating by $\pm\pi/2$, so that each portion of the grating is in antiphase. If we assume that each of the elements in the grating is of equal width, then it is clear that the flux appearing in the zeroth order will be entirely cancelled out by destructive interference. By reference to the previous discussion, the width of each element is now denoted by δ, and the overall period of the grating, d, is equal to twice the element width, i.e. 2δ. Such a phase grating may be implemented by a glass plate, one of whose surfaces has been etched to form a period structure that produces a relative phase delay of π radians. This is illustrated in Figure 11.13.

If we apply a grating period of d, then most of the preceding analysis still applies. That is to say, we may still apply the grating equation, Eq. (11.14) and calculate the dispersion and resolving power in the same way. However, the transmission efficiency with respect to order is markedly different. It is the convolution expression, Eq. (11.19) that has changed, since this expression is the Fourier Transform of a 'step function', as opposed to a top hat function. The theoretical grating efficiency is shown in Figure 11.14. By effectively discarding the zeroth order, the efficiency of all odd orders is increased by a factor of two. Thus, the efficiency of each first diffraction order is increased to 40.4%, as opposed to 20.2%.

11.3.6 Impact of Varying Angle of Incidence

Hitherto, we have pursued our analysis of diffraction grating assuming that the incoming wave is normal to the surface of the diffraction grating, as in Figure 11.10. However, in practice, the incident beam will be at some angle, θ, with respect to the surface normal of the grating and the diffracted beam will be at some other angle, ϕ, with respect to the normal. This is illustrated in Figure 11.15.

The diffraction analysis proceeds as for the normal incidence scenario, except that the phase of the incident beam varies linearly across the grating. The form of the grating equation is as per Eq. (11.14) except with the introduction of an extra term describing the incident angle, θ:

$$d(\sin\phi - \sin\theta) = m\lambda \tag{11.20}$$

Not surprisingly, the zeroth order is characterised by equal diffracted and incident angles. In addition, however, a non-zero angle of incidence destroys the symmetry between positive and negative orders. Thus, one should expect that the diffraction angles for order 1 and order −1 would be different. The dispersion of the

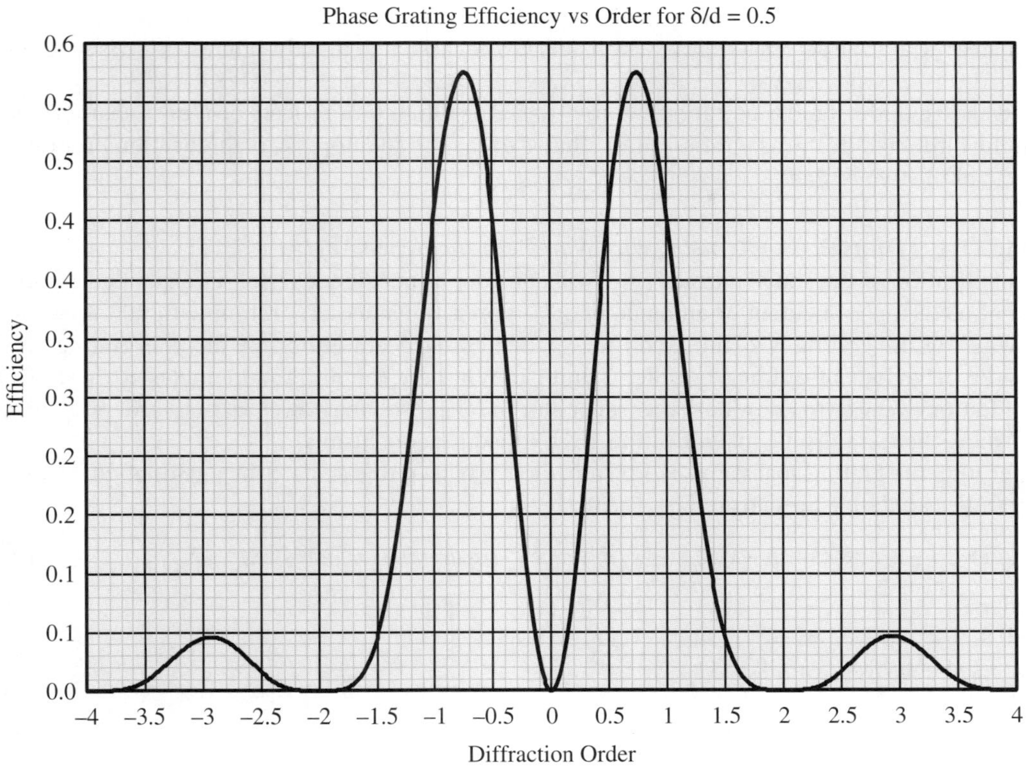

Figure 11.14 Phase grating efficiency.

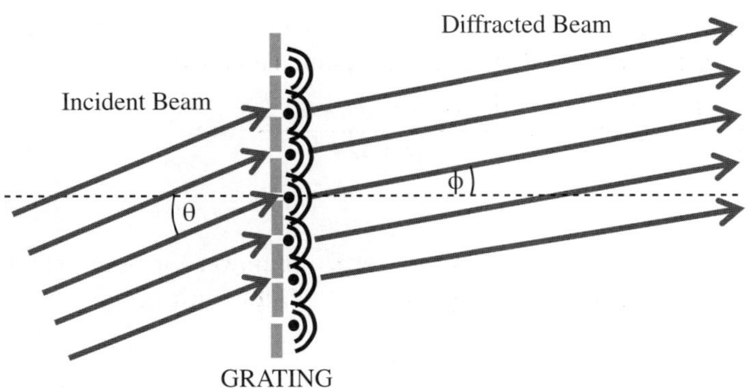

Figure 11.15 Diffraction for non-zero angle of incidence.

grating is now given by:

$$\beta = \frac{m}{d\cos\phi} = \frac{\sin\phi - \sin\theta}{\lambda\cos\phi} \tag{11.21}$$

In addition, the resolution of the grating is similar in form to that described by Eq. (11.18):

$$R = \frac{w(\sin\phi - \sin\theta)}{\lambda} \tag{11.22}$$

It is important to understand in the diffraction analysis presented here, that it is assumed the incident beam uniformly illuminates the grating across its entire width. It is further assumed that this illumination is spatially coherent.

Finally, we need also to understand that, for non-zero orders, the diffraction process produces anamorphic magnification in the diffracted beam, as per Eq. (11.8). That is to say the anamorphic magnification produced is equal to the ratio of the two cosines.

Worked Example 11.4 ***Transmission Grating with Non-Zero Incidence Angle***

To illustrate the application of the above narrative, we will consider a specific scenario. An expanded helium neon laser beam at 632.8 nm is incident at an angle of 30° upon a transmission grating with 600 lines per millimetre. The width of the grating is 20 mm and the expanded laser beam uniformly fills this grating. We are interested in the behaviour of diffraction order 1 and −1. What are the diffraction angles at these two wavelengths? In addition, we wish to know both the dispersion and the resolving power for both orders. Finally, we also desire to know the anamorphic magnification, M, of the diffracted beams.

For 600 lines per mm, the grating spacing, d, is 1.67 μm. First, we need to calculate the angles of diffraction for the two orders. We can derive these from Eq. (11.20):

$$d(\sin\phi - \sin\theta) = \pm\lambda \text{ and } (\sin\phi - 0.5) = \pm\lambda/d \text{ hence}: (\sin\phi - 0.5) = \pm 0.6328/1.67$$

For the two orders:

m = 1: $\sin\phi = 0.879$; $\phi = 61.55°$ **m = −1:** $\sin\phi = 0.121$; $\phi = 6.94°$

The dispersion may be derived from Eq. (11.21):

$$\beta = \frac{m}{d\cos\phi} \quad \beta = \frac{1}{1600\cos(61.55)} \text{ (m = 1) } \beta = \frac{-1}{1600\cos(6.94)} \text{ (m = −1)}$$

This gives:

m = 1: $\beta = 0.001\,31$ rad per nm or 0.0752° per nm.

m = −1: $\beta = -0.000\,63$ rad per nm or −0.0361° per nm.

It is important to note the sign reversal on the −1th order. The orientation of the dispersion for this order is opposite to that of the first order. The resolving power for both orders may be calculated from Eq. (11.22).

$$R = \frac{w(\sin\phi - \sin\theta)}{\lambda}; R = \frac{20000(0.8792 - 0.5)}{0.6328} \text{ (m = 1); } R = \frac{20000(0.1208 - 0.5)}{0.6328} \text{ (m = 1);}$$

This gives:

m = 1: R = 11 985; **m = −1:** R = −11 985;

The sign of the resolving power is not important and it is clear that the resolving power for both orders is identical. It now simply remains to calculate the anamorphic magnification, M, for both orders. From Eq. (11.8) we have:

$$M = \frac{\cos\phi}{\cos\theta} \quad M = \frac{0.476}{0.866} = 0.55 \text{ (m = 1); } M = \frac{0.993}{0.866} = 1.146$$

m = 1: $M = 0.55$; **m = −1 :** $M = 1.146$;

The magnification for the two orders is different. For the m = 1 order, the beam size has been constricted, whereas for m = −1 the beam has been expanded.

11.3.7 Reflection Gratings

We have illustrated the analysis of diffraction gratings by introducing transmission gratings. The principle of their operation is clear to understand. However, in practice, the majority of gratings used in real applications are reflection gratings. To understand how this might work, we can substitute the linear array of slits

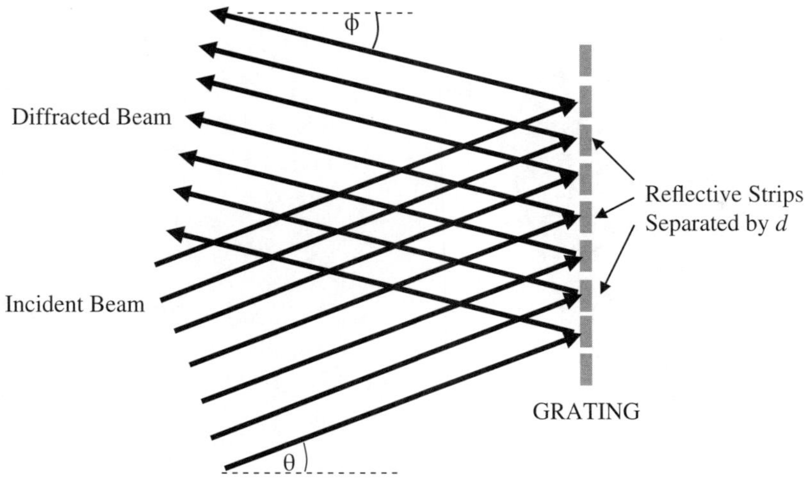

Figure 11.16 Operation of reflective grating.

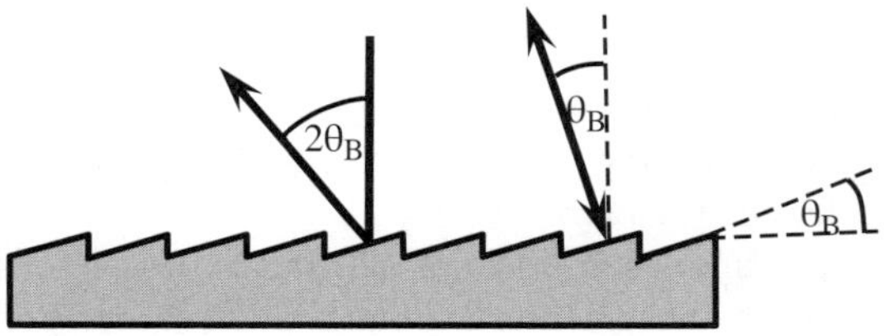

Figure 11.17 Blazed diffraction grating.

in Figure 11.15, with a linear array of reflective strips each separated by a distance, *d*. This is illustrated in Figure 11.16.

Analysis of the reflection grating then proceeds entirely as for the transmission grating, except one needs to be careful about the sign convention for the incident and diffracted angles. That is to say, Eqs. (11.20) to (11.22) apply equally to reflection and transmission gratings. For a reflection grating, the angles of incidence and diffraction are equal for zeroth order.

In Figure 11.16, the reflective grating is represented by a series of reflective strips. However, this is not a practical proposition as such an arrangement is very inefficient in delivering light into non-zeroth orders. The logic of this is similar to that pertaining to transmission gratings and outlined previously. As a consequence of this, the most common form of (reflective) grating is the ***blazed grating***. Each flat reflective strip is replaced by an inclined facet that is tilted at such an angle to direct the light towards a specific order at the design wavelength. As such, the form of the grating is described by a sawtooth profile with the width of each tooth approximately equal to the grating spacing. This is illustrated in Figure 11.17.

The angle, θ_B, which each step makes with respect to the surface of the grating is referred to as the ***blaze angle***. Two potential scenarios are illustrated which seek to take advantage of the grating blaze. In the first case, light is incident normally to the plane of the grating and the desired order is arranged to be diffracted at twice the blaze angle. The second case occurs where the incident and diffracted angles are equal, but not opposite, and both incident and diffracted angles are equal to the blaze angle. This is known as the ***Littrow configuration***. The Littrow configuration is particularly useful, as it allows direct retro-reflection for a non-zeroth order. In addition, the anamorphic magnification in the Littrow configuration is unity. It is a configuration that is widely used in many instrument designs, including tunable lasers. This arrangement is shown in Figure 11.18.

In describing a blazed grating, the convention is to describe the grating in terms of a blaze wavelength, λ_B, rather than a blaze angle, θ_B. This blaze wavelength is usually defined in terms of the Littrow condition. That

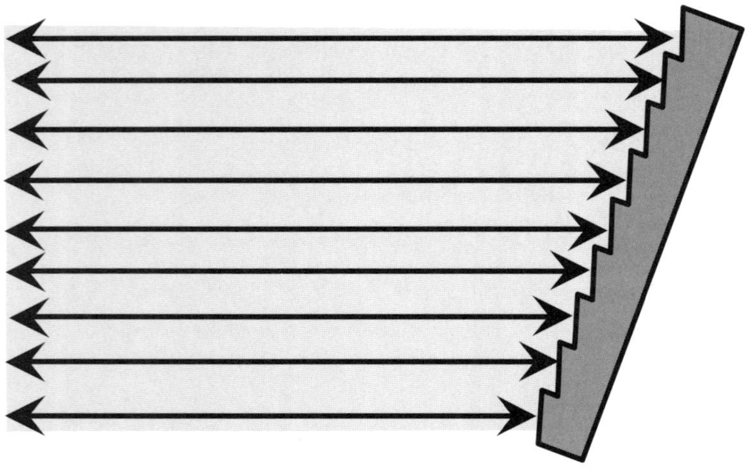

Figure 11.18 Diffraction grating in Littrow configuration.

is to say, the Littrow condition is fulfilled at the blaze wavelength for some specific diffraction order, e.g. 1.

$$2d \sin\theta_B = m\lambda \tag{11.23}$$

For example, a grating is described has having 900 lines per mm and a blaze wavelength of 500 nm. The grating spacing, d, is 1111 nm and, assuming the blaze wavelength is defined in first order, then one may calculate the blaze angle as 13°.

The simplest model that one might present of a blazed grating is to describe each facet as delivering a linear ramp in phase, with the length of the ramp approximately equal to the grating spacing. This will produce an efficiency envelope function similar to that in Figure 11.12. However, each facet represents the full grating spacing, rather than half, as depicted in Figure 11.12. As previously argued, this would lead to zero efficiency at all grating orders, except the zeroth. By contrast, for a blazed grating, the whole pattern has been tilted to align the maximum with the blaze angle. As a consequence, at the blaze wavelength (only) all other orders, in theory, disappear.

$$I(convolve) = \frac{2[1 - \cos(dk(\sin\varphi - \sin\theta_B))]}{(k(\sin\varphi - \sin\theta_B))^2} \tag{11.24}$$

Assuming the Littrow condition, we may use Eq. (11.24) to express more generally the diffraction efficiency versus wavelength. This gives a useful expression for the efficiency which depends only upon the ratio of the wavelength to the blaze wavelength.

$$I(convolve) = \frac{2[1 - \cos(2\pi(1 - \lambda_B/\lambda))]}{(2\pi(1 - \lambda_B/\lambda))^2} \tag{11.25}$$

This expression is illustrated graphically in Figure 11.19.

The efficiency curve falls off more steeply at shorter wavelengths. Indeed, it is clear that the series of zero minima at shorter wavelengths reflect the higher diffraction orders. At the same time, the form of the curve for higher order and shorter wavelengths is broadly the same as for the nominal curve. That is to say, for the curve above, the blaze wavelength is 600 nm in first order, then the efficiency curve for second order diffraction would be identical if an effective blaze wavelength of 300 nm is used instead of 600 nm. Of course, the plot in Figure (11.19) assumes perfect reflectivity and does not take account of absorption losses.

Worked Example 11.5 ***Blazed Grating***

A grating has 1000 lines per mm and is blazed for 500 nm (Littrow) in first order and is deployed in the Littrow configuration.

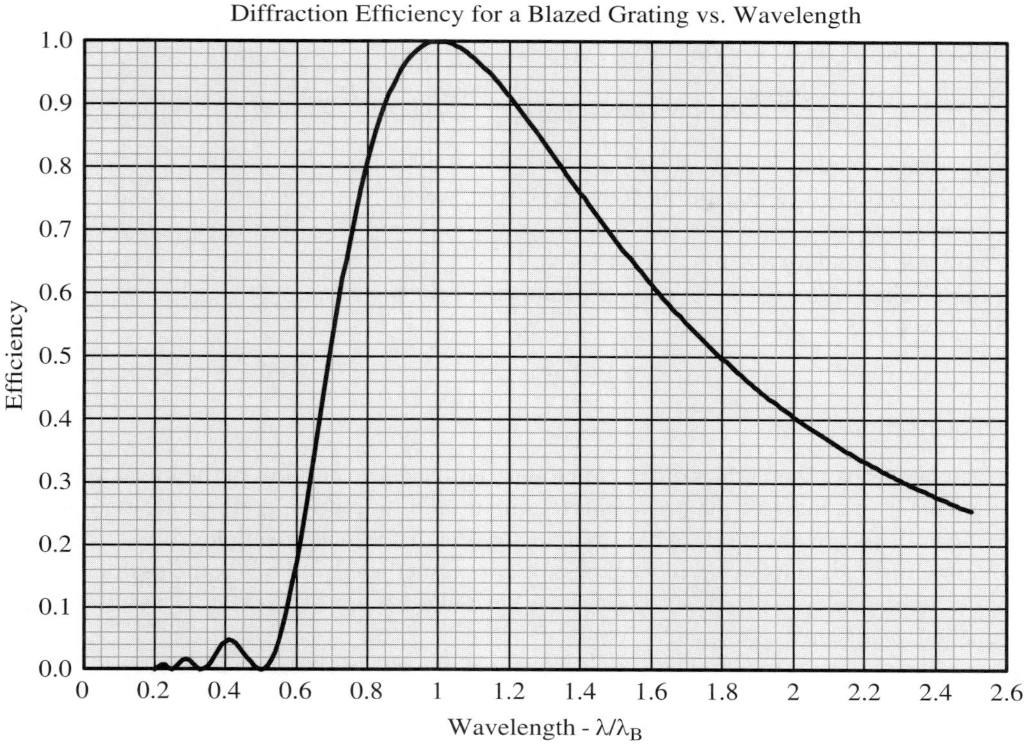

Figure 11.19 Generic efficiency curve for a blazed diffraction grating.

1. What is the blaze angle of the grating?
2. What is the dispersion of the grating around 600 nm in this configuration?
3. What is the diffraction angle for light at 400 nm and 600 nm relative to the 500 nm (Littrow) light?
4. Estimate the anamorphic magnification and the diffraction efficiency at 400 nm and 600 nm

1. The grating spacing is 1 μm or 1000 nm (1000 lines per mm). The relevant wavelength is 500 nm and the diffraction order is one. From Eq. (11.23) we have:

   $$2d\sin\theta_B = m\lambda \text{ and } 2\times 1000\times\sin\theta_B = 1\times 500 \text{ and } \sin\theta_B = 0.25\ \theta_B = 14.48°$$

   The blaze angle is 14.48°
2. The dispersion may be derived from Eq. (11.21). We must, however, be a little careful as to the sign convention for the angles. In the context of Eq. (11.21), the diffracted and incident angles are equal, but opposite. Therefore, we may calculate the dispersion as follows:

   $$\beta = \frac{\sin\varphi - \sin\theta}{\lambda\cos\varphi} = \frac{-0.25-0.25}{500\times 0.9682} = -0.00103\ \textbf{\textit{Dispersion is}} - \textbf{\textit{0.001 03 rad nm}}^{-1}\ \textbf{\textit{or 005 932°nm}}^{-1}$$

   The dispersion is, of course, negative. As the wavelength increases, the diffraction angle moves away from the position of the zeroth (normal reflection) order.
3. The diffraction angles may be derived from Eq. (11.20). In the case of the Littrow arrangement, the order is actually −1, rather than 1, as the sense of the diffraction is away from the zeroth order.

   $$d(\sin\phi - \sin\theta) = -m\lambda\ 1000\times(\sin\varphi_1 - 0.25) = -400 \text{ and } 1000\times(\sin\varphi_2 - 0.5) = -600$$

   This gives: $\sin\phi_1 = -0.15$ and $\sin\phi_2 = -0.35$ or $\phi_1 = -8.63°$ and $\phi_2 = -20.49°$
   Comparing this to the Littrow angle we have: $\boldsymbol{\phi_1 = 5.85°}$ ***(400 nm) and*** $\boldsymbol{\phi_2 = -6.01°}$ ***(600 nm)***

This solution is entirely consistent with the dispersion derived in part 2.

4. The magnification is given by Eq. (11.8).

$$M = \frac{\cos\phi}{\cos\theta}M = \frac{0.9887}{0.9682} = 1.021\ (400\ \text{nm});\ M = \frac{0.9368}{0.9682} = 0.967(600\ \text{nm})$$

The anamorphic magnification is 1.021 at 400 nm and 0.967 at 600 nm.
To estimate the diffraction efficiency for the 400 and 600 nm wavelength, we need simply substitute the relevant λ_B/λ values into Eq. (11.25).
For 400 nm, $\lambda_B/\lambda = 1.25$ and for 600 nm, $\lambda_B/\lambda = 0.833$
Substituting these values into Eq. (11.25) gives the following efficiencies:
$\boldsymbol{I_{400} = 0.875; I_{600} = 0.875.}$

11.3.8 Impact of Polarisation

The analysis presented here is based on scalar diffraction theory and takes no account of the impact of polarisation. This is a significant weakness of the preceding treatment and diffraction gratings are significantly sensitive to polarisation. Of course, the merit of the scalar analysis is that it makes complex diffraction problems tractable and, notwithstanding its inherent weakness, it does nonetheless provide some useful insights. A full and rigorous diffraction analysis of a (blazed) grating is beyond the scope of this text. Nevertheless, one can outline some qualitative insights.

Figure 11.20 shows a blazed grating together with an indication of the two polarisation orientations. Viewed in the direction of p polarisation transverse magnetic (TM), the sharp step in the blazed grating may be seen as a barrier to the movement of charge carriers. By contrast, in the direction of s polarisation transverse electric (TE), charge carriers have an un-interrupted conduction path. As such, one might view a blazed grating as being somewhat analogous to a wire grid polariser. It is clear that the p polarisation will be preferentially reflected when compared to the s polarisation. This is illustrated in Figure 11.21 which shows a grating efficiency plot vs wavelength for the two polarisations. The diffraction efficiency of s polarisation is clearly inferior to that of p polarisation. In some respects, the diffraction grating may be seen as a polarisation device.

The data in Figure 11.21 are for a blaze angle of 19°. Perhaps not surprisingly, the difference in behaviour between the two polarisations is accentuated for large blaze angles. By contrast, for low blaze angles the difference in efficiency between the s and p polarisations is slight.

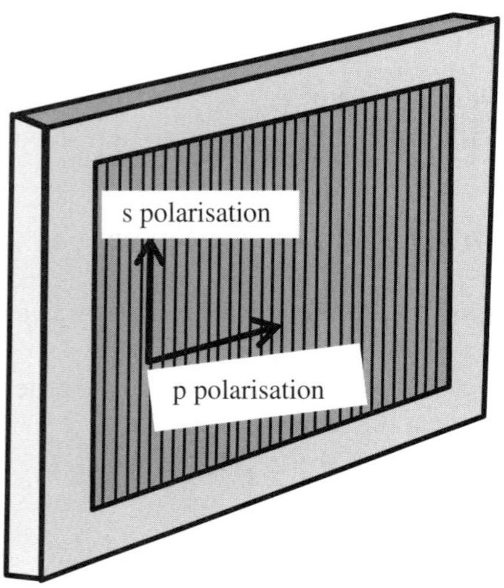

Figure 11.20 Blazed grating showing polarisation orientations.

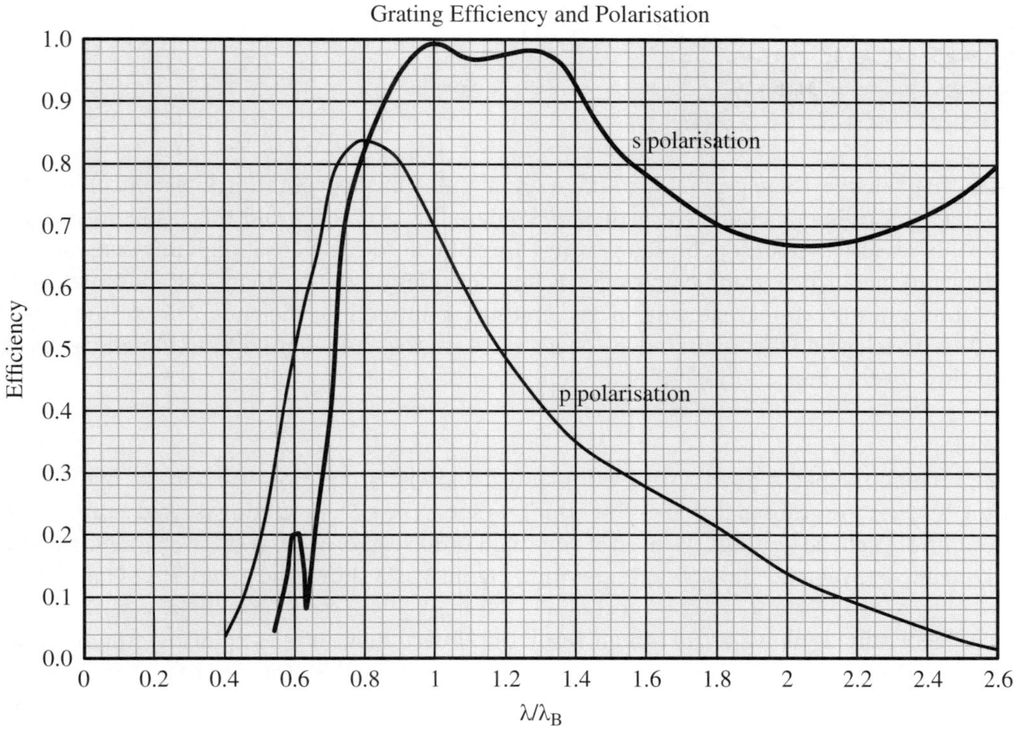

Figure 11.21 Grating efficiency for two polarisation directions ($\theta_B = 19°$). Source: *Courtesy Newport Corporation*. Permission to use granted by Newport Corporation. All rights reserved.

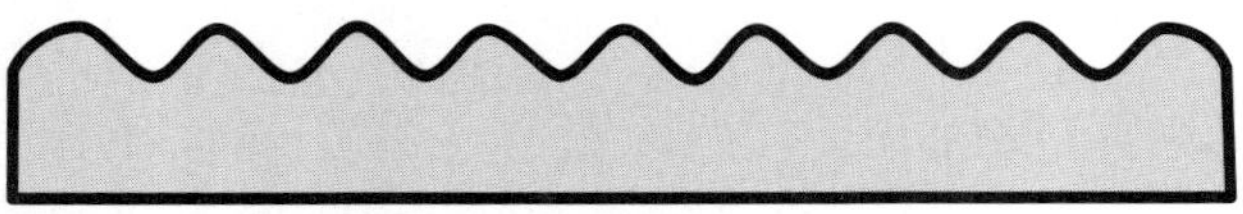

Figure 11.22 Holographic grating profile.

11.3.9 Other Grating Types

11.3.9.1 Holographic Gratings

A holographic grating replaces the sharp and discontinuous step profile of a blazed grating with a smooth sinusoidal profile. This is illustrated in Figure 11.22.

Holographic gratings are created by a photolithographic process involving the interference of two intersecting and mutually coherent laser beams. One clear difference between the holographic profile in Figure 11.22 and that of the blazed profile in Figure 11.17 is the lack of sharp features. Although not predicted by scalar diffraction theory, these sharp features contribute sharply to light scattering. For critical applications, where low levels of background light are important, holographic gratings are preferred.

There is another important advantage of holographic gratings. Since the grating is produced by an interference process between two laser beams, the sinusoidal profile is created with extremely high fidelity. However, blazed gratings are produced mechanically using a (diamond) ruling machine. Sequential machining of equally spaced lines requires conversion of the machine's natural rotary motion into an even linear progression. Traditionally, this conversion is accomplished through a progression of gears. However, this gear train has a tendency to introduce very small periodic errors in the ruling machine's linear progression. This additional periodicity produces faint diffraction maxima close to the main order. These additional diffraction maxima are

Figure 11.23 Echelle grating.

referred to as ***'ghosts'***. In principle, these problems could be ameliorated by incorporating linear encoders and positional feedback into the ruling machine. Nevertheless, holographic gratings avoid this problem entirely and are favoured for this reason. In addition, blazed gratings are difficult to replicate at high line densities, so there is also a tendency to favour holographic gratings for line densities of greater than 1000 lines per mm. That said, the holographic grating cannot match the blazed grating for diffraction efficiency. As a consequence, blazed gratings tend to be selected when working at longer wavelengths and where diffraction efficiency is important, assuming low level light scattering is not an impediment.

11.3.9.2 Echelle Grating

An ***echelle grating*** is, in many respects, geometrically similar to a blazed grating. However, instead of the incident light striking the 'step' of the grating, it strikes the 'riser'. As such, the echelle grating is used at a shallow incidence angle. In addition, echelle gratings are used at very high order, e.g. $m = 40$, and are characterised by a low line density, typically between 30 and 300 lines per mm. Figure 11.23 shows the geometry of an echelle grating.

Echelle gratings are generally used as part of a high-resolution instrument, often exploring quite narrow portions of the spectrum. As discussed earlier, they tend to be used at very high diffraction orders and disambiguation between the different orders is an issue. A technique called ***cross dispersion*** is often used whereby a second, low dispersion grating, is used to provide order sorting dispersion in an orthogonal plane. As a component favoured in high resolution spectroscopy, their application echoes that of the etalon. In both cases, for a given diffraction angle, there are a large number of transmission peaks or orders separated a free spectral range. If we assume that the grating is operating in the Littrow condition, then by rearranging the grating equation we get:

$$1/\lambda = m/2d\sin\theta_B \text{ and } \upsilon = m/2d\sin\theta_B \text{ hence } \Delta\upsilon = 1/2d\sin\theta_B \tag{11.26}$$

As can be seen from Eq. (11.26), the free spectral range, or $\Delta\upsilon$, is equivalent to the inverse of the optical path difference accrued between reflections off adjacent steps. This free spectral range represents the difference (in wavenumbers) between two adjacent orders for a specific diffraction angle. The form of Eq. (11.26) is very similar to the free spectral range of the etalon which is expressed in terms of the inverse of the path difference between reflections at the two plane mirror surfaces. For an echelle grating, θ_B is high, 60–70° being not untypical. Considering an echelle grating with a groove density of 30 lines per mm and a blaze angle of 70°, then the free spectral range is about 160 cm^{-1}. For a wavelength of 2000 nm, this represents an interval of 63.9 nm.

As outlined previously, echelle gratings tend to be used in high resolution applications. Gratings as large as 400 mm may be deployed. In this case, resolving powers of several hundred thousand are achievable. If we revisit Eq. (11.22) describing the resolving power of a grating, it is clear that the resolving power is given by the ratio of the reflective optical path difference across the grating, divided by the wavelength. This is a useful concept in describing the resolving power of a dispersive instrument.

11.3.9.3 Concave Gratings – The Rowland Grating

Hitherto, we have considered a diffraction grating as being arrayed on a plane surface. Implicit in this treatment is the assumption that the incoming light is collimated to produce a plane wave input. Similarly, in order to exploit and image the output from the diffraction grating, it must be further assumed that some form of focusing optics must be incorporated to translate the diffraction angle into an imaged point or line. As will

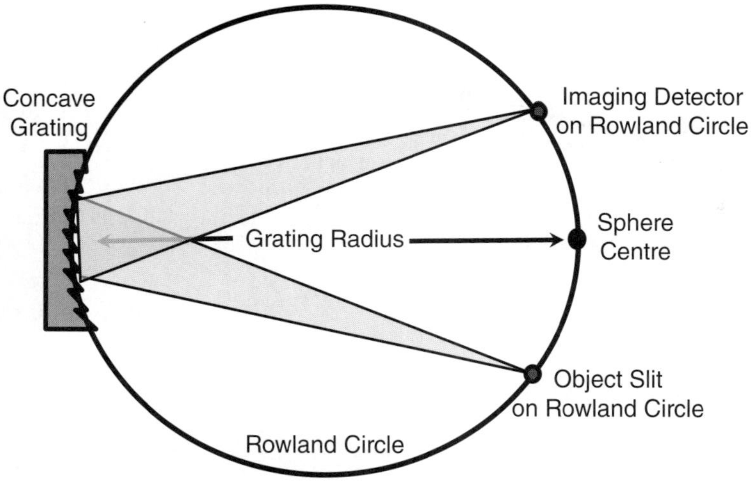

Figure 11.24 Rowland grating arrangement.

be seen when we come to review spectroscopic instruments in a subsequent chapter, this is by far the most common arrangement. However, it is quite possible to rule gratings on a curved surface.

One example of this is the ***Rowland grating*** arrangement which is based on a diffraction grating ruled on a spherical surface. One may now regard the system as an imaging system with a real object and a real image. In this case, the object comprises a slit located near the centre of the spherical grating. Diffracted light from the spherical grating is then imaged at a slit formed close to the centre of the sphere. In practice, both object and image slits lie at different points along a circle whose ***diameter*** is equal to the grating sphere radius; the grating itself also lies along this surface. The circle is referred to as the ***Rowland Circle***. This arrangement is shown in Figure 11.24.

It is important to emphasise that the Rowland circle has a ***diameter*** that is equal to the ***radius*** of the concave grating. The Rowland grating arrangement allows for formation of a dispersed slit image without the need for collimation and focusing optics. The most significant point about this arrangement is that the imaged slit is perfectly in focus, provided both object (slit) and image (detector) lie on the Rowland circle. This can be justified by aberration theory. The Rowland circle allows formation of an unaberrated image for the tangential rays. It does not represent the Petzval surface. There is sagittal aberration (astigmatism), but at the tangential focus, any blurring occurs along the slit, and not significant in a traditional (*non imaging*) spectrometer.

Location of the imaged slit for a particular wavelength is straightforward to determine. The vertex of the grating is represented as lying on the opposite side of the Rowland circle to the sphere centre. The line joining the sphere centre and its vertex represents the grating axis. The angle between this axis and a line jointing the object and the sphere vertex represents the incident angle. The diffraction angle may then be calculated in the same way as for a plane grating. The diffraction angle is then represented as the angle between the grating axis and a line joining the image and the sphere vertex. In this way, the position of the diffracted image may be determined.

11.3.9.4 Grisms

A ***grism*** is a combination of a grating and a prism. It is implemented at a transmissive grating. Under normal circumstances, a transmissive grating will deflect light for all orders except the zeroth. However, by incorporating the grating onto one facet of a prism, this diffractive deflection is counteracted by the deflection provided by the prism refraction. At one specific design wavelength, the refraction will precisely cancel the diffractive deflection and the light will be undeviated. This is of great utility in the design of compact spectroscopic instruments. The principle is illustrated in Figure 11.25.

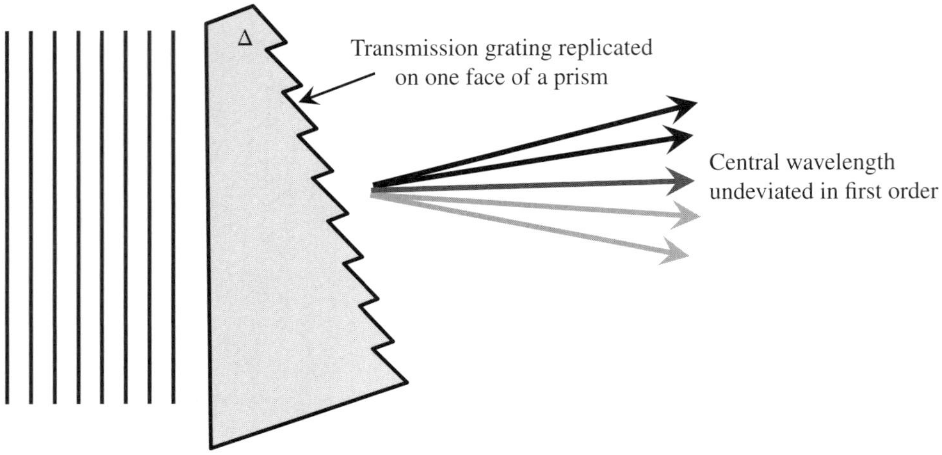

Figure 11.25 Grating prism or grism.

The new characteristic we have introduced is the prism angle, Δ. Otherwise, the analysis of the grism proceeds in the same way as for a general transmissive grating. However, the effective incidence angle, θ is changed by the refractive effect of the prism. We will assume that the light is normally incident upon the first surface of the prism depicted in Figure 11.25. Although internally, the light is incident at the second diffractive surface at an angle, $-\Delta$, externally, the sine of this incidence angle, θ, is actually $-n\sin\Delta$, where n is the refractive index of the grism material. If the grating spacing is d and the angle of diffraction is ϕ, then the grating equation is modified to:

$$d(\sin\phi + n\sin\Delta) = m\lambda \tag{11.27}$$

For the case of an undeviated ray, at the design wavelength, λ_0, then it is clear that the compensating diffractive deviation, ϕ, must be equal to $-\Delta$. Therefore, we have:

$$d(n-1)\sin\Delta = m\lambda_0 \tag{11.28}$$

Equation (11.28) thus defines the required prism angle, Δ, for a given grating spacing and design wavelength. It might be preferable to cast Eq. (11.27) in terms of the deviation angle, δ, where $\delta = \phi + \Delta$.

$$d(\sin(\delta - \Delta) + n\sin\Delta) = m\lambda \tag{11.29}$$

The dispersion of the grism is straightforward to calculate:

$$\beta = \frac{d(\sin(\delta-\Delta) + n\sin\Delta)}{\cos(\delta-\Delta)\lambda} \text{ and } \beta = \frac{(n-1)\tan\Delta}{\lambda_0} \text{ for } \lambda = \lambda_0 \tag{11.30}$$

Similarly, the resolving power of the grism is given by:

$$R = \frac{w(n-1)\sin\Delta}{\lambda_0} \tag{11.31}$$

Worked Example 11.6 ***Design of Visible Grating Prism***

A grism is to be designed for an undeviated central wavelength of 500 nm in first order. It is to operate over the wavelength range of 400–600 nm. The line density is 400 lines mm^{-1} and it may be assumed that the refractive index is 1.52 over the whole operating wavelength region. The width of the grism is 12 mm.

1. What is the angle of the grism?
2. What is the dispersion at the central wavelength of 500 nm?
3. What is the resolving power of the grism?

4. What is the deviation of the diffracted beam at 400 and 600 nm?
   1. The angle of the grism may be determined from Eq. (11.28). The wavelength, λ_0 is 500 nm and the groove spacing, d, is 2.5 μm (400 lines mm^{-1}) and the refractive index, n, is 1.52:

      $$d(n-1)\sin\Delta = m\lambda_0 \text{ and } 2500 \times 0.52 \times \sin\Delta = 500 \text{ Hence } \sin\Delta = 0.385.$$

      Therefore, the prism angle, Δ, is equal to 22.6°.
   2. The dispersion of the prism is simply given by Eq. (11.30):

      $$\beta = \frac{(n-1)\tan\Delta}{\lambda_0} = \frac{0.52 \times \tan(22.6)}{500} \text{ Hence } \beta = 0.000\,443 \text{ rad nm}^{-1}.$$

      The dispersion is 0.025° nm^{-1}.
   3. The resolving power of the grism is derived directly from Eq. (11.31):

      $$R = \frac{w(n-1)\sin\Delta}{\lambda_0} = \frac{12000 \times 0.52 \times 0.385}{0.5} = 4800$$

      The resolving power of the grism is 4800.
   4. The deviation for the two wavelengths may be calculated from Eq. (11.29):

      $$d(\sin(\delta - \Delta) + n\sin\Delta) = m\lambda$$

      For the two wavelengths this gives:

      $$(\sin(\delta_{400} - 22.6) + 1.52 \times \sin(22.6)) = \frac{400}{2500} \text{ and } (\sin(\delta_{600} - 22.6) + 1.52 \times \sin(22.6)) = \frac{600}{2500}$$

      Therefore:

      $$\sin(\delta_{400} - 22.6) = -0.425 \text{ and } \sin(\delta_{600} - 22.6) = -0.360$$

      This gives:
      $\delta_{400} = -2.51°$ and $\delta_{600} = 1.51°$

11.4 Diffractive Optics

One might imagine a diffraction grating emulating the effects of a prism in producing a given deviation for a specific wavelength and diffraction order. In terms of deployment in spectroscopic instruments the parallel is especially clear with diffraction gratings acting as a direct replacement for prisms. However, it is also possible to use periodic structures to emulate more complex components. For a specific area on a grating, the deflection produced by the grating is dependent upon the local groove spacing. If that groove spacing varies across the surface of the component, then the angular deviation may also be made to vary. In this way, for example, one can construct a diffractive lens, if the spatial frequency of the grooves is made to be proportional to the distance from the nominal centre of the component. A diffractive lens is illustrated schematically in Figure 11.26.

The diffractive lens will usually operate in some specific order, e.g. 1. Each facet of the grating is tilted, as shown in Figure 11.26. As for blazing in a diffraction grating, this tilt acts to direct the light, at the design wavelength, $\lambda_{0,}$ exclusively into the design order. The effective blaze condition is met when the optical path difference produced by the step is an integral multiple of the design wavelength. For the first order, this optical path difference is equal to the wavelength and, if the step height is denoted by δ s and the lens refractive index by n, then this step height is given by:

$$\delta s = \frac{\lambda_0}{n-1} \tag{11.32}$$

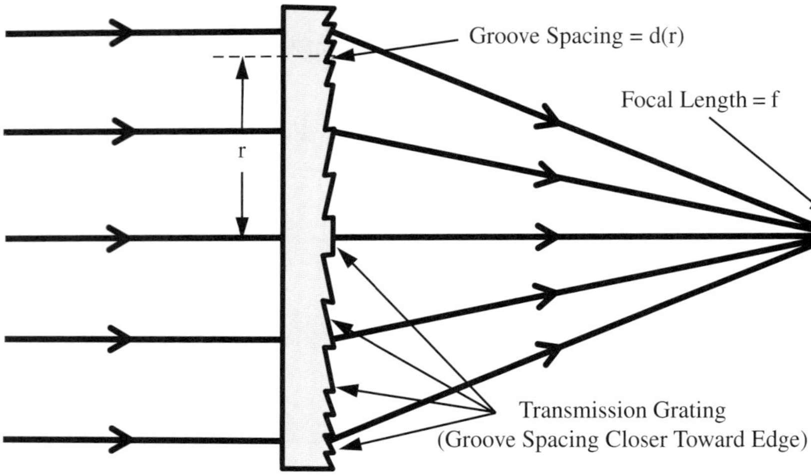

Figure 11.26 Diffractive lens.

As to the focusing power of the lens, if, for simplicity we assume that the lens is working in the paraxial regime, then we can relate the diffracted angle, θ, directly to the variable groove spacing, *d(r)*, and the design wavelength, λ_0:

$$\theta = \frac{m\lambda_0}{d(r)} \tag{11.33}$$

Maintaining the paraxial assumption, we can relate the angle of diffraction to the lens focal length, f, and the lens radial position, r:

$$\theta = \frac{m\lambda_0}{d(r)} = \frac{r}{f} \text{ and } d(r) = \frac{m\lambda_0}{r}f \tag{11.34}$$

We can use Eq. (11.34) to calculate the position of the nth diffraction groove, $r(n)$:

$$r(n) = (2fm\lambda_0)^{1/2}n^{1/2} \tag{11.35}$$

Quite obviously, a diffractive lens is not achromatic. It is clear from Eq. (11.33) that, as the wavelength increases the angular deviation increases in proportion. Hence the effective power of a diffractive lens increases proportionately with the wavelength. This increase in focusing power with wavelength is the reverse of the dispersion generated by refractive materials. That is to say, ***diffractive optics produce anomalous dispersion***. Furthermore, the effect is large, as we can demonstrate by computing the effective Abbe number of a diffractive lens:

$$'AbbeNumber' = \frac{\lambda_D}{\lambda_F - \lambda_C} = -3.452 \tag{11.36}$$

This large anomalous dispersion can occasionally be useful in certain optical designs.

11.5 Grating Fabrication

11.5.1 Ruled Gratings

Ruled gratings are produced by precision single point diamond machining. Based on an extremely rigid machine construction and precision slides and spindles, these machines are capable of generating surfaces

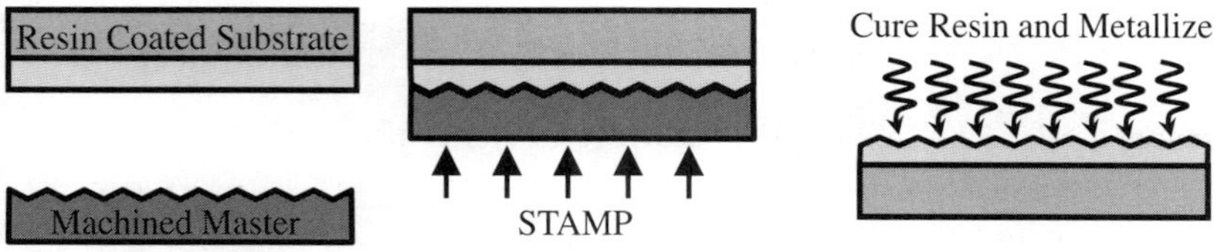

Figure 11.27 Ruled grating replication.

to an optical finish and accuracy. Of necessity, particularly for large gratings, the machining process is very slow. Therefore, this process cannot be used for high volume low cost grating components. In this case, a ***replication process*** must be used. Diamond machining is used to create a high value master which is then used to 'press' a grating in a thin film resin material. This resin is then cured and coated with aluminium or gold to create a reflective surface. The replication scheme is illustrated in Figure 11.27.

11.5.2 Holographic Gratings

By contrast, holographic gratings are formed by a photolithographic process. This process exploits techniques generally used in the semiconductor industry. A substrate coated with an optically sensitive material, a photoresist, is exposed to two angled and coherent laser beams. Interference of these two beams creates a sinusoidally varying flux across the surface of the photoresist. Following exposure, the light sensitive photoresists is treated with a chemical which etches the surface of the resist at a rate dependent upon the optical exposure. As a result, an etched sinusoidal grating is created. This process is shown in Figure 11.28.

The period, d of the grating so created is determined by the wavelength, λ, of the laser beam and the angle of their intersection, θ. By simple trigonometry, this spacing is given by:

$$d = \frac{\lambda}{2\sin(\theta/2)} \tag{11.37}$$

If, for example, the wavelength were 488 nm and the angle 30°, then the period of the grating would be about 943 nm or about 1060 lines per mm. It is a feature of a holographic grating that if one of the laser beams used in creating the grating were to illuminate the finished grating, then the missing beam would effectively be re-created.

The holographic grating process described here represents the formation of a physical 3-dimensional structure by photoetching. However, holographic gratings can also be imprinted as a planar structure by producing a periodic variation in the refractive index. There is a class of materials, photochromic materials, which change their refractive and absorptive properties on exposure to light. One example of this is (ammonium) dichromate absorbed into gelatin films. This produces a refractive index change on exposure to light and can be used to produce a transmitting phase grating. This type of grating is referred to as a ***volume phase hologram (VPH).***

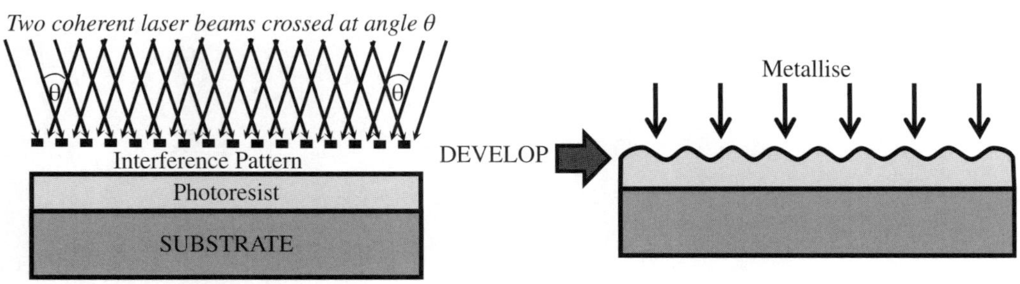

Figure 11.28 Fabrication of a holographic grating.

Further Reading

Born, M. and Wolf, E. (1999). *Principles of Optics*, 7e. Cambridge: Cambridge University Press. ISBN: 0-521-642221.

Hecht, E. (2017). *Optics*, 5e. Harlow: Pearson Education. ISBN: 978-0-1339-7722-6.

Lipson, A., Lipson, S.G., and Lipson, H. (2011). *Optical Physics*. Cambridge: Cambridge University Press. ISBN: 978-0-521-49345-1.

Saleh, B.E.A. and Teich, M.C. (2007). *Fundamentals of Photonics*, 2e. New York: Wiley. ISBN: 978-0-471-35832-9.

Smith, F.G. and Thompson, J.H. (1989). *Optics*, 2e. New York: Wiley. ISBN: 0-471-91538-1.

12

Lasers and Laser Applications

12.1 Introduction

In Chapter 6, in our analysis of diffraction we introduced the idea of coherence and touched upon the analysis of laser beam propagation. In this chapter we will look at these topics in a little more detail, together with a brief description of the underlying operational principles of laser sources. The intention, however, is to provide the applications-focused practitioner practical insights into the use of laser sources; this is not intended as a specialised text on laser devices.

In understanding laser emission, we are fundamentally concerned with the interaction of light with matter. The acronym '***Laser***' stands for Light Amplification by Stimulated Emission of Radiation. The process of optical absorption is a familiar one and is enumerated by Beer's law, wherein the optical flux in an absorbing medium decays exponentially with distance. Einstein originally reasoned that there must be a countervailing process to absorption, namely stimulated emission. Thus, in principle, under certain very specific conditions, one should be able to see exponential amplification of flux in a medium, as opposed to exponential attenuation. However, the term, 'Laser', as it stands, is something of a misnomer. A laser device is, in reality, an oscillator, rather than a mere amplifier. Just as an electronic oscillator relies on amplification ***plus*** positive feedback, then so does the laser. Amplification is provided by the optical medium and feedback via mirrors or something similar.

As outlined, we must begin by understanding the interaction of light with matter, particularly in the quantum mechanical model. In the quantum mechanical model the energy possessed by atoms, molecules, and crystals is partitioned or quantised into discrete values, corresponding to specific states, or energy levels. At the same time, the energy associated with an electromagnetic wave is also quantised into distinct packets called photons. A photon may interact with a pair of energy levels in a medium by, for example, promoting the medium from the lower energy level to the upper energy level; the photon is absorbed. Conversely, a photon may interact with a medium that is already in the upper energy state. Here, the medium is 'demoted' to the lower energy state and an extra photon is ejected. This is the process of stimulated emission. There is a third process. This is the process of spontaneous emission. Here, a photon is spontaneously emitted from a medium in the upper energy state, whereupon the medium reverts to its lower energy state. Figure 12.1 illustrates these three processes.

It is clear from Figure 12.1 that absorption and stimulated emission are competing processes. Therefore, the question arises what are the circumstances under which amplification, as opposed to absorption, is observed in a medium? Since absorption and stimulated emission are related process, it is reasonable to suggest that their effective cross sections are identical. That is to say an atom in the excited state has an equal probability of producing an extra photon by stimulated emission as an atom in the lower energy level has of annihilating a photon by absorption. The condition for amplification to occur, therefore, is that there must be more atoms/molecules in the upper energy level than in the lower energy level. This condition is referred to as a ***population inversion*** and is a departure from the 'default condition' of normal matter.

Optical Engineering Science, First Edition. Stephen Rolt.

Companion website: www.wiley.com/go/Rolt/opt-eng-sci

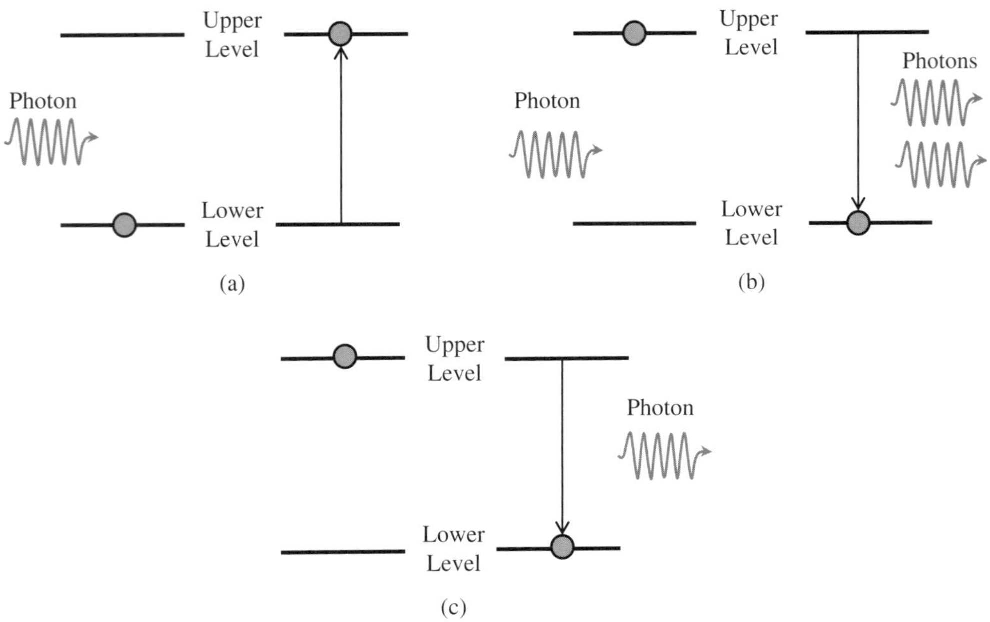

Figure 12.1 (a) Absorption. (b) Stimulated emission. (c) Spontaneous emission.

In general, higher energy levels in a medium will be less populated than lower energy levels. Indeed, for a material in thermal equilibrium, the ratio of the population of any two levels is dictated by the temperature. This ratio is given by the Boltzmann factor:

$$\frac{P_{upper}}{P_{lower}} = e^{-\frac{\Delta E}{kT}} \tag{12.1}$$

ΔE is the energy difference between the two levels; k is the Boltzmann constant and T the absolute temperature.

As T tends to infinity, the ratio of the populations tends to unity. However, the population of the upper state never exceeds that of the lower energy level. Therefore, a population inversion is symptomatic of a material that is not in thermal equilibrium. As a consequence, achievement of a population inversion is by no means easy. Indeed, as implied by the form of Eq. (12.1), a population inversion is sometimes seen as a manifestation of a negative absolute temperature.

Historically, stimulated emission was first demonstrated in the microwave region with the construction of the Ammonia Maser by Townes, Gordon, and Zeiger in 1953. At the time, the device was referred to as a ***MASER*** (**M**icrowave **A**mplification by **S**timulated **E**mission of **R**adiation). Following the work of Schawlow and Townes, Theodore Maiman finally demonstrated laser action in the visible in 1960 with the construction of the Ruby laser. For a brief period, for historical reasons, these optical devices were referred to as 'optical masers'.

Following, their discovery, there was a brief period where lasers were considered a curiosity – an invention 'looking for an application'. However, it was not long before a wide range of applications was opened up. Indeed the current focus of laser development is very much directed towards applications. The first, and perhaps most obvious application area, relates to the exceptionally high flux densities realised by laser sources. The flux density of conventional sources is fundamentally restricted by the thermal equilibrium associated with the source material. That is to say, it is impossible to exceed the blackbody radiance produced by a source at some nominal, but realistic absolute temperature. On the other hand, as we have seen, the fundamental principle underlying laser action defies the notion of thermal equilibrium. Therefore, exceptionally high radiances may be achieved by laser sources. Such radiances are substantially in excess of those pertaining to blackbody emission at any reasonable or practicable temperature. This enables a range of materials processing applications, such as the cutting and welding and shaping of a wide variety of materials, particularly refractory materials.

However, lasers are particularly characterised by their fundamental property as an oscillator. That is to say, they are to be represented as an ideal single frequency (optical frequency) oscillator with a unique and deterministic phase relationship established at all times and in all spatial positions. This is, however, very much of an ideal representation. Nevertheless, this property of coherence is critical to a range of laser applications. The property of phase coherence across a wavefront assures the directionality of the laser beam. As our discussion on Gaussian beam propagation in Chapter 6 revealed, beam spreading or divergence is restricted by the coherence as indicated by the M^2 value or number of independent, incoherent modes. Thus, the directional fidelity of the laser beam promotes a range of applications related to directional alignment, e.g. surveying, dimensional metrology, and so on. We must not, of course, ignore the spectral purity of the laser source. On this account, the laser finds many uses in the area of optical metrology, in interferometry, and laser spectroscopy.

12.2 Stimulated Emission Schemes

12.2.1 General

As we discovered earlier, the greatest difficulty is incurred in establishing a scheme for producing a population inversion. Natural processes tend to establish a status quo whereby lower energy levels are more heavily populated. Creating a population inversion involves injecting energy into a medium in such a way that specific energy levels are excited preferentially. Schemes for this are many and varied and depend upon the materials involved. To illustrate the process a little more thoroughly it is necessary to turn to real examples.

12.2.2 Stimulated Emission in Ruby

As outlined, the first (optical) laser was the ruby laser. Ruby consists of an aluminium oxide crystal (sapphire) that has been doped with a small (fraction of a percent) quantity of chromium. Typically, in a ruby laser, this proportion is around 0.05%. It is this chromium impurity that is the principal actor in the stimulated emission process. Ruby is what is referred to as a ***three level system***. In the case of the ruby laser, the upper laser level is ***pumped*** by optical absorption from the ground state by a powerful (conventional) lamp. Chromium atoms in this state then decay rapidly and non-radiatively to an intermediate third level. A population inversion is then produced between this level and the ground state. For this to happen, the pumping process must be rapid, as must the decay to the third level. To preserve the population inversion, the radiative decay from the third level to the ground state (the laser transition) must be comparatively slow. This process is shown in Figure 12.2.

The natural time constants associated with non-radiative or radiative decay are important. For population inversion to be possible, the time constant τ_1 should be less than the time constant τ_2. Very strong radiative transitions in atoms and other media naturally tend to have very short decay times. As a consequence, in many cases, laser transitions are associated with somewhat weaker optical transitions.

The original ruby laser was pumped by means of a powerful pulsed xenon flashlamp which produced high flux emission in the visible and ultraviolet. Normally, such flashlamps are in the form of a relatively long cylindrical tube with high voltage electrodes at either end and filled to high pressure with xenon. However, the case of the original ruby laser, the flashlamp was in the form of a helical coil wrapped around a cylindrical ruby rod. In many current implementations, a linear flashlamp is substituted, with both flashlamp and laser rod positioned at the foci of an elliptical reflective cavity. In the original device, the polished ends of the ruby road were provided with a metallic reflective coating. One of these mirror surfaces was 'half silvered' or partially reflective. A schematic of the ruby laser is shown in Figure 12.3.

The ruby laser is an example of a solid state laser and the original laser operated in the pulsed mode, rather than operating continuously. In the case of the original ruby laser, the pulse width was of the order of 1 ms. It

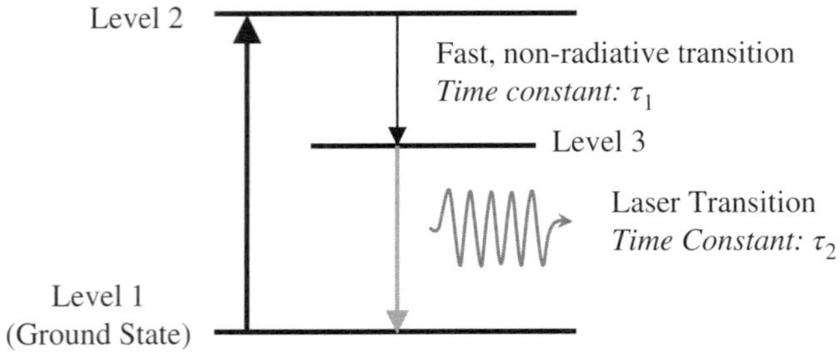

Figure 12.2 Three level laser scheme.

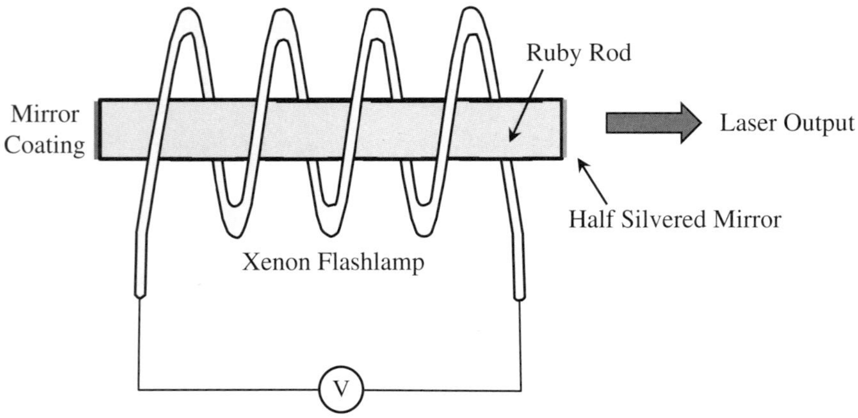

Figure 12.3 Schematic of Ruby laser.

is substantially easier to maintain the high pumping power input required to produce a population inversion when operating in this pulsed mode. Nevertheless, it is possible to operate the ruby laser in a continuous fashion and this was demonstrated within a year of the original invention. For practical reasons associated with pumping intensity, it is easier to demonstrate laser action in a pulsed mode than in a continuously operating fashion. As a consequence, for certain laser types only operation in the pulsed mode is possible.

12.2.3 Stimulated Emission in Neon

Following on from the discovery of the ruby laser, the next laser device to be demonstrated was the helium-neon laser. Unlike the ruby laser, which is a solid state laser, the helium neon laser is a gas laser. The lasing medium, in this case, is a gaseous mixture of helium and neon and the energy input is via an electrical discharge within the medium. Stimulated emission is observed on a 632.8 nm transition in neon. Although neon is, in effect, the active constituent, the helium buffer plays a prominent role. As in the ruby laser, the active constituent, neon is the minority component, with a helium to neon ratio, typically, of 7 : 1. Creation of the population inversion is based upon a adventitious circumstance. The $1s^12s^1\,{}^1S$ and ${}^3S$ excited states in atomic helium are metastable. That is to say, they cannot rapidly decay to the ground state. As a result, with the injection of energy into the medium via an electrical discharge, a large population is built up in this metastable state. The fortuitous circumstance alluded to is that specific energy levels in neon lie very close to the energy levels of these metastable states. These are the energy levels associated with the outer shell 4s and 5s electrons in neon. The proximity of these energy levels to those of metastable helium results in a resonantly enhanced exchange of energy between the helium and neon atoms. As a result of the fortuitous coincidence of energy levels, when helium and neon atoms collide, there is a very high probability of the neon atoms being

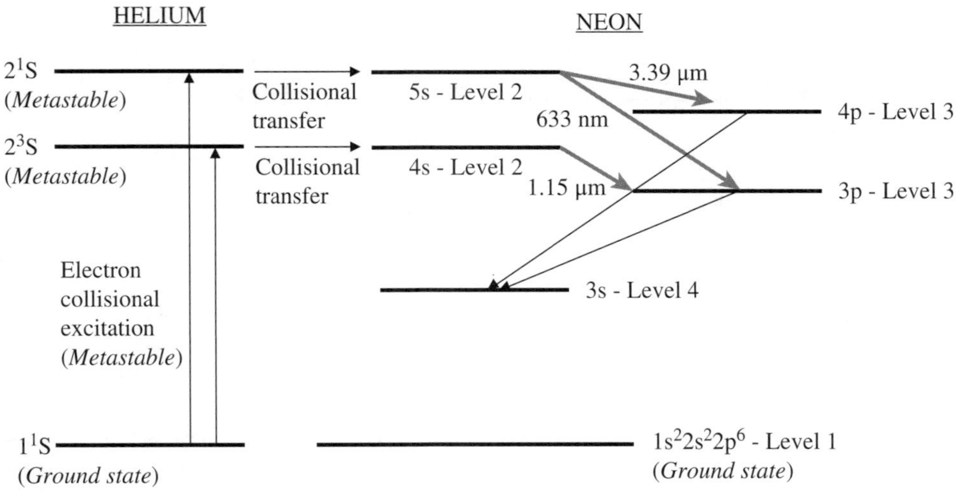

Figure 12.4 The helium neon pumping scheme.

promoted to the 4s or 5s excited states. In consequence, there is preferential pumping of the 4s and 5s states when compared to the lower energy 4p and 3p states. A population inversion is therefore created.

The helium neon laser is an example of a ***four level laser***. Laser action takes place between the 5s and 4p states (3.39 μm), the 5s and 3p states (632.8 nm) and the 4s and 3p states (1.15 μm). The 632.8 nm transition is, of course, the most widely applied. These lower energy levels are short lived, decaying rapidly to a lower energy fourth level, hence the 'four level laser' label, so a population inversion is readily maintained. This scheme is illustrated in Figure 12.4.

The lower energy states in neon, e.g. 3p, are split into a number of relatively closely spaced levels. Although, for example, the 632.8 nm helium neon laser transition is well known, other transitions terminating on the 3p state are available. This includes the 543 nm transition which forms the foundation of the so-called 'GreNe laser'.

Excitation of the lasing medium takes place in the stable positive column of a helium neon electrical discharge. Operation of a four level laser and maintenance of the population inversion is contingent upon the rapid deactivation of the third level to the fourth level. However, in the case of the HeNe laser, the fourth energy level, the 3s level, is, like its corresponding level in Helium, metastable. Population can build up in this level and optical absorption can re-promote atoms back to the third level, partially negating the population inversion. Thus, build-up of population in the 3s level can create a 'bottleneck' and it is desirable to ameliorate this effect. This is done by implementing the positive column of the discharge in a narrow bore capillary tube. Metastable atoms rapidly diffuse to the walls of the capillary and are de-activated there.

The single pass gain of a HeNe laser is relatively low – only a few percent. Therefore the losses incurred in the feedback process – i.e. the mirrors – must be very small. Therefore, multilayer dielectric mirrors, as described in Chapter 10, are used to form the cavity of the laser. The design of a typical HeNe laser is illustrated schematically in Figure 12.5.

One important distinction between the pulsed ruby laser and the helium neon laser is that the gain is very much lower, only a few percent. In the design shown in Figure 12.5, the gas tube is sealed at either end with Brewster angle windows. We might recall from Chapter 8 that Fresnel reflection is entirely eliminated at the Brewster angle for one polarisation (p polarisation). This is especially significant in the context of low gain, as the impact of Fresnel reflection at four interfaces would be sufficient to impede any laser oscillation. In the design shown, there are two mirrors external to the gas envelope. Both mirrors are multilayer dielectric mirrors of the type encountered in Chapter 10. One is nominally 100% reflecting and the other is a partially transmitting but high reflectance (e.g. 99%) mirror. The design illustrated is representative of earlier systems.

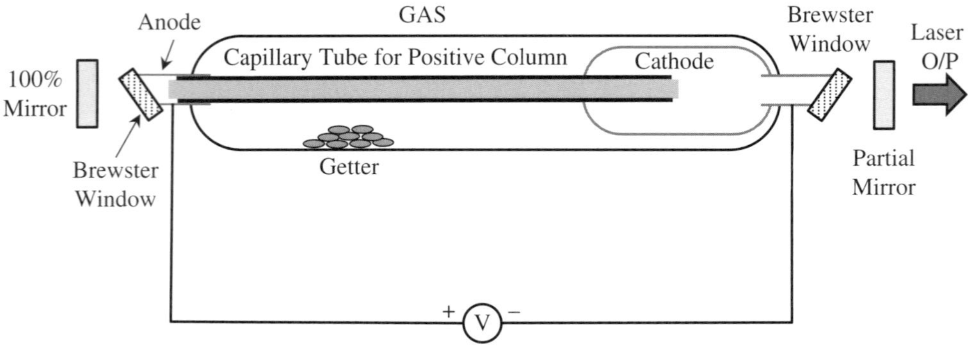

Figure 12.5 Helium neon laser.

In many modern lasers of this type, the external mirrors and Brewster windows are replaced by sealed mirrors physically attached to the glass envelope. A getter is provided to ensure the longevity of the laser tube by removing contaminants that might otherwise build up.

12.2.4 Stimulated Emission in Semiconductors

Semiconductor lasers make up an extremely technologically important class of laser devices. A wide range of light emitting, ***direct bandgap*** semiconductor materials support laser action. These include gallium arsenide, gallium nitride, indium phosphide, and so on, but ***not*** silicon or germanium. Operation of these lasers may be understood in terms of the band structure of these materials. All these materials are necessarily crystalline materials and the energy level structure of the charge carriers is driven by the periodic potential of the crystal lattice. Without going specifically into details, electrons are allocated to two types of state, namely the conduction band or the valence band. The former band is 'free' and the latter band is 'bound'. The critical fact is that these two states are separated by an energy gap – the band gap.

In the conduction band, charge carriers, in many respects, behave as kinetic particles, with increasing momentum associated (quadratically) with increasing energy. Perversely, the valence band displays the opposite behaviour. That is to say, increasing momentum is associated with ***decreasing*** energy. Population of both valence and conductions bands is dependent upon the presence of impurities or dopants in the material. For example, the presence of impurity atoms inclined to 'accept' electrons, or acceptors, will lead to depletion at the top of the valence band. Conversely, the presence of atoms disposed to donate electrons (donors) will preferentially populate the lower part of the conduction band. The former type of material, with depleted energy levels at the top of the valence band is known as ***p-type*** (positive) material. Where electron donors are present, the material is known as ***n-type*** material.

In a semiconductor laser, the two types of material, n-type and p-type are brought together at a ***semiconductor junction***. This junction is essential for the operation of the semiconductor laser. A population inversion is created by driving an electric current through the junction. This promotes electrons to the conduction band which then decay to fill the bottom of the conduction band. A population inversion is then created between the bottom of the conduction band and the top of the valence band. This is illustrated in Figure 12.6.

As illustrated in Figure 12.6, electrical excitation at a pn semiconductor junction promotes electrons to the conduction band which are de-excited by collision to form a pool of electrons at the bottom of the conduction band. This should be compared to the top of the valance band which is somewhat depleted. As a consequence, a population inversion is established. It is important to emphasise that Figure 12.6 illustrates a direct band gap material. In an indirect band gap material, such as silicon, the bottom of the conduction band and the peak of the valence band are offset in terms of momentum. Since a photon has zero momentum, only a vertical transition, as shown in the energy diagram in Figure 12.6 is possible. Therefore an indirect bandgap material cannot be a light emitter; this is only possible for direct bandgap materials.

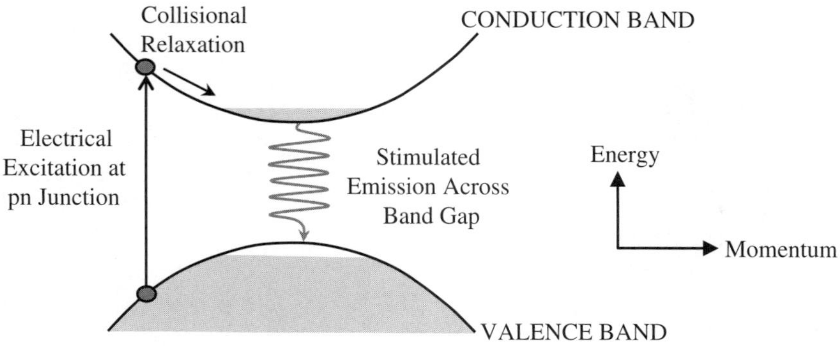

Figure 12.6 Stimulated emission in a semiconductor laser.

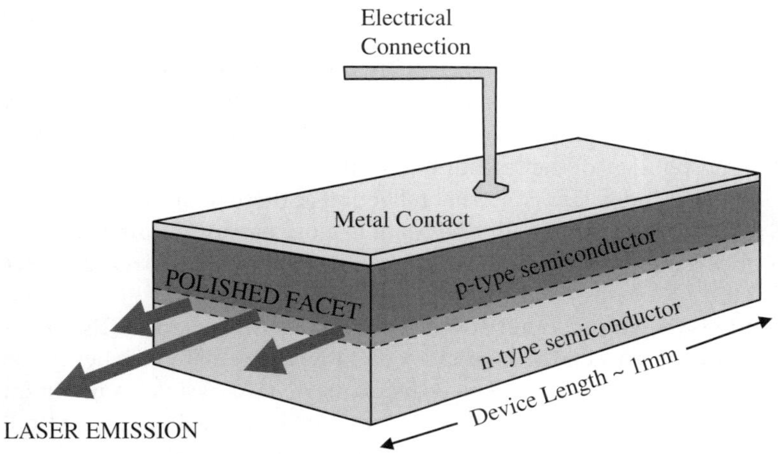

Figure 12.7 Simplified sketch of semiconductor laser.

Of course, establishing amplification alone is insufficient to produce laser emission; feedback is also necessary. To understand how this might proceed, a basic semiconductor laser is illustrated in Figure 12.7.

The first point to note is the size and geometry of the device. Most commonly, the laser is implemented as a small block of material, as sketched, referred to as a '***laser chip***'. The length of the lasing medium is of the order of 1 mm and often only a fraction of a millimetre. Feedback is obtained in a number of ways. The simplest scheme, as employed in the earlier lasers simply uses the polished or cleaved ends of the laser to act as partial mirrors. Many of the semiconductor laser materials, such as gallium arsenide are, by their nature, high index materials with a refractive index in the range of 3–4. As a result, the Fresnel reflections at the air interfaces are relatively large.

Of course, the scheme shown in Figure 12.7 is grossly oversimplified. One specific problem with the device, as shown, is that the gain region is very narrow. In fact, amplification only occurs in the so called thin depletion layer which marks the junction between the p type and n type materials. The thickness of this layer is of the order of a micron or a fraction of a micron. As the amplified beam propagates between the two ends of the laser chip, diffraction has a tendency to spread the beam. As a result, only a small portion of the propagating beam overlaps the active region of the device and, consequentially, the efficiency of the amplification process is substantially compromised.

The picture presented by Figure 12.7 and the subsequent narrative reflects the early development of the semiconductor laser. These devices were inefficient and were only capable of operating under restricted conditions, such as cryogenic cooling. Some contemporary laser sources, such as lead salt lasers can only operate

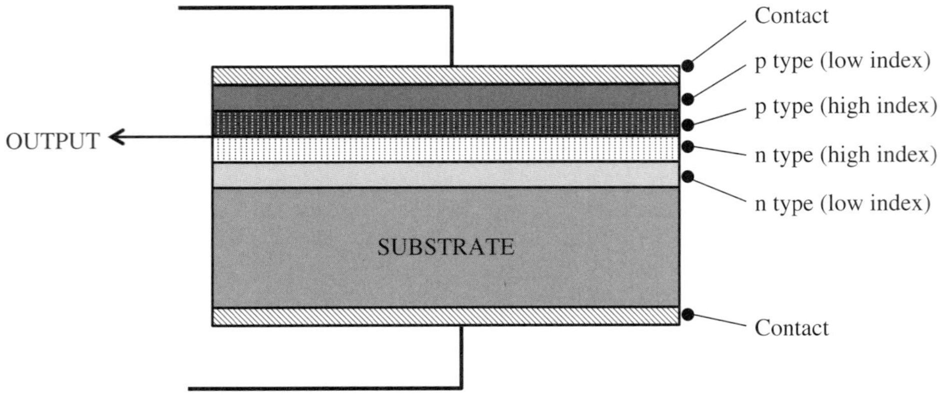

Figure 12.8 Double heterostructure laser.

when cooled. More generally, to address this gain problem, the so called '***double hetero-structure laser***' was developed. Instead of being formed from a single pn junction in a common material, the laser chip is based on a pn junction formed from two complementary materials. Most significantly, these two materials have significantly different refractive indices. The morphology of the structure is arranged such that the higher index active region is surrounded by material of lower index. Total internal reflection at these interfaces has a tendency to counteract the natural effects of diffraction. As a result, the beam is more narrowly confined to the active region and the amplification process is rendered much more efficient. The double hetero-structure laser is illustrated in cross section in Figure 12.8.

As before, the active part of the structure is the pn junction between the two high index layers. This junction is effectively wrapped by the two junctions between the high index and low index materials, hence the term ***double*** heterostructure. One example of a system of this type is the GaAs/AlGaAs system. Aluminium Gallium Arsenide (with variable aluminium/gallium stoichiometry – $Al_xGa_{1-x}As$) is the high index material and Gallium Arsenide is the low index material. These lasers operate around 850 nm.

12.3 Laser Cavities

12.3.1 Background

Of paramount importance to the operation of a laser is the feedback provided by the cavity that surrounds it. The cavity may be thought of as a resonator that promotes oscillation in the presence of amplification. In common with classical resonators, such as tensioned strings or organ pipes, these cavities have a number of different modes. Like the organ pipe example, a laser cavity will have a number of ***longitudinal modes***. The laser cavity effectively acts as an etalon which produces a number of longitudinal modes separated by the free spectral range. From this consideration, it is clear that a laser with a long cavity, such as a helium neon laser, will produce modes that are spaced relatively closely. On the other hand, a semiconductor laser, which has a very short cavity, will produce modes that are widely separated. Depending upon the spectral width associated with the amplification process, a number of cavity modes will be excited. In general, these modes will bear no phase relationship to each other. However, if a single mode is excited, then a unique single frequency oscillation will be produced. This oscillation thus possesses the property of temporal coherence, with the fidelity of the oscillator's phase maintained over very wide path differences and only limited by the very small, but finite linewidth of the oscillator.

In addition, the laser may also have a number of different transverse modes. In our treatment of laser beam propagation in Chapter 6, we introduced the concept of the Gaussian beam. The classical Gaussian

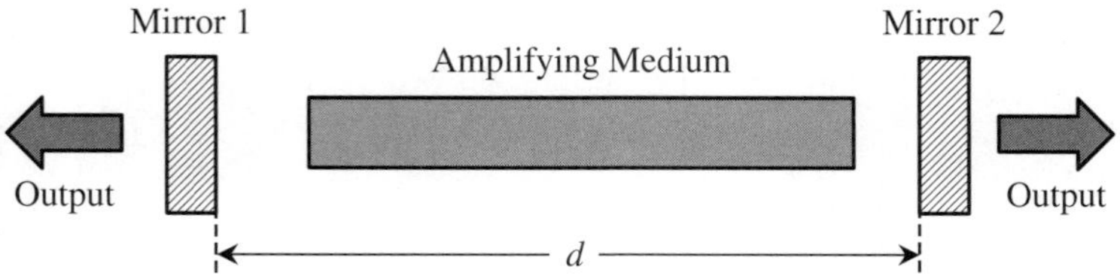

Figure 12.9 Generalised representation of a laser cavity.

beam may be regarded as a single transverse mode. However, other 'higher order' modes may be created, as might be described by the Gauss-Hermite series of polynomials. In practice, these higher order modes tend to extend further laterally than the lower order modes. Controlling the number of transverse modes amounts to restricting the lateral extent of the gain profile. For example, in the helium neon laser, the size of the capillary bore that restricts the positive column determines the lateral extent of the gain region. As understood from Chapter 6, the number of transverse modes is effectively described by the M^2 parameter. This helped us semi-quantitatively to understand the propagation of a multimode laser beam. In fact, the M^2 parameter originates with the laser device itself and describes the number of phase independent modes.

12.3.2 Longitudinal Modes

We will now examine the question of longitudinal modes in a little more detail. As previously suggested, the laser cavity may be represented as an etalon, producing a series of resonant modes separated by the free spectral range. From Chapter 10, this free spectral range is given by:

$$\Delta\upsilon = \frac{\Delta\lambda}{\lambda^2} = \frac{1}{2nd} \tag{12.2}$$

$\Delta\upsilon$ is the spacing in wavenumbers; n the cavity refractive index; d the cavity spacing

This cavity is quite simple to visualise in terms of the general operation of a laser oscillator; this is shown in Figure 12.9.

Most significantly, the laser itself will have a finite bandwidth or gain profile on to which a comb of possible resonant frequencies is superimposed. Of course, the 'teeth' of this comb are separated by the free spectral range. This is illustrated in Figure 12.10.

Worked Example 12.1 ***Longitudinal Modes in Helium Neon Laser***

To understand the operation of longitudinal modes, it is worthwhile to turn to a practical example. A helium neon laser has a cavity length of 250 mm. The laser wavelength is 632.8 nm. What is the spacing of the longitudinal modes in nm?

First, we may assume that the refractive index of the gaseous medium is close to unity. Therefore we may revise Eq. (12.1) to read:

$$\frac{\Delta\lambda}{\lambda^2} = \frac{1}{2d} \text{ and } \Delta\lambda = \frac{\lambda^2}{2d}$$

Substituting the values (in nm) into the above equation:

$$\Delta\lambda = \frac{632.8^2}{2\times 2.5x10^8} = 8.01x10^{-4} \text{ nm}$$

The longitudinal modes are separated by 8.01×10^{-4} nm.

Taken alone, it is by no means straightforward to understand the significance of the above result. To put this into context, it might be useful to express the mode separation in terms of the overall width of the laser gain profile, as depicted in Figure 12.10. The dominant factor determining the width of the helium neon laser

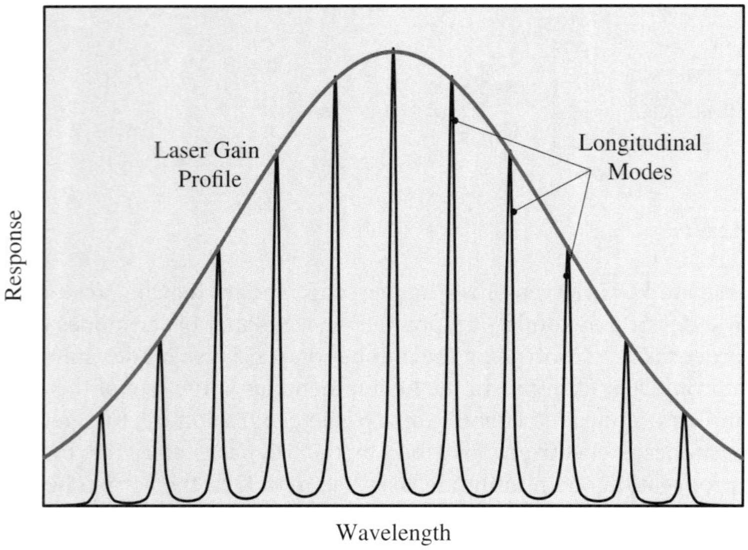

Figure 12.10 Laser gain profile and longitudinal modes.

transition is the Doppler broadening of the atomic neon transition. Each individual atom is travelling at a different thermal velocity with respect to the observer. As a result each neon atom experiences a different Doppler shift and this, by statistical summation, leads to a broadened profile across a population of atoms. This process is referred to as ***inhomogeneous broadening***. This occurs where atoms in a particular energy state experience a variable shift in output wavelength. By contrast, ***homogeneous broadening*** occurs where all atoms in a particular energy state experience identical broadening. For example the natural (radiative or non-radiative) decay of an energy level contributes to broadening – the more rapid the decay, then the greater the broadening.

In the case of the foregoing example, the Doppler broadening of the neon line is Gaussian on account of the random thermal velocity distribution. The width of the line is thus proportional to the average atomic velocity, and hence the square root of the temperature. The linewidth full width half maximum (FWHM) is given by:

$$\Delta\lambda = \lambda_0\sqrt{\frac{8kT\ln 2}{Mc^2}} \tag{12.3}$$

λ_0 is the central wavelength; k the Boltzmann constant; T the absolute temperature; M the mass of the (neon) atom.

From the above, the FWHM of the neon laser line is 2×10^{-3} nm. Thus the mode spacing is about 40% of the FWHM. As such, the laser is, in principle, capable of supporting 3–4 longitudinal modes. As soon as laser action is initiated on one mode, the stimulated emission process will act to deplete the population of atoms associated with that mode's wavelength. Effectively, the stimulated emission 'burns a hole' in the gain profile shown in Figure 12.10. At equilibrium, the residual amplification associated with that mode must be just sufficient to cover losses. Depending on the design, cavity losses and pumping levels more than one mode may be supported. It is, however, possible, by careful design to produce a helium neon laser that will operate on a single mode.

The position for a semiconductor laser is rather different. Although the longitudinal mode spacing is very much greater, this is substantially compensated by the much greater intrinsic gain bandwidth. Typically, the mode spacing might be about a nanometre or a fraction of a nanometre for near infrared devices. This might compare to a gain bandwidth of the order of 20 nm for a semiconductor laser. Excitation of a single mode is not straightforward. However, techniques are available and these are discussed very briefly a little later.

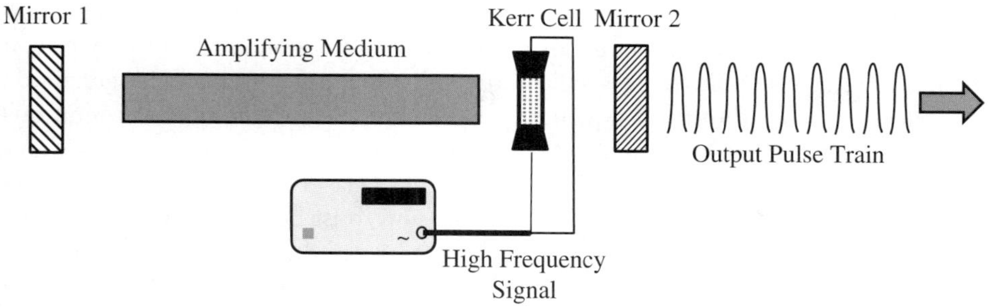

Figure 12.11 Active mode locking.

12.3.3 Longitudinal Mode Phase Relationship – Mode Locking

To this point, we have assumed that the phase of the longitudinal modes are randomly disposed. As a consequence, most generally, where large number of modes are excited, these will sum randomly. However, it is possible to establish a coherent phase relationship between the modes by imposing a time varying phase disturbance on the cavity that is matched to the round trip time of the cavity. That is to say, the imposed phase disturbance is sinusoidal with the same frequency as the frequency separation of the modes themselves. The result of this is that the different modes interfere coherently, producing a train of very short pulses whose separation is the same as that of the cavity round trip time. This process is known as mode locking.

Mode locking may be produced under two broad schemes. First there is ***passive mode locking***, which uses saturable absorption to produce mode locking. Saturable absorption is a non-linear effect whereby the absorption is reduced at higher optical fluxes. Second, ***active mode locking*** employs direct modulation of the amplitude or phase of the intra-cavity light by means of an optically active material. One can imagine an optically active material as a birefringent material whose birefringence is proportional to some externally applied electric field. The frequency of the applied electric field is the same as the round trip time; an active mode locking scheme is shown in Figure 12.11.

As depicted in Figure 12.11 the Kerr Cell is the electro-optic device that provides the sinusoidal phase change. With the phase change established, then the coherent amplitudes of all the different modes will sum across the gain bandwidth of the laser. This process is as set out in Eq. (12.4).

$$A(t) = \sum_{p=0}^{p=N} e^{i(\omega_0 + (\pi c/nd)p)t} \tag{12.4}$$

ω_0 is the nominal angular frequency of the laser oscillation

Equation (12.4) is simply a series of oscillating modes separated in frequency by c/(2×n×d). One might imagine these phase related modes 'beating' with respect to each other with a frequency equal to the mode spacing. The width of each pulse thus created is determined by the Fourier transform of the gain envelope. For convenience, the gain envelope of the laser might be described by a Gaussian profile with a frequency FWHM equal to Δf. Of course, the Fourier transform of a Gaussian is another Gaussian and so each pulse in the train is represented by a Gaussian temporal profile with an FWHM, $\Delta\tau$ given by:

$$\Delta\tau = \frac{2(\ln 2)^2}{\pi \Delta f} \tag{12.5}$$

For the previous example of the HeNe laser, the pulse width is approximately 500 ps. For solid state lasers and semiconductor lasers, where the gain bandwidth is considerably higher, the pulse widths that can be achieved are very much less. In particular, in the case of solid state lasers, the bandwidth extends across nanometres or tens of nanometres. As a consequence, these lasers are used in the generation of picosecond or sub-picosecond pulses for which there are a wide range of useful applications.

Another more recent development of mode locking is in supercontinuum lasers and ***frequency comb generation***. For a typical solid state laser the active mode locking frequency, as per Figure 12.11 might be of the order of a few hundred megahertz. When the laser is locked, the individual mode frequencies generated are spaced by this interval. However, the ***absolute*** frequency is indeterminate. Under special conditions where the laser emission is exceptionally broad, then the absolute frequency is itself equal to an integer multiple of the locking frequency. For this to occur, the frequency range of the laser emission must be at least an octave. That is to say, the highest frequency component of the emission must be at least a factor of two greater than the lowest frequency emission. This condition cannot be achieved directly in a practical laser device. Rather, the generation of broad or supercontinuum emission relies on frequency sum generation in non-linear optical materials. Non-linear optical materials possess polarizability that is dependent upon the applied electric field. The non-linear polarizability results in the creation of sum and difference frequencies, leading to an overall broadening of the output. The significance of frequency comb generation is that a suite of optical frequencies may be generated that are locked to the fundamental standard frequency (currently in the microwave region). As such, frequency comb generation has applications in precision metrology and may offer the possibility of moving the current 9.19 GHz (Caesium) frequency standard from the microwave to the optical domain. Frequency comb generation is thus an extremely powerful technique for the precise calibration of optical frequencies/wavelengths.

12.3.4 Q Switching

Mode locking exploits our ability to externally control the phase or amplitude of light in a laser cavity. Another widely used application of amplitude control is ***'Q switching'*** or quality switching. Q switching amounts to a deliberate and transient attempt to degrade the resonance quality, or Q factor of the laser oscillator. In all laser systems, the pumping process drives the population inversion. However, as the level of stimulated emission in the cavity increases, this process, in itself, serves to deplete the population inversion. After a brief period, an equilibrium is established, whereby the static population inversion is just sufficient in order to provide sufficient amplification to overcome cavity losses. Ultimately, the laser output is determined by the efficacy and the vigour of the pumping process.

Q switching is applied specifically to naturally transient or pulsed laser systems where the instantaneous level of pumping is especially high. At the beginning of the transient laser pumping cycle, absorption is introduced into the laser cavity. This absorption is sufficient to supress amplification in the cavity and thus to inhibit laser action. During this period, the population inversion is being continually augmented by the powerful pumping process. Moreover, since there is no stimulated emission to deactivate the upper level, the population inversion increases to a level that is orders of magnitude greater than would have been created by a continuous process. At some critical point in the pumping sequence, the absorption is removed and the 'Q switch' opened. Since the population inversion created is so large and the amplification so intense, the laser energy is released in a short, but giant pulse. One could regard the 'Q switching' process as an energy storage scheme, whereby energy is stored over hundreds of microseconds during pumping and released in a pulse lasting a few or tens of nanoseconds. Most typically, a Q switched laser will be a solid state laser. With a pulse energy perhaps measured in joules and a pulse width of nanoseconds, peak powers of several hundred Megawatts or more are possible. This opens many applications in non-linear optics and materials processing.

As for mode locking, implementation of Q switching can either be passive or active. In the case of passive Q switching, a saturable absorber is used. For active mode locking the electro-optic effect is exploited to produce a fast optical switch. Application of an electrical field in a crystal produces an index difference and phase delay between two orthogonal polarisation directions. This effect is known as the ***Pockels effect***. In practice, an external electric field is used to create a temporary 'quarter wave plate' between two polarisers. As a consequence, this ***Pockels cell*** blocks the transmission of light within the laser cavity. When the electric field is removed, then transmission is resumed. Operation of a Q switched laser is shown in Figure 12.12.

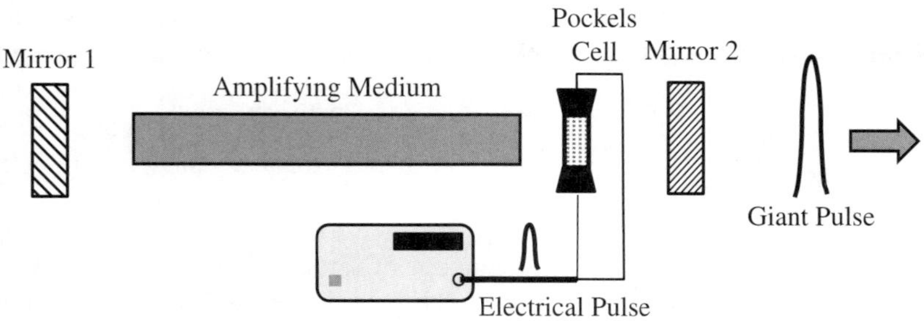

Figure 12.12 Q switched laser.

12.3.5 Distributed Feedback

In semiconductor lasers there is an additional feedback mechanism (aside from cavity mirrors) and this is distributed feedback (DFB). For double heterostructure lasers, as we have seen, there exist refractive index boundaries within the laser chip that effectively guide the laser beam and reduce loss. In a ***DFB laser***, a periodic grating structure is etched into one of these boundaries. As a result, there is continuous interaction between the two counter-propagating waves within the cavity. Moreover, this interaction is selective or resonant. That is to say, the coupling between the two waves is greatest when the period of the grating matches that of the wavelength of the light. Sometimes this concept is used in conjunction with a more conventional cavity resonator, thereby selecting one specific longitudinal mode whose frequency is aligned with that of the DFB grating. In any case, the effect is to create a single frequency source with a narrow linewidth.

12.3.6 Ring Lasers

Hitherto, we have presented the resonant laser cavity as a simple linear space bounded by two mirrors. However, this is not the only geometric structure capable of supporting optical resonance. One such structure is the ***ring laser***. Feedback is provided by continuous circulation of two counter-propagating beams and not by reciprocating reflection. Typically, a ring cavity might be formed by the deployment of two or three mirrors. This is illustrated in Figure 12.13.

The key feature of the ring laser is the two counter-propagating beams which produce two outputs. These two outputs are phase related and frequency shifted. This frequency shift, where the beams are induced to

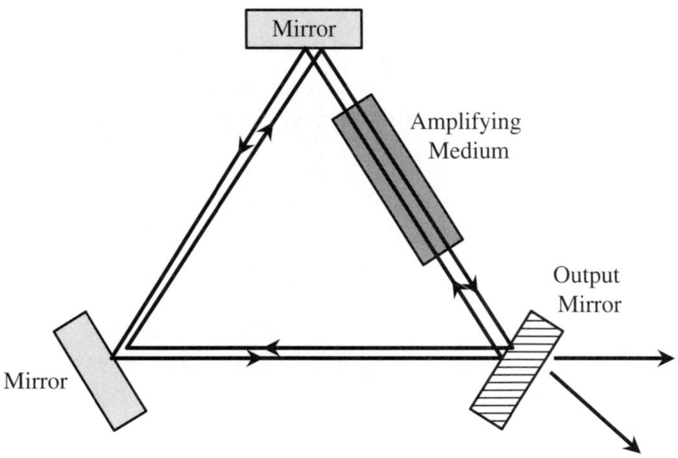

Figure 12.13 Ring laser.

interfere, produces a beat frequency that is directly related to the angular velocity of the ring device. This is by virtue of the so-called ***Sagnac effect*** and, hence, the ring laser may be used as a precisions gyroscopic sensor to measure angular velocity.

12.3.7 Transverse Modes

The previous discussion around longitudinal modes relates to the concept of temporal coherence. That is to say, the number of longitudinal modes excited directly affects the fidelity of the phase relationship of a beam at two separate times, but at the same point. By contrast, the number of transverse modes impacts the phase relationship across a wavefront at a specific time, i.e. spatial coherence. To understand the generation of transverse modes in a laser cavity, we must first understand the concept of a stable cavity.

Thus far, we have presented a simplistic picture of a laser cavity, with two parallel plane mirrors. However, for a laser, such as a helium neon laser, with a low gain per pass, the maintenance of a stable beam profile over many cavity round trips is essential to its operation. In the case of two parallel plane mirrors, there will be a tendency for the laser beam to diverge over time. Figure 12.14 presents a more realistic description of a laser cavity, with two curved mirrors with radii, R_1 and R_2, separated by a distance d.

To understand the stable resonator geometry, we must postulate that there exists a specific ray geometry, as defined by ray height and angle, which is stable over multiple passes through the above cavity. In order to investigate this scenario, we must return to the matrix ray analysis presented at the beginning of the book. Before proceeded further, it is necessary to re-iterate the sign convention used. A positive radius of curvature is always associated with a mirror sag that is directed towards the right. In other words, in Figure 12.14, R_1 is positive and R_2 is negative. Positive displacement is towards the right. With this in mind, we may present the overall system matrix for one round trip, starting at Mirror 1 and returning there:

$$M = \begin{bmatrix} 1 - \frac{4d}{R_1} + \frac{2d}{R_2} - \frac{4d^2}{R_1R_2} & -2d - \frac{2d^2}{R_2} \\ \frac{2}{R_1} - \frac{2}{R_2} + \frac{4d}{R_1R_2} & 1 + \frac{2d}{R_2} \end{bmatrix} \tag{12.6}$$

Instability will be produced by a beam that successively expands with each cavity round trip. Conversely, if this does not occur, then the cavity might be said to be stable. To analyse this further, we postulate that this expansion may be described by some constant factor, λ, that acts upon the vector describing the ray height and angle:

$$\begin{bmatrix} A & B \\ C & D \end{bmatrix} \begin{bmatrix} y \\ \theta \end{bmatrix} = \lambda \begin{bmatrix} y \\ \theta \end{bmatrix}$$

For those familiar with the mathematics, the vector is referred to as the ***eigenvector*** and the multiplier as the ***eigenvalue***. The eigenvalue describes the round trip expansion factor and its value is given by the following

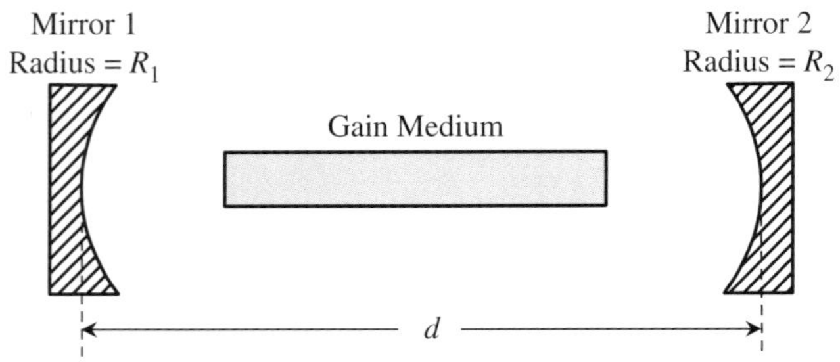

Figure 12.14 Stable resonator geometry.

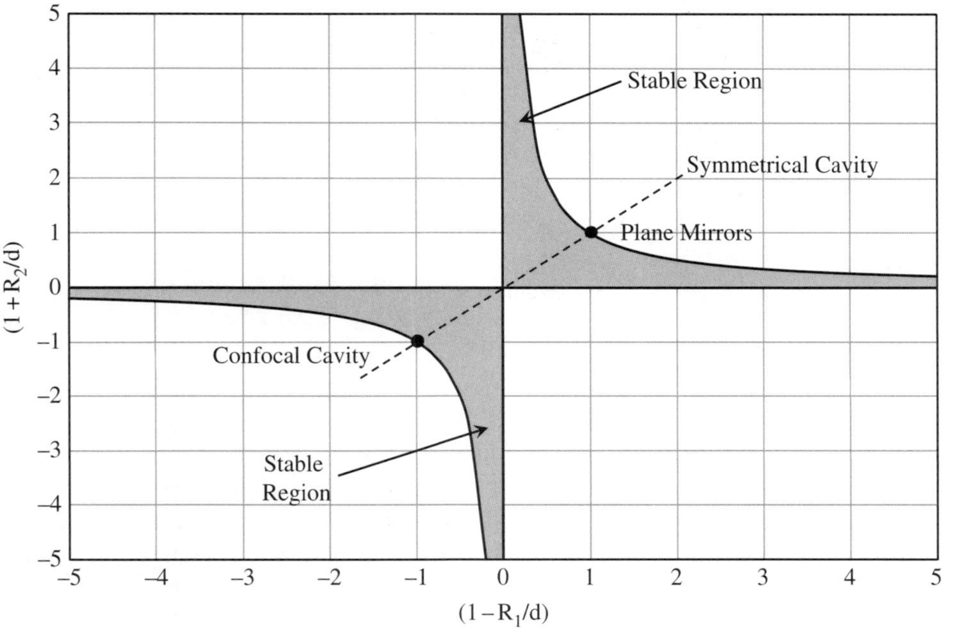

Figure 12.15 Laser cavity stability.

quadratic equation:

$$\lambda^2 - 2\beta\lambda + 1 = 0 \text{ where } \beta = \left[1 - 2\left(1 - \frac{d}{R_1}\right)\left(1 + \frac{d}{R_2}\right)\right] \quad (12.7)$$

For the beam not to expand successively it follows that the modulus of λ should be less than one. We can understand the implications of this if we set out the solution for the above quadratic.

$$\lambda = \beta \pm \sqrt{\beta^2 - 1} \quad (12.8)$$

For the modulus of λ to be less than one, then the modulus of β must also be less than one. This latter condition implies an imaginary component to the solution. The presence of this imaginary term suggests that the beam size, instead of expanding exponentially, oscillates with successive cavity round trips.

$$-1 <= \left[1 - 2\left(1 - \frac{d}{R_1}\right)\left(1 + \frac{d}{R_2}\right)\right] <= 1 \text{ and } 0 <= \left(1 - \frac{d}{R_1}\right)\left(1 + \frac{d}{R_2}\right) <= 1 \quad (12.9)$$

We should emphasise, once again, the sign convention adopted in Figure 12.14 and, indeed, throughout the book. For Mirror 1, a positive value represents a concave surface, whereas for Mirror 2 a positive value represents a convex surface. Perhaps unsurprisingly, Eq. (12.9) suggests that concave surfaces ***tend*** to confer greater stability. A plane mirror, plane mirror combination, which has hitherto been chosen to illustrate a generic laser cavity, is just stable. For a symmetrical concave cavity, then each radius of curvature must be greater than half the cavity length (*confocal arrangement*). Conditions for cavity stability are sketched out in Figure 12.15 using terms derived from Eq.(12.9).

12.3.8 Gaussian Beam Propagation in a Laser Cavity

To examine the behaviour of a resonant cavity a little more realistically, we might want to include the effects of diffraction. To this end, we reintroduce our treatment of Gaussian beam propagation we encountered in Chapter 6. First, it would be useful to define the problem somewhat more clearly. From the perspective of

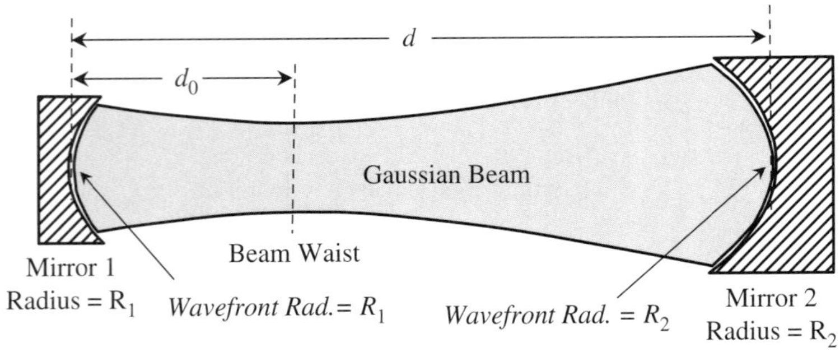

Figure 12.16 Gaussian beam and cavity geometry.

Gaussian beam propagation, it would be reasonable to propose that the wavefront curvature at each mirror is equal to the radius of curvature of the mirror itself. This would certainly give rise to a stable and self-perpetuating beam configuration within the cavity. Figure 12.16 sketches the geometry.

The wavefront radii for a Gaussian beam are governed by Eq. (6.36). Initially, we will assume that the mirror radii are known and we wish to determine the location of the beam waist, d_0 and the Rayleigh distance of the beam, Z_R. We have the following two relationships:

$$R_1 = d_0 + \frac{Z_R^2}{d_0} \quad R_2 = d_0 - d + \frac{Z_R^2}{(d_0 - d)}$$

The position of the beam waist is given by:

$$d_0 = \frac{(d + R_2)}{(2d - R_1 + R_2)} d \tag{12.10}$$

The Rayleigh distance may be computed from the following relationship:

$$Z_R^2 = \frac{(R_1 - R_2)d - d^2}{4} + \frac{(R_1 d + R_2 d)^2 (d^2 - R_1 d + R_2 d)}{4(2d^2 - R_1 d + R_2 d)^2} \tag{12.11}$$

For a symmetrical cavity, where $R_2 = -R_1$, then Eq. (12.11) becomes rather more simple, with only the first term in the expression applying. We can see how this analysis might proceed from a worked example.

Worked Example 12.2 ***Helium Neon Laser Beam***

A helium neon laser has a symmetrical cavity formed by two mirrors separated by 300 mm. We wish to create a beam with a beam waist size, w_0, of 0.5 mm. What should the radius of curvature of the mirrors be to give the necessary beam waist size?

In this example, we are effectively given the Rayleigh distance, and we must use this to calculate the mirror radii. We know that the cavity is symmetrical, with each mirror having equal and opposite curvatures. Therefore, $R_1 = R$ and $R_2 = -R$. Therefore, Eq. (12.11) may be greatly simplified:

$$Z_R^2 = \frac{Rd}{2} \frac{-d^2}{4} \tag{12.12}$$

We are informed that the beam waist size should be 0.5 mm and the wavelength of the HeNe laser is 632.8 nm. This can be used to calculate the Rayleigh distance:

$$Z_R^2 = \frac{\pi w_0^2}{\lambda} = \frac{3.142 \times 0.5^2}{6.328x10^{-4}} = 1241mm$$

We know that the Rayleigh distance is 1241 mm. It is then straightforward to calculate R by re-arranging Eq. (12.11).

$$R = \frac{2Z_R^2}{d} + \frac{d}{2}$$

The radius of curvature is 10 420 mm or 10.42 m.

As suggested by this example, stable cavity mirrors tend to have quite large radii in proportion to the cavity length. Formation of a stable cavity is more critical for low gain systems such as HeNe lasers where many cavity round trips are required to develop sufficient amplification. This consideration does make such systems difficult to align, as significant mirror tilt will cause the beam to be diverted from the critical gain region after a few passes.

Promotion of single mode operation may be understood in terms of the Gaussian-Hermite polynomials presented in Chapter 6. The wavefront curvature of a set of these polynomials is identical for that of the underlying 'zeroth order' Gaussian. As the mode number increases, then the overall size of the mode increases. It is possible, therefore, to preferentially attenuate higher order modes by introducing an aperture at some point in the system. The aperture is of a size that it attenuates higher order modes to such an extent that there is no nett amplification on a single pass. Conversely, the attenuation of the underlying Gaussian mode is sufficiently small that nett amplification is preserved.

12.4 Taxonomy of Lasers

12.4.1 General

Since the discovery of the laser in 1960, the number of different types of laser has multiplied prolifically. Laser action has observed in an exceptionally wide range of materials, from diamond to edible jelly. Without exhaustively describing all types of lasers we will attempt now to categorise the broad range of laser systems. Laser systems may be usefully categorised by a limited range of parameters, as set out below:

1. Material type – (gas, chemical, solid state, semiconductor, dye, fibre)
2. Pulsed operation or continuous wave (cw)
3. Power (or energy per pulse)
4. Wavelength of operation
5. Tunable or fixed wavelength
6. Other (mode locked, Q switched, etc.)

12.4.2 Categorisation

12.4.2.1 Gas Lasers

Gas lasers are typically energised by electrical discharge and examples include helium neon lasers, argon ion lasers, carbon dioxide lasers, excimer lasers, etc. Operating wavelengths range from the vacuum ultraviolet to the mid-far infrared. Although many gas lasers produce only modest output powers, they are intrinsically scalable. Carbon dioxide lasers, for example, are capable of delivering many tens of kilowatts of continuous power. One specific practical problem faced by all high power laser devices is the management of a troublesome by product, namely thermal energy. In gas lasers particularly, forced convection offers an efficient mechanism for heat transfer.

12.4.2.2 Solid State Lasers

In contrast to gas lasers, most typically, solid state lasers are optically pumped. For example, the ruby laser (the first laser) is pumped by a flashlamp source. Although ruby lasers are still in use, other materials, such as

Nd:YAG (Neodymium: Yttrium Aluminium Garnet) are more prevalent. They rely on low level atomic doping of a crystal or glassy material with, for example, chromium doping in a ruby laser or neodymium in a Nd:YAG laser. Generally, dopants tend to be rare earth or transition metals. Many of these lasers are, nowadays, pumped by compact laser sources, such as arrays of semiconductor lasers. Output wavelength range is from the visible to the near infrared, with the majority in the infrared.

12.4.2.3 Fibre Lasers

Fibre lasers represent a subcategory of solid state lasers. Instead of the dopant being incorporated into a solid rod, it is directly added to the core of an optical fibre. Optical pumping is then provided by a shorter wavelength laser. Feedback may, as usual, be provided by mirrors. However, most usually some form of structure internal to the fibre is used. This takes the form of a periodic modulation of the fibre structure or index along the length of the fibre to create resonantly enhanced reflection. Such a structure may be likened, in function, to a multilayer dielectric mirror, as described in Chapter 10 and are referred to as ***fibre Bragg Gratings***. Alternatively, by forming a fibre into a loop, a fibre laser may be deployed in a ring laser geometry. Fibre lasers are based around rare-earth dopants, such as erbium or yttrium and cover the near infrared to mid infrared spectral range.

12.4.2.4 Semiconductor Lasers

Semiconductor lasers are based on direct bandgap materials which restricts the range of material choice. For practical purposes, the majority of such laser sources are based on III-V compounds, such as gallium arsenide and indium phosphide. The recent addition of sources based upon gallium nitride has extended their range of operation into the ultraviolet. Gallium nitride is a wide bandgap material and the wavelength of operation is ultimately fixed by the material bandgap. With this in mind, operating wavelengths extend from the ultraviolet to the near infrared. However, this wavelength range is extended further into the infrared by ***quantum cascade lasers***, which exploit inter-band transitions within the conduction band.

Semiconductor lasers have found wide application as telecommunication sources. More recently, by careful attention to morphology and thermal design high power versions have become available. In particular, semiconductor processing techniques enable the creation of arrays of such sources. Such devices are used in materials processing and for the optical pumping of other laser systems. As outlined earlier, the favoured geometry is the double heterostructure laser where the optical cavity is directed along the length of the semiconductor junction and the current flow is perpendicular to this. However, another recent development is the emergence of the ***Vertical Cavity Surface Emitting Laser*** (***VCSEL***). In this case, the cavity is aligned with the current flow and perpendicular to the semiconductor junction structure. Because the semiconductor junction geometry dictates that the gain region is necessarily short, high levels of feedback must be applied to offset the lower overall gain. Therefore, multilayer dielectric mirrors must be provided at the surface. In terms of the semiconductor processing geometry, emission is perpendicular to the surface of the underlying semiconductor wafer; for the conventional heterostructure laser emission is in the plane of the wafer and devices are created by cutting up or 'dicing' the wafer. In contrast, for the VCSEL a high density array of emitters may be fabricated, for example, in a grid structure. This opens up a wide range of applications, for instance in communications.

12.4.2.5 Chemical Lasers

Chemical lasers create excited state populations by virtue of a chemical reaction. It is clear that many chemical reactions, particularly combustion, are accompanied by optical emission through the generation of excited species in the reaction products. In specific cases, this process leads to the creation of a population inversion. Examples include the chemical iodine oxygen laser at 1.32 μm and the hydrogen fluoride and deuterium fluoride lasers at 2.8 and 3.8 μm respectively. Since chemical energy represents an extremely compact and scalable energy storage mechanism, chemical lasers are capable of releasing very significant quantities of energy. As such, they are favoured in military applications, such as directed energy weapons and also in some industrial applications.

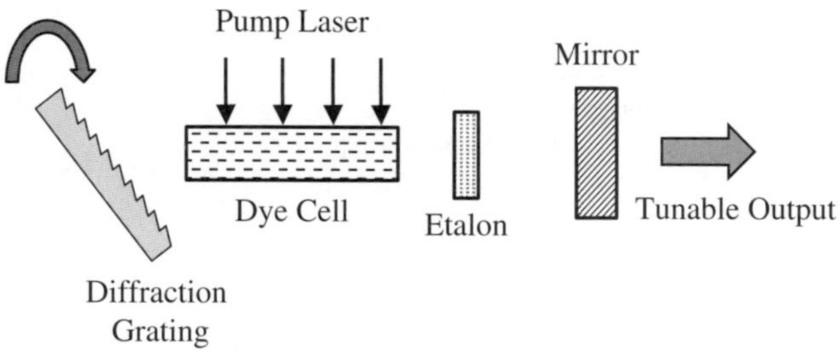

Figure 12.17 Dye laser schematic.

12.4.2.6 Dye Lasers

Dye lasers are optically pumped lasers that are in some way similar in principle to solid state lasers. However, instead of using a solid material doped with an atomic substance, they use organic dyes, such as Rhodamine 6G dissolved in some solvent, such as methanol. A particular feature of dye lasers is their wide gain spectrum encompassing a width of a few tens of nanometres, typically. In order to isolate the lasing at one wavelength some form of dispersive feedback much be used whereby feedback is restricted to one wavelength. Moreover, this feedback can be controlled externally so the wavelength of operation (within a specific range) may be selected. That is to say, the laser itself is tunable. The dispersive feedback might take the form of a prism or diffraction grating for coarse tuning and an etalon for fine tuning. Figure 12.17 illustrates the set-up of a dye laser.

Tuning of the laser is achieved by rotation of the dispersive elements, such as the diffraction grating or etalon. Dye coverage is relatively restrictive, from the near ultraviolet to the very near infrared. Furthermore, coverage of a relatively wide wavelength ranges requires the substitution of a significant number of different dyes. Moreover, in addition to the manifest practical difficulties of handling liquid dyes, many of these dyes present further difficulties on account of their toxicity and carcinogenic properties. Therefore, there has been a tendency, in more recent years, for tunable dye lasers to be displaced by widely tunable solid state lasers, such as the Ti:Sapphire (650–1100 nm) and Alexandrite (700–820 nm) lasers.

12.4.2.7 Optical Parametric Oscillators and Non-linear Devices

As previously outlined, laser sources represent a departure from the norms of optical emission bounded by thermal equilibrium. In consequence, exceptionally high flux densities are available. As a result, the electric fields associated with this emission become comparable to the underlying electric fields that bind the constituent components of matter. In our treatment of polarisation and birefringence, we built up a model in which local oscillating dipoles were created whose magnitude was linearly proportional to the applied optical field. However, for flux densities applicable in laser applications, this simple linear approximation breaks down and we must apply non-linear terms to describe the behaviour of these dipoles. In crystal materials, in practice, it is sufficient simply to include an additional quadratic term to describe the induced polarisation. This additional quadratic term is set out in the revised expression for polarisation in Eq. (12.13).

$$\mathbf{P} = \varepsilon_0(\chi^{(1)}\mathbf{E} + \chi^{(2)}\mathbf{E}^2) \tag{12.13}$$

We now apply an oscillating electric field with two frequency components, ω_1 and ω_2.

$$\mathbf{E} = \mathbf{E}_1 e^{\pm i\omega_1 t} + \mathbf{E}_2 e^{\pm i\omega_2 t}$$

The resultant polarisation is given by:

$$\mathbf{P} = \varepsilon_0\chi^{(1)}(\mathbf{E}_1 e^{\pm i\omega_1 t} + \mathbf{E}_2 e^{\pm i\omega_2 t}) + \varepsilon_0\chi^{(2)}(\mathbf{E}_1^2 e^{\pm i2\omega_1 t} + \mathbf{E}_2^2 e^{\pm i2\omega_2 t} + 2\mathbf{E}_1\mathbf{E}_2(e^{\pm i(\omega_1+\omega_2)t} + e^{\pm i(\omega_1-\omega_2)t})) \tag{12.14}$$

The important point about the form of Eq. (12.14) is that the non-linearity has created a number of different frequency terms. In addition to two terms of double the original frequency ($2\omega_1$ and $2\omega_2$), there are also sum

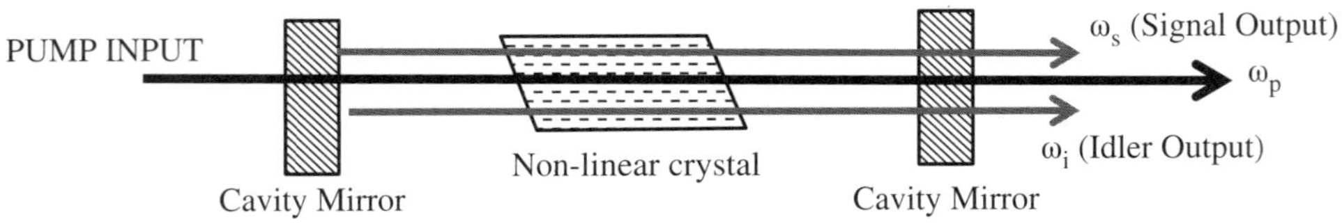

Figure 12.18 Parametric oscillator.

and difference frequencies present ($\omega_1 + \omega_2$ and $\omega_1 - \omega_2$). This process forms the basis of ***frequency doubling*** or ***second harmonic generation,*** where incident laser radiation is partially converted in to light of twice the original frequency. For example, the crystal, potassium di-hydrogen phosphate (KDP) is used commercially to convert infrared Nd:YAG at 1064 nm into green emission at 532 nm.

Sum and difference frequency generation is used in ***optical parametric oscillators (OPOs).*** A crystal is fed with laser light at some pump frequency, ω_p. Second order non-linearity in the crystal results in the output of two frequencies, referred to as the signal and idler frequencies, ω_s and ω_i. These frequencies must sum to ω_p.

$$\omega_p = \omega_s + \omega_i \tag{12.15}$$

Parametric oscillators are especially useful because, within reason, any combination of signal and idler frequencies may be generated, provided they conform to Eq. (12.15). A simplified arrangement is illustrated in Figure 12.18. The crystal is confined within a cavity and the frequency generated is determined by a variety of wavelength selection mechanisms in the cavity. Of course, in practice generation of useful output is dependent upon selecting crystals with a high second order susceptibility and good transmission in the spectral region of interest. Examples of materials include barium borate (BBO), lithium borate (LBO) and periodically poled lithium niobate (PPLN).

The ability to oscillate over a wide frequency range opens up a wealth of applications, particularly in the provision of widely tunable sources. To a significant extent, they have taken over from dye lasers in many spectroscopic applications. A more detailed description of OPOs and second harmonic generation is beyond the scope of this text and forms part of the very broad field of ***non-linear optics***.

12.4.2.8 Other Lasers

The Gas Dynamic Laser uses rapid expansion of supersonic gas mixtures to provide rapid cooling that generates transient non-equilibrium energy level distributions in the flowing gas. In some special cases, this process produces a population inversion between specific states. Most frequently encountered is the CO_2 Gas Dynamic Laser, operating around 10.6 μm and the CO laser at around 5 μm. These devices yield exceptionally high flux levels, in the range of tens to hundreds of Kilowatts. As a consequence, they tend to be featured in military or research applications.

The ***Free Electron Laser*** is an electron beam device offering a continuous and wide range of emission wavelengths. A linear accelerator provides a high energy relativistic electron beam which passes between a periodic array of magnets known as ***wigglers***. This causes oscillation of the electron beam in a direction perpendicular to the nominal direction of the beam. This oscillating electron beam can naturally be thought of as an oscillating dipole and produces optical emission at the oscillating frequency. Careful design of the wiggler assembly ensures that this optical emission then extracts further energy from the oscillating electron beam, resulting in amplification. Naturally, an optical cavity is used to extract the laser emission. Adjusting the spatial period of the wiggler and the energy of the electron beam changes the output wavelength. A free electron laser is, of course, by its nature, a large fixed facility and is used in research applications where tunability across a wide range of wavelengths is at a premium.

The ***Raman Laser*** or amplifier relies on the ***stimulated Raman effect***. In the Raman effect, light is inelastically scattered by lattice vibrations in a solid or molecular vibrations in a molecule. If the vibrational angular frequency is ω_v and the photon angular frequency is ω_p, then the sum and difference frequencies are produced. The ***Stokes frequency*** is at $\omega_p - \omega_v$ and the ***Anti-Stokes frequency*** at $\omega_p + \omega_v$. The important point is that at sufficiently high photon flux, the scattering process can be stimulated and that light at the Stokes or Anti-Stokes frequencies can undergo amplification. For example, the vibrational frequency of the hydrogen molecule is about 1.25×10^{14} Hz. The argon fluoride excimer laser at 193 nm may be shifted to 210 nm (Stokes) or 179 nm (Anti-Stokes) using a high pressure hydrogen gas cell. Raman fibre lasers use the frequency shift ($\sim 1.3 \times 10^{13}$ Hz) produced by fundamental vibrations in silica to shift the wavelength of near infrared (e.g. 1064 nm) pump radiation. Generally, Raman lasers are used where high power laser output needs to be converted to a different wavelength for a specific application.

12.4.3 Temporal Characteristics

Many lasers are intrinsically pulsed because of the nature of their pumping scheme. In some laser systems the level of pump power required to maintain a population inversion is so high that this process can only be accomplished for a short duration. This is the case for some solid state lasers, in particular. In the case of ultraviolet excimer lasers, which are high pressure gas discharge lasers, pumping can only be maintained for several tens of nanoseconds before electrical discharge instabilities terminate the useful pumping process. Moreover, in general, the achievement of continuous operation of a laser is more challenging than pulsed operation. Intrinsically, pulsed laser systems have pulse widths ranging from tens of nanoseconds to ten milliseconds or so. Intrinsic pulse repetition rates for commercial pulsed laser systems can vary from a few Hertz to several tens of KHz for Copper vapour lasers.

In addition to these intrinsically pulsed systems, as previously outline there are Q-switched systems which store energy over a comparatively long for release in a giant pulse of a few nanoseconds. Mode locking produces pulses with a width of picoseconds or even femtoseconds. Where the achievement of ultrashort pulses (i.e. femtosecond) is at a premium, pulse compression is possible using non-linear optics. The repetition rate of these pulses is dictated by the cavity round trip time and is of the order of a few hundred MHz.

12.4.4 Power

For continuous lasers laser power or flux is the useful denominator. This ranges from microwatts to megawatts. High power is at a premium for materials processing applications involving the joining or removing of material. Many systems, such as the helium neon laser and the argon ion laser have very low efficiency, typically a fraction of a percent. By contrast, the efficiency of the carbon dioxide laser and semiconductor lasers is very high, of the order of tens of percent. To appreciate the significance of power in laser systems it is useful to compare the brightness or radiance of an ordinary laser source and compare this with conventional sources. Take for example, a high power cw Nd:YAG laser, with an output flux of 30 W. Assuming single mode operation at 1064 nm, the effective étendue is simply given by the square of the wavelength, or 3.56×10^{-12} m^2 sr. This gives a radiance of 8.44×10^{12} Wm^{-2} sr^{-1}. This is consistent with a blackbody temperature of about 110 000 K. It is therefore easy to appreciate why lasers should have a key role in materials processing applications.

Hitherto, we have considered continuous lasers. Pulse systems such as Nd:YAG or Nd:Glass are capable of delivering many joules, even hundreds of joules in pulses lasting a few nanoseconds. With such systems instantaneous powers of hundreds of Megawatts are possible, extending to GigaWatts for larger systems. With non-linear pulse compression, these nanosecond pulses may be reduced to picosecond durations or less. In which case, powers of Terawatts or even petawatts are possible. For these systems, the effective temperatures are in the range of tens of millions of degrees. These are equivalent to temperatures found at the centre of the sun and such laser systems are used in ***inertial confinement fusion*** experiments which seek to artificially generate and control nuclear fusion.

12.5 List of Laser Types

Before proceeding to discuss briefly the major applications of lasers, we will tabulate some of the principal laser sources together with their main attributes. Figures for power and energy are indicative only and are intended to provide a guide to maximum values.

12.5.1 Gas Lasers

Table 12.1 sets out a basic list of the common gas laser types.

12.5.2 Solid State Lasers

Solid state lasers form a very extensive group of lasers with many commercial embodiments. The principal members of this class are set out in Table 12.2. Historically, these lasers were optically pumped using (xenon) flashlamps. Nowadays, they are predominantly pumped by other lasers, e.g. semiconductor lasers.

12.5.3 Semiconductor Lasers

Semiconductor lasers are predominantly based on an active material incorporating one or more of the Group III–V compound semiconductors. These are set out in Table 12.3.

12.5.4 Chemical Lasers

There are two principal types of chemical laser, namely the hydrogen fluoride laser and the chemical iodine laser. The deuterium fluoride laser is essentially the same laser as the hydrogen fluoride laser, except the

Table 12.1 Common gas laser types.

Laser	Wavelength (nm)	cw Pulsed	Power/energy	Application
Excimer (Exciplex)	157 (F_2), 175 (ArCl), 193 (ArF), 222 (KrCl), 248 (KrF), XeBr (283), XeCl (305), XeF (351)	Pulsed (few ns)	Up to several J	Materials processing, research, semiconductor lithography, photochemistry and medical (eye tissue ablation)
Helium cadmium	325 and 442	Cw	Hundreds of mW	Photochemistry, spectroscopy, microscopy
Nitrogen	337	Pulsed (10s of ns)	Few mJ	Research, photochemistry, laser pumping
Argon ion	351–529	Cw	Tens of W	Research, entertainment, laser pumping
Krypton ion	416–799	Cw	Few W	Research, entertainment, medical
Copper vapour	511 and 578	Pulsed (10s of ns)	Few mJ @ 10s KHz rep. rate	Research, materials processing, dye laser pumping, high-speed photography
Gold vapour	627	Pulsed (10s of ns)	Few mJ @ 10s KHz rep. rate	Medical
Helium neon	632.8	Cw	<50 mW	Metrology, alignment, surveying
Carbon monoxide	5000	Both	<1 KW	Materials processing
Carbon dioxide	9400 and 10 600	Both	Over 100 KW	Materials processing

Table 12.2 List of solid state lasers.

Laser	Wavelength (nm)	cw Pulsed	Power/energy	Application
Ti:Sapphire	650–1100	Both	10s of Watts	Spectroscopy, metrology, research, LIDAR
Ruby	693	Pulsed	Several joules	Research and medical
Alexandrite	700–820	Both	Several watts	Spectroscopy, research
Yb Doped Fibre	1000	Cw	Kilowatts	Telecoms – materials processing
Nd:YLF	1047 and 1053	Both	10s of Watts	Materials processing and research
Nd:Glass	1062	Pulsed	Kilojoules	Research (fusion)
Nd:YAG	1064	Both	Kilowatts	Materials processing and research
Nd:YVO_4	1064	Pulsed	10s of Watts	Materials processing and research
Er Doped Fibre	1530–1560	Cw	Several watts	Telecoms – fibre amplifiers
Th:YAG	1900	Pulsed	>100 W (average)	Medical
Cr:ZnSe	2000–3000	Pulsed	>10 W (average)	Military
Ho:YAG	2100	Pulsed	>100 W (average)	Medical
Colour Centre Lasers	2300–3200	Cw	Milliwatts	Research, spectroscopy
Er:YAG	2940	Pulsed	10s of Watts	Medical

Table 12.3 Semiconductor lasers.

Laser	Wavelength (nm)	cw Pulsed	Power/energy	Application
AlGaN	250–400	cw	mW	Research
GaN	400	cw	mW	High capacity optical discs and research
InGaN	400–500	cw	mW	Optical displays and research
AlGaInP	635–780	cw	Tens of mW	Display, laser pointers, optical discs
AlGaAs	700–1100 W	cw	KWs (in large arrays or 'bars')	Materials processing, laser pumping, telecoms
InGaAsP	1000–2100	cw	W	Optical fibre communications, laser pumping, medical
PbSnSe	8000–30 000	cw	mWs (Cooled)	Research, spectroscopy
PbSnTe	7000–30 000	cw	mWs (Cooled)	Research, spectroscopy
PbSSe	4000–8000	cw	mWs (Cooled)	Research, spectroscopy
PbGeTe	3000–7000	cw	mWs (Cooled)	Research, spectroscopy
PbCdS	2000–4000	cw	mWs (Cooled)	Research, spectroscopy
Quantum cascade	2500–250 000	cw	mWs (Cooled)	Research, spectroscopy

presence of a heavier atom (deuterium) shifts the relevant laser transition to longer wavelengths. These lasers are summarised in Table 12.4.

12.5.5 Dye Lasers

Dye lasers are optically pumped and the resulting dye fluorescence produces optical gain. As the process is molecular fluorescence, the exciting wavelength must be shorter than the output. Argon ion and copper

Table 12.4 Chemical lasers.

Laser	Wavelength (nm)	cw Pulsed	Power/energy	Application
HF	2700–2900	cw	MW	Military
DF	3600–4200	cw	MW	Military
Iodine	1315	cw	100 s KW	Military

Table 12.5 Tuning range for selected dyes.

Dye	Wavelength (nm)	Dye	Wavelength (nm)
p-Terphenyl	320–365	Rhodamine 6G	555–580
Polyphenyl 1	365–410	Pyromethane 597	565–625
QUI	370–400	Rhodamine B	580–610
Stilbene 3	410–460	DCM	600–670
Coumarin 120	420–465	Rhodamine 101	605–665
Coumarin 47	430–495	Pyridin 1	660–720
Coumarin 102	450–500	Pyridin 2	665–755
Coumarin 307	480–535	Styryl 8	710–770
Coumarin 153	520–570	Styryl 8	785–845
Fluorescein 27	540–590		

vapour lasers are useful pump sources, with ultraviolet sources, such as nitrogen and excimer lasers preferred for shorter wavelength applications. Dye lasers may be run in the continuous or pulsed modes of operation, yielding powers in excess of several tens of watts. Although the useful range of dyes is largely restricted to the visible, with some limited extension into the near ultraviolet and near infrared, this spectral range may be further extended by the use of second harmonic generation and other non-linear techniques. There is a tendency for dye lasers to be replaced by more convenient and compact solid state solutions, such as OPOs. The tuning range of a limited number of laser dyes is set out in Table 12.5. The range enumerated is indicative only, as the actual tuning range is dependent upon the exciting wavelength.

12.5.6 Other Lasers

Table 12.6 sets out limited characteristics for lasers not falling into the previous categories.

Table 12.6 'Other Lasers'.

Laser	Wavelength (nm)	cw Pulsed	Power/energy	Application
Optical parametric oscillator	400–20 000	Both	Watts	Spectroscopy, metrology, research
Fibre Raman laser	800–2200	Both	10s of watts	Laser pumping, telecommunications, research
Free electron laser	$0.1–10^6$	cw	100 s KW	Research, materials processing, military

12.6 Laser Applications

12.6.1 General

The early history of the laser was dominated by a clear sense of the remarkable potential of the technology combined with serious difficulties in realising practical or commercial applications. As such, in those times, the laser was dubbed 'an invention looking for an application'. Indeed, there was a pervasive mood, perhaps fuelled by contemporary science fiction, that saw the laser merely as a source of 'death rays'. However, those times are now quite remote and laser technology is embedded into many parts of everyday life, from entertainment and telecommunications to manufacturing industry and surveying. The breadth of these applications is so diverse, nowadays, that it is not possible to describe them in any great detail in this short space. For greater coverage of specific topics, the reader is encouraged to consult more specialised texts. This section will cover, in outline the broad categories of contemporary laser applications.

The most obvious feature of a laser beam in terms of potential applications is its directed power. Of course, it is not the raw power, as such, that defines its utility. Very considerable levels of power may be derived from conventional, e.g. blackbody or other luminous sources. It is, however, (spatial) coherence that truly underpins applications involving directed power. Aside from applications requiring high flux, spatial coherence confers upon the laser beam its remarkable property of directionality. This opens up a wealth of areas in dimensional metrology and allows the laser to be used in many routine surveying and alignment applications.

If spatial coherence confers the property of directionality upon the laser, it is temporal coherence that defines its property as an oscillator. From this are derived many applications in metrology, communications, spectroscopy, and scientific research in general.

12.6.2 Materials Processing

In Section 12.4.4, we were made aware of the exceptionally high brightness associated with laser sources. The corresponding blackbody temperatures, even for relatively modest laser fluxes, are well in excess of melting points or boiling points of even the most refractory materials. For example an early application of the Ruby laser was the laser drilling of holes in diamond to created diamond dies for the drawing of hard metals. Of course, the ruby laser is generally a pulsed laser and the pulse width is of critical importance in materials applications. For example, a pulse energy of 1J and a pulse with of 30 ns corresponds to a peak power in excess of 30 MW. Furthermore, when interacting with a solid material, the pulse width determines the depth to which any heat might penetrate.

The majority of materials processing applications are subtractive in nature, where material is being removed. This might include laser cutting and laser drilling. However, there are also a number of niche applications involving material addition, such as ***photolytic*** or ***pyrolytic deposition***. That is to say processes that use either a light induced or thermally induced mechanism to break down a precursor compound to deposit, for example, a metal conductor. In addition there are processes involving the alteration of material, such as photo-polymerisation as used in 3D printing or the laser induced modification of refractive index to create patterned waveguides. Finally, the fourth broad area of applications is that of joining, i.e. welding.

As suggested earlier the impact of thermal absorption of laser radiation at a material interface is not only dependent upon the laser flux, but also the interaction time. For a pulsed laser, the interaction time is represented by the pulse width. Alternately, for cw emission, the interaction time is dictated by the time which a laser beam dwells at a specific surface location. In either case, the interaction time determines the penetration depth into the material. For a material of thermal diffusivity, κ, and a material interaction time of τ, then the characteristic thermal penetration depth, δ, is given by:

$$\delta = \sqrt{\kappa\tau} \tag{12.16}$$

*The thermal diffusivity, κ, is the thermal conductivity divided by the heat capacity per unit **volume**.*

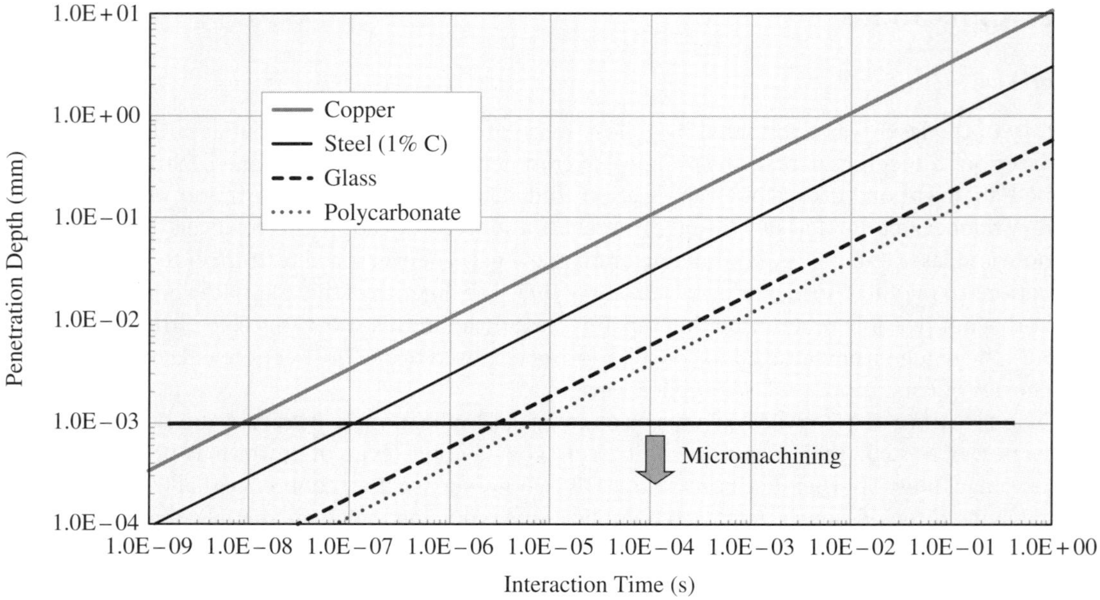

Figure 12.19 Laser penetration depth vs. interaction time.

For example, for a Q switched ruby laser with a pulse width of 15 ns, then the penetration depth in copper is equal to just over a micron. In this instance, the laser is able to remove small quantities of material in a controlled fashion. This is illustrated in Figure 12.19, which shows thermal penetration depth as a function of interaction time for a few selected materials. With the laser energy deposited in such a small depth of material, material is removed by rapid vaporisation, generally accompanied by a shock wave in a process referred to as ***ablation***. Applications involving the controlled removal of small quantities of materials are referred to as micromachining.

For longer interaction times – tens of milliseconds or seconds, penetration depths of several millimetres result. Thus, generally, short interaction times favour materials removal, e.g. drilling and cutting, whilst longer interaction times favour welding applications and so on. Overall, materials processing applications may be grouped by interaction time and irradiance. This is further illustrated in Figure 12.20 which is a graphical representation of applications as bounded by interaction time on the one hand and irradiance on the other.

Laser glazing refers to the rapid melting and solidification of a surface layer of metal to provide a thin semi-amorphous coating that is resistant to crack propagation and fatigue. There is a distinction between penetration welding and conduction welding. Conduction welding relies entirely upon conduction to convey heat through the thickness of the metallic workpiece, as per Eq. (12.16). It tends to be associated with small scale processes, e.g. spot welding. By contrast, penetration welding relies on higher irradiance levels and thermo-mechanical deformation to form a 'keyhole' of molten material extending through the depth of the material. Laser hardening is equivalent to 'case hardening' where, for example carbon is infused into the surface of a metal (steel). Laser hardening may be area selective, unlike conventional processes, and is used, for example in the hardening of vehicle crankshafts. Laser alloying is a similar process except a dissimilar metal is fused into the surface of a metal to create an alloyed surface layer.

All these processes are well established, generally requiring high fluxes. Other, more esoteric applications fall into the materials processing category. One example is a 3D printing process that relies on photo-polymerisation of a liquid precursor. Laser light is focused into a bath of the precursor compound, and due to non-linear (two photon) absorption of the light, polymerisation only occurs at the focal point. By

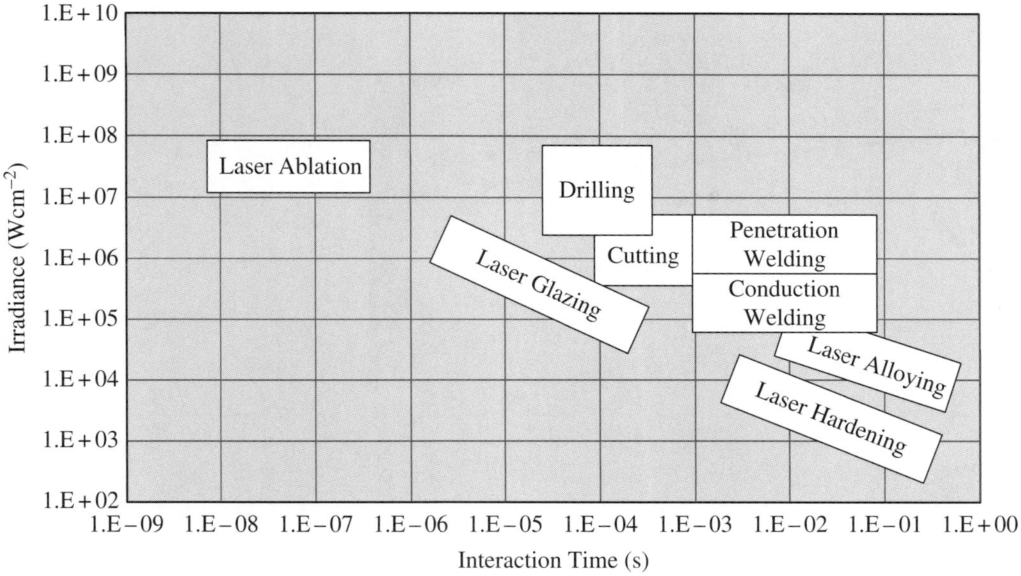

Figure 12.20 Chart of laser materials processing applications.

modulating the flux of the laser and translating it in three dimensions, a three dimensional solid structure may be (slowly) built up.

12.6.3 Lithography

In Chapter 6, we learned that the resolving power of an optical system is dependent upon the wavelength of illumination and the numerical aperture of the system. As a result of this consideration, lasers, particularly short wavelength lasers have a significant role in pattern replication. This is particularly true in the replication of semiconductor circuits by lithography. The dramatic and rapid expansion of integrated circuit functionality, as expressed by ***Moore's Law***, has been facilitated by the ability optically to replicate features as small as a few tens of nanometres. This has, in practice, meant the deployment of ultraviolet excimer lasers, with wavelengths as short as 157 nm (F_2). In this case, the laser is effectively used as a flood lamp to illuminate a patterned mask. The object mask is then imaged onto the semiconductor wafer with a high performance, high numerical aperture lens. Design of this lens is greatly facilitated by the nominally monochromatic nature of the laser illumination.

12.6.4 Medical Applications

The laser is widely used in surgery and dentistry as a surgical tool. One could perhaps regard medical applications as a subset of materials processing applications. For the most part, lasers are used in cutting or cauterisation and rely on absorption of laser radiation in tissue. Medical lasers tend to fall within a specific wavelength range in the near- to mid-infrared, with wavelengths around 2000 nm predominating. This is largely driven by the absorption properties of water and so wavelengths in the infrared where water absorbs most strongly are generally most favoured. A specific exception to this is laser eye surgery. In this case, (pulsed) ultraviolet excimer laser radiation is used to ablate corneal tissue and shape the external envelope of the eye. This effectively adjusts the 'sag' of the eye to adjust the focal power of the eye and to minimise aberrations, such as astigmatism.

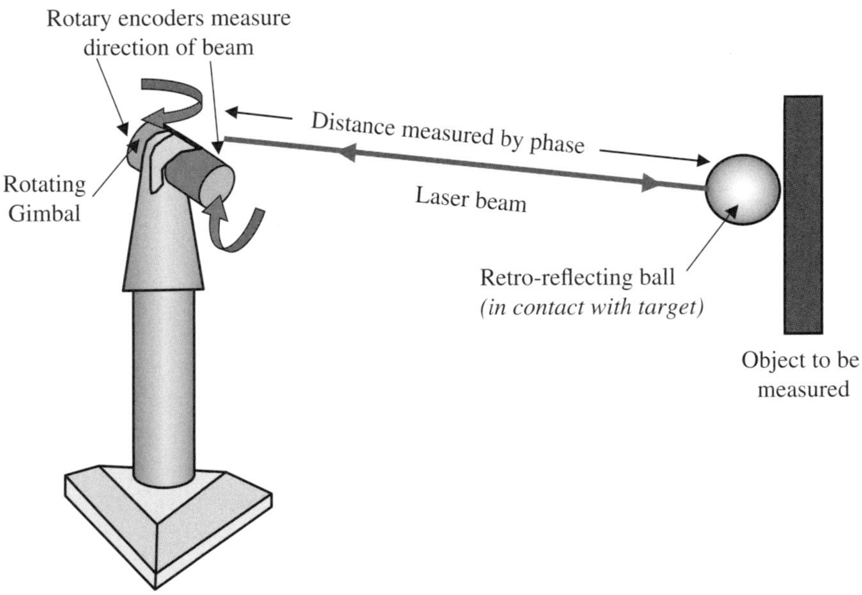

Figure 12.21 Laser tracking – 3D coordinate metrology.

12.6.5 Surveying and Dimensional Metrology

Surveying and associated applications depend upon the directionality of a laser beam. At a basic level, the use of lasers in surveying is ubiquitous with the widespread availability of 'laser levels' for both domestic and industrial use. This ranges from the relatively mundane in the alignment of timber in sawmills for the optimum extraction of sawn timber to critical alignment tasks in tunnelling and other major civil engineering projects. Indeed, this type of application was a feature of early development, with the deployment of laser rangefinders for military applications. Of course, the directional properties of the laser are readily exploited for weapons guidance.

Lasers also find use in rather more sophisticated surveying and metrology applications. For example, ***laser trackers*** are used for extremely accurate three dimensional co-ordinate measurements. These instruments track the position of a corner cube retro-reflector ('cats eye'), contained within a mechanically referencing solid sphere. A ***corner cube***, or cat's eye, is a truncated cube encompassing one of its eight corners and incorporating three mutually perpendicular and reflective surfaces. Reflection off all three surfaces will automatically reverse the path of any ray. The laser tracker set up is illustrated in Figure 12.21.

A rotating gimbal directs a laser beam onto the corner cube target which retro-reflects the laser beam. The measurement head detects the returning laser beam and 'tracks' the position of the retroreflector such that the rotating gimbal always maintains the laser beam in alignment with the corner cube wherever it is. In this application the retro-reflector is withdrawn manually from a known position close to the laser tracker and then placed in contact with the object to be measured. During this interval, the reflector is 'tracked'. In this, way the distance to the object may be computed from the phase of the reflected signal, exploiting the laser's temporal coherence. In addition, the orientation of the gimbal is known (from rotary encoders) and so the 3D position of the object can be determined. Laser trackers are useful in the precision measurement of large engineering artefacts, e.g. aircraft in assembly. A related instrument is the so-called 'laser radar'. Instead of using a retroreflector, this instrument relies on diffuse reflectance from the object target itself. The reflected signal is much weaker and the distance measurement is carried by modulating the laser flux and determining the relative phase of this carrier wave.

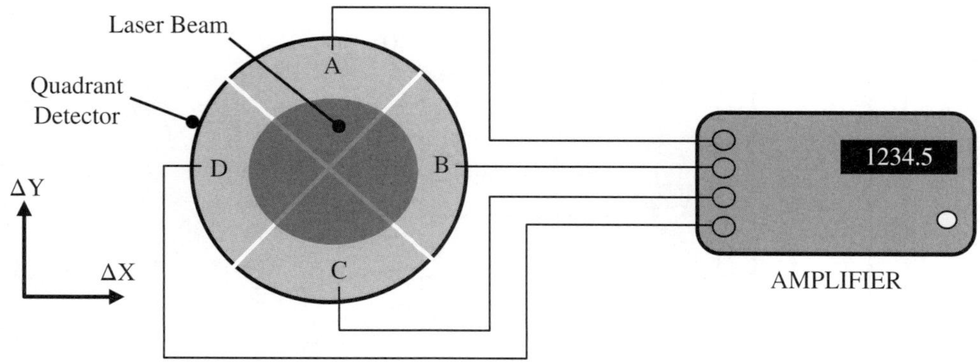

Figure 12.22 Quadrant detector.

A ***laser triangulation gauge*** is a general purpose instrument for measuring the vertical profile of surfaces. The triangular geometry is effected by splitting a laser beam into two arms and arranging them to converge on a single point. This point may be recognised as the vertex of a triangle. The separate origin of the two beams, as they emerge from the sensor head, represent the base of the triangle. The sensor head is then arranged to traverse linearly across a surface to be measured. A camera then views two laser beams as the scatter from the surface. The separation of these two beams is an indication of the surface height.

12.6.6 Alignment

The directional properties of the laser find use in a wide variety of industrial and laboratory applications. For example, lasers are particularly useful in the shaft alignment of machine tools and other rotating machinery. This relies upon the deviation of a reflected beam from a rotating shaft or spindle and the precise adjustment or elimination of any wobble in the reflected beam. Any deviation or wobble can be automatically detected to submicron precision. This may be achieved using a charge coupled device (CCD) detector (digital camera) or a ***quadrant detector***.

The quadrant detector consists of a circular photodiode sensor split into four separate quadrants with a small gap between the segments. Each quadrant produces a separate signal that is proportional to the total flux falling upon that segment. If a laser beam falling on this detector is perfectly centrated, then the output from each sensor will be equal. Deviation from this condition may be taken as a measure of misalignment. The principle is illustrated in Figure 12.22.

Precise sensing of the alignment of the laser beam in two dimensions, X and Y may be derived from the output from the quadrants labelled A, B, C, and D. This may be expressed as follows:

$$\Delta X \propto \frac{A - C}{A + B + C + D} \qquad \Delta Y \propto \frac{B - D}{A + B + C + D} \tag{12.17}$$

In addition to such specific alignment tasks, visible lasers are sometimes used as a proxy alignment marker for non-visible lasers, particularly infrared lasers. That is to say, a co-aligned helium neon laser may provide a useful marker for the output of a Nd:YAG or CO_2 laser.

One important feature of the laser in alignment applications is its ***pointing stability***. In practice, thermal effects, particularly when a laser system is switched on and warming up, causes the alignment of the laser to drift. As such, the pointing stability of a laser is an important and descriptive parameter.

This narrative is intended to provide some sense of the utility of lasers in alignment applications. It is, however, recognised that the field is exceptionally diverse and specialist texts are recommended for the interested reader.

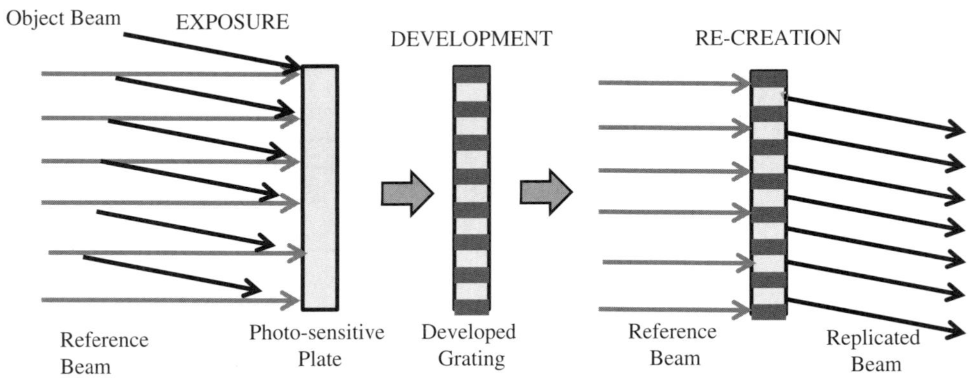

Figure 12.23 Underlying principle of holography.

12.6.7 Interferometry and Holography

The application of lasers in interferometry and holography is dependent upon their coherence properties. Interferometry is based upon the amplitude division of a beam, usually by a beam splitter, and subsequent recombination after one or both beams has probed an optical surface or inhomogeneous medium. Recombination of the two beams leads to interference, producing contrasting light and dark fringes. From the disposition of this pattern of fringes information about the geometrical distribution of the relative phase of the two beams may be gleaned. This information may be put to a variety of purposes, including the precise determination of the shape of optical surfaces, the refractive index distribution of inhomogeneous media, and so on. Interferometry is able to discriminate the relative optical path to a precision of a few nanometres or better. Of course, interferometry was practised before the advent of the laser. Spectral line sources, such as those derived from the mercury lamp, were used. However, lack of temporal and spatial coherence greatly reduced the utility of interferometry. In particular, interference could only be generated if the relative path length of the interfering beams was less than the (very short) coherence length of the lamp. Not surprisingly, this consideration greatly compromised experimental set ups. Interferometric techniques will be considered in more detail when we cover these instruments later in the text.

Holography is a technique for wavefront re-creation using interferometry. The spatial and temporal coherence of the laser beam allows the re-creation of a three dimensional wavefront both in phase and amplitude. As a result, the re-created wavefront represents a full 3D replication of the original object field. The technique relies on the interference of a plane parallel reference beam with light scattered from the object. Illumination of the object is generated from the same source as the reference beam. To visualise how the process works, one can image the light scattered from the object as a superposition of a large number of plane waves. We now select a particular plane wave and imagine it interfering with the reference beam at a light sensitive or photographic plate. This generates a sinusoidal interference pattern that is recorded by the imaging plate as a phase or amplitude diffraction grating. Following recording of the diffraction grating by the imaging plate, the reference beam is then re-applied. As a result, the original wavefront from the object is re-created in the first diffraction order. When presented with a multiplicity of individual plane waves, the more complex diffraction pattern will recreate the original pattern of individual plane waves. This underlying principle illustrated in Figure 12.23.

As with interferometry, holography is, in principle, possible with conventional optical sources, but immeasurably more difficult. Of course, the discovery of holography preceded that of the laser by some 12 years.

12.6.8 Spectroscopy

Application of lasers in spectroscopy is predicated upon the property of the laser as a very narrowband oscillator. When used in these applications, the linewidth of the laser is measured in MHz or even KHz. This represents and extremely high Q factor when one bears in mind that the base frequency of a laser oscillator is in the range of hundreds of THz. Another key feature of (some) lasers is the ability to tune the wavelength of emission and so to probe the structure of matter by precision spectrometry. Precision measurement of atomic and molecular structure is often hampered by the finite linewidth of optical transitions, particularly as caused by the Doppler effect. From the early advent of the tunable laser, various schemes were put in place for circumventing this difficulty. This included two photon spectroscopy, that could, under specific circumstances, obviate the impact of Doppler line broadening. More recently the use of cold atom trapping by lasers, using 'photon recoil' to slow and trap thermal atoms, has extended the possibilities of line narrowed spectroscopy.

On a more practical level, tunable lasers, particularly those derived from compact semi-conductor lasers, can be used in the spectroscopic monitoring of atmospheric constituents, e.g. of industrial contamination or pollution and so on. With a pulsed laser system monitoring backscattered light from the atmosphere, the spatial distribution of contamination may be derived by analysing the temporal distribution of the backscattered light. This relies of the same principle as RADAR to discriminate the temporal signal and the technique is referred to as ***LIDAR*** (Light Detection and Ranging).

As discussed earlier, the discovery of frequency comb generation opens the possibility of locking an optical frequency to the ultimate (currently microwave) frequency standard. Thus the derived optical frequency may be established to the same precision as the underlying standard. This opens the possibility of the frequency standard being shifted from the microwave to the optical domain in the near future.

12.6.9 Data Recording

Perhaps, numerically, the most widespread deployment of laser systems is in data recording. Optical disc recording is a universal feature in a wide range of domestic and commercial data storage applications. Early systems were based around an extremely compact AlGaAs semiconductor laser system operating at 780 nm. Data was recorded on an aluminised polymer disc into which a spiral pattern of pits had been etched. These pits represented the digital data and were approximately one quarter of a wavelength deep (~100 nm). As the disc was spun, these pit and surrounding plateau regions produced a time varying response of high and low reflection. This response was recorded by a sensor on the CD head and decoded and presented as data.

The advantage of optical data recording is the density of data storage, which is ultimately governed by diffraction and hence the wavelength of light. It is therefore not surprising that more recent years have seen the deployment of shorter wavelength sources. In particular, the development of GaN-based semiconductor lasers has facilitated the introduction of 'Blue-ray' technology based upon a 405 nm source. All things being equal, it is clear that the density of data storage, i.e. bits per unit area, is inversely proportional to the square of the wavelength of the source.

It is possible also to write data using a laser head. In this case, the write laser here has sufficient power, when focused, to change the morphology and hence reflectivity of a specially formulated metal film. In this case, the data is written on an entirely planar metal film. When the disc is read, data patterning is determined entirely by the local film morphology. In all cases, the reflective layer on an optical disc is protected by a thin layer of polymer or resin.

12.6.10 Telecommunications

As early as the mid to late 1980s the bulk of core 'backbone' telecommunications has been provided by the transmission of laser light through optical fibres. In this instance, the laser acts as a very high frequency carrier wave, upon which digital data is impressed. Indeed, the technology that it displaced, the microwave link, was hampered by the comparatively low bandwidth available from a source with a carrier frequency of a few GHz. Of course, this compares very unfavourably with the bandwidth available from an optical source running at several hundred TeraHertz. ***Wavelength division multiplexing*** (WDM) allows several different wavelengths to be sent down the same length of fibre and data rates as high as 10s of Terabits/s are possible.

Initially, the major barrier to the deployment of optical fibre technology was the transmission of the optical fibres themselves. The transmission and guidance of light through tens or hundreds of kilometres presented serious difficulties particularly through scattering and material absorption. However, Kao and Hockham at Standard Telecommunication Laboratories realised, in 1966, that these problems were not insuperable. Rayleigh or atomic scattering that predominated at shorter wavelengths could be controlled by selection of longer wavelengths. In addition, they realised that the application of semiconductor processing technology to optical fibre development could reduce impurity absorption to such an extent that transmission over exceptionally long distances might become a reality. Indeed, current long-haul technology employs a wavelength of around 1500 nm where the attenuation of optical fibre is as low at 0.2 dB K^{-1}. For example, if 1 W of flux is injected into a 300 km long fibre, then 1 μW (adequate) would be detected at the far end. For exceptionally long transmission links, e.g. *trans*-oceanic links, regenerators must be established at intervals. These regenerators detect the weak signal and, after amplification, impress it on another high power laser source to continue the journey along the fibre link.

Further Reading

Milonni, P.W. and Eberly, J.H. (1988). *Lasers*. New York: Wiley. ISBN: 0-471-62731-3.
Silvast, W.T. (1996). *Laser Fundamentals*. Cambridge: Cambridge University Press. ISBN: 0-521-55424-1.
Yariv, A. (1989). *Quantum Electronics*, 3e. New York: Wiley. ISBN: 978-0-471-60997-1.

13

Optical Fibres and Waveguides

13.1 Introduction

Optical fibres are a key element in the communications infrastructure that so dominates modern life. The utility of optical fibres is based on the guiding effect produced by total internal reflection of light in a cylindrical optical medium. Although seen as a modern invention, optical fibres have a rather longer history than might be anticipated. In the nineteenth century, Lord Kelvin created optical fibres by attaching a stub of molten glass to an arrow and firing the arrow from a crossbow to draw a thin strand of glass. In some ways, this echoes modern fibre production methods which rely on the very rapid drawing of fibres from a heated boule or preform of specially prepared glass. Of course, at that time, optical fibres were very much a scientific curiosity. It was not until the 1950s that useful applications started to emerge. However, these applications were relatively narrow in scope and focused on the use of the guiding properties of fibres in delivering illumination, particularly in medical applications.

It was in the early 1960s that the utility of optical fibre in communications began to be appreciated. At that time, there was a significant heritage within the telecommunications industry of delivering voice and data over a microwave link. Essentially, the microwave radiation acted as a carrier wave to deliver data at some rate. Naturally, it is clear that the maximum rate at which the data can be delivered is dependent upon the bandwidth of the (microwave) carrier. Thus, by substituting an optical carrier with a frequency of hundreds of Tera-Hertz for a microwave carrier at a few Giga-Hertz, a massive increase in the data transmission rate should, in principle, be possible. This argument was first advanced by Charles Kao and George Hockham working at Standard Telecommunication Laboratories in England in 1966. Analysis of the optical fibre as a waveguide was built upon the extensive scientific heritage invested in the study of microwave waveguides and many parallels were drawn. Kao and Hockham fully appreciated that the most significant barrier to the deployment of optical fibres in long-distance telecommunications was their high attenuation. Ultimately, the insight provided by Kao and Hockham was that this obstacle was not insurmountable, given adequate resources. Indeed, this proved to be the case.

In understanding optical fibres, we are presented with the dichotomy that underpins the understanding of all optics. On the one hand, it is very useful to understand the working of an optical fibre in terms of total internal reflection. This presents a geometrical optics view of the propagation of light within an optical fibre. As might be expected, this approach is most acceptable for large diameter optical fibres. Where the diameter of an optical fibre approaches the wavelength of light, then geometrical optics is entirely inadequate to describe light propagation. To describe the propagation of light more generally in an optical fibre, one must revert to the description provided by the wave equation and its solutions. Unlike the solution to the wave equation in free space, the presence of sharply defined material boundaries in an optical fibre imposes specific boundary conditions on these solutions. As a result, for an optical fibre, there are a strictly finite number of independent solutions to the wave equation. These solutions are known as ***modes***. The number of modes that are supported in an optical fibre is dependent principally upon the size of the fibre and also the refractive index contrast. Quite naturally, a larger fibre will tend to support more modes than a smaller fibre. At the opposite extreme,

Optical Engineering Science, First Edition. Stephen Rolt.

Companion website: www.wiley.com/go/Rolt/opt-eng-sci

a fibre sufficiently small might be capable of supporting only one mode. Such a fibre is called a ***single mode fibre***. By contrast, fibres capable of supporting many modes are referred to as ***multimode fibres***.

Thus far we have restricted the discussion to optical ***fibres***. These are structures defined entirely by an extended cylindrical geometry. However, other geometries are used for the guidance of light, most notably those with square or rectangular geometries. These structures are described under the generic label of ***waveguides***. At the same time, an optical fibre may be strictly defined as a specific type of waveguide. On the whole, these rectangular waveguides tend to be deployed as part of highly miniaturised and integrated structures in miniature semiconductor devices.

At this point, we will analyse fibre and waveguide propagation with a simple geometrical analysis. This will be extended to cover the analysis of fibre modes using a wave-based model.

13.2 Geometrical Description of Fibre Propagation

13.2.1 Step Index Fibre

The geometrical description of optical fibre propagation is most applicable to multimode fibres. In this model, retention of light within the optical fibre is entirely due to 100% reflection at the interface between the fibre and a lower index medium. Initially, we might suppose that this boundary exists between the glass of the fibre and the surrounding air. However, most usually, an optical fibre comprises a ***core*** region surrounded by a solid ***cladding*** layer of lower refractive index. The range of angles over which rays successfully propagate along the fibre is determined by the refractive index contrast between the core and cladding layer. The rays must always strike the interface with an incident angle that is greater than the critical angle. By convention, the maximum cone angle, relative to the fibre axis, for propagating rays, is described in terms of the numerical aperture of the fibre. Fibre propagation is illustrated in Figure 13.1

It is clear that the incident angle at the interface, Δ, must be greater than the critical angle. If the core refractive index is n_0 and that of the cladding is n_1, then we have the following expression for the minimum interface angle:

$$\sin(\Delta) > n_1/n_0$$

We can express this instead in terms of the cone angle, ϕ, that the ray makes with respect to the fibre axis.

$$\cos(\phi) > n_1/n_0 \quad \text{and} \quad \sin(\phi) < \sqrt{1 - \left(\frac{n_1}{n_0}\right)^2}$$

We can now finally express this in terms of the numerical aperture:

$$\sin(\theta) < n_0\sqrt{1 - \left(\frac{n_1}{n_0}\right)^2} \quad \text{Hence} \quad NA < \sqrt{n_0^2 - n_1^2} \tag{13.1}$$

As previously advised there is a tendency, especially for larger core fibres, to specify them in terms of their numerical aperture. Fibres and waveguide structures where the difference between core and cladding index is low are referred to as weakly guiding structures. In practice, this condition applies to most optical fibres. For a silica fibre, a typical value for n_0 might be 1.47 and for n_1 1.45. This gives a numerical aperture of about 0.24.

In the foregoing analysis, there is an implicit assumption that the fibre is characterised by a sharp and well defined boundary between two distinct materials. At the boundary between these materials there is a step change in the refractive index. This type of fibre is therefore known as a ***step index fibre***. However, there are types of fibres where there is no discontinuity in the index at a specific interface, but rather a smooth gradation in the index from the centre of the fibre outwards.

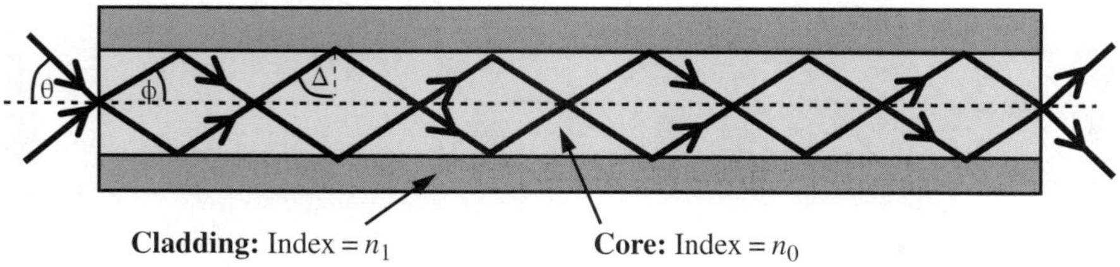

Figure 13.1 Fibre propagation.

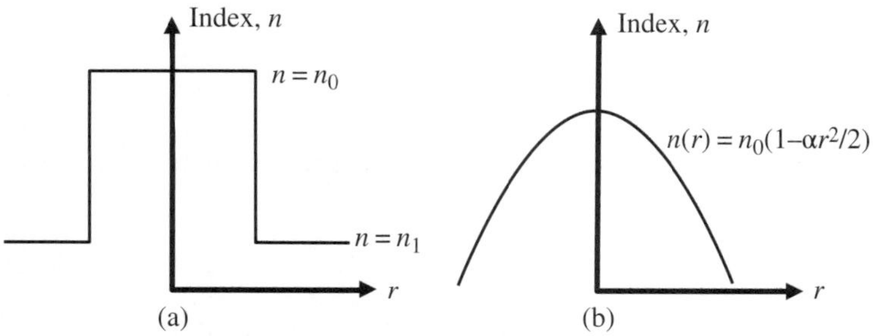

Figure 13.2 (a) Step index fibre, (b) Graded index fibre.

13.2.2 Graded Index Optics

13.2.2.1 Graded Index Fibres

As previously advised, there are fibres where the refractive index of the fibre varies smoothly from the centre of the fibre outwards. Such fibres are referred to as ***graded index fibres***. In line with the behaviour of step index fibres, the refractive index must decline from the centre outwards, in order to confine the light within the fibre. The most common refractive index profile in a graded index fibre is the so called quadratic index profile. This is illustrated in Figure 13.2, comparing the quadratic index profile with the step index profile.

Figure 13.2 plots the refractive index, n, as a function of the radial distance from the centre of the fibre. The index profile is defined by n_0, the base index and the coefficient, α. That is to say, the refractive index profile of the fibre is defined by the following relationship:

$$n(r) = n_0\left(1 - \frac{\alpha}{2}r^2\right) \tag{13.2}$$

The significance of the quadratic index profile is that, in the paraxial approximation, the fibre acts as a series of positive lenses. The fact that the index, and the optical path declines further away from the centre of the fibre, confers positive focal power upon the fibre. In fact, the graded index fibre may be modelled as a series of positive lenses of infinitesimal thickness. If the thickness of such an element is Δz, then its power, $(1/f)$, is given by α. This result may be derived with reference to Fermat's principle. Reverting to a matrix formulation we may described the incremental change in angle produced by one such lens element:

$$\begin{bmatrix} y + \Delta y \\ \theta + \Delta\theta \end{bmatrix} = \begin{bmatrix} 1 & 0 \\ -\alpha\Delta z & 1 \end{bmatrix} \begin{bmatrix} y \\ \theta \end{bmatrix} \tag{13.3}$$

Thus, we can use Eq. (13.3) to formulate the incremental change in the angle, $\Delta\theta$:

$$\Delta\theta = -\alpha y \Delta z \quad \text{and} \quad \frac{d\theta}{dz} = -\alpha y$$

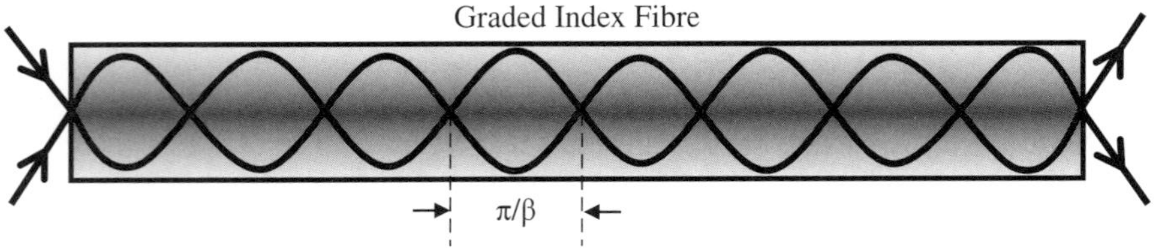

Figure 13.3 Periodic propagation in a graded index fibre.

In the paraxial approximation, the angle, θ, may be represented as the first derivative of the height, y, with respect to the distance, z. Hence:

$$\frac{d^2y}{dz^2} = -\alpha y \tag{13.4}$$

Since Eq. (13.4) is in the form of a simple harmonic oscillator equation, it is clear that a ray will propagate along a sinusoidal path:

$$y = y_0 \sin \beta z \quad \text{where } \beta = \sqrt{\alpha} \tag{13.5}$$

This process is illustrated in Figure 13.3 which shows ray propagation through a quadratic index fibre:

As Figure 13.3 illustrates, rays propagate along a sinusoidal path with a period equal to $2\pi/\beta$ and defined by a maximum ray height, y_0. It is straightforward to calculate the angle of the ray as it propagates along the fibre:

$$\theta = y_0\beta \cos \beta z \tag{13.6}$$

We might, in the paraxial approximation, wish to express Eq. (13.6) in terms of numerical aperture rather than angle:

$$NA = n_0y_0\beta \cos \beta z \quad \text{or} \quad NA = NA_0 \cos \beta z \quad \text{where} \quad NA_0 = n_0y_0\beta \tag{13.7}$$

One extension of the paraxial approximation is that any variation in the refractive index across the profile is smaller than n_0. This assumption is implicit in the derivation of Eq. (13.7). However, the implication of Eq. (13.7) is clear – the larger the numerical aperture, then the larger the maximum ray height, y_0. Algebraically, the choice of numerical aperture and hence maximum ray height seems unrestricted. However, the graded index profile shown in Figure 13.2b cannot extend indefinitely; there must be a limit to the refractive index contrast, Δn, that the fibre can support. Since every infinitesimal refractive index boundary obeys Snell's law, the product of the sine of the incident angle and the refractive index must always be constant. If, using the nomenclature of Figure 13.1, we label the angle of incidence, Δ, and the ray angle with respect to the fibre axis, θ, we have:

$$n(r) \sin \Delta = Constant \quad \text{and} \quad n(r) \cos \theta = Constant \tag{13.8}$$

We will assume that the initial (maximum) ray angle, at the centre of the fibre, is θ_0 and the refractive index at the centre is n_0. Thereafter, we will assume that, at the maximum height, θ, is equal to zero and $n(r)$ is equal to $n_0 - \Delta n$:

$$n_0 \cos \theta_0 = n_0 - \Delta n \quad \text{and} \quad n_0^2 - NA_0^2 = (n_0 - \Delta n)^2 \quad \text{where } NA_0 = n_0 \sin\theta_0. \tag{13.9}$$

Finally:

$$NA_0 = \sqrt{2n_0\Delta n - (\Delta n)^2} \tag{13.10}$$

It can be seen that Eq. (13.10) is equivalent to Eq. (13.1), if we substitute $n_1 = n_0 - \Delta n$. If we now substitute Eq. (13.10) into Eq. (13.7), which relates the numerical aperture to the maximum ray height, we may determine

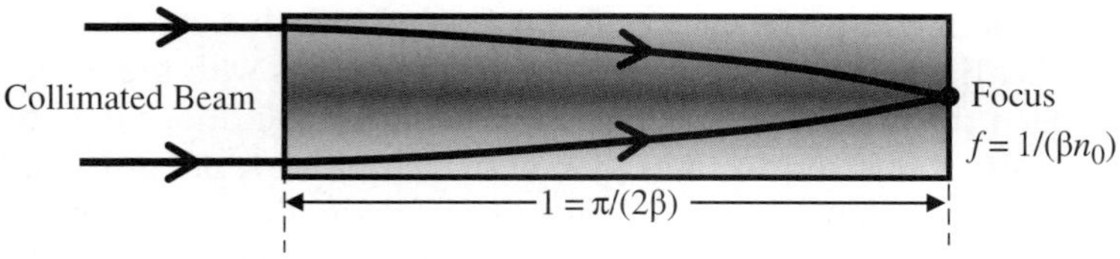

Figure 13.4 Ray paths in a focusing GRIN lens.

this height in terms of the fibre parameters.

$$y_0 = \frac{NA_0}{\beta n_0} = \sqrt{\frac{2n_0\Delta n - (\Delta n)^2}{\alpha n_0^2}} \tag{13.11}$$

If we now apply the paraxial approximation, assuming $\Delta n \ll n_0$, then:

$$y_0 = \sqrt{\frac{2\Delta n}{\alpha n_0}} \tag{13.12}$$

Not surprisingly, Eq. (13.12) expresses the radius at which, according to Eq, (13.2), the refractive index difference of Δn is actually attained.

For the most part, graded index fibres are used in multimode fibres. In multimode applications, the utility of graded index fibres lies in their quadratic index profile. In the geometrical optics narrative, all rays depicted in Figure 13.3 progress with the same optical path along the fibre, as per Fermat's principle. This conclusion is unaffected by the maximum height, y_0 of the ray path. As a consequence, all rays take the same ***time*** to progress along the fibre. This conclusion does not pertain to step-index fibres. Naturally more oblique rays travel a longer distance and take a longer time to progress along the fibre. In the language of physical, as opposed to geometrical, optics, this is referred to as ***modal dispersion***. As a result, a short pulse of light is likely to be temporally spread out and this compromises the data carrying capacity of the fibre.

13.2.2.2 Gradient Index Optics

At this point, we take a very brief detour from our discussion of optical fibres and consider the related topic of ***gradient index (GRIN) lenses***. In essence, as for a gradient index fibre, a gradient index lens is formed by creating a quadratic index profile in a glass material, as depicted in Figure 13.2b. However, in the case of a gradient index lens, the scale is rather different, with the diameter likely to be of the order of a few millimetres, as opposed to tens or hundreds of microns. A typical application for a gradient index lens is focusing light into a fibre or collimating light originating from a fibre. As such, a GRIN lens will mostly, but not exclusively, operate at the infinite conjugate. If we now proceed according to the previous paraxial analysis for the graded index fibre, we seek solutions of the form of $\sin(\beta z)$ or $\cos(\beta z)$ for the ray height as a function of propagation distance, z. For a simple focusing lens, the solution follows a 'quarter wave' progression along the length of the lens, l. This is depicted in Figure 13.4.

We can now determine the focal length of the GRIN lens. If we follow the analysis set out in Eq. (13.3) and determine the ray height at the focal plane for a specific paraxial field angle, then the focal length is simply given by:

$$f = \frac{1}{\beta n_0} \quad \text{or} \quad f = \frac{1}{\sqrt{\alpha} n_0} \tag{13.13}$$

In the light of the 'quarter wave' pattern depicted in Figure 13.4, the length of the GRIN lens is not equivalent to the focal length. The focal length is, of course, further modified by the base refractive index of the lens, n_0. The length of the GRIN lens, l, is then given by:

$$l = \frac{\pi}{2\beta} = \frac{1}{2\sqrt{\alpha_0}} \tag{13.14}$$

It is interesting at this point to compare the aberration performance of a GRIN lens with that of a conventional lens. In this analysis, we will assume that the lens is acting as a simple focusing lens with the object located at the infinite conjugate and the pupil at the first GRIN lens interface. As fibre applications tend to be largely in a substantially on-axis configuration, it is spherical aberration of the lens that we are primarily interested in. To determine this, we must calculate the optical path as a function of the initial ray height, y_0. Furthermore, if we are to calculate the third order aberrations, then this calculation must be performed to fourth order in ray height. First, the ray height, $r(z)$, as a function of distance, z, must be set out, as follows:

$$r(z) = y_0 \cos(\beta z) \tag{13.15}$$

It remains only to calculate the optical path and thence the optical path difference. The optical path, s, is given by:

$$s = \int_{z=0}^{z=\pi/(2\beta)} \sqrt{1 + \left(\frac{dr}{dz}\right)^2}\, n(z)dz \quad \text{and} \quad n(z) = n_0\left(1 - \frac{\alpha y_0^2}{2}\cos^2(\beta z)\right) \tag{13.16}$$

Since, in evaluating the third order spherical aberration term, we are only interested in terms of up to fourth order in r, we may use the binomial theorem to provide a useful approximation to Eq. (13.16). In addition, we may use Eq. (13.5) to substitute for α:

$$s \approx \int_{z=0}^{z=\pi/(2\beta)} n_0\left(1 + \frac{1}{2}\left(\frac{dr}{dz}\right)^2 - \frac{1}{8}\left(\frac{dr}{dz}\right)^4\right)\left(1 - \frac{\beta^2 y_0^2}{2}\cos^2(\beta z)\right)dz \tag{13.17}$$

We can now substitute for the differential expressions in Eq. (13.17):

$$s \approx \int_{z=0}^{z=\pi/(2\beta)} n_0\left(1 + \frac{\beta^2 y_0^2}{2}\sin^2(\beta z) - \frac{\beta^4 y_0^4}{8}\sin^4(\beta z)\right)\left(1 - \frac{\beta^2 y_0^2}{2}\cos^2(\beta z)\right)dz \tag{13.18}$$

Performing the integration, we obtain the following:

$$s \approx n_0\left(1 - \frac{5\beta^4 y_0^4}{64}\right)\left(\frac{\pi}{2\beta}\right) \tag{13.19}$$

Note that the only terms are the constant term and the quartic term in y_0. The absence of any quadratic term in Eq. ((13.19)) indicates that we are at the paraxial focus. We are now able to set out the spherical aberration, expressed as an optical path difference, in terms of the aperture and the parameter, β:

$$K_{SA} = -\frac{5\pi n_0 \beta^3 y_0^4}{128} \tag{13.20}$$

We could more usefully express Eq. (13.20) in terms of the lens focal length, f, rather than the parameter, β. Using Eq. (13.13), we get:

$$K_{SA}(GRIN) = -\frac{5\pi}{128 n_0^2}\left(\frac{y_0^4}{f^3}\right) \tag{13.21}$$

At this point, it is useful to compare the performance of a GRIN lens with that of a best form singlet. Comparing Eq. (13.21) with Eq. (4.34) from Chapter 4, we have:

$$K_{SA}(singlet) = -\frac{4n^2 + n}{32(n-1)^2(n+2)}\left(\frac{y_0^4}{f^3}\right) \tag{13.22}$$

Comparing Eqs. (13.21) and (13.22) for both n_0 and n equal to 1.5, we find:

$$K_{SA}(GRIN) = -0.054\left(\frac{y_0^4}{f^3}\right) \quad \text{and} \quad K_{SA}(singlet) = -0.375\left(\frac{y_0^4}{f^3}\right)$$

The GRIN lens thus reduces the spherical aberration by a factor of approximately 7. As such, not only does a GRIN lens provide a useful, compact focusing or collimation lens for optical fibre coupling, it also has significantly better aberration performance than the comparable conventional lens. This improvement may be thought of as a direct consequence of distributing the refractive power of the lens across an infinite series of individual lenses of infinitesimal individual power. Furthermore, the aberration of a GRIN lens may be further improved by tailoring the refractive index profile, $n(r)$. By analogy to an aspheric lens, if an additional quartic term is added to the refractive index profile, the spherical aberration performance may be further optimised:

$$n(r) = n_0\left(1 - \frac{\alpha}{2}r^2\right) + n_2 r^4 \qquad (13.23)$$

Worked Example 13.1 ***GRIN Lens***

A GRIN lens is to be used to focus collimated light into an optical fibre. The lens is 12 mm long. What is the maximum allowable lens diameter that ensures diffraction limited focusing at a wavelength of 633 nm for an on-axis beam. What numerical aperture does this correspond to when the focused beam enters the fibre? The base refractive index of the GRIN lens material is 1.52.

We are told that the length of the fibre is 12 mm. We first need to calculate the focal length, f, from Eqs. (13.13) and (13.14):

$$f = \frac{2}{\pi n_0} l = \frac{2}{\pi \times 1.52} \times 12 \quad \boldsymbol{f = 5.026\ mm}$$

From Eq. (13.21), we know the spherical aberration:

$$K_{SA}(GRIN) = -\frac{5\pi}{128 n_0^2}\left(\frac{y_0^4}{f^3}\right)$$

However, in order to compute the wavefront error with respect to the Maréchal criterion, we need to know the rms wavefront error. For spherical aberration, this is determined by recourse to the relevant Zernike function:

$$\Phi_{RMS} = \frac{K_{SA}}{\sqrt{5} \times 6} = \frac{\sqrt{5}\pi}{6 \times 128 n_0^2}\left(\frac{y_0^4}{f^3}\right)$$

For diffraction limited performance to be attained, according to the Maréchal criterion, the following condition must apply:

$$\Phi_{RMS} < \sqrt{0.2}\frac{\lambda}{2\pi} \quad \text{and} \quad \boldsymbol{\Phi_{RMS} < 45.05\ nm} \text{ for 633 nm input.}$$

Therefore, for the diffraction limited performance to apply:

$$\frac{\sqrt{5}\pi}{6 \times 128 n_0^2}\left(\frac{y_0^4}{f^3}\right) < 45.05nm$$

Substituting all relevant values (in mm), we get:

$$\frac{\sqrt{5}\pi}{6 \times 128 \times 1.52^2}\left(\frac{y_0^4}{5.026^3}\right) < 45.05x10^{-6} \quad \text{and} \quad y_0 < 1.10\ \text{mm}$$

The maximum allowable lens diameter is thus 2.2 mm.

The maximum permitted numerical aperture may be expressed in terms of the maximum ray height, y_0 and the focal length f:

$$NA_{\max} = \frac{y_0}{f} = \frac{1.10}{5.026} = 0.218$$

The maximum numerical aperture is thus 0.218.

13.2.3 Fibre Bend Radius

We now return to the more general geometrical treatment of fibre optics. Hitherto, we have assumed that the optical fibre under consideration is perfectly straight. However, to some degree, the practical utility of optical fibres is enhanced by their flexibility or ability to bend. Therefore, we are forced to consider the impact of the mechanical bending of optical fibres upon light propagation. It is inevitable that, to some degree, the impact of fibre bending is deleterious to the containment of light by total internal reflection. A rigorous evaluation of the effect of bending can only be obtained through consideration of the wave properties of light. However, from the geometrical perspective, fibre bending serves to further restrict the numerical aperture over which total internal reflection may be preserved. This is illustrated in Figure 13.5

It is clear from Figure 13.5, that the angle of reflection on the inside of the bend, θ, is different to that on the outside, φ. If we consider the fibre radius, r, and the bend radius, R, we may establish a clear geometrical relationship between the two angles, as set out in Figure 13.6.

From inspection of Figure 13.6 and basic trigonometry, we can establish the following:

$$\frac{\sin\phi}{\sin\theta} = \frac{R-r}{R+r} \tag{13.24}$$

Equation (13.24) establishes the relationship between the angles on the inside and the outside bend of the fibre. To determine the impact on propagation, we are more interested in the angle, Δ, that might be attributable to the incident angle in the absence of fibre bending. Taking Δ as the intermediate angle between θ and φ, it is reasonable to suppose:

$$\frac{\sin\phi}{\sin\Delta} = 1 - \frac{r}{R} \tag{13.25}$$

The maximum numerical aperture attainable, NA', will be defined by the product of the base refractive index, n_0 and the cosine of the angle, Δ. Similarly, the original numerical aperture, NA_0, as defined by the critical angle, will be given by the product of the refractive index, n_0 and the cosine of the angle, φ. This gives

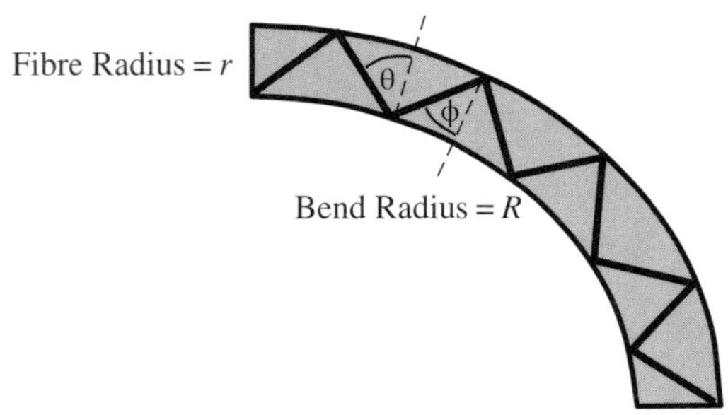

Figure 13.5 Impact of fibre bend radius.

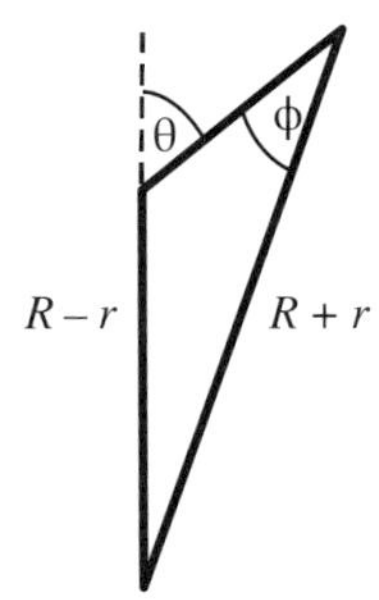

Figure 13.6 Geometry of fibre bending.

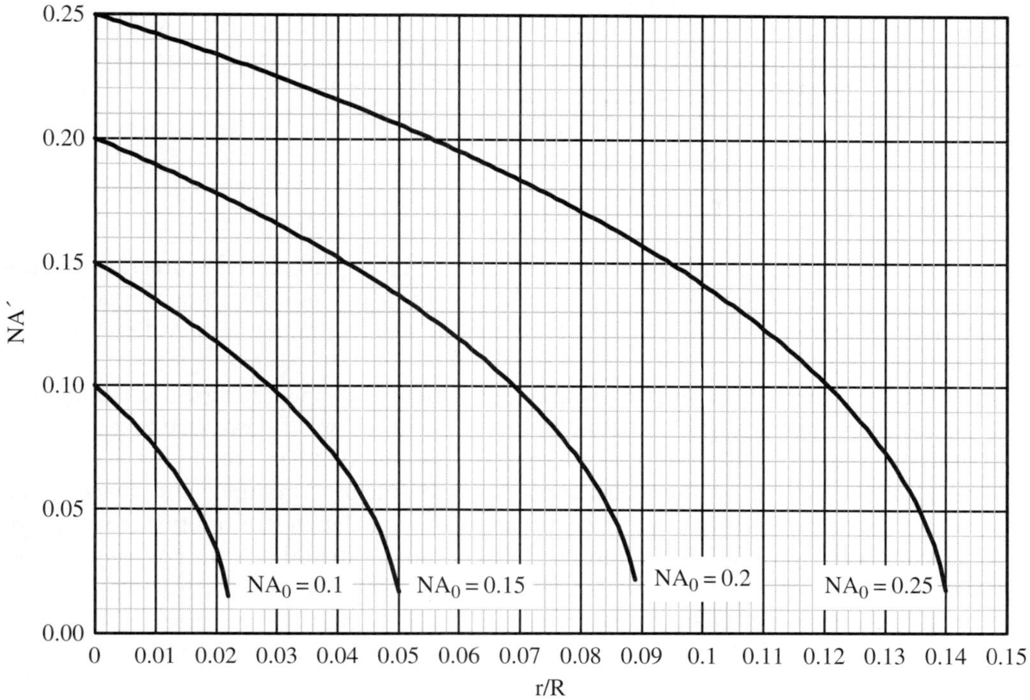

Figure 13.7 Geometrical effect of fibre bending on numerical aperture ($n_0 = 1.5$).

the relationship between the two numerical apertures as:

$$NA' = \sqrt{\frac{NA_0^2 - r/(Rn_0^2)}{1 - r/R}} \tag{13.26}$$

It is clear from Eq. (13.26), that the reduction in the maximum numerical aperture created by fibre bending is more severe for fibres with a lower intrinsic numerical aperture, NA_0. This is sketched out in Figure 13.7, which shows the reduction in numerical aperture produced by fibre bending for various intrinsic fibre numerical apertures. In all cases, it is assumed that the base fibre refractive index is 1.5. For the lowest numerical aperture fibre, a bend radius of one hundred times the fibre radius produces a significant effect. In practice, in the wave description of fibre propagation, fibre bending leads to loss, by virtue of light escaping from the waveguide. For the loss to be acceptable, a ***minimum bend radius*** is imposed, which might be 50–100 times the core diameter or more. As outlined in the introduction, the study of optical fibre propagation had its roots in microwave waveguide analysis. In the case of microwave waveguides, in some cases, minimum bend radii of several hundred metres were contemplated!

13.3 Waveguides and Modes

13.3.1 Simple Description – Slab Modes

We will present here a very simplified description of ***waveguides*** and ***modes***. To this end, we will establish the geometry of a waveguide in one dimension only. That is to say, the waveguide is represented by a planar structure consisting of a high refractive index slab surrounded by slabs of material with a lower refractive index. The purpose of this description is that it conveys many of the salient principles of waveguide propagation

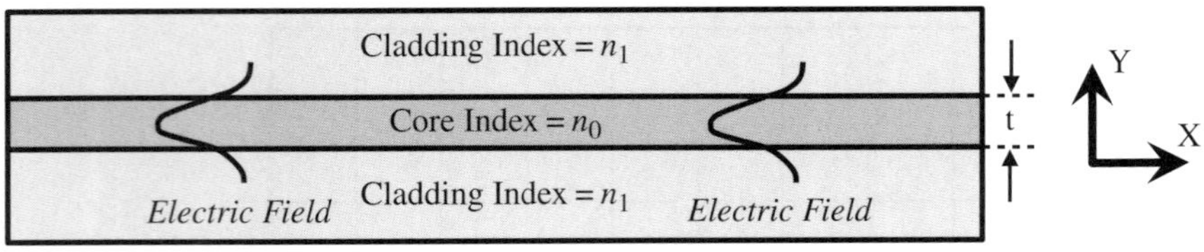

Figure 13.8 Slab waveguide.

whilst presenting a more tractable analysis. In this model, propagation takes place in the z direction and the structure of the waveguide is defined along the y axis. The structure extends to 'infinity' along the x ordinate. Furthermore, for all practical purposes, the extent of the two cladding slabs is infinite in the y direction. The geometry is illustrated in Figure 13.8.

To determine the electric field distribution within the waveguide structure, we must solve the wave equation within the separate media. Most specifically, we must find separate solutions for the two cladding slabs and for the core slab. At the two interfaces, the solutions must obey specific boundary conditions. As we will see, polarisation plays an important part in the realisation of these boundary conditions and hence in the characteristics of the modes. As such, in this instance, we will look for two independent solutions to the wave equation, one in which the electric field is oriented in the x direction and the other with the electric field oriented in the y direction.

To each region, we will apply the wave equation:

$$\frac{\partial^2 \mathbf{E}}{\partial x^2} + \frac{\partial^2 \mathbf{E}}{\partial y^2} + \frac{\partial^2 \mathbf{E}}{\partial z^2} = \frac{n^2}{c^2}\frac{\partial^2 \mathbf{E}}{\partial t^2} \tag{13.27}$$

First, we might look for a solution with a specific temporal frequency, ω. As far as the spatial dependence is concerned, we can effectively ignore the dependence in the x direction, as the boundary extends to infinity in this direction. In the z, propagation direction, it is reasonable to assume the solution propagates as a wave with some propagation constant, β. Furthermore, it is important to recognise that the propagation constant must be identical for both the core and cladding solutions. In summary, the proposed solution might be expressed in the following form:

$$\mathbf{E}(x, y, z, t) = \mathbf{E_0}F(y)e^{i(\beta z - \omega t)} \tag{13.28}$$

It is the form of $F(y)$ that is of principal concern. Substituting Eq. (13.28) into Eq. (13.27) we get:

$$\frac{\partial^2 F(y)}{\partial y^2} = (\beta^2 - n^2k_0^2)F(y) \quad \text{where } k_0 = \omega/c \tag{13.29}$$

Equation (13.29) yields solutions that are either sinusoidal or exponential, depending on the sign of the bracketed expression on the RHS of Eq. (13.29). If this sign is negative, then the solutions are sinusoidal or exponential if the sign is positive. Reviewing the geometrical form of the waveguide in Figure 13.8, it is reasonable to suppose that the solution within the core is sinusoidal and that outside the core is (decaying) exponential. So, we have the following solutions for $F(y)$ in the core and in the cladding respectively:

$$F_0(y) = E_0\cos(\alpha_0 y) \quad \text{or} \quad F_0(y) = E_0\sin(\alpha_0 y) \quad \textit{core}; F_1(y) = E_1 e^{-\alpha_1 y} \ \textit{cladding} \tag{13.30}$$

From Eq. (13.29), the two coefficients, α_0 and α_1 must be related to the propagation constant, β, and the refractive indices, n_0 and n_1, according to (13.29).

$$\alpha_0 = \sqrt{(n_0^2k_0^2 - \beta^2)} \qquad \alpha_1 = \sqrt{(\beta^2 - n_1^2k_0^2)} \tag{13.31}$$

The boundary conditions applicable at the interface between the two solutions defined by $F_0(y)$ and $F_1(y)$ depend upon the state of polarisation of the light. We will define two polarisations, TE, with the electric field parallel to the interfaces, i.e. in the x direction and TM with the electric field perpendicular to the interfaces.

In both cases, the magnetic field is continuous across the interface. However, for TE polarisation it is the electric field, **E**, that is continuous across the boundary, whereas for TM polarisation, the electric ***displacement***, **D**, is constant. For the TE polarisation, the boundary conditions may be expressed as:

$$F_0(y_{\text{int erface}}) = F_1(y_{\text{int erface}}) \quad \text{and} \quad F'_0(y_{\text{int erface}}) = F'_1(y_{\text{int erface}}) \quad (\text{TE}) \tag{13.32}$$

Similarly for TM polarisation, the condition is:

$$n_0F_0(y_{\text{int erface}}) = n_1F_1(y_{\text{int erface}}) \quad \text{and} \quad \frac{F'_0(y_{\text{int erface}})}{n_0} = \frac{F'_1(y_{\text{int erface}})}{n_1} \quad (\text{TM}) \tag{13.33}$$

If the thickness of the core waveguide is t, then the two boundary conditions may, for 'even' solutions of the form $\cos(\alpha_0 y)$, be summarised as:

$$(\text{TE polarisation}): \ \alpha_0 \tan\left(\frac{\alpha_0 t}{2}\right) = \alpha_1 \qquad (\text{TM polarisation}): \ \frac{\alpha_0}{n_0^2} \tan\left(\frac{\alpha_0 t}{2}\right) = \frac{\alpha_1}{n_1^2} \tag{13.34a}$$

Similarly for 'odd' solutions, of the form $\sin(\alpha_0 y)$, the same boundary conditions may be summarised as:

$$(\text{TE polarisation}): \ \alpha_0 \cot\left(\frac{\alpha_0 t}{2}\right) = -\alpha_1 \qquad (\text{TM polarisation}): \ \frac{\alpha_0}{n_0^2} \cot\left(\frac{\alpha_0 t}{2}\right) = -\frac{\alpha_1}{n_1^2} \tag{13.34b}$$

The boundary conditions are thus defined by the tangent or cotangent of the thickness. Since this function is periodic, there is every prospect that there will be multiple solutions to this equation. As alluded to earlier these solutions are referred to as ***modes***. However, under specific conditions, which depend on the thickness, t, there can only be one solution. Under such circumstances, the waveguide is said to operate as a single mode waveguide. There are no conditions for which no solutions may be found. As the thickness of the slab becomes smaller and smaller, eventually α_1 and the tangent expression in Eqs. (13.34a) and (13.34b) must both tend to zero. Nevertheless, the equality required by Eqs. (13.34a) and (13.34b) will always be satisfied. For a further (second) solution to exist, then, at a minimum, then Eqs. (13.34a) and (13.34b) must be satisfied for the next zero value for $\tan(\alpha_0 t/2)$ or $\cot(\alpha_0 t/2)$. In fact, it is quite apparent that the lowest order mode must be symmetrical. Hence for the next (anti-symmetrical) mode to exist, the cotangent must be set to zero. Thus, the following condition must apply:

$$\frac{\alpha_0 t}{2} = \frac{\pi}{2} \quad \text{and} \quad \alpha_1 = 0 \ \text{Since } \alpha_1 = 0, \text{then (from Eq. [13.31])} \ \alpha_0 = k_0\sqrt{(n_0^2 - n_1^2)} \tag{13.35}$$

As a condition for single mode propagation, we thus have the following condition which applies to both polarisations:

$$\frac{k_0 t\sqrt{n_0^2 - n_1^2}}{2} < \frac{\pi}{2} \tag{13.36}$$

At this point we introduce a generalised parameter, V, the ***normalised frequency***, which is defined as follows:

$$V = k_0 t\sqrt{n_0^2 - n_1^2} \ \text{and, in this instance, for single mode propagation}: \ V < \pi \tag{13.37}$$

The implication of Eq. (13.37) is that there exists a ***cut-off wavelength***, λ_c, above which only single mode propagation is allowed. If the critical value of the normalised frequency is labelled V_c, then the cut-off wavelength is defined by:

$$\lambda_c = \frac{2\pi}{V_c}\sqrt{(n_0^2 - n_1^2)}t \tag{13.38}$$

Equation (13.38) is fundamental to the consideration of single mode propagation in an optical fibre or waveguide. To illustrate this further, we will briefly consider an example based on a simple slab waveguide.

Table 13.1 Description of slab modes.

Polarisation	Mode	n_{eff}	α_0	α_0
TE	Low order	1.4539535	0.055172	0.107149
	High order	1.4512075	0.104984	0.059188
TM	Low order	1.4539499	0.055268	0.107099
	High order	1.4512017	0.105064	0.059046

Worked Example 13.2 ***Cut off Wavelength of a Slab Waveguide***
A slab waveguide has a central core thickness of 2.5 μm. The refractive index of the core is 1.50 and that of the cladding is 1.48. What is the cut-off wavelength for this waveguide?

From Eq. (13.37), we know that $V_c = \pi$. Therefore it is clear from Eq. (13.38):

$$\lambda_c = 2 \times \sqrt{(n_0^2 - n_1^2)}t = 2 \times 0.244 \times 2.5$$

The cut-off wavelength is thus 1220 nm. For all wavelengths above this value, the waveguide will be single moded.

13.3.2 Propagation Velocity and Dispersion

Solution of the wave equation for a fibre or waveguide involves the determination of the propagation constant, β. Each solution or mode will have its own unique propagation constant. Of course, the propagation constant determines the phase velocity of the mode. Since each mode will tend to have a different propagation constant, then different modes will also tend to travel at a different velocity. However, in specific cases, as dictated by symmetry, some modes have identical propagation constants; these modes are referred to as ***degenerate***. In any case, the difference in propagation velocities accorded to different modes is referred to as ***modal dispersion***.

We will attempt to illustrate this problem, by constructing a slab waveguide to propagate light at 633 nm. This waveguide will be 'multimode' in the sense that it is capable of supporting two modes (for each polarisation). Our slab waveguide has a core that is 4 μm thick and with a refractive index of 1.455. The refractive index of the cladding is 1.45. Of course, the solutions are defined by their unique propagation constant, β. However, we designate each mode according to its ***effective index***, n_{eff}, which is defined in the following way:

$$n_{eff} = \frac{\beta}{k_0} \tag{13.39}$$

The effective index is useful in describing the ***phase velocity***, v_p, of the mode as it propagates along the fibre. The phase velocity is simply the speed of light divided by the index. One has to be cautious in interpreting this velocity as relating to the velocity at which information flows along the fibre. This description is reserved for the ***group velocity.*** The phase velocity is simply given by the angular frequency, ω, divided by the wavevector, k, whereas the group velocity is the differential of the angular frequency with respect to the wavevector:

$$\textit{phase velocity}:\ v_p = \frac{\omega}{\beta} \qquad \textit{group velocity}:\ v_g = \frac{d\omega}{d\beta} \tag{13.40}$$

We now turn to the solutions for our particular waveguide. There are four solutions to consider – a low order and a high order mode for both the TE and TM polarisations. Relevant values are given in Table 13.1 for the key parameters.

It is quite apparent that there are significant differences between the effective indices of the higher and lower order modes for each polarisation. So, in this instance, the modal dispersion is significant, amounting

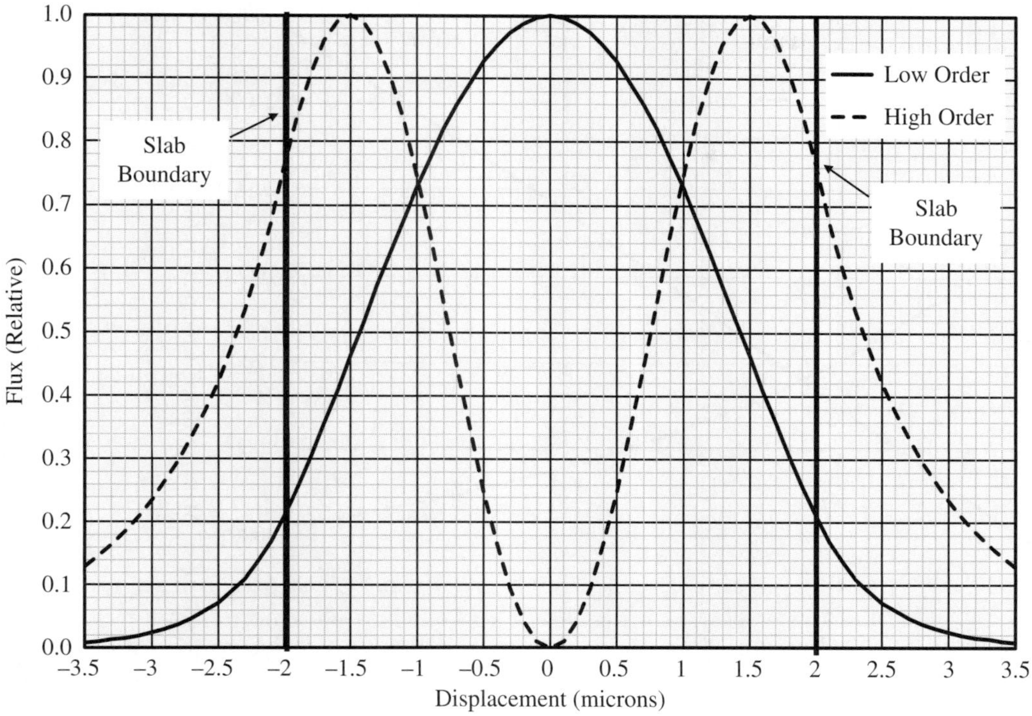

Figure 13.9 Slab waveguide (weakly guided).

to an effective index difference of about 0.0027. If this slab were representative of a 10 km fibre, then the modal dispersion would be equivalent to a propagation delay of 90 ns between the two modes. In the context of optical fibre communication, this is a significant impediment. The difference between the two polarisations is more subtle. For the low order modes, the effective index difference is approximately 3.6×10^{-6}. Again, if we consider the slab as representing a 10 km long optical fibre, this dispersion would amount to about 120 ps. Whilst a relative delay of this magnitude might not seem significant, in the context of high bit rate communications, e.g. 40 Gbits s^{-1}, it cannot be ignored. As we shall see a little later, ***polarisation mode dispersion*** is of significance in high bandwidth communications.

It would be useful to examine the modal solutions in a little more detail at this point. The two TE modes are illustrated in Figure 13.9 which also shows the boundary between the core and cladding regions.

Figure 13.9 plots the normalised flux, i.e. the square of the electric field, against displacement in the y direction. The low order mode is 'symmetric' with just one maximum, whereas the higher order mode is antisymmetric and has two maxima. In viewing Figure 13.9, it must be remembered that the ***square*** of the field is plotted, so whilst the higher order mode appears symmetric, when plotted as electric field it is nonetheless antisymmetric. One feature clearly shown in Figure 13.9 is the penetration of the wave into the cladding region. This is particularly true for the higher order mode. We will return to this theme presently.

Thus far, we have considered the dispersion arising from the difference in effective index between the two modes and the polarisation state. In addition, we must consider the ***chromatic dispersion*** produced by a variation in index with wavelength. At first, we might view this simply as a function of the changing material properties (refractive index) with wavelength. That is to say, in the absence of anomalous dispersion, we might expect the effective index of a mode to diminish with wavelength. Thus, the anticipated behaviour of the waveguide is that the propagation velocity will increase with wavelength. We must bear in mind, however that the effective index of a mode is dependent upon the core and cladding indices, n_0 and n_1. Nevertheless, as far as the properties of the two materials are concerned, the dispersion seen should be normal.

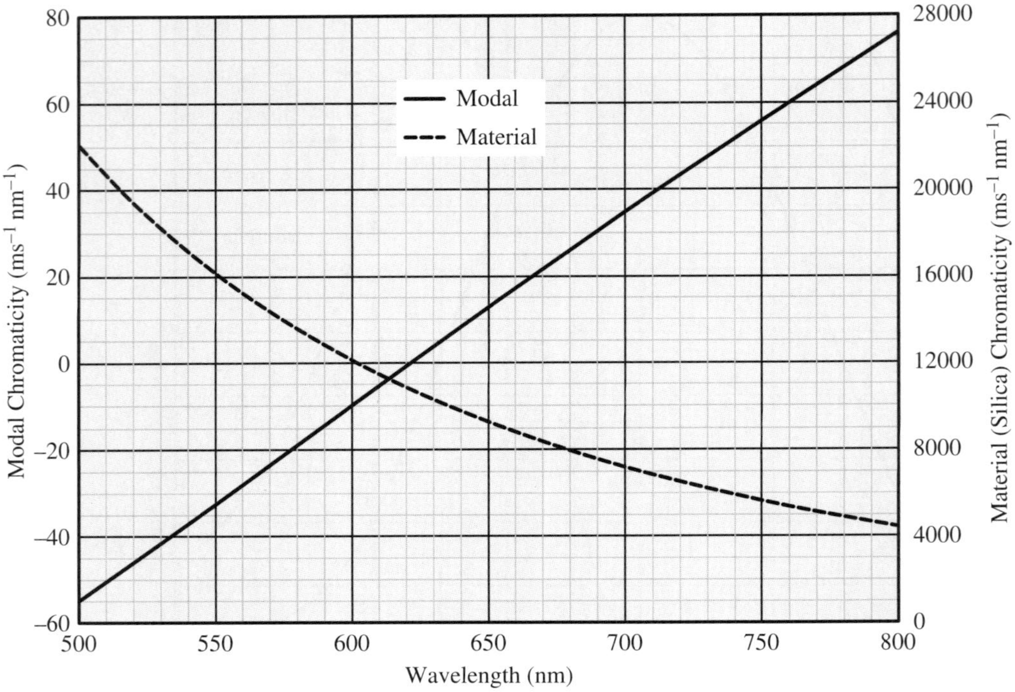

Figure 13.10 Modal chromaticity for example waveguide.

Whilst the dispersion of fibre materials, such as silica, or chalcogenide glasses, is normal, this consideration tends to be restricted to dispersion of the ***phase velocity***. As far as transmission of information in an optical fibre is concerned, it is the ***group velocity*** dispersion that is important. For silica, group velocity dispersion changes from positive to negative at around 1.2–1.3 μm. This is important for telecommunications applications.

Thus far, we have considered the impact of material dispersion. There is, however, a significant omission in the preceding narrative. The implication of Eq. (13.31) is that β, and hence the effective index, is likely to change with wavelength, even in the absence of any change in refractive index of the core and cladding. Most significantly, in some instances, the group velocity, as opposed to the phase velocity tends to ***reduce*** with wavelength, as a result of this modal effect. As such, this effect is equivalent to anomalous dispersion. This is important, as this effect may be used, particularly in single mode fibres, to adjust for the impact of material dispersion. In this way, it is possible to engineer an optical fibre that has zero chromatic dispersion at a specific wavelength. Such fibres are referred to as ***dispersion flattened*** fibres.

To illustrate this effect, we will return to our example and, specifically, the low order TE mode. This modal chromaticity depends upon wavelength. It is possible to plot the modal chromaticity, expressed in metres per second per nanometre, as a function of wavelength. This is shown in Figure 13.10 which plots the modal chromaticity between 500 and 800 nm for this waveguide.

As Figure 13.10 illustrates, the effect of modal chromaticity is quite modest. This is further illustrated by the comparison presented for a typical material chromaticity. In this case, the material properties of fused silica are presented and the chromaticity computed on the simple basis of propagation in a continuous medium. Clearly, in this instance, the impact of material effects is very much larger. However, on progression to longer wavelengths and different waveguide designs, the relative impact of material properties declines when compared to modal effects. Thus for telecoms fibres operating between 1.3 and 1.5 μm, the two effects can become comparable.

Elimination of chromatic dispersion in a fibre is useful. Dispersion in an optical fibre leads to a degradation or blurring of high frequency information, as expressed by differential propagation delay of fast optical signals in a long fibre. Overcoming chromatic dispersion permits dispersion free performance to be extended over a wider wavelength band, thus further increasing available bandwidth. In practice, the change in effective index of a mode is a complex function of the waveguide structure and material properties. As outlined previously, the effective index of a mode is dependent upon both the core index, n_0 and the cladding index, n_1. Both these parameters change with wavelength. Despite this, it is possible to engineer fibres and waveguides to control chromatic dispersion.

13.3.3 Strong and Weakly Guiding Structures

This cladding penetration seen in Figure 13.9 is a feature of a ***weakly guiding*** structure, as is the case for that particular structure. Weakly guiding structures are associated with a low refractive index contrast between the two layers, as is the case of the previous example, where the contrast is 0.005. For ***strongly guiding*** structures, the index contrast is much higher and penetration into the cladding layer much less marked. Of course, the simplest possible waveguide or fibre structure, notably a strand of glass, is, by definition, a strongly guiding structure. The core/cladding interface is marked by the glass air boundary. Any penetration into the 'cladding' in this instance is represented as propagation in air. This wave is referred to as an ***evanescent wave*** and, if any high index material is brought very close to this interface, radiation from this evanescent zone may be directly sampled. This phenomenon is known as frustrated internal reflection.

We now introduce a strongly guiding structure by selecting a core index of 1.76 (e.g. Al_2O_3) and a cladding index of 1.45 (e.g. silica). In this instance, the waveguide core is 0.6 μm thick and the operating wavelength is 1.55 μm. At this wavelength, the structure is a single mode waveguide. The effective index of the single mode at 1.55 μm for this waveguide is 1.6287. The mode structure is shown in Figure 13.11.

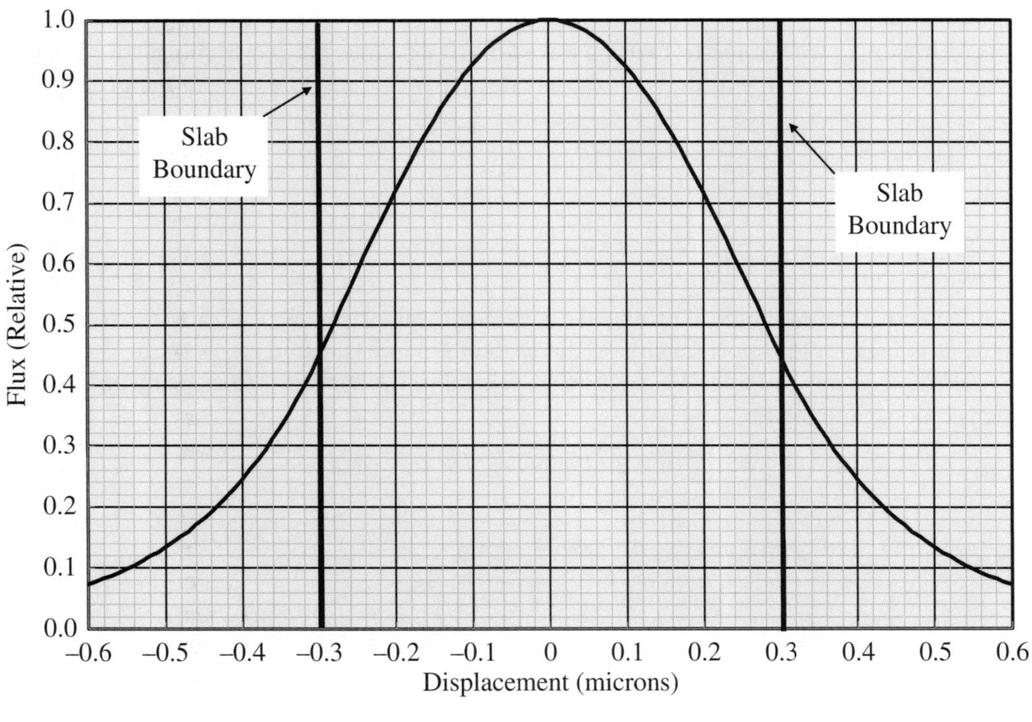

Figure 13.11 Strongly guided waveguide.

The penetration into the cladding is much less marked when compared to the weakly guiding structure. The excursion into the cladding, in this instance, is only of the order of 0.3 μ.

13.4 Single Mode Optical Fibres

13.4.1 Basic Analysis

At this point we leave behind our illustrative discussion of slab modes and turn to the analysis of more representative structures, in this case, optical fibres and, in particular, single mode optical fibres. Such fibres are of exceptional practical interest in the configuration of long haul telecommunications networks. The analysis of these waveguides draws heavily on the preceding discussion, as many aspects of their behaviour may be understood in terms of the rather simpler slab analysis. Single mode optical fibres are, in practice, weakly guiding structures, with an index contrast of 0.005 being typical. Although we must now extend our analysis to an extra dimension, the cylindrical symmetry of an optical fibre does somewhat simplify the analysis. The geometry of an optical fibre is illustrated in Figure 13.12.

In the model the fibre is characterised by its core radius, a, and the refractive index of its core (n_0) and its cladding (n_1). Analysis proceeds in the same way, as per the slab analysis, in that we are seeking separate solutions for core and cladding regions, reconciling these solutions by boundary conditions at the interface. As for the slab waveguide both core and cladding solutions must be defined by the same propagation constant, β, which describes propagation in the z direction. An additional simplification may be made under the assumption that we are interested in single mode propagation and that the single mode is also radially symmetric. The solutions we are looking for thus depend only upon r, the radial distance from the fibre core and the sinusoidal propagation in the z direction. Thus, we might expect that the solutions in both core and cladding are of the form:

$$\mathbf{E}(r,z,t) = \mathbf{E}_0 F(r) e^{i(\beta z - \omega t)} \tag{13.41a}$$

To determine the form of $F(r)$, we need to re-cast the wave Eq. (13.27), in terms of cylindrical co-ordinates. Furthermore, we may describe both modes in terms of the effective index, n_{eff}, rather than the propagation constant, β:

$$\text{core:}\ \frac{1}{r}\frac{\partial}{\partial r}\left(r\frac{\partial F_{core}(r)}{\partial r}\right) = k_0^2(n_{eff}^2 - n_0^2)F_{core}(r) \tag{13.41b}$$

$$\text{cladding:}\ \frac{1}{r}\frac{\partial}{\partial r}\left(r\frac{\partial F_{clad}(r)}{\partial r}\right) = k_0^2(n_{eff}^2 - n_1^2)F_{clad}(r) \tag{13.41c}$$

It is clear that the effective index should lie somewhere between the extremes of n_0 and n_1. Hence, the effective 'eigenvalue' or constant of proportionality on the RHS of Eq. (13.41a) is negative in the case of the core and positive in the case of the cladding. We can further generalise the above expressions by introducing the following dimensionless parameters:

$$U^2 = a^2k_0^2(n_0^2 - n_{eff}^2) \quad \text{and} \quad W^2 = a^2k_0^2(n_{eff}^2 - n_1^2)\ \textit{a is the fibre core radius} \tag{13.42a}$$

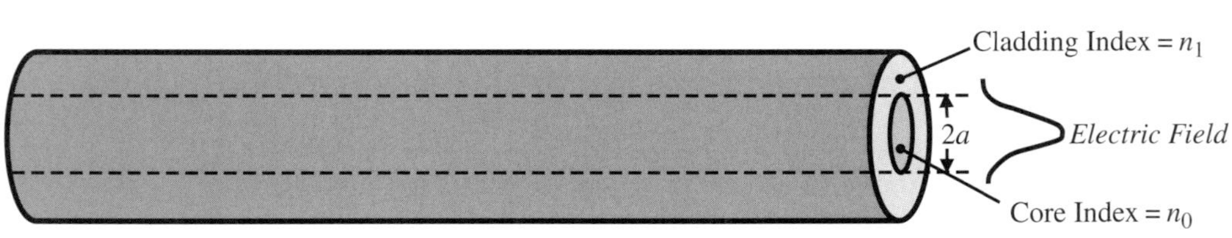

Figure 13.12 Optical fibre model.

At this point we also introduce the normalised frequency parameter, V_a, which, by analogy with the slab mode treatment, is given by:

$$V_a^2 = a^2 k_0^2 (n_0^2 - n_1^2) \quad \text{and} \quad V_a^2 = U^2 + W^2 \tag{13.42b}$$

The solution for the core yields a Bessel function of the first kind. For the lowest order mode, this is a Bessel function of order zero, $J_0(r)$. In the case of the cladding, the solution is represented by a modified Bessel function of the second kind and zeroth order (for the lowest order mode), $K_0(r)$. These solutions may be written as:

$$\text{core: } F_{core}(r) = A_{core} J_0 \left(U\frac{r}{a} \right); \text{ cladding: } F_{clad}(r) = A_{clad} K_0 \left(W\frac{r}{a} \right) \tag{13.43}$$

To determine the effective index, it is necessary to apply the boundary conditions. In the case of a weakly guiding structure, such as a single mode fibre, we may use the weakly approximation, i.e. $n_0 - n_1 << n_0$. In this case we may apply the boundary conditions as:

$$F_{core}(a) = F_{clad}(a) \quad \text{and} \quad F'_{core}(a) = F'_{clad}(a) \tag{13.44}$$

As the derivative of the zeroth order Bessel functions are equal to the first order functions, the boundary conditions may be rewritten as:

$$U\frac{J_1(U)}{J_0(U)} = W\frac{K_1(W)}{K_0(W)} \tag{13.45}$$

Equation (13.45), taken together with Eqs. (13.42a) and (13.42b) is sufficient to unambiguously determine both U and W in terms of V_a. For the fibre to support just one single mode, V_a must be less than the critical value, V_c, which is 2.405 for this cylindrical geometry. We can derive the cut-off wavelength, λ_c, from this:

$$\lambda_c = \frac{2\pi}{2.405}\sqrt{(n_0^2 - n_1^2)}a \quad \text{or} \quad \lambda_c = 2.613\sqrt{(n_0^2 - n_1^2)}a \tag{13.46}$$

Worked Example 13.3 ***Single Mode Fibre***

A single mode fibre is to work at 1.55 μm. It consists of a core with a refractive index of 1.46 and a cladding index of 1.455. The core diameter is 8 μm. First, we would like to calculate the cut-off wavelength. Thereafter, we will attempt to solve Eq. (13.45) numerically, to compute the electric field distribution of the single mode at 1.55 μm.

First, the cut-off wavelength is straightforward to calculate from Eq. (13.46):

$$\lambda_c = 2.613\sqrt{(1.46^2 - 1.455^2)} \times 4 = 1.26$$

The cut-off wavelength of the fibre is 1.26 μm.

As the cut-off wavelength is 1.26 μm, the 1.55 μm operating wavelength is clearly within the single mode regime. At this wavelength, the value of V_a is 1.958. We can use this to (numerically) calculate the values of U and W from the boundary conditions. From these calculations, we determine that U is equal to 1.513 and V is 1.241. We now have sufficient information to plot the mode distribution in the fibre. This is shown in Figure 13.13, which shows the distribution of the flux associated with the mode, i.e. the square of the amplitude.

As with the slab modes, there is significant penetration of the mode into the cladding. In addition to the modal distribution itself, Figure 13.13 shows the best fit Gaussian curve. The flux distribution is modelled in terms of a beam waist, w_0, as used in Gaussian beam propagation analysis:

$$\Phi = \Phi_0 e^{-\frac{r^2}{2w_0^2}} \tag{13.47}$$

In this particular instance, the beam waist size is 4.98 μm, close to that of the core size. Whilst the distribution does not correspond exactly to the fitted Gaussian, use of a Gaussian fit to model the distribution of a fibre mode is useful, as will be seen later. This beam distribution can be used to model propagation of the

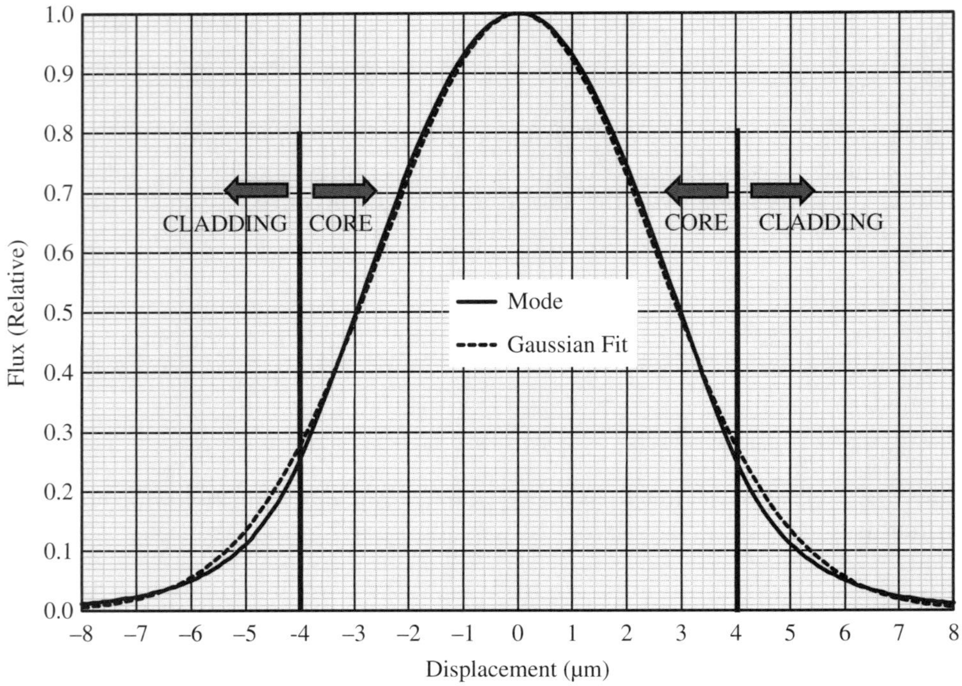

Figure 13.13 Flux distribution in single mode fibre.

beam as it emerges from the fibre, in line with the Gaussian beam modelling previously described. In addition, Gaussian analysis can be used to analyse the coupling of (laser) light into a single mode fibre. The advantage of Gaussian analysis lies in its utility in facilitating practical calculations, rather than its fidelity in replicating the modal distribution. In general, the real distribution has considerably more flux in the wings of the distribution when set against the comparable Gaussian.

13.4.2 Generic Analysis of Single Mode Fibres

Following on from an analysis of the conditions imposed by Eqs. (13.42a), (13.42b), and (13.45), we can provide a generic analysis of the dependence of U and W on V. Furthermore, if we also fit the resultant mode distribution to the Gaussian illustrated in Figure 13.13, we can quantify the physical mode size in terms of the normalised frequency parameter, V_a. Figure 13.14 shows a plot of U and W against V for values of V lying between 1 and 4.

Figure 13.15 shows a plot of characteristic Gaussian beam size, w_0, against the normalised frequency parameter. The beam size is expressed as a fraction of the core radius size, a.

Moreover, we can fit the plots in Figures 13.14 and 13.15 to a useful empirical curve in order to quantify all sets of parameters against the normalised frequency parameter. For the relationship between W and V, the following useful empirical function applies:

$$W = -0.6237 + 1.0471V + 2.8057e^{-V/0.345} - 0.7353e^{-V/1.475} \tag{13.48}$$

The value of U is then simply derived from Eqs. (13.42a) and (13.42b). We can further quantify the Gaussian beam size and express it as a ratio of the core radius, a:

$$\frac{w_0}{a} = 1.0906 + 6.6109V^{-1} - 5.5363V^{-0.8} \tag{13.49}$$

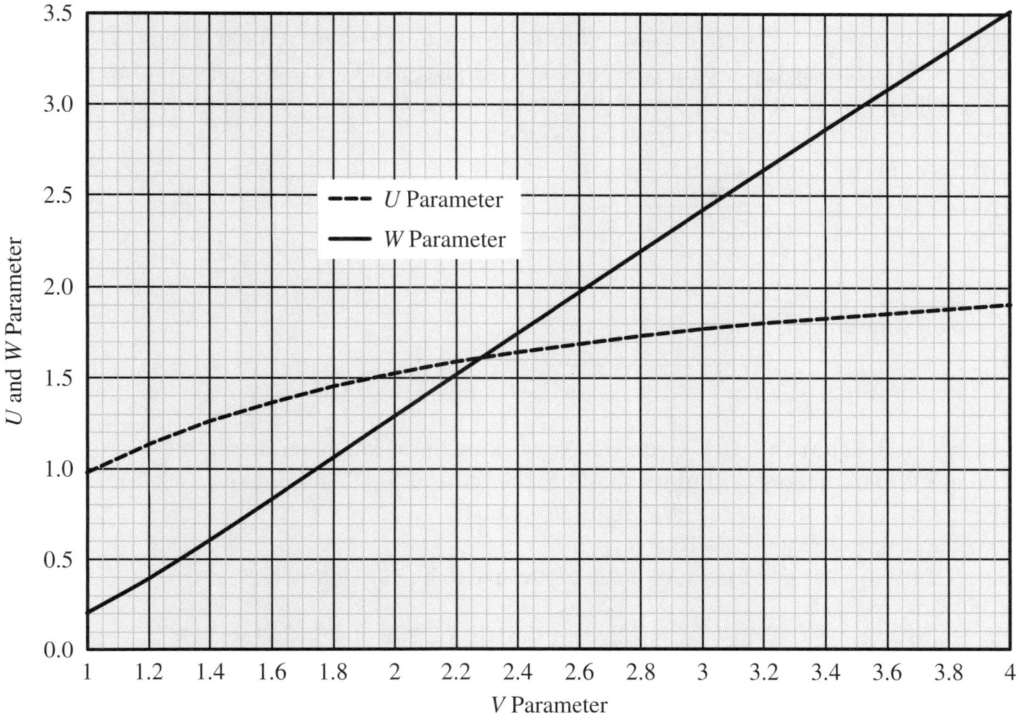

Figure 13.14 Dependence of U and W parameters on normalised frequency parameter, V.

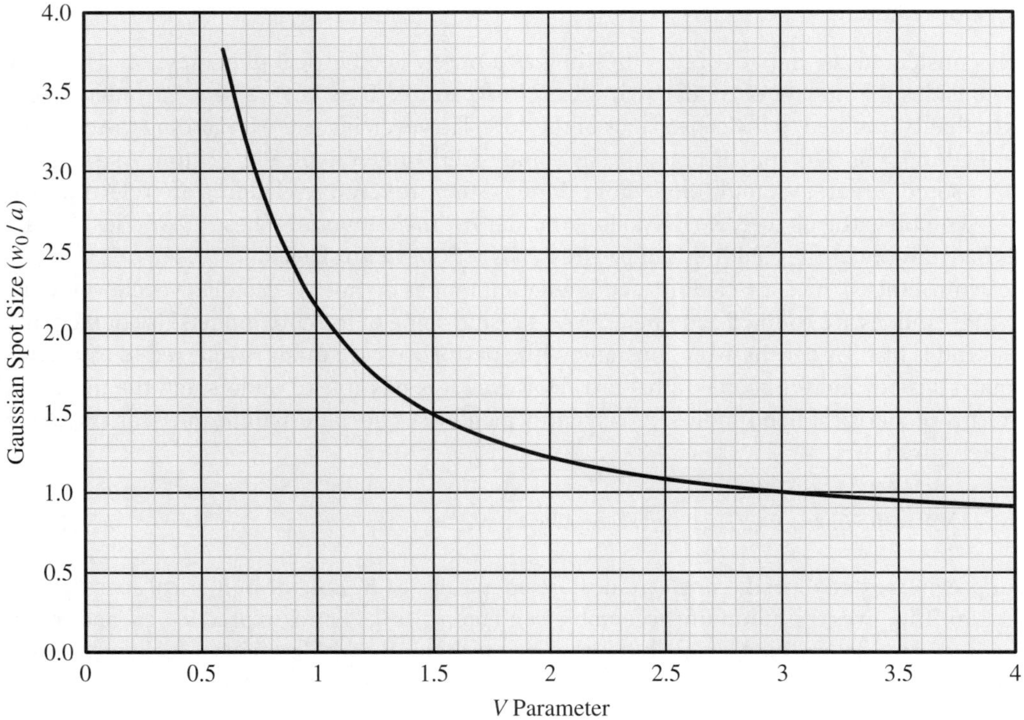

Figure 13.15 Gaussian beam size, w_0, vs, normalised frequency parameter, V.

Worked Example 13.4 ***Single Mode Fibre Mode Size***
A single mode fibre is to operate at 1.52 μm. Its core radius is 4.5 μm and the core refractive index is 1.46 and the cladding index is 1.455. Calculate the V parameter at this wavelength and calculate the cut-off wavelength for single mode propagation. In addition, we would like to know the U and W values to define the mode distribution and we also wish to determine the Gaussian Mode size.

First, we wish to determine V_a – from Eq. (13.44):

$$V_a = \frac{2\pi}{\lambda}\sqrt{(n_0^2 - n_1^2)}a = \frac{2\pi}{1.52}\sqrt{(1.46^2 - 1.455^2)}4.5 = 2.246$$

The V_a parameter is 2.246

To calculate the cut-off wavelength, λ_c we refer to Eq. (13.46):

$$\lambda_c = 2.613\sqrt{(n_0^2 - n_1^2)}a = 2.613 \times \sqrt{(1.46^2 - 1.455^2)} \times 4.5 = 1.42$$

The cut-off wavelength is 1.42 μm and, at 1.52 μm we are working in the single mode regime.

To calculate the U and W parameters we simply substitute the value of V_a (2.246) into Eq. (13.48). This gives W as 1.5716. Applying Eqs. (13.42a) and (13.42b), yields U – 1.6042.

U is equal to 1.6042 and W is equal to 1.5716.

Finally, we use Eq. (13.49) to calculate the ratio of the mode size to be core diameter:

$$\frac{w_0}{a} = 1.0906 + 6.6109V^{-1} - 5.5363V^{-0.8} = 1.0906 + 6.6109 \times 2.246^{-1} - 5.5363 \times 2.246^{-0.8}$$

This gives the ratio as 1.1362 and the Gaussian mode size is 5.11 μm

The Gaussian mode size is 5.11 μm.

13.4.3 Impact of Fibre Bending

We had previously considered the impact of fibre bending as a purely geometrical phenomenon. However, it is necessary to consider the impact of modal analysis on this picture. Thus far, we have considered only propagation of light that is confined to the central axis of the core. However, instead of regarding the cladding as a semi-infinite medium, the cladding, in practice has some real diameter, typically 125 μm. As such, the cladding behaves, to some extent, as a multimode fibre. Modes supported in the cladding are referred to as ***cladding modes***. To complete the picture, we must also consider ***radiation modes***. These are modes that are not confined to either the core or the cladding and progressively leak out into space.

Under normal circumstances, the core, cladding, and radiation modes do not interact. However, where the fibre is bent, then these modes are coupled in some way. As a result, some of the core mode(s) is coupled into the radiation mode and is lost. Under these circumstances, the perturbed core mode is referred to as a ***leaky*** mode. Detailed analysis is beyond the scope of this text. However, it is possible to understand the impact of fibre bending in terms of a critical bend radius, r_c, beyond which losses become significant. This threshold value is approximately given by:

$$r_c = \frac{3n_0^2}{4\pi NA^3}\lambda \;\; NA \textit{ is the fibre numerical aperture where}:\; NA = \sqrt{n_0^2 - n_1^2} \qquad (13.50)$$

In the case of our previous example of the single mode fibre operating at 1.52 μm, the numerical aperture was 0.1207 and n_0 was 1.46. This gives the minimum acceptable bend radius as 0.44 mm. In practice, the minimum acceptable bend radius is not determined by loss but rather by mechanical considerations. Mechanical stress in the fibre creates undesirable polarisation sensitivity and may also lead to mechanical damage or failure. As an approximate rule, the minimum bend radius should not be less than 50–100 times the outer fibre diameter. For a 125 μm single mode fibre (including cladding) the bend radius should not be less than 6–12 mm.

13.5 Optical Fibre Materials

13.5.1 General

In the majority of cases, the implementation of optical fibres is based on fused silica as the material of choice. This allows transmission from the ultraviolet to the near infrared (~250–2000 nm). To modify the refractive index profile and to create cladding layers, small quantities of dopants such as germanium, arsenic, and fluorine may be added. Otherwise, polymer fibres offer a low-cost solution with greater restrictions on the passband. Ultraviolet transmission is markedly inferior to that of conventional glass with transmission below 400 nm unfavourable. Molecular vibrations also restrict transmission in the infr-red, although this can be ameliorated by substituting fluorine for hydrogen in hydrocarbon polymers. For applications in the mid infrared and beyond, chalcogenide glasses are used. These materials are deployed in specific niche applications, as these glasses are difficult to process and handle. Nevertheless, they do find significant application in infrared laser beam delivery and in chemical sensors.

13.5.2 Attenuation

Long-haul telecommunications has been the defining commercial application for the development of optical fibres. As alluded to in the introduction, it was optical transmission, or lack of, that presented the most significant barrier to the adoption of the technology. Whilst at the outset, the advantage of information transmission over an optical frequency carrier was clear, light absorption over kilometres of fibre seemed to present an impenetrable obstacle. At the time, the benchmark comparison was with conventional coaxial electric capable which carry data with an attenuation of a few dB/km.

Attenuation in silica arises from three potential mechanisms. Firstly there is Rayleigh or molecular scattering, by which a disordered material with a significant polarisability must inevitably re-radiate or scatter. In line with Rayleigh scattering, this is a short wavelength phenomenon, with a scattering cross section proportional to the inverse fourth power of the wavelength. Secondly, incorporation of metal ion impurities causes significant absorption particularly in the visible. The distinctive intense colouration of gemstones and the production of coloured glasses is a testament to this. Finally, the incorporation of small quantities of water causes significant absorption in the infrared. Since, by virtue of the underlying physics, Rayleigh scattering is a fundamental attribute of a material with a refractive index greater than one, it is impossible to ameliorate this source of attenuation. However, both ion and water absorption can be substantially reduced by advances in materials processing. It was the application of semiconductor processing technologies to fibre production that proved decisive. Silicon with impurity levels around a few parts per billion or less was demanded and delivered with advanced purification techniques, such as zone refining. Typically, the refined silicon might be converted into silane (SiH_4) and deposited as silica by a thermal chemical vapour deposition process.

In modern fibre applications, attenuation is substantially dictated by Rayleigh scattering, lattice and residual water absorption. The former dominates at shorter wavelength and the latter increases towards the near/mid infrared. As a result, there exists a wavelength region at which the attenuation is at a minimum. This is around 1.55 μm, where the attenuation is approximately 0.2 dB km^{-1}. To understand the significance of this, one might imagine an optical fibre link 100 km long. The attenuation amounts to 20 dB, equivalent to an attenuation by a factor of 100 over this distance. Or alternatively, the flux is reduced by a factor of two for every 15 km of propagation along the fibre. Figure 13.16 shows the typical attenuation of a silica optical fibre, as a function of wavelength. The minimum absorption region is referred to as the '***telecoms window***'.

Very early fibre communication systems used a wavelength of around 850 nm, based largely on available laser sources. Naturally, this wavelength is not close to the minimum attenuation of silica fibres. However, the next generation of systems operated at about 1.3 μm, the first 'window', giving attenuation of around 0.3 dB km^{-1}. Current systems almost entirely use the second telecom window at 1.55 μm, close to the absolute minimum attenuation of 0.2 dB km^{-1}.

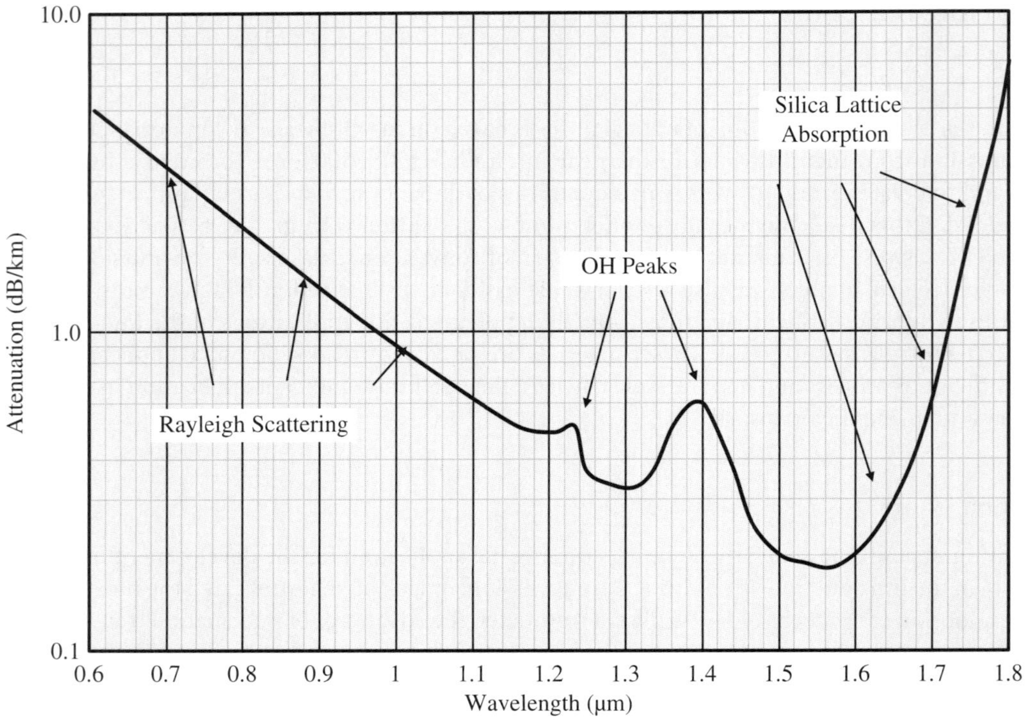

Figure 13.16 Silica fibre attenuation.

13.5.3 Fibre Dispersion

Group velocity dispersion is a significant issue in optical fibre communications. For a given group velocity dispersion, the information bandwidth that the fibre can transmit is inversely proportional to the fibre length and the bandwidth of the laser source. Of course, in the case of a perfectly efficient optical carrier wave, the laser bandwidth must at least match the information bandwidth. Therefore, fundamentally, group velocity dispersion limits the data carrying capacity of an optical fibre. Historically, the first 1.3 μm window for silica fibres corresponded with the point at which the group velocity dispersion is zero for silica. When transmission technology migrated to the 1.5–1.6 μm region, movement away from the zero dispersion point was compensated by redesign of the index profile of the fibre, particularly in the cladding layer. As per the earlier discussion on modal chromaticity, this had the effect of adjusting the total group dispersion to compensate for the change in wavelength. Material group velocity dispersion for silica is shown in Figure 13.17.

13.6 Coupling of Light into Fibres

13.6.1 General

In terms of coupling a conventional light source into an optical fibre, this is governed by the étendue of the fibre and the radiance of the source. The étendue is the product of the fibre solid angle and area. In fact, in a multimode fibre the étendue is directly proportional to the number of modes supported by the fibre. If one imagines the single mode emerging from a fibre is characterised by some Gaussian width, w_0, we can calculate the étendue, G. The principle is based on application of the overlap integral discussed in the next section to

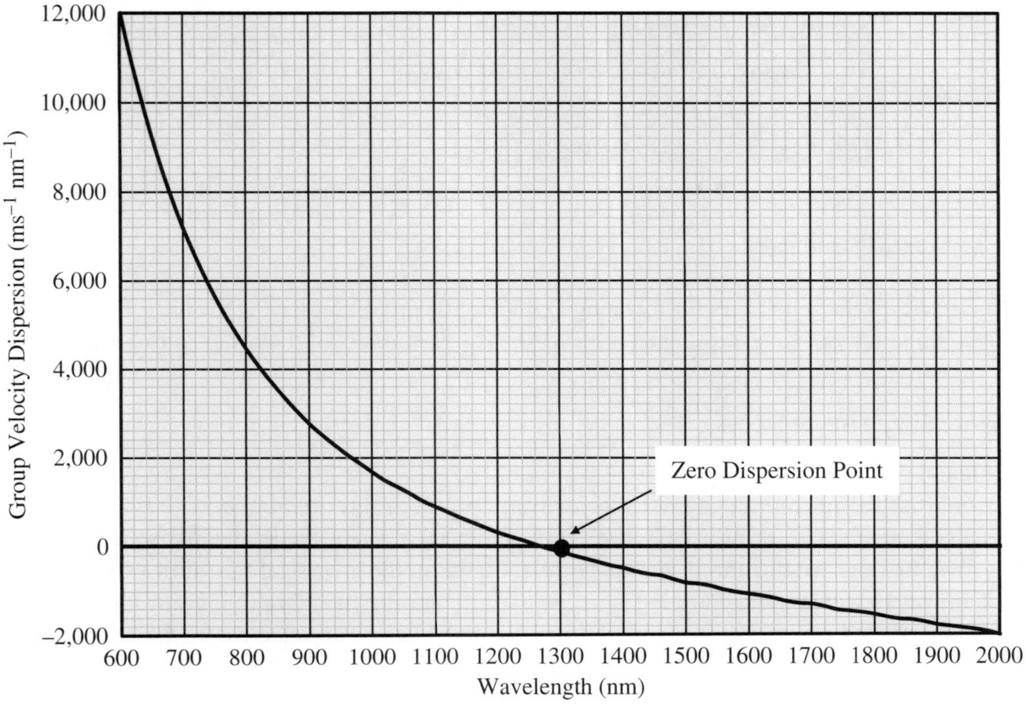

Figure 13.17 Group velocity dispersion in silica.

an input of even radiance.

$$G = \frac{\pi w_0^2}{2} \times \pi NA_0^2 \quad NA_0 \textit{ is the far field numerical aperture}$$

However, the numerical aperture is a function of the beam width, w_0:

$$G = \frac{\pi w_0^2}{2} \times \pi \times \frac{\lambda^2}{\pi^2 w_0^2} = \frac{\lambda^2}{2} \text{ since } NA = \frac{\lambda}{\pi w_0} \tag{13.51}$$

Assuming the étendue of each mode in a multimode fibre is the same as in Eq. (13.51), the number of modes supported in a multimode fibre should be equivalent to the étendue of the fibre divided by the étendue of a single mode. It is clear, therefore, that the number of modes supported is proportional to the square of the reduced frequency parameter, V_a. If N is the number of modes supported, then the following applies:

$$N = \frac{V_a^2}{2} \qquad \text{for } V_a >> 1 \tag{13.52}$$

If we now consider the coupling of light into a fibre from a source whose numerical aperture is matched to that of the fibre, then, for a multimode fibre, where N is large, a geometrical description suffices. This is illustrated in Figure 13.18.

In Figure 13.18, a small defocus, Δz, is applied. If the defocus is sufficiently large then a proportion of the light will fall outside the fibre core and not be collected. Using simple geometry, it is possible to calculate the full width half maximum focal range, Δf. This focal range is equal to the distance between the two points at which the geometrical coupling is reduced to one half. This range is given by:

$$\Delta f = \frac{2\sqrt{2}}{NA} a \quad a \textit{ is the fibre radius} \tag{13.53}$$

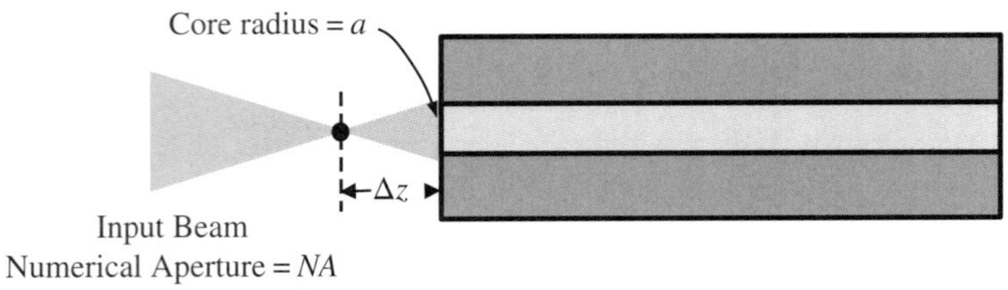

Figure 13.18 Coupling into a multimode fibre.

Equation (13.52) gives the axial sensitivity for focusing and for a 125 μm diameter, 0.2*NA* fibre, this is about 0.9 mm. Naturally the sensitivity of the coupling to transverse displacement is of the order of the fibre size.

13.6.2 Coupling into Single Mode Fibres

13.6.2.1 Overlap Integral

It must be emphasised that the preceding geometrical analysis applies to multimode fibres. If one considers coupling of a Gaussian beam into a fibre, intuitively, the depth of focus is given by the Rayleigh distance which is proportional to the inverse square of the numerical aperture, rather than the inverse of the numerical aperture. In fact, the convenient way to view modal coupling is to consider the modes in a fibre (including cladding and radiation modes) as comprising an orthonormal set of functions. Any arbitrary electric field distribution at the input of the fibre may be expressed as a linear combination of these modes. On the assumption that these modes form an orthonormal set, then the coupling efficiency is determined by the overlap integral between the input field distribution and the specific mode distribution. If the electric field distribution is across the input of the fibre is represented by $E(x,y)$ and the modal distribution by $M(x,y)$, the power coupling coefficient, C, is given by:

$$C = \frac{\left[\int E(x,y)M(x,y)dxdy\right]\left[\int E(x,y)M(x,y)dxdy\right]\times}{\left[\int E(x,y)E\times(x,y)dxdy\right]\left[\int M(x,y)M\times(x,y)dxdy\right]} \tag{13.54}$$

Although Eq. (13.54) is of general application to all fibre modes, it is particularly useful for determining coupling into single mode fibres. Furthermore, this analysis is made considerably more tractable by assuming a Gaussian beam profile. This provides further justification for our earlier attempts to fit a single mode profile to a Gaussian distribution.

13.6.2.2 Coupling of Gaussian Beams into Single Mode Fibres

By substituting two Gaussian profiles into Eq. (13.54) for the input beam and the mode distribution, it is possible to produce a simple analytical expression for the coupling coefficient, C. The Gaussian beam radius, w, is defined as in Eq. (13.47) and we define the input beam radius as w_b and the mode radius as w_m.

$$C = \frac{4}{[w_b/w_m + w_m/w_b]^2} \tag{13.55}$$

Not surprisingly the expression for coupling is symmetric with respect to the ratio of the two radii. Quite obviously, if the radii are equal, then the coupling is 100%. Where the ratio is 2 : 1, the coupling falls to 64%. Equation (13.54) only applies for perfect alignment, where the Gaussian beam waist is perfectly aligned centrally to the core of the fibre. There are three ways in which this alignment may be compromised and we will deal with each successively:

1. Axial misalignment in z (fibre not at beam waist)
2. Lateral offset
3. Tilt (axis of Gaussian beam not parallel to fibre axis)

Figure 13.19 Fibre coupling and offset beam.

In considering axial misalignment, we must be careful in applying Eq. (13.54). A full description of a Gaussian beam (away from its waist) must include the imaginary (phase) component associated with wavefront curvature. As such, the lateral dependence of the Gaussian beam amplitude is given by:

$$A = A_0 e^{-\frac{r^2}{w^2(z)} + i\frac{k_0 r^2}{2R(z)}} \tag{13.56}$$

w(z) is the Gaussian beam size and R(z) is the wavefront radius. Both are described as a function of z, the distance from the beam waist.

If we now assume that the mode size of the fibre is the same as that of the beam waist, w_0, then we may perform the overlap integral to calculate the coupling:

$$C = \frac{4}{[w(z)/w_0 + w_0/w(z)]^2 + (w_0^2 w^2(z) k_0^2)/4R^2(z)}$$

The above expression may be simplified by substituting for $w(z)$ and $R(z)$ and expressing the coupling coefficient entirely in terms of the Rayleigh distance, Z_R and the axial displacement, z:

$$C = \frac{4Z_R^2}{4Z_R^2 + z^2} \tag{13.57}$$

In the case of single mode coupling, it is the Rayleigh distance that impacts the depth of focus. Since the Rayleigh distance is inversely proportional to the square of the numerical aperture, this conclusion stands in contrast to that for multimode coupling, as exemplified in Eq. (13.53).

Having dealt with axial displacement, we now wish to consider the impact of lateral displacement on coupling efficiency. The geometry is shown in Figure 13.19, showing a small lateral offset of Δx. It is assumed that the laser beam size, w_0, is the same as that of the fibre and that the beam waist is at the fibre input.

It is straightforward to perform the offset coupling calculation using the overlap integral in Eq. (13.54) and this gives:

$$C = e^{-\frac{(\Delta x)^2}{w_0^2}} \tag{13.58}$$

A similar calculation may be performed for axial tilt of the incoming beam. If we define the axial tilt angle in terms of a numerical aperture offset, ΔNA and if the effective numerical aperture of the mode is NA_0 then:

$$C = e^{-\frac{(\Delta NA)^2}{NA_0^2}} \text{ where } NA_0 = \frac{\lambda}{\pi w_0} \tag{13.59}$$

Worked Example 13.5 ***Single Mode Fibre Coupling***

A laser beam with a wavelength of 0.85 μm is focused into a single mode optical fibre. However, the size of the beam is mismatched. The focused spot has a Gaussian beam radius of 2 μm, whereas the size of the mode is 5 μm. What is the coupling coefficient for this laser beam? Following this experiment, the laser beam is re-sized to match the fibre mode size. We wish to determine the alignment sensitivity of this new arrangement. What misalignments are required to produce a 20% reduction in the coupling over the optimum in the following cases:

1. For axial misalignment (in z)

2. For lateral misalignment (in x or y)
3. For tilt misalignment (in θ or NA).

In the first case, for the mismatched fibre, the coupling coefficient is given by Eq. (13.55):

$$C = \frac{4}{[w_b/w_m + w_m/w_b]^2} = \frac{4}{[2/5 + 5/2]^2} = \frac{4}{[2.9]^2} = 0.476$$

The coupling coefficient is 47.6%.

Coupling loss is often expressed in decibels or 10×log10(C). In this case, ***the loss is 3.2 dB.***

i) Axial misalignment

With a loss of 20% (~1 dB), we know that the coupling coefficient is 0.8. We can calculate the axial displacement required to produce this coupling loss from Eq. (13.57):

$$C = \frac{1}{1 + (z/2Z_R)^2} = 0.8 \qquad \text{Therefore:} \ (z/2Z_R)^2 = 0.25 \ \text{and} \ (z/Z_R) = 1$$

We know that the (matched) mode size is 5 μm and that the wavelength is 0.85 μm. Accordingly, the Rayleigh distance, Z_R is given by:

$$Z_R = \frac{\pi w_0^2}{\lambda} = \frac{\pi 5^2}{0.85} = 92.4$$

The axial misalignment sensitivity for 20% coupling loss is 92.4 μm.

ii) Lateral misalignment

The coupling with respect to lateral alignment is given by Eq. (13.58) and setting the coupling to 0.8:

$$C = e^{-\frac{(\Delta x)^2}{w_0^2}} = 0.8 \ \text{and} \ \frac{(\Delta x)^2}{w_0^2} = 0.22 \ \text{or} \ \Delta x = 0.47 \times w_0.$$

We know w_0 is 5 μm giving Δx as 2.36 μm.

The lateral misalignment sensitivity is 2.36 μm.

iii) Angular misalignment

To determine the angular sensitivity of the coupling we first need to calculate the fibre numerical aperture, NA_0:

$$NA_0 = \frac{\lambda}{\pi w_0} = \frac{0.85}{\pi 5} = 0.054$$

To determine the sensitivity, we simply apply Eq. (13.59):

$$C = e^{-\frac{(\Delta NA)^2}{NA_0^2}} = 0.8 \ \text{and} \ \Delta NA = 0.47 \times NA_0 \ \text{or} \ 0.47 \times 0.054, \text{giving} \ \Delta NA \ \text{as} \ 0.0256.$$

The sensitivity in numerical aperture is 0.0256 corresponding to an ***angular sensitivity of 1.46°***.

13.7 Fibre Splicing and Connection

The previous analysis clearly articulated the sensitivity of fibres, particularly single mode fibres, to misalignment. This is especially troublesome where fibres are to be joined or connected. Lateral displacements are particularly problematical, as the previous exercise demonstrated, with lateral sensitivities of a few microns.

Before fibres can be spliced or connected they must be stripped of any protective coating. The optical fibre itself is invariably coated with a polymer formulation. This serves as a general protective coating and as a mechanical inhibitor to crack propagation. In addition, there are usually other protective sleeves or, in some circumstances, armouring. After stripping, the fibre ends are precisely cleaved and, if necessary, polished, to create perpendicular faces. Splicing is usually accomplished by a fusion process employing an electric arc. The ends of the fibre are precisely aligned in a precision jig and the fusion process is enacted. The fusion process needs to be very well controlled to avoid inter-mixing of the core and cladding regions.

As well as joining by splicing, fibres may be connected using mechanical connectors. As with splicing, the fibres must first be stripped and cleaved. Broadly, the two fibres to be connected are precisely and centrally located within a cylinder and cylindrical socket. Both cylinder and socket are manufactured to high precision and the fit of the cylinder into the mating socket is exceptionally tight. In this way, the lateral alignment tolerance can be maintained. Fresnel reflection losses at the fibre – air interface can cause undesirable losses. In this case, interstices within the connector may be filled with a refractive index matching fluid or gel to reduce these losses.

Another way of improving coupling between fibres is the introduction of 'lensed' fibres. A lens is introduced at the end of the fibre by a fusion or other process. This can be used to create a narrow collimated beam at the exit of the fibre, effectively increasing the Gaussian beam width and reducing sensitivity to misalignment.

13.8 Fibre Splitters, Combiners, and Couplers

There are occasions when light from one fibre needs to be split into more than one channel. This task is carried out by a ***fibre splitter***. Conversely, it is sometimes useful to combine light from more than one fibre into a single fibre; this is done by a ***combiner.*** Alternatively, light from two or more fibres might need to be combined into two or more outputs using a ***fibre coupler***. Coupling between different fibres or waveguides is either achieved by local fusing of fibre cores or by the fabrication of a specially designed waveguide structure. For the most part, these devices are ***passive***, where the configuration of optical inputs and outputs is fixed. However, by manipulating the refractive properties of waveguide structures, either thermally, or electro-optically, ***active*** devices may be created. As such, it is possible to control or switch the light path from one fibre to another. This is illustrated in Figures 13.20a–c.

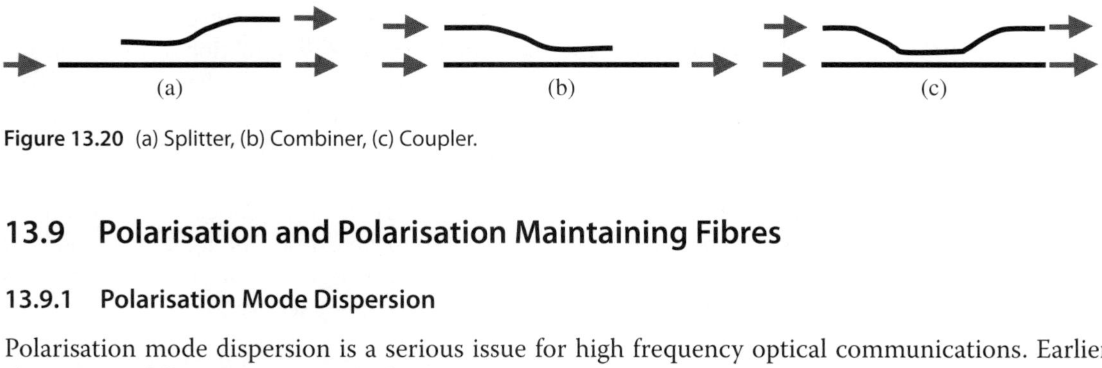

Figure 13.20 (a) Splitter, (b) Combiner, (c) Coupler.

13.9 Polarisation and Polarisation Maintaining Fibres

13.9.1 Polarisation Mode Dispersion

Polarisation mode dispersion is a serious issue for high frequency optical communications. Earlier we saw the impact of fibre geometry on polarisation mode dispersion. Strictly, this effect is dependent upon the existence of geometrical asymmetries in a fibre or waveguide. For a standard cylindrical optical fibre, there is no such asymmetry, so, at least nominally, modes associated with different polarisations are entirely degenerate. According to this reasoning, a standard optical fibre should not be susceptible to polarisation effects. However, for long fibre lengths small random effects, such as strain due to bending, manufacturing defects, inhomogeneity, and the presence of electric and magnetic fields conspire to break this degeneracy in a chaotic way. Therefore, in practice, an optical fibre acts as a series of waveplates of random strength and orientation. Hence, a polarised beam launched into an optical fibre will become progressively more randomly polarised as it progresses along the fibre. Such random fluctuations in polarisation form an important source of noise in very high speed optical fibre communications.

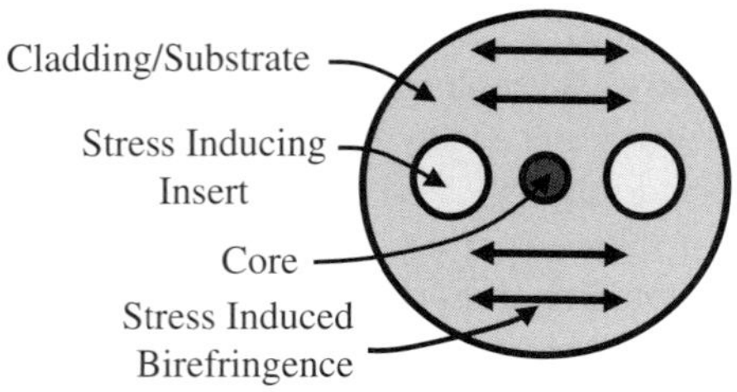

Figure 13.21 Polarisation maintaining fibre preform.

13.9.2 Polarisation Maintaining Fibre

In some applications, particularly in the delivery of polarised light to (micro) optical devices, it is necessary to maintain some particular polarisation orientation. As discussed previously, standard optical fibre has a propensity for randomising polarisation. To overcome this problem, special fibres, known as polarisation maintaining fibres, are used. In this case, the natural polarisation mode degeneracy is deliberately broken down by introducing significant asymmetry into the fibre. This asymmetry might be brought about by stressing the fibre core preferentially in some direction to created aligned stress induced birefringence. Alternatively, the geometry of the core region may be changed to remove its circular symmetry. For reasonably short lengths of fibre, the polarisation state of the transmitted light will be directly related to that of the incident light.

An example of a glass preform used to generate a polarisation maintaining fibre is shown in Figure 13.21. Two holes in the preform accommodate two cylindrical rods which, by virtue of differential expansion induce stress in the preform which is preserved in the drawn fibre.

13.10 Focal Ratio Degradation

Focal ratio degradation is a phenomenon in multimode fibre that leads to an apparent increase in the étendue of light propagating through a fibre. This is of particular significance where larger fibres are used as light conduits in instrument applications, for example, in spectroscopy. The input light might be defined by a specific focal ratio or numerical aperture ***that is less than the numerical aperture of the fibre itself***. That is to say, the angular distribution is characterised by a hard aperture stop to produce illumination of some specific focal ratio, e.g. f#10. This input, by virtue of the overlap integral cited in Eq. (13.54), will be split into a large number of modes, each with its own amplitude and ***phase***. If this phase relationship is maintained through the fibre, then the focal ratio will be preserved. However, as with polarisation mode dispersion, real fibres have imperfections which degrade the phase coherence between the modes. As with the presence of aberrations in an optical system, there is an ineluctable tendency for the étendue to increase. This tendency is exacerbated by fibre bending and longer fibre lengths and conspires to reduce the focal ratio of the light at the output. Naturally, in the limit of long fibre length, the output light will evenly fill the fibre étendue and the focal ratio at the output will be entirely determined by the numerical aperture of the fibre itself.

13.11 Periodic Structures in Fibres

13.11.1 Photonic Crystal Fibres and Holey Fibres

Instead of using index contrast in two contiguous and continuous structures, light guidance can be produced by the introduction of periodic structures into an otherwise continuous material. The underlying principle behind photonic crystal fibres echoes the electronic behaviour of periodic crystalline materials. Just as a periodic lattice has a tendency to exclude certain electron energy states in a crystal band gap, the same applies to

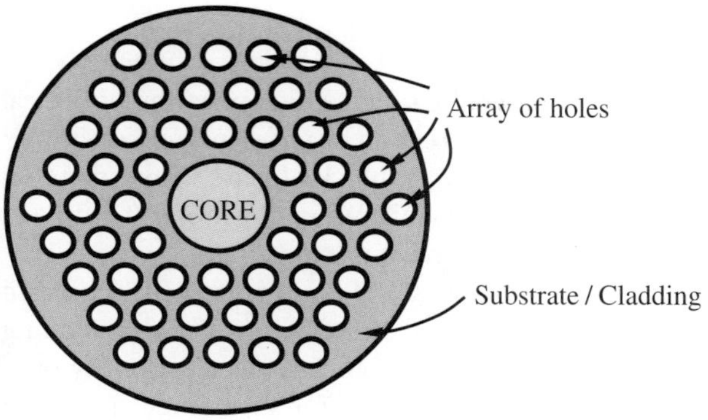

Figure 13.22 Photonic crystal fibre cross section.

light in a periodic refractive medium. That is to say, in such a periodic structure, light between certain specific wavelengths is not transmitted.

A ***holey fibre*** uses the effect of the introduction of these periodic structures, usually holes in a continuous material, to substantially modify mode dispersion effects of a conventional cladding material. In a ***photonic crystal fibre***, the lower index cladding is replaced by a periodic structure (of holes) that does not permit transmission over a specific range of wavelengths. In this case, it is possible for light to be confined in a lower index core, or even in air. Such structures permit the transmission of exceptionally high irradiance levels (i.e. laser power) within the compact core of a fibre. The structure of a photonic crystal fibre is illustrated schematically in Figure 13.22.

13.11.2 Fibre Bragg Gratings

Hitherto, we have been concerned with periodic structures in the lateral plane of the fibre. Another important group of structures embed periodic structures in the axial direction of the fibre. These are referred to as ***Fibre Bragg Grating***s and usually take the form of periodic variations in the core index or size. Their role lies in the creation of narrowband filters or mirror structures within a fibre. In effect, they may be thought of as a stack of quarter wave interference films. Although the index or size variations are very small, by extending the structure over several millimetres, amounting to several thousand periods, very high narrowband reflection may be attained. As well as functioning as narrowband mirrors in fibre lasers, these structures can be used to provide controlled dispersion in a fibre system. They are fabricated by periodic illumination of a fibre using crossed laser beams. This periodic illumination can either be used to initiate periodic etching of the fibre or, at high powers, to create small permanent refractive index changes in the core of the fibre. This process is illustrated in Figure 13.23.

Figure 13.23 Creation of fibre Bragg grating (*Period of grating is* $\lambda/2\sin\theta$).

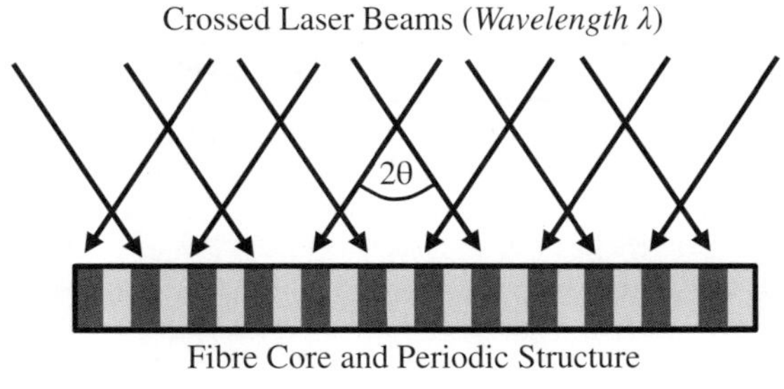

13.12 Fibre Manufacture

As outlined earlier, material purity is an exceptionally critical aspect in the successful application of optical fibre technology. The process of manufacturing optical fibres relies on the creation of a ***fibre preform*** followed by a fibre drawing process. The fibre preform is essentially a large cylindrical block of glass whose material geometry entirely reflects that of the fibre, but at a much larger geometrical scale. When the preform is drawn, the initial geometry (on a scale of centimetres) is replicated on a scale of tens of microns.

The preform is manufactured by a chemical vapour deposition process (CVD), a process widely used in the semiconductor industry. Manufacture is started using a silica tube as a substrate onto which the cladding and core layers may be deposited. This deposition process uses the thermal breakdown of exceptionally pure reactant gases, such as silane, germane, phosphine, and arsine to create the correct refractive index profile in the cladding and the core. These layers are deposited onto the inside of the silica tube, heating the tube with gas burners; the tube is rotated throughout on a lathe spindle. Following the deposition process, the hollow tube is transformed into a solid preform by a controlled collapse process. After this the fibre is drawn from the heated preform. The process is illustrated in Figure 13.24.

For the drawing process, the fibre preform is fed slowly into a furnace and the fibre strand withdrawn at speed by a capstan. Control of the fibre diameter is the critical parameter in the drawing process. This is achieved by using a non-contact sensor to continuously monitor the diameter of the drawn fibre. A feedback loop is used to maintain the fibre diameter at the desired value. That is to say, if the fibre diameter starts to increase beyond the desired value, the capstan is commanded to draw more rapidly and vice-versa in the event of a reduced fibre diameter. Production of the fibre is a continuous process and the fibre is wound onto a drum directly following the drawing process. Before the fibre is wound onto the drum, it is coated with a UV curable polymer coating, for mechanical protection.

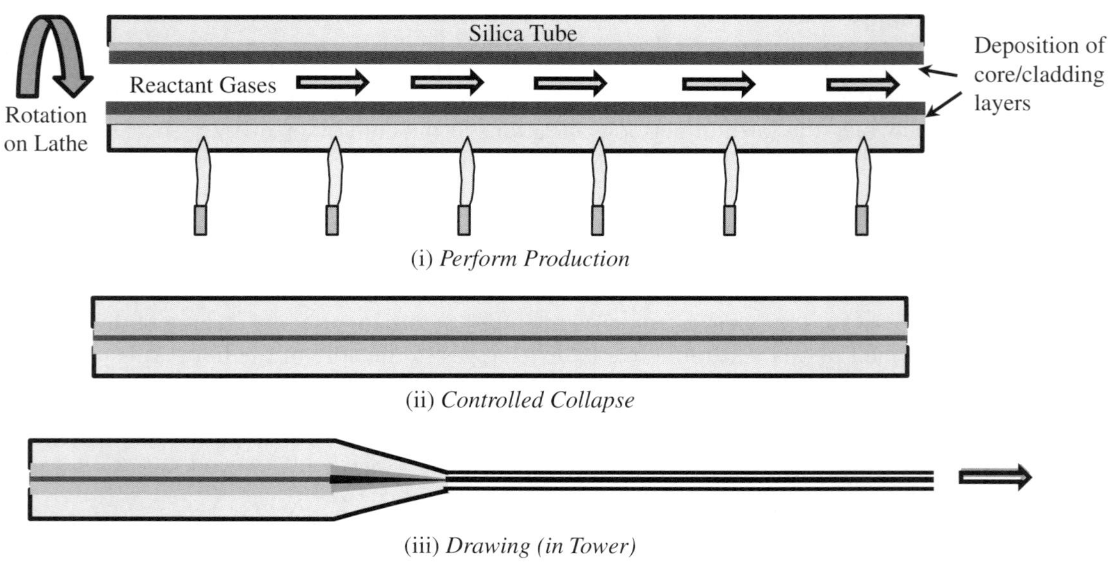

Figure 13.24 Optical fibre manufacture.

13.13 Fibre Applications

It is beyond the scope of this text to detail the very many applications of optical fibre technology. Economically, it is fair to say that fibre applications are dominated by their ubiquity it telecommunications applications. Broadly, the scope of optical fibre applications may be partitioned as follows:

1. Data transport (telecoms)
2. Compact sensors – chemical etc.
3. Delivery of light to and from inaccessible or hazardous locations
4. Instrument miniaturisation
5. Laboratory applications

Fibres are widely used in spectroscopy and lend themselves to use in, for example, the detection of gaseous contaminants. Fibres have particular sensitivity to gases or contaminants adsorbed onto the surface of a fibre. An early type of fibre sensor used the scattering caused by oil droplets formed on the surface of a fibre to detect low levels of contamination of sea water by oil. Chemical sensing by optical fibres is an exceptionally wide field.

Optical fibres have obvious widespread application in internal examination in medicine. They are used in visual inspection of hazardous locations, such as inside a nuclear reactor. They are widely used in large scientific instrumentation programmes to deliver light in a flexible or structured way, particularly from locations that are difficult to access.

Many optical instruments, such as interferometers, may be replicated in fibre form. In replacing bulky conventional optics with optical fibre equivalents, perhaps also using graded index optics, a considerable reduction in scale might be afforded.

Further Reading

Adams, M.J. (1981). *An Introduction to Optical Waveguides*. New York: Wiley. ISBN: 0-471-27969-2.

Grattan, K.T.V. and Meggitt, B.T. (1995). *Optical Fiber Sensor Technology*. London: Chapman & Hall. ISBN: 0-412-59210-X.

Koshiba, N. (1992). *Optical Waveguide Analysis*. New York: McGraw-Hill. ISBN: 0-07-035368-9.

Saleh, B.E.A. and Teich, M.C. (2007). *Fundamentals of Photonics*, 2e. New York: Wiley. ISBN: 978-0-471-35832-9.

14

Detectors

14.1 Introduction

Much of the effort in designing an optical system is concerned with the manipulation of light for some specific purpose, e.g. imaging. A large portion of classical optics and optical engineering falls within this very broad arena, particularly when concerned with aberration and image formation. In addition, we must also be concerned with object illumination through the study of radiometry. However, no system design can be complete without some consideration of how the output of an optical system might be utilised. This is crucial, particularly in a modern context, as the useful presentation of an image is generally in the form of data which may be manipulated or processed in some useful fashion. Of course, historically, the ultimate detector was universally the human eye. More latterly, the human eye was replaced by photographic media in specific applications.

It is very apparent nowadays that in a very large number of commercial and scientific applications, these traditional aspects have been superseded. Electronic detectors, particularly ***pixelated detectors*** for imaging, are ubiquitous in instrument and device design. As such, this text will largely focus on electro-optical ***detectors*** used in the conversion of light into electrical signals.

An electronic detector is a ***transponder***, whose purpose is to convert one form of energy (optical) into another (electrical). For the most part, detectors fall into two broad categories, as defined by their end application. They may either be used for measurement, or for imaging, or perhaps a combination of the two. Where detectors are used for measurement, the issue of calibration is especially salient, particularly in absolute radiometric measurements. For imaging applications, geometrically structured or pixelated electronic detectors dominate.

Physically, the majority of optical detectors operate with regard to the quantum nature of matter and light. For example, a single photon interacts with a semiconductor material and elevates a charge carrier from the valence band to the conduction band. This principle underlies the operation of the majority of current devices. A limited range of devices rely on the absorption of radiation and its conversion to heat which can then be sensed electronically. Such devices are inevitably less sensitive when compared to purely electronic devices. However, they do have the benefit of wider spectral coverage, particularly into the mid and far infrared where there are no or few materials available for purely electronic detection. In addition, since they convert light directly into heat, thermal detectors have the additional benefit of conferring direct absolute radiometric calibration.

14.2 Detector Types

14.2.1 Photomultiplier Tubes

14.2.1.1 General Operating Principle

A photomultiplier tube (PMT) is a highly sensitive optical detection device that is based on the photoelectric effect. It is well known that conductive materials are capable of ejecting charge carriers into space through

Optical Engineering Science, First Edition. Stephen Rolt.

Companion website: www.wiley.com/go/Rolt/opt-eng-sci

the absorption of a single photon. For this to happen, the energy of the incoming photon must exceed some threshold energy, known as the ***work function***. Like the interaction of light with semiconductor materials, this is a quantum effect, for the elucidation of which Albert Einstein famously won his Nobel Prize. Of course, metals have very high work functions and would not be much practical use in detectors, outside the deep ultraviolet. In practice, materials used in PMTs tend to be based on alkali metals with their low work functions or some compound semiconductor materials. This extends the usefulness of PMTs across the visible and into the near infrared, ***but no further***.

The PMT is a vacuum device with the photosensitive cathode enclosed within a high vacuum envelope. Electrons are ejected from the cathode when light is incident upon it. In principle, one electron should be ejected from the cathode for every photon absorbed with an energy that is in excess of the work function. However, in practice, only some proportion, η, is ejected. This proportion is known as the quantum efficiency. After ejection from the cathode, electrons are swept away by an externally imposed electric field and directed to a series of specially sensitised surfaces, known as dynodes. These surfaces perform the electron multiplication function. When an electron strikes one of these surfaces (at speed), two or three electrons are ejected. This multiplication effect is compounded by a series of many (e.g. 12) dynodes, each successively multiplying the number of electrons that eventually reach the anode. Depending upon the acceleration voltage used, usually in the range of 1 to 3 kV, multiplication factors of over 10^6 are possible. A reactive getter is incorporated within the vacuum envelope to maintain the vacuum and to scavenge outgassing products from the inside of the tube.

Figure 14.1 shows a sketch of a PMT, illustrating its operational principle. The voltage between each dynode is maintained by dividing the potential from a high voltage power supplier across a resistor chain. The voltage across each dynode gap is nominally equal, although relative voltages may be adjusted somewhat to optimise performance. By its nature, the photomultiplier tube, or PMT, is a current detection device. That is to say, an optical signal is sensed by virtue of the electrical current that emerges from the anode, as shown in Figure 14.1. The current generated is dictated by the input flux incident on the photocathode, the quantum efficiency of the photocathode and the multiplication ratio, M. A transimpedance amplifier then converts this current into a (digital) voltage, as determined by the nominal impedance of the amplifier.

In many respects, the PMT represents a legacy detection technology. Low cost and physically robust solid state detectors have replaced PMTs in the majority of applications. However, as will be seen later, their sensitivity at low signal levels, in terms of noise performance, remains unmatched.

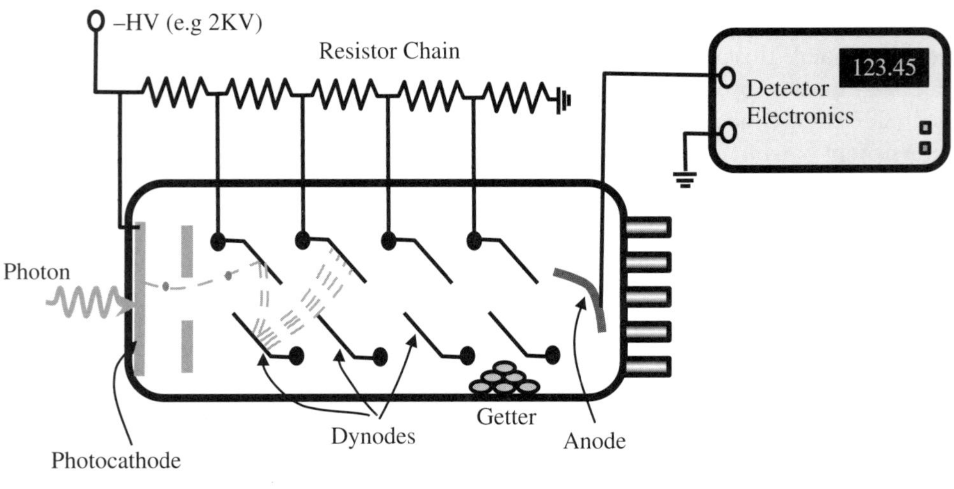

Figure 14.1 Photomultiplier tube.

14.2.1.2 Dynode Multiplication

Each dynode provides electron multiplication according to some ratio, δ, the ***secondary emission ratio***. Naturally, if there are n dynodes, the system multiplication factor, M, would be given by:

$$M = \delta^n \tag{14.1}$$

Stage multiplication factors of about 3 are typical, so for a 12 dynode device, the overall multiplication factor, M, would be about 5×10^5. Of course, the secondary emission ratio is itself dependent upon the dynode voltage. This dependence upon voltage can be modelled by a power law dependency, with the exponent, α, being close to one (i.e. approximately linear). Hence if the overall voltage across the PMT is V, then the secondary emission ratio and tube multiplication factor is given by:

$$\delta \propto V^{\alpha} \qquad M \propto V^{\alpha n} \tag{14.2}$$

The consequence of Eq. (14.2) is that one might expect a very sharp dependence of the overall electron multiplication, M, on applied voltage. That is to say, a very small drift in the applied voltage will produce a disproportionately large change in the output signal.

14.2.1.3 Spectral Sensitivity

The spectral sensitivity of a PMT is determined by the photocathode coating material. For the most part, the sensitivity of PMTs is restricted to the ultraviolet to visible spectral range. However, the use of III–V semiconductor materials, such as gallium arsenide, extends sensitivity to the near infrared. On the whole, the quantum efficiency of photocathode materials is rather low, with quantum efficiencies typically between 1 and 10%. This compares rather unfavourably with solid state detector technology. Figure 14.2 shows a plot of photocathode sensitivity versus wavelength for a number of photocathode materials.

Figure 14.2 also displays lines representing quantum efficiency as a function of wavelength. Some materials show no sensitivity within the visible, or, in particular, no overlap with the natural solar spectrum. Such

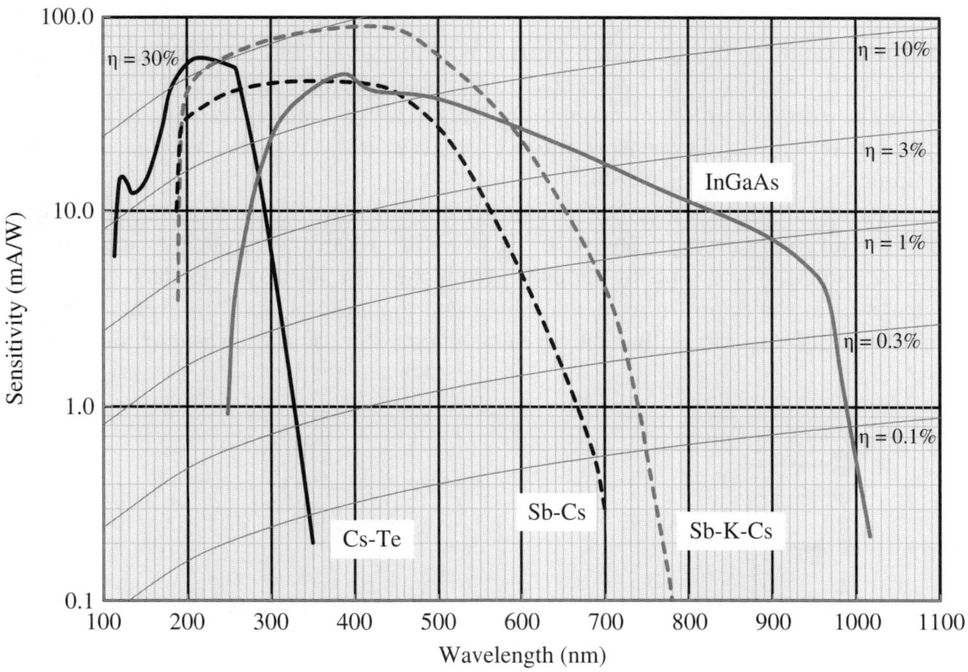

Figure 14.2 Sensitivity of some photocathode materials.

materials are referred to as 'solar blind'. All materials display some form of 'cut-off' in sensitivity at long wavelengths. The cut off wavelength is dictated by the material work function; a photon must have sufficient energy to overcome this barrier. For the device as a whole, the overall sensitivity, S, (in amps per watt) is dependent upon the quantum efficiency, η, the overall electron multiplication, M, and the wavelength of the incident light. The sensitivity may be represented by:

$$S = \frac{e\lambda}{hc}\eta M \tag{14.3}$$

e is the charge on the electron, c the speed of light, and h Planck's constant.

14.2.1.4 Dark Current

The photoelectric effect operates by a photon providing sufficient energy to eject an electron from a material into free space. However, this energy may also be supplied thermally or by other means. That is to say, even where no light is incident upon the photocathode, a small electric current will be produced at the anode. This is the so-called '***dark current***'. This dark current is problematical, even if it can be removed by subtraction, as it also adds to the device noise.

Thermionic emission is of particular concern. Such thermionic emission may originate either from the photocathode itself or any of the dynodes. This process is illustrated in Figure 14.3.

Whilst the work energy function, ψ, typically amounts to a few electron volts, thermal energy at room temperature is equivalent to about a fortieth of an electron volt. Clearly, the levels of thermionic emission are small. Nevertheless, with electron multiplication, thermionic emission can make a significant contribution, particularly when signal levels are low. Only the 'tail' of the thermal energy distribution can make a contribution to thermionic emission and its magnitude, as a function of absolute temperature, T, may be modelled by a Arrhenius type relationship, with the activation energy given by the work function. The dark current, I_D is given by:

$$I_D = AT^{5/4}e^{-e\psi/kT} \tag{14.4}$$

e is the charge on the electron and k the Boltzmann constant.

As with any Arrhenius type relation, the change in dark current with temperature is exceptionally rapid. For example, where the work function is equivalent to 2 eV, reduction in the operating temperature from 20 to 0 °C reduces the level of emission by over two orders of magnitude. Thus, relatively modest cooling substantially reduces the dark current. For materials with a large workfunction, such as the ultraviolet sensitive, 'solar blind' materials, then the thermionic dark current is negligible.

There are other sources of dark current, including field emission from the cathode at high voltage and ohmic conduction within the PMT – an internal resistance of $10^{12}\,\Omega$ will still produce a significant dark current. In addition, external ionising radiation and ionising radiation from the tube itself creates dark current by ejection of multiple electrons from the photocathode. Such high energy photons result in events that produce very many electrons at the cathode, rather than single electron emission, as is the case for the 'normal' photo-electric effect.

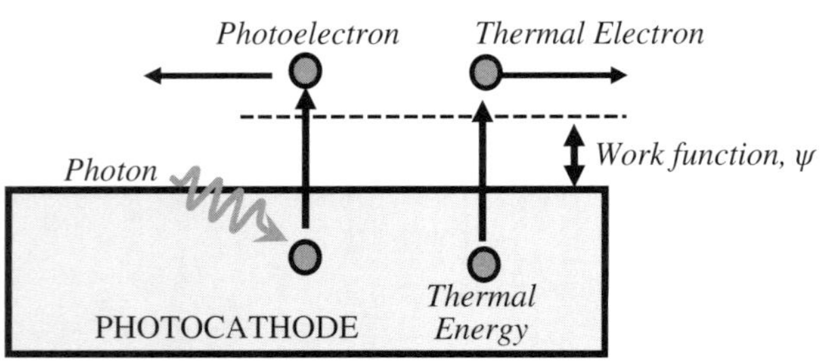

Figure 14.3 Photo-emission and thermionic emission.

14.2.1.5 Linearity

The predominant application domain for PMTs is in optical measurement, as opposed to imaging. In many cases, specifically for absolute radiometry, knowledge of the absolute sensitivity is of critical importance. In any case, even where measurements are relative, ***detector linearity*** is a highly desirable characteristic. That is to say, the output current should be strictly proportional to the input optical flux. For the most part, this is true for PMTs. Each photon incident upon the detector will, on average produce a specific electric charge at the output, so, in principle, the photomultiplier current is proportional to the optical flux. However, there are processes that lead towards ***detector saturation***, or at least a compromise in the output linearity. First, the photocathode and dynodes have some finite 'ohmic' impedance and drawing current reduces the driving voltage seen at each stage. In addition, this driving voltage is further reduced by voltage drop across the dynodes that is produced by drawing significant current through the dynode resistors. In practice, linearity may be maintained for currents up to a few tens of micro-amps in the case of a DC (continuous) measurement. For pulsed measurements, linearity may be improved by providing capacitive charge storage across the dynodes, particularly the latter stages. In this case linearity may be preserved up to tens of milliamps.

14.2.1.6 Photon Counting

As a result of the exceptional sensitivity of PMTs, single photon counting is possible. That is to say, in view of the very high electron multiplication factor ($>10^6$), a single photocathode emission event may be detected as a single electrical pulse at the output. As will be revealed later, this single pulse substantially swamps any random detector noise. In many respects, PMTs are unique in this capability and this opens up a range of scientific applications.

14.2.2 Photodiodes

14.2.2.1 General Operating Principle

A photodiode is very much the reverse of a light emitting diode (LED) or semiconductor laser. Photodiodes are formed at a semiconductor junction between p type and n type materials. Unlike in the case of LEDs, it is not necessary to employ a direct bandgap material. Silicon is, of course, a direct bandgap material and so cannot be used in LEDs. However, in contrast it is perhaps the most commonly employed photo-sensor material. A photodiode works by the creation of an electron hole pair close the p-n junction. In some respects, there is an analogy with the operation of a PMT. Instead of the work function being the operative activation energy, the absorbed photon must elevate an electron across the semiconductor band gap. Therefore, it is the material bandgap, as opposed to workfunction, that determines the spectral sensitivity of the material. In the case of silicon, the band gap is 1.14 eV. This corresponds to a wavelength of about 1100 nm. Therefore, a silicon photodiode will have useful sensitivity for wavelengths shorter than this. As a consequence, silicon is a useful detector material across the ultraviolet and visible spectral range and into the near infrared.

Figure 14.4 shows the general layout of a p-n photodiode. Naturally, Figure 14.4 represents a simplified representation of a photodiode. A photon is incident upon the detector and is absorbed within the so-called depletion zone that exists at the boundary between the n-type and p-type material. However, at the air interface, an anti-reflection coating is provided. This is of particular practical significance. Most semiconductor materials have especially high refractive indices, typically between three and four. Without the provision of an anti-reflection layer, Fresnel reflection losses would be unacceptably high. Physically, the device consists of a junction between two semiconductors, one n-type where the negative charges or electrons are mobile and one p-type layer where the positive charges or 'holes' are mobile. Interdiffusion of charge carriers between the two regions creates the depletion zone where the preponderance of particular charge carriers is equalised. The depletion zone or layer is not an engineered layer, rather a local modification brought about bythe physical processes involved.

This depletion zone is marked by a naturally occurring 'potential wall' that blocks further diffusion of electrons into the p-type layer and holes into the n-type layer. The potential wall is created by space charge effects

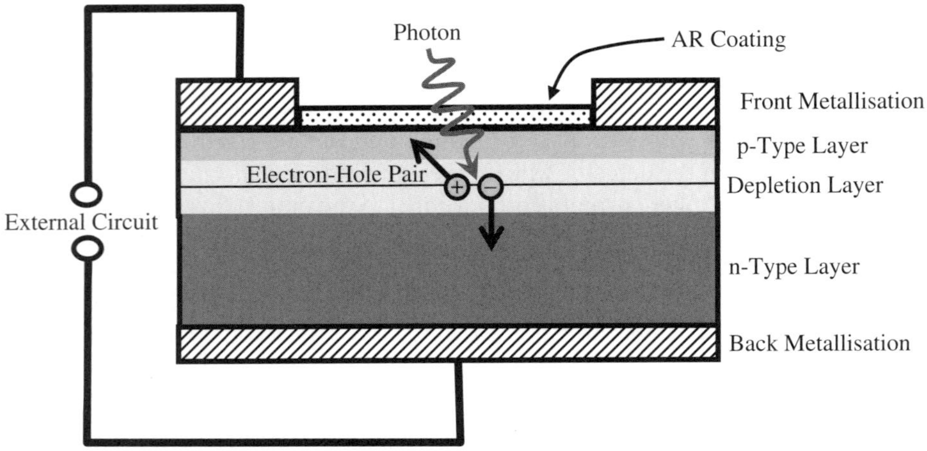

Figure 14.4 Operational principle of p-n photodiode.

produced by diffusion of charge carriers. When a photon creates an electron-hole pair in this region, the potential gradient serves to sweep the electron back towards the n-type layer and the hole towards the p-type layer, generating current in the external circuit. It is important to recognise the role of the semiconductor junction in this process. Outside this depletion region, where the potential gradient is small, any charge pair created will be largely immobile and the pair will be annihilated by recombination. For the minority carriers, e.g. electrons in a p-type material, this annihilation process is especially rapid. In this case, no external current will be produced. For current to flow in the external circuit, the minority carrier must reach its 'host' material. This process can only occur for carriers generated at or close to the depletion layer. Therefore, from the perspective of quantum efficiency, it is desirable to make the depletion zone as wide as possible.

A photodiode, in common with the PMT, is a current source. It can be operated without bias, i.e. no voltage applied by the external circuit. However, it is often operated with a reverse bias, with a negative voltage applied to the p-type terminal. The effect of reverse bias is to increase the width of the depletion layer. This increases the sensitivity by effectively extending the active volume over which charge carriers are preserved. It also reduces the capacitance of the junction, thus reducing the response time of any detection circuitry. Furthermore, reverse biasing of the detector also increases the range over which the detector provides a linear response. Unfortunately, reverse biasing increases the dark current, thus low noise applications tend to favour the absence of bias.

As outlined, extending the depletion layer thickness increases the quantum efficiency of the device. One way of further enhancing this process is to sandwich a layer of neutral or intrinsic semiconductor material between the n-type and p-type materials. The intrinsic material is effectively 'natural' or undoped semiconductor and serves to increase the effective thickness of the depletion layer. This is the so-called ***p-i-n detector***, or p-type – intrinsic – n-type detector. The layout of a p-i-n detector is shown in Figure 14.5.

It is important to appreciate that, unlike in the p-n junction, the intrinsic layer is an engineered layer, as opposed to the depletion layer that is created by physical processes. As highlighted, the thicker intrinsic layer allows photo-induced charge to be collected from a greater volume, thus increasing efficiency. In addition, the junction capacitance is reduced and, as such, the device has a shorter response time.

14.2.2.2 Sensitivity

By far the most commonly used photodiode material is silicon. Its spectral sensitivity is determined by the width of the bandgap and, as suggested previously, its sensitivity diminishes to zero at about 1100 nm. At shorter wavelengths, especially in the ultraviolet, a higher proportion of the input radiation is absorbed at a greater distance from the junction and the depletion layer. This is in consequence of the higher absorption

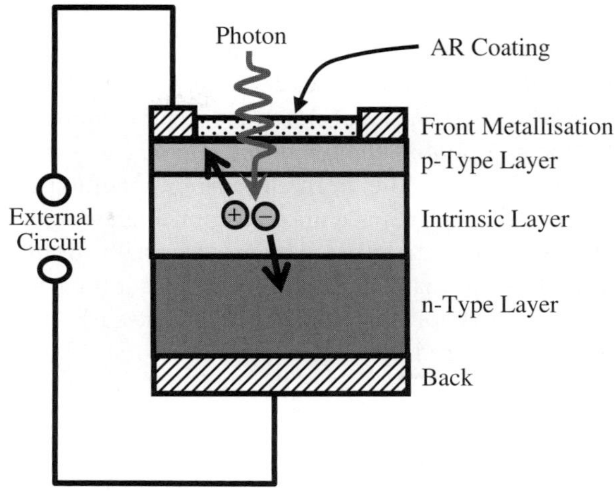

Figure 14.5 Layout of p-i-n detector.

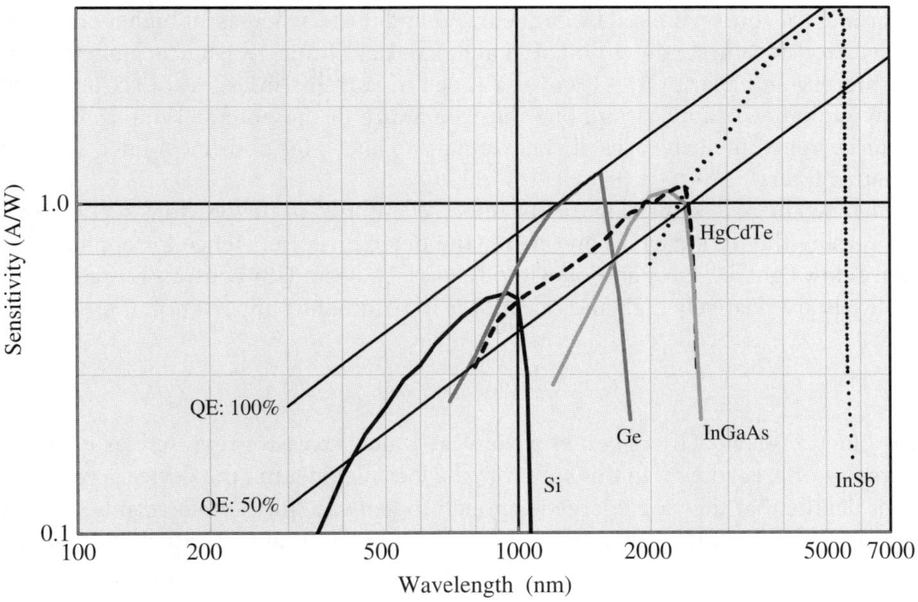

Figure 14.6 Sensitivity of photodiode materials.

coefficient at shorter wavelengths. Substitution of germanium for silicon, with its narrower bandgap, extends coverage further into the infrared. Further infrared coverage is provided by III–V compounds, such as indium gallium arsenide or II–VI compounds such as mercury cadmium telluride. Sensitivity curves for a variety of semiconductor photodiodes is shown in Figure 14.6. When compared to the sensitivity of a photocathode material in a PMT, the quantum efficiency of a photo-diode is much greater, almost approaching unity on occasion. Ternary semiconductors, such as indium gallium arsenide are especially useful as the stoichiometry of these materials may be adjusted to tailor the bandgap and spectral sensitivity for a specific application. To illustrate this further, indium gallium arsenide may be represented formulaically as $In_xGa_{1-x}As$ and mercury cadmium telluride as $Hg_xCd_{1-x}Te$. Adjusting the value of x in each formula changes the width of the bandgap. For example, in the case of mercury cadmium telluride, the bandgap varies from zero (metallic) to about 1.5 eV, as x is varied from 0 to 1.

14.2.2.3 Dark Current

The production of dark current in photodiodes is, in many respects a similar process to that in PMTs. A thermal diffusion current flows across a p-n junction when a small reverse bias is applied. For this to happen, charge carriers must surmount a potential equivalent to that of the bandgap. As such, the process is similar to that at a photocathode, in that charge carriers must acquire sufficient activation energy to surmount the potential barrier. Therefore, the current quantitatively follows an Arrhenius type dependence upon temperature. Furthermore, since the activation energies involved are significantly lower for photodiodes, dark current effects are potentially more severe, especially for infrared sensors with low bandgaps and activation energies. Therefore, for many infrared applications it is customary to cool detectors to cryogenic temperatures. In line with an activation energy dependence, the dark current is reduced by a factor of two for every 10 °C reduction in temperature. The primary motivation for reducing the dark current is to reduce the additional noise contribution that is inherent in the subtraction of any background. This aspect will be considered in more detail when we cover the topic of detector noise.

14.2.2.4 Linearity

Photodiodes are highly linear detectors and, as such, find application in calibrated radiometric measurements. At the low current end, linearity is, of course, limited by dark current and noise, whereas, at higher currents, there is a tendency for the detector to saturate due to internal and external 'ohmic' resistance. Nonetheless, photodiodes are capable of preserving linearity to $\pm 1\%$ over a range of currents in excess of six orders of magnitude. Of course, in any radiometric application, one must be aware of the potential implications of non-linearity at high detector currents. In such cases, it is customary to check for non-linear effects by the judicious insertion of attenuating filters.

Linearity and speed are enhanced by reverse biasing the detector. As outlined in the previous section, this has the adverse effect of increasing the dark current and hence the detector noise. Hence for applications involving low noise detection at low light levels, operation without bias is preferred. By contrast for radiometric applications where light levels are relatively high and linearity is of paramount importance, then reverse biasing is more suitable.

14.2.2.5 Breakdown

As with any semiconductor junction device, there comes a point at which excessive reverse bias causes a catastrophic increase in current – or breakdown. In this scenario, the local field within the device accelerates charge carriers to a sufficient degree that they are energetic enough to generate further electron hole pairs by collision. There comes a point at which the magnitude of the charge carrier 'amplification' so induced becomes sufficient to produce a runaway chain reaction. Breakdown voltages of a few tens of volts might be typical.

It would be useful at this stage to illustrate the form of photodiode current dependence upon the applied voltage. Where reverse bias is applied and before the onset of breakdown, the current reaches some saturation level, as the bias voltage is increased. This saturation current level is the so-called ***diffusion current***. For forward biasing, the (forward) current rapidly increases with voltage, with the device simply performing its function as an electronic diode or rectification device. Excluding the breakdown region, the voltage dependence of current, I, upon the bias voltage, V, is given by the following expression:

$$I = I_D\left(e^{eV/kT} - 1\right) - I_p \tag{14.5}$$

I_D is the diffusion current; I_p the photoinduced current; k the Boltzmann constant; and T the absolute temperature.

The dependence of current on bias voltage is illustrated in Figure 14.7 for a range of different photocurrents. The breakdown voltage in Figure 14.7 is labelled as V_B. In practice, the separation of the forward bias region and the breakdown voltage is much larger than (literally) suggested by Figure 14.7.

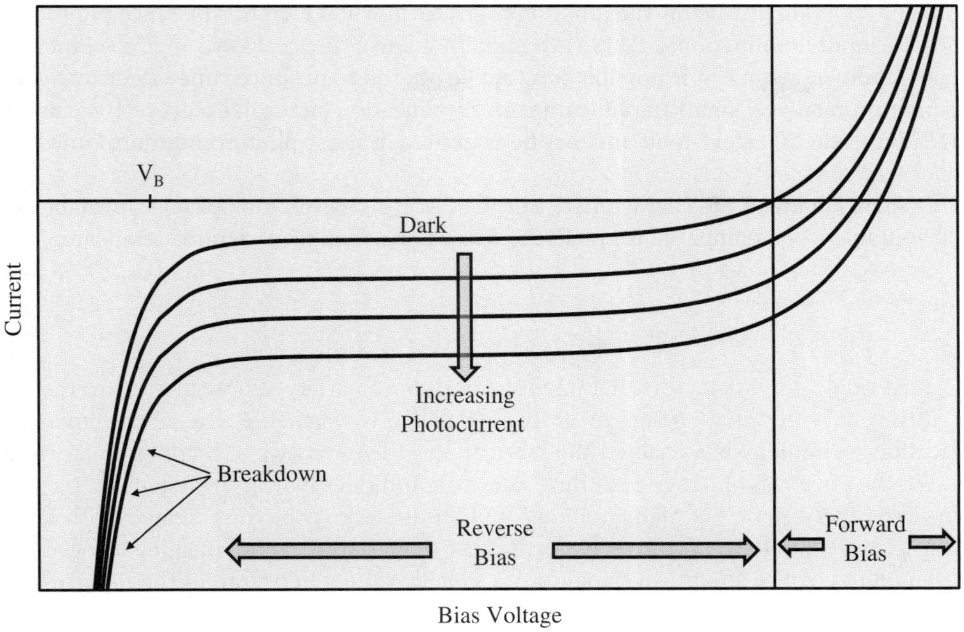

Figure 14.7 Effect of bias voltage on photodiode current.

14.2.3 Avalanche Photodiode

In discussing the phenomenon of diode breakdown, we introduced the concept of (runaway) generation of charge carriers through collisional excitation at high applied electric field. In the so-called ***Avalanche Photodiode*** (***APD***), this phenomenon is exploited in a controlled fashion to amplify the photo-generated current. That is to say, a high reverse bias electric field is applied to a suitably designed junction in such a way as to amplify any photo-induced charge by energetic collisions. In many respects, the APD is the solid state equivalent of the PMT. Both devices use collisional generation of charge to amplify the initial creation of a single charge carrier. However, the amplification produced by an APD is very much lower than that of a PMT. Typically, gains in APDs amount to no more than a hundred, substantially in contrast to that of the PMT. Operation of an APD is shown schematically in Figure 14.8.

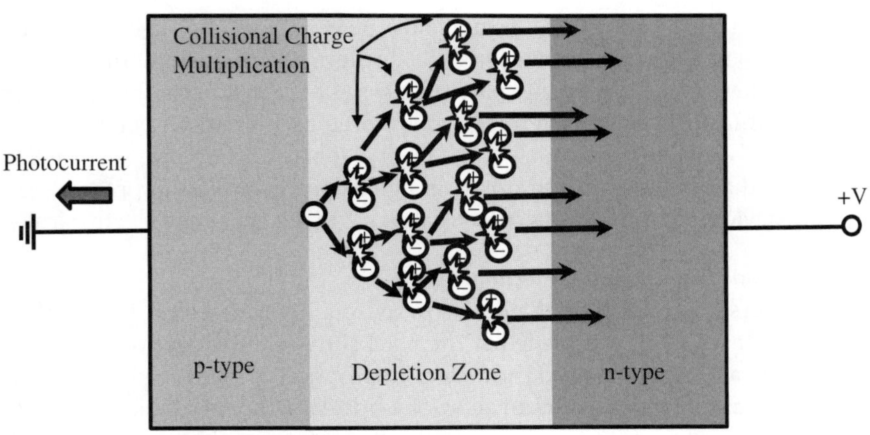

Figure 14.8 Operation of an avalanche photodiode.

In normal operation, APDs operate just below the junction breakdown voltage. In this instance photocurrent is directly related to the input illumination; an APD is nominally a linear device. However, if the junction is biased just above the breakdown voltage, it is possible for a single photon to produce runaway charge generation. Instead of producing a relatively small amplification, an amplification factor in excess of 10^{12} can be produced. This mode is the so-called 'Geiger Mode' and may be employed in single photon counting. However, detection is not linear.

As alluded to, the APD shares many of the characteristics of the PMT. However, the noise performance of the APD is rather less favourable when compared to the PMT. This will be discussed in more detail later.

14.2.4 Array Detectors

14.2.4.1 Introduction

Hitherto, we have described single point detectors that might be useful in a variety of optical measurements. By themselves, however, they have no innate image recording capability. Nevertheless, the advancement of integrated circuit fabrication technology has enabled the fabrication of large arrays of highly miniaturised photodiode structures. For the purposes of image recording, these photodiodes may be arranged in a rectangular array of photosensitive ***pixels***. Arrays of many millions of photodiodes may be thus created, with array spacings of a few microns. This technology, of course, underpins the operation of digital imaging devices.

These devices are designed to acquire an optical signal over a specific time, referred to as the ***integration time***, as defined by the ***frame rate***. This terminology, originating from the motion picture industry, describes the number of separate images acquired in a specific time, usually one second. As such, for each pixel, the detector integrates the total amount of signal received over the specified time, converting it into stored electrical ***charge***.

14.2.4.2 Charged Coupled Devices

The difficulty in the development of array detectors lies not so much with the photodetectors themselves, but rather the provision of means to faithfully monitor the output of many individual detectors in a limited time frame. There are a number of schemes for achieving this. Perhaps the most established technology is the ***charge coupled device*** or ***CCD*** technology. Each of the many photodiodes has a dedicated contiguous capacitor structure that stores output from its photodiode during a specific timeframe. When this specific timeframe has elapsed, all the individual photodiode charges are sequentially 'clocked out'. This is done by successively transferring charge from one pixel capacitor to its neighbour. This charge transfer is accomplished by successively applying a bias voltage between each contiguous capacitor, which operate as semiconductor MOS (metal oxide semiconductor) charge storage cells. As such, application of a controlling 'gate voltage' switches stored charge from one 'storage bucket' to the next.

The CCD process is illustrated schematically in Figure 14.9. Individual pixel sensors and their associated storage capacitors shift (slowly) stored charge to its neighbouring capacitor immediately below. This may be envisioned as a ***vertical shift register***. These are offloaded into particular series of capacitors at the bottom of the sensor, which function as a ***horizontal shift register***. In turn charge stored in this register is rapidly and sequentially shifted towards the output for reading. Eventually each pixel will be transferred to the 'outside world' where it is read by an analogue to digital converter and stored as digital data. In effect, this process converts the spatial signal of the imaging device into a sequential, temporal signal that can be read electronically.

14.2.4.3 CMOS (Complementary Metal Oxide Semiconductor) Technology

An alternative scheme for array detectors is provided by the ***active pixel*** or ***CMOS detector***. In this class of devices, amplification and readout of each pixel is provided at the pixel site itself. Readout of each pixel is accomplished by directly interrogating an individual pixel through an appropriately configured network of interconnections. For example, readout may be provided by an array of vertical and horizontal interconnections. Addressing one specific pair of horizontal and vertical connections provides access to a specific pixel. A schematic diagram is shown in Figure 14.10.

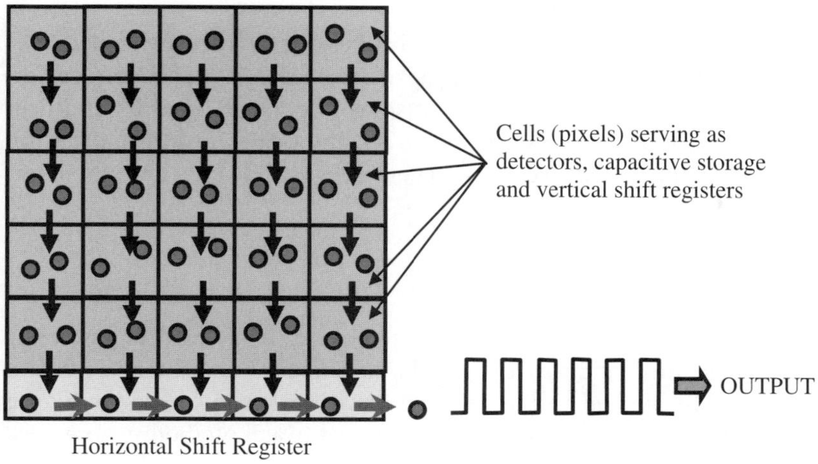

Figure 14.9 Operation of a CCD device.

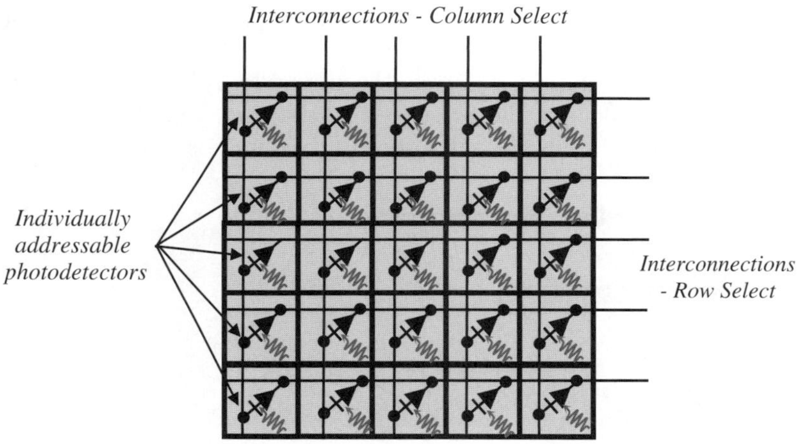

Figure 14.10 Active pixel or CMOS detector.

Figure 14.10 is greatly simplified, showing just photodiodes arrayed between the two lines of interconnects. In practice the photodiode would be replaced by a photodiode circuit, including the photodiode itself plus amplification and other signal processing electronics.

14.2.4.4 Sensitivity

The sensitivity of array detectors mirrors that of the underlying semiconductor technology. For example, silicon technology has a sensitivity that extends from 300 to 1100 nm. Use of compound semiconductors, such as indium gallium arsenide and mercury cadmium telluride, extend performance into the infrared, as illustrated in Figure 14.6. In common with their discrete counterparts, detection quantum efficiencies are high.

14.2.4.5 Dark Current

As array detectors are based upon underlying semiconductor photodetector technology, the same considerations apply to dark current as apply to the corresponding discrete devices. However, as pixelated detectors work by integration of a signal over a specific time frame, the phenomenon is often described by dark count or signal, rather current. However, most pertinently, for low signal applications, where noise levels are critical, the dark count may be substantially reduced by cooling the detector, as for discrete devices.

14.2.4.6 Linearity

In principle, as for their discrete counterparts, array detectors are highly linear. In that sense, the detectors find use not only in general imaging applications, but in spatially resolved radiometric measurements. The important feature of array detectors is their propensity to saturate at some integrated signal level. This is usually expressed as a '***well depth***' – the number of electrons that can be accommodated in a MOS capacitor without the charge 'overflowing'. Provided the number of electrons generated is significantly lower than this well depth, then the device will be substantially linear.

14.2.5 Photoconductive Detectors

Photoconductive detectors or photoresistors are also semiconductor-based detectors, but rely on photo-generated charge carriers to change the conductivity of a homogenous slab of semiconductor material. As such, a forward voltage must be applied to the material and the change in conductivity is detected as an increase in the resistive current generated. This stands in contrast to a photodiode where there is no requirement to apply an external voltage in order to produce a photocurrent. Furthermore, operation of a photodiode is based upon the provision of a semiconductor junction.

The morphology of a photoconductive detector is illustrated in Figure 14.11. A set of 'interdigitated' electrodes is provided on the surface of a homogenous semiconductor slab. A selection of photoconductive sensor materials is set out in Table 14.1, together with their operating wavelength range.

Photoresistors are less sensitive than photodiodes and are markedly less linear. Traditionally, they have found application in low-cost sensors designed to detect the presence of light. For example, cadmium sulfide detectors have found widespread use in visible applications, although they are less favoured due to

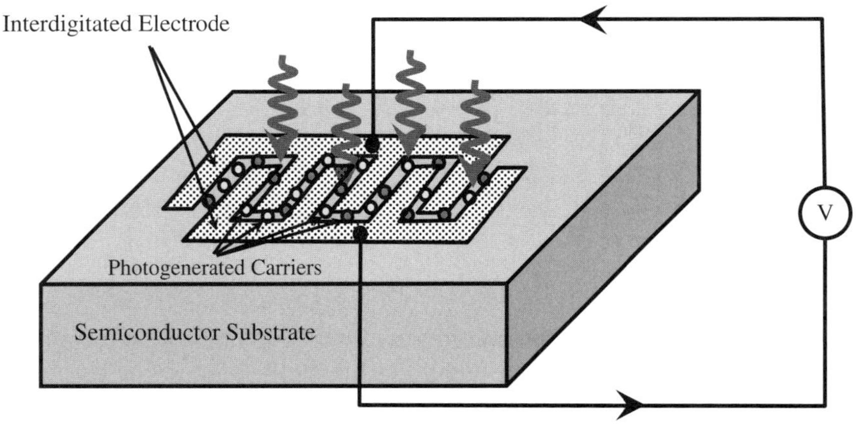

Figure 14.11 Photoconductive detector.

Table 14.1 Some photoconductive materials.

Material	Wavelength range (μm)	Material	Wavelength range (μm)
CdS	0.42–0.78	Si:Ga	2–20
PbS	1.0–2.5	Si:As	3–25
PbSe	1.5–4.5	Ge:Cu	5–30
InSb	1.0–5.5	Ge:Hg	4–14
Si:In	1.0–8.0	Ge:Au	2–9

environmental concerns regarding cadmium toxicity. Otherwise, photoconductive sensors tend to be based on narrow bandgap materials. Their long wavelength cut off is, of course, determined by the bandgap. For mid infrared applications, the commercial field is dominated by Lead Sulfide or Lead Selenide devices. For longer wavelengths, doped silicon and doped germanium, particularly Ge:Cu is the material of choice. These materials are referred to as ***extrinsic*** as opposed to ***intrinsic*** semiconductors. That is to say, the conductivity and photoconductivity is based on the creation of shallow energy levels accessible for charge generation to long wavelength, low energy photons. Such long-wavelength detectors require cooling (to liquid nitrogen temperatures) for operation. Photoconductive sensors tend to be slow.

14.2.6 Bolometers

Hitherto, we have considered electronic detectors which convert incident light into an electrical signal. By contrast, a ***bolometer*** absorbs optical radiation, converting it into heat, analysing the induced temperature rise and converting that into optical power. Measurement of the induced temperature rise may be recorded by a thermocouple, pyroelectric detector, or by sensing changes in electrical resistance. As a consequence, the detector provides a more direct link to radiometric measurements of flux, with the ability to convert detector output directly into an absolute measurement in watts. As was outlined in the section on radiometry, substitution of a calibrated ohmic heating source within the bolometer allows direct calibration of the bolometer. That is to say, if absorption of optical radiation produces a measured temperature rise of 2.5 °C, one needs simply to provide an (adjustable) electrical heating source that provides the same temperature rise. This is the principle of the substitution radiometer. Figure 14.12 shows a sketch of a simple bolometer.

Incoming radiation is absorbed at the absorbing layer and a temperature difference is established across the insulator and between the thermal layer and the thermal ground. This temperature difference is proportional to the radiation flux. Ideally, the absorbing layer should mirror the performance of a black body as closely as possible. Organic, black carbon-based coatings provide good performance over a wide range of wavelengths. More recently, this has included carbon nano-tube-based preparations which provide close to 100% absorption from the visible to infrared spectrum. For harsher applications, such as the monitoring of high power lasers, nano-structured metallic films, such as gold-black, may be used.

As outlined earlier, for most general applications monitoring of the induced temperature difference uses either a resistive thermometer, a thermopile or, for pulsed irradiation a ***pyroelectric detector***. A ***pyroelectric*** material, such as barium titanate or lithium tantalate, is a crystalline material that develops an internal (DC) electric polarisation in response to a temperature change. In such a detector, the crystalline material is sandwiched between two metal electrodes, effectively creating a capacitor. In the case of absorption of pulsed laser radiation, the heating of the crystalline material produces a polarisation within the crystal which is detected by virtue of the charge generated at the electrodes and hence the voltage across them. The voltage so generated is proportional to the incident energy.

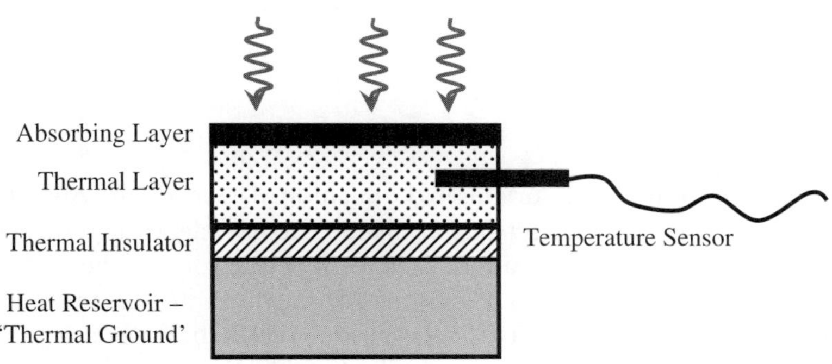

Figure 14.12 Simple bolometer.

Bolometers are useful in general purpose radiometric applications where measurement of absolute flux levels are demanded. However, in comparison to electronic detectors, bolometers are not especially sensitive. Nevertheless, more specifically, bolometers are especially useful for the detection of very long wavelength radiation (>100 μm) where electronic detectors are not readily available. Working at such extreme wavelengths, for example in astronomical applications, detectors must be cooled to cryogenic temperatures. On example of this is the so-called ***superconducting*** bolometer. This device relies on the very rapid change in resistivity produced by the heating of a superconducting material and is the most sensitive type of detector in this wavelength range. A hot electron bolometer relies on the heating of free electrons in a semiconductor material, such as indium antimonide, to produce a change in material resistivity.

14.3 Noise in Detectors

14.3.1 Introduction

Thus far, we have classified the major types of detectors and analysed their sensitivity, linearity, etc. That is to say, our focus has been on the output signal produced by the detector. However, unfortunately the desired signal is inevitably also accompanied by a stochastic contribution that limits the ultimate sensitivity of detection. This random contribution is referred to as noise and may arise from a variety of sources. This narrative is restricted to the treatment of electronic devices.

Of specific interest in this discussion is the ***signal to noise ratio*** (***SNR***). In the definition pursued here, the SNR is the ratio of the mean signal to its standard deviation. However, the reader should be aware that an alternative definition exists related to noise power rather than amplitude. In this case, the SNR is defined as the ratio of the square of the mean of the signal to its variance. The convention adopted here is defined below for a signal whose mean level is μ_S and whose standard deviation is σ_S:

$$SNR = \frac{\mu_S}{\sigma_S} \tag{14.6}$$

Most of the noise sources we will consider produce a specific level of noise ***power*** per unit bandwidth, irrespective of the underlying frequency. Since we are considering noise amplitude in our convention, set out in Eq. (14.6), then the total noise amplitude, σ_S, is proportional to the square root of the frequency interval, Δf.

$$\sigma_S \approx (\Delta f)^{1/2} \tag{14.7}$$

We will consider six distinct sources of noise in the current analysis:

- Shot Noise
- Gain Noise
- Background Noise
- Dark Current Noise
- Johnson Noise
- Pink Noise

The first process to consider is so-called ***shot noise*** and has its origins in the quantum nature of light. Absorption of photons in a detector creates a specific number of electrons or charge carriers, depending on the quantum efficiency. The arrival of a stream of photons and generation of a stream of electrons is inherently a stochastic process governed by Poisson statistics. Amplification in PMTs or APDs further enhances the noise generation process, as the collisional multiplication of charge is itself a stochastic process. In considering the impact of dark current or signal background, we must be aware that any background (including dark current) we may wish to subtract from the signal also includes random shot noise, which cannot be subtracted.

In contrast to the previous noise sources, based on the quantum nature of light, Johnson or thermal noise is caused by random, thermal motion of charge carriers. All these sources of noise are described as ***white noise***. That is to say, the noise power per unit bandwidth is independent of the underlying frequency. On the other hand, the final noise source, ***pink noise***, has a noise power that is inversely proportional to the underlying frequency. This noise is due to natural imperfections in the realisation of electronic circuits and it is sometimes referred to as ***1/f noise*** or ***flicker noise***.

14.3.2 Shot Noise

If we consider a detector upon which a stream of photons is incident. In a time interval, Δt, some number of photons, N, arrives. This is then converted into ηN charge carriers, where η is the quantum efficiency of the detector. However, the creation of these charge carriers is an entirely stochastic process, governed by Poisson statistics, and the standard deviation is given by:

$$\sigma_S = (\eta N)^{1/2} \qquad \text{and } SNR = (\eta N)^{1/2} \tag{14.8}$$

If now the radiometric flux incident upon the detector is Φ, then the number of photons, N, of angular frequency, ω, arriving in a time interval Δt is given by:

$$N = \frac{\Phi \Delta t}{\hbar \omega} \tag{14.9}$$

We can substitute Eq. (14.9) into Eq. (14.8):

$$SNR = \left(\frac{\eta \Phi \Delta t}{\hbar \omega} \right)^{1/2} \tag{14.10}$$

The important point about shot noise is that the noise is proportional to the square root of the signal and the SNR is inversely proportional to the square root of the signal. To understand the impact of signal bandwidth, we can represent the time interval, Δt, as the inverse of some frequency interval, Δf.

$$SNR = \left(\frac{\eta \Phi}{\hbar \omega \Delta f} \right)^{1/2} \tag{14.11}$$

It is clear that the bandwidth dependence of shot noise follows the pattern prescribed by Eq. (14.7). In other words, the noise level is proportional to the square root of the bandwidth.

Worked Example 14.1 ***Laser Beam Shot Noise***

A laser beam of wavelength 530 nm is incident upon a silicon detector. The flux of the laser beam is 1 μW. At this wavelength, the quantum efficiency of the detector is 60%. We wish to monitor the signal across a 1 MHz bandwidth. What is the signal to noise ratio into this bandwidth?

First, the photon energy, $\hbar\omega$, at 530 nm is equal to:

$$\frac{2\pi\hbar c}{\lambda} = \frac{6.626 \times 10^{-34} \times 2.998 \times 10^{8}}{5.3 \times 10^{-7}} = 3.75 \times 10^{-19} \quad \text{Photon energy is } 3.75 \times 10^{-19}\ \text{J}$$

Applying this value to Eq. (14.11) for a flux of 10^{-6} W, a quantum efficiency of 0.6 and a frequency bandwidth of 10^{6} Hz gives:

$$SNR = \left(\frac{0.6 \times 10^{-6}}{3.75 \times 10^{-19} \times 10^{6}} \right)^{1/2} = 1265$$

The signal to noise ratio is 1265.

14.3.3 Gain Noise

PMTs and APDs produce gain. However, the gain process itself adds additional noise. This additional source of noise is ***gain noise***. For example, in a PMT, each dynode stage multiplies each incident electron by some factor, ô, e.g. 3.2. However, in fact, each multiplication event produces an integer number of electrons, e.g. 1, 2, 3, etc., as dictated by Poisson statistics. The additional noise produced by this process, over and above the un-amplified shot noise, is accounted for by an ***excess noise factor***, F. Therefore, to account for the effect of gain noise, we must re-model Eq. (14.11) to give the revised SNR:

$$SNR = \left(\frac{\eta\Phi}{F\hbar\omega\Delta f}\right)^{1/2} \tag{14.12}$$

For PMTs, the excess noise factor, F is generally low, around 2. In the case of APDs, the excess noise factor is usually somewhat higher, around a factor of 10. In this regard, APDs are somewhat more noisy than PMTs. Of course, the effect of amplification is to produce inferior noise performance when compared to pure shot noise. The utility of amplification lies in those scenarios where there is some other prevailing source of noise, e.g. thermal noise, that is not amplified by the PMT or APD. In this case, the signal is boosted without adding proportionately to the ***overall*** noise level.

14.3.4 Background Noise

If one is making an optical flux measurement in the presence of scattered light, it is possible, provided the background light is at a constant level, to account for this by subtracting the background light level. Unfortunately, however, the background light not only contributes to the deterministic DC signal, but it also contributes to the shot noise. Before attempting to analyse the impact of this, we must first seek to understand how the introduction of multiple independent noise sources contributes to the overall noise level. In summing the contributions from independent stochastic variables the overall system behaviour is described by summing the variance of all the individual sources:

$$\sigma_{system}^2 = \sigma_1^2 + \sigma_2^2 + \sigma_3^2 + \sigma_4^2 + \ldots\ldots \tag{14.13}$$

If we now have two independent noise sources, one arising from the signal flux, Φ_{sig} and the other originating from the background light, Φ_{back}, then it is possible to use Eqs. (14.12) and (14.13) to determine the overall SNR.

$$SNR = \left(\frac{\eta\Phi_{sig}}{F\hbar\omega\Delta f}\right)^{1/2} \frac{1}{\sqrt{1 + \Phi_{back}/\Phi_{sig}}} \tag{14.14}$$

For measurements at low signal fluxes, it is clearly important to minimise any background light falling on the detector, as far as possible. This becomes especially important for mid to far infrared radiation where thermal radiation from the surroundings can be significant. Therefore, in order to minimise the background light, the environment around the detector must be cooled. Indeed, in such cryogenic optical systems one has a 'cold stop' to substantially reduce the background radiation originating from outside the system étendue. If the sensitivity of the detector, in this context, is defined by a signal that is equal to the background, it is easy to see that reducing the temperature of the thermal background radiation would increase sensitivity.

We can illustrate this more quantitatively by considering an InSb photodiode with a diameter of 1 cm. It is monitoring an optical signal with a wavelength of 5 μm. By calculating the spectral irradiance of blackbody radiation at a given wavelength, it is possible to calculate the spectral flux arriving at the detector at a specific wavelength. This may then be weighted by the sensitivity of the InSb, as illustrated in Figure 14.6. By integrating over the sensitivity range of the detector and comparing this to the signal generated at 5 μm, it is possible to determine, for any temperature, the sensitivity weighted flux arriving at the detector. This is illustrated in Figure 14.13.

Of course the sensitivity will vary according to detector size and other factors. However, Figure 14.13 does illustrate the utility of cooling the surrounding environment, in that the effect of even modest cooling is quite dramatic. Cooling from 300 to 100 K reduces the background by some eight orders of magnitude.

14.3.5 Dark Current

Even in the complete absence of any background light, an electronic detector will produce some finite dark current. In terms of the effect it produces on the signal and on the noise, this dark current may be viewed as an effective background flux, Φ_{dark}. That is to say, it is the effective optical flux that would contribute a current equivalent to the dark current. Thereafter, this 'dark flux' may be treated in the same way as any background flux and inserted in Eq. (14.14) to determine its contribution to the overall SNR. As with background light, there is every incentive to minimise the dark current in order to maximise the SNR. This is especially true at low signal levels.

As expressed in Eq. (14.4), there is a tendency for the dark current in both PMTs and photodiodes to follow an Arrhenius type relationship. Broadly, the dependence of detector sensitivity as a function of temperature will follow a similar pattern to that set out in Figure 14.13. Therefore, as is the case for thermal emission, cooling of the detector has a disproportionate effect on the dark current and the dark noise. This is especially true for infrared detectors, where the effective activation energy, as determined by the bandgap, is low. For PMTs, modest cooling, e.g. using 'dry ice' is effective. For precision measurements in the infrared cooling to cryogenic temperatures is often necessary.

14.3.6 Johnson Noise

14.3.6.1 General

Hitherto, all the sources of noise we have considered are related, in some way, to the quantum nature of matter and light. By contrast, ***Johnson noise*** or ***thermal noise*** is caused by the random, thermal motion of charge carriers. Broadly speaking, the randomised collective motion of electrons within a resistor is thermodynamically assigned an energy of ½kT for each degree of freedom. As a result, a randomly fluctuating voltage appears

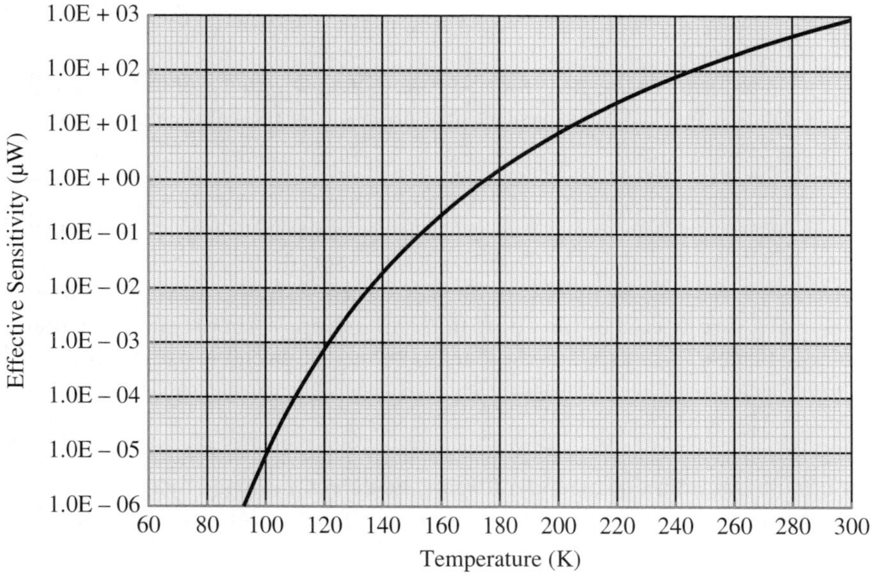

Figure 14.13 Sensitivity of InSb detector vs background temperature.

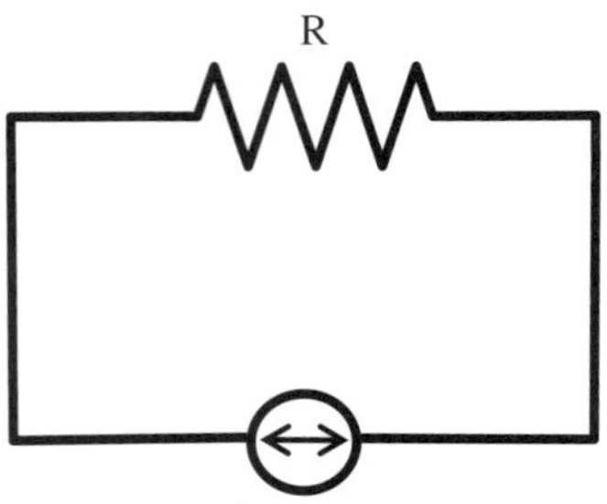

Figure 14.14 Equivalent circuit for Johnson noise.

across the resistor, accompanied, of course, by a randomly varying current. In the context of detector noise, it is the randomly varying current that is significant. There is a direct equivalence between the randomly varying current produced by thermal noise in a photomultiplier circuit and a randomly varying optical signal. Johnson noise is 'white noise', whose power per unit bandwidth is independent of frequency. It can be shown, for frequencies significantly less than the electron collisional frequency that the noise power per unit bandwidth, P, is equal to:

$$P = 4kT \tag{14.15}$$

k is the Boltzmann constant and T is the absolute temperature.

The rms current, I_{rms}, and voltage, V_{rms}, attributable to a resistor of resistance, R, is given by:

$$I_{rms} = \sqrt{\frac{4kT}{R}\Delta f} \quad \text{and} \quad V_{rms} = \sqrt{4kTR\Delta f} \tag{14.16}$$

Δf is the frequency bandwidth.

The equivalent circuit is shown in Figure 14.14.

Any circuit used to detect photocurrent will inevitably possess some measure of resistance and will contribute some noise, Φ_{rms}, to the measured optical flux. It is possible to calculate this noise flux from Eq. (14.16):

$$\Phi_{rms} = \left(\sqrt{\frac{4kT}{RG^2e^2\eta^2}\Delta f}\right)\hbar\omega \tag{14.17}$$

It is useful also to present Eq. (14.17) in terms of the steady 'background flux', Φ_{eff} that would produce this level of noise signal. In other words, at what level of flux does the shot noise attain the value denominated in Eq. (14.17)? This value is set out in Eq. (14.18).

$$\Phi_{eff} = \sqrt{\frac{4kT\hbar\omega}{RG^2e^2\eta F}} \tag{14.18}$$

It is clear that the effective noise flux is reduced by increasing the resistance, R, to the maximum possible value. However, in practice the value of R is restricted by the desire for some specific response time, τ. Usually, the detector and associated circuitry will have some associated capacitance and the circuit resistance is limited by its RC time constant:

$$\tau = RC \tag{14.19}$$

The resistance of the detector circuit is ultimately dependent upon the detector capacitance, as a smaller capacitance is associated with a higher resistance for a given response time. The effect of gain on the effective noise flux is of especial importance. Equation (14.17) clearly illustrates the impact of the gain, G, on the effective noise flux. With the quadratic term in the denominator, the effect of introducing gain is to multiply the effective resistance by a factor equal to the square of the gain. Of course, the noise is temperature dependent, increasing as the square root of the absolute temperature. As such, in contrast to background noise and dark current, the temperature dependence is not dramatic. There is therefore much less scope for reducing its magnitude by cooling. As Eq. ((14.18)) illustrates, control of Johnson noise is afforded by optimising the input amplifier resistance and, most particularly, by exploiting PMT or APD gain.

Worked Example 14.2 ***Photomultiplier Sensitivity***

A photomultiplier tube is designed to monitor very weak emission from the hydrogen Balmer beta line at 486.1 nm. The amplifier circuit has an impedance of 50 KΩ and the signal bandwidth is 1 MHz. The temperature of the PMT is 300 K, the gain is 10^5, the noise factor is 2.2 and the quantum efficiency is 0.15. What is the effective noise signal for this detection system for the wavelength of interest?

The photon energy, as defined by $\hbar\omega$, is 4.09×10^{-19} J. Substituting all relevant values into Eq. (14.17), we get:

$$\Phi_{rms} = \sqrt{\frac{4 \times (1.38 \times 10^{-23}) \times 300}{10^{10} \times (5 \times 10^4) \times (2.56 \times 10^{-38}) \times 0.15^2} 10^6} \times (4.09 \times 10^{-19}) = 9.81 \times 10^{-14}\text{ W}.$$

This noise signal level corresponds to a flux of about 240 000 photons per second. Supposing the signal were 10^6 photons per second, then a reasonable signal to noise ratio is afforded at a photon arrival rate that is equivalent to the system bandwidth (1 MHz). Therefore it will, in principle, be possible to detect single photons in this arrangement. This illustrates the utility of gain in PMTs and APDs. The equivalent effective noise flux in this example (from Eq. (14.18)) is about 40 nW.

14.3.6.2 Johnson Noise in Array Detectors

Array detectors effectively integrate the signal arising from a single pixel in some integration time. The Johnson noise associated with the readout circuitry is the so-called ***read noise***. As the signal in such a detector is an integrated signal, the noise amounts to an uncertainty in the integrated charge produced. As such, the read noise is usually denominated in multiples of charge, most notably in 'electrons'. For example, the read noise may be quoted as 10 electrons.

Read noise in array detectors is presented as a fixed amount of noise that is independent of the signal level (c.f. shot noise, background noise, etc.). However, read noise is effectively a manifestation of Johnson noise. The process of reading an accumulated signal may be understood in terms of collecting charge from a capacitor via an amplifier with a specific input resistance. Choice of resistor value determines the level of noise per unit bandwidth – the lower the resistor, the lower the noise (current) per unit bandwidth. However, the choice of resistance also determines the bandwidth over which the signal is processed. The lower the resistance, the greater the bandwidth. Ultimately, the integrated noise depends only upon the circuit capacitance. The equivalent circuit is shown in Figure 14.15. We can explore the noise behaviour of this circuit by making the rather simplistic assumption that the circuit time constant, τ, is equal to the inverse of the angular frequency bandwidth, $\Delta\omega$. Furthermore we wish to express the noise in terms of charge rather than current:

$$q_{rms}^2 = I_{rms}^2 \tau^2 \quad \text{and} \quad \Delta f = \frac{\Delta\omega}{2\pi} = \frac{1}{2\pi\tau}$$

from Eq. (14.16):

$$q_{rms}^2 = \frac{2kT}{\pi R}\tau$$

However, we know that the time constant τ, is equal to the product RC. Therefore:

$$q_{rms}^2 = \frac{2CkT}{\pi}$$

By carrying out a more rigorous analysis, the noise charge is given by the following expression:

$$q_{rms}^2 = CkT \tag{14.20}$$

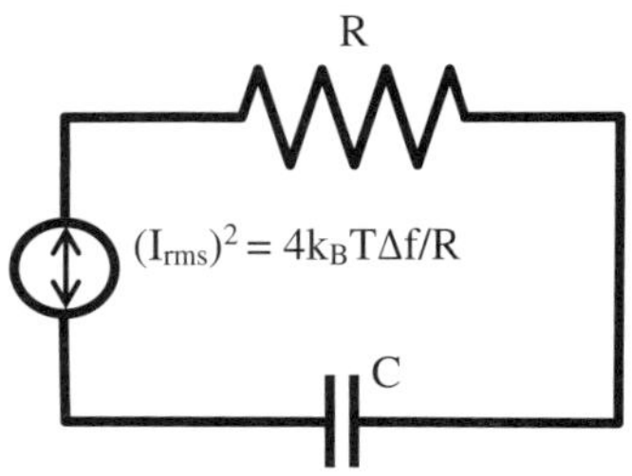

Figure 14.15 Equivalent read circuit for array detector pixel.

Equation (14.20) reveals that a fixed amount of charge-related noise is generated during the read process. This read noise is dependent only upon the capacitance of the cell and its absolute temperature.

Worked Example 14.3 ***Read Noise in a Representative Pixel***
We will now imagine a representative charge storage cell in an array detector as a 10 μm × 10 μm capacitor whose dielectric comprises a 100 nm layer of silica, whose relative permittivity, ε, is 3.8. The ambient temperature is 300 K. We now wish to estimate the read noise attributable to the cell, expressing the answer in equivalent electrons.

Before we can calculate the read noise, we need to derive the capacitance of the cell. The capacitance of a simple plate capacitor of area, A, and thickness, t, is given by:

$$C = \frac{A}{t}\varepsilon\varepsilon_0$$

where ε_0 is the permittivity of free space.

Substituting the values given we have:

$$C = \frac{10^{-5} \times 10^{-5} \times}{10^{-7}} 3.8 \times (8.85 \times 10^{-12}) = 3.36 \times 10^{-14} F$$

The capacitance of the cell is 3.36×10^{-14} F and we may substitute this value into Eq. (14.20):

$$q_{rms}^2 = Ck_BT = (3.36 \times 10^{-14}) \times (1.38 \times 10^{-23}) \times 300 = 1.39 \times 10^{-34} C^2$$

The square root of the above expression then yields the rms charge:

The rms noise charge is 1.18×10^{-17} C. **This may be expressed as a read noise of 74 electrons.**

This exercise illustrates the fundamental impact of material properties on the noise performance of array detectors. We can extend this analysis a little to account for the underlying signal to noise ratio of a pixelated detector. An individual cell has a fundamental 'well capacity' of electrons that it can accommodate before 'overflowing'. In considering the cell as a simple storage capacitor, we can express its storage limit as being defined by a critical electric field E_c above which the dielectric will break down. Hence, the maximum permissible voltage, V_{max}, across the storage capacitor is given by:

$$V_{\max} = E_c t$$

The charge storage capacity, or the signal in this case, is simply given by the product of the voltage and capacitance:

$$q = CE_c t$$

The signal to noise ratio is then give by:

$$\frac{q}{q_{rms}} = \frac{CE_c t}{\sqrt{CkT}} = \sqrt{\frac{C}{kT}} E_c t \quad \text{this gives}: \; SNR = \sqrt{\frac{\varepsilon\varepsilon_0 At}{kT}} E_c \tag{14.21}$$

Equation (14.21) clearly suggests that the signal to noise ratio of a pixel that is associated with the read noise scales fundamentally as the square root of the volume of material (At). As the signal to noise ratio is such a critical performance factor, there is an incentive to consider means of improving this figure. As the analysis of photomultiplier sensitivity revealed, introducing gain into a detector acts as a powerful means for improving sensitivity. It is relatively straightforward to quantify this effect. If the SNR without gain is labelled as SNR_0, then the enhancement is given by:

$$\frac{SNR}{SNR_0} = \sqrt{\frac{G^2}{F}} \tag{14.22}$$

G is the gain and F is the excess noise factor.

Technologically, introducing gain into an array detector is a substantially more challenging proposition than is the case for discrete devices such as APDs. ***Electron multiplying charge coupled devices*** (EMCCDs) successively multiply charge from individual pixels during the frame transfer process. This can be thought of as an impact ionisation process that takes place at each charge transfer stage. Whilst the gain at each stage is low, the large number of shifts that take place before the final read process ensures a high overall gain, e.g. 1000. An older competing technology uses a separate ***image intensifying tube***. The image intensifying tube may be considered as an array version of a photomultiplier tube. However, instead of the output electrons being detected as an electrical signal, they strike a phosphor and are converted back into light. Light emerging from the phosphor replicates the original incident light pattern, but at higher flux. Coupling this device with a CCD gives the image intensifying charge coupled device or IICCD.

14.3.7 Pink or 'Flicker' Noise

The final source of noise we need to consider is so-called ***pink noise*** or ***flicker noise***. This noise source is, like Johnson noise, electronic in origin. It is a product of the physical imperfections of real electronic circuits and devices. Whereas Johnson noise is a fundamental attribute and determined absolutely by the underlying physics, pink noise is highly variable and much more difficult to quantify, depending on device imperfections. Unlike all the other noise sources, pink noise is not 'colourless'. That is to say, the noise per unit bandwidth depends upon frequency. In fact, the noise power per unit bandwidth is proportional to the inverse of the frequency. If, in this context, the power per unit bandwidth is represented by the square of the signal current per unit bandwidth, we may define pink noise is the following manner:

$$I_{rms}^2 \propto \frac{1}{f}\Delta f \tag{14.23}$$

One consequence of Eq. (14.23) is that pink noise contains the same power in each decade of frequency. Integrating Eq. (14.23) with respect to frequency produces a logarithmic relationship. Therefore, for example, one might expect the same level of power to contained in the interval between 1 and 10 Hz as between 10 and 100 Hz. In practice, pink noise occurs in conjunction with other (colourless) sources of noise. Therefore, one might expect that the noise spectrum is defined by Eq. (14.24):

$$I_{rms}^2 = \left(\frac{A}{f} + B\right)\Delta f \tag{14.24}$$

The value B in Eq. (14.24) is termed the '***noise floor***' and the frequency at which the two contributions are equal is the so-called ***corner frequency***. This is illustrated in Figure 14.16.

Previously, we have discussed how to minimise sources of noise, such as those arising from background noise and dark current noise, as well as thermal noise. As we are presented with an additional source of noise, the question arises as to how this might be minimised. However, there is a distinct and rather unpleasant feature of pink noise, namely its propensity for defying the utility of signal averaging. Under normal circumstances, the process of averaging random noise effectively reduces the signal bandwidth in inverse proportion to the sampling time. Therefore, for a constant signal, increasing the sampling time, for example, by a factor of four,

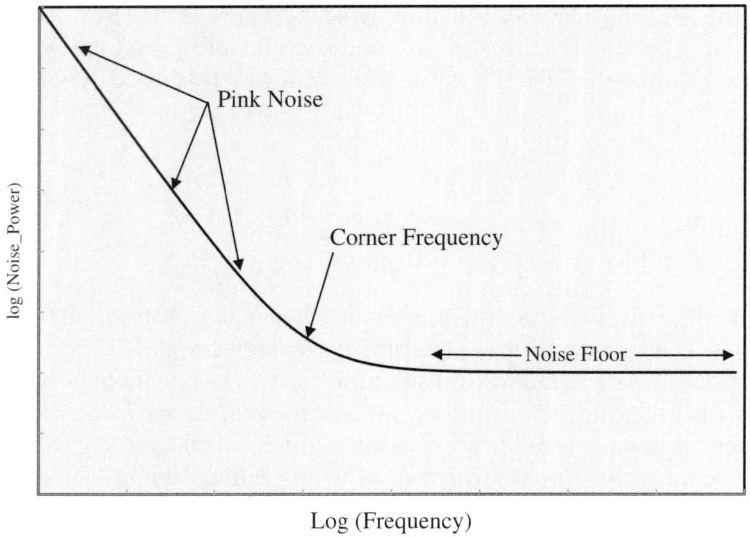

Figure 14.16 Frequency dependence of pink noise.

will reduce the noise level by a factor of 2. However, in the case of pink noise, not only is the bandwidth being reduced but the part of the frequency spectrum being sample is shifted to lower frequencies. On the one hand, restriction of the band width improves SNR, whereas the shift to lower frequencies degrades it.

An insight can be gained from inspection of Figure 14.16. It is clear that if we wish to minimise the impact of pink noise, by preference we should be operating in the region of the noise floor. One example of a typical scenario might be flux measurement with a photodiode. Highly sensitive measurements could, in principle be made by averaging the signal over long periods of time, in effect sampling the noise over a very small bandwidth. However, in the case of a nominally steady or DC flux, the contribution from pink noise might be prohibitive. Therefore, one useful strategy might be to convert the DC flux to an AC flux with a frequency within the noise floor. This is accomplished by 'chopping' the optical signal with an optical chopper to produce a signal with a frequency of a few tens or hundreds of Hertz. This strategy is common in sensitive laboratory measurements. The AC output from the photodiode is amplified by a lock-in amplifier that detects only the AC signal at the requisite frequency, using a reference signal derived from the optical chopper. The optical chopper itself consists of a vaned rotor that interrupts the optical beam. This arrangement is shown in Figure 14.17.

14.3.8 Combining Multiple Noise Sources

Thus far, we have dealt with the multiple sources of noise as single entities. We must now pose the question as to how the different sources of noise might be combined. ***Provided each individual noise source is uncorrelated*** to the others, then we simply sum the squares of all the individual noise sources. We will resort to the expedient of quantifying each source of noise in terms of the equivalent input flux required to generate the same level of noise as shot noise, e.g. as set out in Eq. (14.18). We can therefore express the aggregate SNR as follows:

$$SNR = \left(\frac{\eta\Phi_{sig}^2}{F\hbar\omega\Delta f}\right)^{1/2}\frac{1}{\sqrt{\Phi_{sig}+\Phi_{back}+\Phi_{dark}+\Phi_{thermal}+\Phi_{pink}}} \tag{14.25}$$

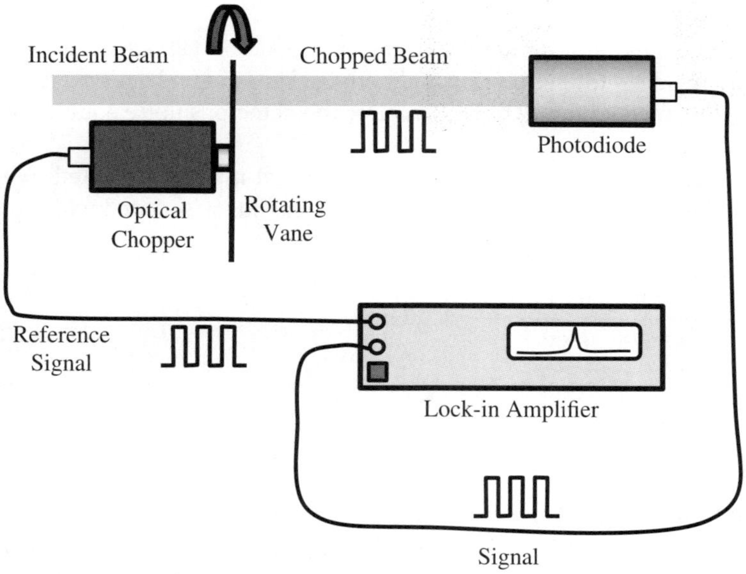

Figure 14.17 Optical measurement with optical chopper and lock-in amplifier.

14.3.9 Detector Sensitivity

The detector sensitivity provides a measure of the minimum optical signal that can realistically be detected by a specific device. This is captured by an arbitrary but useful signal level where the SNR is equal to one. If we consider shot noise alone, then the derivation is relatively trivial:

$$\Phi_{shot} = \frac{F\hbar\omega\Delta f}{\eta} \tag{14.26}$$

For example, the minimum signal level for a 1 Hz bandwidth is equivalent to the flux of one photon per second, as modified by the detector quantum efficiency and excess noise factor (if any). In the case of a photodiode with a quantum efficiency of 0.5 viewing a 500 nm beam, then the minimum signal flux is about 8×10^{-13} W.

In practice, the minimum detectable signal is governed by dark current, thermal, or pink noise, etc. In this case, we may ascribe an effective 'noise flux', Φ_{noise}, for all these sources. In this case, we can modify Eq. (14.25) to give:

$$SNR = \left(\frac{\eta\Phi_{sig}^2}{F\hbar\omega\Delta f}\right)^{1/2} \frac{1}{\sqrt{\Phi_{sig} + \Phi_{noise}}} \tag{14.27}$$

We are specifically interested in the limiting case where the signal is very small. Therefore we can make the following approximation:

$$SNR \approx \left(\frac{\eta\Phi_{sig}^2}{F\hbar\omega\Phi_{noise}\Delta f}\right)^{1/2} \tag{14.28}$$

If we set the SNR to one and extract the signal level consistent with this, we get:

$$\frac{\Phi_{sig}}{(\Delta f)^{1/2}} = \left(\frac{F\hbar\omega\Phi_{noise}}{\eta}\right)^{1/2} \tag{14.29}$$

Equation (14.29) gives the sensitivity in terms of the power per square root of the frequency interval. This is known as the ***noise equivalent power*** or the ***NEP*** and is expressed in Watts per root Hertz. NEP is used as a generic figure of merit by manufacturers to describe the sensitivity of detectors.

Worked Example 14.4 ***NEP of a Photodiode***

A simple photodiode is used to detect light at 500 nm. It has a quantum efficiency of 0.5 and it is to be used in a circuit with an input impedance of 50 KΩ. The detector has no gain and both background and dark currents are negligible. Furthermore we may ignore the impact of pink noise. Estimate the detector's noise equivalent power at an ambient temperature of 300 K.

We need only consider thermal noise in this analysis; background and dark current may be neglected. We may assume the gain and the excess noise figure to be unity, so the effective optical flux associated with the thermal noise is given by Eq. (14.18):

$$\Phi_{eff} = \sqrt{\frac{4kT\hbar\omega}{RG^2e^2\eta F}} = \sqrt{\frac{4(1.38\times10^{-23})\times300(3.98\times10^{-19})}{50000(2.56\times10^{-38})\times0.5}} = 3.2\text{ mW}$$

Applying this to Eq. (14.29), we get:

$$NEP = \left(\frac{F\hbar\omega\Phi_{noise}}{\eta}\right)^{1/2} = \left(\frac{(3.98\times10^{-19})\times(3.2\times10^{-3})}{0.5}\right)^{1/2} = 5.1\times10^{-11}\text{ WHz}^{-1/2}$$

The noise equivalent power is 5.1×10^{-11} ***WHz***$^{-1/2}$.

In the previous example we defined a reasonable input impedance for the amplifier. However, as outlined previously, the value of this resistance is not entirely independent of the detector characteristics. For a given response time, τ, we might expect the input resistance of the amplifier to reduce with increasing detector capacitance and, hence detector area. Furthermore, both dark and background currents have a tendency to scale with detector area. Therefore, one might expect the noise equivalent power to increase with detector area, A. As a consequence, it is customary to introduce a figure of merit to describe the noise performance of a detector. This parameter is known as the specific detectivity, D^*, and is defined as follows:

$$D^* = \frac{\sqrt{A}}{NEP} \tag{14.30}$$

A higher specific detectivity is associated with superior detector performance. Both noise equivalent power and specific detectivity are widely quoted for commercial devices.

14.4 Radiometry and Detectors

The practical use of detectors is inextricably linked with the study of radiometry. Absolute measurement of radiometric flux and radiance is dependent upon the use of traceably calibrated detectors. This topic was introduced in Chapter 7. Of particular importance in the practical application of detectors is the use of simple radiometric methods to estimate signal and noise levels in experimental and instrumental scenarios. For example, it is very straightforward to estimate the flux, Φ, arriving at a detector in an imaging system in terms of the radiance of the source, L, the throughput of the optical system, ξ, and the étendue of the system, G.

$$\Phi = \xi GL$$

It is very straightforward to calculate the étendue associated with a detector in an optical system. Whether the detector element is a single pixel or group of pixels in an array device, or a discrete detector, then it will have a specific area, A. The étendue is simple equal to the product of the area A and the solid angle associated with the system aperture.

$$G = A\times\pi NA^2$$

NA is the system numerical aperture.

Therefore, in any particular scenario we can use the source radiance and system and detector characteristics to calculate the flux arriving at the detector. From the flux arriving at the detector and the detector quantum efficiency, dark current, etc. we can finally calculate the signal and SNR. In many practical instances in instrument design, the SNR will be an important requirement and must be met.

Worked Example 14.5 ***SNR in a Thermal Camera***

To understand the practical application of radiometry, it is useful to demonstrate it using a plausible scenario. We wish to design a thermal camera system to monitor a nominally blackbody source with a temperature of 315 K. A bandpass filter restricts the transmitted wavelength to between 4.5 and 5.0 μm, with a system throughput of 75% in this range. The detector is an InSb detector with a pixel size of 10 μm and a quantum efficiency of 80% in the desired range. In this specific case, it may be assumed that the background and dark current are zero and the system noise is defined entirely by a read noise of 50 electrons. Finally, the camera has an aperture of f#3 and we may assume a detector integration time of 5 ms. Determine the signal to noise ratio of this system.

First, we need to calculate the radiance emitted by the source. We are told the source is a blackbody source at 315 K, so, initially, we calculate the spectral radiance, L_λ, at the mid-wavelength of 4.75 μm. The spectral radiance is given by Eq. (7.9):

$$L_\lambda(\lambda, T) = \frac{2hc^2}{\lambda^5} \frac{1}{e^{hc/\lambda kT} - 1}$$

This gives a spectral radiance of 3.21×10^6 W m^{-3} sr^{-1} or 3.21 W m^{-2} sr^{-1} μm^{-1}. The radiance into the 0.5 μm filter bandwidth is simply half the latter figure or 1.605 W m^{-2} sr^{-1}. In order to calculate the flux arriving at a single pixel, we need to calculate the relevant étendue, *G*. We are told that the aperture of the system is f#3, or a numerical aperture of 0.17 and the 10 μm pixels have an area of 10^{-10} m^2. The étendue, *G* is given by:

$$G = A \times \pi NA^2 = 10^{-10} \times 3.141 \times 0.17^2 = 1.96 \times 10^{-11}$$

The étendue of a single pixel is thus $8.72 \times 10–^{12}$ m^2 sr and given the throughput of 75% and the previously calculated radiance, it is possible to calculate the flux arriving at a single pixel:

$$\Phi = \xi GL = 0.75 \times (8.72 \times 10^{-12}) \times 1.605 = 1.05 \times 10^{-11}\ \text{W}$$

The flux per pixel is thus 1.05×10^{-11} W and the energy arriving at the detector in 5 ms is thus 5.25×10^{14} J. The photon energy associated with the mid-wavelength of 4.75 μm is 4.13×10^{-20} J. For a quantum efficiency of 80% the total number of charge carriers generated would be:

$$0.8(5.25 \times 10^{14})\text{J}/4.13 \times 10^{-20}\ \text{J} = 940\,000 \text{ electrons.}$$

The rms shot noise is simply the square root of the above figure or 970 electrons. The read noise of 50 electrons is small compared to be shot noise, but by adding the two contributions (by rss) we get a total noise of 971 electrons. Finally the SNR is given by:

$$\text{SNR} = 940\,000/971 = 968$$

The signal to noise ratio is 968.

This exercise illustrates the power of simple radiometric calculations in evaluating the performance of a design at its inception.

14.5 Array Detectors in Instrumentation

14.5.1 Flat Fielding of Array Detectors

We dealt with the general radiometric calibration of detectors in Chapter 7. The absolute radiometric calibration of discrete detectors largely relies on the provision of calibrated radiometric sources, whose irradiance or radiance is known. However, there are aspects of radiometric calibration that are peculiar to pixelated detectors. Modern array detectors are complex devices provided with many million pixels. Although commercial applications of such devices are not excessively demanding, complex instrumentation programmes require the provision of calibrated detectors. That is to say, each pixel in a detector must be either relatively or absolutely calibrated for sensitivity. It is inevitable in a real manufacturing process that no two pixels will be absolutely identical in terms of performance. Furthermore, as a result of defects in the manufacturing process, it is inevitable that there will be a few pixels that do not function at all. These are referred to as '***dead pixels***'.

To calibrate the relative sensitivity of an array detector, it must be presented with a pool of uniform irradiance across its surface. This is most usually accomplished by means of an arrangement which incorporates an integrating sphere. The process is known as ***flat fielding***.

14.5.2 Image Centroiding

A pixelated detector is, geometrically, a precision component. That is to say, the location and size of individual pixels is constrained to a very high precision. This property is exploited in the application of array detectors in precision metrology. As such, it is possible to locate an image on the surface of a detector to a very high precision. This process is referred to as ***image centroiding***. Most frequently, this is applied to the location of point images, such as those related to the imaging of stars in astronomical instrumentation. The extent of such a point image will be characterised by its point spread function. In terms of locating the centre of this image, it is advantageous that the point spread function covers several pixels. Figure 14.18 illustrated the process schematically.

A number of algorithms exist for image centroiding. Perhaps the most simple to apply is the procedure that effectively locates the centre of gravity of the image. That is to say, each pixel location, (x_i, y_i), is weighted by the pixel flux, Φ_i. The centroid location, (x_0, y_0), is then given by:

$$x_0 = \frac{\sum x_i \Phi_i}{\sum \Phi_i} \qquad y_0 = \frac{\sum y_i \Phi_i}{\sum \Phi_i} \tag{14.31}$$

The precision of centroid location is substantially less than one pixel, most typically 0.1 pixels or less. In practice, any centroiding algorithm should be capable of dealing with any background illumination which may produce a significant systematic error. In addition to dealing with systematic error, the impact of detector noise produces a random positioning error. It is possible to estimate the contribution of this error by adding noise contributions in an rss fashion, applying them to Eq. (14.31), or similar centroiding algorithm.

A centroiding function may also be provided by a semi-discrete segmented detector. One such detector is the so-called quadrant detector which consists of a circular photodiode segmented into four quadrants. Such detectors are most useful in alignment applications, tracking the offset of an alignment laser beam. This application is covered in more detail in Chapter 12.

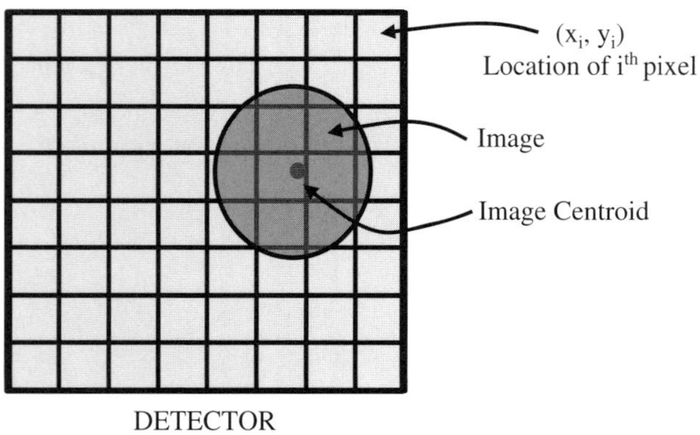

Figure 14.18 Image centroiding.

14.5.3 Array Detectors and MTF

Array detectors, by their nature, do not provide a smooth and continuous representation of an image. ***Nyquist sampling*** dictates that the maximum effective spatial frequency that can be resolved by a pixelated detector is governed by a spatial wavelength equivalent to twice the pixel spacing. That is to say, for a pixel spacing of 5 μm, the maximum spatial frequency that can be satisfactorily observed is 100 mm^{-1}. At spatial frequencies that are higher than this, then ***aliasing*** effects will be observed. The aliasing effect amounts to the generation of a 'beat frequency' between the signal spatial frequency, f_s and the pixel spatial frequency, f_p. The aliasing spatial frequency, f_a, is given by:

$$f_a = f_s - nf_p \tag{14.32}$$

where n is an integer.

As previously outlined, the modulation transfer function or MTF of a system is defined by the reduction in contrast ratio produced by the optical system. It is possible to calculate the effective MTF of the detector resulting purely from the effect of the pixels. Naturally, the MTF should tend to unity for the lowest spatial frequency. It is straightforward to prove by integration that the relevant MTF is defined by the sinc function:

$$MTF = \frac{\sin(\pi f_s/f_p)}{(\pi f_s/f_p)} \tag{14.33}$$

Equation (14.33) is illustrated graphically in Figure 14.19. As Figure 14.19 shows, the contrast or MTF is zero when the spatial frequency of the signal is identical to that of the detector. At the Nyquist frequency, or half the pixel frequency, the MTF is equal to $2/\pi$ or 0.637. The MTF of the detector is an important part of the overall system performance budget. As outlined in Chapter 6, the MTF of a system is equal to the product of all the individual sub-system MTFs; this includes that of the detector.

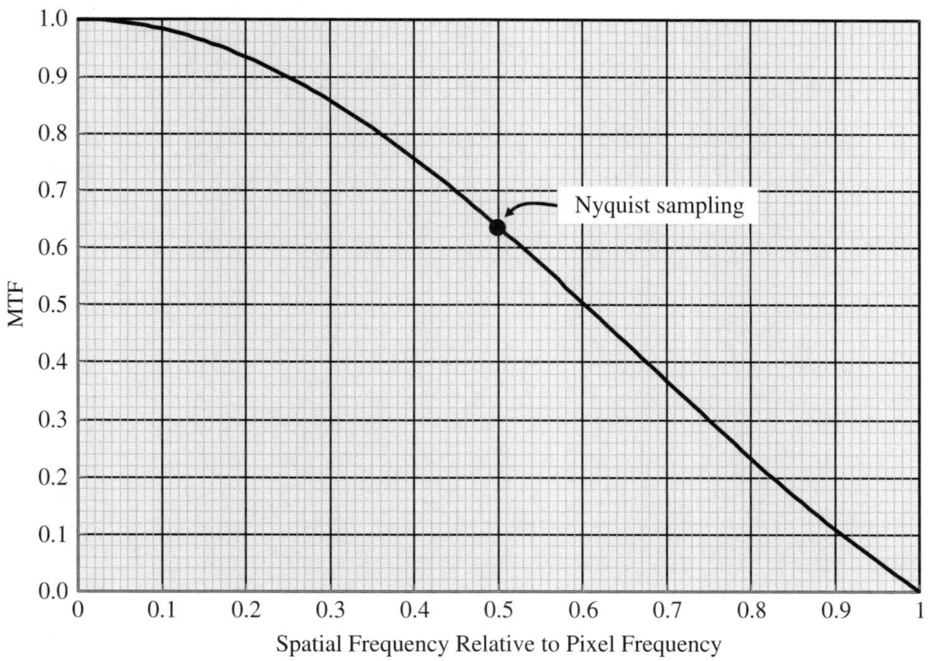

Figure 14.19 MTF of pixelated detector illustrating Nyquist sampling.

Further Reading

Bass, M. and Mahajan, V.N. (2010). *Handbook of Optics*, 3e. New York: McGraw Hill. ISBN: 978-0-07-149889-0.
Derniak, E.L. and Crowe, D.G. (1984). *Optical Radiation Detectors*. New York: Wiley. ISBN: 978-0-471-89797-2.
Kingston, R.H. (1995). *Optical Sources, Detectors and Systems: Fundamentals and Applications*. London: Academic Press. ISBN: 978-0-124-08655-5.
Saleh, B.E.A. and Teich, M.C. (2007). *Fundamentals of Photonics*, 2e. New York: Wiley. ISBN: 978-0-471-35832-9.

15

Optical Instrumentation – Imaging Devices

15.1 Introduction

In this chapter we will examine in a little detail the design of imaging devices, such as telescopes, microscopes, and cameras. The focus will be specifically on the optical design rather than other aspects, such as detectors and the mechanical mounting of optical components. Historically, design of imaging systems has been underpinned by the fundamental notion that the human eye is the only available optical sensor. As such, all instruments were originally designed to relay any image to the eye, usually via a special purpose adaptor, or eyepiece. This might apply to the telescope or the microscope, where an eyepiece of broadly similar design might be used. However, with the advent of photographic media and, more recently, digital sensors, the urgency of this demand has receded somewhat. That is not to say that the design and fabrication of the eyepiece, for example, is wholly unimportant in the current context.

Notwithstanding the continuing demand for 'eye friendly' optics in consumer products, recent developments in sensing media have radically altered the design envelope for imaging optics. For example, the spatial resolution of pixelated detectors is far superior to that of traditional photographic media. High resolution 35 mm slide media might produce a Modulation Transfer Function (MTF) of 0.5 at 40 cycles per mm, whereas a digital detector with a 5 μm pixel size would replicate this performance at about 120 cycles per mm. It is clear, therefore, for a specific angular resolution requirement, the effective focal length of a digital design would be a fraction of that for a traditional design. This fundamental change of length scale not only has implications for product miniaturisation, but also for the realisation of performance metrics.

In this chapter we will consider the design of eyepieces, telescopes, microscopes, and cameras from a more fundamental perspective. Although modern computer aided design tools remove much of the labour from the design process, an understanding of the underlying principles is of great benefit. Underpinning the optimisation of all imaging devices is the desire to minimise all aberrations, particularly third order aberrations. Hitherto, in our treatment of aberrations, we have only considered very simple building blocks, such as mirrors, singlet lenses, and achromatic doublet lenses. Only in the most benign applications are these simple elements adequate. In most practical applications, a large number of optical elements is obligatory in order to provide sufficient degrees of freedom to control aberrations adequately. More specifically, for 'fast', i.e. high numerical aperture, and wide angle systems, the requirement to correct higher order aberrations becomes more pressing. As such, the number of design constraints multiplies considerably and, with this, the number of surfaces required for optimisation.

Some basic design principles have already been set out. In the design of microscopes and telescopes, where the field angles are significantly smaller than the numerical aperture, there is a clear 'hierarchy' of aberrations. The most pre-eminent is spherical aberration, followed by coma, field curvature, and astigmatism. As a consequence, aplanatic elements, which have no spherical aberration or coma feature strongly in such designs. The picture is less straightforward for camera designs where the control of aberrations is complicated by the large field angles involved. Although optical design may be understood, to some degree, by a few elementary principles, elaborate designs feature large numbers of surfaces whose optimisation cannot be related in such a

Optical Engineering Science, First Edition. Stephen Rolt.

Companion website: www.wiley.com/go/Rolt/opt-eng-sci

simple way. As with the study of the game of chess, the variable space is so extensive that, for complex designs, any study based exclusively on first principles has limited tractability. Therefore, optical design, in such cases relies on a library of 'prior art' that is optimised to specific applications.

Another important factor to recognise is that all systems are optimised to operate at a specific conjugate ratio. With the exception of relay lens design, the majority of applications are designed to operate at the infinite conjugate. The contradiction of the Helmholtz equation and the Abbe sine rule implies that substantial correction of aberrations can only be maintained at one conjugate ratio.

15.2 The Design of Eyepieces

15.2.1 Underlying Principles

The function of an eyepiece is to accept light from an intermediate focal plane, usually the focus of an objective lens, and relay it to the eye at the infinite conjugate. Most particularly, the design of an eyepiece cannot be readily understood without comprehending the paraxial properties of the human eye. In setting out some reasonable parameters here, it must be understood that, as human attributes, they are significantly variable. On average, the eye's effective focal length is about 17 mm with a pupil size as large as 8 mm for a dark-adapted eye and as small as 2 mm under bright illumination. This means that the effective numerical aperture of the human eye varies between 0.06 and 0.24 or f#8.5 and f#2.1. This is important, since, ideally, the eyepiece aperture should match that of the eye that it is illuminating. Moreover, the exit pupil of the eyepiece should not be larger than that of the eye itself; any light falling outside the pupil of the eye is thus wasted. In practice, eyepieces are designed with a pupil size of 3–6 mm in mind. As such, the eyepiece geometry is well constrained. Typically, all lenses within an eyepiece are constrained within a standard sized barrel, e.g. 1.25 in. or 31.75 mm.

Although the exit pupil is nominally defined by the human iris, it must be remembered that the ***system*** pupil may be defined elsewhere. For a telescope system, the entrance pupil might be defined by a circular mirror or lens aperture. The limiting aperture is then the smallest of the telescope and eyepiece combination. Ideally, they should be matched. If the telescope lens produces an f#6 beam, then for a 6 mm iris, the focal length of the eyepiece should be 36 mm. If, as is likely, the eyepiece focal length is shorter, then f#6 remains the limiting aperture, not that of the eyepiece. Short focal length, high-magnification eyepieces equate to a large aperture input. The situation is markedly different for a microscope eyepiece. Here, the aperture of the microscope objective is always the limiting aperture and, as projected on the iris, is less than a tenth of the iris diameter.

As important as the size of the exit pupil of the eyepiece is its location. Quite naturally, the exit pupil should coincide with the pupil of the eye. However, ideally, the exit pupil should be a reasonable distance from the last mechanical surface of the eyepiece, e.g. 10–20 mm. This distance is referred to as the ***eye relief*** of the eyepiece. Eye relief allows for the accommodation of eyelashes or even spectacle lenses between the eyepiece and the eye.

In terms of the paraxial treatment of an eyepiece, as expressed by the cardinal point locations, the input focal plane is coincident with the first focal point of the eyepiece and the second focal plane is approximately co-located with the second focal plane. This is, of course, contingent upon the input pupil, usually located at the objective, being far (compared to the eyepiece focal length) from the input focal plane. This is illustrated in Figure 15.1

Another aspect of eyepiece design not apparent from Figure 15.1 is the requirement, in certain specific applications, for an intermediate focal plane to be located within the eyepiece tube. The purpose of this might be to accommodate a reticle for dimensional metrology. The effective focal length of the eyepiece is a fundamental parameter of paramount importance. The focal length might fall in the range from 8 to 30 mm, equivalent to a magnification of between ×5.3 and ×20 for a 160 mm tube length.

Thus far, we have considered only the paraxial properties of the eyepiece. There are a number of further aspects that we need to consider before we can understand how to control aberrations. Eyepieces can be

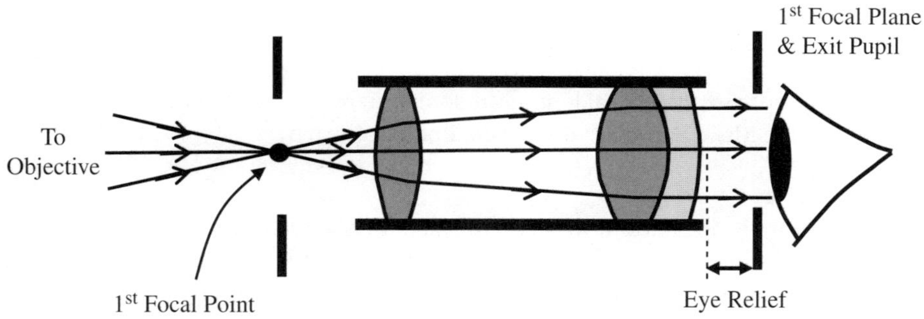

Figure 15.1 Paraxial layout of eyepiece.

nominally designed to provide wide angle viewing, e.g. 60°. This angle of view, the ***apparent field angle***, is the field angle as denominated at the eyepiece itself, rather than at the original object. So, potentially, in view of the wide angles involved, all third order aberrations may contribute significantly. However, it must be understood that, at any one time, the human eye cannot survey this whole field. High acuity vision for the human eye is only reserved for a small field of a few degrees around the central viewing point. Within this restricted field of view, resolution is approximately 1 arcminute and this limitation necessarily drives optical quality requirements for eyepiece design. Surveying of the full field is accomplished by the eye ranging across it. Therefore, field curvature may not be a significant problem. As such, a greater emphasis is placed on image quality in the central field. This is in stark contrast to other imaging systems, such as cameras, where the designer must be equally conscious of performance across the entire field. Whereas field curvature has less prominence in eyepiece design, on the other hand, astigmatism must be considered more carefully. The other aspect that is characteristic of the eyepiece is its relatively short focal length. As outlined in Chapter 2, the magnification of an eyepiece is given by a standard tube length (e.g. 160 mm) divided by the eyepiece focal length. Eyepiece focal lengths may typically fall into the range from 10 to 30 mm.

The first aberration that we need to consider is chromatic aberration. Elementary calculations, based on simple lens elements, indicate that uncorrected chromatic aberration predominates over spherical aberration for typical materials for numerical apertures less than about 0.25. For a 6 mm pupil, this corresponds to a focal length of 12 mm or greater, or an eyepiece magnification of about 14 and less. Hence, chromatic aberration is the primary concern in most practical applications. However, chromatic aberration manifests itself in the form of transverse and longitudinal aberration. The former depends upon pupil size, but not object size, whereas, conversely, the latter depends upon object size, but is independent of pupil size. Essentially, the ratio of the two effects depends upon the ratio of the eyepiece's numerical aperture and field angle. For eyepieces with significant field angles, particularly those will low magnification and longer focal lengths, then transverse chromatic aberration tends to predominate. There is a tendency, in general, for transverse aberration to be the primary concern. Therefore, particularly in more basic eyepiece designs, there is a tendency for the effective focal length of the eyepiece to be colour corrected, as opposed to the focal point locations.

15.2.2 Simple Eyepiece Designs – Huygens and Ramsden Eyepieces

As captured by the preceding discussion, the most elementary designs focus on correction of chromatic aberration. Such designs can operate only over modest fields and their aperture is necessarily restricted. Although these eyepieces feature in a range of low cost consumer designs, their performance is insufficient for even moderately demanding applications. However, examining their design is useful in understanding the correction of chromatic aberration in eyepieces. The ***Huygens eyepiece*** and the ***Ramsden eyepiece*** consist of two simple thin lenses, of the same material, of focal length, f_1 and f_2, separated by half the sum of their focal lengths. That is to say, if d represents the lens separation, it is given by:

$$d = (f_1 + f_2)/2 \tag{15.1}$$

If the given focal lengths f_1 and f_2, represent the focal lengths at some reference wavelength, then at some other wavelengths their revised focal lengths, f_1' and f_2', are approximately given by:

$$f_1' = (1 + \Delta)f_1 f_2' = (1 + \Delta)f_2 \tag{15.2}$$

To assess the impact of lateral chromatic aberration, we need to calculate the effective focal length of the combined system. This may be done by simple matrix analysis which the reader might wish to replicate. The effective system focal length, f, at the reference wavelength is then given by:

$$f = \frac{f_1 f_2}{f_1 + f_2 - d} \tag{15.3}$$

Substituting Eqs. (15.1) and (15.2) into Eq. (15.3), we may determine the effective focal length at any other wavelength:

$$f' = \frac{f_1 f_2 (1 + \Delta)^2}{(f_1 + f_2)(1 + \Delta) - (f_1 + f_2)/2} = \frac{2f_1 f_2}{(f_1 + f_2)}\left(\frac{(1 + \Delta)^2}{(1 + 2\Delta)}\right) \tag{15.4}$$

Assuming Δ is small – it is typically of the order of 1%, then we may approximate Eq. (15.4), as follows:

$$f' \approx \frac{2f_1 f_2}{(f_1 + f_2)}(1 + 5\Delta^2) \tag{15.5}$$

It is clear from Eq. (15.5) that we have isolated the design from any linear dependence on the material dispersion. Any remaining dependence is quadratic. For a Δ equal to about 1%, we have reduced the impact of dispersion from 1% to 0.05%, which is significant. However, this process has only impacted the transverse chromatic aberration, in the form of the effective focal length. If we repeat the matrix analysis to determine the location of focal points we will find that these are significantly impacted by chromatic aberration. Indeed, the shift of the focal point combined with a constant effective focal length implies that the location of the principal planes is significantly affected by dispersion.

In the case of the Ramsden eyepiece, the two element focal lengths are identical. As such, the effective focal length is equal to the focal length of each element. Principal points and focal points are located at the individual lenses themselves, so this design affords no eye relief. Some eye relief may be provided by sacrificing chromatic performance. By contrast, in the Huygens design, the two lenses have different powers. The eye lens (that closest to the eye) is of lower power than the first lens and the ratio of the two lens powers may typically be of the order of a factor of two. This design produces an intermediate focal point between the two lenses, which is useful if one wishes to incorporate a reticle. In the Huygens design, the second focal point location is shifted away from the eyelens and is located between the lenses. It is thus impossible to properly optimise the location of the exit pupil with this design. Figure 15.2 illustrates the paraxial characteristics of these two eyepieces.

15.2.3 Kellner Eyepiece

Apart from the difficulty with inferior eye relief, the major disadvantage of the Ramsden and Huygens eyepieces is their lack of proper colour correction. Substitution of one of the singlet lenses in the basic design with a doublet lens, allows the correction of both transverse and longitudinal colour. Such a modified design is known as the ***Kellner eyepiece***. The design originated in the mid-nineteenth century, when material choices were limited and when there was no reliable technology for producing antireflection coatings. Practical application of the original design was troublesome, particularly with regard to uncontrolled 'ghosting' from internal Fresnel reflections. However, modern Kellner ***type*** designs which use the basic singlet doublet arrangement do provide useful performance at moderate cost.

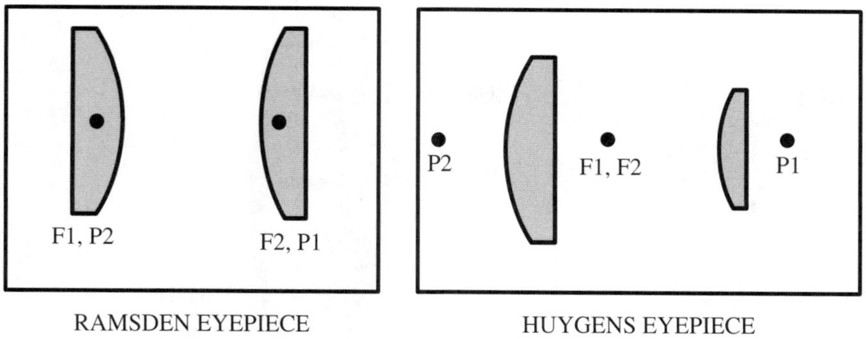

Figure 15.2 Cardinal points of Ramsden and Huygens eyepieces.

Table 15.1 Optimised Kellner design (dimensions in mm).

Surface	Description	Radius	Thickness	Material
1	Entrance pupil (eye)	Infinity	15.0	Air
2	Eye lens – 1st surface	22.091	5.5	SF5
3	Eye lens – 2nd surface	12.479	2.0	BK7
4	Eye lens – 3rd surface	−44.076	23.09	Air
5	Field lens – 1st surface	26.953	3.5	BK7
6	Field lens – 2nd surface	Infinity	4.454	Air
7	**Image**	**−29.0**		

We may illustrate the principles by considering an eyepiece design with a focal length of 30 mm. This would correspond to an effective magnification of around ×5. An eye relief of 15 mm is required together with a 30° field of view. Beyond consideration of chromatic effects, off-axis aberrations, particularly astigmatism, feature significantly. The effective focal length of 30 mm must apply to two wavelengths for colour correction. In addition, the first focal point location is determined by the eye relief requirement and the second focal point, wherever that is, must be collocated for two wavelengths. These four constraints are set against four degrees of freedom, namely the focal power of the singlet and doublet, their separation, and the residual dispersion of the doublet. In principle, it is possible to determine the paraxial prescription algebraically from these considerations. This analysis, applying the thin lens approximation, suggests the eye lens should have a focal length equal to the effective focal length (30 mm) and be designed as a standard achromatic doublet. Separation of the two lenses should be equal to the effective focal length (30 mm) and the focal length of the second lens equal to the square of the effective focal length divided by the focal length minus the eye relief. Adjusting the shape of the doublet and, to a lesser extent, the singlet then serves to minimise other aberrations. However, in this case, different optimisation priorities pertain to this configuration as opposed to a simple achromatic doublet designed for the infinite conjugate.

Whilst this analysis provides a measure of conceptual understanding of the problem, the restricted geometry of eyepiece designs tends to accentuate the impact of lens thickness. In practice, therefore, such designs are now optimised with ray tracing tools, as will be seen in a later chapter. In the meantime, Table 15.1 shows the optimised prescription for our eyepiece design.

The curved surface at the image reflects the latitude we have afforded for field curvature. Over the ±15° field, the curvature amounts to an accommodation of about one dioptre in the eye's focusing power, which is very

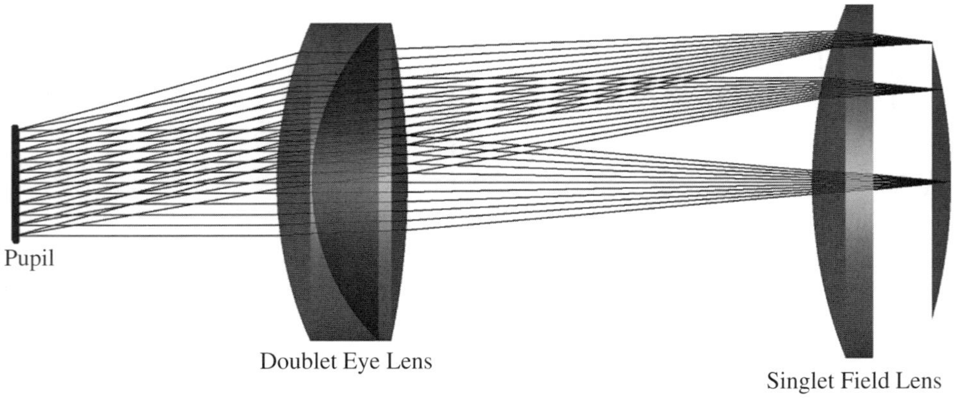

Figure 15.3 Layout of optimised Kellner design.

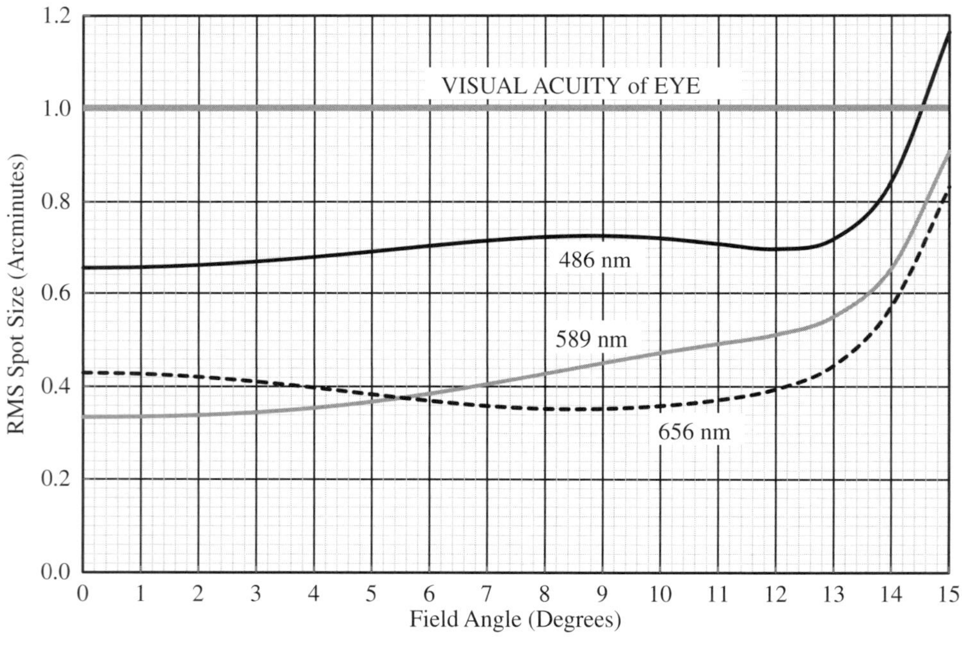

Figure 15.4 Performance of modified Kellner eyepiece.

modest. Figure 15.3 shows the layout of the design. The performance of the design is illustrated in Figure 15.4 which shows the rms spot size, denominated by angle, as a function of field angle.

15.2.4 Plössl Eyepiece

The previous example of the Kellner eyepiece represented a relatively undemanding application. As the physical pupil size is ineluctably fixed by the human iris, decreasing eyepiece focal length and hence increasing magnification inevitably increase the numerical aperture of the eyepiece. This will inevitably exacerbate aberrations. However, in addition, the relative proportion of the desirable eye relief to the focal length also increases. Accommodating both image quality and eye relief make design especially challenging for high magnification eyepieces. Furthermore, high specification designs feature large field angles, further compounding difficulties. The multiplication of design constraints is inevitably accommodated by the introduction of more degrees of

Table 15.2 Plössl eyepiece prescription.

Surface	Description	Radius	Thickness	Material
1	Entrance pupil (eye)	Infinity	15.0	Air
2	Eye lens – 1st surface	−82.382	4.0	SF8
3	Eye lens – 2nd surface	46.136	8.5	SK4
4	Eye lens – 3rd surface	−25.508	1.60	Air
5	Field lens – 1st surface	25.508	8.5	SK4
6	Field lens – 2nd surface	−43.136	4.0	SF8
7	Field lens – 3rd surface	82.382	21.98	Air
8	**Image**	**−52.42**		

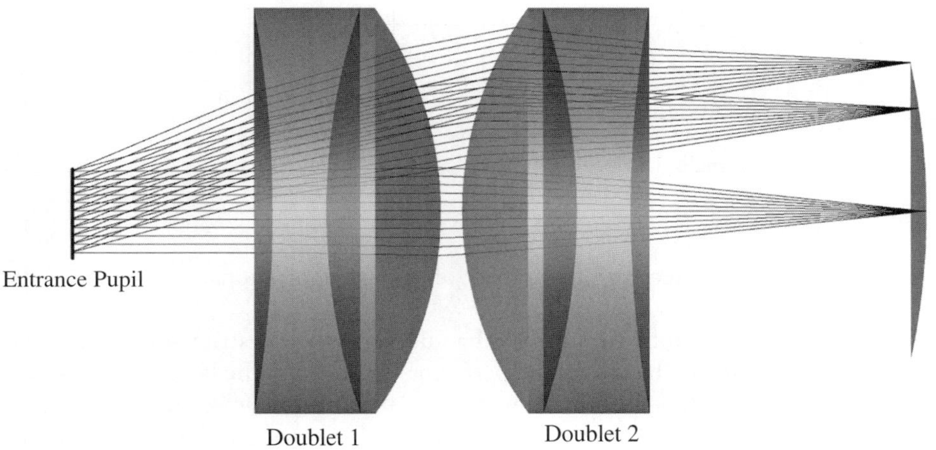

Figure 15.5 Plössl eyepiece layout.

freedom, i.e. more surfaces and elements. The simplest extension to the Kellner eyepiece is a symmetrical four element design, known as the ***Plössl eyepiece***. This consists of two symmetrically arranged achromatic doublets. Table 15.2 shows the prescription for an illustrative design for a symmetrical Plössl eyepiece. In this case, the same specifications apply as for the Kellner design, except for a substantially increased field angle of ±22.5° (45° FOV). Figure 15.5 shows the layout of the design example.

The Plössl eyepiece does provide an incremental increase in image quality over the Kellner design. This is illustrated in Figure 15.6 which shows RMS spot size of the design versus field angle.

Comparison of Figures 15.4 and 15.6 clearly shows an improvement in the spot size, particularly considering the larger field angles. Astigmatism and coma feature significantly in residual aberration. However, analysis of the wavefront error reveals a significant presence of higher order aberration terms.

15.2.5 More Complex Designs

In introducing this topic we have introduced a few simple designs that illustrate both the design principles involved and the historical evolution of eyepiece design. Nevertheless, the Kellner and Plössl designs or variants thereon do feature in modern applications where performance requirements are relatively modest and where cost is a factor. More sophisticated designs feature an increasing field of view (>60°) as a salient requirement. In addition, adequate eye relief, particularly for short focal length eyepieces is a further challenge.

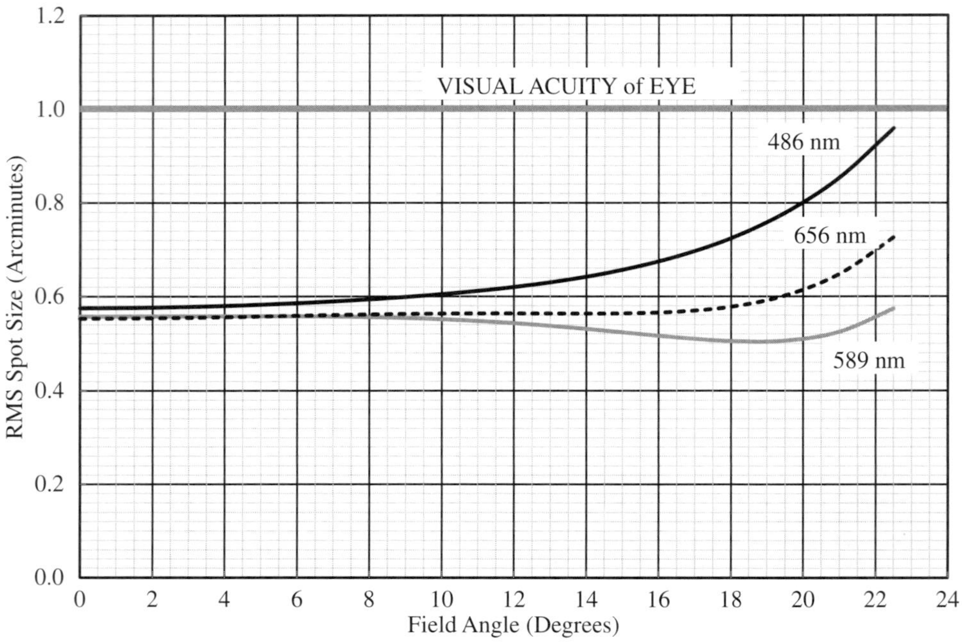

Figure 15.6 Performance of Plössl eyepiece.

Furthermore, shorter focal lengths further exacerbate the impact of on axis aberrations, by increasing the numerical aperture.

Historically, eyepiece design was constrained by two specific handicaps. First, a restricted range of glass types was available to the designer to optimise the chromatic performance. Second, the lack of high performance optical coatings made reflections from optical surfaces particularly troublesome and this militated against the adoption of designs with a large number of optical surfaces. This constraint has been substantially removed and cost is the only predominating factor in the complexity of eyepiece design. Inevitably, high performance is achieved by increasing the number of optical elements, permitting more degrees of freedom in the design. For the designs previously introduced all elements were positive. As such, these simple designs inevitably have significant Petzval curvature. Therefore, inevitably, most sophisticated designs feature elements with negative power to achieve a flatter field.

As indicated previously, complex, multi-element designs rely, to some extent, on modifications to a 'library' of existing designs, rather than a simple process of design from first principles. Optimisation, where higher order aberrations are present, is substantially a 'non-linear' problem, where a large number of interactions between variables make optimisation an inherently complex process. Of course, traditionally, this problem was tackled with abstruse high order aberration analysis techniques and by useful general principles, such as the Abbe sine law. However, these difficulties have been largely overcome, with modern computational power. Ray tracing packages allow for the rapid optimisation of highly complex designs with a large number of variables.

Refinements to the basic three-element Kellner design feature a reversal in the layout, with the achromatic doublet featuring as the field lens and the singlet as the eye lens. These are the so-called König and RKE (*Rank-Kellner Eyepiece*) designs. Another useful four element design is the ***orthoscopic*** or ***Abbe eyepiece***. In this case, the eye lens is a simple plano-convex singlet, followed by a triplet lens. The term orthoscopic refers to the eyepiece's low distortion. An incremental improvement to the Plössl eyepiece inserts an additional singlet lens between the two doublets. This improvement is the ***Erfle eyepiece***. These designs may be adapted and

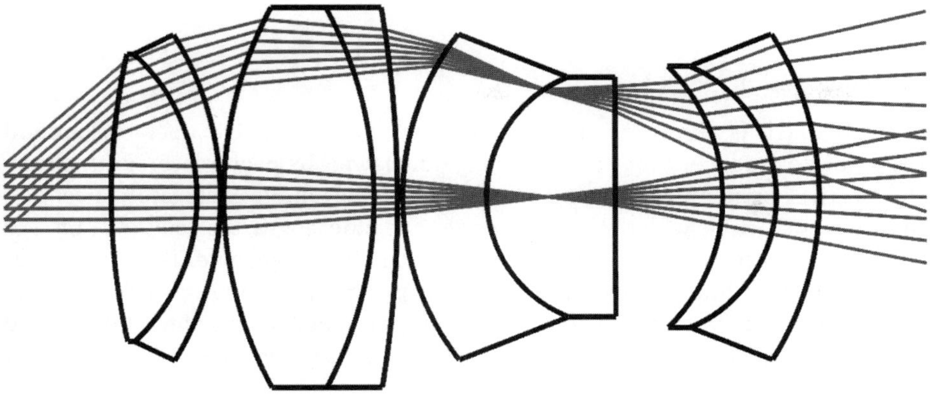

Figure 15.7 Modified Nägler eyepiece.

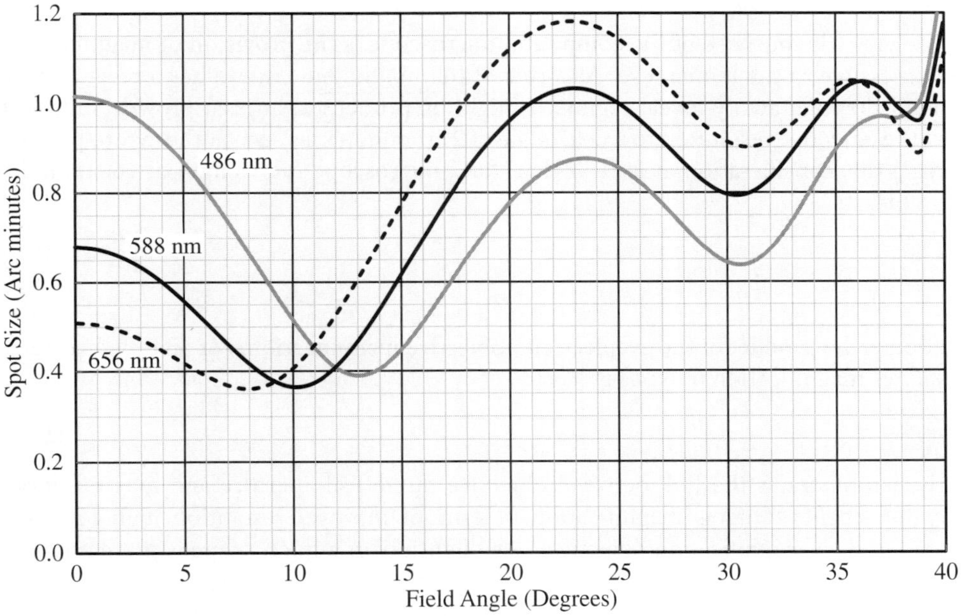

Figure 15.8 Performance of modified Nägler eyepiece.

variants may introduce additional lens elements. An example of a more modern, complex design is the ***Nägler eyepiece.*** This consists of a doublet field lens with negative power, followed by a large group of positive lenses. Up to eight lens elements may feature in the design. The design of the field lens helps to reduce the overall Petzval sum. Furthermore, spreading the refractive power over a relatively large number of elements helps to further reduce aberrations. Nägler eyepieces are specifically designed for high performance over a very wide field; field angles in excess of 80° are possible. In addition, they can be designed to provide excellent eye relief. Figure 15.7 shows an example of a modified Nägler design with eight elements. This design is for a ×10 eyepiece with a focal length of 16 mm and an eye relief of 16 mm, with a maximum field angle of ±40°.

Figure 15.8 illustrates the performance of the eyepiece graphically, confirming the improvement in performance.

15.3 Microscope Objectives

15.3.1 Background to Objective Design

A microscope objective is a compound lens with a very short focal length usually designed to operate at around the infinite conjugate. Its purpose is to produce an intermediate image for viewing by an eyepiece, or other relay lens system. Although such objectives are nominally designed to function at the infinite conjugate, convention dictates that the image distance is compatible with a standard microscope tube length, typically 160 mm. For some specific applications, certain objectives ***are*** designed to work at the infinite conjugate. These objectives are referred to as ***infinity corrected***. The ultimate purpose of a microscope objective is to resolve the smallest possible detail. Unlike the eyepiece, the microscope objective is generally designed to give diffraction limited performance across its field of view. As such, the resolution of a microscope objective is driven by its numerical aperture, as given by the classical Rayleigh criterion formula:

$$\Delta x = \frac{0.61\lambda}{NA} \tag{15.6}$$

λ is the wavelength and NA the numerical aperture

In the case of ocular viewing, the purpose of the compound microscope is the provision of magnification to make this resolution accessible to the human eye. As a rule of thumb, the human eye has a resolution of about 1 arc minute around the high acuity foveal region. It might be prudent to at least provide sufficient magnification to convert the objective resolution to a field angle of 2 arcminutes, as seen at the eye, to provide adequate margin. Applying this consideration, it is clear that, for a microscope tube length of 160 mm, the magnification should be at least:

$$M_{system} > \frac{153 \times NA}{\lambda} \tag{15.7}$$

λ is in microns

The system magnification is the product of the eyepiece and objective magnification. For an eyepiece magnification of ×5, the objective magnification must be at least:

$$M_{objective} > \frac{30.5 \times NA}{\lambda} \tag{15.8}$$

Equation (15.8) clearly demonstrates the link between objective numerical aperture and magnification. Hence, generally, to reap the benefits of higher magnification, in terms of improved resolution, higher numerical apertures are essential in high magnification objectives. As the utility of a microscope is pre-eminently driven by its resolution, there is a significant premium for maximising the numerical aperture of an objective. An extreme example of this is in the design of the oil immersion objective. In this design, both objective and object are immersed in oil, or some other high refractive index fluid. As a consequence, numerical apertures in excess of one, e.g. 1.3 are achievable. Given that we have defined two key attributes of a microscope objective, namely high numerical aperture and high magnification, we can characterise a microscope objective as a diffraction limited lens of high numerical aperture and short effective focal length.

If the field angle is defined by the field stop of the eyepiece, then the field seen by the objective is equal to the eyepiece field divided by its magnification. For an eyepiece magnification of ×5 and a viewing field of ±15°, then the objective will see a field angle of only ±3°. Thus, we can further view the objective as a lens of high numerical aperture, but low field angle. As such, the behaviour of a microscope objective with respect to third order aberrations follows the discussion framed in Chapter 4 concerning the 'hierarchy' of aberration. The correction of on-axis aberrations, such as spherical aberration is the most salient task facing the designer. Of the off-axis aberrations, only coma is a particular concern. Therefore, the design of microscope objectives is significantly informed by the narrative in Chapter 4, particularly concerning the value of aplanatic designs that eliminate both spherical aberration and coma. Indeed, Chapter 4 introduced the example of the aplanatic hyperhemisphere, which is a key element in the design of a high-power objective. In this case, the hyperhemisphere is introduced as the primary aplanatic component at the object location, with additional power provided by adding meniscus lenses.

However, the preceding discussion omits the significant impact of chromatic aberration. To quantify this, we should imagine an objective fabricated from a material with an Abbe number given by V_D. Furthermore, the focal length of the objective is f and the numerical aperture, NA. The wavefront error caused by chromatic defocus (difference between the C/F and D wavelengths) is then given by:

$$\Phi = \pm\frac{NA^2}{V_D 8\sqrt{3}}f \tag{15.9}$$

For the system to be diffraction limited:

$$\frac{NA^2}{V_D 8\sqrt{3}}f < \frac{\lambda}{14} \text{ and } V_D > \frac{7NA^2 f}{4\sqrt{3}\lambda} \tag{15.10}$$

If we attempt to capture the relationship between focal length and magnification (via. Eq. (15.8)), we arrive at the following inequality defining the minimum Abbe number:

$$V_D > 33 \times D \times NA \textit{ where D is denominated in mm.} V_D > 5300 \times NA \text{ (for } D = 160) \tag{15.11}$$

For most practical materials, V_D falls in the range from 25 to 100. Clearly, an uncorrected objective is not a tenable design. Furthermore, as one might expect, the problem becomes more severe as the numerical aperture is increased. Whilst we have established that the correction of primary colour is imperative, we need to examine the impact of secondary colour. Equation 4.58 in Chapter 4 established the focal shift due to secondary colour, in an achromatic doublet as expressed by the partial dispersions of the two glasses involved.

$$\Delta f = \frac{P_2 - P_1}{V_1 - V_2}f \tag{15.12}$$

P_1 and P_2 are the partial dispersions of the glasses and V_1 and V_2 are the Abbe numbers

Expressing Eq. (15.12) as a minimum condition for satisfactory performance, as per Eq. (15.11), we may set out the requirement for secondary colour.

$$\frac{V_1 - V_2}{P_2 - P_1} > 5300 \times NA \tag{15.13}$$

For the main series of glasses there is a clear linear relationship between the Abbe number and the partial dispersion:

$$\frac{\Delta V}{\Delta P} = 2000 \tag{15.14}$$

Taken together, Eqs. (15.13) and (15.14) set clear limits on the numerical aperture for simple correction of primary colour.

$$NA < 0.38 \tag{15.15}$$

Therefore, there is a clear need for secondary colour to be corrected in higher numerical aperture and hence higher magnification objectives. So called '***apochromatic***' designs, incorporating fluorite elements are a feature of such high-specification microscope objectives.

Another important aspect of microscope objective design, again set out in Chapter 4, is the general use of microscope cover slides with optical microscopes. Microscopes are designed to work with a thin piece of glass, the '***cover slip***', to protect the specimen. For high numerical aperture objectives, the spherical aberration produced by a flat piece of glass is proportional to its thickness and the fourth power of the numerical aperture. In practice the aberration produced is sufficient to compromise diffraction-limited performance. Therefore, microscope objectives are designed specifically to compensate for this added spherical aberration. Quite clearly, as the aberration is proportional to the thickness, objectives are designed for a specific standard thickness of cover slip. The most common standard thickness, particularly for biological specimens is 0.17 mm.

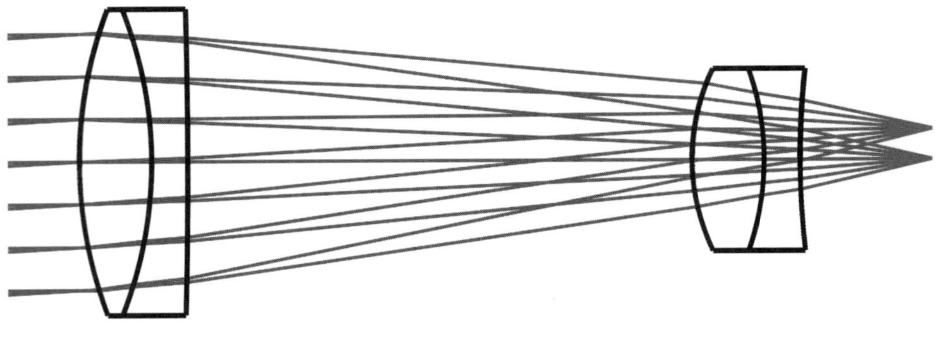

Figure 15.9 Simple ×10 microscope objective.

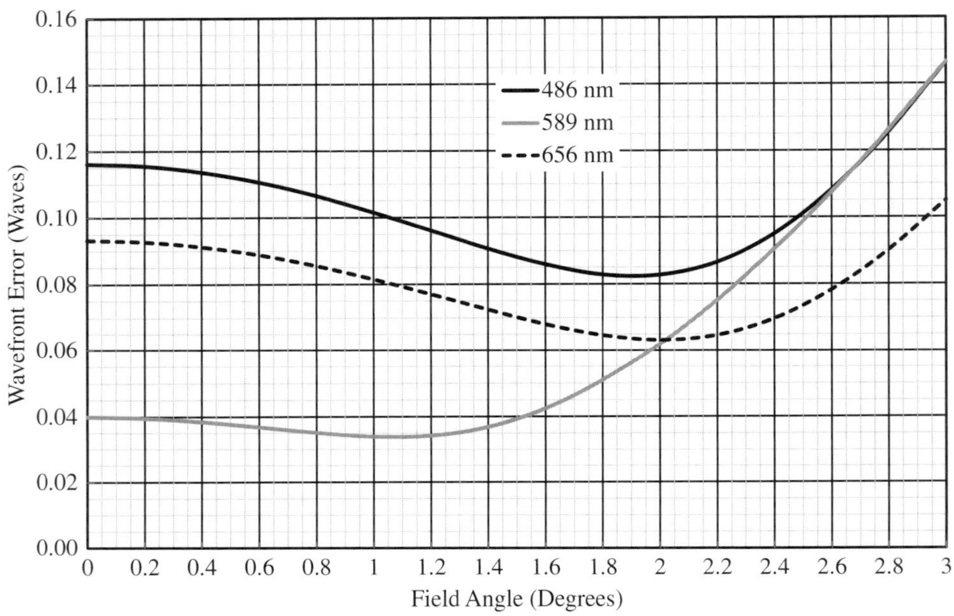

Figure 15.10 Wavefront error performance of simple ×10 microscope objective.

15.3.2 Design of Microscope Objectives

For the lowest possible magnifications and numerical apertures, the simplest microscope objective is an achromatic doublet. This is adequate for very low magnifications and numerical apertures. Somewhat more effective than this approach is the incorporation of an additional doublet. Not only does this provide extra degrees of freedom in the design but, by sharing the power between more surfaces, higher order aberrations are further restricted. Figure 15.9 shows an example design for a ×10 objective with a numerical aperture of 0.2.

In general, in optimising a design for visible wavelengths, it is customary to optimise for three representative wavelengths spanning the visible region. A popular convention is to use the C, D, and F lines at 486, 589, and 656 nm respectively. The simple design exhibits outlined provided close to diffraction limited performance across the 3° field, especially for the central wavelength of 589 nm. This is illustrated in Figure 15.10.

Higher magnification objectives are based on an aplanatic design, often featuring a hyperhemisphere as the first element. Meniscus lenses are incorporated to add power to the system whilst preserving the aplanatic character of the design. The related exercises in Chapter 4 covered only the monochromatic aberrations. To the basic aplanatic design, therefore, must be added appropriate colour correction. In addition, we must include

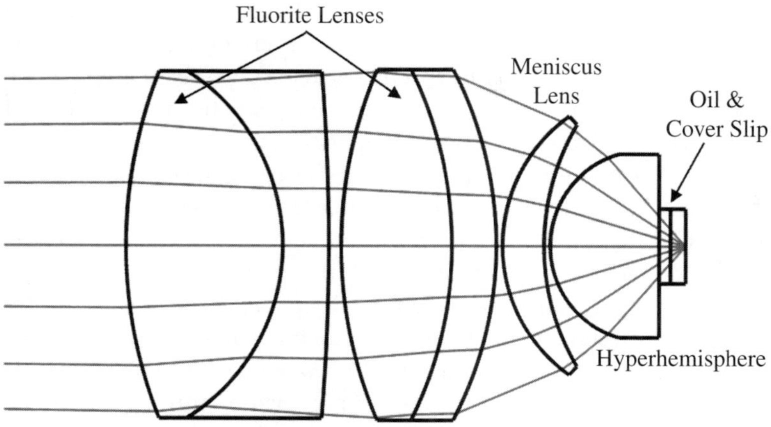

Figure 15.11 ×100 Microscope objective.

the cover slip in the design and, for high numerical aperture oil immersion objectives, a specified thickness of oil also forms an integral part of the designs; this is assumed typically to be around 0.14 mm.

It is worthwhile, at this point, to discuss more fully the utility of the aplanatic hyperhemisphere in objective design. The aplanatic hyperhemisphere is especially useful not only in eliminating third order spherical aberration and coma, but also providing perfect imaging on axis regardless of numerical aperture. As an example, in the design of a ×100 objective, a SF66 ($n = 1.923$) hyperhemisphere will, ***on its own***, yield substantially diffraction-limited performance for numerical apertures up to 0.9. Of course the hyperhemisphere does not, in itself, produce an image at the correct conjugate. If one assumes that the image is to be located at the infinite conjugate, then the hyperhemisphere produces an intermediate object whose effective numerical aperture has been reduced by a factor equal to the square of the refractive index. So, in the preceding example, the effective numerical aperture has been reduced from 0.9 to about 0.24. This effect may be enhanced by addition of further meniscus lenses, with each addition reducing the effective numerical aperture by a factor equal to the refractive index. Hence, the design of the succeeding optical train becomes more tractable and less demanding with a lower numerical aperture; the effective field angle will, of course, increase. This is illustrated in Figure 15.11.

Upon the succeeding optical train will necessarily fall the entire burden of colour correction. As discussed earlier, an achromatic design does not provide adequate colour correction in high-magnification objectives. Fluorite or calcium fluoride optics feature in all high-specification designs. This is because the fluoride group of materials lies outside the 'main series' of glass characteristics and do not follow the behaviour indicated in Eq. (15.14). Although the aplanatic hyperhemisphere does provide good correction for higher order aberrations, nonetheless a large number of surfaces is needed to provide full correction in high-performance objectives.

Another critical feature of high-performance microscope objectives is their sensitivity to alignment, particularly lateral alignment offset. The effect of small lateral misalignments in optical elements is to produce off axis type aberrations, such as coma, for central field locations. As such, objects often implement some form of (factory) alignment adjustment to compensate.

15.4 Telescopes

15.4.1 Introduction

Apart from the obvious size distinction, telescopes share many of the design imperatives of microscope objectives. In the case of telescopes, in general, the field angle is even more restricted than that of the microscope objective, often amounting to no more than a fraction of a degree. Moreover, although operating at infinite conjugate ratio, it is the object, rather than the image that is located at infinity. Therefore, it is the angular resolution of the telescope objective that is the critical requirement and this is determined by the size of the objective. This lies in contrast to the microscope objective where the performance is determined by spatial resolution and hence the numerical aperture of the objective. From a design perspective, this creates a more benign environment, with the premium on high system numerical aperture sharply reduced. As a consequence, with a smaller numerical aperture and more restricted field, it is possible to design a telescope system with relatively few surfaces whilst maintaining diffraction limited performance.

Whilst the design environment for the telescope might be relatively benign, the premium on objective size indicates that the challenges lie primarily in the engineering, rather than the design. However, for terrestrial observations, an important restriction relates to the fundamental angular resolution of a telescope system. The atmosphere produces a stochastic contribution to system wavefront error. As a rule of thumb, at ***visible wavelengths*** and for ***highly stable*** atmospheric conditions (at night), vertical propagation through a nominal 11 km atmospheric thickness contributes to a Strehl ratio of 0.8 for an aperture size of 100 mm. That is to say, an aperture size of 100 mm might be deemed to produce 'diffraction limited performance'; thereafter, in terms of system resolution, the utility of further increases in aperture size is sharply diminished.

Historically, from the perspective of astronomical optics, the value of large system apertures was invested primarily in photometric performance, rather than optical resolution. That is to say, according to this perspective, the primary role of the telescope is to act as a 'light bucket'; the provision of greater étendue enables the detection of fainter sources. However, this rationale has changed substantially in more recent years. Firstly, a significant number of systems, for example, the Hubble Space Telescope, have been designed for the space environment. Here the impediment of atmospheric mediation has been entirely removed. This consideration also applies, to a significant extent, to the increasing number of Earth observation systems in low Earth orbit. Although atmospheric effects do degrade performance to an extent, this is much less marked than for comparable terrestrial applications. In addition, in terrestrial applications, technological advances have enabled the compensation of atmospheric propagation effects through ***adaptive optics***. The study of adaptive optics lies beyond the scope of this book. Broadly, it involves the monitoring of wavefront error across the pupil with a wavefront sensor and then compensating this error by means of a conjugated deformable surface, such as a deformable mirror. In any case, the fruition of such technologies has stimulated the development of larger terrestrial telescope systems with diffraction limited performance, in more recent years.

It must be further emphasised that the hierarchy of aberrations applies to the design of telescope systems. First, one should correct for spherical aberration, then coma, and then field curvature or astigmatism. In analysing the telescope system, we are simply considering an optical system with a long focal length, or plate scale, delivering light to some focal plane or surface. Subsequent viewing of this image plane by eyepiece or further instrumentation optics is not considered here.

15.4.2 Refracting Telescopes

If one ignores the radiometric aspects of telescope design, then in terms of resolution, a useful benchmark indicator, for traditional instruments, is the performance of a system with an aperture of 100 mm. The most simple optical system imaginable comprises a single achromatic doublet. This lens is substantially corrected for spherical aberration and coma as well as primary chromatic aberration. For a given telescope aperture, third order spherical aberration scales with the inverse third power of the focal length. Similarly, chromatic aberration, both primary and secondary, scales inversely with the focal length. Therefore, from an optical perspective, the longest possible focal length is desirable. However, from an engineering perspective, a compact design is preferred, with a shorter focal length preferred. As such, practical design is a classic engineering compromise.

In older refractive designs, an f#10 aperture may have been representative; in more recent times, somewhat faster designs are preferred. Nevertheless, it is useful to consider a 100 mm aperture f#10 design, and consider the magnitude of the different aberrations. Uncorrected chromatic aberration, for an Abbe number of 60 would produce as much as $\pm 3\,\mu m$ rms defocus wavefront error. As far as secondary colour is concerned, for 'main series' glasses, the defocus error might be about one-thirtieth of this or about 100 nm rms. This is not quite diffraction limited and there is some utility in correction for secondary colour. This is particularly true for 'faster' designs and, as such, some 'high end' amateur telescopes do employ triplet lenses incorporating one fluorite element. Higher order (i.e. fifth order) spherical aberration, by comparison is negligible. Sphero-chromatism has a larger impact but is less significant than secondary colour.

By far the most salient objection to the use of refracting objectives in large telescopes is their inherent lack of scalability. As a transmitting optic, a glass lens must necessarily be held by mounting at the periphery of the optic. For larger optics, this poses serious mechanical challenges, requiring prohibitively large lens thicknesses to provide the necessary rigidity. This difficulty does not apply to mirror optics where mounting support may be distributed evenly across the optics. Therefore, larger telescopes almost exclusively employ mirror optics, where, in addition to the advantages of scalability, the concern about chromatic effects is entirely removed. There are some exceptions to this general rule. For solar observations, especially of the solar corona, refracting telescopes are preferred because these inherently produce lower levels of optical scattering which might otherwise swamp the observational signal. In addition, lens optics may be used in combination with mirror optics to provide aberration compensation, rather than optical power. These systems are referred to as ***catadioptric systems***.

15.4.3 Reflecting Telescopes

15.4.3.1 Introduction

For the most part, for astronomical and remote applications, reflecting telescope solutions are preferred. First, chromatic dispersion is absent and second reflecting designs are inherently scalable. Initially, we will consider only ***on-axis designs***, where the chief ray consistently follows a common axial path when passing from one reflecting surface to the next. For a system with several mirror surfaces, primary, secondary, and tertiary etc., it is inevitable that succeeding surfaces will engender some obscuration of the light path. Most usually, the system stop is defined by the first or primary mirror. Almost inevitably, by design, subsequent mirror surfaces are smaller. The effect of this is to produce an annular pupil shape with a small central obscuration. Apart from the small reduction in étendue, this is not, in itself, a problem. Otherwise, the only tangible impact of pupil obscuration is a subtle amendment of the Airy diffraction pattern. However, should it be necessary, this problem may be avoided in off axis systems, where mirror elements are tilted to avoid pupil obscuration. This comes at the cost of increased design complexity, with a greater requirement for compensating off axis aberrations.

15.4.3.2 Simple Reflecting Telescopes

The most basic reflecting telescope designs use a single reflecting mirror to deliver optical power. The most familiar example is the ***Newtonian*** telescope which uses a parabolic primary mirror as the objective; light is diverted to the eyepiece or camera via a 45° mirror. However, in terms of physical implementation, the geometry of this simple design is somewhat inconvenient. A wholly axial design may be preferred, where a secondary mirror is used to retroreflect light back through a conveniently engineered aperture in the primary mirror. This basic design is referred to as a ***Cassegrain*** system. These two basic designs are illustrated in Figure 15.12.

The use of a parabolic primary mirror confers perfectly unaberrated image formation for on-axis field points. For this simple system, the most significant uncorrected aberration is coma. To understand a little more about the underlying principles of telescope design, it would be useful, at this point to quantify this aberration. If the

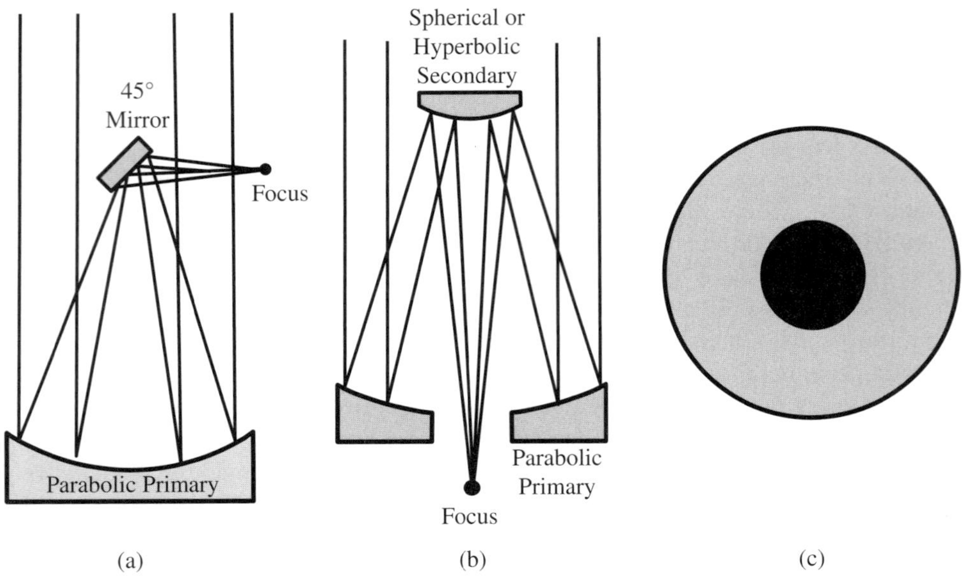

Figure 15.12 (a) Newtonian layout. (b) Cassegrain layout. (c) Pupil obscuration.

maximum field angle is denoted by θ_0, the primary (pupil) semi-diameter is r_0, then the rms wavefront error associated with coma is given by:

$$\Phi_{rms} = \frac{r_0^3\theta_0}{\sqrt{72}R^2} \tag{15.16}$$

R is the radius of the primary mirror

For the system to be diffraction limited at the extreme field, by virtue of the Maréchal criterion, the maximum field angle must obey the following inequality:

$$\frac{r_0^3\theta_0}{\sqrt{72}R^2} < \frac{\lambda}{14} \text{ and } \theta_0 < \sqrt{2}\frac{12\lambda f^2}{7r_0^3} \tag{15.17}$$

f = focal length = R/2

For a fixed focal ratio, e.g. f#8, the maximum field angle varies inversely with the system focal length and aperture. In the case of an f#8 system with a primary mirror diameter of 200 mm, the maximum field angle is about 0.21°. For wider fields and for larger systems, then further correction would be needed. This is more especially true for those systems not degraded by atmospheric propagation. A further measure of the utility of these simple systems is a measure of the number of lines, N, that might be resolved across the entire field. As we have now established the maximum field for diffraction limited resolution, we simply need to divide the total field, $2\theta_0$, by the diffraction limited resolution:

$$\Delta\theta = \frac{0.61\lambda}{r_0} \text{ and } N = \frac{2\theta_0}{\Delta\theta} = \frac{7.95f^2}{r_0^2} \text{ or } N = 31.8(f\#)^2 \tag{15.18}$$

Hence the maximum number of resolvable lines is simply proportional to the square of the f number. Although increasing the f number, in principle, improves the system resolution, it does come at the cost of reduced system étendue. Hence, as with systems engineering in general, the design process, in practice, reflects an arbitration between seemingly conflicting goals. The resolution metric directly relates to the granularity or information content of the final image. For example, in the case of an f#8 system, the resolution, N, would

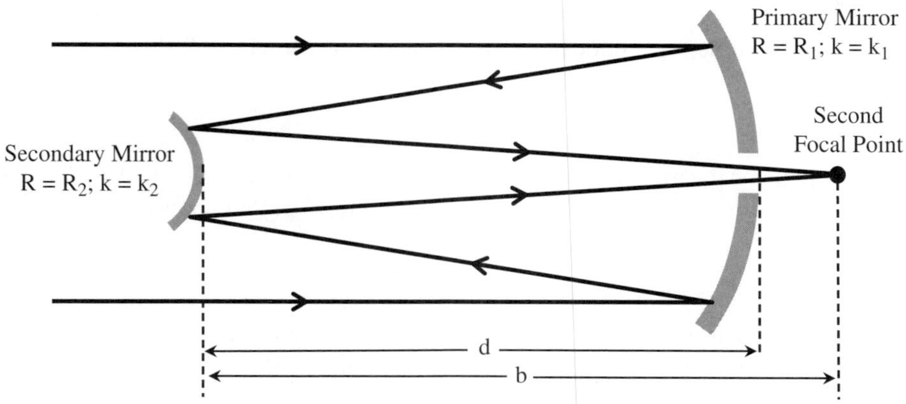

Figure 15.13 Ritchey-Chrétien telescope.

amount to about 2050 lines. If one assumes that this is to be sampled by a digital camera, then, for Nyquist sampling, at least 4000 pixels are required across the field. Depending upon format, this might be represented by a 4000 × 3000 pixel detector, or a 12 MPixel camera. This discussion clearly illustrates that increasing performance and resolution in telescope design must necessarily be accompanied by a proportional increase in the capacity of the detection system.

15.4.3.3 Ritchey-Chrétien Telescope

The incorporation of two curved mirrors into a reflecting telescope design increases the degrees of freedom available to the designer. In terms of the first order optical parameters, a two-mirror design enables the definition of a compact system with a very long focal length. For some astronomical systems, a long focal length is essential to provide a large plate scale, enabling the resolution of objects of very small angular size. In a Newtonian or simple Cassegrain reflector, the length of the system envelope is determined by the focal length, which is directly related to the primary mirror radius. To illustrate this, the focal length of the Hubble Space Telescope is 56.6 m. Quite apart from other design considerations, to realise this effective plate scale in a single mirror system would create an instrument of quite excessive length.

At this point we introduce the Ritchey-Chrétien design which consists of a conic primary mirror and a conic secondary mirror. The design is illustrated in Figure 15.13 showing a system with a primary mirror of radius R_1 and a secondary mirror of radius R_2. The two mirrors are separated by a distance, d, with a distance b separating the second focal point at the secondary mirror. This distance is sometimes referred to as the back focal length, although, strictly, in this instance, the surface that is most proximate to the focal plane is the primary mirror itself.

To establish the first order parameters, we need to consider the matrix for the primary mirror, the translation matrix for the mirror separation (minus d) and the matrix for the secondary mirror. For consistency, we should then trace back to the original reference at the primary mirror. The relevant system matrix is given by:

$$M = \begin{bmatrix} \left(1 + \frac{4d}{R_1} - \frac{2d}{R_2} - \frac{4d^2}{R_1R_2}\right) & \left(2d + \frac{2d^2}{R_2}\right) \\ \left(\frac{2}{R_1} - \frac{2}{R_2} - \frac{4d}{R_1R_2}\right) & \left(1 - \frac{2d}{R_2}\right) \end{bmatrix} \tag{15.19}$$

The effective focal length is given by:

$$f = \frac{R_1R_2}{2(R_1 - R_2 + 2d)} \tag{15.20}$$

It should be noted that for both mirrors, positive curvature consists a sag that lies in the positive direction – in the case of Figure 15.13, to the right. As such, both mirrors in Figure 15.13 have negative curvature.

The effectiveness of the design in contracting the system length is defined by the so called secondary magnification, M_2. This is defined as the ratio of the system focal length to that of the primary ($-R_1/2$).

$$M_2 = \frac{-R_2}{(R_1 - R_2 + 2d)} \tag{15.21}$$

We can also derive the radii from the focal length and the mirror separation, d, and the so-called back focal length, b, which, itself may be derived from the system matrix as the second focal point location:

$$R_1 = \frac{2df}{b-f}\ R_2 = \frac{2db}{d+b-f} \tag{15.22}$$

However, perhaps the most significant feature of the Ritchey-Chrétien design is its ability to restrict aberration over a wider field. In terms of the hierarchy of aberrations, the Newtonian type telescope effectively dealt with spherical aberration with its parabolic primary mirror. By adding a second mirror, we are, in principle, able to correct for the next candidate in the hierarchy, namely coma. This we do by independently adjusting the conic constant of the primary (k_1) and that of the secondary (k_2). As such, we are able to provide aberration correction over a wider field. Of course, with two mirrors, we are still unable to correct for off-axis field curvature and astigmatism. Nevertheless, this represents a significant advance. By analogy with our previous discussion for the single mirror telescope, the resolution, in terms of the number of lines resolved across the field, will be proportional to the cube of the focal ratio. This will increase substantially the granularity of the final image, or enable the use of lower focal ratios and increased radiometric performance.

As highlighted previously, whilst such telescopes present formidable engineering challenges, understanding the basic design analysis is relatively elementary. If we assume that the primary mirror represents the input pupil, by applying the stop shift equations to the secondary mirror, we are able to calculate the system spherical aberration and coma. Furthermore, if we include the impact of the conic surfaces, via the two conic constants, k_1 and k_2 we can, by simple algebraic manipulation, determine the two constants. The spherical aberration contributions of the primary and secondary mirrors are given by:

$$K_{1SA} = -(1+k_1)\frac{r_0^4}{4R_1^3}\ K_{2SA} = \left[k_2 + \left(\frac{R_1+2d-2R_2}{R_1+2d}\right)^2\right]\frac{(R_1+2d)^4 r_0^4}{4R_2^3 R_1^4} \tag{15.23}$$

In the case of coma, contributions arise from both the primary and secondary mirrors in the usual way. However, the secondary mirror also contributes coma by virtue of the transformation of spherical aberration via the stop shift effect. Overall, coma contributions are given by:

$$K_{1CO} = -\frac{r_0^3\theta}{R_1^2} \tag{15.24}$$

$$K_{2CO} = -\left[\left(\frac{R_1+2d-2R_2}{R_1+2d}\right)\right]\frac{(R_1+2d)^2}{R_2^2R_1^2}r_0^3\theta + \left[k_2 + \left(\frac{R_1+2d-2R_2}{R_1+2d}\right)^2\right]\frac{d(R_1+2d)^3}{R_1^3R_2^3}r_0^3\theta \tag{15.25}$$

The conic constant, k_2 may be uniquely determined by the requirement that the coma should be zero. Eliminating common factor and equating to zero, we obtain:

$$1 + \left[\left(\frac{R_1+2d-2R_2}{R_1+2d}\right)\right]\frac{(R_1+2d)^2}{R_2^2} - \left[k_2 + \left(\frac{R_1+2d-2R_2}{R_1+2d}\right)^2\right]\frac{d(R_1+2d)^3}{R_1R_2^3} = 0 \tag{15.26}$$

Furthermore, we may simplify this expression for the coma by expressing it in terms of the secondary magnification, M_2 and the 'back focal length', b.

$$1 - \frac{M_2^2-1}{M_2^2} + \left[k_2 + \left(\frac{M_2+1}{M_2-1}\right)^2\right]\frac{d(M_2-1)^3}{2M_2^2(dM_2+b)} = 0 \tag{15.27}$$

And

$$k_2 = -\frac{2M_2^2(dM_2 + b)}{d(M_2 - 1)^3} + \frac{2(M_2 + 1)(dM_2 + b)}{d(M_2 - 1)^2} - \left(\frac{M_2 + 1}{M_2 - 1}\right)^2 \tag{15.28}$$

Finally:

$$k_2 = -1 - \frac{2}{(M_2 - 1)^3}\left[M_2(2M_2 - 1) + \frac{b}{d}\right] \tag{15.29}$$

Having calculated the second conic constant, we may now set the spherical aberration to zero and determine the first conic constant.

$$K_{SA} = -(1 + k_1)\frac{r_0^4}{4R_1^3} + \left[k_2 + \left(\frac{R_1 + 2d - 2R_2}{R_1 + 2d}\right)^2\right]\frac{(R_1 + 2d)^4 r_0^4}{4R_2^3 R_1^4} \tag{15.30}$$

Eliminating common factors and setting the spherical aberration to zero, we get:

$$-(1 + k_1) + \left[k_2 + \left(\frac{R_1 + 2d - 2R_2}{R_1 + 2d}\right)^2\right]\frac{(R_1 + 2d)^4}{R_2^3 R_1} = 0 \tag{15.31}$$

Once more, we substitute R_1 and R_2 for the secondary magnification and back focal length, obtaining:

$$-(1 + k_1) + \left[k_2 + \left(\frac{M_2 + 1}{M_2 - 1}\right)^2\right]\frac{(M_2 - 1)^3 b}{M_2^3(dM_2 + b)} = 0 \tag{15.32}$$

Substituting k_2 we obtain:

$$-(1 + k_1) + \left[-\frac{2M_2^2}{d} + \frac{2(M_2^2 - 1)}{d}\right]\frac{b}{M_2^3} = 0 \tag{15.33}$$

Finally, rearranging, this gives:

$$k_1 = -1 - \frac{2b}{dM_2^3} \tag{15.34}$$

It is clear from this analysis, that both primary and secondary mirrors have a conic constant that is less than −1. That is to say, both surfaces are hyperbolic in cross section. In practice, for most compact telescope designs, the secondary magnification is considerably greater than one. Therefore, to a degree, the primary mirror shape is approximately parabolic.

Worked Example 15.1 ***Hubble Space Telescope***

At this point, we can demonstrate the application of this analysis to a real system, namely the Hubble Space Telescope. The telescope is a classic Ritchey-Chrétien design. The system focal length, as informed by the imaging plate scale requirement is 57.6 m. Practical scaling requirements determine the mirror separation of 4.9067 m and the 'back focal length' of 6.4063 m. This is sufficient information to determine the optical prescription for the telescope mirrors. First, we wish to determine the two mirror radii:

$$R_1 = \frac{2df}{b - f} \quad R_2 = \frac{2db}{d + b - f}$$

Substituting the relevant values:

$$R_1 = \frac{2 \times 4.9067 \times 57.6}{6.4063 - 57.6} = -11.04 \quad R_2 = \frac{2 \times 4.9067 \times 6.4063}{6.4063 + 6.4063 - 57.6} = -1.358$$

The primary mirror radius is − 11.04 m and the secondary mirror radius is − 1.358 m.

We may now calculate the secondary magnification:

$$M_2 = \frac{-R_2}{(R_1 - R_2 + 2d)} = \frac{1.358}{-11.04 + 1.358 + (2 \times 4.907)} = 10.435$$

Having calculated the secondary magnification, we may calculate the two conic constants:

$$k_1 = -1 - \frac{2b}{dM_2^3} = -1 - \frac{2 \times 6.4063}{4.9067 \times 10.435^3} = -1.0023$$

$$k_2 = -1 - \frac{2}{(M_2 - 1)^3}\left[M_2(2M_2 - 1) + \frac{b}{d}\right] = -1 - \frac{2}{(10.435 - 1)^3}\left[10.435(2 \times 10.435 - 1) + \frac{6.4063}{4.9067}\right]$$

$$k_2 = -1.49685$$

The conic constant of the primary mirror is – 1.0023 and that of the secondary – 1.49685.

These values, as calculated, are very close to the ***design*** values for the telescope. Famously, due to an error in the metrology set up, the manufactured primary mirror had a conic constant that was significantly different from the intended design value. This error resulted in significant image degradation and had to be compensated by addition of corrective optics in the camera system.

15.4.3.4 Three Mirror Anastigmat

Elimination of coma by the addition of a second corrective mirror provides a significant improvement in performance over nominally single mirror telescopes. However, this does not represent full correction of all third order aberrations. Addition of a third mirror should, in principle, enable the correction of all the third order Gauss–Seidel aberrations. This is the basis of the so-called ***three mirror anastigmat*** or ***TMA***. That this should be necessary, particularly in astronomical instrumentation, is a testament to technological improvements that have taken place in recent decades. In the so-called hierarchy of aberrations, field curvature and astigmatism are the least prominent. It is only the significant amelioration of atmospheric effects, either by adaptive optics or exo-atmospheric deployment, that has enabled full use to be made of the superior resolution afforded by larger aperture telescopes. With this in mind, in order to obtain diffraction-limited performance over a significant field of view, additional correction must be provided. Again, extending the previous analysis, addition of the third mirror further enhances the resolution/étendue metric.

To illustrate the points discussed, it would be useful to assess the shortcoming of a Ritchey-Chrétien design, as embodied in the Hubble Space Telescope. As a simple illustration, we may estimate the impact of field curvature, assuming this is described by the Petzval curvature. Of course, this is not intended as an accurate calculation of wavefront error, but merely as an ***estimate*** of the magnitude of field curvature/astigmatism. The Petzval radius of the Hubble Telescope is 1.55 m. In calculating the Petzval radius, we need to be exceptionally careful about sign convention. For all our descriptions of ray tracing, we have adopted a universal common reference frame to denominate the sign of surface sag and ray propagation. However, where the direction of ray propagation is reversed, on single mirror reflection, the sign of ***even*** aberrations, such as spherical aberration, astigmatism, and field curvature (not coma) is reversed. Therefore, the Petzval curvature of a Ritchey-Chrétien telescope is given by:

$$1/R_{PETZ} = 1/R_1 - 1/R_2 \tag{15.35}$$

The field of view of the telescope is ±0.03°, representing ±30 mm for a 57.6 m focal length. Therefore, the defocus at the edge of the field, for a field curvature of 1.55 m, is about 0.29 mm or ± 0.145 mm. The diameter of the telescope primary (the entrance pupil) is 2.4 m and for a 57.6 m focal length, the numerical aperture is about 0.021. The maximum wavefront error associated with this defocus is about 9 nm rms. Clearly, this is well within the diffraction limit and will not impact imaging performance. However, looking towards the next generation, the James Webb Space Telescope, this has a larger field of view, at ±0.04° and a much larger aperture at 6.5 m. As field curvature scales with the square of the aperture and field, pursuing a two mirror design

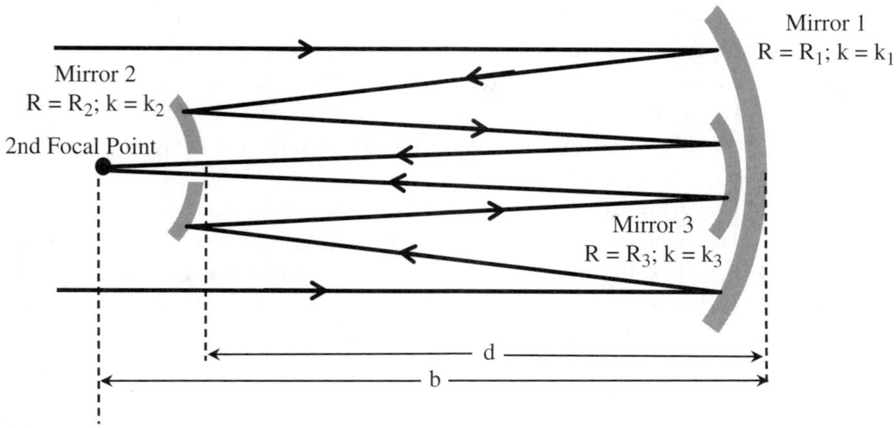

Figure 15.14 Three mirror anastigmat.

with this instrument would produce a wavefront error over 12 times larger, or around 115 nm rms. However, in practice, the figure is likely to be larger than this when one specifically calculates the field curvature and astigmatism separately.

In a three mirror anastigmat, we now incorporate an extra conic mirror to control the third aberration, astigmatism. The reason that we do not need to consider field curvature in this analysis is that the first order design is constrained to eliminate Petzval curvature altogether. The design is illustrated in Figure 15.14. There are two mirror separations to consider in this design, between the first and second mirrors and between the second and third. However, to make the analysis a little more tractable, we will assume that both distances are identical and denoted by the symbol d. As in the Ritchey Chrétien design, the so-called 'back focal length', b, is the distance from the third mirror to the focus. In terms of the first order design of the telescope the analysis is quite straightforward. The curvature of the three mirror is determined by three constraints: the system focal length, the location of the second focal plane, and zero Petzval curvature. Instead of describing the mirrors by their respective radii, we describe them in terms of their curvatures, c_1, c_2, c_3. It is very straightforward to analyse the system in first order with matrix analysis and we may formalise the three constraints as follows:

$$\text{Zero Petzval Curvature}:\ c_1 - c_2 + c_3 = 0 \tag{15.36a}$$

$$\text{System Focal Length}:\ 4d(2c_1^2 - 2c_1c_2 + c_2^2 + 2dc_1c_2(c_2 - c_1)) = -1/f \tag{15.36b}$$

$$\text{Second Focal Point Location}:\ 1 + 4dc_1 - 2dc_2 - 4d^2c_1c_2 = -b/f \tag{15.36c}$$

Manipulation of the above set of equations presents a quadratic type equation with, in general, two possible solutions. The quadratic equation is presented in terms of the curvature, c_1 of the first surface and is set out below:

$$[4d^2((f/b) - 1)]c_1^2 + [2d(2 + 2(f/b) + (d/b))]c_1 + [2 + (d/b) + (f/b) + (b/f)] = 0 \tag{15.37}$$

And

$$c_2 = \frac{1 + b/f + 4dc_1}{2d + 4d^2c_1} \tag{15.38}$$

Finally:

$$c_3 = c_2 - c_1 \tag{15.39}$$

The above equations wholly define the system in first order fixing the three mirror radii. Each of the three surfaces are conic surfaces with conic constants, k_1, k_2, and k_3. We may determine the value of these constants by setting the system spherical aberration, coma, and astigmatism to zero. There is no need to analyse the field curvature, as this will be automatically set to zero when the astigmatism is zero, given the zero Petzval curvature. The approach is broadly similar to the Ritchey-Chrétien analysis except with three unknowns. For the spherical aberration coma and astigmatism, the different stop shift factors may be determined from the matrix elements at each mirror surface. A set of three simultaneous equations results, that may be solved for the three conic constants.

It would be useful, here, to set out the procedure for defining these simultaneous equations. Broadly, the approach is to determine, for each mirror, its contribution to the global spherical aberration, coma, and astigmatism. This may be done by deriving the ray tracing matrix for each surface, ***before being refracted from that surface***. The radius of the nth surface is R^n and its conic constant, k_n and the relevant matrix elements are A_n, B_n, C_n, and D_n. Aberration contributions for each surface are listed below, with r_0 representing the pupil radius and θ the field angle:

$$\frac{\Phi_{SA}}{r_0^4} = \sum_{n=1}^{n=N} -\left[k_n + \left(1 + \frac{C_n R_n}{A_n}\right)^2\right] \frac{A_n^4}{4R_n^3} \times |M| \tag{15.40}$$

$$\frac{\Phi_{CO}}{r_0^3\theta} = \sum_{n=1}^{n=N} -\left[k_n + \left(1 + \frac{C_n R_n}{A_n}\right)^2\right] \frac{A_n^3 B_n}{R_n^3} \times |M| - \left[\left(1 + \frac{C_n R_n}{A_n}\right)\right] \frac{A_n^2}{R_n^2} \tag{15.41}$$

$$\frac{\Phi_{AS}}{r_0^2\theta^2} = \sum_{n=1}^{n=N} -\left[k_n + \left(1 + \frac{C_n R_n}{A_n}\right)^2\right] \frac{A_n^2 B_n^2}{2R_n^3} \times |M| - \left[\left(1 + \frac{C_n R_n}{A_n}\right)\right] \frac{A_n B_n}{R_n^2} - \frac{1}{2R_n} \times |M| \tag{15.42}$$

The sign of each mirrors contribution depends upon the determinant of the ray tracing matrix, $|M|$. In practice, it reverses from mirror surface to mirror surface and depends upon the direction of propagation.

Worked Example 15.2 ***TMA Design***

We will illustrate the design process by adapting the Hubble design and introducing an extra mirror. All other parameters will remain unchanged. The desired focal length will remain 57.6 m, the separation, *d*, 4.9067 m and the 'back focal length', *b*, 6.4063 m. Since the first order parameters are derived from the quadratic equation, there are two solutions for the mirror radii, as set out below:

R_1 (m)	R_2 (m)	R_3 (m)
−11.534	−2.479	−3.158
−5.622	−0.910	−1.086

We are faced with a choice between two solution sets. In comparing the two, one might conclude that the solution with the lower curvatures might be best, as surfaces with high curvature might introduce higher order aberration. We therefore select the first of the two solutions and, as previously indicated, sum the spherical aberration, coma, and astigmatism across all three surfaces, setting them to zero. This produces a set of linear equations for the three conic constants, as indicated below:

$$\begin{bmatrix} -0.000163 & 8.122E-6 & -1.214E-6 \\ 0 & 0.00107 & 0.00128 \\ 0 & 0.0176 & -0.168 \end{bmatrix} \begin{bmatrix} k_1 \\ k_2 \\ k_3 \end{bmatrix} = \begin{bmatrix} 0.000134 \\ -0.00419 \\ 0.0397 \end{bmatrix}$$

This gives the solutions: $\boldsymbol{k_1 = -0.98219; k_2 = 3.23249; k_3 = -0.57524.}$

This analytical solution is very close to that derived from optimisation using ray tracing analysis. Over a very wide field of 0.2° the wavefront error is very low, of the order of a few nanometres for a 6 m diameter primary mirror.

Data relating to the Hubble Space Telescope and James Webb Space Telescope and used in these examples is courtesy of the National Aeronautics and Space Administration.

15.4.3.5 Quad Mirror Anastigmat

In a further extension to the TMA design, a fourth conic mirror may be added to the system to produce a ***Quad Mirror Anastigmat, QMA***. From a first order perspective, the addition of an extra surface adds an extra degree of freedom to, for example, locate the pupil at a specific conjugate. This might be used to create a design with a telecentric output. The four constraints addressed by the four surfaces are the system focal length, focal point location, zero Petzval curvature, and pupil location. Furthermore, by adjusting all four conics, it is possible to produce a design that has no third order aberrations, ***including distortion***. Determination of the four radii of curvature and the four conic constants is a slightly more elaborate process than for the TMA, but proceeds broadly along the same lines.

Thus far, all these mirror designs have been analysed under the assumption of a clear axial symmetry. That is to say, the designs presented are all 'on-axis' designs. The weakness of this geometry is that is results in portions of the beam being vignetted by some of the mirrors, producing an 'obscured' pupil. Therefore, to remedy this, the more complex TMA and QMA designs are 'off-axis' or have off-axis elements. Without the underlying axial symmetry, these designs are more complex to analyse. However, although the fundamental symmetry that underpins Gauss–Seidel aberration theory has been destroyed, it is nonetheless possible to analyse the system on the basis of revised assumptions. In the off-axis scenario, each mirror or optic produces Gauss–Seidel type aberrations. However, the field symmetry of each aberration type is displaced, according to the local tilt. The result of this is that there are well defined ***nodes*** at which a particular Gauss–Seidel aberration vanishes. The critical distinction when compared to classical aberration theory is that these nodes are located at off-axis field points. Furthermore, the number of nodes is dictated by the azimuthal symmetry order. For example, for spherical aberration, there are no nodes (constant optical path difference [OPD] over all field points), for coma, there is one node, and for astigmatism there are two nodes. This forms the basis of ***nodal aberration theory***. Its details lie beyond the scope of this text. For the interested reader, a useful reference is provided at the end of the Chapter.

15.4.4 Catadioptric Systems

In catadioptric telescope systems, both mirrors and lenses or transmissive optics are combined. Most generally, mirrors provide the 'heavy lifting' or focusing power of the instrument and glass optics are used for aberration correction. The most well-known example of a catadioptric telescope system is the Schmidt Camera. In the Schmidt design, the parabolic primary mirror is dispensed with, and replaced with a spherical mirror. Aberration correction is provided by a specially shaped glass plate that is located before the primary mirror. Essentially, the glass plate has little focusing power, but is shaped to convey a significant fourth order form element to provide spherical aberration correction. The utility of this type of design lies in the historical difficulty of polishing non-spherical mirrors, such as parabolic mirrors. Of course, the creation of a fourth order form for aberration correction is not, in itself a non-trivial task. Originally, creation of the correction plate was realised by a simple but highly effective technique. A glass plate was deformed under vacuum to create a fourth order profile and subsequently polished flat. Removal of the vacuum then preserved the inverse of the original deformation. The Schmidt telescope system is illustrated in Figure 15.15.

There are numerous variants of this catadioptric scheme, with the essential principle of refractive elements providing correction rather than power. A simpler large meniscus lens element substituted for the Schmidt corrector is the basis of the Maksutov system. The Modified Dall-Kirkham is a variant of the Ritchey Chrétien telescope. Here, the conic secondary is substituted by a spherical mirror, which, naturally, is easier to fabricate and test. Correction of off-axis aberrations is provided by a group of lenses sited towards the focal plane. One useful combination is a doublet consisting of a positive and negative lens of equal power. This combination

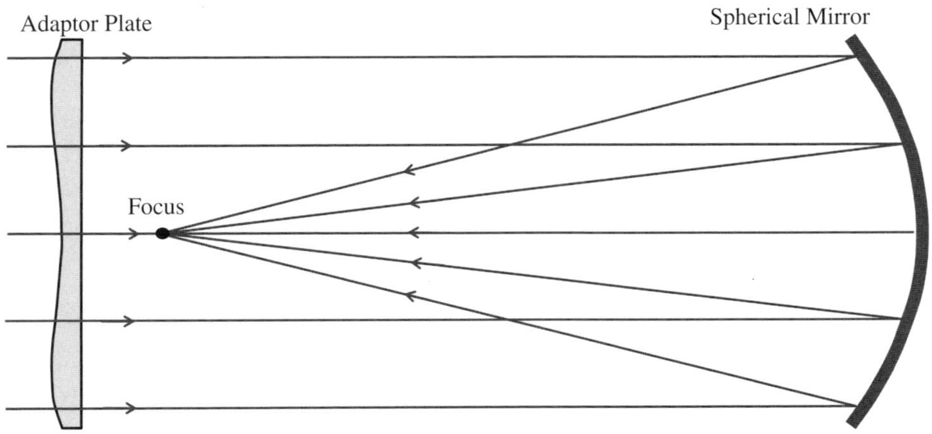

Figure 15.15 Schmidt camera system (sag of adaptor plate greatly exaggerated).

provides no optical power but does introduce (potentially correcting) spherical aberration. One might regard it as a substitute for an aspheric plate.

15.5 Camera Systems

15.5.1 Introduction

A camera is essentially a wide field, wide aperture optical imaging system. Its task is to provide high resolution across a wide field and its utility might be framed in terms of the number of lines it can resolve across the field. Traditionally, the camera has been associated with the use of photographic media. However, image sensing nowadays is almost exclusively provided by pixelated digital media. To provide a contextual comparison between historical and digital media, the MTF of high resolution black and white film falls to 0.5 at a spatial frequency of 50 cycles per mm; resolution of colour film is inferior. This might be compared to Nyquist sampling for a 10 μm pixel digital detector. On this basis, a single frame of 35 mm film might compare to a 8 MPixel digital camera. This emphasises the paramount importance of resolution, rather than wavefront error in defining the utility of a design. As such, it is the system MTF that is most usually used to describe lens performance.

Application of camera systems extend beyond the simple and direct capturing of images. Often, they form part of a more extensive optical system, especially in scientific and technical applications. For example, they may provide imaging in microscopic, telescopic, or spectroscopic applications. The salient feature that is common to all these applications is the requirement to deliver high resolution across a wide field and with a low focal ratio. From a design perspective, the challenges inherent in camera design are severe. The prominence of all third order aberrations is broadly equivalent. There is no hierarchy of aberrations to simplify the design task.

In order to set the scene, we might establish some useful parameters. In terms of useful standards, the old 35 mm camera format provides guidance as to field angles that might be encountered in camera design. The format consisted of a 36 × 24 mm rectangular image plane for a 'standard' focal length of 50 mm. The extreme corners of this field represent a departure of ±21.6 mm from the central field location or ± 23.4°, giving a total field of 46.8°. In general, cameras are not diffraction limited systems and the system aperture is dictated by étendue and light gathering capacity, rather than resolution. In the context of capturing images in limited light levels and in a limited acquisition time, high numerical aperture is always preferred. As such, high numerical aperture lenses, with a high étendue facilitated rapid exposure or image acquisition and were therefore referred to as 'fast lenses'. Conversely, low numerical aperture lenses were regarded as 'slow'. Perhaps, in a more modern

Table 15.3 Detector formats.

Format	Field (mm – H×V)	Comments
IMAX	69.6 × 48.5	One of many large cinematographic formats
120	56 × 56, 56 × 84	Medium format photographic
35 mm	36 × 24	Compact photographic and high-end digital
16 mm	10.26 × 7.49	Legacy amateur cinematographic
1.5″	18.7 × 14.0	Digital sensor
4/3″	17.3 × 13.0	Digital sensor
1″	12.8 × 9.6	Compact digital sensor
2/3″	8.8 × 6.6	Compact digital sensor
1/3.2″	4.54 × 3.42	Mobile phone cameras

context, particularly in scientific applications, it is detector signal to noise ratio that is the critical parameter. Taken together, a merit function defining the utility (and complexity and cost) of a camera lens would be the system étendue divided by the area of a single resolution element.

A numerical aperture of 0.1 (f#5) represents a relatively undemanding goal, whereas, by contrast, a numerical aperture of 0.5 (f#1), is significantly challenging. By way of comparison, a marginal ray corresponding to a numerical aperture of 0.4 (f#1.25), subtends an angle of about 24°, equivalent to the maximum field angle in a typical camera. For the slower, e.g. f#5, lenses, then the extreme fields are usually greater than the marginal ray angles. Therefore, correction of off-axis aberrations becomes of primary interest, as opposed to dealing with on-axis aberrations.

Another important aspect of modern digital technology is the impact of miniaturisation. Scaling of the recording media from photographic emulsion to imaging chip results in a potential reduction in scale by a factor of 3 to 5. Accordingly, first order parameters, such as focal length are scaled by a similar factor. Therefore, all things being equal, for a given field angle and numerical aperture, the wavefront error is reduced in proportion. In some respects, this lightens the load of the optical designer, although the reduced étendue of each resolution element must be compensated by increased detector efficiency. It is clear from the preceding arguments that camera scaling is largely dictated by camera geometry. It would be useful, at this point to illustrate this with some examples of common detector formats, both current and historical. This is set out in Table 15.3.

Overall, the most common format ratio is either 3 : 2 or 4 : 3 (H×V), although less common variants exist. Curiously, the size of digital sensors is denominated by the diameter (in inches) of the equivalent legacy image intensifying tube; this is significantly larger than the size of the chip itself. Comparison of a typical compact digital sensor (1″) and the ubiquitous 35 mm film format suggests a geometrical scaling of 3 : 1. Thus, a 'standard' 50 mm focal length 35 mm camera lens would equate to a 17 mm focal length in the corresponding compact digital camera. As far as the mobile phone camera is concerned, the corresponding focal length would be about 6 mm.

It is important to emphasise, as highlighted earlier, that, in general, the geometric spot size of a camera lens is its defining performance characteristic, rather than its wavefront error. This resolution may also be defined by the camera's MTF as a function of spatial frequency. For a digital detector, Nyquist sampling is equivalent to a spatial wavelength equal to two pixel widths. In theory, as set out in Chapter 14, the MTF at this spatial frequency is 0.637. It is reasonable to suppose that a camera lens designed for use with such a detector would match this MTF at the spatial frequency in question. A lower MTF would significantly degrade system performance and a higher MTF would face diminishing returns for the inevitable added cost and complexity.

The focus on spatial resolution, rather than wavefront error also affects the depth of focus. In a diffraction-limited system, the depth of focus is inversely proportional to the ***square*** of the numerical aperture. For non-diffraction camera systems with a spot radius of Δr, the depth of focus, Δf, is inversely proportional to the numerical aperture, NA:

$$\Delta f \approx \frac{\Delta r}{NA} \tag{15.43}$$

As an example, for a (35 mm) camera with an f#3 aperture (NA 0.16) and a resolution of 20 μm, the depth of field would be of the order of 0.12 mm in the image plane. In terms of the impact in object space, for a nominal object distance of infinity, the depth of focus in object space would extend from 20 m to infinity for a 50 mm focal length lens. Of course, for a diffraction limited lens, the depth of focus would be rather less.

15.5.2 Simple Camera Lenses

Perhaps the simplest form of lens imaginable is the singlet. Early camera designs were very slow and it is not surprising that field curvature and astigmatism were the principal concerns for a flat field. With a specific focal power in mind, the Petzval curvature is a fixed quantity. However, it is possible to balance the tangential and sagittal aberrations by giving them an equal and opposite sign. This can only be done, for a single lens, by shifting the stop away from the lens, for astigmatism and field curvature for a lens placed at the stop are of the same sign. Thereafter, balancing and optimising these two aberrations may be accomplished by adjusting the shape factor of the lens. Of course, it cannot be eliminated on account of the non-zero Petzval curvature. Optimum performance is achieved for a meniscus lens. Taking a specific example, the design of a 100 mm focal length f#10 lens with a field angle of 10°. Assuming a refractive index of 1.52 and with the stop placed 20 mm from the lens, the optimum shape factor is about 4.5; this is consistent with a meniscus form. This analysis is entirely based upon third order theory and the use of the stop shift equations. This very simple optimisation is very much in line with the design of the earliest camera lenses. Once again, it must be emphasised, although modern design proceeds by computer-based optimisation, it is enormously beneficial to be fully aware of the underlying principles.

In this preceding analysis, we have ignored chromatic effects. In common with field curvature and astigmatism, chromatic aberration follows a second order dependence on numerical aperture. Thereafter, the relative significance of field curvature and chromatic aberration is given by the product of the square of the field angle and the Abbe number. This merely affords an estimate of the relative magnitude of the two aberrations, indicating that, for an Abbe number of 64 (BK7), the two effects might be comparable for a field angle of 7°. This indicates that the next significant improvement is to be obtained by eliminating chromatic effects. The next refinement substituted the single meniscus lens with a doublet of similar form. However, since they were not optimised for coma or spherical aberration, these simple lenses were very slow.

Many of these very simple historical designs predated the development of photographic media and were incorporated into camera obscura and eyeglass design. The first lens specifically designed for photographic media was the Petzval portrait lens and was designed with some measure of mathematical rigour, as opposed to using trial and error. This design specifically sought to increase the speed of the lens. However, the field size was rather more limited. The lens is, of course, named after its inventor, Joseph Petzval. However, despite this, and perhaps allowing for the relatively modest field, the lens does not have zero Petzval curvature. In essence, it consists of two achromatic doublets arranged symmetrically about the stop. Although radically different from preceding designs, it does emphasise the significance of the stop location, as we illustrated with the simple meniscus lens design.

The Petzval lens introduced an important element in camera lens design. The symmetry about the stop looks forward to more modern designs which follow the same general principle. As a consequence of the dependence of the Gauss-Seidel aberrations on stop shift, there is a tendency for those aberrations which have an odd power dependence on field angle, such as distortion and coma, to be cancelled out. The weakness of the design, of course, is that it possesses significant Petzval curvature. Therefore, it is impossible to

Table 15.4 Cooke triplet paraxial design variables and constraints.

Variable	Constraint
Lens 1 power	Focal length λ_1
Lens 2 power	Focal length λ_2
Lens 3 power	Focal point λ_1
Separation lens 1 – lens 2	Focal point λ_2
Separation lens 2 – lens 3	Zero Petzval curvature

eliminate field curvature and astigmatism and this represents a significant impediment to its use in wider angle systems.

15.5.3 Advanced Designs

15.5.3.1 Cooke Triplet

Much of the preceding narrative relates to a theoretical understanding of lens design in the development of the modern camera lens. What must also be emphasised are the severe restrictions that pertained to the choice and quality of optical glasses. It is more modern developments that have provided the designer with a wide choice of optical materials. Furthermore, the utility of multi-element designs was historically restricted by the lack of (anti-reflection) coating options. Multiplicity of surfaces inevitably led to image contrast reduction through the chaotic summation of various Fresnel reflections. With this in mind, the ***Cooke Triplet*** represents the first modern camera lens. Although referred to as the Cooke Triplet, it was actually first developed (in 1894) by Harold Taylor, who was employed by Cooke and Sons of York. Unlike the Petzval lens, it is aplanatic and specifically designed to have zero Petzval curvature.

In its simplest form, the Cooke Triplet consists of three separate singlet lenses. Most usually, it consists of two positive 'crown glass' lenses, surrounding a negative flint lens. From this very simple recipe it is possible to correct all third order aberrations (with the exception of distortion). Furthermore, the design is very straightforward and the underlying principles easy to grasp. In essence, analysis of the design proceeds in a very similar manner to the three mirror anastigmat. As such, the Cooke triplet is, itself, an example of an anastigmatic lens design. Initially, one considers only the paraxial design of the system. With this in mind, in terms of paraxial design variables, there are five, namely three lens powers, and two lens separations. It is customary to fix the back focal length – the distance from the last surface to the focal point – to some specific value, usually expressed as a fraction of the focal length. Viewed in terms of paraxial constraints there are also five to consider. Firstly, there is a system focal length that must be fulfilled for two wavelengths and a focal point location which must also be identical for two wavelengths. Finally, there must be zero Petzval curvature. These fundamental design considerations are summarised in Table 15.4.

Following this process, the paraxial design is finalised with the three lens powers and separations now fixed. We now need to eliminate the three Gauss–Seidel aberrations, spherical aberration, coma, and astigmatism. In the case of the Three Mirror Anastigmat, we did this by varying the three conic constants. Of course, we could repeat this strategy for this lens design, by introducing conic prescriptions to three of the lens surfaces. However, in the case of the Cooke Triplet, as spherical surfaces are much easier to manufacture, controlling aberrations is accomplished by simply varying the shape factor of each of the three lenses and taking into account any stop shift effects in the analysis.

This presents a clear and logical progression of the design process for this relatively simple lens. What is presented here is a thin lens third order analysis. Of course, as stated many times previously, final optimisation inevitably proceeds using ray tracing software. However, not only does this process provide a useful initial starting point for computer based optimisation, more significantly it does provide the designer with ***understanding***.

Worked Example 15.3 ***Cooke Triplet***

At this point, we follow through the preceding discussion with a design example. We are to create a Cooke triplet with a design focal length of 50 mm, with a back focal length of about 40 mm. The design is to use two glasses, PSK53A for the front and back positive lenses and SF2 for the middle diverging lens. The design is to be optimised for two wavelengths, the F and C wavelengths at 486 and 653 nm, respectively. In the description provided here, the solutions ware derived through spreadsheet analysis, although this process could be more automated. Firstly, we arbitrarily select two lens focal lengths for the first and second lenses and we will label them f_{d1} and f_{d2}, as these are the focal lengths at the 589 nm D wavelength. There is no need to choose the third focal length as this is given by the zero Petzval curvature condition:

$$\frac{1}{n_{d1}f_{d1}} + \frac{1}{n_{d2}f_{d2}} + \frac{1}{n_{d3}f_{d3}} = 0 \quad \mathrm{n_{d1}, n_{d2}, and\ n_{d3}\ are\ the\ refractive\ indices\ at\ 589\ nm.}$$

The relevant refractive indices are tabulated below:

Lens index values

Lens	Material	n_D	n_C	n_F
Lens 1	PSK53A	1.61791	1.61503	1.62478
Lens 2	N-SF2	1.64752	1.6421	1.66125
Lens 3	PSK53A	1.61791	1.61503	1.62478

Having selected f_{d1} and f_{d2}, the focal lengths of all three lenses may be computed at all wavelengths. With some simple matrix analysis, it is possible to force the focal length at the C and F wavelengths to be 50 mm, and so determine the two thicknesses, t_1 and t_2. In fact, for given conditions, two solutions are produced, as the underlying equation is quadratic. From this analysis, the following solution is computed, with the focal lengths and separations listed.

First order parameters for triplet (values in mm)

Lens	Focal length	Separation
Lens 1	25.000	8.555
Lens 2	−11.422	9.099
Lens 3	21.752	41.0

In this instance, the stop is placed at the first lens. Adjusting the shape of each lens, spherical aberration, coma, and astigmatism are set to zero, using the basic equations and stop shift equations from Chapter 4. The resulting lens shape factors are set out below:

Triplet lens prescriptions

Lens	Focal length	Shape factor
Lens 1	25.000	1.153
Lens 2	−11.422	−0.279
Lens 3	21.752	−0.5

Of course, this analysis is a thin lens analysis and forms the basis for computer optimisation. As such, each lens must be ascribed a reasonable thickness, paying particular regard to mechanical integrity. Thereafter, this full optimisation process, which accounts for lens thicknesses and higher order aberrations, produces a relatively modest change in the lens prescription. The final computer generated optimisation is tabulated for comparison.

Optimised triplet prescription (values in mm)

Lens	Focal length	Shape factor	Thickness	Separation
Lens 1	21.098	0.994	1.75	6.743
Lens 2	−10.954	−0.253	1.00	8.522
Lens 3	23.916	−0.412	4.00	34.3

The general layout is shown in the figures here for a f#5 aperture and a 20° field of view. The spot size as a function of field is shown in the Figure. For most of the field the spot size is less than 4 μm. Over the whole field, the spot size is less than 15 μm, giving a resolution of about 1250 lines over the whole field.

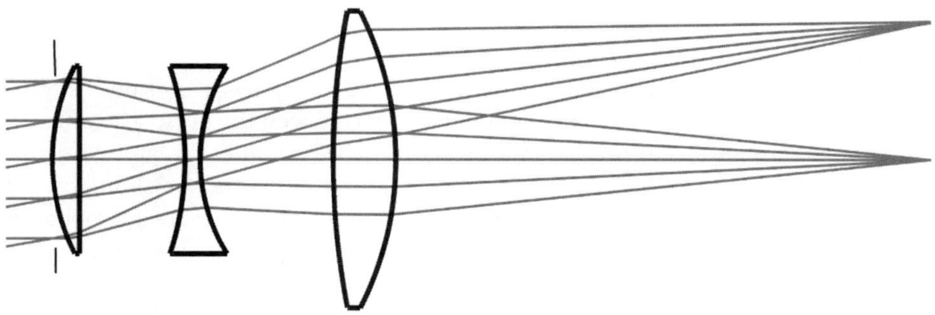

Sketch of Optimised Triplet

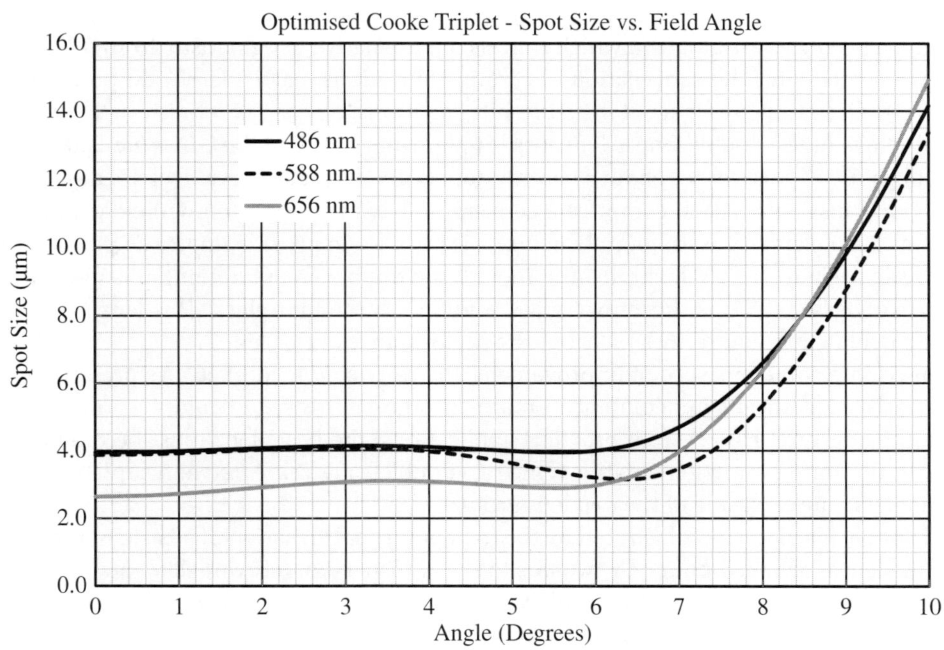

Optimised Triplet Performance

15.5.3.2 Variations on the Cooke Triplet

The preceding analysis of the Cooke Triplet provides some understanding of the processes involved in formulating a simple optical design. First, we sketch out an initial design that is derived from the fundamental principles outlined in this book. Thereafter, detailed optimisation proceeds by computer-based optimisations. This process will be discussed more fully later in the text. The Cooke triplet itself formed the basis for more sophisticated designs. In the ***Tessar lens***, the final lens in the triplet is replaced by a doublet lens. As with many imaging lens systems, further improvement in performance is obtained by moving the stop to a more central location between the lens elements, rather than at the first element. A further development is to split both the front and rear elements of the triplet into doublets. This modification produces the ***Heliar lens***. These lenses provide good performance at modest apertures, e.g. f#4.5 with covering a field of up to 40°. Although relatively 'slow' by modern standards, they do retain the advantage of simplicity and, consequentially, economy. Furthermore, historically, before the advent of reliable, low cost anti-reflection coating, the limited number of lens groups in the triplet design, ameliorated contrast-reducing reflections. As well as applications in photographic imaging, such simple lenses also have applications in other imaging areas, such as projection and imaging enlarging.

15.5.3.3 Double Gauss Lens

The triplet lens and its derivatives suffer from the disadvantage of being relatively slow. Another basic anastigmatic design is the ***Double Gauss Lens***. In Chapter 4, we introduced the air spaced aplanatic achromat. Solution of the thin lens equations to produce zero spherical aberration, coma, and chromatic aberration yielded a quadratic equation and hence two independent solutions. The first, or Fraunhofer solution, produces the classic 'off the shelf' achromatic doublet, used, for example, in refracting telescopes. The alternative solution is the ***Gauss lens*** and is more meniscus like and less planar in form. This is the basis of the double Gauss lens which is based on two Gauss doublets arranged symmetrically about a central stop.

Broadly, each lens may be optimised to have zero spherical aberration, chromatic aberration, and Petzval curvature. There is residual coma and astigmatism in each lens group. To some degree, the symmetry of the two lens groups produces coma of opposite sign enabling the coma to be cancelled out in the system. The stop shifted coma produces astigmatism which cancels out the native 'astigmatism' from the two lens groups. Overall symmetry facilitates the cancellation of certain aberrations, particularly 'odd powered' aberrations, such as coma and distortion, by virtue of the stop shift equations.

The core structure of the double Gauss lens, first implemented as the ***Clark Double Gauss lens*** is a symmetric assembly of two Gauss lenses about a central stop. Both meniscus curvatures are directed towards the central stop. The Gauss lenses each comprise a converging low dispersion (crown) element and a diverging high dispersion (flint) element. With four elements, it is possible to provide improved performance over the basic triplet design. Figure 15.16 illustrates the most basic double Gauss design, a 50 mm focal length f#3 design, using two glasses, BK7 as the 'crown glass' and SF2 as the flint glass. This lens has been computer optimised for a field of 20°. Figure 15.17 sets out the performance of this lens.

Although providing some measure of improvement, the restricted field of 20°, as presented in the simple example, is clearly a significant impediment in many instances. This field could, in principle, be extended by restricting the aperture. However, in practical applications, the basic lens must be modified to improve performance. In the simplest implementations, the two doublets are cemented and supplemented by at least two further singlets, one each side of the two lens groups.

A modified double Gauss lens is illustrated by the (computer) optimisation of a basic modified design. Two cemented Gauss lenses are surrounded by two positive, low dispersion (crown) elements. In this simple optimisation process, only two glasses are used, with N-LAK34 forming the 'crown' elements and SF2 forming the flint elements. The optimised prescription is given in Table 15.5.

The lens is designed for a 50 mm focal length and a comparatively fast aperture of f#2.5. Figure 15.18 shows a layout of the optimised design.

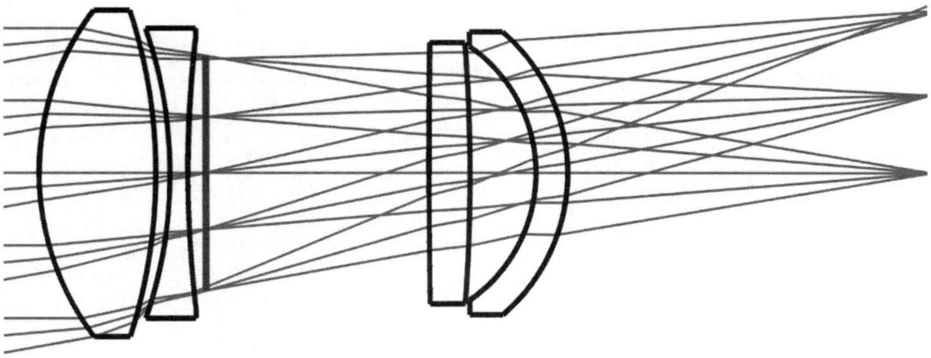

Figure 15.16 Basic Gauss doublet.

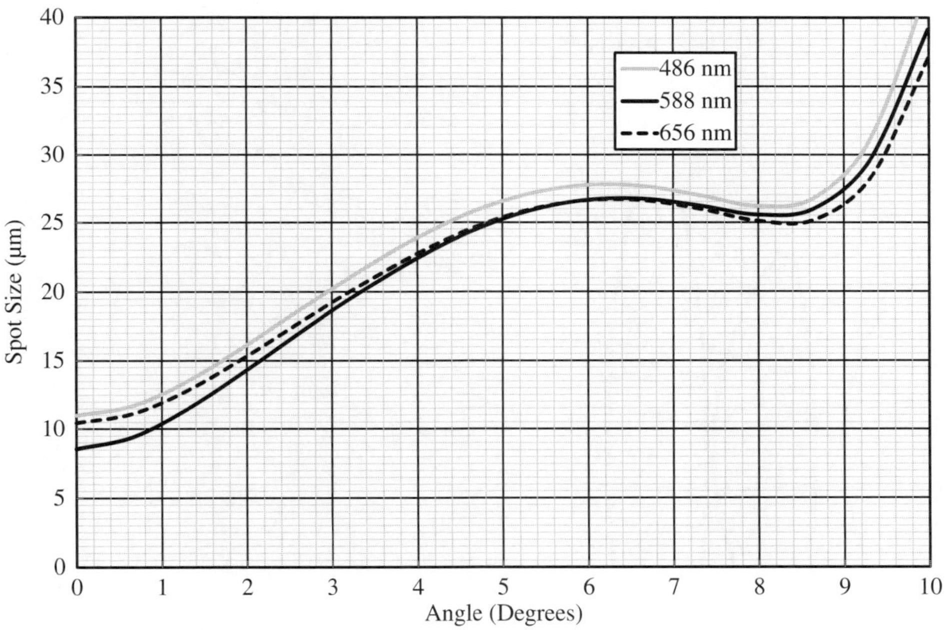

Figure 15.17 Performance of simple Gauss lens.

Addition of the two extra lenses yields a substantial increase in performance. The lens is designed to cover a field of 46.8° which entirely covers the 24 × 36 mm 35 mm format for a 50 mm focal length lens. The improvement in performance is illustrated in Figure 15.19.

This optimisation process is somewhat idealised, as some care is taken to accommodate all fields without vignetting at any surface. In practice, this ethos compels one to significantly increase the size of some elements, adding cost and weight to the design. As always, design is a compromise between seemingly conflicting priorities. As such, many wide field, high numerical aperture designs accept some vignetting for extreme fields.

This lens has been designed as a 50 mm focal length lens for a 35 mm aperture. As such, this represents either the format of a legacy 35 mm camera or a large format digital camera. It would be instructive, at this point to adapt this design for a compact digital camera. The camera is to use a 1″ sensor (12.8 × 9.6 mm) and we wish to preserve the same horizontal field angle in the new design. This gives a focal length of 17.8 mm, suggesting all dimensions be scaled by a factor of 0.355. This is very straightforward to do. Figure 15.20 shows an MTF plot of such a lens. The MTF is plotted for specific fields, but incorporates chromatic dispersion and

Table 15.5 Modified Gauss prescription.

Surface #	Comment	Material	Thickness (mm)	Radius (mm)
1	Singlet	N-LAK34	5.1	42.66
2		Air Gap	0.54	162.167
3	Gauss doublet	N-LAK34	5.48	21.45
4		SF2	2.84	42.077
5		Air Gap	4.62	14.556
6	STOP	Air Gap	7.71	
7	Gauss doublet	SF2	2.18	−20.819
8		N-LAK34	8.2	70.984
9		Air Gap	0.54	−25.816
10	Singlet	N-LAK34	8.02	97.735
11		Air Gap	27.23	−96.955
12	Image plane			

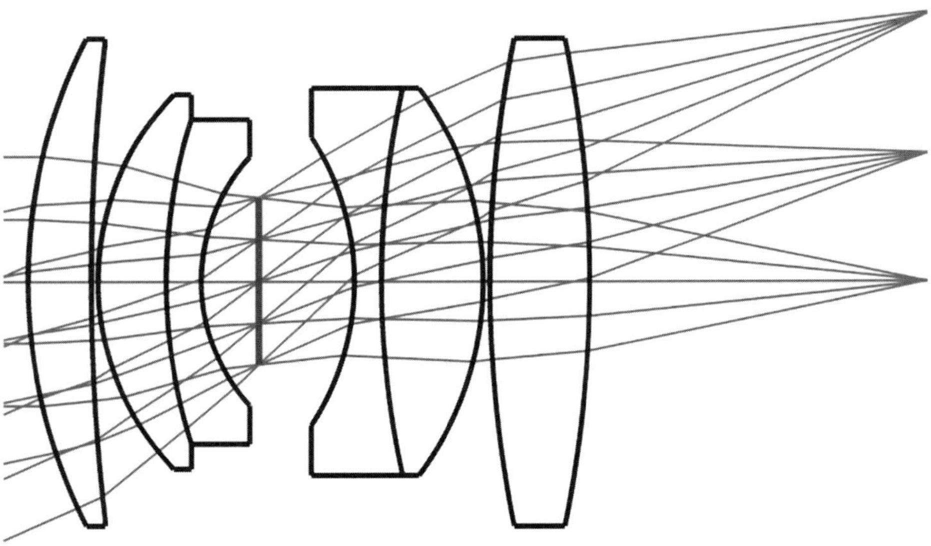

Figure 15.18 Optimised modified Gauss lens.

averages tangential and sagittal MTF. For an average field of 13.5°, the MTF falls to 0.5 at a spatial frequency of 53 cycles per mm. This corresponds to Nyquist sampling for a pixel size of 9.4 μm. However, in a digital camera, with a three colour RGB filter, effectively only half of the pixels (the 'green pixels') provide a proxy for image contrast. Therefore, Nyquist sampling for this lens would actually be equivalent to a pixel size of 6.67 μm - dividing by square root of two. This is equivalent to a 2.8 MPixel sensor. In practice, the detector would have a greater resolution than this. The balance of detector and lens performance is dictated primarily by economics. Incremental performance in detector capability is generally more economic to deliver than incremental improvements in lens performance.

The Double Gauss Lens and its derivatives are ubiquitous in modern imaging lens applications. These lenses offer high performance, with apertures of f#1 and less and a field of view in excess of 40°. Although the Double

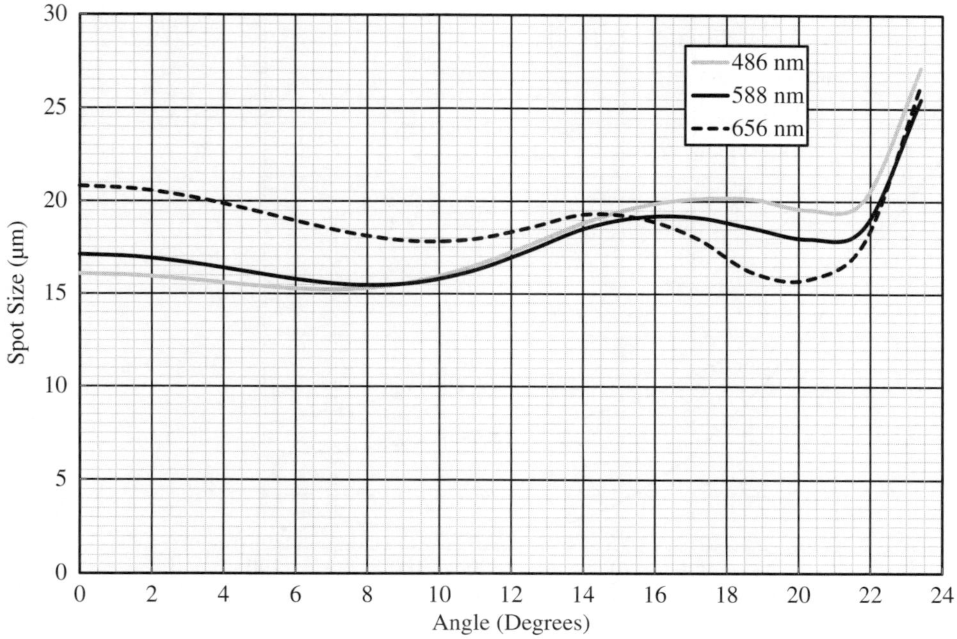

Figure 15.19 Modified double Gauss performance.

Gauss Lens is described as anastigmatic, correction of the classical third order aberrations is insufficient for apertures as large as f#1. As such, control of higher order aberrations is essential. Furthermore, even for the analysis of third order aberrations, accounting for finite lens thickness is essential; adjustment of lens thicknesses forms a central part of the lens optimisation process. In principle, it is possible to contemplate analytical treatment of higher order aberrations and this was formally carried out. However, this analytical process has fallen out of favour with the advent of computer aided optimisation. Nevertheless, as with the design process in general, understanding the principles underpinning aberration control is useful before proceeding to detailed optimisation.

15.5.3.4 Zoom Lenses

In many imaging applications one may wish to alter the plate scale at will to produce a focused image with controllable magnification. Since the imaging plate scale is synonymous with the effective focal length, this amounts to the creation of a variable focal length lens. Such a lens is reconfigured by moving elements or groups of elements within a system. If an independent mechanical adjustment must be made to maintain focus at the image plane, then such a system is referred to as a ***varifocal lens***. If, on the other hand, the system offers automatic focus compensation, such a system is referred to as a ***zoom lens***.

In a zoom lens, one group of lenses is moved to provide focal length adjustment, whilst a second group is adjusted to provide focus compensation. If the two groups are moved independently, but in a co-ordinated fashion, then such a system is called a ***mechanically compensated zoom.*** The relationship between the movement of the two groups is often highly non-linear and requires a relatively complex mechanical design involving cams, cogs, and linkage mechanisms. On the other hand, if it is possible to design a system where both groups move approximately the same distance, then the two groups may be linked and only one mechanical motion is sufficient to provide the zoom function and the focus. Such a system is referred to as an ***optically compensated zoom***.

The design of a zoom lens is necessarily complex, involving a very large number of optical elements. One useful way to consider the design of a zoom lens is to split the lens up into groups whose first order paraxial

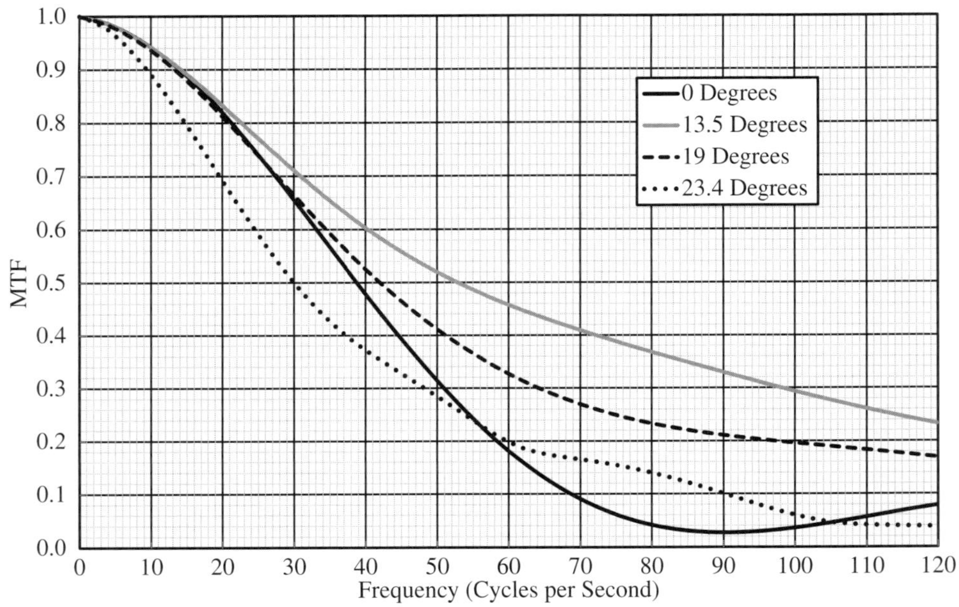

Figure 15.20 MTF of compact double Gauss lens.

behaviour may be understood readily. Each of these groups naturally contains a number of elements for the control of chromatic and other aberrations. As such, a zoom lens contains a large number of elements, often in excess of 15. Only since the development of reliable anti-reflection coatings and the availability of a wide range of high quality optical glasses has the manufacture of zoom lenses become a practical proposition. Of course, the deployment of compact zoom type lenses has become ubiquitous with the development of digital camera technology. However, most usually, digital cameras employ a separate and independent focusing process based on digital image (sharpness) processing. Strictly, one might therefore consider a digital zoom as a varifocal lens, rather than a zoom lens.

To illustrate the (paraxial) design of a zoom lens, it is useful to consider one specific category of zoom lens. In this example, the basic zoom lens consists of four groups. The first three groups comprise an afocal system whose purpose is simply to provide adjustable magnification. Adjustment of the relative position of two of these three groups provides both the afocal function and variable magnification. The fourth and final group then provides the ultimate focusing function. To maintain a constant lens speed, the stop is located close to this final group. The basic design is sketched in Figure 15.21, in simple paraxial format. In this design, the first group is fixed and the second and third groups are translated in such a way as to maintain the afocal character of the system. Most usually, the stop is located at the final lens group, so that the aperture of the lens is preserved during zooming.

In analysing the zoom lens, it is the ratio of the focal powers of the first three lens groups that is critical in the analysis. We simply assume that the focal power of the first lens is unity and the second and third lenses have focal powers, P_2 and P_3 respectively. Thereafter we must adjust the separation between the first and second lens groups (d_1) and between the second and third groups (d_2) to maintain the afocal condition. The application of this co-ordinated movement produces a magnification, M, in the diameter of the collimated beam. If the focal length of the final lens is f_4, then the effective focal length, f, of the system is given by:

$$f = f_4/M \tag{15.44}$$

Simple matrix analysis enables the derivation of the magnification produced by adjustment of the first thickness, d_1 and the compensating thickness d_2 required to maintain collimation. The magnification is given by

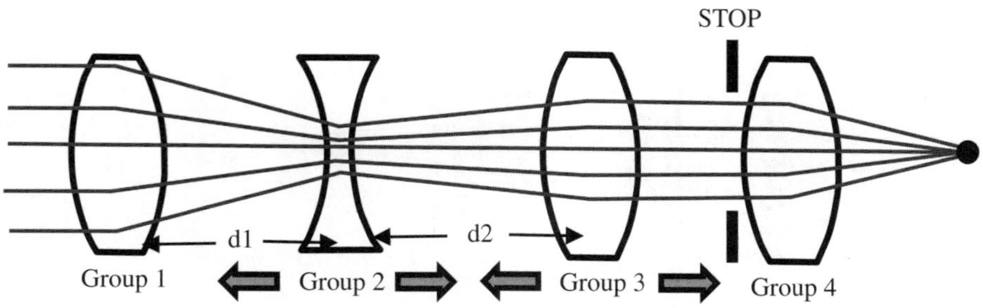

Figure 15.21 General layout of a zoom lens.

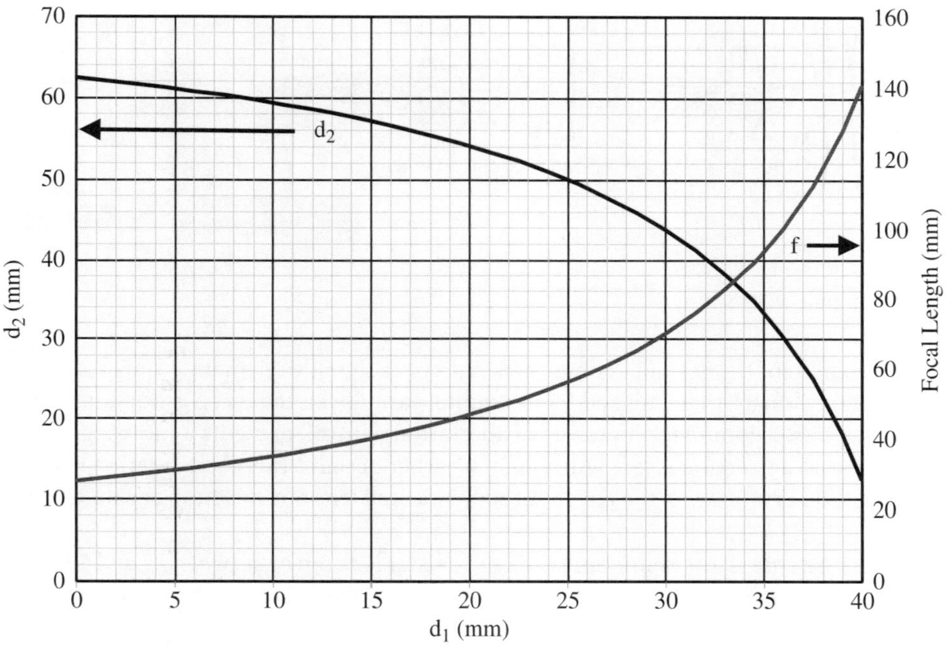

Figure 15.22 Paraxial analysis of zoom lens performance.

Eq. (15.45).

$$M = \frac{P_1(d_1 - 1) - 1}{P_2} \tag{15.45}$$

$$d_2 = \frac{(P_1 + P_2)(d_1 - 1) - 1}{P_2(1 + P_1(d_1 - 1))} \tag{15.46}$$

All values are referenced to the power or focal length of the first lens group. We can illustrate this paraxial analysis with a simple example. In this example, the focal length of both the first and final groups is 75 mm. The second group is represented as a diverging group with a focal length of −25 mm whilst the third group is positive with a focal length of 100 mm. The system performance is depicted in Figure 15.22 which shows both the second displacement, d_2 and the system focal length, f, as a function of the first displacement, d_1.

In this design, the focal length ranges from about 25 to about 130 mm. Of course, each of the paraxial lens groups must be converted into a group of several elements with, at the very least, some achromatic capability. That is to say, within each group a range of different glass types will be found. As a consequence, a zoom lens is an inherently complex system with many elements therein. In addition, the necessary separations between the

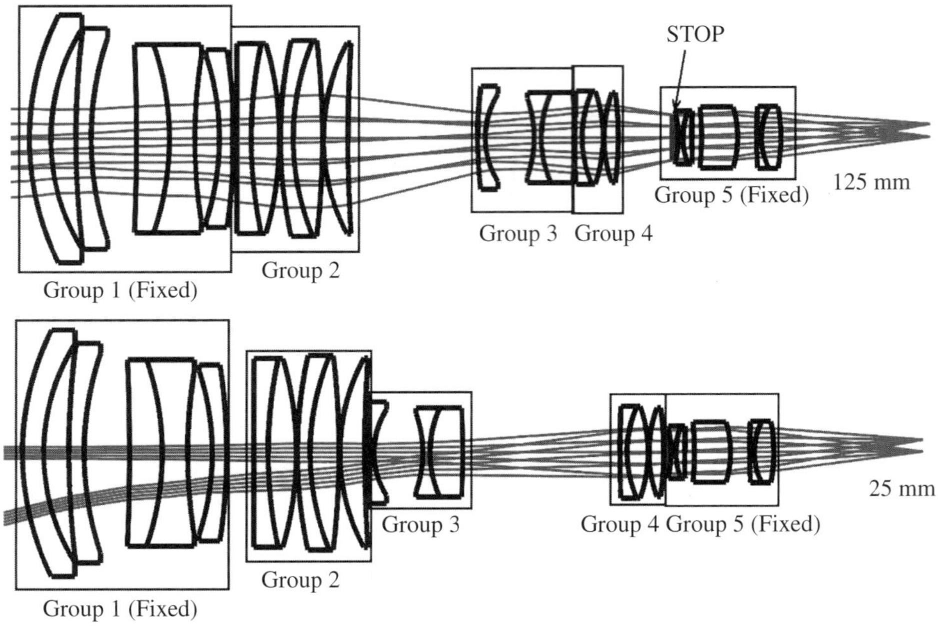

Figure 15.23 Mechanically compensated zoom lens.

groups (d_1 and d_2) adds to the length of the design, as does the length of each multi-element group. As such, a zoom lens tends to be considerably longer than its fixed focus counterpart. Given that each group must function over a range of conjugate ratios, a zoom lens, in terms of aberration control, presents a more significant design challenge when compared to a fixed focus lens. Although in earlier designs, this consideration resulted in the acceptance of compromised performance, the advent of sophisticated lens optimisation capabilities has largely ameliorated this effect.

Figure 15.23 shows the design of a cinematographic lens with an adjustable zoom of between 25 and 125 mm. In this instance, there are five, as opposed to four, groups, with three of these moving independently. The diagram serves to illustrate the complexity of such a lens with a total of 21 lens elements. However, the broad principles outlined are maintained, with a broadly collimated beam focused by a fixed final lens group where the stop is located.

In the preceding discussion, we have considered a mechanically compensated zoom lens with two separate mechanical movements. In an ***optically compensated zoom lens***, adjustment is accomplished by the identical displacement of two or more separate groups using a co-ordinated mechanical movement. In practice, the

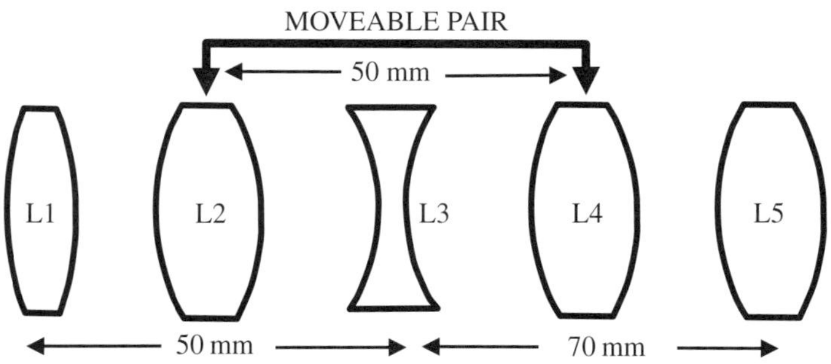

Figure 15.24 Paraxial outline of optically compensated zoom lens.

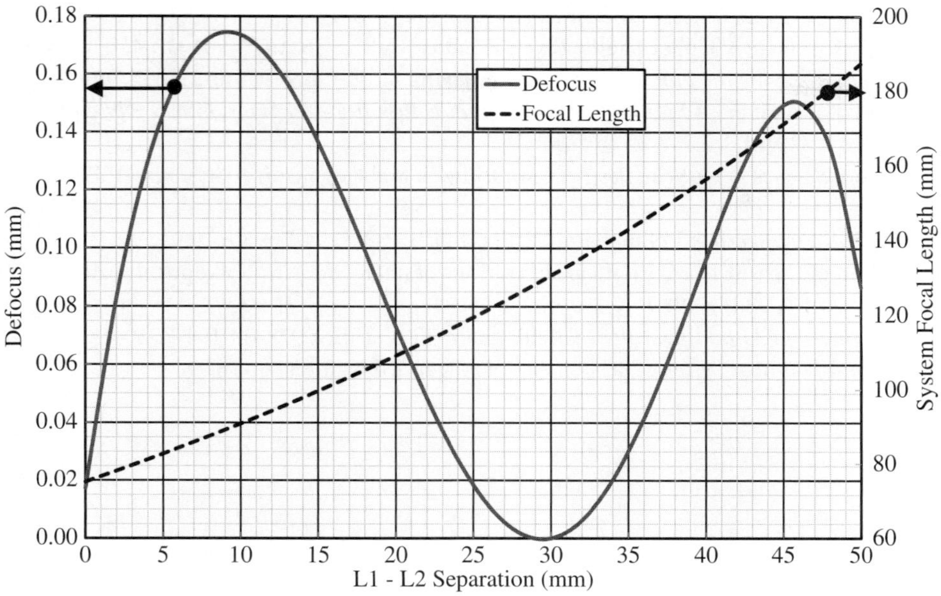

Figure 15.25 Paraxial behaviour of optically compensated zoom lens.

location of the focal point must be maintained to within some nominal depth of focus. Optimisation of an optically compensated system introduces further complexities, when compared to mechanical compensation and details are beyond the scope of this text. Even at the paraxial level, computer optimisation is needed. A basic example is illustrated in Figure 15.24, with five paraxial lenses, three of which are fixed, with the other two moving together.

In this example, the focal lengths of the five paraxial lenses from L1 to L5 are 117.1, 38.2, 588, 31.97, and 62.5 mm respectively. The separation of L1 and L2 may be varied nominally between 0 and 50 mm; thereafter all distances are fixed. The paraxial behaviour of this system is illustrated in Figure 15.25.

Further Reading

Allen, L., Angel, R., Mangus, J.D. et al., *The Hubble Space Telescope, Optical Systems Failure Report*, National Aeronautics and Space Administration Report **NASA-TM-104343** (1990).

Bass, M. and Mahajan, V.N. (2010). *Handbook of Optics*, 3e. New York: McGraw-Hill. ISBN: 978-0-07-149889-0.

Conrady, A.E. (1992). *Applied Optics and Optical Design*. Mineola: Dover. ISBN: 978-0486670072.

Dereniak, E.L. and Dereniak, T.D. (2008). *Geometrical and Trigonometrical Optics*. Cambridge: Cambridge University Press. ISBN: 978-0-521-88746-5.

Ditteon, R. (1997). *Modern Geometrical Optics*. New York: Wiley. ISBN: 0-471-16922-6.

Hecht, E. (2017). *Optics*, 5e. Harlow: Pearson Education. ISBN: 978-0-1339-7722-6.

Kidger, M.J. (2001). *Fundamental Optical Design*. Bellingham: SPIE. ISBN: 0-8194-3915-0.

Kidger, M.J. (2004). *Intermediate Optical Design*. Bellingham: SPIE. ISBN: 978-0-8194-5217-7.

Kingslake, R. (1983). *Optical System Design*. Orlando: Academic Press. ISBN: 978-0124121973.

Kingslake, R. and Johnson, R.B. (2010). *Lens Design Fundamentals*, 2e. Orlando: Academic Press. ISBN: 978-0123743015.

Laikin, M. (2012). *Lens Design*, 4e. Boca Raton: CRC Press. ISBN: 978-1-4665-1702-8.

Levi, L. (1980). *Applied Optics*, vol. 2. New York: Wiley. ISBN: 0-471-05054-7.

Mandler, W., *Design Of Basic Double Gauss Lenses,* Proc. SPIE **237**, 222 pp. (1980).
Nussbaum, A. (1998). *Optical System Design*. Upper Saddle River: Prentice Hall. ISBN: 0-13-901042-4.
Riedl, M.J. (2009). *Optical Design: Applying the Fundamentals*. Bellingham: SPIE. ISBN: 978-0-8194-7799-6.
Rolt, S., Calcines, A., Lomanowski, B.A. et al., *A four mirror anastigmat collimator design for optical payload calibration,* Proc. SPIE, **9904**, 4 U (2016).
Shannon, R.R. (1997). *The Art and Science of Optical Design*. Cambridge: Cambridge University Press. ISBN: 978-0521454148.
Smith, W.J. (2007). *Modern Optical Engineering*. Bellingham: SPIE. ISBN: 978-0-8194-7096-6.
Thompson, K. (2005). Description of the third-order optical aberrations of near-circular pupil optical systems without symmetry. *J. Opt. Soc. Am. A* 22 (7): 1389.
Walker, B.H. (2009). *Optical Engineering Fundamentals,* 2e. Bellingham: SPIE. ISBN: 978-0-8194-7540-4.
Welford, W.T. (1986). *Aberrations of Optical Systems*. Bristol: Adam Hilger. ISBN: 0-85274-564-8.

16

Interferometers and Related Instruments

16.1 Introduction

Hitherto, our narrative has been almost exclusively focused on how the underlying principles of optics and engineering impact optical system design. It has largely been taken for granted that all the optical surfaces that populate the finished design will be fabricated with absolute fidelity. In this case, system performance may be derived entirely from the analyses previously described without the inconvenience of having to measure or verify that performance. Of course, in practice, this is absolutely not the case. Manufacture of optical components, lenses and mirrors, etc., must be fabricated at finite cost in finite time. In consequence, imperfections must be accepted. Furthermore, the same considerations apply to system integration, so that small misalignments between optical components must also be contemplated. Therefore, it is generally imperative to verify system performance by measurement and analysis.

Wavefront error is a critical performance metric for an optical system. Furthermore, the determination of wavefront error across an aperture is central to the formalised analysis of an optical system. Key to the measurement of wavefront error is the comparison of the phase of the measured wavefront across the pupil and the phase of a nominally flat reference wavefront. This phase measurement is accomplished by the interference of the measured and reference wavefronts, converting the phase difference into a spatially or temporally varying amplitude or flux that can be measured with a detector. This process is the foundation of the technique of interferometry.

16.2 Background

16.2.1 Fringes and Fringe Visibility

For a phase difference to be translated by interference into a palpable flux variation, measured and reference wavefronts must exhibit some degree of mutual coherence. A reference beam is usually created by division of amplitude – diverting a portion of a collimated beam by using a beam splitter. One beam passes through the optical system under test, whilst the other is preserved as a reference. ***Interferometry*** exploits the interference produced when these beams are recombined and any phase differences between them translated into spatially varying irradiance. If this spatially varying phase difference across a circular pupil is described as $\Phi(x, y)$, then the variation in irradiance produced by interference with the reference may be given by:

$$I(x,y) = A(x,y) \times A^*(x,y) = A_0^2(1 + e^{i\Phi(x,y)}) \times (1 + e^{-i\Phi(x,y)}) = 2A_0^2[1 + \cos(\Phi(x,y))] \tag{16.1}$$

With their high degrees of spatial and temporal coherence, laser systems have greatly enhanced the development of practical interferometers. Of course, a high degree of coherence is not strictly essential and the science of interferometry antedates the development of the laser by a considerable margin. However, if the temporal coherence is low, then fringe visibility will only be preserved if the optical path difference between

Optical Engineering Science, First Edition. Stephen Rolt.

Companion website: www.wiley.com/go/Rolt/opt-eng-sci

the measurement and reference beams is less than the restricted coherence length. Thereafter, the fringe visibility, or the contrast between the light and dark fringes, diminishes as the first order correlation function. Thus, Eq. (16.1) represents an idealised scenario. More generally, lack of coherence reduces the visibility of the fringes and this is captured by the fringe visibility, V, which represents the fringe contrast, or the difference between the maximum and minimum irradiances divided by their sum:

$$V = \left[\frac{I_{max} - I_{min}}{I_{max} + I_{min}}\right] \tag{16.2}$$

Taking into account the fringe visibility, Eq. (16.1) now becomes:

$$I(x, y) = 2A_0^2[1 + V\cos(\Phi(x, y))] \tag{16.3}$$

The visibility itself is defined by the first order correlation function we introduced in the chapters covering diffraction and laser devices. Correlation or coherence may be analysed with respect to differences in time or position (spatial or temporal). Many, but not all, interferometers use sources with well-defined spatial coherence. With this in mind, interferometers employing low coherence sources, such as mercury lamps, must ensure that the optical path difference between the two beams is as small as possible. For a Doppler broadened atomic line, such as mercury, the coherence length is of the order of a few centimetres. Beyond this path difference, the coherence falls off significantly. Thus, whilst this lack of coherence considerably complicates and constrains the design of an interferometer it does not render such a design impossible.

In this analysis, the initial amplitudes of each beam, A_0, are identical. By virtue of the sinusoidal term in Eq. (16.1), the interference process has a tendency to produce alternating bands or fringes of light and dark where there is systematic variation of phase between the two beams. Of course, this systematic difference in phase between the two beams translates directly into wavefront error, assuming that the reference beam is entirely 'flat'. Figure 16.1 presents an illustration of how this might work in practice. A collimated beam is split by means of a beam splitter and subsequently re-combined with the probe beam after it has passed through the optical system.

Figure 16.1 illustrates the general form of the interferogram with a pattern of alternating light and dark fringes. It is by no means straightforward to deconvolute the interferogram into a map of the phase across the beam or the pupil. Firstly, and most obviously, it is clear from Eq. (16.1) that there is an ambiguity in the phase difference with regard to integer multiples of 2π. That is to say an apparent phase difference Φ could be legitimately interpreted as a phase difference of $2\pi n + \Phi$, where n is an integer. In principle, this could be dealt with under the assumption that the form of the wavefront is continuous across the pupil and 'stitching' a wavefront map across the pupil on the basis of this assumption. However, a further ambiguity arises. The form of Eq. (16.1) is such that the measured irradiance of the interferogram is independent of the sign of the phase difference. That is to say a phase difference of $+\Phi$ is indistinguishable from a phase difference of $-\Phi$. In principle, this difficulty may be circumvented by taking two interferograms where the relative phase of the reference beam is shifted by 90°. In other words, the relative phase of the two interferograms is in

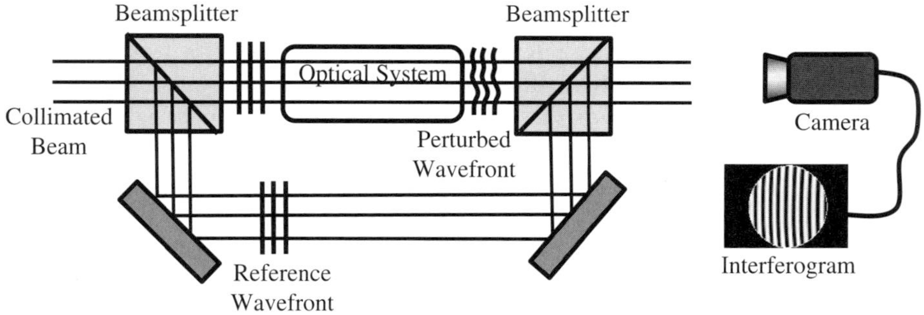

Figure 16.1 Basic principle of interferometry.

quadrature. In practice, conversion of an interferogram – a 2D image – into a wavefront map requires the capture of multiple interferograms. This is especially true, since, in practice, the fringe visibility, V, is an extra parameter that needs to be ascertained.

16.2.2 Data Processing and Wavefront Mapping

Just as the invention of the laser has had a significant impact on the design of practical interferometry systems, the introduction of digital imaging and computing power is of especial importance. Historically, the analysis of interferograms has been, to a degree, interpretive. That is to say, an interferogram is presented simply as an image, viewed either directly by eye or as a photograph and qualitative or semi-quantitative information derived by inspection of the fringe pattern. With the availability of digital technology, the fringe pattern can be converted directly into a continuous phase map across the pupil. At each pixel in the interferogram image, the irradiance is analysed and compared to the maximum and minimum values in the fringe pattern and the phase difference derived. As alluded to earlier, decoding the phase information from the interferogram requires the capture of at least two interferograms with phase offsets between the two beams. In practice, this is accomplished by a process referred to as ***phase shifting*** whereby the absolute phase of the reference beam is varied over one cycle.

16.3 Classical Interferometers

16.3.1 The Fizeau Interferometer

The ***Fizeau Interferometer*** is an arrangement generally used for the testing of optical components or optical surfaces. In this set up, the reference beam is provided by a precision reference surface that has a form that is close to that of the surface to be measured. For example, it may be used to test a flat surface, in which case, the reference is provided by a precision flat. Where a lens surface is to be tested, the reference is provided by a precision sphere known as a ***test plate***. Lens manufacturers tend to keep a large stock of precision reference plates to test specific lens radii used frequently in their manufactured components.

A Fizeau interferometer provides a collimated or spherical beam that strikes both the reference and test surface which lie in close proximity. Thereafter, the two reflected beams are diverted by a beam splitter for fringe viewing by camera. This set up is illustrated in Figure 16.2.

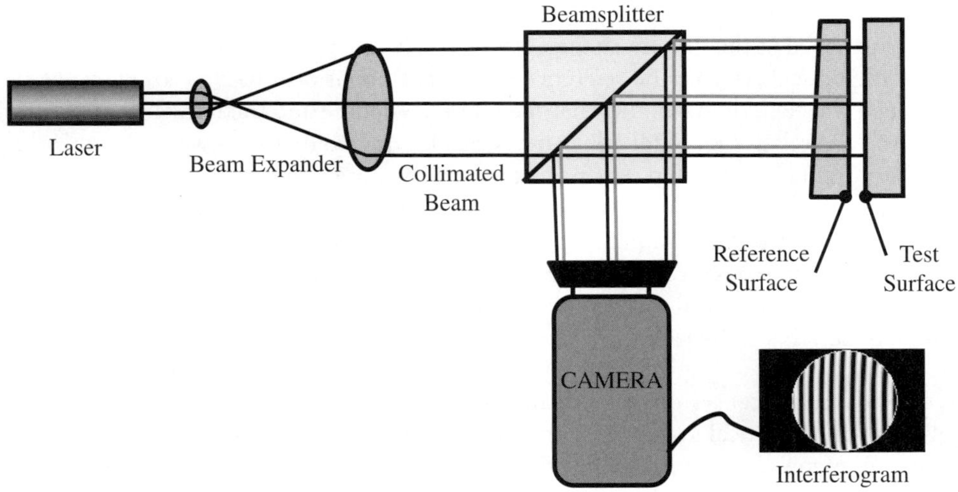

Figure 16.2 Fizeau interferometer.

In Figure 16.2, a planar set up is shown with a test and a reference flat. By inserting a lens in the parallel beam after the beamsplitter, a spherical wavefront may be created enabling the testing of spherical surfaces. An important feature of the Fizeau interferometer is that both test and reference beams share a ***common path***. Firstly, the absolute phase difference between the two beams is small and sensitivity to source coherence is diminished. Secondly, and most importantly, is that other components inserted into the optical path, such as the beamsplitter and any focussing lenses, will have the same impact on the optical path of the test and reference beams. As such, the measurement is insensitive to wavefront errors added by other components in the interferometer as these contributions are identical for both beams. Therefore, this will not affect the interferogram which depends upon the phase ***difference*** between the two beams.

The Fizeau test is based upon the comparison of a ***nominally perfect*** reference surface with the test surface. This reference is a high value precision artefact with a fidelity of form (with respect to the nominal sphere or plane) of up to $\lambda/100$ peak to valley. Of course, the whole measurement is entirely dependent upon the fidelity of this reference and its creation is a topic that will be taken up a little later. In the meantime, inspection of Figure 16.2 suggests that the optical path difference of the wavefront is double that of the relative form error of the two surfaces. Therefore, one full fringe on the interferogram, e.g. from dark fringe to dark fringe, although it represents an optical path difference of one wavelength, only represents a relative difference in surface form of half a wavelength. Before the advent of computational analysis of interferograms, it was customary to introduce a tilt between the two surfaces. Assuming both surface have the same form, this would produce a series of straight, evenly spaced fringes. Small errors in the form of the test piece would therefore be seen as a deviation in the straightness of individual fringes. By inspection, the maximum peak to valley deviation in the straightness of these fringes (in 'fringes') could then be determined. As a result, by virtue of this historical precedent, there is a tendency to denominate surface form error as a peak to valley figure, rather than the more optically meaningful rms deviation. Of course, the latter value cannot be determined without recourse to digital image processing and computational algorithms.

16.3.2 The Twyman Green Interferometer

For the more general testing of optical systems, as opposed to individual components and surfaces, other arrangements may be used. A particularly popular set up is the so-called ***Twyman-Green Interferometer***. In this configuration, the initial wavefront division takes place at a beam splitter rather than at a reference surface close to the test piece. The reference beam is then retroreflected by a precision plane mirror for combination with the test beam. This arrangement is shown in Figure 16.3.

In the arrangement illustrated, the output is in the form of a collimated beam. The optical system under test, e.g. a camera lens is being tested at the infinite conjugate with a reference sphere placed with its centre at the second focal point of the optical system. The reference sphere is a precision sphere whose form error is some very small fraction of a wavelength. Assuming that the reference sphere and the reference mirror do not contribute to system error, the relative path difference measured is equal to twice the wavefront error of the optical system. This is because the test beam passes twice through the optical system.

Other test arrangements are possible. For example, the collimated output may be used to test a flat optical surface directly. In addition, although the illustration in Figure 16.3 depicts a collimated beam, a focusing lens may be inserted so that a spherical mirror can be tested. Furthermore, the focal point of this focusing lens may be aligned to the focal point of an optical system under test and the collimated output retroreflected by a precision flat mirror. In effect, this is the reverse of the arrangement in Figure 16.3, with optical system inverted.

Where the wavefront error of an optical system is being measured, it is important that the camera image is conjugate to the pupil plane. In an optical system, it is the optical path difference referenced to the pupil plane that is the significant parameter in any analysis. It is therefore important that the fringes are viewed at the correct conjugate location. The camera system must be so designed as to focus at this particular location. Furthermore, the camera, should be designed to image the pupil with minimal distortion.

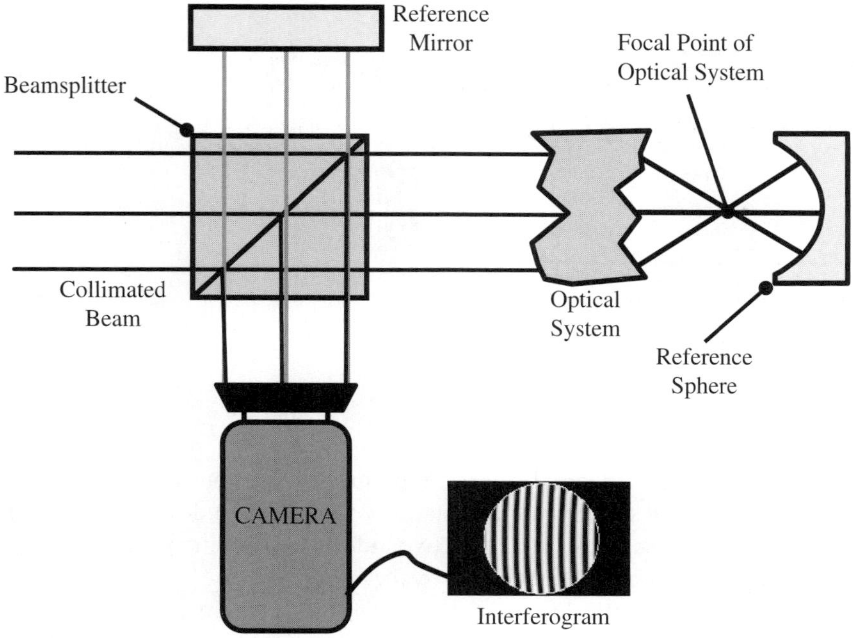

Figure 16.3 Twyman-Green interferometer.

Where measuring systems or components, the optical components within the interferometer contribute to the perceived optical path difference. Unlike the Fizeau interferometer, the reference and test beams in the Twyman-Green arrangement do not follow a common path for much of the optical path length. These can lead to 'non-common path errors' where optical path perturbations are added to one beam but not to the other. For example, a focusing lens impacts the test path, but not the reference path, whereas the reference mirror only affects the reference beam. These errors, as attributed to the interferometer optics are systematic and may be removed by a process of calibration. Calibration is effected by use of a precision artefact, such as a plane or sphere with very low or well characterised form error. The systematic wavefront deviation is simply subtracted as a background. However, this procedure is predicated upon the assumption that all wavefront contributions are additive. This assumption holds provided each ray samples the same portion of the pupil at all surfaces for both measurement and calibration scenarios. Removal of this assumption produces what are referred to as ***retrace errors*** where wavefront error contributions at different surfaces interact in a non-linear fashion. In practice, these errors are negligible if the wavefront error or path difference of the test path is low. That is to say, when testing a system or component, the engineer must strive to ensure that the interferogram contains few fringes. In other words, the set up must be designed to produce a '***null interferogram***'. Sometimes this requires a great deal of imagination and ingenuity, particularly when testing non-spherical optics.

For both the Fizeau and Twyman-Green interferometers, the measurement is essentially a ***double pass*** measurement. After accounting for any calibration offsets, the wavefront error or form error is equal to half the optical path difference derived from the interferogram.

16.3.3 Mach-Zehnder Interferometer

In the previous two classical arrangements, the interferogram is produced in double pass. By contrast, the ***Mach-Zehnder Interferometer*** is a single pass arrangement. In the Mach-Zehnder interferometer, the divided test and reference beam are recombined at a second beam splitter. The setup is illustrated in Figure 16.4.

The Mach-Zehnder Interferometer is a very flexible arrangement and is often used for the sensitive characterisation of refractive media, e.g. gases. For example, if the system under test were a gas cell, by counting

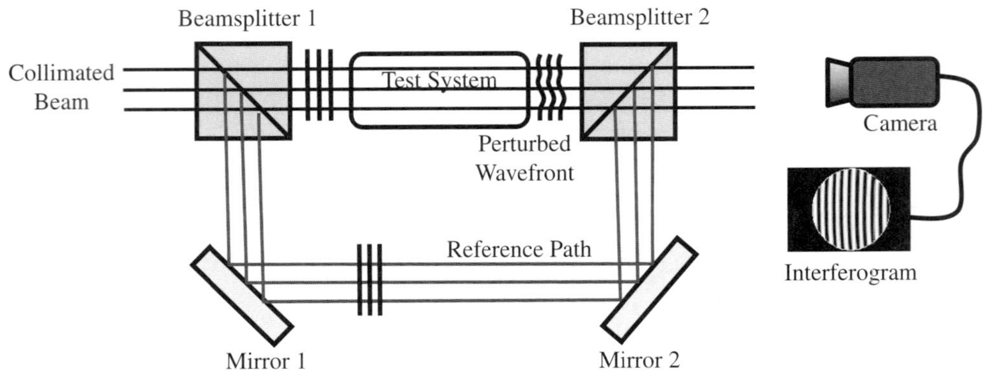

Figure 16.4 Mach-Zehnder interferometer.

fringes, it is possible to determine accurately the path length contribution produced when gas is introduced into the cell (from vacuum). This provides a very sensitive determination of refractive index. It can also be used to visualise a range of phenomena that cause density or refractive index variation in extended media. This might include the viewing of turbulence or convection currents in air.

Interferometer systems may be implemented in optical fibre or waveguide structures. In this scenario, the function of the beamsplitter may be replicated by a fibre or waveguide splitter or combiner. Implementation of an interferometer system in single mode fibre structures serves to transform phase differences into modulated flux output. More specifically, in the case of the Mach-Zehnder interferometer, a waveguide structure is created to replicate the optical paths shown in Figure 16.4, featuring one splitter and one combiner. The refractive index of one of the paths or 'arms' of the interferometer may be varied by application of an electric field, producing a modulation in the relative phase of the two arms. This produces a (high frequency) modulation in the output after the combiner and is the basis of Mach-Zehnder modulators that are of key importance in optical fibre communication networks. Refractive index modulation is based on the electro-optic effect which occurs in certain crystalline materials, such as lithium niobate or gallium arsenide.

16.3.4 Lateral Shear Interferometer

Hitherto, the arrangements that we have described are designed for the testing of wavefronts with little departure from a spherical or planar geometry. For such systems, great care is taken in adapting these classical arrangements to produce a null interferogram, i.e. one with only a few fringes of wavefront departure. This presents a great challenge in the measurement of components or systems with large aspheric departure. The problem is especially acute, as modern design and manufacturing techniques facilitates the use and production of ever more exotically shaped or freeform surfaces. One particular solution to this issue is the ***Lateral Shear Interferometer***. This instrument directly characterises a wavefront by splitting it into two laterally offset beams. This is accomplished by means of a ***shear plate***, a flat glass plate of appropriate thickness whose flat surfaces have been polished to exceptionally high fidelity. Furthermore, since one of the two beams must pass through the glass, the material must be exceptionally uniform in terms of its refractive index and have low stress induced birefringence. The arrangement is shown in Figure 16.5.

In essence, the Lateral Shear Interferometer compares the phase of a wavefront at two positions which are laterally displaced from each other. As such, the instrument essentially measures the gradient of the wavefront at any specific point. For example, a tilted wavefront would produce a null interferogram as the gradient of the wavefront is constant across the wavefront. By the same token, the interferogram of a spherical wavefront should produce a set of uniformly spaced fringes. More generally, if the phase of a wavefront at a specific point in space is described by the function, $\Phi(x, y)$, then for a lateral displacement in x of Δx, the phase difference

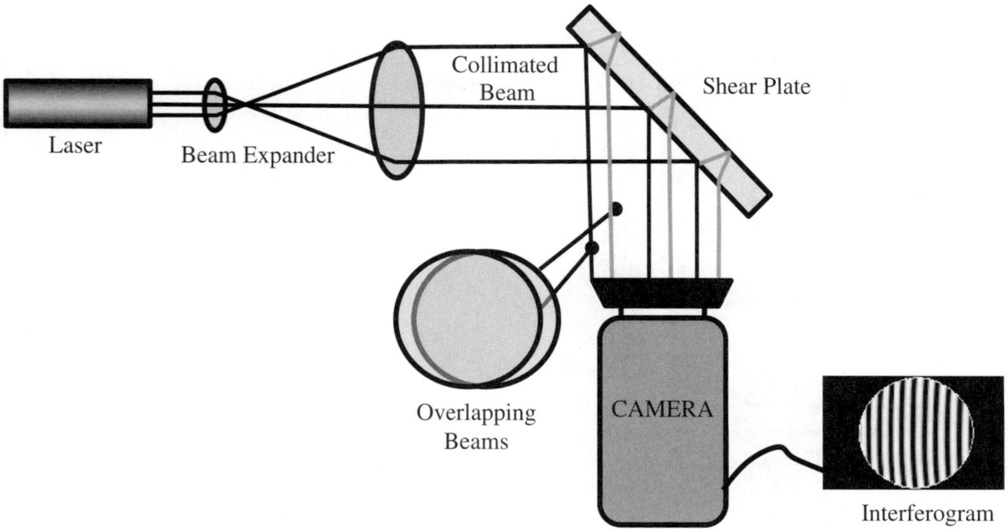

Figure 16.5 Lateral shear interferometer.

is described by:

$$\Delta\Phi(x,y) = \Phi(x+\Delta x, y) - \Phi(x,y) \tag{16.4}$$

If the displacement of the wavefront, Δx, is small and the curvature of the wavefront is low, then Eq. (16.4) may re-cast in linear form, with the difference in phase being linearly proportional to the local wavefront slope.

$$\Delta\Phi(x,y) = \frac{\partial\Phi(x,y)}{\partial x}\Delta x \tag{16.5}$$

By integration, Eq. (16.5) may be used to help derive the wavefront error across the whole pupil. However, since the measurement for a relative beam displacement in the x direction only provides the gradient in that direction, a separate measurement must be made for a displacement in the y direction. Assuming some constant phase offset ***and gradient*** at the centre of the pupil, e.g. zero, the wavefront error, $\Delta\Phi(x, y)$, across the entire pupil may be mapped. Since, in this instance, the ***absolute*** phase difference measured is of no significance, then the technique is insensitive to the absolute tilt or gradient of the wavefront.

Of course, this linear approach is predicated upon the assumption that the wavefront displacements and curvatures are small. Where this is not the case, the linear approximation breaks down and higher order terms (in Δx) need to be considered. Under the assumption that the wavefront error may be described by a well behaved continuous function in x and y there are a number of mathematical approaches that will facilitate extraction of the wavefront error. For example, it may be assumed that the wavefront error is capable, within prescribed limits, of being represented in terms of a series of Zernike polynomials, up to some specific order. The two sets of measured phase differences (in Δx and Δy) may be decomposed themselves into their relevant Zernike components. For the set of Zernike polynomials, and given Δx and Δy, it is possible to determine a linear relationship (in the form of a matrix) between those polynomials representing the wavefront and those representing the measured difference. Given knowledge of this linear relationship (derived mathematically given Δx and Δy), the wavefront error across the pupil may be derived.

16.3.5 White Light Interferometer

The so called ***White Light Interferometer*** does not, strictly speaking, belong to the realm of classical interferometers. As the name suggests and, unlike conventional interferometers, the instrument uses broadband,

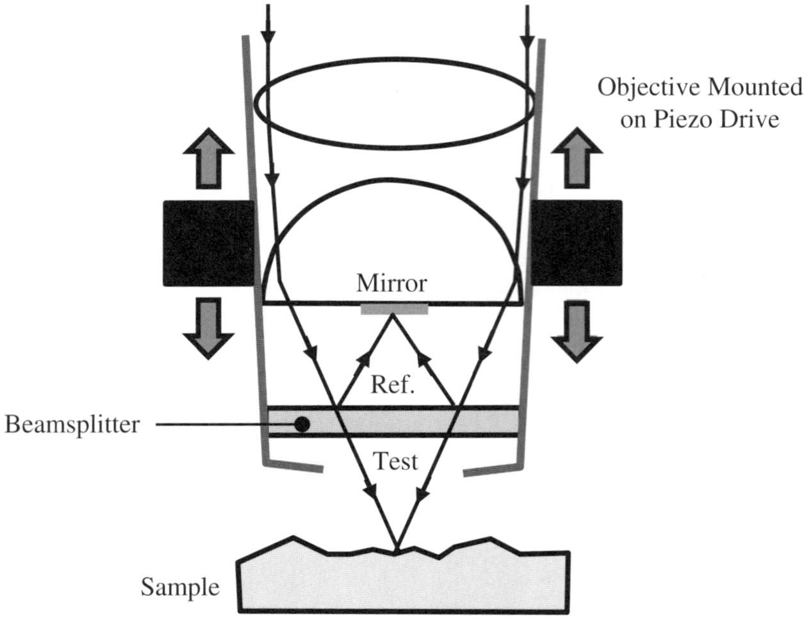

Figure 16.6 Mirau objective.

as opposed to coherent and narrowband, sources. Naturally, interference is only possible for the smallest path differences. The White Light Interferometer is used as a microscope with a specially devised objective providing the necessary reference beam by division of amplitude at a beamsplitter. This type of objective is known as a Mirau objective and is illustrated in Figure 16.6.

An objective lens focuses light from a broadband source onto the sample, as illustrated. A beamsplitter divides the illumination into two paths – a test path and a reference path, as shown in Figure 16.6. The reference path is reflected off a mirror and re-joins the test path thereafter. As meaningful interference, for a broadband source, can only be observed where the reference and test beams have a very small path difference, the design of the objective is such that the two paths are equivalent. Furthermore, as illustrated in Figure 16.6, the relative paths of the two beams may be adjusted precisely by moving the objective relative to the sample with a piezo drive. This facilitates adjustment of the path difference to a precision of better than a nanometre.

Where the path lengths are very similar, fringes are observed in the final image, as captured digitally. As the objective height is adjusted by the piezo drive, these fringes are observed to move across the captured image. Using image processing techniques, the behaviour of these fringes under scanning may be used to build up a picture of the object relief to nanometre precision.

At this point, we might care to analyse the formation of the white light fringes in a little more detail. To illustrate the formation of the fringes, we can model the broadband flux as an idealised Gaussian distribution with respect to its spatial frequency, k. The source has a maximum spectral flux, $\Phi(k)$, at some spatial frequency, k_0 and the width of the broadband emission is defined by Δk:

$$\Phi(k) = \Phi_0 e^{-[(k-k_0)/\Delta k]^2} \text{ and } A(k) = A_0 e^{-[(k-k_0)/\sqrt{2}\Delta k]^2} \tag{16.6}$$

We introduce a relative shift of Δx in the path between the test and reference beams. For each specific spatial frequency, k, this will modulate the output flux by interference according to Eq. (16.1). If we represent the flux following interference as $\mathrm{I}(k, \Delta)$, then, for a specific spatial frequency, this is given by:

$$I(k, \Delta) = 2\Phi(k)(1 + \cos(k\Delta x)) \tag{16.7}$$

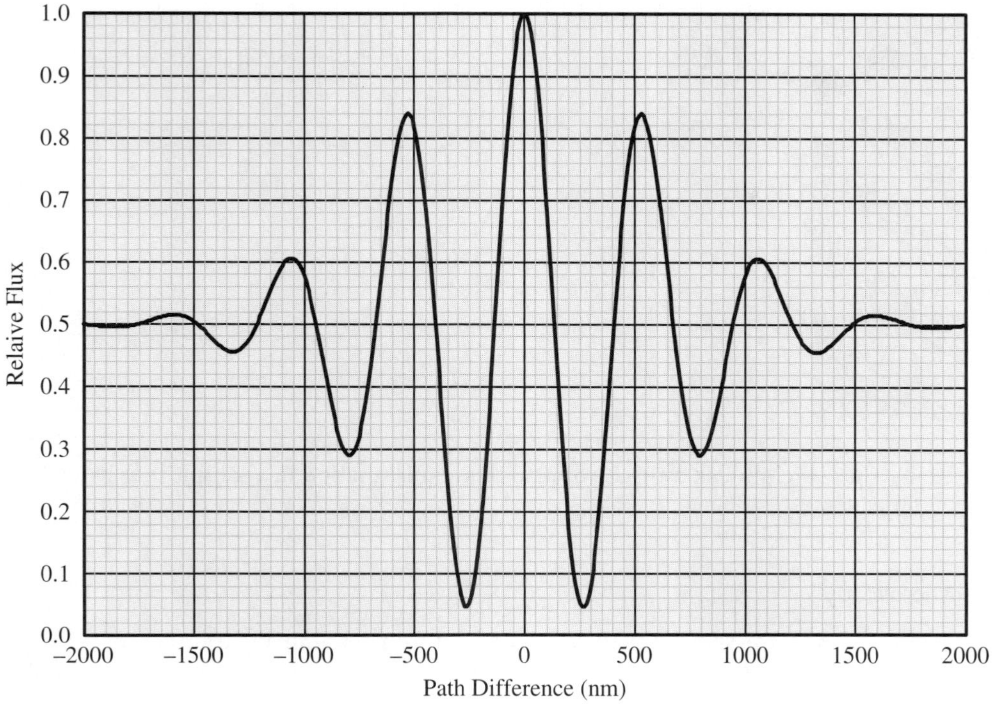

Figure 16.7 Modelled white light fringes.

We simply need to integrate the above expression with respect to k to obtain the total integrated flux, $I(\Delta)$:

$$I(\Delta) = 2\int \Phi_0 e^{-[(k-k_0)/\Delta k]^2}(1 + \cos(k\Delta x))dk \tag{16.8}$$

Integrating the above expression gives:

$$I(\Delta) = 2\Phi_0 \lfloor 1 + e^{-[(\Delta k\Delta x)/2]^2} \cos(k_0 \Delta x) \rfloor \tag{16.9}$$

In terms of the visibility of the fringes at the detector, any expression for the effective spectral flux of the illumination must take into account the spectral sensitivity of the detector. Of course, the analysis pursued here is somewhat idealised, but we may propose a reasonable model of a 'white light source' in terms of a maximum spectral flux at 540 nm and a Δk value that is 20% of the k_0 value. The results of this modelling are shown in Figure 16.7.

Figure 16.7 confirms that the fringes are only visible over a very restricted range of path differences. In a practical instrument, with a pixelated detector, the behaviour shown in Figure 16.7 would be available on a pixel by pixel basis. The objective would be scanned in height over some range, e.g. 5 μm and, for each pixel, the data would be analysed to determine the height at which the signal is maximised, as per Figure 16.7. The product of this analysis is a very precise two dimensional height map of the object under inspection. For example, the White Light Interferometer may be used to measure the surface roughness of polished surfaces, to nanometre or subnanometre level.

Figure 16.8 shows an interferogram for a diamond machined aluminium surface. The regularity of the fringes is disturbed by the morphology of the machined surface.

The White Light Interferometer is one specific application of interferometry in microscopy. It is fundamentally a metrological application providing quantitative information about the surface in question. There are other techniques where interferometry is applied to microscopy to improve contrast in inherently low

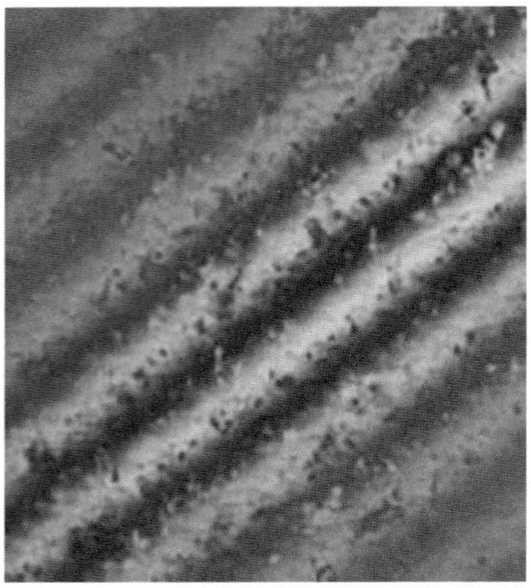

Figure 16.8 White light interferogram of diamond machined Al surface.

contrast objects. Essentially, these techniques translate optical phase information in a sample into irradiance fluctuations for enhanced imaging.

16.3.6 Interference Microscopy

Phase contrast microscopy enhances the sensitivity to scattered light from a nominally transparent (e.g. biological) sample. In this instance, the sample is illuminated by an annular pupil and any light outside this annulus must have been scattered by the sample itself. A phase shift plate is then used to shift the relative phases of these two regions (e.g. by 90°) such that the scattered light interferes constructively with the transmitted light. Another plate is used to control the amplitude of the two regions to maximise diffraction efficiency. In this example, where constructive interference is engendered, regions of the sample corresponding to the highest levels of scattering appear to be bright against a dark background. Alternatively, this scenario may be reversed by adjusting the relative phases to produce destructive interference. In this case regions of scattering appear dark against a light background.

In the ***Nomarski Microscope***, phase contrast between two orthogonal polarisations is converted into irradiance contrast by interference. Illumination is polarised and this linearly polarised light is itself split into two orthogonal and coherent components at +45° and −45° by a Wollaston prism. Both components are imaged at the (transparent) object by condensing optics and then recombined at a second Wollaston prism. If there is no relative phase offset between the two polarisations, then the original polarisation state will be re-created and any variation will change the polarisation state at the output. Indeed, if a second (crossed) polariser is placed at the output then, where there is no phase disturbance, no light will be transmitted. Small differences in the relative phase of the two polarisations as they pass through the sample will then produce large variations in contrast.

16.3.7 Vibration Free Interferometry

In interferometer arrangements where there are significant non-common path lengths, such as the Twyman-Green interferometer, any random variability in that path length, on the order of a fraction of a wavelength, will significantly degrade fringe visibility. There are two prime causes of this path length variability. Firstly, small vibrations in optical mounts will produce changes in relative path lengths. Secondly, over long path lengths in air, variations in air density caused by thermal currents and air motion will produce significant phase disturbances. As a result, interferometric measurements are by no means straightforward and great care must be taken to avoid these path length disturbances. Therefore, it is customary to mount interferometric set-ups in special stable laboratory settings on optical tables that are vibrationally isolated

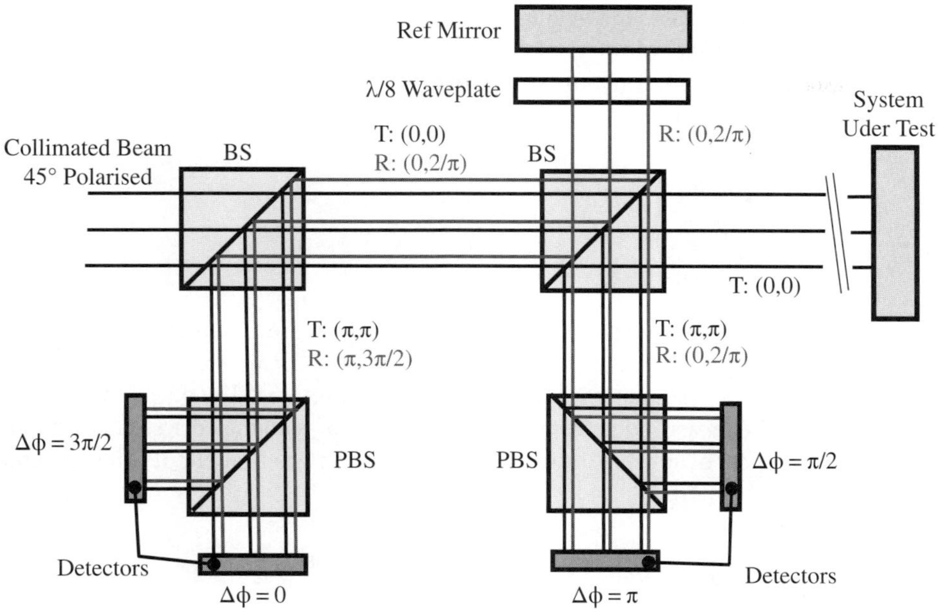

Figure 16.9 'Vibration Free' interferometer.

from floor vibration. In addition, great care must be taken to ensure that the laboratory is thermally stable in order to minimise low level air turbulence.

As we will see, the decoding of meaningful phase information from fringe data requires measurement of at least three separate sets of fringe data. Application of varying phase shifts between these separate measurements is generally accomplished sequentially. Thus, even if the fringe image acquisition time is very short, significant time elapses between each measurement. As a consequence, the actual phase shift between each measurement is substantially compromised by random phase shifts produced by any instability in the test arrangement. For all the precautions that might be taken in mitigating the impact of vibration and air currents, there may be instances where these effects cannot be ameliorated sufficiently. Vibration Free Interferometry overcomes these problems by acquiring a number of phase shifted fringe images, e.g. four, ***simultaneously***.

There are a number of interferometric instruments and tests that exploit polarisation to produce variable phase shifts. One such example is shown in Figure 16.9.

In the example shown in Figure 16.9, the input collimated beam, derived from a laser source is polarised at 45°. This polarisation state may be decomposed into two orthogonal and in phase polarisation components. The arrangement consists of two 'normal' beamsplitters, labelled 'BS' and two polarising beam splitters, labelled 'PBS', to split these two components at the detectors. As usual, the reference beam is diverted at a beamsplitter. Interposed in the reference path is a $\lambda/8$ waveplate which, after a double pass, imposes a $\lambda/4$ or $\pi/2$ phase difference between the two polarisations. This is indicated in Figure 16.9, by the label 'R: (0, $\pi/2$)' to indicate the phases of the two components in the reference beam. By contrast, the test beam is not modified and its polarisation state is indicated by the label 'T: (0, 0)'. At this point, it is useful to examine the phase impact of reflection at the beam splitter. From one direction the effective index changes from low to high, producing a 'hard reflection'. From the other direction, from high index to low index, the reflection is 'soft'. There is a relative phase difference of 180° between these reflections. The consequence of this is that a further relative phase shift π radians is introduced between the test and reference paths in one of the arms. Thus, the arrangement in Figure 16.9 is able simultaneously to produce four separate interferograms incorporating different and equally spaced phase differences.

The arrangement shown in Figure 16.9 illustrates the operating principle clearly, although, in practice, it is a little cumbersome, requiring accurate alignment of the four separate detectors. It is possible to integrate this arrangement to generate these images on a common detector, providing a more compact and useful instrument. Further details are available in the literature.

16.4 Calibration

16.4.1 Introduction

We have previously alluded to the important role of calibration artefacts, sphere and flats, that are figured to some very small fraction of a wavelength, representing a few nanometres of form error. These artefacts are critical in the removal of background or systematic errors in experimental set ups. The question arises as to how such surfaces, themselves, may be measured precisely in the presence of recognised systematic error. Such surfaces are characterised by a process of absolute interferometry and the measurement of spherical and planar surfaces will be considered here.

16.4.2 Calibration and Characterisation of Reference Spheres

Calibration of a reference sphere involves at least three measurements in three separate configurations. For example, we might assume that the arrangement follows a standard Twyman-Green set up in which the interferometer contributes some small but unknown wavefront error to each measurement. The three measurements may be categorised as follows:

i. Interferometer focus at centre of reference sphere
ii. Position as i) except sphere rotated by 180° about optical axis
iii. Interferometer at 'cat's eye' position on surface of sphere

The measurement scheme is illustrated in Figure 16.10.

To understand the principle of the measurement we may assume that the wavefront error associated with the sphere may be split into even and odd functions, $E_1(x, y)$ and $O_1(x, y)$ respectively. For the even functions, the form is preserved on rotation through 180°; for the odd functions, the form is reversed. Similarly for the instrument contribution, this may also be split into even and odd functions, $E_0(x, y)$ and $O_0(x, y)$. It is to be understood that these contributions take into account the double passage through the interferometer. Assuming that the reference sphere only contributes a small amount of wavefront error, the retro-reflected

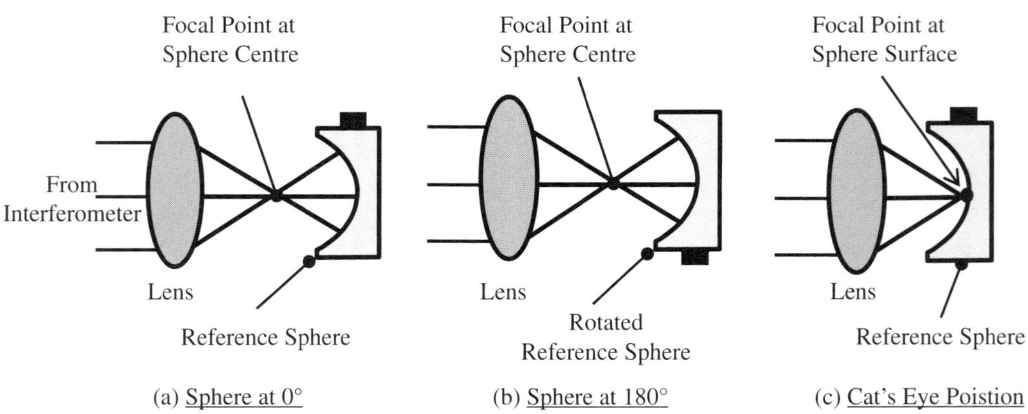

Figure 16.10 Absolute form measurement of reference sphere.

beam passes through the same part of the interferometer, where the interferometer focus is at the sphere centre. Under these conditions, the total system wavefront error may be computed as the sum of the interferometer and reference sphere contributions. This analysis is straightforward to implement for the first two configurations – both even and odd functions simply sum for the two passages. However, for the cat's eye measurement, the retro-reflected beam samples a portion of the pupil that is rotated by 180° about the centre. Therefore the even contribution is preserved, whereas the odd contribution is cancelled. If we label the even and odd measurements for each of the three scenarios as $E_a(x, y)$, $O_a(x, y)$, etc., then the following equations apply:

$$E_a(x,y) = E_0(x,y) + E_1(x,y) \quad O_a(x,y) = O_0(x,y) + O_1(x,y) \tag{16.10a}$$

$$E_b(x,y) = E_0(x,y) + E_1(x,y) \quad O_b(x,y) = O_0(x,y) - O_1(x,y) \tag{16.10b}$$

$$E_c(x,y) = E_0(x,y) \quad O_c(x,y) = 0 \tag{16.10c}$$

This gives:

$$E_1(x,y) = E_a(x,y) - E_c(x,y) \quad O_1(x,y) = (E_a(x,y) - E_b(x,y))/2 \tag{16.11a}$$

$$E_0(x,y) = E_c(x,y) \quad O_0(x,y) = (E_a(x,y) + E_b(x,y))/2 \tag{16.11b}$$

Once calibrated in this way, the reference sphere may be directly used with an interferometer instrument to calculate the systematic instrument background for subtraction in further measurements.

16.4.3 Characterisation and Calibration of Reference Flats

Characterisation and calibration of reference flats proceeds by the so-called ***3 flats method***. In this test, three flats, A, B, C are compared in a Fizeau arrangement in three separate tests. The basic arrangement for these tests is illustrated in Figure 16.11.

In the test configurations shown, we define our coordinate system as shown and we assume that the 'standard' orientation of each surface is that presented by the surface on the right. We further assume that the surface on the left has been placed by rotating that particular flat about the y axis, as shown. If we represent the measured wavefront error in the three scenarios as $\Phi_1(x, y)$, $\Phi_2(x, y)$ and $\Phi_3(x, y)$, then the ***symmetric*** (about y) contribution of each flat, A(*x*, *y*), B(*x*, *y*) and C(*x*, *y*) may be represented as follows.

$$\Phi_1(x,y) = A(x,y) + B(x,y) \tag{16.12a}$$

$$\Phi_2(x,y) = A(x,y) + C(x,y) \tag{16.12b}$$

$$\Phi_3(x,y) = B(x,y) + C(x,y) \tag{16.12c}$$

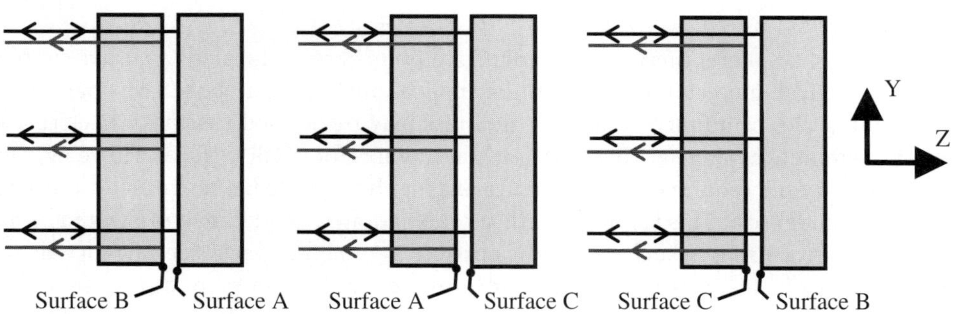

Figure 16.11 The three flat test.

The reader should note the sign of the terms of the right hand side of Eqs. (16.12a, 16.12b, and 16.12c). Although the Fizeau interferometer compares the form of two surfaces by subtraction, one of the two surfaces has been flipped with respect to the other. In the presented scenario, for those contributions to surface form that are symmetrical about the y axis, then the process simply inverts the surface, hence the form of Eqs. (16.12a, 16.12b, and 16.12c). ***For surface form contributions symmetrical in y***, then these contributions are simply given by:

$$A(x,y) = (\Phi_1(x,y) + \Phi_2(x,y) - \Phi_3(x,y))/2 \tag{16.13a}$$

$$B(x,y) = (\Phi_1(x,y) - \Phi_2(x,y) + \Phi_3(x,y))/2 \tag{16.13b}$$

$$C(x,y) = (-\Phi_1(x,y) + \Phi_2(x,y) + \Phi_3(x,y))/2 \tag{16.13c}$$

If one analyses all terms as Zernike components, then y symmetry encompasses all terms with an azimuthal component of the form $\cos(n\theta)$ where n is even and those terms of the form $\sin(n\theta)$ where n is odd. To elucidate Zernike components of the form $\cos(n\theta)$ where n is odd, then each measurement should be performed with one of the two flats rotated by 180°. Analysis proceeds as per Eqs. (16.13a)–(16.13c). Similarly, for components of the form $\sin(n\theta)$ where n is even, then each measurement should be performed with one of the two flats rotated by 90°.

16.5 Interferometry and Null Tests

16.5.1 Introduction

The interferometric testing of standard shapes, such as spherical and planar surfaces represents no especial challenge. The generation of spherical or planar wavefronts in such tests is inherently straightforward. However, the testing of aspherical surfaces does create significant difficulties. For reasons indicated earlier, the test wavefront must, to a degree, match that of the surface under test. A null interferogram should be generated with only a few fringes of departure over the clear aperture. Advances in design and manufacturing capability have encouraged the adoption of more exotic optical surfaces in a variety of applications. However, form testing of these surfaces is an essential part of the manufacturing process. A measure of ingenuity is required to formulate a test arrangement to characterise such surfaces. These tests are referred to as ***null tests***.

Some special surface types do lend themselves to generic test arrangements. One group of such surfaces is the conic surface. Conic (mirror) surfaces produce perfect on-axis imaging for a specific conjugate pair whose conjugate parameter is related to the conic parameter of the surface. This applies only where the conic parameter is less than zero and the conjugate parameter, t, is given by the following simple expression:

$$t = \pm\sqrt{-k} \tag{16.14}$$

Of course, for the trivial case of a sphere, then the two conjugate points are co-located. Another simple conic surface that is of great practical interest in many scientific applications is the parabola. For a parabola, one of the ideal conjugate points lies at infinity. A suitable null test may be devised that uses a reference plane mirror to retroreflect the infinite conjugate. This is perfectly acceptable for small surfaces. However, the testing of large telescope mirrors would require the manufacture of a highly accurate flat mirror whose size is equivalent to that of the mirror under test. This is not a practical proposition as its manufacture would be as least as costly as the telescope mirror itself. In such cases, it is possible to design a special lens arrangement, referred to as a ***null lens***, that permits testing to be carried out from the nominal centre of curvature of the parabola or conic. The null lens effectively produces spherical aberration that cancels that generated by the surface.

For surfaces that are more complex, such as even aspheres or truly freeform surfaces lacking axial symmetry, ***then computer generated holograms (CGHs)*** are available to facilitate interferometric testing. CGHs are transmission gratings created by depositing patterned thin film structures on a transparent substrate. Whilst the gratings do produce diffraction into specific orders, zeroth, first, second etc. and the amplitude produced has a distinctive spatial variation, it is rather the differential phase that is produced across the wavefront that is of interest. By careful calculation of the grating pattern, the resulting diffraction produces a wavefront whose shape is tailored to that of the surface in question. The technique is extremely flexible and may be applied, within reason, to virtually any surface. The principal disadvantage of the CGH is that of cost; a tailor made CGH must be designed and manufactured to suit a specific surface.

16.5.2 Testing of Conics

The simple test for conic surfaces proceeds as a double pass measurement. To illustrate this, we will exemplify the process by considering the testing of a simple parabola. The focus of the (e.g. Twyman-Green) interferometer is placed at the focus of the parabola. A collimated beam then emerges from the parabola which is reflected back to the parabola by a precision flat. The parabola then refocuses the beam back to the interferometer focus. If the measurement is an 'on-axis' measurement, it is inevitable from the geometry that some portion of the beam or pupil is obscured by one of the co-axial surfaces. Most typically, these tests are employed for telescope mirrors with a central aperture, producing an annular, as opposed to circular, pupil. In addition, because the measurement is a double pass measurement, the phase difference generated is twice that of the corresponding single pass measurement. Indeed, the wavefront error produced is actually four times the form error of the parabolic mirror. The arrangement is illustrated in Figure 16.12.

Figure 16.12 applies specifically to the testing of parabolic mirrors. However, by extension, any conic surface may be tackled. Specifically, for surfaces with a conic constant of less than zero (but not -1) a precision sphere should be substituted for the flat and should be located with its centre at the appropriate focus. That is to say, one focus of the conic should be at the sphere centre and the other at the interferometer focus. For concave surfaces and for $-1 < k < 0$ (ellipsoidal surfaces) both foci are real. Similarly, for convex hyperboloidal surfaces ($k < -1$) both foci are also real. Otherwise, the mirror foci are virtual and effectively lie behind the mirror surface.

In this narrative, we have dealt with conic surfaces with a negative conic parameter. Where the conic parameter is positive, the form of the surface describes an oblate spheroid or ellipsoid. However, in this instance, the effective conic axis, on which the two foci lie, is rotated by 90°. This considerably complicates access to the two foci. Nonetheless, access may be provided by an off-axis arrangement. As with the equivalent prolate spheroid test, the interferometer focus is placed at one focus and the centre of a precision sphere at the other. This arrangement is shown in Figure 16.13.

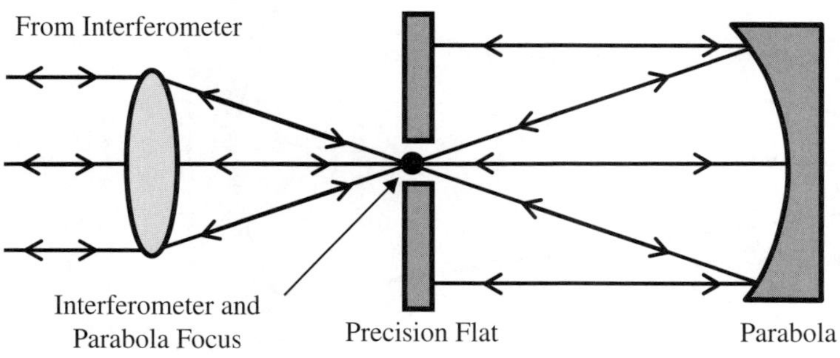

Figure 16.12 Interferometric testing of a paraboloidal mirror.

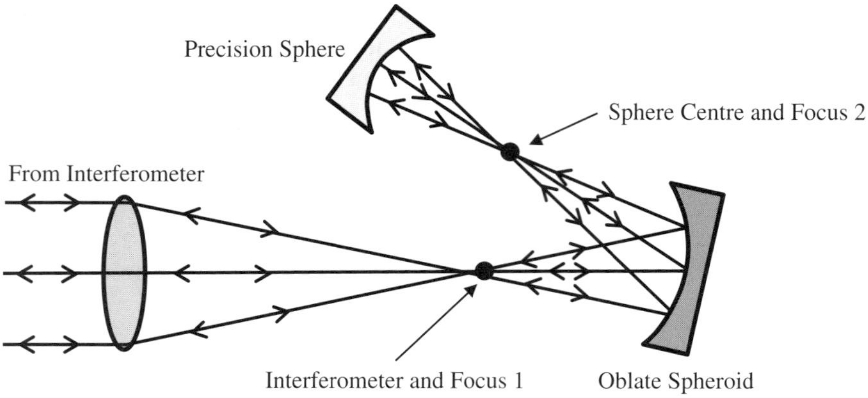

Figure 16.13 Oblate spheroid test.

All these tests may be used to test more generic, freeform surfaces, provided that the departure of the surface from the nominal does not amount to more than a few fringes. Of course, the ability to test surfaces that are nominally conic, as opposed to spherical extends the range of surfaces that can be tested.

16.5.3 Null Lens Tests

The tests thus described are useful tests for characterising relatively small components. Where larger surfaces are to be tested, then it is possible to design an optical system to cancel out the on-axis aberrations produced by the surface. As such, for concave surfaces, these so-called ***null lenses*** are very compact, even for test surfaces several metres in diameter. Of course, this consideration does not apply to convex surfaces. In consequence, large convex surfaces are generally quite troublesome to measure and there is a tendency to avoid substantially large convex surfaces in most optical designs.

For an on-axis test of a conic surface, the simplest scenario imaginable is where a singlet lens is used to correct the third order spherical aberration produced by the conic surface. This forms the backdrop to the ***Ross null test***. In this test, the interferometer focus is effectively located close to the centre of the conic. The arrangement is shown in Figure 16.14.

We commence the analysis by computing the third order spherical aberration attributable to a conic surface of base radius, R, with a conic constant of k, and illuminated by a test beam with a numerical aperture of NA. From Chapter 5, Eq. (5.7) the contribution of the conic is:

$$\Phi_{Mirror} = -\frac{kR}{4}NA^4 \tag{16.15}$$

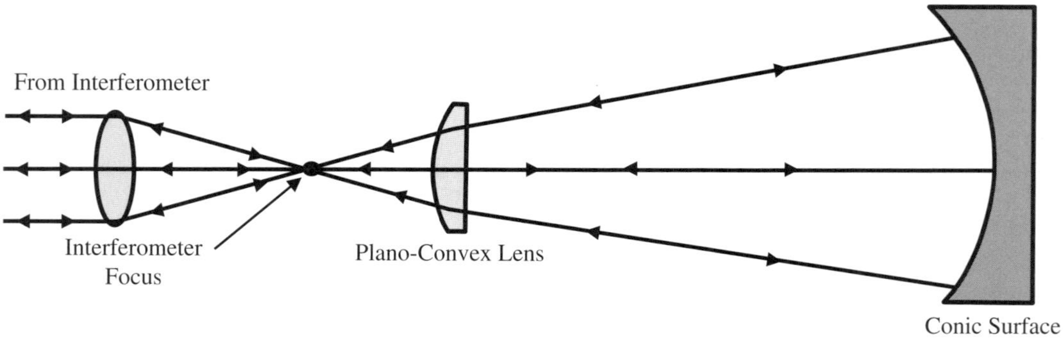

Figure 16.14 Ross null test.

For a plano-convex lens in the orientation described, the shape factor is 1 and, for a given focal length f, we may assume that the conjugate parameter, t is adjustable. From Chapter 4, the contribution of the plano lens to the spherical aberration is given by:

$$\Phi_{Lens} = -\left(\frac{1}{32f^3}\right)\left[\left(\frac{n}{n-1}\right)^2 - \left(\frac{n}{n+2}\right)t^2 + \frac{(n+2)}{n(n-1)^2}\left[1 + 2\left[\frac{n^2-1}{n+2}\right]t\right]^2\right] r^4 \quad (16.16)$$

Equation (16.6) expresses the spherical aberration in terms of the pupil radius, r, as viewed at the lens. However, Eq. (16.15) is cast in terms of the numerical aperture seen by the mirror, or that following the lens. We can express r in terms of NA thus:

$$r = \frac{2fNA}{1-t} \text{ and } r^4 = \frac{16f^4NA^4}{(1-t)^4}$$

Thus ***and allowing for a double pass*** through the lens:

$$\Phi_{Lens} = -\left[\left(\frac{n}{n-1}\right)^2 - \left(\frac{n}{n+2}\right)t^2 + \frac{(n+2)}{n(n-1)^2}\left[1 + 2\left[\frac{n^2-1}{n+2}\right]t\right]^2\right]\frac{fNA^4}{(1-t)^4} \quad (16.17)$$

Since the two contributions from the length and mirror must sum to zero, we arrive at the following equation to solve for the focal length in terms of the conjugate parameter.

$$\frac{kR}{f} = -\left[\left(\frac{n}{n-1}\right)^2 - \left(\frac{n}{n+2}\right)t^2 + \frac{(n+2)}{n(n-1)^2}\left[1 + 2\left[\frac{n^2-1}{n+2}\right]t\right]^2\right]\frac{4}{(1-t)^4} \quad (16.18)$$

All this analysis is based on the thin lens approximation; detailed (computer-based) analysis would account for a finite lens thickness.

Worked Example 16.1 A 500 mm diameter ellipsoidal telescope mirror is to be tested using a Ross null lens. The base radius of the mirror is 2400 mm and its conic constant is −0.75. We are told that the conjugate parameter is 3.5. What is the focal length of the lens and how far should the lens be from the interferometer focus, assuming its refractive index is 1.52? We have only considered third order spherical aberration. Estimate the contribution of uncorrected fifth order aberration.

From Eq. (16.18) we find:

$$\frac{kR}{f} = -11.74$$

Substituting the values of k and R, we obtain the focal length of 153.4 mm. It is a simple matter to calculate the distance of the lens from the interferometer focus, or the object distance from the following relation:

$$u = \frac{2f}{1+t} \quad u = 68.2 \text{ mm}.$$

The lens focal length is 153.4 and the distance from lens to interferometer focus is 68.2 mm.

In this analysis, we ignored terms of sixth order in the mirror sag. The relevant sixth order contribution, as a function of radial distance r, base radius, R and conic constant, k is given by:

$$z(6) = \frac{(1+k)^2}{16R^5}r^6$$

What we are really concerned with is the difference in sag when compared to the best fit sphere and this is given by:

$$\Delta z(6) = \frac{(1+k)^2 - 1}{16R^5}r^6 = \frac{k^2 + 2k}{16R^5}r^6$$

If the maximum radius or semi-diameter is given by r_0, then the relevant Zernike polynomial contribution may be used to calculate the rms value:

$$\Delta z_{rms}(6) = \frac{k^2 + 2k}{320\sqrt{7}R^5} r_0^6$$

Substituting $r_0 = 250$ mm, $R = 2400$ mm and $k = -0.75$, we get:

$$\boldsymbol{\Delta z_{rms}(6) = 3.6\ nm}$$

Although this value does seem small, in the context of a precision measurement the systematic error entailed may be significant. Clearly, this method is restricted in application for smaller mirrors; for the characterisation of significantly larger mirrors of this type, a more elaborate test arrangement may be called for.

Another test that is comparable to the Ross test is the ***Couder test***. Instead of employing a single lens, the Couder test employs two lenses of equal and opposite power deployed close to the mirror's centre of curvature. Whilst the arrangement adds no optical power, it does produce spherical aberration that corrects for that invested in the test mirror. As the foregoing exercise suggests, characterisation of larger optics requires the use of more sophisticated arrangements correcting higher order aberrations. Such Null lenses employ several lenses or mirrors and are often designed to correct many orders of on axis aberrations. A relatively simple example is the ***Offner null test*** which employs a positive lens, as in the Ross test and another positive lens, ***the field lens***, which is located close to the focus of the first lens. This is effective in controlling higher order aberrations.

A specially designed null lens consisting of several mirrors and a field lens was used to characterise the primary mirror for the Hubble Space Telescope during the manufacturing process. Revisiting the earlier exercise, the spherical aberration correction provided by the Ross test depends upon the distance from the interferometer focus to the plano lens. Indeed, small offsets in this distance add a proportional contribution to the residual spherical aberration. Unfortunately, in the Hubble design, due to an error in the alignment, one of the mirror separations was set incorrectly in the null lens. The effect, as with the simple Ross lens displacement, was to add a small amount of spherical aberration to the measurement in proportion to the displacement. As the manufacturing process was designed to minimise the measured form error, a significant amount of spherical aberration was imprinted in the primary mirror. The profound impact of this manufacturing error on the Hubble Telescope performance had to be corrected at great expense in a subsequent Space Shuttle mission.

16.5.4 Computer Generated Holograms

CGHs are essentially tailor made transmission gratings that produce a far field distribution such that a given test surface will focus the light onto a perfect spot. They are especially useful for testing unorthodox surfaces that are difficult to test otherwise. Intuitively, one can think of their generation in the following manner. If one places the focus of a laser beam at the nominal centre of curvature of the test surface, then, by collimating the reflected beam, one may produce a wavefront whose deviation from planarity is dictated by the spherical departure of the original surface. If one then imagines this wavefront interfering with a plane wavefront, then a series of interference fringes is produced. This series of fringes may then be replicated and, when illuminated by a collimated beam, would tend to recreate the original wavefront distribution emanating from the test surface.

In practice, a CGH is designed to create a tilted beam. In other words, a spherical wavefront would be re-created by a CGH consisting of a series of equally spaced lines. Typically, the test wavefront is created by the first diffraction order. The zeroth order may be used to create a reference beam to be reflected from a reference sphere in a Fizeau arrangement.

Figure 16.15 sketches out a Fizeau test using a CGH. A collimated beam strikes the CGH and, in this example, the first order is used to define the test beam and the zeroth order defines the reference beam. A lens then focuses both beams and relays them onto the test surface and reference sphere which are juxtaposed in a

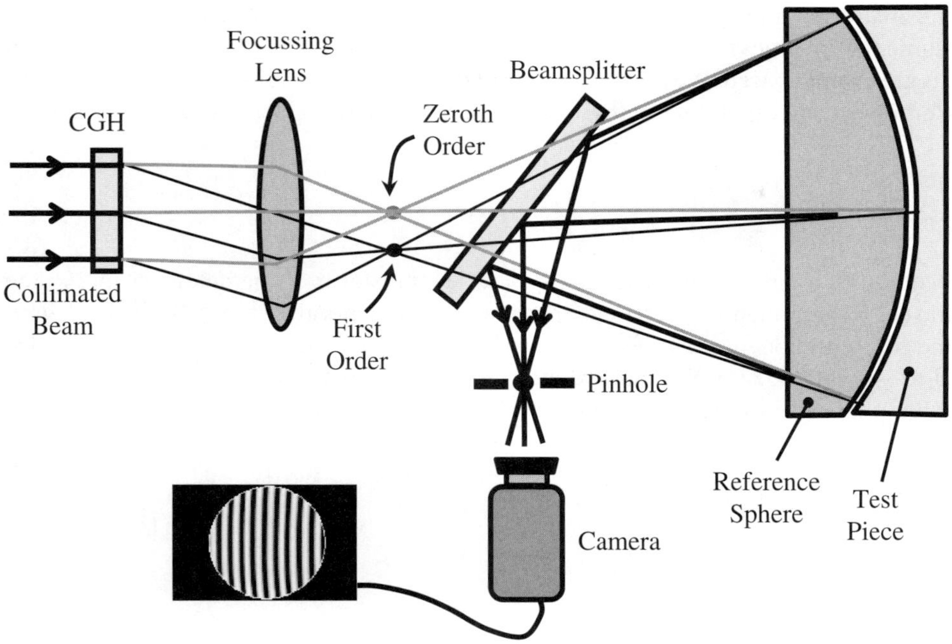

Figure 16.15 Computer generated hologram Fizeau test.

classic Fizeau arrangement. The test surface and reference sphere are conjugate to the CGH which defines the entrance pupil. The first order beam as reflected by the test surface, and the zeroth order beam as reflected by the reference sphere are focused at a common point. Thereafter, the reflected beams are monitored by a camera (via. a beamsplitter) which, itself, is also conjugate with the test and reference surface. The fringe pattern observed defines the departure of the test surface from its design prescription. Of course, the reference surface will not only reflect the zeroth order beam, but it will also reflect the first order beam. Similarly, the test piece will also reflect the zeroth order beam. Other diffraction orders will also be created and reflected. These 'extraneous' orders are displaced at the focus and may be removed by insertion of a pinhole aperture, allowing transmission of the proper test and reference beams only.

Correct alignment of the reference sphere and test piece is essential for the elimination of systematic errors. It is therefore customary to incorporate into the CGH a number of fiduciary markers that can be projected onto the reference sphere and test piece. These are in the form of diffracted cross hairs or similar distinctive location markers.

16.6 Interferometry and Phase Shifting

Digital imaging and processing techniques enable the extraction of real quantitative phase information from an interferogram. In order to unambiguously measure the phase at a specific location at least three independent interferograms must be acquired each with a different and known phase offset between the test and reference beams. For each pixel in the interferogram, the only information that is available is the measured flux level and the phase must be extracted from this information. There are a variety of techniques for providing this phase offset, for example, exploiting polarisation sensitivity and using variable wave retarders to provide the offset. In many compact commercial devices, this phase shifting is simply accomplished by fine adjustment of a reference mirror using piezoelectric actuators. The overall process is referred to as ***phase shifting interferometry***.

If we now consider the analysis of a phase shifting procedure incorporating three independent measurements, then we may assume that the measurements are made at relative offsets of 0°, 120° and 240°. That is to say, the measurements are evenly spaced. We are seeking to determine the phase offset of the bright fringe and the three flux measurements corresponding to these offsets are Φ_1, Φ_2, Φ_3. The phase angle, ϕ, is given by:

$$\tan\phi = \frac{\sqrt{3}\Phi_2 - \sqrt{3}\Phi_3}{2\Phi_1 - \Phi_2 - \Phi_3} \tag{16.19}$$

In effect, Eq. (16.19) computes a discrete, if rather sparse Fourier transform of the data, calculating the amplitude of the $\sin\theta$ and $\cos\theta$ components. The ratio of these two components gives $\tan\phi$. By paying regard to the sign of the $\sin\theta$ and $\cos\theta$ components, the value of ϕ may be determined unambiguously over the range $-\pi < \phi < \pi$ (as opposed to $-\pi/2 < \phi < \pi/2$). We may extend this analysis, to a more general measurement involving N equally spaced phase measurements. As previously, we designate the flux values as Φ_1, Φ_2,…, Φ_{N-1}, Φ_N.

$$\tan\phi = \frac{\sum_{i=0}^{i=N-1} \sin(2\pi i/N)\Phi_{i+1}}{\sum_{i=0}^{i=N-1} \cos(2\pi i/N)\Phi_{i+1}} \tag{16.20}$$

Equation (16.20) may be applied to four, five, and six measurement scenarios:

$$\tan\phi = \frac{(\Phi_2 - \Phi_4)}{(\Phi_1 - \Phi_3)} \textit{ four measurements} \tag{16.21a}$$

$$\tan\phi = \frac{\sqrt{(10+2\sqrt{5})}(\Phi_2 - \Phi_4) + \sqrt{(10-2\sqrt{5})}(\Phi_3 - \Phi_5)}{\Phi_1 + \sqrt{(6-2\sqrt{5})}(\Phi_2 + \Phi_4) - \sqrt{(6+2\sqrt{5})}(\Phi_3 + \Phi_5)} \textit{ five measurements} \tag{16.21b}$$

$$\tan\phi = \frac{\sqrt{3}\Phi_2 + \sqrt{3}\Phi_3 - \sqrt{3}\Phi_5 - \sqrt{3}\Phi_6}{2\Phi_1 + \Phi_2 - \Phi_3 - 2\Phi_4 - \Phi_5 + \Phi_6} \textit{ six measurements} \tag{16.21c}$$

In terms of the overall interferogram, we are presented with a set of discrete phase shift values, one for each pixel. Taken as an isolated data point, there is no way for a specific measurement to discriminate between phase shifts offset by integer multiples of the wavelength. That is to say, where an apparent phase shift of ϕ is measured, this measurement could also be satisfied by a phase shift of $2\pi n + \phi$, where n is an arbitrary integer. ***Stitching*** the individual pixelated phase measurements to produce a phase map can only be carried out under the assumption that the phase shifts between pixels is small and the wavefront error across the pupil may be represented as a smooth continuous function.

16.7 Miscellaneous Characterisation Techniques

16.7.1 Introduction

There are a number of optical techniques which, although not based on interferometry, perform many of the wavefront characterisation and optical surface characterisation tasks often attributed to interferometry. These include instruments such as the Shack-Hartmann wavefront sensor and a variety of techniques employing fringe projection and knife edge aperturing. What these techniques have in common is the ability to convert subtle disturbances in a wavefront into some form of measurable contrast variation in a projected optical image.

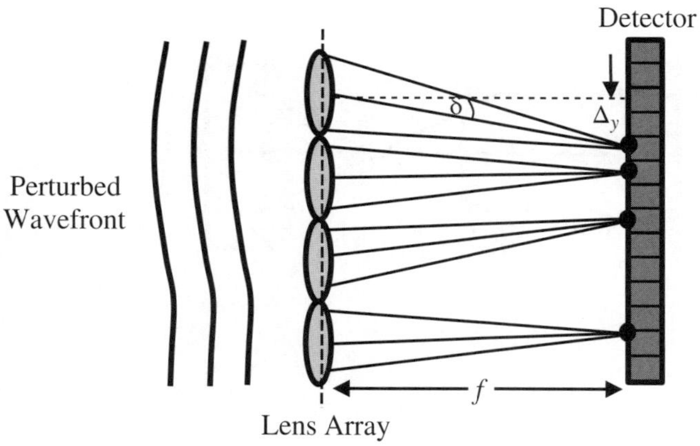

Figure 16.16 Shack-Hartmann wavefront sensor.

16.7.2 Shack-Hartmann Sensor

The ***Shack-Hartmann Sensor*** is used to characterise the planarity of a nominally plane wave. If one imagines the plane wave to be sampled at a location conjugate to the pupil, then this provides a measure of the wavefront error. The planarity of the wave is characterised by measurement of the local slope of the wavefront, rather than by comparison with a reference wavefront. This measurement is conducted by breaking the pupil into a large number of sub-pupils by means of a 2D array of circular lenslets. Each individual lenslet then focuses the beam onto an array detector. Generally, since each micro-lens is small, typically with a diameter of less than a millimetre, then the numerical aperture is correspondingly small. As a result, each focused spot is essentially diffraction limited. Naturally, the pixelated detector is able to measure the centroid location of each spot to high precision. The displacement of the spot centroid (in two dimensions) divided by the lenslet focal length determines the slope angle of the wavefront at that particular sub-aperture location. The principle is illustrated in Figure 16.16.

If, as illustrated, the focal length of each lenslet is f and the observed shift in the spot centroid is Δy, then the local wavefront slope δ is given by:

$$\delta = \Delta y/f \tag{16.22}$$

By 'stitching' measured local wavefront gradients, it is possible to recreate a wavefront map across an entire pupil. A typical Shack-Hartmann sensor might have an overall diameter of 10 mm with lenslets of around 100 μm in diameter. A lenslet focal length of a few millimetres is typical. With a centroid resolution of the order of 0.1 μm or less, then local wavefront slopes of the order of 10 μrad may be resolved. Indeed, one may model the random centroid location uncertainty in terms of the pixel size and the detector signal-to-noise ratio. This suggests a resolution of the order of a few tens of nanometres or less in the peak to valley wavefront error. In practice, a Shack-Hartmann sensor can measure wavefront error to an accuracy of a few nanometres rms. In certain instances, the accuracy may approach that of a conventional interferometer, although this is not generally the case.

A Shack-Hartmann sensor is generally deployed in an arrangement analogous to that of an interferometer. Illumination may be provided by a collimator narrowband source, e.g. a laser or light emitting diode (LED) and introduced into the optical path by means of a beamsplitter. Thereafter, the collimated beam may be used to test planar optics, or it can be focused by a lens to create a spherical wavefront. This arrangement is shown in Figure 16.17.

A Shack-Hartmann Sensor may be used to characterise the wavefront error of a system or the form of an optical surface. Since the sensor can only measure the offset of each focused spot relative to some nominal position, the sensor must first be calibrated. This, as for interferometers, is done by using a calibrated

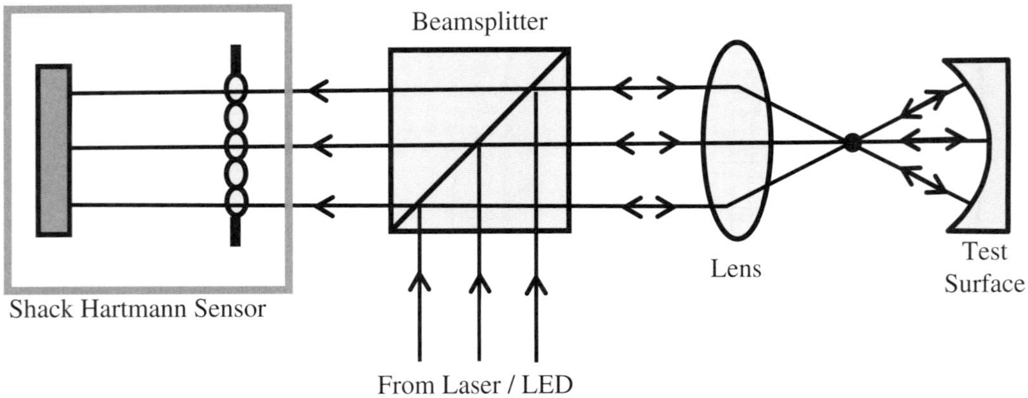

Figure 16.17 Deployment of Shack-Hartmann sensor.

reference artefact such as a sphere or plane. The effects of any optics in the Shack-Hartmann system, such as lenses is taken into account by this calibration process, provided the wavefront departure of the test system is small. Otherwise, retrace errors must be accounted for. Since the detector cannot, in any meaningful sense, determine the absolute position of the focused spots, it is insensitive to any absolute tilt of the wavefront. To calibrate the absolute tilt of the wavefront, a retro-reflecting corner cube may be inserted into the collimated beam path and the resultant centroid positions preserved and used as a reference in subsequent measurements.

The spatial resolution of the sensor is dictated by the number of lenslets in the array and this tends to be of the order of 100 × 100 or less. As such, the resolution afforded is rather less than that of the comparable interferometer. In addition, accuracy of the Shack-Hartmann sensor is not as great as the interferometer. However, operation of the sensor is largely immune to vibration. The effect of vibration is to add some noise to the centroiding process, whereas, in an interferometer, a small amount of vibration will completely compromise fringe visibility. A Shack-Hartmann sensor may therefore be deployed in a wide range of adverse environments.

16.7.3 Knife Edge Tests

The ***Foucault knife edge test*** is a remarkably simple test to determine the axial location of a focused beam and can be used in the form testing of optical mirrors. In the test, a sharply defined edge is placed close to the focal position of a beam. The knife edge is then translated laterally such that it is placed in a central position, obscuring half the beam. The projected beam is then viewed in a far field position. As the knife edge is translated axially through the focus, the character of the far field image changes. Either side of the focal position, the projected illumination takes the form of an illuminated semicircle on one side with darkness on the other. The handedness of this distribution changes on progression from one side of the focus to the other. Close to the focal position, the illumination becomes uniform, with significant diffraction effects evident either side of the focal position. The test is very sensitive, with the focal position determined as the knife edge location where the far field illumination is the most uniform.

Naturally, the ease of implementing the Foucault test has been greatly facilitated by the incorporation of laser sources. Figure 16.18 illustrates a typical Foucault test arrangement, showing the change in the projected illumination as the knife edge traverses the focal position.

The arrangement shown in Figure 16.18 is used in the accurate determination of the location of a focal point. A variation of this arrangement may be used to test the fidelity of a spherical mirror. If the screen shown in Figure 16.18 is replaced with a mirror whose centre lies at the focal point of the beam and at the location of the knife edge, then the reflected illumination may be viewed in the far field by a beamsplitter. With the knife edge located at the focus of both the beam and its reflection, the far field illumination should appear uniform.

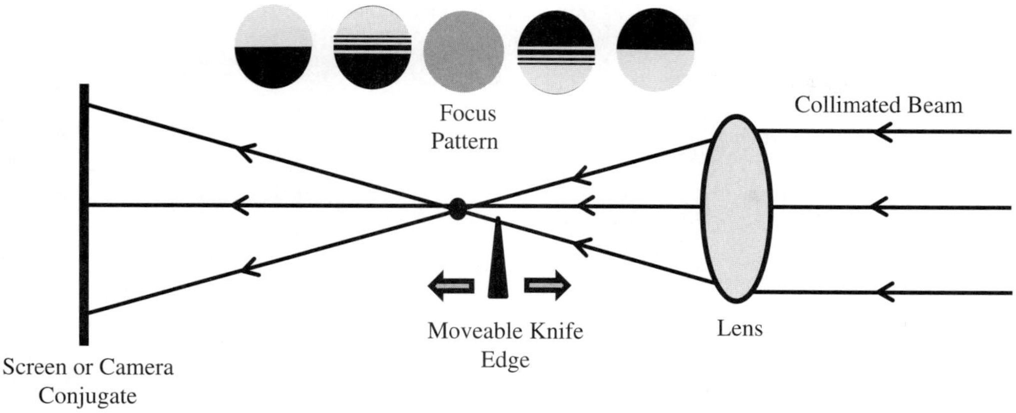

Figure 16.18 Foucault knife edge test.

However, the knife edge obscuration is so sensitive to small lateral deviations in the ray path, that any very small perturbations of the mirror's form will be revealed as distinct variations in the contrast of the far field image. This test of mirror form is, however, qualitative, although very useful. Nonetheless, its sensitivity is almost equivalent to that of interferometry.

A variant of the Foucault knife edge test is the ***Schlieren test***. This also exploits the sensitivity of the deployment of a knife edge at an optical focus. The Schlieren test is designed to translate small variations in optical path into palpable variations in irradiance at some other conjugate plane. In particular, the Schlieren test is designed to image very small variations in the density of transparent media, such as those produced by shock waves or thermal convection currents. As such, the test finds use in a variety of engineering applications. The scene of interest is illuminated by a collimated beam which is then focused onto a knife edge, as in the Foucault test. Subsequently, the original scene is imaged at some other conjugate. Very small refractive deviations are translated into significant increments in knife edge obscuration and presented as changes in contrast at the imaged conjugate.

In all these applications, the knife edge may be replaced by a coarse, transparent grating. This grating is known as a ***Ronchi grating***. The grating consists of parallel bars of metallisation a few tens of microns wide imprinted on a glass substrate, interspersed with transmissive regions. The advantage of this approach is that the single knife edge is effectively replaced by multiple knife edges enhancing the efficiency of the differential obscuration. In a further modification, two Ronchi gratings are deployed at conjugate focal points. Between these two conjugate points, at the nominal pupil location, the object of interest is located. If the two gratings are offset such that the transmissive regions of one overlap the reflective bars of the other, then transmission is blocked and the viewed image will be entirely dark under stable conditions. Any small instability in the optical path will thus be very efficiently converted into output illumination at the image plane.

16.7.4 Fringe Projection Techniques

The application of digital imaging technology to the analysis of images has facilitated the derivation of quantitative and accurate data from stored images. We have witnessed this in the application of digital imaging in the Shack-Hartmann sensor, whereby spot centroids may be located to a small fraction of a microscopic pixel width. The application of these analytical techniques to the analysis of image geometry comes under the general heading of ***photogrammetry***. One such application is, in common with the applications hitherto discussed, applied to the analysis of surface form. This is the so-called '***fringe projection***' technique. In this technique, a series of alternating light and dark bars are optically projected onto a surface and are subsequently deflected by the geometry of the surface under test. Accurate and quantitative analysis of the imaged geometry of these fringes provides information about the form of the surface.

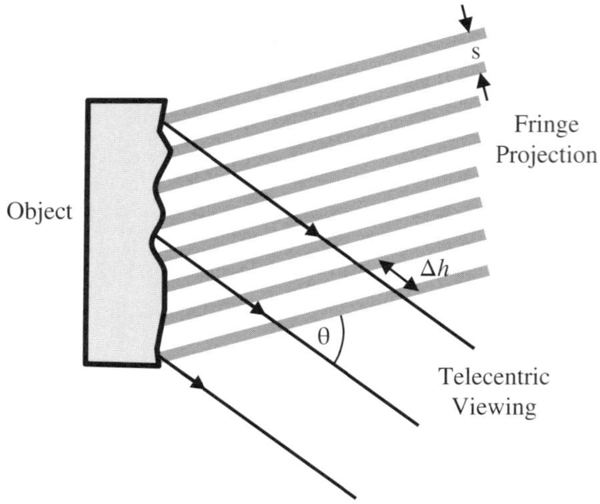

Figure 16.19 Fringe projection.

A set of parallel fringes is projected onto a surface and then viewed at some angle, θ. The fringes could be generated by the interference of two overlapping and coherent beams or by the simple projection of a patterned mask. It must be imagined that the surface has some deviation in form from the planar and that this deviation is converted into a corresponding deviation in the spacing of the fringes. The arrangement is shown in Figure 16.19

In many respects, analysis of fringe projection is analogous to interferogram analysis. A set of (as viewed) parallel fringes represents a planar surface. If the separation of the projected fringes perpendicular to the axial direction is s and the viewing angle is θ, then the surface contour interval, Δh, of the viewed fringes in the observation direction is given by:

$$\Delta h = s/\sin\theta \tag{16.23}$$

Equation (16.23) suggests that the observed fringes simply mark a contour map of the test surface. Digital imaging enables the accurate quantitative characterisation of these contours to produce a height map of the surface. Furthermore, the projected fringes themselves may be modified by using ***spatial light modulators*** (as ***liquid crystal displays***) as the source of the fringes. Strictly speaking, Eq. (16.23) relies on the elimination of perspective in analysing an image. That is to say, fringe spacing may also depend upon the (variable) proximity of different parts of the test piece to the viewing camera. As such, practical implementation of fringe projection often relies upon telecentric lenses, which are widely used in metrology. These lenses convert a specific lateral offset in an object's lateral displacement to a constant image displacement, irrespective of the object distance. Greatest sensitivity is, of course, afforded when the angle, θ, is as close as possible to 90°. However, in practice, the choice of angle is dependent upon the range of angles present in the test object; excessive viewing angles lead to obscuration of some parts of the object.

The technique of fringe projection is generally applied to the 3D characterisation of surfaces with a relatively large dynamic range. That is to say, the resolution does not match that provided by interferometry. Furthermore, fringe projection is applied to surfaces that can scatter light reasonably efficiently; specular surfaces do not support this method. ***Fringe reflection*** provides an extension to fringe projection, enabling the accurate characterisation of reflective surfaces. As with fringe projection, a fringe pattern is projected onto the test surface. However, in this instance, the projected fringe pattern is not viewed at the conjugate corresponding test object, but at some other remote location. The result of this is that the observed fringes are significantly displaced by small tilts in the test object. This enhanced sensitivity is itself complemented by the sensitivity of the pixelated detector in locating geometrical features, such as fringes. Where the fringes are located to a precision compatible with the detector noise performance, measurement of mirror form to a precision comparable to that of interferometry is possible.

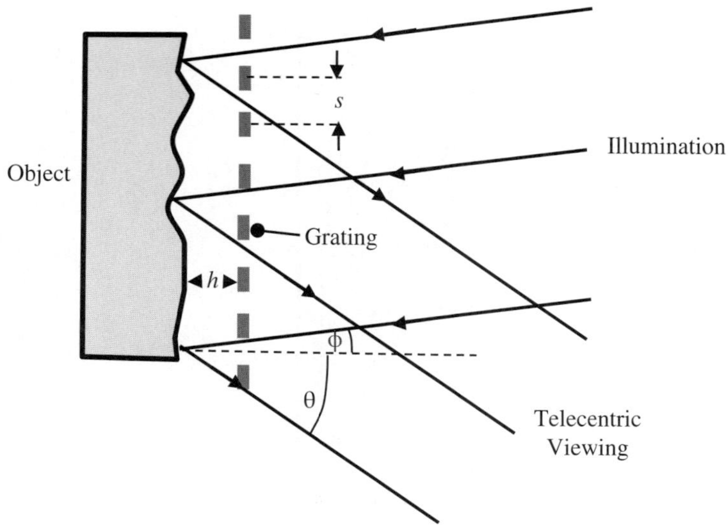

Figure 16.20 Shadow Moiré technique.

For all fringe methods, greater precision is afforded for small fringe spacings. Ultimately, however, the use of increasingly fine fringes may compromise fringe contrast. There are a variety of techniques that permit the use of finer fringes by detecting the 'beat pattern' between two sets of fine fringes with slightly differing spatial frequencies. These techniques come under the general heading of ***Moiré fringe*** methods. There are many variations of this approach and a comprehensive listing is beyond the scope of this text. One specific example is the so called shadow moiré technique, where a transmission grating is placed in front of the surface of interest. The beat pattern arises from the interference between the grating pattern projected onto the surface and the image subsequently viewed through the grating. The arrangement is illustrated in Figure 16.20.

If the surface is illuminated at an angle of θ and viewed at an angle of ϕ, then, for a grating spacing of s, then the height increment, Δh, for each moiré fringe is given by:

$$\Delta h = \frac{s}{\tan\theta + \tan\phi} \tag{16.24}$$

Careful calibration is essential to facilitate quantitative characterisation. For accurate calibration of mirror surface form, precision reference surfaces, such as spheres and planes are used, as in interferometry. In other cases, for more general measurements using fringe projection, then precision reference artefacts are used.

16.7.5 Scanning Pentaprism Test

Whilst interferometric testing is the primary procedure for testing large concave mirrors, the unfortunate experience of the Hubble Space Telescope primary mirror test has demonstrated the value of supplementary or corroborative tests. One such test is the scanning pentaprism test which is particularly useful in testing mirrors of approximately parabolic form. A pentaprism has the useful property of deflecting a beam by 90° regardless of the tilt of the prism. A laser beam is launched perpendicularly to the mirror axis, and is deflected by the pentaprism to produce an axially aligned laser beam. The pentaprism is then scanned across the mirror, producing a beam that can be scanned across the pupil whilst maintaining its axial alignment. A detector is then placed at the mirror focus. For a perfect parabola, the laser beam will maintain a constant position at the detector as the prism is scanned across the aperture.

The arrangement is illustrated in Figure 16.21. A specific prism position corresponds to a particular location within the mirror aperture. Deviation of the laser beam at the detector is a measure of the deflection of the mirror surface from the nominal at that aperture location. Translation of the mirror across the mirror aperture

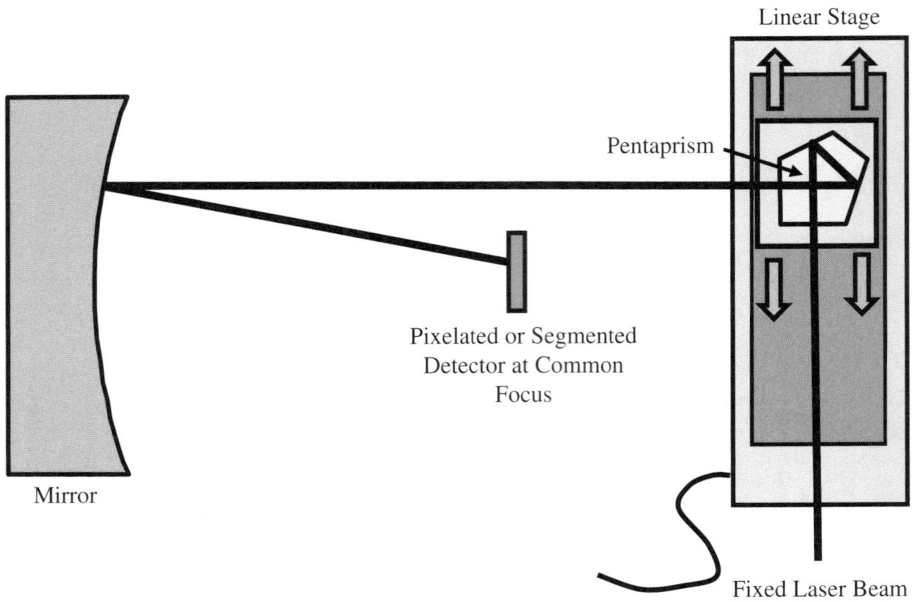

Figure 16.21 Scanning pentaprism test.

is accomplished by a linear stage, as illustrated. Critical to the understanding of the method is an appreciation of the uncertainties introduced by the operation of the linear stage. As the mirror is translated, there may be some angular yaw of the prism produced by mechanical imperfections of the stage. However, this does not impact the angular deflection in a plane containing the scan axis and the mirror axis. This is determined by the prism angles alone and the constant 90° deflection is a fundamental attribute of the pentaprism. Similarly, any pitch in the prism has no effect upon the direction of the outgoing beam. However, any 'roll' of the prism as it progresses along the scan axis will be converted into out-of-plane deflection of the laser beam. Therefore, any deflections in this direction are ignored and not analysed. As such, the only useful data comprises components of deflection in the plane of the fixed laser beam and mirror axis.

The in-plane deflection can be measured with great sensitivity by centroiding the laser spot at the detector. This, of course, provides a measure of the local surface slope error to a fraction of a microradian. These local slope errors may be stitched together to provide a map of the mirror form error. To accomplish a full mapping of the surface, a number of different linear scans must be arranged in some pattern. For example, for a circular mirror without a central aperture, a series of radial scans may suffice. Otherwise, a series of parallel linear scans may be arrayed in two orthogonal directions in grid fashion. At any grid point, from the two orthogonal scans, the tilt error may be defined in the two orthogonal orientations. Overall, from such data, a form error resolution of a few nanometres is possible with this technique.

16.7.6 Confocal Gauge

In ***confocal microscopy***, an illuminated pinhole is imaged onto a surface by a microscope objective and the scattered or reflected light gathered and imaged onto the same pinhole. This back-scattered light may then be diverted by means of a beam splitter and sampled. To build up a picture of the surface, the test piece must be scanned in two dimensions with respect to the focusing objective. In practice, the pinhole is often replaced with an optical fibre. The basic arrangement is shown in Figure 16.22.

In the context of the evaluation of surface form, the efficiency of fibre coupling is depends upon the height of the sample surface with respect microscope focus. Provision of a third (axial) scanning axis enables topographic information to be gathered on the basis that maximum detector signal occurs where the surface is

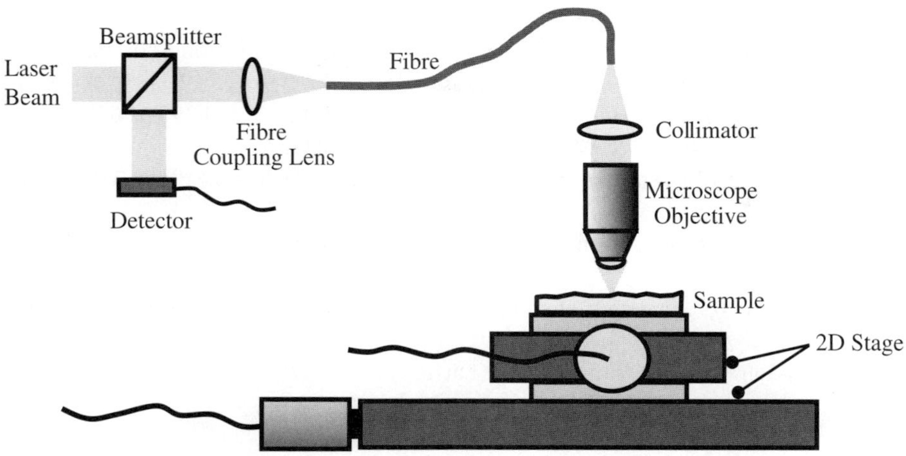

Figure 16.22 Confocal microscopy.

located at the objective focus. This is the basic principle of the ***confocal gauge***. The difficulty with this method is that the data collection is inherently serial in character with a single detector monitoring the scattered signal over the not insignificant time required to perform a 2D scan at reasonable resolution. This is further complicated by any additional vertical scanning that might be required to elucidate surface topography. It is possible to overcome the latter difficulty in a number of ways. The ***confocal length aberration*** gauge employs a microscope objective that is (deliberately) poorly corrected for chromatic aberration. White light is fed into the optical fibre and, because of the chromatic aberration of the objective, only scattered light at one wavelength is optimally re-focused onto the fibre. The single detector is replaced by a spectrometer and the peak signal wavelength is recorded. This peak wavelength effectively acts as a 'proxy' for the surface height. Of course, the system must be calibrated with a precision artefact in order to convert this wavelength proxy into a real surface height.

Notwithstanding its inherent slow speed, confocal measurement is particularly useful in the characterisation of discontinuous surfaces. Interferometry, with the exception of white light interferometry, operates under the assumption that the surface under investigation is continuous. Therefore, confocal microscopy is particularly useful in the characterisation of segmented surfaces, for example, faceted mirrors, or any surface with 'steps'.

Another particular advantage of confocal microscopy is its improvement in resolution over conventional microscopy. Theoretically, it offers a resolution enhancement of $\sqrt{2}$ over conventional microscopy. If the imaged spot of the confocal system is modelled as a Gaussian distribution of width Δx, then the projected fibre aperture may also be modelled in the same way. As such, the overall sensitivity function, $\Phi(x)$, is represented as the product of two Gaussian functions (one for the illumination at the object and one for the fibre input):

$$\Phi(x) = e^{-(x/\Delta x)^2} \times e^{-(x/\Delta x)^2} = e^{-(\sqrt{2}x/\Delta x)^2} \tag{16.25}$$

Further Reading

Brock, N., Hayes, J., Kimbrough, B. et al. (2005). Dynamic interferometry. *Proc. SPIE* 5875: 0F.

Burge, J.H., Zhao, C., Dubin, M. et al. (2010). Measurement of aspheric mirror segments using Fizeau interferometry with CGH correction. *Proc. SPIE* 7739 (02).

Damião, A.J., Origo, F.D., Destro, M.A.F. et al. (2003). Optical surfaces flatness measurements using the three flat method. *Ann. Opt.* 5.

Evans, C.J. and Kestner, R.N. (1996). Test optics error removal. *Appl. Opt.* 35 (7): 1015.

Goodwin, E.P. and Wyant, J.C. (2006). *Field Guide to Optical Interferometric Testing*. Bellingham: SPIE. ISBN: 978-0-819-46510-8.

Hariharan, P. (2003). *Optical Interferometry*, 2e. Cambridge, MA: Academic Press. ISBN: 978-0-123-11630-7.

Malacara, D. (2007). *Optical Shop Testing*, 3e. New York: Wiley. ISBN: 978-0-471-48404-2.

Malacara, D., Servín, M., and Malacara, Z. (2005). *Interferogram Analysis for Optical Testing*, 2e. Boca Raton: CRC Press. ISBN: 1-57444-682-7.

Rolt, S. and Kirby, A.K. (2011). Flexible null test for form measurement of highly astigmatic surfaces. *Appl. Opt.* 50: 5473.

Rolt, S., Kirby, A.K., and Robertson, D.J. (2010). Metrology of complex astigmatic surfaces for astronomical optics. *Proc. SPIE* 7739: 77390R.

Wittek, S. (2013). Reaching accuracies of lambda/100 with the three-flat-test. *Proc. SPIE* 8788: 2L.

17

Spectrometers and Related Instruments

17.1 Introduction

In this chapter we will analyse in a little detail the design of ***spectrometers*** and related instruments. The function of a spectrometer is to extract spectral information in an optical signal and present in a format suitable for observation ***and measurement***. Most usually, in contemporary instruments, the end detector is a pixelated detector rather than the human eye. Traditionally, an instrument designed to provide spectrally dispersed data probing specific properties for subsequent analysis is denominated a spectrometer; a system adapted for simple recording of an optical spectrum is known as a ***spectroscope.*** A ***spectrograph*** transforms incoming light into spatially dispersed illumination. This spatially dispersed illumination is then presented at a pixelated detector, or more traditionally, at a photographic plate, for subsequent analysis. In practice, the boundaries between these terms are somewhat fluid and they are often used interchangeably.

Spectral information is, of course, of immense practical and scientific consequence in a wide range of applications. This ranges from the study of astronomical sources to the spectroscopic evaluation of trace gas contamination. As with imaging devices, the introduction of compact pixelated sensors has revolutionised the development of compact instruments.

For the majority of instruments, spectrometer design is based around the exploitation of dispersive components. In Chapter 11, we introduced and analysed dispersive elements, both diffractive and refractive. Modern designs are, for the most part, exclusively based upon diffractive components, such as gratings. Prisms, as dispersive devices, do not feature in modern instruments. A typical instrument features a collimated beam derived from an illuminated slit and presented to a diffraction grating. This parallel beam is then angularly dispersed by the grating in a direction perpendicular to the slit object. The dispersed, collimated beam is then imaged at some focal plane by a lens or mirror. As such, the slit object is recreated by the imaging optics. However, because of the grating dispersion, the location of this object within the focal plane is dependent upon the illumination wavelength.

In a spectrometer, typically, a pixelated detector is located at the focal plane and captures the spectrally dispersed illumination. The orientation of the grating, itself, is fixed. In a ***monochromator***, a matching image slit is placed at the output focal plane allowing transmission of a single wavelength for recording by a single, discrete, photodetector. Tuning is achieved by rotation of the grating.

Although the analysis of dispersive components is central to the design of spectroscopic instruments, other topics covered are also important. The design of instruments to function in low light levels, such as those deployed in astronomical and other scientific applications, requires a clear understanding of photometry and detector performance. In addition, an understanding of the use and performance of optical filters is critical to discrimination between the different diffraction orders produced by gratings.

Optical Engineering Science, First Edition. Stephen Rolt.

Companion website: www.wiley.com/go/Rolt/opt-eng-sci

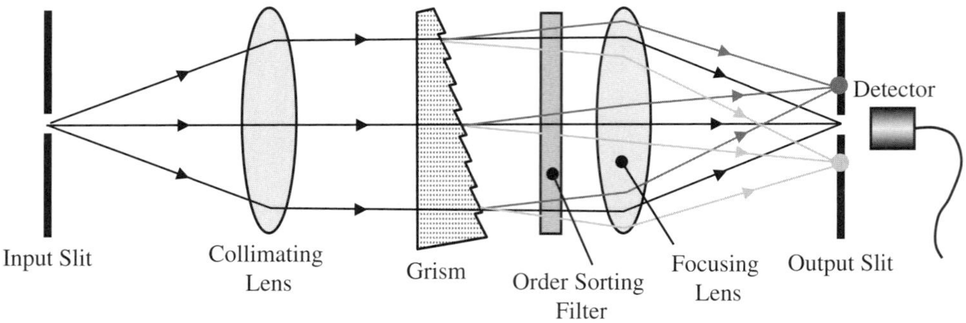

Figure 17.1 General layout of a monochromator.

17.2 Basic Spectrometer Designs

17.2.1 Introduction

To illustrate the basic architecture of a spectrometer, we will introduce some basic designs for simple grating systems. In the coverage of dispersive components, key performance metrics, such as resolution, were entirely dependent upon the grating properties. However, when these components are integrated into optical systems, other factors, such as image quality (impact of aberrations) and object slit width are also of importance. Whilst from a spectral resolution perspective it is desirable to make the object slit as narrow as possible, this naturally compromises the optical signal and hence the signal-to-noise ratio.

17.2.2 Grating Spectrometers and Order Sorting

Figure 17.1 provides a schematic illustration of a generic spectrometer design, featuring input and output slits, grating, and collimation and focusing optics. In addition to these essential components, an ***order sorting filter*** is also added. This filter is essential to remove the inherent ambiguity produced by the different diffraction orders. For example, first order diffraction of 700 nm light is indistinguishable from second order diffraction at 350 nm; both will follow an identical path in the instrument. Therefore, if we are interested only in the 700 nm light, the second order contribution at 350 nm must be removed. This is accomplished by an order sorting filter, usually in the form of a long pass filter, which transmits longer wavelengths whilst blocking those below a certain cut-off wavelength.

Figure 17.1 shows a transmissive arrangement with a grism used as the dispersive component. Of course, a reflective grating may be substituted for the grism and reflective mirrors for the two lenses. The arrangement shown in Figure 17.1 is a monochromator configuration, where a single wavelength is transmitted through the output slit where it is sampled by a single point detector. Otherwise, as shown in Figure 17.2, an array detector may be substituted for the slit allowing a range of wavelengths to be sampled simultaneously. In this instance, some care must be taken in selecting the order sorting filter. The filter characteristics must be such that all wavelengths are transmitted in the instrument passband whilst also rejecting spurious diffraction orders. In the monochromator arrangement more flexibility is permitted. As different transmission wavelengths are selected, different order sorting filters may be substituted. Most usually, this is done by arranging a selection of filters in a rotatable filter wheel.

17.2.3 Czerny Turner Monochromator

17.2.3.1 Basic Design

The ***Czerny Turner Monochromator*** provides a useful example illustrating, in more detail, the design of a dispersive instrument. As such, the analysis presented here is intended to explore more generally the underlying

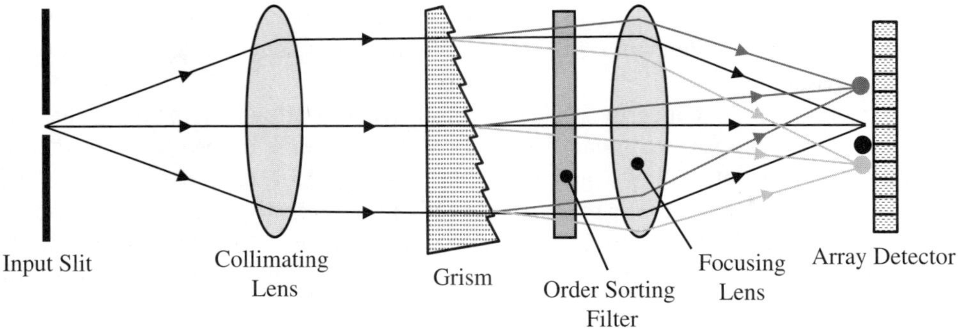

Figure 17.2 General layout of a spectrometer.

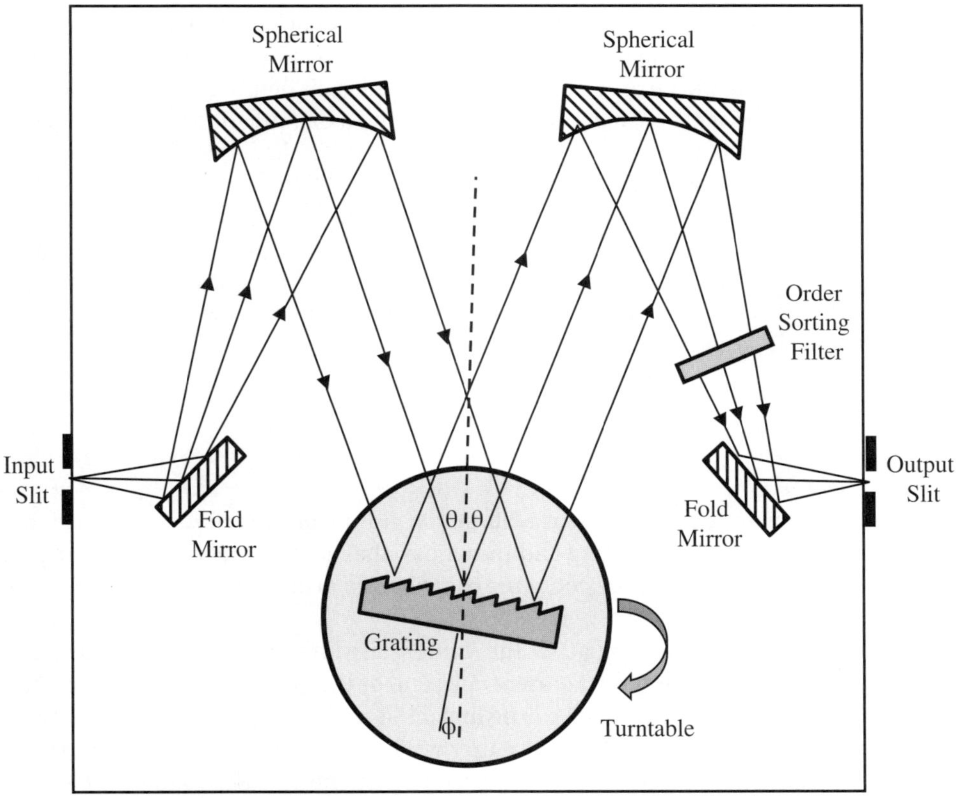

Figure 17.3 Czerny-Turner monochromator.

choices in designing such an instrument. Only reflective optics are used throughout the design and this eliminates any effects due to chromatic aberration. In most commercial instruments, collimation and focusing is provided by spherical mirrors, directing light to and from fixed slits. A reflective grating is mounted on a rotating platform. Rotation of this platform directs a different wavelength from the grating onto the output slit.

The basic layout of the Czerny Turner Monochromator is shown in Figure 17.3.

The entrance pupil of the instrument is assumed to be located at the grating. As such, the size of the grating, as the limiting aperture, determines the size of the pupil. The overall size of the instrument tends to be specified by the system focal length, as determined by the radii of the two mirrors. As will be seen later, the focal length plays an important role in the instrument's resolving power in practical systems. In addition, the system aperture, as expressed by the focal ratio, f#, determines the system étendue and hence the optical flux for a given spectral radiance.

As far as the dispersive characteristics are concerned, it is the characteristic grating half angle, θ, that is of central importance. This half angle expresses the angular divergence of the two 'arms' of the monochromator. The rotation angle of the grating, φ, then determines which wavelength is selected for transmission. Of course, if φ were zero, then the grating is effectively acting as a plane mirror and only the zeroth order will be transmitted. With the geometry shown in Figure 17.3 in mind, the grating incidence angle is $\theta + \varphi$ and the diffracted angle is $\theta - \varphi$. From the basic grating equation, it is possible to establish the condition for transmission to occur:

$$d(\sin(\theta + \phi) - \sin(\theta - \phi)) = m\lambda \text{ and } 2d\cos\theta\sin\phi = m\lambda \tag{17.1}$$

d is the grating spacing, m the order and λ the wavelength.

Equation (17.1) shows that the wavelength transmitted is linearly proportional to the sine of the grating angle. Naturally, for a non-zeroth order, the grating tilt must be biased in one direction. This breaks the apparent symmetry shown in Figure 17.3; the significance of this will be discussed later. Before the advent of digital data processing, it was considered convenient to rotate the turntable using a combination of a linear screw thread and a ***sine bar***. An arm or bar, terminated by a ball, projects along a line aligned to the centre of the turntable. A plane surface attached to the leadscrew then pushes the sine bar as the leadscrew progresses. This produces a turntable rotation angle whose sine is proportional to the leadscrew displacement.

The dispersion is straightforward to calculate from Eq. (17.1). Furthermore, in the context of the whole instrument, it might be useful to present the dispersion as differential ***displacement*** (at the slit) with respect to wavelength, as opposed to a differential angle. If the focal length of the instrument (i.e. the mirrors) is f, then the dispersion, δ, at the output slit is given by:

$$\delta = \frac{dx}{d\lambda} = \left[\frac{2\tan\phi}{1 + \tan\theta\tan\phi}\right]\left(\frac{f}{\lambda}\right) \tag{17.2}$$

17.2.3.2 Resolution

In Chapter 11 we derived an expression for the resolution of a diffraction grating ***in isolation***. We learned that the resolution is proportional to the width of the grating and the sine of the incident and diffraction angles. However, in a spectrometer design, we must take into account the contribution made by all parts of the system, not just the grating itself.

The most obvious additional factor relates to the impact of the (finite) slit width. In effect, the resolution is dictated by the convolution of the slit function and the Fourier transform of the grating, as imaged at the slit by the instrument optics. Clearly, for the slit width to have little impact, it must be significantly smaller that the diffraction pattern of the grating. The grating diffraction pattern may be represented as a sinc function whose width is inversely proportional to the system numerical aperture and proportional to the wavelength. For example, in an instrument described as 'f#4', having a numerical aperture of 0.125, this limiting slit width would be 2 μm for a wavelength of 500 nm. This is clearly an exceptionally small slit width and, in most practical applications, the slit width is likely to be substantially larger than this.

The most useful expression of the instrument resolution is the profile that would be recorded when a very narrowband source (atomic line or laser) is scanned by the instrument. Where the slit width is the limiting factor, the ***slit function*** would adopt a triangular profile as the instrument is tuned across the line by rotating the grating. For a slit width of a, (both input and output), the slit function, f(x), reaches zero when the image of the input slit at the output slit is displaced by one full slit width in either direction. As such, the slit function may be expressed as:

$$f(x) = \frac{a + x}{a} \quad (-\mathrm{a} < \mathrm{x} \leq 0); f(x) = \frac{a - x}{a} \quad 0 < \mathrm{x} \leq \mathrm{a} \text{ otherwise } \mathrm{f(x)} = 0 \tag{17.3}$$

Conversely, where the grating width is the limiting factor, the slit function would adopt a sinc profile. Assuming that the grating width is defined by a numerical aperture, *NA*, then the form of the grating diffraction envelope ***as imaged at the slit*** is given by:

$$f(x) = \left[\frac{\sin(2\pi NAx/\lambda)}{(2\pi NAx/\lambda)}\right]^2 \tag{17.4}$$

It is most natural and useful to express the slit function in terms of wavelength increment, $\Delta\lambda$, rather than displacement, x. The relationship between the two is expressed by the dispersion, as given in Eq. (17.2). Where slit width is the limiting factor, the resolution is determined by the condition whereby one wavelength is effectively displaced by one full slit width with respect to the adjacent wavelength:

$$\Delta\lambda = \frac{\lambda a(1+\tan\theta\tan\phi)}{2f\tan\phi} \text{ and } R = \frac{\lambda}{\Delta\lambda} = \frac{2f\tan\phi}{a(1+\tan\theta\tan\phi)} \tag{17.5}$$

In the grating limited scenario, from Chapter 11, the resolution is given by:

$$R = \frac{\lambda}{\Delta\lambda} = \frac{w\sin(\theta+\phi) - w\sin(\theta-\phi)}{\lambda} = \frac{2w\cos\theta\sin\phi}{\lambda} \tag{17.6}$$

Equation (17.6) establishes the resolution as being equivalent to the path difference (in waves) between rays striking the opposite edges of the grating. The slit width at which the grating and slit width contributions are identical is given by:

$$a = \frac{\lambda f}{w\cos(\theta-\phi)} \text{ or } a = \frac{\lambda}{2NA} \quad \textit{where NA is the numerical aperture} \tag{17.7}$$

Where the slit width is between these two extremes, then the slit function is a convolution of the two profiles. This is illustrated in Figure 17.4 which shows the variation in slit function for three different slit widths, namely, ×4, ×1, and ×0.25. These slit widths are referenced to the value set out in Eq. (17.7). That is to say, the ×1 slit width corresponds to that width where grating and slit width resolutions are identical. As such, and as expected, the ×4 slit function exhibits a triangular profile, whereas the ×0.25 function follows a sinc profile.

In this analysis of resolution, we have focused on the view at the instrument slit. When we consider the analysis of resolution when viewed at the grating, the effect of increasing the slit size is to reduce the effective size of the diffraction pattern of the slit at the grating. From this perspective, the effective size of the grating sampled is reduced and the resolution is correspondingly diminished. In this sense, the resolution can still be thought of as a path difference, in waves, between two extreme ray paths.

17.2.3.3 Aberrations

Having examined the impact of slit width, one might reasonably expect the resolution to be determined by the grating size as the slit width is reduced. However, one important consideration has been omitted. The examination of the grating contribution presented here is essentially a diffraction limited analysis. For a real system, then aberrations will have a significant impact on performance. In particular, in the Czerny-Turner monochromator, an off-axis mirror system is used to collimate and focus the beams. Therefore, any off-axis aberrations must be considered carefully. Initially, in this analysis, we consider the simplest configuration where the collimating and focusing mirrors are spherical surfaces placed in a symmetrical arrangement.

In the analysis of aberrations, we make the assumption that the 'in-plane', off-axis angle dominates, i.e. the θ in Figure 17.3 and that any contribution in the 'sagittal direction', due to the finite ***height*** of the slits, may be ignored. In practice, this sets a finite limit on the slit height before those 'sagittal' aberrations become unacceptable. Furthermore, because of the system geometry, it must be assumed that the effective collimator tilt angle, $\theta/2$ is significantly larger than the numerical aperture. As a consequence, it is the off-axis aberrations associated with the folded geometry that might be expected to dominate, as opposed to the on-axis aberrations.

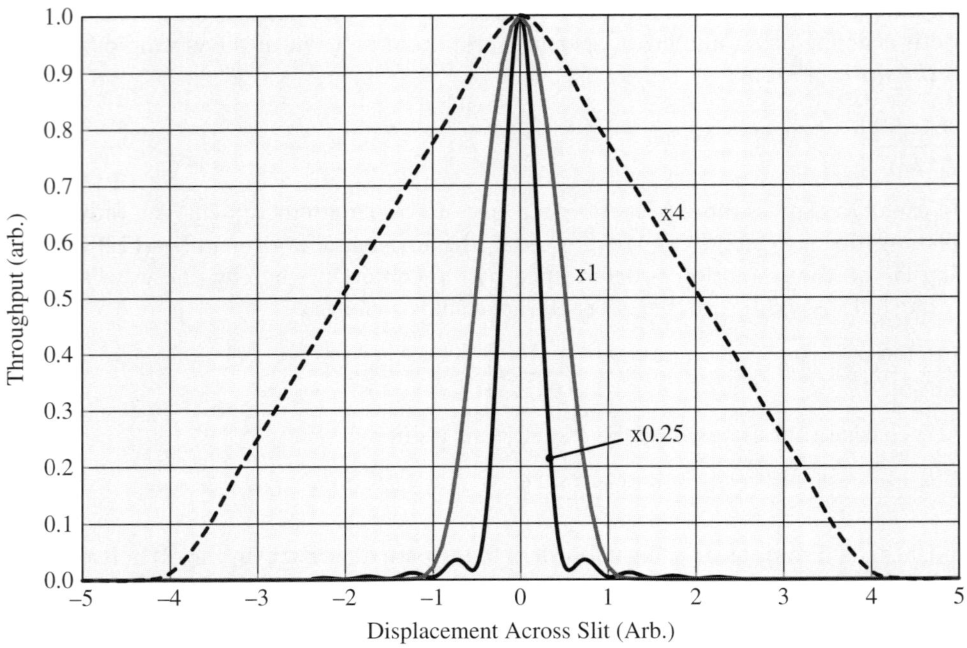

Figure 17.4 Slit function for varying slit widths.

With these assumptions in mind, it might be evident that astigmatism and field curvature should dominate. However, in imaging a linear slit we are substantially unconcerned about transverse aberrations along the length of the slit; only transverse aberrations perpendicular to the slit degrade resolution. As a consequence, we are substantially unconcerned about astigmatism and field curvature. As such, the output slit needs to be placed at the tangential focus. Any sagittal defocus, no matter how great, is simply resolved along the direction of the slit and does not affect instrumental resolution.

Of all the third order aberrations, it is coma that is the most interesting. The two mirrors, collimating and focusing both contribute to coma. In the arrangement sketched in Figure 17.3, the layout is symmetrical, with the off-axis angles the same, or rather equal and opposite. At first sight, therefore, their nett contribution to coma should be zero, as each mirror contributes an equal and opposite amount. Indeed, this would be true for the specific case of zeroth order diffraction where the grating acts as a mirror. Otherwise, as outlined in Chapter 11, the grating produces anamorphic magnification which distorts the pupil, transforming a circular pupil into an elliptical pupil. The effect of this transformation is to scale the coma and to apply that scaling to one mirror only. As a consequence, there is residual coma for a symmetrical system.

The anamorphic magnification produced is equal to the ratio of the cosine of the incident and diffracted angles. For the symmetrical Czerny-Turner system, then the anamorphic magnification, M, may be expressed as:

$$M = \frac{1 + \tan\theta \tan\phi}{1 - \tan\theta \tan\phi} \text{ and } M \approx 1 + 2\tan\theta \tan\phi \tag{17.8}$$

In calculating the coma produced by each mirror, the off-axis angle of each mirror amounts to $\theta/2$, and we further assume that mirrors are parabolic in form, so there is no contribution from stop shifted spherical aberration. If the radius of each mirror is R and its numerical aperture is NA, then for the zeroth order scenario, the rms coma produced by each is given by:

$$\Phi_{rms}(coma) = \frac{NA^3}{12\sqrt{2}} R\theta \tag{17.9}$$

The impact of the anamorphic magnification is, for example, to transfer the pupil co-ordinates in the y direction only. This has the effect of transforming the coma, as expressed by the Zernike 7 (Noll Convention) polynomial into an admixture of Zernike 7 and Zernike 9 (trefoil) terms. If the original rms coma is described by the parameter $Z7$, then the revised third order terms, $Z7'$ and $Z9'$ may be expressed by:

$$Z7' = \left(\frac{3M^3 + M}{4}\right) Z7 \text{ and } Z9\prime = \left(\frac{3M - 3M^3}{4}\right) Z7 \tag{17.10}$$

Substituting the approximation in Eq. (17.8) and further assuming that M does not differ greatly from one, we obtain:

$$Z7\prime \approx (1 + 5\tan\theta\tan\phi)Z7 \text{ and } Z9\prime \approx 3\tan\theta\tan\phi Z7 \tag{17.11}$$

The implication of Eq. (17.11) is that, for a symmetrical system the two coma contributions will not cancel and we are left with a residual coma described by Eq. (17.12)

$$\Phi_{rms}(coma) = \frac{5NA^3}{12\sqrt{2}} R\theta \tan\theta \tan\phi \tag{17.12}$$

In practice, a simple Czerny-Turner monochromator is deliberately designed to be asymmetric, so the off-axis angles for the collimating and focusing optics differ. It is thus possible to balance out the coma arising from the anamorphic magnification term. However, as applied in Eq. (17.11), the residual coma changes broadly linearly with wavelength. Therefore, it is only possible to apply this correction for one wavelength.

The analysis hitherto presented assumes the use of spherical surfaces. However, the substitution of off-axis conics, particularly off-axis parabolas removes any off-axis aberration. Of course, there is a penalty for the use of non-spherical surfaces in terms of component manufacturing and cost. Nonetheless, in more recent years, this option has become increasingly attractive for high performance instruments. With the substitution of off-axis parabolas, the slits themselves lie upon the parabolic axis. Therefore, the centre of the slit corresponds to an on-axis scenario. As the parabola provides perfect control of spherical aberration, the principal concern is with off-axis aberrations exhibited at the extreme ends of the slit. This consideration and its impact upon resolution sets the boundary upon slit height. Once more, symmetry dictates that contribution to coma from the collimating and focussing mirrors are equal and opposite. Therefore, it is field curvature and astigmatism that are the principal concerns. The rms wavefront error produced by coma arising from the finite slit height may be derived from Eq. (17.9). The precise balance of field curvature and astigmatism depend upon the location of the pupil and manipulation of the stop shift equations. However, we may obtain a broad notion of the impact of these aberrations by calculating the Petzval curvature. The Petzval curvature generated by the two mirrors is $2/R$ and this may be used to calculate the defocus produced at either end of the slit and the wavefront error attributable to it. If the height of the slit is h, the focal shift, Δf, at either end of the slits is given by:

$$\Delta f = \frac{h^2}{2R} \text{ or } \Delta f = \frac{h^2}{4f} \quad \textit{where f is the instrument focal length} \tag{17.13}$$

Expressing Eq. (17.13) as a defocus rms wavefront error, Φ_{rms} we get:

$$\Phi_{rms} = \frac{h^2}{16\sqrt{3}f} NA^2 \quad \textit{NA is the system numerical aperture} \tag{17.14}$$

If the system is to be diffraction limited, then there is a significant restriction on the slit height. Taking a typical instrument, with a focal length of 300 mm and a numerical aperture of 0.125, then Eq. (17.14) suggests that the slit height should be no more than 5 mm to fulfil the Maréchal criterion at a wavelength of 550 nm. In practice, this may be extended somewhat, e.g. to 10 or 20 mm with the sacrifice of some resolution. However, the scope for such increases in slit height is strictly limited. Another useful insight provided by Eq. (17.14) is an understanding of the impact of scaling. Equation (17.14) suggests that, if diffraction limited performance is to be maintained, then the slit height will scale as the square root of the instrument focal length.

Worked Example 17.1 A symmetric Czerny monochromator, with a numerical aperture of 0.125 and focal length of 300 mm, is designed to operate in first order. A grating with 800 lines mm^{-1} is deployed and the symmetric monochromator angle is 20°. Assuming a slit width if 50 μm calculate the resolution at a wavelength of 550 nm.

The first point to note is that the slit width is substantially larger than the diffraction limited width associated with a f#4 beam at 550 nm. Therefore, we must use Eq. (17.5) to calculate the resolution. Firstly, we must determine the grating rotation angle, φ, from Eq. (17.1).

$$2d\cos\theta\sin\phi = m\lambda \; \mathrm{d} = 1250\,\mathrm{nm};\, \mathrm{m} = 1;\, \lambda = 550\,\mathrm{nm};\, \cos(20) = 0.9397$$

$$\sin\phi = \frac{550}{2 \times 1250 \times 0.9397} = 0.2341 \text{ and } \phi = 13.54°$$

$$R = \frac{2f\tan\phi}{a(1 - \tan\theta\tan\phi)} = \frac{2 \times 300 \times 0.2408}{0.05 \times (1 - 0.364 \times 0.2408)} = 3167$$

The instrument resolution is 3167.

17.2.3.4 Flux and Throughput

Determination of the flux passing through a spectrometer requires some thought, especially if the source is a broadband source with some spectral radiance. It is clear that the area of the slit and the system aperture establish the étendue of the system. However, the slit width also impacts the spectral bandwidth sampled. Therefore, for a continuum source, the flux that emerges at the output slit will be proportional to the square of the slit width, rather than the slit width itself. The flux is governed also by the angular dispersion, δ, of the system, which translates the slit width into an effective spectral bandpass, $\Delta\lambda$. For a focal length of f and a slit width of a, the effective bandpass is given by:

$$\Delta\lambda = \delta f a \tag{17.15}$$

The system étendue, G, is proportional to the area of the slit, or the product of the height, h, and width, w, and is also proportional to the square of the numerical aperture, NA.

$$G = \pi a h N A^2 \tag{17.16}$$

From our previous analysis of radiometry, the flux passing through the system is given by the product of the system étendue, G, the spectral radiance at the slit, L_λ and the bandwidth, $\Delta\lambda$. However, this must be modified by the system throughput of the system, ξ, which takes into account the diffraction efficiency of the grating and absorption of the mirrors:

$$\Phi = \xi L_\lambda G \Delta\lambda \text{ and } \Phi = \xi\pi\delta f h a^2 N A^2 L_\lambda \tag{17.17}$$

The important point is that the output flux is proportional to the square of the slit width. The picture, however, changes for a narrowband source, such as a laser or spectral line, where the intrinsic linewidth is considerably smaller that the system resolution. In this case, the radiometry is defined by the input radiance, L, as opposed to the spectral radiance. Therefore, the output flux is proportional to the slit width, as opposed to the square of the slit width:

$$\Phi = \xi\pi h a N A^2 L \tag{17.18}$$

In many applications, one is dealing with comparatively weak sources and the determination and optimisation of output flux is essential to delivering adequate signal-to-noise performance.

17.2.3.5 Instrument Scaling

At this stage, we pause to consider the impact of initial instrument requirements upon instrument scaling. As with much of the preceding discussion, although we are explicitly analysing the Czerny-Turner instrument, this is, in reality, a vehicle for understanding the behaviour of spectroscopic instruments in general. Science-driven requirements will identify a spectral range across which the instrument is required to perform and, critically, a spectral resolution that applies across that range. In addition, the light gathering capacity of the instrument will be constrained by flux requirements and identified in the form of the system étendue. In practice, the spectrometer will, in itself, be a sub-system in a larger overall system comprising other subsystems. Preceding subsystems will substantially constrain parameters such as the instrument étendue. For example, an astronomical telescope might precede a spectrometer. To a large extent, it is the étendue of this subsystem that constrains the étendue of the spectrometer. The étendue of the telescope might be determined by the minimum angular object size and the diameter of the telescope mirror. Therefore, in this specific instance, larger telescopes automatically correspond to larger downstream spectroscopic instruments.

Initially, we might define, as a requirement, both the sub-system étendue, G, and the resolution, R. In this instance, we are interested in some luminous object imaged at the slit, so a representative object area, as presented to the slit, is the square of the slit width. The system étendue is therefore given by:

$$G = \pi a^2 NA^2 \tag{17.19}$$

If we now assume that it is the slit width that determines the resolution, then the slit width is constrained by the following expression:

$$a = \frac{2f \tan\phi}{R(1 - \tan\theta \tan\phi)} \tag{17.20}$$

It is possible to substitute the grating width, w, in Eq. (17.21), incorporating the system numerical aperture, NA, at the exit slit.

$$a = \frac{w \cos\theta \sin\phi}{RNA} \tag{17.21}$$

We may now substitute Eq. (17.21) into Eq. (17.19) to obtain a revised expression for the étendue, expressed in terms of the grating width:

$$G = \frac{\pi w^2 \cos^2\theta \sin^2\phi}{R^2} \tag{17.22}$$

Equation (17.22) sets the area of the grating in terms of the required resolving power and the system étendue:

$$w^2 = \frac{GR^2}{\pi \cos^2\theta \sin^2\varphi} \tag{17.23}$$

Hence, in the case that the slit width determines the resolution, then the area of the grating is proportional to the étendue and the square of the resolution. This confirms the notion that the size of a spectrograph instrument inherently follows the size of any 'downstream' sub-systems systems. Furthermore, the angular term in the denominator suggests that the effective resolution 'efficiency' is increased by tilting the grating as far as possible. That is to say, if the instrument size is to be minimised, it is preferable to maximise the grating tilt angle ϕ as far as possible.

This analysis applies where the slit width determines the resolution. Where the resolution is to be diffraction limited, this inevitably constrains the system étendue. Equation (17.7), which prescribes the limit on the slit width, suggests, not surprisingly, that the limiting étendue is of the order of the square of the wavelength. This limit, of course, applies to downstream instrumentation.

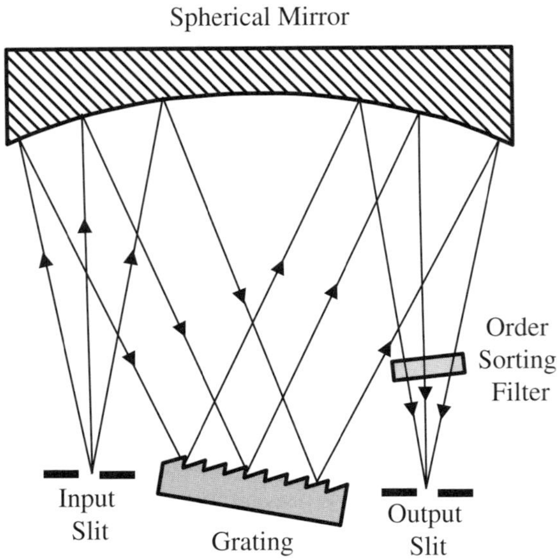

Figure 17.5 Fastie-Ebert spectrometer.

Worked Example 17.2 ***Extremely Large Telescope Spectrometer Scaling***
The E-ELT telescope has a primary mirror 38 m in diameter which defines the system entrance pupil. An area of sky 0.1 arcseconds square is to be sampled and presented to the input of a visible spectrometer. If, in the context of a Czerny-Turner monochromator, we assume a grating tilt, φ, of 20° and an arm angle, θ, of 10°, estimate the size of the instrument, given a required resolution of 12 000.

Firstly, we need to calculate the system étendue, G. The angle, 0.1 arcseconds, corresponds to 0.485 μrad.

$$G = (4.85 \times 10^{-7})^2 \times \pi \times 19^2 = 2.668 \times 10^{-10}\ \mathrm{m}^2 \text{ or } 2.668 \times 10^{-4}\ \mathrm{mm}^2.$$

We can estimate the size of the grating from Eq. (17.22):

$$w^2 = \frac{GR^2}{\pi \cos^2\theta \sin^2\phi} = \frac{(2.668 \times 10^{-10}) \times 12000^2}{\pi \times 0.985^2 \times 0.342^2} = 0.1078\mathrm{m}^2$$

The area of the grating is approximately 0.1078 m^2 or 0.328 m × 0.328 m. This is an very large grating and, if we assume f#4 optics, corresponds to an instrument with a focal length of over 1.3 m. This illustrates the clear relation between the scaling of spectrometers and the optical size of downstream sub-systems.

17.2.4 Fastie-Ebert Spectrometer

The previous detailed analysis of the Czerny-Turner Monochromator served to illustrate a number of important principles that are germane to all configurations covered here. The Fastie-Ebert Spectrometer is a simple, low-cost design which replaces the two mirrors in the Czerny-Turner configuration with a single common mirror. The basic arrangement is illustrated in Figure 17.5.

The most salient feature of the Fastie-Ebert design is its robustness and simplicity of construction. However, unlike the Czerny-Turner design, it does suffer significantly from uncorrected off-axis aberrations. It therefore does not feature in high specification instruments.

17.2.5 Offner Spectrometer

The ***Offner Spectrometer*** is based upon the classical ***Offner Relay*** which consists of a single large concave spherical mirror and a small convex spherical mirror whose centres of curvature are co-located. The Offner Relay functions as a 1:1 imaging system and, if the radius of curvature of the convex mirror is half that of the

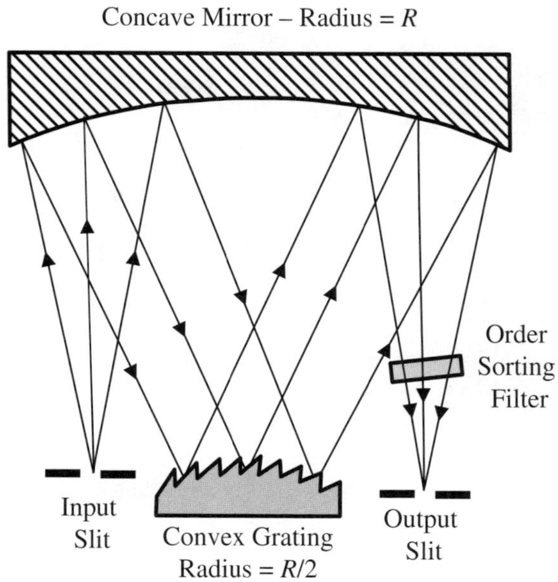

Figure 17.6 Offner spectrometer.

larger concave mirror, then the Petzval curvature will be zero. As with the arrangement in the Fastie-Ebert spectrometer, the large concave mirror is sampled twice and a flat field results. For an input pupil located at infinity, this simple relay provides full third order correction. The inherent symmetry of the system eliminates coma and spherical aberration and astigmatism is corrected by virtue of the pupil location and the co-location of the sphere centres. With astigmatism eliminated and a zero Petzval sum, the field curvature is also removed. In the spectrometer design, the concave mirror is replaced by a concave grating. This scenario is similar to that of the simple concave Rowland Grating covered in Chapter 11, where the object and image slits must lie on the Rowland circle to minimise aberrations. The effect of the Offner Relay arrangement is to flatten the Rowland circle. The basic arrangement is shown in Figure 17.6.

As depicted, the system is entirely symmetric and, in this configuration, the aberration performance is extremely robust. The only significant aberration is higher order astigmatism that follows a fourth order angular dependence. Even this can be tolerated by locating the slits at the tangential focus. However, as with the Czerny-Turner instrument, the effect of diffraction into the non-zeroth order is to remove the symmetry. In practice, although one continuous concave mirror is shown in Figure 17.5, most generally this is split into two separate mirrors. Generally the principle of colocation of the mirror centres of curvature is preserved. Otherwise, curvatures of the two separate mirrors may be adjusted in order to cancel out aberrations, especially coma for a specific wavelength or range of wavelengths.

17.2.6 Imaging Spectrometers

17.2.6.1 Introduction

Hitherto, we have considered a monochromator or spectrometer in terms of a single point dispersive measurement with output flux recorded by a discrete detector. Dispersion takes place in a direction that is orthogonal to the line of the linear slit. In this scenario, the additional length of the slit, over, and above the resolution defining slit width, is simply to provide extra signal. However, it is possible to use the length of the slit to provide an extra layer of data in the form of spatially resolved information. Naturally, recording of the image is facilitated by a pixelated detector. Full use is now made of the additional discrimination provided by a 2D detector array. The slit image is aligned to one axis of the 2D detector, and this is referred to as the ***spatial direction***; this contains spatial information from the object. Dispersion of the slit occurs in a direction that

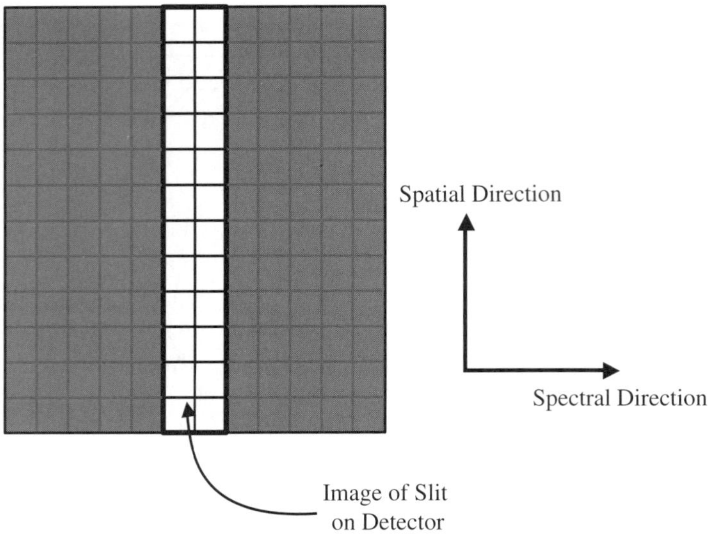

Figure 17.7 Image of slit at detector.

is perpendicular to the slit and is projected upon the other axis of the detector; this direction is known as the ***spectral direction***. Such an instrument is referred to as an ***Imaging Spectrometer***.

By further manipulation, it is possible to map a 2D object onto a linear slit to produce a 3D map with spatially resolved spectral information. The process of gathering this rich pixelated spectral data is referred to as ***hyperspectral imaging***. This may be used to provide, for example, a 2D map of atmospheric contamination by producing an individual spectrum for each imaged point for an extended object. We will describe some of the schemes used for this geometrical mapping a little later. In the meantime, we must consider that the spatial information incident upon the slit is encoded along the length of the slit only.

At this point, it is useful to examine the impact of pixel size on resolution. As was revealed in the coverage of optical detectors, the impact of pixels may be revealed through consideration of their contribution to the system modulation transfer function (MTF). The concept of Nyquist sampling was introduced, providing a useful rule of thumb for maximum pixel size. Accordingly, the pixel size should be half of the nominal resolution. Most importantly, this consideration applies to the spectral direction as well as the spatial direction. Therefore, it is customary for the slit to be imaged across some number of pixels, e.g. two, to ensure that the finite pixel width does not significantly degrade resolution; that consideration also applies to all spectrometers using pixelated detectors, not just imaging spectrometers.

The general scheme is illustrated in Figure 17.7, which illustrates the slit location for one specific wavelength and designed specifically to fulfil Nyquist sampling.

17.2.6.2 Spectrometer Architecture

The spectrometer consists of three distinct subsystems. Firstly, there is the collimator, together with the slit assembly which provides collimated illumination for the second sub-system, the grating. The system pupil is located at the grating assembly and diffracted light from the grating is focused by the camera subsystem which includes the detector. Ideally, the grating is arranged at an angle close to the Littrow condition. This would reduce the impact of anamorphic magnification on the system. In practice, for a reflective grating, the incident and diffracted paths must be separated sufficiently, so the precise Littrow condition cannot be attained in this instance. To account for this architecture, for some central wavelength, we might describe the diffracted beam path in terms of zeroth order reflection with an incident and reflected angle of θ. Diffraction is accounted for a tilt, φ, of the grating from the zeroth order condition. This analysis is identical to that presented for the Czerny-Turner monochromator. This general architecture is shown in Figure 17.8.

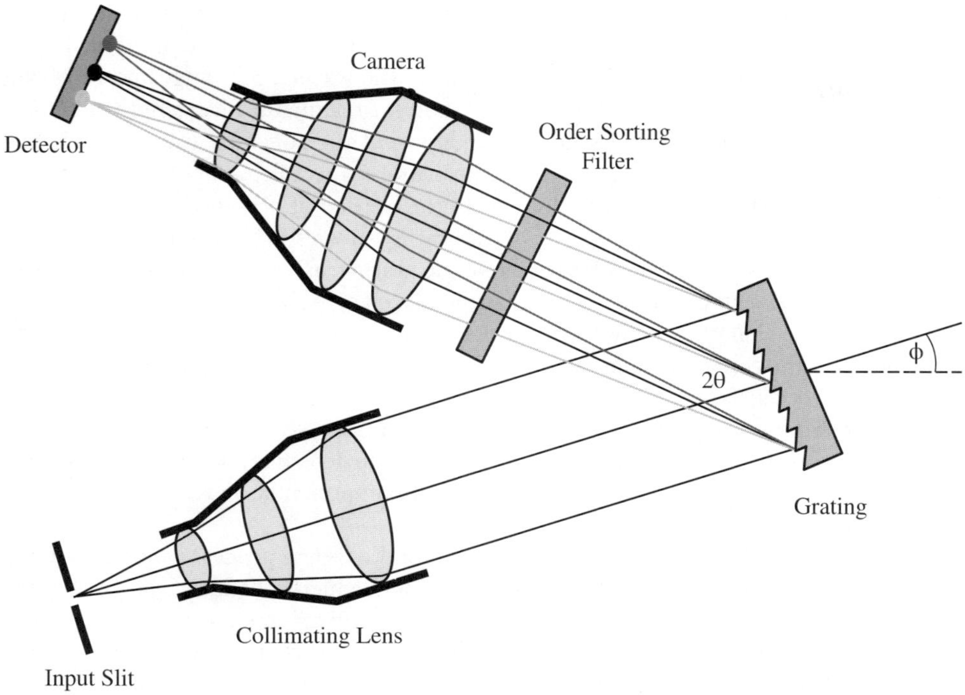

Figure 17.8 Layout of imaging spectrometer.

The input characteristics of the instrument are dictated by the system étendue which is a function of the underlying input imaging subsystem specification. However, what is clear from Figure 17.8 is that the étendue at the camera has been substantially increased by virtue of the grating dispersion. That is to say, the dispersion process has significantly extended the field. As a consequence, design of the camera is rather more challenging than that of the collimating lens. Furthermore, as the preceding discussion emphasised, the slit must be imaged onto a specific number of detector pixels, typically two. In most applications, the width of the slit is likely to be larger than the pixel width and this consideration demands that the camera demagnifies the slit image. By virtue of the Lagrange invariant, the ineluctable consequence of this is that the camera lens is considerably 'faster' than the collimator. This places further demands on the cameral design.

As suggested earlier, it is desirable to restrict the effective incidence and reflected angle, θ, as far as possible. It is clear from Figure 17.8 that this is dictated by the need to separate the camera and collimating optics. In practice, some extra margin must be added to allow for mechanical mounting. Naturally, the size of the camera is also influenced by the ranged of diffracted angles and hence the wavelength range covered by the instrument. Therefore, the instrument wavelength range affects the arm angle, θ.

17.2.6.3 Spectrometer Design

From an optical perspective, there are a limited number of critical requirements that inform the design process. These might be summarised as:

- Wavelength range
- Resolving power
- Étendue
- Image quality
- Source (spectral) radiance
- Signal-to-noise ratio

The initial design phase is very much a 'paper exercise', in which the fundamental design attributes of the instrument sub-systems are sketched out. This would involve establishing the focal length and numerical aperture of the collimator and camera and selecting the diffraction grating. The wavelength range and resolving power together with the system étendue broadly set the grating size and the collimator focal length and aperture. Selection of grating dispersion should enable the accommodation of the specified wavelength range within a reasonable angular range. This makes design of the camera more tractable. A field angle range of ±15° might be acceptable for the camera; larger field angles add to the burden of preserving image quality. With this information, it is possible to specify the line density of the grating, given some reasonable grating tilt angle, θ, e.g. 30°. In addition, we must also select the blaze angle of the grating, assuming a ruled, as opposed to holographic grating is to be used. The assumption here is that the blaze angle should be chosen to deliver maximum efficiency at the central wavelength. Finally, the pixel size of the detector sets the focal length of the camera. As such, the focal length of the camera is then established by the magnification required to image the slit across the appropriate number of pixels.

In practice, choice of critical components, such as gratings and detectors, is restricted. Compromise must inevitably be accepted, as gratings are available only with certain specific line densities and blaze angles. Similarly, the pixel size of detectors is also constrained. For these critical components, there are only limited opportunities for customisation.

Of course, once the outline design has been established, completion of the design process would inevitably involve the use of ray tracing software. This is especially true of the camera which is defined by its comparatively wide field of view and its high numerical aperture. The collimator and camera could be either a reflective or transmissive design. In the case of a reflective design, the chief difficulty is the requirement for the mirror surfaces to be 'off-axis' to prevent obscuration of the beam path. This compromises the designer's ability to produce high image quality with simple, especially spherical, surfaces. On the other hand, use of transmissive optics is complicated by the need to preserve achromatic performance. Furthermore, since spectroscopic instruments are often required to operate in wavelength regimes outside the visible, choice of suitable glass materials is often more restricted. For example, in the ultraviolet, one is restricted to fused silica and the alkali fluorides, such as calcium and barium fluoride.

Worked Example 17.3 ***Spectroscope Design***

At this point, it would be useful to amplify these basic principles with a simple example. Our task is to design a spectrometer to cover the visible wavelength range from 450 to 700 nm. The required resolution is 3200 and it is to operate in the first order. Input to the spectrometer is from a telescope with an aperture of 4 m and we wish to resolve spatially objects that subtend an angle of 2 arcseconds. The 'arm' angle is 15° and the range of diffracted angles, which defines the camera field, is ±15°. Finally, we are to use an array detector with a pixel size of 25 μm.

Our first task is to establish the scale of the instrument. To estimate the grating size we need to know the étendue, G, and the resolution, R. The latter we know to be 3200. The étendue, G, may be calculated from the field angle, Δ, and the mirror diameter, D, from the following:

$$G = \pi D^2 \Delta^2/4; \quad \mathrm{D} = 4\,\mathrm{m}; \quad 9.7\,\mu\mathrm{rad}; \quad G = \pi 4^2 (9.7x10^{-6})^2/4 \text{ and } \mathrm{G} = 1.18 \times 10^{-9}\,\mathrm{m}^2.$$

From Eq. (17.23) we have:

$$w^2 = \frac{\pi G R^2}{\cos^2\theta \sin^2\phi}$$

As previously outlined, we make a reasonable estimate for θ (15°) and φ (30°) and this gives:

$\boldsymbol{w = 404\ mm}$.

The width of the beam emerging from the collimator and 'covering' the grating depends upon the angle of the grating with respect to the collimated beam. In this case, it is assumed to be 15°, giving a ***collimated beam diameter of 390 mm***. To make the collimator design reasonably tractable, we choose a relatively 'slow' implementation, e.g. f#4. This gives a ***collimator focal length of 1560 mm***. The system étendue is preserved through the Lagrange invariant and, with a collimated beam diameter of 390 mm, approximately one-tenth of the telescope mirror diameter, the angular width of the slit must be about 10 times that of the original object. This gives a slit width of 20.5 arcseconds or 99.5 μrad, corresponding to a physical ***slit width of 155 μm***.

We may now turn to the camera paraxial design. According to the Nyquist sampling criterion, the slit, as imaged at the detector should correspond to two pixel widths or 2 × 25 μm or 50 μm. The camera should provide a magnification equal to 50/155 or 0.32. As such, the ***camera focal length should be 503 mm*** with an ***aperture of approximately f#1.3***. Taken along with the extended field angle for the camera, this illustrates the greater challenge that is invested in the camera design.

The grating characteristics must now be established. We fix the 'arm angle', θ, at 15°. However, both the tilt angle, φ, and the grating spacing, d, must be calculated. At the 'central' wavelength, λ_0, the grating equation results in an expression with the same form as Eq. (17.1):

$$2d\cos\theta\sin\phi = m\lambda_0$$

It is the range of diffracted angles (±15°) that effectively defines the grating angle. If the shorter wavelength (450 nm) is labelled, λ_1 and the longer wavelength, λ_2, then the following equations apply:

$$d(\sin(\theta+\phi)-\sin(\theta-\phi+\alpha)) = m\lambda_1 \text{ and } d(\sin(\theta+\phi)-\sin(\theta-\phi-\alpha)) = m\lambda_2 \qquad \textit{for positive } m$$

This equation yields a tilt angle of 31.29°. The grating spacing d is 603.79 nm or 1656 lines mm^{-1}. In practice, depending upon the availability of commercial gratings, a grating with 1800 lines mm^{-1} may be selected. The 'central' wavelength, where the ***angular deviation*** is between the two extremes, is 605.77 nm. Note, in wavelength terms, this is not halfway between the two extremes (600 nm). Finally, to complete the picture, we need to calculate the blaze angle for the grating. A reasonable basis to calculate this is to assume that the grating is blazed to deliver maximum efficiency for the central wavelength. In fact, the blaze angle is equal to the tilt angle of 31.29°. Gratings are often specified in terms of a blaze wavelength, λ_B, for the Littrow condition. The blaze wavelength is determined by the blaze angle,θ_B, according to:

$$m\lambda_B = 2d\sin\theta_B \quad \lambda_B = 2\times 603.79\times\sin(31.29) = 627.2\ \text{nm}.$$

17.2.6.4 Flux and Throughput

Somewhat surprisingly, in view of the relative complexity of the instrument, it is rather straightforward to calculate the flux incident upon each detector pixel. Each pixel has a well-defined étendue, the product of its area and the solid angle described by the camera aperture. The flux incident upon the pixel is simply given by the product of the source spectral radiance, L_λ, the pixel étendue, G_{pix}, the effective bandwidth of the instrument, $\Delta\lambda$, and the system throughput, ξ, as given by Eq. (17.17). In calculating the effective bandwidth for a single pixel, one must remember that a single pixel width may not correspond to the width of the imaged slit. For example, it may be half that width, in the case of Nyquist sampling. In this instance, assuming the slit width is larger than the diffractive resolution, the slit function will be a trapezoidal rather than a triangular profile. In the more general case, the effective instrument bandwidth is the integral of the slit function $f(\lambda)$:

$$\Delta\lambda = \int f(\lambda)d\lambda \tag{17.24}$$

A significant portion of the throughput, ξ, is determined by the grating efficiency, which is a strong function of wavelength and polarisation state. A brief account of grating efficiency and its dependence upon polarisation and wavelength was given in Chapter 11. We must also consider the impact of the collimator and camera on throughput. If both sub-assemblies are transmissive, then we must consider the impact of Fresnel losses. If the surfaces are uncoated, losses of the order of 4% per surface or 8% per component must be contemplated. However, this scenario is rather unlikely, and it is to be expected that the surfaces will be anti-reflection coated in some way. Nevertheless, despite this, losses of up to 1% per surface must be budgeted for. For an all mirror design, larger losses are likely, except, perhaps in the infrared. Relatively high losses of >10% per surface apply for metal films in the visible and near infrared. All these losses, taken with the grating efficiency must be factored into any computation of the throughput.

17.2.6.5 Straylight and Ghosts

In the analysis of the imaging spectrometer design the narrative has focused exclusively upon ***sequential ray tracing*** analysis. This is true for the bulk of the analysis of optical systems presented in this text. That is to say, the behaviour of light as it passes through a system is entirely deterministic as it progresses, in sequence, from one surface to the next. However, the practical designer is often exercised about the issue of ***straylight***, whereby light scattered stochastically from optical and mechanical surfaces provides an undesirable background level of illumination. For sensitive measurements at low signal levels, this background illumination might be of critical importance, especially if this background level is greater than the detector dark current. In this instance it will form the dominant noise source. Straylight also has the propensity to degrade contrast in imaging applications.

Due to the random nature of the scattering process, the light path from object to detector is not deterministic or sequential. In an instrument with many surfaces this problem is in no way amenable to analytical solution. Therefore analysis of straylight is exclusively the domain of computer modelling, simulating the stochastic behaviour of scattering through ***non-sequential ray tracing*** analysis. This topic is considered in a little more detail later in the book. The important point about straylight analysis is that it must consider the contribution made by mechanical mounts and other mechanical surfaces, as well as the optical surfaces themselves.

Each surface is modelled according to some general description of its scattering behaviour. Some aspects of this were covered in the treatment of radiometry. The geometry of scattering may be modelled as Lambertian or by some more subtle scattering distribution defined by the surface's BRDF (Bi-directional reflection distribution function). In the context of a spectrometer design, we are specifically interested in light of a specific wavelength that ends up in the 'wrong place'. As well as generic scattering from mirrors and optical mounts and other surfaces, we are particularly concerned with the behaviour of the grating. The fate of diffracted light that misses the detector cannot simply be ignored. That light will inevitably strike other surfaces and scatter around the inside of the spectrometer and may eventually reach the detector. In addition, the grating is itself a significant contributor to low level, but significant, scattering. Much interest, therefore, is focused on the coating of internal and mechanical surfaces. There are many proprietary 'black' coatings specifically designed for optical applications, which seek to reduce internal scattering to a minimum. Any non-sequential ray tracing model that seeks to analyse straylight must account for the scattering properties of such coatings as well as the complex internal geometry of the instrument, embracing both optical and mechanical surfaces.

Ghosting refers to the impact of undesired specular reflections leading to the formation of auxiliary images. In spectroscopy, the most notorious of these are the so-called 'grating ghosts'. Grating ghosts are a feature of mechanically ruled gratings and not holographic gratings and are associated with small, subtle periodic errors in the machine that rules the gratings. Since these errors are subtle, these ghosts are faint. Traditionally, grating ghosts have been identified in atomic spectra as weak features produced by intense spectral lines. In addition to these ghosts, in any design, we must consider very carefully the fate of undesired diffractive orders. We can, by no means, assume that they simply disappear. For example, they could undergo (Fresnel) reflection at the order sorting filter, as shown in Figure 17.8. Thereafter, they could strike the grating again and be diverted once more into different diffractive orders. These reflections could produce 'ghosts' at the detector and such multiple reflections must be carefully analysed. Otherwise, the generation of ghosts is an issue associated with Fresnel reflections in transmissive optics. Quite apart from the elimination of chromaticity, the adoption of an all mirror design in the collimator and camera eliminates such ghosts.

17.2.6.6 2D Object Conditioning

In the majority of practical applications, we are interested in a physical object defined by a 2D field. Unfortunately, the intrinsic field of a spectrometer is that of the one-dimensional slit. There are a number of ways

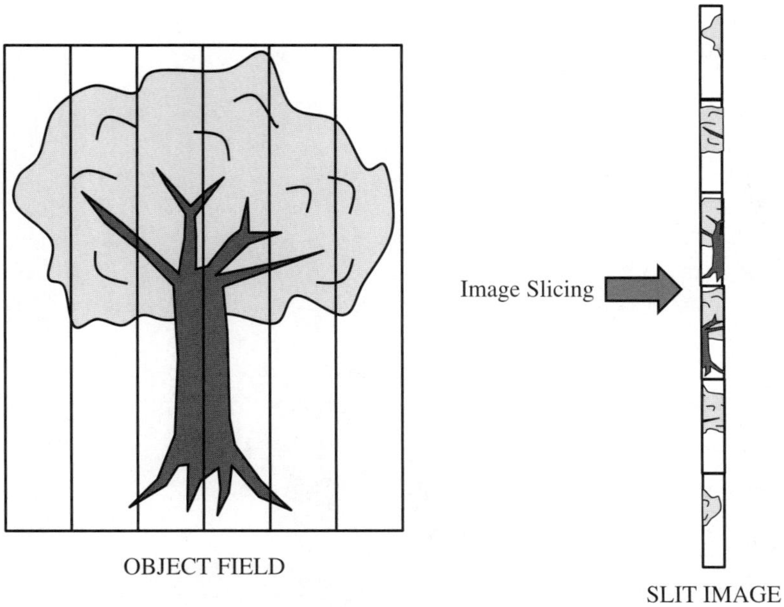

Figure 17.9 Principle of image slicing.

of resolving this difficulty. These schemes fall into two broad categories. Firstly, there are techniques that partition or slice a 2D field and re-arrange these slices along a one-dimensional slit. An example of this is the ***integral field unit*** (IFU) which uses segmented mirrors to slice a square or rectangular field and then re-arrange these slices along a linear slit. This scheme is shown in Figure 17.9. Alternatively, this object field rearrangement may be accomplished by a 2D array of optical fibres placed at the original input field. The output end of these fibres can then be configured along the nominal or virtual slit of the spectrometer. Of course, the resolution of the original object is limited by the number of pixels arrayed along the length of the spectrometer slit. For example, if the number of spatial pixels along the length of the slit is 2500, then this would correspond to a square input field of only 50×50 pixels. Generally, this type of technique is used where the granularity of physical objects in the original object field is relatively spare. An example of this might be in astronomical applications where the input field consists of a relatively restricted configuration of discrete objects.

Otherwise, for more densely configured object fields, it is possible to scan the object field across the slit. That is to say, at any particular time, only a linear strip is sampled within the object field. Over some period of time, the strip within the object field that is projected upon the slit is scanned in a direction that is orthogonal to the slit. This scanning procedure could, of course, be effected by a rotating, scanning mirror, as if often the case. Another technique of scanning that is popular in aerospace applications is the so-called ***pushbroom scanner***. In this instance, the scanning process is produced by the motion of the viewing platform, e.g. satellite or aircraft. Naturally, the slit is oriented orthogonally to the direction of relative motion. As the object field, for instance the Earth's surface, moves with respect to the instrument, a different linear portion of the object field is presented at the slit. This arrangement is illustrated in Figure 17.10.

Figure 17.10 represents a typical application of a push broom scanner, based upon an Earth observation satellite. The imaging optics consists of a telescope that images a strip or swath of the Earth's surface onto the slit of a spectrometer. Orbital motion effectively scans this slit across the surface of the earth. Although spatial resolution is significantly impacted by the image quality provided by the imaging optics and the spectrometer, it is also impacted by the effective integration time of the detector. For example, for an orbital velocity of $7600\,\mathrm{ms^{-1}}$, typical of low Earth orbit, and a detector integration time of 50 ms, this corresponds to a movement of about 400 m. Clearly, any attempt to improve the imaging resolution beyond this value will lead to

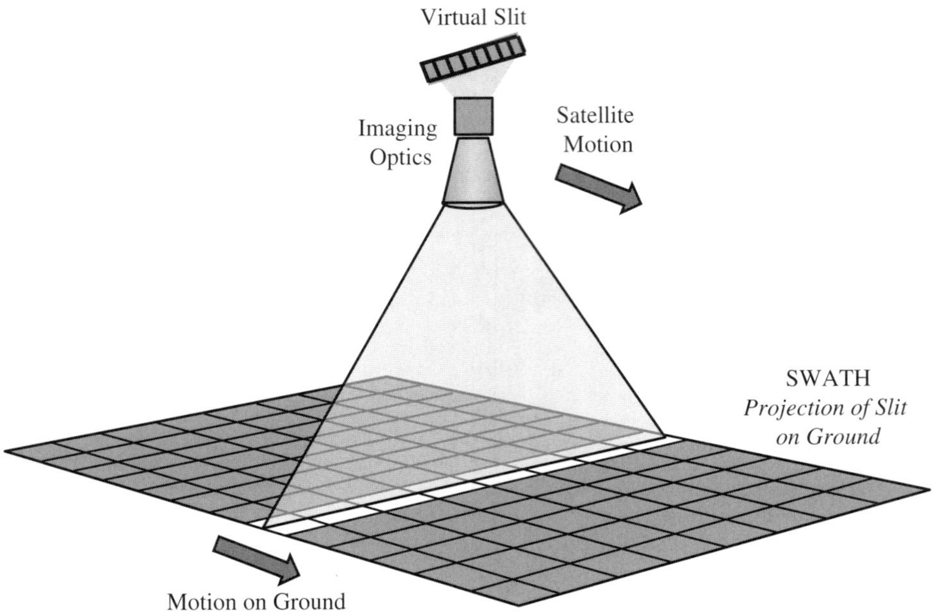

Figure 17.10 Operation of a push broom scanner.

diminishing returns in terms of the overall system resolution. Effectively, this displacement ultimately governs the size of a spatial pixel, as projected at the Earth's surface. This type of application also illustrates how straightforward it is to estimate the flux incident on each detector pixel, according to Eqs. (17.17) and (17.24). We start out with the well-known spectral irradiance of solar illumination and, from some understanding of the reflectivity/BRDF of some portion of the Earth's surface, we may derive the emerging spectral radiance at some wavelength of interest. The known slit function, system throughput, and pixel étendue may be used to estimate the flux at each pixel. Thereafter, it is possible to estimate the number of charge carriers generated at each pixel during an integration period. As the signal-to-noise ratio is often an important system requirement, this calculation forms an important part in establishing fundamental design constraints for the system.

17.2.7 Echelle Spectrometers

Equation (17.23) offers a general insight into instrument scaling. It clearly illustrates the dominant impact of system spectral resolution upon instrument size. For certain applications, particularly those probing the structure of spectral lines and closely packed spectral features, resolutions of many tens of thousands are required. With the relatively straightforward designs thus far outlined, delivering such resolutions in any reasonably configured instrument becomes impractical. In Chapter 11, we introduced echelle gratings, low density gratings specifically engineered to work at very high diffraction orders, e.g. 50–100. Moreover, unlike standard ***echelette***, gratings, they are optimised to work over multiple (high) diffraction orders. They are characterised by very high Littrow angles, e.g. 60°–70° and involve reflection off the grating 'riser' as opposed to 'step', for the conventional grating. High resolution is ultimately conferred by the very high grating tilt angles used, according to the logic of Eq. (17.23). This geometry, as previously outlined, is inevitably associated with the use of coarse gratings and operation at high diffraction orders. However, whilst this approach confers higher resolution, it renders the inherent ambiguity in distinguishing the different diffraction orders even more acute.

In common with other high dispersion components, such as Fabry-Perot etalons, the echelle grating character may be denominated by its free spectral range (FSR). The FSR is the interval (in wavenumbers) between adjacent orders. From Eq. (17.1), in the Czerny-Turner geometry, the FSR is given by:

$$FSR = \frac{1}{2d\cos\theta\sin\phi} \tag{17.25}$$

For an echelle grating with 50 lines mm^{-1} ($d = 0.02$ mm) and $\theta = 15°$, $\varphi = 70°$, then the FSR corresponds to 550 cm^{-1}. For a near infrared wavelength around 1 μm, this interval corresponds to about 55 nm. A common technique employed in high resolution instruments to overcome this ambiguity is ***cross dispersion***. Essentially, this involves the addition of a further dispersive sub-system with the axis of dispersion at some angle, usually orthogonal, to that of the echelle dispersion. This sub-system may employ a conventional grating, or, less commonly, a prism. Where a grating is used as the cross disperser, this is optimised to operate in a single order, unlike the echelle grating. In this way, the ambiguity is removed, as successive echelle orders are removed by displacing them along the second axis. The resulting diffraction pattern is displayed on an array detector. We must, of course, be careful to ensure that the FSR of the cross dispersion is not some integer multiple of the echelle FSR. In this instance, the ambiguity might be retained for some specific orders. Clearly, in this case, we are using the additional detector dimensionality to enhance the density of spectral information. By doing this, in the case of an imaging spectrometer, we are attenuating the available information bandwidth for analysing spatial information.

The principle is illustrated in Figure 17.11. In this example, we have a grating of 20 lines mm^{-1} operating in the spectral region from 2 to 3 μm. For simplicity, it is assumed that both echelle and cross dispersion grating are operating in the Littrow configuration at the central wavelength of 2.5 μm. The echelle grating is blazed at about 60°, with the central wavelength of 2.5 μm corresponding to order number 35 under the Littrow condition. The cross dispersion grating is blazed at 30° in first order for the central wavelength, giving a line density of 400 lines mm^{-1}. For each order and each wavelength, Figure 17.11 plots both the dispersion along the y-axis (echelle grating) and the x-axis (cross disperser) and illustrates the dispersion for discrete wavelengths between 2.0 and 2.9 μm. The echelle grating diffraction order is also clearly shown.

In practice, the angular displacements in the two directions will be resolved into displacements at the detector, following imaging by the camera optics. Thus, the pattern seen in Figure 17.11 is representative of what would be seen at the array detector. Although Figure 17.11 replicates the echelle spectrometer behaviour at specific wavelengths, the lines marked out in Figure (17.11) reveal the continuous spectrum that would be observed for the various orders. This illustrates the way in which cross dispersion clearly separates out the different orders.

17.2.8 Double and Triple Spectrometers

As previously alluded to, the analysis of straylight of optical systems in general and spectrometers in particular is often neglected in standard texts. As a guideline, in a typical monochromator design 'out of band' background light levels are typically four to five orders of magnitude lower than the nominal signal. For many applications, this is perfectly adequate. However, in specific instances this performance is not acceptable. One example is that of Raman spectroscopy. Typically, Raman scattering is triggered by a high-power laser and one is looking for a spectral feature arising from the scattering process; this feature is very close to the single frequency laser line. The difference in wavenumber of the scattered and pump radiation corresponds to the wavenumber of vibrational spectral feature of interest. Unfortunately Raman scattering is notoriously inefficient, with scattering levels of around 10^{-6} of the original pump. Furthermore, detailed straylight analysis reveals that the scattered background level is not constant across the monochromator wavelength range. On the contrary, for an input dominated by a single spectral feature, scattered levels are very much higher at wavelengths close to that feature. As a consequence, in a standard monochromator design, the Raman feature is likely to be swamped by laser source.

To overcome this problem, monochromators can be placed in tandem, with the output slit of one monochromator acting as the input slit of the next monochromator in the chain. The most common arrangement is that

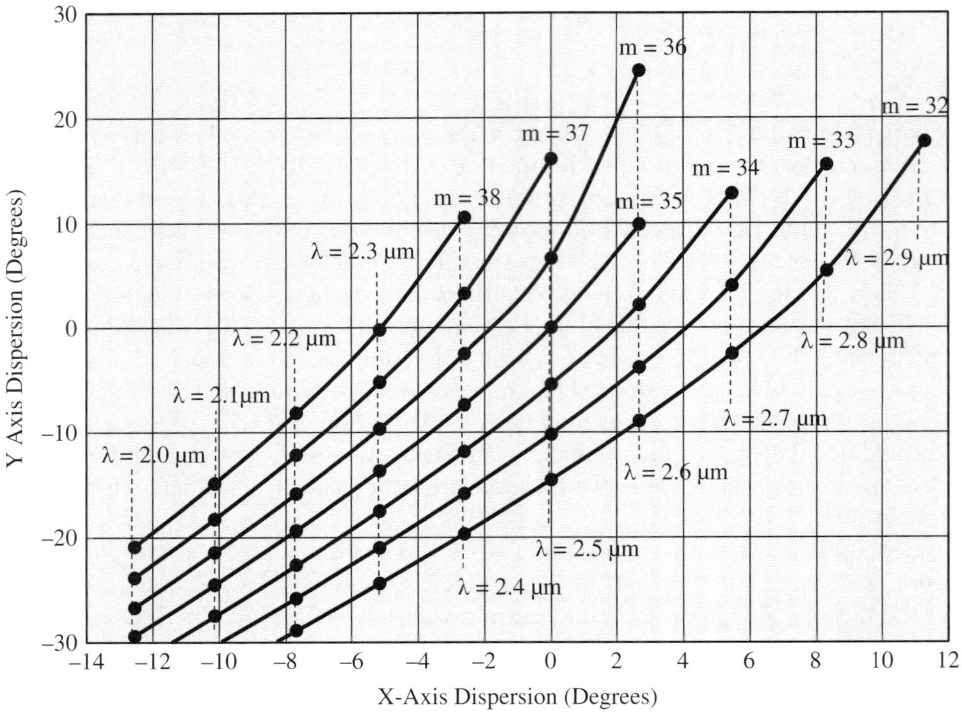

Figure 17.11 Cross dispersion in an Echelle grating spectrometer.

of the ***double monochromator***. In this example, the two monochromators are tuned to a common wavelength. The slit function of the combined instrument, is a convolution of the two sub-system slit functions, $f_1(\lambda), f_2(\lambda)$.

The effect of this convolution process is to reduce the contribution from stray light to below 10^{-6} for a double spectrometer and further still for a triple instrument. Although both instruments are most usually identical, by virtue of symmetry, they may be arranged in such a way that their two dispersions are either additive or subtractive. With additive dispersion, the effective combined slit function is narrower than for subtractive dispersion and higher resolution is obtained.

17.3 Time Domain Spectrometry

17.3.1 Fourier Transform Spectrometry

'Traditional' spectroscopy involves the analysis of the optical signal in the frequency (wavelength) domain. However, the phase of an optical wave may be probed to capture the time dependent amplitude of a wave. This may be converted to the frequency or wavelength domain by the expedient of the Fourier transform. In practice, this time dependent phase information is captured by an analysis of the phase variation of a plane wave along its propagation direction. This is accomplished by interferometry whereby a reference plane wave is created by amplitude division, e.g. at a beam splitter. Phase information as a function of axial displacement is extracted by varying the relative path lengths of test and reference beams. This is generally accomplished by a retroreflector mounted on a linear stage. This process is referred to as ***Fourier Transform Spectrometry*** and is most widely applied to infrared spectrometry.

The arrangement is shown in Figure 17.12. Overall, the set-up is similar to that of the Michelson interferometer, whereby a 45° beamsplitter divides the input beam, creating a reference path via a fixed mirror. The

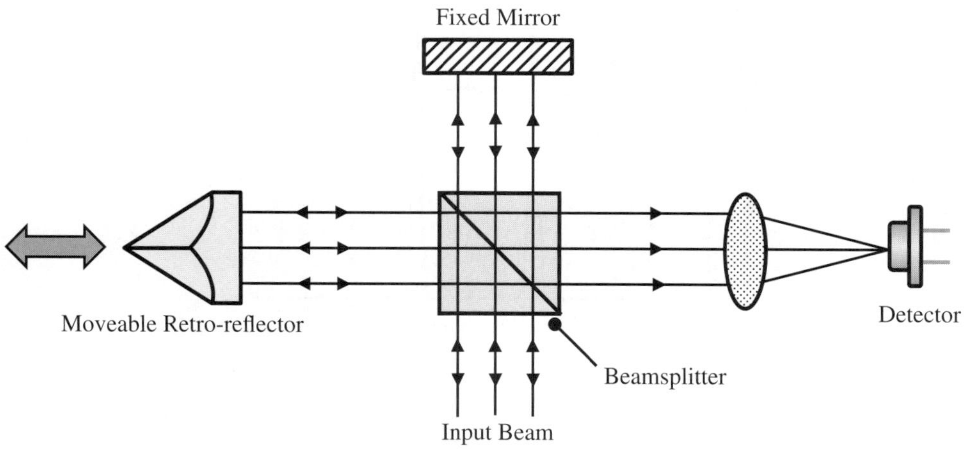

Figure 17.12 Fourier transform spectrometer.

other portion of the beam is diverted to a movable retro-reflector positioned on a linear stage. As the stage and retro-reflector move along a straight path, the relative optical path length of the two beam is changed. The resulting interference pattern is observed at the detector.

It is perfectly clear that, for a single frequency, such as a stabilised laser, the signal at the detector would describe a perfect sinusoidal dependence upon retro-reflector displacement. Indeed, in terms of retro-reflector displacement, the effective wavelength of the detector signal will be one-half of the actual laser wavelength. More generally, the detector flux as a function of reflector displacement is given by the Fourier transform of the original signal. For example, two closely spaced spectral lines, such as the sodium 'D' lines will show a spectrometer signal where a beat pattern is observed corresponding to the line spacing. In addition, the extent over which fringe behaviour is observed is determined by the spectral width of the signal. For example, a very narrow line, such as a laser line, will show the fringe pattern over an extensive range of displacements. Conversely, as in the white light interferometer described in the previous chapter, broad band emission only produces fringing over a narrow displacement range.

More rigorously, the flux observed at the detector, as a function of the reflector displacement, $\Phi(\delta)$, is given by the following integral:

$$\Phi(\delta) = \int A^2(k)(1 + \cos(2k\delta))dk \tag{17.26}$$

A(k) is the input wave amplitude as a function of the wavevector, k.

The flux is effectively given by the Fourier transform of the square of the amplitude. Therefore it is possible to extract the amplitude as a function of wavenumber (wavelength) by performing a Fourier transform on the flux as a function of displacement. This is the basis of Fourier Transform Spectroscopy.

The principle is further illustrated in Figure 17.13, which shows a Fourier transform spectrum of two closely spaced lines and the extraction of the original spectrum.

Not only does Figure 17.13 reveal the 'beating' between the two adjacent lines, but the width of the individual lines is itself characterised by an envelope whereby the fringe visibility diminishes away from the zero path difference condition. Quite naturally, a narrow spectral line corresponds to a broad envelope function and vice versa. In this instance, both line and envelope functions have been modelled as a Gaussian distribution.

One important attribute of the Fourier transform spectrometer is its resolving power. The resolving power is determined by the length of scan of the instrument. For example, an instrument with a scan length of 2 m would have a resolution of 4×10^6 at 1.0 μm. This is equal to the scan path difference (2× the mirror displacement) divided by the wavelength. As such, the instrument has applications in high resolution spectroscopy, particularly in National Measurement Institutions. For example, such instruments are, nowadays,

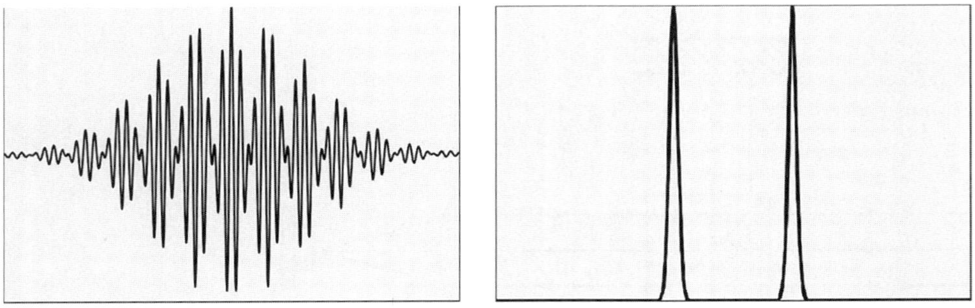

Fourier Transform | Spectrum

Figure 17.13 Fourier transform spectrograph of two closely spaced lines.

indispensable in providing a highly accurate 'atlas' of spectral lines, particularly atomic lines, with measurement uncertainties of the order of 10^{-4} nm. The association of the instrument resolving power with optical path difference is in some way connected to the resolving power of a grating. In the case of a grating, operating in the diffraction limit, the resolving power is proportional to the optical path difference between the two extreme ends of the grating, as opposed to the path difference ascribed to retroreflector movement.

As far as commercial instruments are concerned, Fourier transform spectroscopy is largely applied to infrared instruments. Signal-to-noise performance is critical at low signal levels. For infrared instruments, particularly where the detector and instrument is not cooled, background emission and dark current make an important contribution to the noise level. The two traces shown in Figure 17.13 may also be considered to represent a Fourier transform spectrograph and a conventional (grating based) instrument. Comparing the two traces and assuming them each to be characterised by a number, n, of data points gathered over an equivalent period of time. Assuming the same signal level may be applied to both traces and an equivalent background noise, when one analyses the Fourier transform trace and converts it into a spectrum, the noise is diminished by a factor equivalent to $\sqrt{n}$, compared to the conventional spectrum. This is because, in the Fourier transform instrument, we are making use of the input signal over the full range of n points, rather than a restricted range, equivalent to the linewidth, in the conventional instrument. In fact, the signal-to-noise enhancement is equivalent to the square root of the ratio of the number of data points to the linewidth. This circumstance is known as ***Fellgett's advantage***. This does not, however, apply where shot noise is the dominant mechanism, hence the greater application of Fourier transform instruments in the background limited infrared.

17.3.2 Wavemeters

A ***wavemeter*** is a compact Fourier transform device that is specifically applied to the precision measurement of the wavelength of a single frequency source, such as a laser or a tunable laser. Absolute calibration of the instrument is accomplished by provision of an internal calibration source, most usually a stabilised laser source. Typically, for visible applications, a stabilised helium-neon source is used. Effectively, the calibration source shares a common path with the source under test, or, alternatively, shares the same retroreflector geometry.

Calibration uncertainty is then determined by the residual uncertainty in the calibration source wavelength and, to a lesser extent, by the instrument resolving power, as dictated by the mirror scan length. In principle, the centration uncertainty of a single frequency signal will be much lower than the inverse of the resolving power. That is to say, an instrument with a resolving power of 10^6 will be capable of finding the centre of a line to an accuracy that is superior to 1 part in 10^6. The extent to which this is possible is dictated by the signal-to-noise ratio and is strongly dependent upon the signal level. Quite obviously, high levels of precision are possible when monitoring the signal arising from laser sources. Otherwise, the measurement uncertainty is dictated by the fidelity of the calibration laser source. This source is usually a frequency stabilised atomic laser source, such as the helium-neon laser, whereby the oscillation frequency is actively locked to the centre of the doppler broadened atomic line to ~1 part in 10^7 or better. Higher precision is obtained by locking the laser line to an external and fundamental absorption feature, such as an iodine absorption line.

A variant of Fourier transform wavemeter is the ***Fizeau wavemeter***. This design is based upon a Fizeau interferometer which views the interferogram of two slightly inclined planar surfaces or an optical wedge. The interferogram thus produced yields a uniform series of fringes which are detected by a camera and the image digitised. Comparison of the pattern produced by the test beam and that produced by a calibration standard yields the wavelength of the test beam. The advantage of this configuration is that it eliminates the use of moving parts and is compatible with a compact layout. Precision is, however, somewhat compromised.

Further Reading

Bazalgette Courrèges-Lacoste, G., Sallusti, M., Bulsa, G. et al. (2017). The Copernicus sentinel 4 mission: a geostationary imaging UVN spectrometer for air quality monitoring. *Proc. SPIE* 10423: 07.

Chandler, G.C. (1968). Optimization of a 4-m asymmetric Czerny-Turner spectrograph. *J. Opt. Soc. Am.* 58 (7): 895.

Closs, M.F., Ferruit, P., Lobb, D.R. et al. (2008). The integral field unit on the James Webb space telescope's near-infrared spectrometer. *Proc. SPIE* 7010: 701011.

Content, R. (1998). Advanced image slicers for integral field spectroscopy with UKIRT & Gemini. *Proc. SPIE* 3354-21: 187.

Eversberg, T. and Vollmann, K. (2015). *Spectroscopic Instrumentation: Fundamentals and Guidelines for Astronomers*. Berlin: Springer. ISBN: 978-3-662-44534-1.

Hollas, J.M. (1998). *High Resolution Spectroscopy*, 2e. New York: Wiley Blackwell. ISBN: 978-0-471-97421-5.

Julien, C. (1980). A triple monochromator used as a spectrometer for Raman scattering. *J. Opt.* 11: 257.

Pavia, D. and Lampman, G. (2014). *Introduction to Spectroscopy*, 5e. Pacific Grove: Brooks Cole. ISBN: 978-1-285-46012-3.

Prieto-Blanco, X., Montero-Orille, C., Couce, B. et al. (2006). Analytical design of an Offner imaging spectrometer. *Opt. Express* 14 (20): 9156.

Ramsay Howat, S.K., Rolt, S., Sharples, R. et al. (2007). Calibration of the KMOS Multi-object Integral Field Spectrometer. In: *Proceedings of the ESO Instrument Calibration Workshop held in Garching 23–26 January 2007* (eds. A. Kaufer and F. Kerber). Berlin: Springer. ISBN: 978-364-209566-5.

Smith, B.C. (2011). *Fundamentals of Fourier Transform Infrared Spectroscopy*, 2e. Boca Raton: CRC. ISBN: 978-1-420-06929-7.

18

Optical Design

18.1 Introduction

18.1.1 Background

In this chapter, we shall discuss optical design on a rather more practical footing. Hitherto, we have been concerned with the principles that underpin optical design. Moreover, this narrative has, to a large extent, been entirely preoccupied with system performance. Important as performance requirements such as wavefront error and throughput are, there are many more practical concerns to take into account. Cost is quite obviously a critical factor in any design. This cost must not only include material costs, e.g. the cost of using more exotic glasses, but must also take into account manufacturing difficulties which, of course, add to the manufacturing costs. Another salient practical concern is that of instrument space and mass. A compact and light design is often a great asset, adding significantly to convenience in consumer applications; it is invariably essential in many aerospace applications. In addition, having fixed the optical prescription in a design, the practical issue of mounting the components cannot be ignored. Straylight is another issue that is often neglected and was briefly introduced in our examination of spectrograph instruments.

During the course of this chapter, we will very briefly sketch out the place of the optical design process within the overall context of more general systems engineering. It is essential for the optical designer to understand the constraints that lie outside the narrow confines of his or her specialisation. However, brevity precludes anything more than a very brief outline. Naturally, the focus of the chapter will be the optical design process in general and, more specifically, optical modelling software.

18.1.2 Tolerancing

Having produced a workable design, an essential stage in the design process is ***tolerancing.*** This exercise establishes whether it is possible to manufacture and assemble the system at a reasonable cost. This process must account for uncertainties in the manufacture of components and optical surfaces, for example, the impact of the inevitable departure from the prescribed shape, or form error. Furthermore, due account must be taken of the uncertainties in the placement of components during the alignment process. Naturally, this work focuses predominantly on optical design. However, it is very clear that the mechanical design of a system, particularly with regard to component placement, has an exceptionally strong linkage to the ultimate optical performance. An optical designer is also interested in thermal aspects of the design, particularly where the system must operate over a wide temperature range. As well as impacting component position and alignment through thermal expansion, focal power of transmissive components is affected via the temperature-dependent refractive index. In situations where a wide operating temperature range is mandated, designers go to great lengths to produce an ***athermal*** design, particular in regard to the temperature dependence of the focal points. All these factors must be accounted for in any tolerancing exercise.

The tolerancing exercise, in general, plays substantially to the strengths of computer simulation. Component characteristics and spacings may be perturbed at random to simulate manufacturing and alignment errors and

Optical Engineering Science, First Edition. Stephen Rolt.

Companion website: www.wiley.com/go/Rolt/opt-eng-sci

the impact on optical performance assessed. To provide a realistic simulation many different combinations must be analysed; this is not really tractable with traditional analytical techniques.

18.1.3 Design Process

Before we turn to these practical aspects, we must first consider the initial design process. The first and most important task for the designer is to interpret and understand the system requirements and convert these into an outline design where the first order parameters, such as the system focal points, cardinal point locations are approximately established. However, the process of understanding the underlying requirements and converting them into clear and achievable specifications is essentially a process of negotiation between the designer and the end user or customer. Where the initial requirements are transparently not achievable in combination, this must be clearly articulated in any discussions between the designer and end customer. In many cases, this process is very straightforward; otherwise the definition of a coherent set of requirements inevitably involves some negotiation and compromise.

This early phase of the design process involves sketching out a design perhaps following the approach highlighted in some of the exercises in this text. Thereafter, according to modern practice, the subsequent detailed design process inevitably involves the use of ray tracing software. Of course, the underlying design process is informed by an understanding of the basic principles of optical design. Nonetheless, the overwhelming processing power of modern computers allows the rapid optimisation of very complex designs that would otherwise be beyond the scope of more traditional analytical techniques. This process will be described in a little more detail later in this chapter. Since optical performance is inextricably linked to mechanical design, this stage of the process is often supplemented by mechanical or thermo-mechanical modelling of the system. This may take the form of ***finite element analysis*** (FEA) whereby the entire physical system is broken down into a set of discrete points and the relevant partial differential equations that govern thermal and mechanical behaviour are solved numerically. This topic will be covered in the succeeding chapter.

18.1.4 Optical Modelling – Outline

18.1.4.1 Sequential Modelling

The basic definition of an optical system in a computer-based optical model is provided by a description of each ***surface*** in spreadsheet form. The important point to recognise is that the description is based on a collection of individual surfaces, rather than components, such as lenses or prisms, etc. A full description of each surface is included in each row of the spreadsheet, providing details of surface thickness (distance to next surface), material, and a description of the form of the surface. This spreadsheet must include, as the first entry, the object location with the image location recorded as the final entry.

Other parameters that describe the system as a whole must also be entered. These include details of the input field, pupil size and location, and the wavelength range under consideration. Analysis of the system then proceeds according to two different modes of operation. Most commonly, the optical model is pursued in the ***sequential mode***. That is to say, light proceeds from one surface to the next in a sequential and deterministic manner. First and foremost, computation is by geometrical ray tracing from the source to the image. This enables the derivation of critical metrics, such as the wavefront error, spot radius, etc. for all field points and wavelengths.

Before the software can be brought into action, an initial design must be sketched out for the program to optimise. It is the definition of this initial starting point that requires the deployment of intuitive and more traditional optical skills highlighted in earlier chapters. In addition, the use of libraries of existing designs for similar systems, e.g. microscope objectives and camera lenses, may further help to identify a useful starting point.

Perhaps the most distinctive feature of optical modelling software is the optimisation process by which a design is refined. This process hinges critically on the definition of a ***merit function***. The merit function encapsulates in a single parameter all desirable (or undesirable) performance metrics for the system under design.

As such, the merit function might include obvious contributors, such as wavefront error for specific wavelengths and fields. However, it may also incorporate mechanical constraints that do not seem, at first sight, to be directly related to optical performance. These might include, for example, constraints on the maximum or minimum lens thickness. A merit function might include a large number of these different parameters all weighted according to their perceived importance, contributing (by root sum square addition) to a single figure of merit. The figure of merit is so devised that a lower merit function corresponds to better performance. Thereafter, the software seeks to minimise the merit function by adjusting certain parameters within the optical prescription that have been set as variable parameters. This is an exceptionally demanding process, in terms of computational resource, as, in advanced optical systems, there are very many variables to be optimised.

18.1.4.2 Non-Sequential Modelling

As well as operating in the sequential mode, ray tracing provides a ***non-sequential mode*** for the modelling of illumination systems, straylight, and ghosting. That is to say, the primary interest is not in imaging at all, but rather the broad irradiance distribution of light as perceived by some detector. As with the sequential mode, surfaces or objects are defined in spreadsheet form. The object from the sequential mode is replaced instead by a source with some defined illumination characteristics, for example a Lambertian emitter in the form of a disc, or a simulated light emitting diode. Similarly, the image is replaced by a simulated detector whose purpose is to record the irradiance distribution over some area.

The distinguishing feature of the non-sequential mode is that there is no deterministic sequential progression of light from one surface or object to the other. A ray leaves one particular surface with its denominated vectorial characteristics. It is then traced to the most proximate surface that it strikes and not to the next surface in the sequence. Moreover, the behaviour of a ray once it strikes a surface is non-deterministic; this is another distinguishing feature. With some probability, it may be absorbed, reflected, or scattered according to some stochastically modelled angular distribution. In sequential modelling attention is paid to those surfaces whose function is exclusively optical. However, in non-sequential modelling ***all*** surfaces must be modelled. This includes mechanical assemblies used to mount and support optics and all enclosures and baffling.

First and foremost, non-sequential modelling is used to assess the degradation in performance produced by straylight and ghosting. However, it also has an essential role in the design of non-imaging illumination systems such as automobile headlamps, where, for example, we are concerned about the uniformity of illumination. As we established in the chapter on imaging optics, the provision of cost effective antireflection coatings has provided an impetus to the development of complex systems with very many optical surfaces. Unlike the basic optical design itself, analysis of image degradation due to scattering and parasitic reflections is not amenable to analytical treatment. By statistical modelling of the path of many millions of rays the irradiance distribution at some surface of interest can be accurately modelled. In this type of mathematically defined, repetitive process, the computer naturally excels.

18.2 Design Philosophy

18.2.1 Introduction

As outlined, before embarking upon the design of an optical system, we must understand the wider design environment, incorporating a raft of issues including manufacturability, cost, reliability, convenience as well as performance. Most importantly, we must be prepared to consider all these issues at the outset of the design process, instead of sequentially during development. This, more efficient, parallel, approach is often referred to as ***concurrent engineering***. Concurrent engineering (CE) is a systematic approach to integrated system development that emphasises the expectation of the customer or end user. This approach embodies cooperation, between all parties, particularly the transparent sharing of relevant information to facilitate effective decision

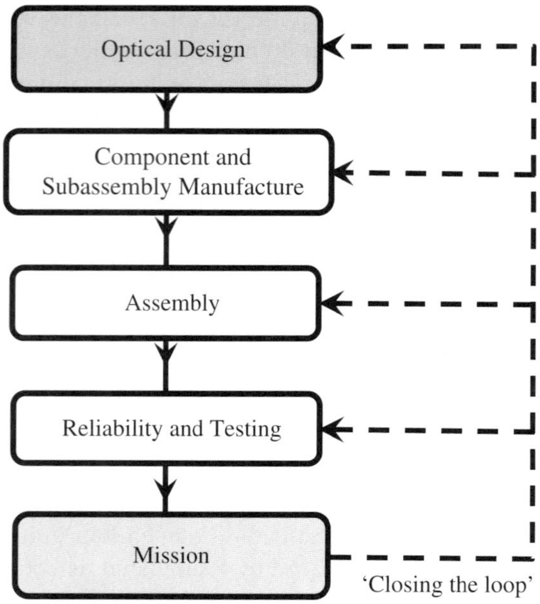

Figure 18.1 Concurrent engineering – 'Closing the Loop'.

Table 18.1 Example of optical system requirements.

Type of requirement	Examples
Optical performance	Pupil size, wavefront error, rms spot size, encircled energy, spectral irradiance, signal-to-noise ratio
Mechanical	Volume envelope, mass, stiffness
Environmental	Temperature range, temperature cycling, thermal and mechanical shock, vibration, humidity, chemical
Other	Cosmetic, exterior finish

making by consensus. Most critically, all perspectives should be considered in parallel, from the outset of the product life-cycle. In addition, this philosophy recognises the wider role of all stakeholders in the process, not only the various optical or mechanical design specialists, but also customers, end users, contractors, etc. This applies as much to consumer products as to the development of complex scientific instruments. Above all, concurrent engineering emphasises the importance of closing the loop between all the key activities in the product or system lifecycle. This is illustrated in Figure 18.1.

18.2.2 Definition of Requirements

Before the design process can begin, inputs from all stakeholders must be consolidated to provide a clearly articulated set of requirements. This cannot be accomplished without transparent communication between all parties. It is essential, that if any doubt exists amongst any of the parties regarding any specification, then this must be resolved at the earliest opportunity. The golden rule is ***'when in doubt ask'***.

Requirements may be broadly divided into optical performance requirements, mechanical requirements, and environmental requirements. Furthermore, the optical requirements must clearly define the wavelength range and the field over which requirements, such as wavefront error, must be met. This is summarised in Table 18.1.

The optical performance requirements need little further elaboration in the context of this text. The mechanical requirements, such as volume envelope and mass are fairly self-evident. Where the usage environment is relatively benign, as in many consumer applications, definition of the environmental conditions is not an especially salient issue. However, for more aggressive conditions, such as those pertaining to aerospace, industrial, and military applications, the role of the environment becomes more prominent. The environmental specifications set out the conditions under which the system is expected to meet its performance requirements. Occasionally, however, the specifications might also indicate environmental conditions that the unit must **survive**, but is not expected to meet performance targets under those conditions. An example might be in satellite applications, an optical payload must be able to withstand the shock and vibration pertaining to launch conditions, without having to meet performance requirements in that environment. More generally, systems performance must not be degraded by exposure to the transport environment, e.g. due to shocks caused by fork-lift truck handling.

Of particular relevance to optical design is the thermal environment. A great deal of importance is attached to reducing the sensitivity of a system to temperature change, particularly with regard to shifts in the focal plane and chief ray alignment, or ***boresight error***. A design whose performance is substantially unaffected by changes in the ambient temperature is referred to as an ***athermal design***. The thermal environment, to a large extent, informs the material choices in optical systems and has to be recognised in every aspect of the design process.

18.2.3 Requirement Partitioning and Budgeting

Most usually, the designer is concerned with the performance of an optical system in its entirety and the requirements are initially articulated at this level. However, the optical system as a whole is generally broken down into separate functional modules. Each module will, to a degree, be designed and modelled independently. As far as the optical design is concerned, it is customary also to establish a full system model (simulation), or 'end-to-end' model, which concatenates the individual modules. For example, in a spectrograph, we might have collimator, grating, and camera subsystems. If the design is to be broken down into these individual elements, then the requirements must also be partitioned into subsystem requirements to reflect this. In addition, the individual susbsystems must interact with each other in such a way as to deliver the end requirements. This necessity is captured by the ***interface requirements***. Interface requirements, for example, might include the locations of entrance and exit pupils of the individual sub-systems, so these may be aligned during system integration.

In this way key, performance attributes, such as wavefront error, are broken down as part of a budget, allocating a specific figure to each module. For the most part, it is clear how individual module budgets impact the global system budget. For example, with wavefront error and rms spot radius it is reasonable to use the root sum of squares (RSS) to derive the system figure. This assumes that the errors associated with these attributes are statistically independent and do not correlate across subsystem elements. On the other hand, for modulation transfer function (MTF) and throughput, then the contribution of the individual sub-systems is multiplicative. This process is illustrated in Figure 18.2.

We have now allocated the system performance requirements amongst the individual sub-systems. However, it is not sufficient to allocate all the budget of a subsystem to its basic design. As such, we must account,

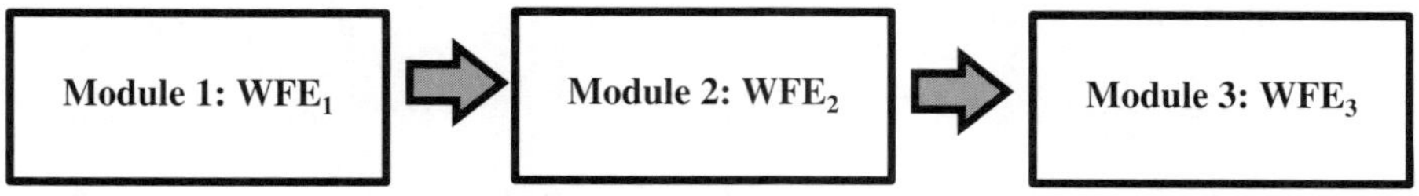

$$WFE_{system} = \sqrt{WFE_1^2 + WFE_2^2 + WFE_3^2}$$

Figure 18.2 Subsystem partitioning of requirements.

Table 18.2 Subsystem error budget.

Description	RMS wavefront error allocated (nm)
Design	100
Manufacture	150
Alignment	80
Total	201.5

not only for the design itself, but we must also allow for component manufacture, e.g. form errors, and also budget for alignment and integration errors. Indeed, these other errors may dominate over that of the initial design. An example of a subsystem budget is shown in Table 18.2.

The manufacturing figure may be further sub-divided down to the component level, to provide the manufacturer with guidance as to individual component form error tolerances. The system and subsystem budget provide an initial ***estimat***e of the most efficient allocation of tolerances, to be refined during the design process. This initial estimate is largely based upon general experience. However, with the advent of modelling software, this process is ultimately placed on a more scientific footing. With this resource, tolerances, such as surface form errors and tilt can be budgeted on a detailed component or surface basis. Although some understanding of manufacturing and alignment capabilities is necessary to inform this process, it is possible to definitively specify individual component tolerances and clearly understand the impact upon ultimate system performance. Before the advent of such capability, the resulting uncertainty in the ultimate system performance had to be resolved by design conservatism. That is to say, in order to ensure system reliability, components had to be over-specified, leading to unnecessary costs and manufacturing difficulties. This is a clear illustration of the ultimate design philosophy of ***optimising*** performance rather than ***maximising*** it. Ultimately, good design practice is directed towards the satisfactory resolution of competing and conflicting demands.

Worked Example 18.1 ***Partitioning of Requirements***

We are tasked to design an imaging spectrometer with diffraction limited performance at a wavelength of 1000 nm. This instrument is to consist of three subsystems: a telescope system, a collimator and a camera subsystem (which includes the grating). The wavefront error budget is to be allocated evenly amongst the three sub-systems. What are the individual sub-system wavefront allocations? Following this analysis, we wish to focus our attention on the collimator sub-system. We have decided on an all mirror design – a three mirror anastigmat (TMA). Our experience tells us that we must allocate as much to manufacturing tolerance as to the design and half as much to the alignment process. Calculate the respective design, manufacturing, and alignment contributions. Finally, we need to specify the individual mirror form error requirements. Assume that the mirror form error represents the sole contribution to the manufacturing allocation and that each of the three mirrors contributes equally.

Firstly, we need to calculate that system wavefront error, Φ_{rms} that will deliver diffraction limited performance at 1000 nm. This is the so-called Maréchal criterion:

$$1 - 4\pi^2 \frac{\Phi_{rms}^2}{\lambda^2} = 0.8 \quad \text{and} \quad \Phi_{rms} = \frac{\lambda}{14.05} \; \Phi_{\mathrm{rms}} = 71.2\ \text{nm rms}.$$

We are told that this global figure is to be allocated evenly amongst all subsystems:

$$\Phi_1^2 + \Phi_2^2 + \Phi_3^2 = 71.2^2 \quad \text{and therefore:} \quad 3\Phi_1^2 = 71.2^2 \quad \text{and} \quad \Phi_{\mathrm{rms}} = 41.1\ \text{nm}.$$

Therefore, we should allocate 41.1 nm rms to each of the three subsystems.

We now turn to the collimator design. We know that the wavefront error to be allocated to this subsystem is 41.1 nm. Furthermore the contribution allocated to manufacturing is the same as the design figure, whereas the alignment allocation is half this:

$$\Phi_{\text{man}} = \Phi_{\text{des}}; \Phi_{\text{ali}} = \Phi_{\text{des}}/2$$

Therefore:

$$\Phi_{des}^2 + \Phi_{man}^2 + \Phi_{ali}^2 = 41.1^2 \quad \text{and} \quad \frac{9}{4}\Phi_{des}^2 = 41.1^2 \; \Phi_{\text{des}} = 2 \times 44.1/3.$$

Therefore, the design and manufacturing allocation for the collimator must be 29.4 nm rms and that for the alignment process 14.7 nm rms.

Finally, we need to assess the impact of mirror form errors, which we are told are the sole contributors to the manufacturing errors. The corresponding allowable wavefront error produced by the three mirrors is given by:

$$\Phi_1^2 + \Phi_2^2 + \Phi_3^2 = 29.4^2 \quad \text{and} \quad 3\Phi_1^2 = 29.4^2 \quad \text{and} \quad \Phi_1 = 17.0 \text{ nm rms}.$$

The rms wavefront error averages optical path differences (OPDs) across the pupil and a 1 nm deviation in a ***mirror*** surface will contribute 2 nm to any path difference. Therefore, the form error, Δ, is half the above figure.

The allowable form error for each mirror is 8.5 nm rms.

18.2.4 Design Process

The design process begins with the proposal of a conceptual solution to a specific problem. This is then refined, in negotiation with the relevant stakeholders into a clearly articulated set of requirements. A very basic initial sketch might follow this, capturing the paraxial requirements of the system and laying out components or component groups in a way compatible with the system volume envelope. After this, the basic optical design may be undertaken, using software tools to maximise image quality and throughput, etc. to ensure the basic performance requirements are met. This process, however, only covers the optical design in isolation. Therefore, throughout this process, aspects of mechanical design, especially component mounting and volume envelope, must be considered. This is especially important, as, in practice, the optical and mechanical aspects will be covered by different specialists; good communication is essential.

Throughout the design process formal design reviews are generally held, involving all stakeholders, to monitor progress and to agree any programme adjustments or changes to requirements. The broad pattern of these reviews is fairly general across all sectors. Usually, the process will start off with a kick-off meeting where the conceptual design is discussed and requirements and contractual matters are agreed between parties. Following completion of the preliminary, outline, design, a ***preliminary design review*** will be held. At this stage, a well-established optical design and some basic elements of the mechanical design is available for detailed examination by all parties. Adjustments to the design and further refinement of the requirements may be agreed at this stage. Subsequently, more detailed optical modelling will follow, including full tolerance modelling and consideration of the impact of straylight and environmental modelling. Detailed mechanical design will also be completed at this stage with the provision of a full set of mechanical drawings. This is followed by a critical design review where any adjustments and agreements result in the final design. This stage might be followed by some prototyping and testing and some formal verification and acceptance process. The overall process is illustrated in Figure 18.3.

18.2.5 Summary of Design Tools

A range of software tools are available to the designer to help construct the detailed design. Although this chapter will focus largely on the optical tools, mechanical and thermal simulation does play a very significant

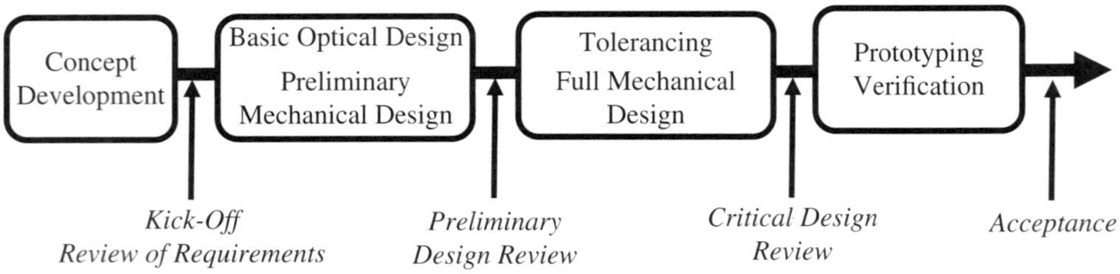

Figure 18.3 Design process.

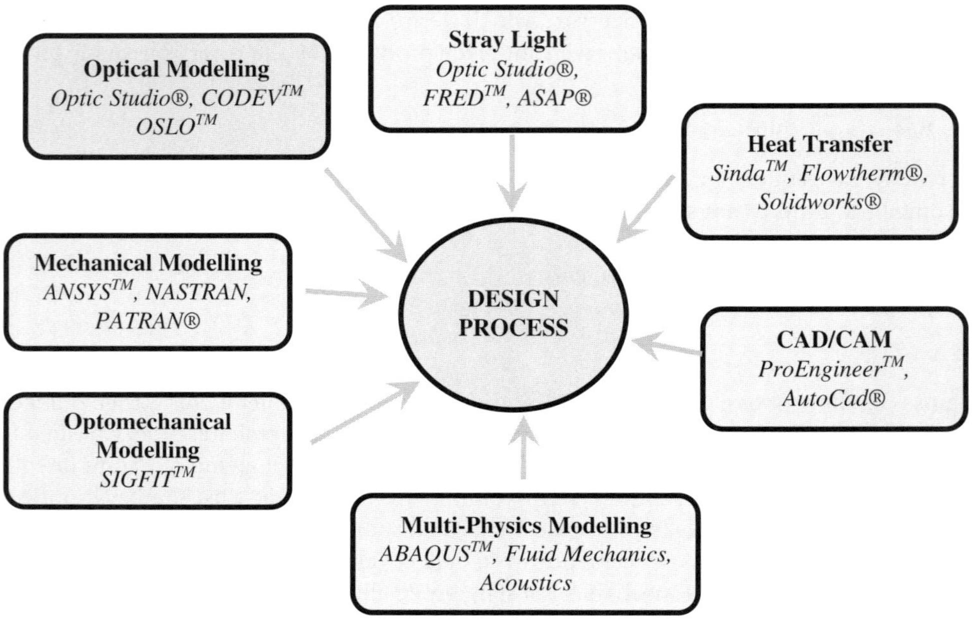

Figure 18.4 Design tools.

role in the development of an optical system. As indicated earlier, optical modelling encompasses the modelling of straylight and illumination through non-sequential modelling as well as conventional optical design. Basic computer-aided design (CAD) tools enable the detailed design of optical mounts and the relative placement of optical components according to the prescription of the optical model. Indeed, the optical models are designed to allow the export of optical surfaces into a format that can be accessed and manipulated by most CAD modelling tools. Support structures, optical benches or breadboards, as well as light-tight enclosures may be configured. Furthermore, just as it is possible to export data from the optical model to the CAD model, the reverse scenario applies. All the mounts and enclosures from the CAD model may be imported into the non-sequential optical model. Since all these surfaces have the potential to contribute to the scattering of straylight, they are modelled as part of the straylight analysis.

Figure 18.4 clearly illustrates how many disciplines contribute to the design of an optical system, beyond the purely optical. As well as basic mechanical modelling, there are a range of different tools that model the impact of the environment upon the system. This is especially critical where the system is to be deployed in an aggressive environment, such as in 'outside plant' or in aerospace and defence applications. For example, the system may be subject to mechanical loads (static or inertial forces) or thermal loads, e.g. deep temperature cycling due to solar loading. All these will bring about mechanical or thermomechanical distortion in functional optical surfaces, or in 'optical bench' type surfaces that support and mechanically integrate the system.

The former will produce image degradation by directly impacting the system wavefront error. The latter will cause misalignment, perhaps also contributing to image degradation. In terms of the application of software tools, it is important to understand that these tools should be used in combination. Any simulation of mechanical or thermomechanical distortion under load cannot be viewed in isolation but must be fed back into the opto-mechanical and optical models.

As with the optical modelling, thermo-mechanical modelling benefits from an understanding of the underlying physics and engineering. Software modelling of thermo-mechanical effects captures the details of a complex system. However, a basic understanding of mechanical distortion and flexure under both mechanical and thermal loads is highly desirable before embarking on the detailed design.

18.3 Optical Design Tools

18.3.1 Introduction

In this section, we will outline in some detail the operation of optical modelling software. Although we have emphasised the importance of working with other software tools, especially those that analyse mechanical and thermal aspects of the design, the focus here will be on optical modelling. As illustrated in Figure 18.4, there are a number of commercial tools available. However, to illustrate how the modelling process works we will follow the evaluation of a very simple design using one specific modelling tool, namely Optic Studio® from the Zemax corporation. This is widely used in the industry and has a wide range of capabilities covering both sequential and non-sequential optics as well as physical optics (diffraction) modelling. The description that follows is intended to provide the reader with a general picture of the use of such powerful tools within the optical design profession. This is important and significant in itself, as the topic is generally neglected. However, it is not intended as a 'training manual' in the use of such software. Tailored courses are available within the industry to help the budding designer to start. However, no training course can ever match that which is gained by day-to-day regular use of the software in practice. Only by such experience can one's initial faltering efforts be transmuted into substantial expertise.

18.3.2 Establishing the Model

18.3.2.1 Lens Data Editor

Our task will be the design and optimisation of a simple achromatic doublet. In this instance, we are working in the sequential mode. First of all, we must establish the prescription for the system. As outlined earlier, the system description is characterised on a surface by surface basis, rather than by the definition of components as single entities. As such, a number of surfaces is laid out in the spreadsheet row by row, with the first row occupied by the object plane and the final row by the image plane. However, it is not possible to start the design process with a 'blank sheet'. Some initial working prescription must be established before the software can get to work. In the case of the achromatic doublet the starting point is fairly clear. In preceding chapters of the book we had examined the design of an achromatic doublet from a thin lens perspective. This forms a useful starting point to populate the model prior to computer optimisation. The same principle applies generally to more complex designs. As such, the efficient use of software tools is predicated upon a deep understanding of the underlying principles.

The prescription information is contained in a spreadsheet, the 'Lens Data Editor', which describes the characteristics of each surface, including shape, thickness, and material composition. Each surface is allocated a row in the spreadsheet and particulars of that surface are entered in the spreadsheet columns. As stated, each surface is assigned a thickness and a material composition. The thickness ascribed is the distance from that surface to the next surface. Similarly, the material designation refers to the material between that surface and the next. The designer can chose from a wide range of materials from an extensive library, covering all commercially available glasses, optical polymers and exotic materials. This library contains details of refractive index, dispersion, and thermal properties over the useful wavelength range for the material. If no material description is entered, Optic Studio will assume a 'default' medium, usually air or vacuum.

A wide variety of surface shapes may be specified, too many to describe fully here. The most common is the 'standard' surface which allows the definition of a spherical or conic shape, as defined by its radius and conic constant. Geometrically, a sequential model has a well-defined optical axis, progressing in the direction of the incremental surface thicknesses. This axis is recognised as the local surface z axis. The shape is so defined that its vertex lies at the local origin (x = 0, y = 0, z = 0). Optic Studio defines the surface form in terms of the local sag, Δz. For instance in the case of a 'standard surface' the sag is given by:

$$\Delta z = \frac{cr^2}{1 + \sqrt{1 - (1+k)c^2r^2}} \quad r^2 = x^2 + y^2 \tag{18.1}$$

c is the curvature (1/radius) and k the conic constant

In the case of the achromatic doublet, all surfaces are spherical, so the standard surface type can be used. More generally, the standard surface type is by far the most common surface type in general use. Most, but not all, surface types are symmetrical about the central axis. In Chapter 5, we introduced the more complex even aspheric surface which is a logical extension of the standard surface to cover even polynomial terms in the radial offset. The surface sag is given by:

$$\Delta z = \frac{cr^2}{1 + \sqrt{1 - (1+k)c^2r^2}} + a_2r^2 + a_4r^4 + a_6r^6 + a_8r^8 + a_{10}r^{10} + \ldots\ldots \tag{18.2}$$

Not all surface types are radially symmetric. These include the 'Zernike Standard Sag' where the surface is defined by Zernike polynomials and the biconic surface which is effectively a standard surface with separate prescriptions along the x and y axis. In Optic Studio, over 70 different surface types are listed. A selection of some of the more commonly used surfaces types is set out in Table 18.3.

The ***co-ordinate break*** surface is worthy of comment. In fact, it is not an optical surface, as such, and has no tangible impact upon ray propagation. It provides a means of tilting or offsetting optical surfaces when building up an off-axis system. All optical surfaces are placed with their vertex at the origin and oriented with the surface normal to the ***local*** z axis. The co-ordinate break surface merely effects a rotational and translational transformation of this ***local co-ordinate system*** with respect to the overall ***global co-ordinate system***. As such, the co-ordinate break surface describes a transformation of the system co-ordinate frame, by specifying rotations about three axes and lateral translation in two directions (x and y).

We now return to a simple achromatic design which featured as a worked example from Chapter 4. We were originally tasked to design a 200 mm focal length achromatic lens using N-BK7 as the 'crown lens' and SF2 as the 'flint lens'. The lens is designed to operate at the infinite conjugate with the object located at infinity. This design was originally analysed according to the thin lens approximation as a Fraunhofer doublet, whereby, as well as ensuring a common focus for two separate wavelengths, both spherical aberration and coma had been eliminated at the central wavelength. The radius values are as listed below, with R_1 and R_2 being the radii for the first N-BK7 lens and R_3 and R_4 the radii for the second SF2 lens.

Solution 1: $R_1 = 121.25$ mm; $R_2 = -81.78$ mm; $R_3 = -81.29$ mm; $R_4 = -281.88$ mm

In the lens data editor we must enter six surfaces. The first surface is the object surface, labelled 'surface 0', followed by the four lens surfaces and finally the image surface. Following on from our previous discussion, all these simple surfaces are captured by the 'standard surface'. To describe each surface, we need only enter its radius of curvature, its thickness and the material used. The first surface is the object surface, whose radius, as a planar surface is assumed to be infinite. For the infinite conjugate, the thickness is obviously also infinite and the material (between the object and first surface) is air or vacuum, so the relevant column is blank.

For the next four lens surfaces, (1–4) we are to enter the radii, as given above, the thickness and the radii. In the initial analysis, we had paid no heed to the lens thickness, as per the thin lens approximation. In some

Table 18.3 Some common surface types.

Surface type	Symmetric	Parameters	Comment
Standard	Yes	Curvature and conic parameter	Most common surface. Models spherical surfaces
Even asphere	Yes	Curvature, conic parameter, and polynomial terms	Used in more elaborate designs. Trade ease of optimisation against increased cost
Biconic	No	Curvature and conic parameter for both x and y axes	Astigmatic surface used occasionally
Toroidal	No	Curvature an conic parameter for y axis; rotation radius R	Similar to Biconic, but defined as conic line in YZ plane which is rotated about centre of curvature in X.
Standard Zernike Sag	No	All even asphere terms plus (rms) Zernike coefficients for up to 232 terms (Noll Convention)	Essentially freeform surface with significant manufacturing difficulty. Useful in off-axis, complex high value designs
Toroid	No		
Co-ordinate break	N/A	Offsets in x and y direction; tilts along x, y, and z axes.	Not really an optical surface. This brings about a transformation in the optical co-ordinate system. Useful for systems with tilts and off-axis components
GRIN lens	Yes	Base refractive index, n_0, and n_{r2} value.	There are a number of different GRIN type surfaces with different parameters to be entered.
Diffraction grating	No	Grating lines per mm. Diffraction order.	Grating lines are parallel to local X axis
Paraxial lens	Yes	Lens focal length	Allows the substitution of a 'perfect lens' Useful in the evaluation and analysis of a design.

more complicated designs, the thicknesses of the glass lens elements are critical parameters in the overall optimisation process. In this instance, this is not the case, and thicknesses are governed solely by mechanical considerations. It must be remembered, since all lens surfaces are defined with their vertex at the local origin, then, in the absence of co-ordinate transformation, the thickness represented in the editor is the ***central thickness***. As a useful rule of thumb, the central thickness of each lens should be at least one tenth of the ***physical*** element diameter ***and*** the edge thickness greater than one twentieth of the element diameter. The lens physical diameter is usually 10–15% larger than the ***clear aperture***, the circle through which all rays will pass. Element sizes are determined by the pupil size and location and by the field. We will consider these definitions in the next section. In the meantime, we are obliged, in the lens data editor, to select the location of the stop or entrance pupil. In this case, the stop is to be placed at the first lens, at surface 1.

In the meantime, we must also define the material columns for the four lens surfaces. Since the material description is applied to the material ***following*** the surface in question, surface 1 is labelled as 'N-BK7' and surface 3 is labelled as 'SF6'. The other surfaces (2 and 4) are left blank, representing air or vacuum. The thickness of surfaces between glass elements must allow a reasonable physical air gap between the glass surfaces. A gap of at least 0.5 mm should be left at the centre, with the air gap at the edge allowing insertion of a physical spacer (e.g. ring). As such, the air gap at the edge might be at least 1.5 mm. For surface 4, the thickness is the gap between the final lens and the image. In this thin lens approximation, of course, this thickness is the focal length of 200 mm. However, it is clear that this value will be modified by the finite lens thickness. Nevertheless, in the meantime, a thickness of 200 mm will be ascribed to the surface, for subsequent adjustment and optimisation by the software.

Table 18.4 Lens data editor spreadsheet.

Surface	Type	Comment	Radius	Thickness	Material	Semi-Dia.	Conic
0	Standard	Object	Infinity	Infinity			0
1	Standard	Lens 1[a)]	121.25 V	9.0	N-BK7	22.54	0
2	Standard		−81.78 V	1.5		22.21	0
3	Standard	Lens 2	−81.29 V	5.0	SF2	21.83	0
4	Standard		−281.88 V	200.0 V		21.63	0
5	Standard	Image	Infinity	—		4.97	0

a) Surface 1 is denoted as the stop or entrance pupil.

Finally, the last surface, labelled number 5, is the image surface. Under the assumption that the image plane is flat, as in the case of a standard pixelated detector or photographic film, and no special provision has been made to accommodate field curvature, then this surface should have an infinite radius of curvature; it has no thickness. The material is, of course, air or vacuum.

Table 18.4 shows a substantially edited version of the Optic Studio lens data editor. Only relevant data columns have been included; columns relevant to surface types other than the standard surface have been omitted. Eight main data columns are shown, covering the surface number, surface type, descriptive comment, radius, thickness, material type, semi-diameter, and conic constant. In all columns, with the exception of the surface number and semi-diameter, the user is expected to enter initial values. The semi-diameter represents the effective semi-diameter of the optic. The user may enter a value which forces the program to ascribe a physical aperture; otherwise, by default the program shows the ***clear aperture*** based upon ray path calculation. In fact, this is an important distinction. The default clear aperture is the portion of the surface's area actually illuminated by the overall field. This is the aperture that would just admit all rays without vignetting. However, as we shall see when we come to consider component manufacturing in Chapter 20, the ***physical aperture*** is invariably larger than the clear aperture. In fact, both the clear aperture and physical aperture may be tabulated in the Lens Data Editor. It is over the clear aperture that the ***optical*** requirements of the surface hold. Specifying a larger physical aperture generates additional 'real estate' to facilitate mounting in the manufacturing process and in the assembly itself. Furthermore, as will be seen when we consider the lens manufacturing process, the grinding and polishing procedure is less reliable close to the edge of a surface. It is therefore inevitable, in any case, that any figuring errors are accentuated close to the physical edge of the lens. As a rule of thumb, the clear aperture tends to be about 85–90% of the physical aperture.

It will be noted that there is a sub-column adjacent to the radius and thickness column. Depending upon the entry in that sub-column, the program is permitted to adjust the adjacent parameters; otherwise they are fixed. In the case of the four lens radii and one of the thicknesses, a 'V' has been entered into the relevant sub-column. This entry denotes that the parameter is used as a variable in the subsequent computer optimisation process. That is to say, the software is permitted to adjust these parameters, and only these parameters, as it attempts to optimise the system performance. All values presented in Table 18.4 represent the prescription ***prior to the optimisation process***, which will be described later.

All values listed in Table 18.4 are given in the 'standard lens units' set in the software which, in this case, is millimetres. The real lens data editor would embrace significantly more columns than in Figure 18.2; what is shown is the relevant subset. As a working value, the physical aperture has been set to 50 mm and this has been used to set reasonable values for the lens thickness, as previously described. All four lens radii are selected as variable parameters under software control during the optimisation process. In addition, the final thickness is also selected as a variable parameter, as the finite lens thicknesses will have moved the focal point by a few millimetres.

18.3.2.2 System Parameters

The lens data editor unambiguously defines the optical prescription. However, we need to define the many and varied system parameters that define the interfaces and the system environment. As a minimum, the following parameters must be defined:

- Aperture size (location defined in lens data editor)
- Wavelengths (up to 12 wavelengths may be defined)
- Field locations (up to 12 field points defined)

In addition to the above it is also possible to delineate the environment – the ambient temperature and pressure. This takes into account not only the temperature coefficient of refractive index for the glass materials but also the refractive properties of the atmosphere which are, of course, dependent upon both temperature and pressure. In addition, there is also provision for describing the polarisation state of the input radiation. This is useful in specific applications. In the current example, polarisation is not accounted for and polarisation is not explicitly analysed.

The aperture size may be defined in a number of different ways. Firstly, and most obviously, the diameter of the actual physical aperture may be set. This is referred to as ***'float by stop size'***. It must be emphasised, that in selecting this option, this does not represent the entrance pupil diameter. Most usually, it is the ***entrance pupil diameter*** that is quantified and, to re-iterate, this is the size of the physical stop as imaged in object space, not the physical size of the stop. Otherwise, the pupil size may be defined through the object space numerical aperture or the image space f# number. In our example, it is the entrance pupil diameter that is set at 45 mm.

Up to 12 wavelengths may be defined. For this simple example, there is provision for the three standard visible wavelengths, F (486 nm), d (588 nm) and C (653 nm), to be included. One wavelength is defined as the primary wavelength. In our example, this wavelength is 588 nm.

As with the wavelengths, up to 12 individual fields may be introduced. These are described by their angular or positional co-ordinates in two dimensions (x and y). One cannot, of course, define the field in terms of positional co-ordinates if the object is located at infinity, as it is in this case. In terms of analysing the performance of a rotationally symmetric system, there is no necessity to define fields along more than one axis. That is to say, performance of a field point displaced by 1° in x will be identical to that displaced by 1° in y. For our simple exercise, we have defined five field points by angle: one central field point (zero displacement), two field points displaced by ±0.7° in x and two field points displaced by ±1.0° in x.

To summarise our system:

Aperture (entrance pupil size): 45 mm *located at first lens surface*

Wavelengths (3): 486 nm, 588 nm (principal), 653 nm.

Field Points (5): {0°, 0°}, {1°, 0°}, {−1°, 0°}, {0.7°, 0°}, {−0.7°, 0°}.

The information provided so far is sufficient for the software to calculate the path of an arbitrary ray through the system. To describe an individual ray unambiguously, we must know its field position, normalised pupil co-ordinates and the wavelength number of the ray. In Optic Studio, the field position is defined in a manner analogous to that of the pupil position. That is to say, the input field co-ordinates are 'normalised' by dividing the real field co-ordinate by the maximum field value. Thus the description of a ray, R, may be generalised as:

$\mathbf{R} = \{\mathbf{H_x}, \mathbf{H_y}, \mathbf{P_x}, \mathbf{P_y}, \mathbf{n_\lambda}\}$; *$H_x$ & H_y are normalised field co-ordinates; P_x & P_y are pupil co-ordinates*

Thereafter, each ray may be traced through all surfaces defined in the lens editor using Snell's law etc. Such calculations can be performed exceptionally rapidly and the results used to analyse the system by computing, for example, average distortion at the image, or wavefront error, etc.

18.3.2.3 Co-ordinates

All Cartesian co-ordinates are referenced to the ***global co-ordinate system***. The global co-ordinate system is the same as the local co-ordinate system for a specific surface which may be selected as a system parameter. The surface sag data for each surface is computed in the local co-ordinate system, with the surface vertex located at the origin and the z axis describing the local optical axis and the nominal direction of ray propagation. In the simple example presented here, there are no co-ordinate transformations, so all six surfaces share the same co-ordinate system. As outlined earlier, co-ordinate transformations are effected by introducing the co-ordinate break surface. For example, one might wish to introduce an off-axis parabola into a system with a 50 mm offset in x. Immediately before the parabola, a co-ordinate break surface is introduced with a 50 mm displacement in x. This places the parabola at an offset of 50 mm with respect to the previous surface. Without such an offset, then the parabola would always be placed with its vertex on axis.

18.3.2.4 Merit Function Editor

Definition of the merit function lies at the heart of the computer optimisation process. As indicated previously, the merit function encapsulates the fidelity of a design within a single number. A low merit function value is synonymous with a high-quality design and the merit function effectively quantifies the extent to which the performance is in conflict with the requirements.

The merit function comprises a list, often very large, of individual operands each weighted according to their importance, as perceived by the designer. These individual operands (there may be hundreds) are then summed according to a root sum square process. For complex systems, there is quite an art in defining a good merit function from the plethora of different operands available. In the simple example highlighted, however, the definition of the merit function is relatively straightforward. It is possible to define a ***'default merit function'***. This reflects the fact that for most (sequential) optical systems, the primary concern is image quality. Therefore, the merit function can be established on the basis of quantifying either the wavefront error or spot size. As established in previous chapters, choice is determined by whether the system is to operate at or close to the diffraction limited regime. Wavefront error is most appropriate for systems reasonably close to diffraction limited performance; otherwise spot size should be selected for the basis of the merit function.

In this case, we wish to design on the basis of wavefront error, as we believe that the ultimate performance of the doublet over the relatively small field angle should be close to diffraction limited. The default merit function, in this instance, consists of a weighted series of operands enumerating the OPD across all wavelengths and fields. In fact, the OPD is calculated on a ray by ray basis, and the wavefront error for each wavelength and field is itself represented by a series of operands presenting the OPD at a number of representative pupil locations. The more pupil locations are represented, the more accurate the wavefront error computation; however, computational speed is reduced. In fact, the pupil is represented by a mesh of points in normalised pupil co-ordinate space whose frequency is defined radially by the number of rings and azimuthally by the number of arms. As such, the default merit function describing OPD contains a substantial number of individual operands.

In our example, in order to produce a more compact merit function for the reader to view, a slightly different approach has been used to describe the wavefront error. One of the very many operand types is the Zernike wavefront operand. Although the underlying computation is the same as for the OPD, the OPD for a specific field and wavelength is fitted to a Zernike polynomial series across the pupil. Furthermore, this approach provides a clearer description of the optimisation process in terms of minimising the underlying third order aberrations. To ensure correction of chromatic aberration, the defocus (Zernike 4) term for the first and third wavelengths (483 and 653 nm) are both included in the merit function and applied to the central field point. Whilst this has the potential to reduce the defocus of the outlying wavelengths to zero, it will not truly optimised defocus for all wavelengths. Therefore, defocus of the central wavelength (588 nm) is also added. In addition, for the optimised, air-spaced doublet, we must minimise the spherical aberration and coma. Spherical aberration is expressed through the appropriate (Zernike 11) term and, since this is not an off-axis aberration, it is also applied to the central field. Coma is expressed through the Zernike 7 polynomial and, as an off-axis aberration, is applied to one of the off-axis fields. Both the spherical aberration and coma operands are applied using the central (588 nm) wavelength. This relatively simple and compact merit function is illustrated in Table 18.5, representing the value of all operands, prior to the optimisation process.

Table 18.5 Merit function.

#	Type	Comment	Surf 1	Surf 2		Target	Weight	Value	Contrib.
1	MNCG	Min centre thick glass	1	4		5.0	0.0001	5.0	0%
2	MNEG	Min edge thick glass	1	4		2.5	0.0001	2.5	0%
3	MNEA	Min edge thick air	1	4		1.5	0.0001	1.5	0%
4	MNCA	Min centre thick air	1	4		0.5	0.0001	0.5	0%
#	Type	Comment		λ#		Target	Weight	Value	Contrib.
5	EFFL	Focal Length		2		200.0	1	200.0	0.0002%
#	Type	Comment	Zern.#	λ#	Field#	Target	Weight	Value	Contrib.
6	ZERN	Defocus	4	1	1	0	1	−0.042	3.01%
7	ZERN	Defocus	4	2	1	0	1	0.186	58.51%
8	ZERN	Defocus	4	3	1	0	1	−0.15	38.48%
9	ZERN	Coma	8	2	2	0	1	−0.0007	0.0007%
10	ZERN	Spherab	11	2	1	0	1	−0.0004	0.0003%

The relevant Zernike functions are at the bottom of the table. All entries contain a column for the desired or target value. For all the Zernike functions representing the defocus and aberrations, we desire these to be as close to zero as possible, so these targets are set to zero. The weight column is an entry whereby we can express the importance of a specific operand. The higher the weighting, then the greater importance we attach to minimising the contribution from that particular parameter. The value column refers to the actual value of the operand as computed by the software. Finally, the contribution column expresses the proportional contribution of that operand to the final merit function.

The fifth operand, EFFL, refers to the system focal length which is targeted to be 200 mm. If this entry were omitted, the software would seek to minimise the wavefront error alone by reducing the numerical aperture to a minimum. For a fixed aperture (45 mm), this would amount to increasing the effective focal length towards infinity. The first four operands do not relate directly to optical performance but control the mechanical thickness of the lens elements. The first two operands control the minimum centre and edge thicknesses of the glass elements which, as previously advised, should be set to 5.0 and 2.5 mm respectively. Minimum centre and edge thicknesses for the air gaps are set to 0.5 and 1.5 mm respectively. For all these operands, provided the minimum criterion is not breached, the contribution for the operand is set to zero.

All the operands in Table 18.5 are summed by a RSS procedure to give a single figure that is used to drive the optimisation process. Although Table 18.5 provides a basic insight into the compilation of a merit function, its purpose is largely to illustrate the process. Most particularly, Optic Studio has a very large and diverse array of different operands that may be used to compile a merit function to optimise complex optical systems with many conflicting requirements.

18.3.3 Analysis

Before we may proceed to system optimisation, some appreciation of the software's analytical capabilities is desirable. All analysis is underpinned by the calculation of large numbers of discrete ray paths through the system. For example, the software generates a large number of discrete rays for a nominated field point, calculating the rms wavefront error for that field point from the OPD of those individual rays. The celerity and efficiency with which this process is accomplished is the hallmark of the modelling tool.

At the most basic, the software is able to compute and present all the critical paraxial parameters that we encountered in the opening chapters of this book. That is to say, the location of all six cardinal points may be presented together with the location and size of the entrance and exit pupils. In Optic Studio, these parameters are laid out in a text file referred to as ***'Prescription Data'***, together, for example, with information about co-ordinate transformations (relative to the global) that apply at each surface.

Other than that, the analytical tools automatically replicate and graphically display the detailed analysis of image quality, etc. that we have encountered throughout this text. Most straightforwardly, the calculation of ray paths may be used to generate a 2D or pseudo-3D diagram that includes both the ray paths themselves and an outline of the optical elements. For each field point, the distribution of rays may be selected by the user. They may be in the form of ray fans with a certain number of rays laid out in the XZ or YZ planes, or as a random grid of points across the entrance pupil.

A large number of the analytical tools help to quantify the system image quality, one of the most salient performance attributes. These, to a significant degree, mirror our analysis of image quality in Chapter 6. At the most elementary level, ignoring diffraction effects, the basic image quality for an individual field point is determined by the geometric spot diagram which is one of the analytical tools. As described in earlier chapters, inspection and interpretation of these diagrams may be used to establish the presence of key aberrations, such as spherical aberration and coma. These same data may be used to generate the transverse aberration fans that were originally introduced in our discussions concerning the third order aberrations. Of course, these transverse ray fans can be either presented for the tangential or sagittal plane. In addition to graphical display, the information may be presented in text format for further analysis.

By default, we consider this analysis as being applied to the image plane. It may equally be applied to any other surface. By applying the geometrical spot diagram for all representative system fields at each surface, we generate the ***footprint diagram***. As such, the footprint diagram delineates the total area that is illuminated by the entire field at each surface. This can be used to calculate the ***clear aperture***, for each surface, which is the aperture that will transmit all system rays without vignetting. It is customary that the ***physical aperture*** be made 10–15% larger than the clear aperture to accommodate component fixturing during the manufacturing process.

In line with analysis presented in earlier chapters, OPD fans may also be computed and presented graphically. The OPD, by accepted convention, is computed by tracing all rays to the image plane and thence to a reference sphere located at the exit pupil whose centre is located on axis at the paraxial image. As with the transverse aberration fans, both tangential and sagittal fans are displayed. Again, these may be used to identify prominent aberration types. Furthermore, OPD information may be presented in 2D form as wavefront maps, for example, illustrating in a false colour plot the OPD variation across a circular pupil. This two dimensional information may be further analysed by decomposing this wavefront error profile into constituent Zernike polynomials.

The MTF is another familiar image quality metric that is computed by Optic Studio. The MTF data is presented as a function of the spatial frequency input. Other computations provided include the calculation of encircled, ensquared, or enslitted energy as well as direct analysis of the principal Gauss-Seidel aberrations.

Although the software tool is based ultimately upon geometrical ray tracing calculations, it does have very significant capabilities in physical optics. In addition to the presentation of the geometrical spot distribution, it can also compute the Huygen's point spread function. Other aspects of diffraction analysis are also provided for, including (Gaussian) physical beam propagation and the analysis of fibre coupling, as presented in Chapter 13.

We have attempted to convey the plethora of analytical tools that are available to the model. However, in this instance, for our simple system, we shall simply illustrate this with a plot of the wavefront error versus field angle for all wavelengths and with a simple ray diagram. These are shown in Figures 18.5 and 18.6 respectively. The wavefront error plot is for the ***pre-optimised system***. As such, it shows the substantial defocus error caused by the addition of the finite lens thicknesses. Indeed, such is the extent of the dominance by simple defocus, there is no observable change in wavefront error with field angle.

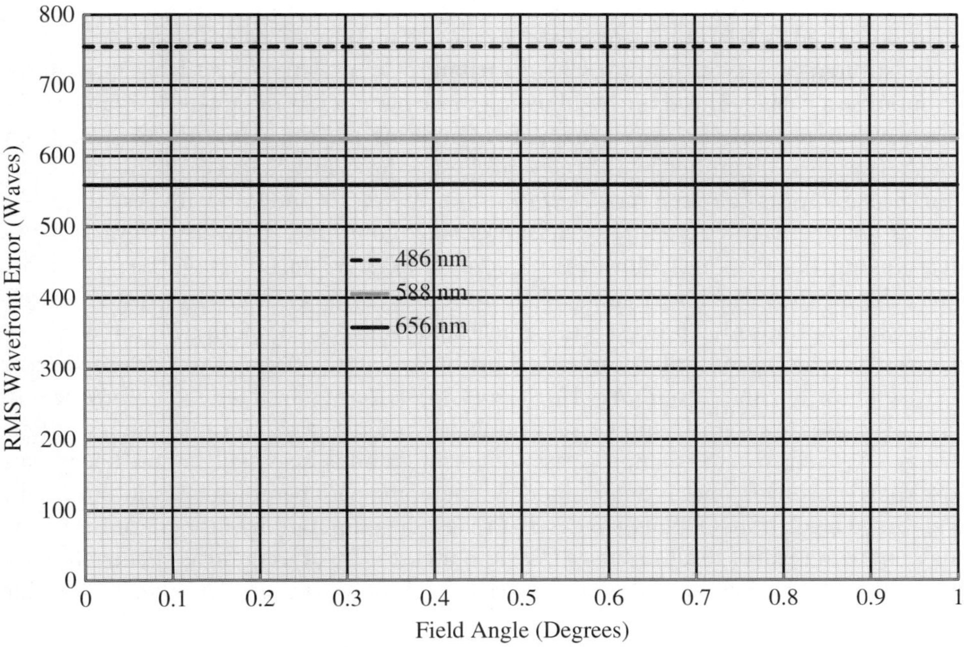

Figure 18.5 Doublet wavefront error vs field angle (before optimisation).

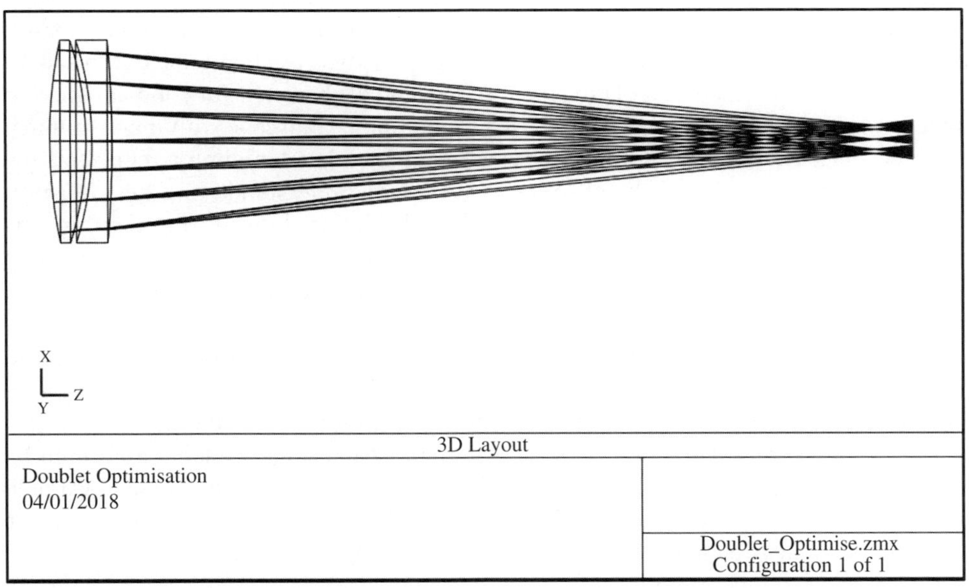

Figure 18.6 Doublet ray trace plot.

18.3.4 Optimisation

The analysis, as previously presented, is very much a passive operation. It merely describes the performance of the system, as currently constituted. However, the most salient attribute of a software tool, such as Optic Studio is its ability to refine or optimise a design. This done by adjusting those parameters designated variable in the lens data editor in such a way as to minimise the merit function. In our example, the merit function effectively describes the wavefront aberration as quantified by the relevant third order Gauss-Seidel contributions.

The basic optimisation process seeks to find a ***local minimum*** of the merit function with respect to the variable parameters. This is not necessarily an entirely trivial process. In our case, there are five variables to be optimised, four curvatures and one thickness. However, in more complex systems there will be many more variables to be adjusted. Each time the variables are adjusted, the (potential complex) merit function is entirely re-computed. As such, the whole optimisation process is extremely demanding on computing resources. Overall, there are two processes by which the local optimisation proceeds. Firstly, there is the ***damped least squares*** method, otherwise known as the ***Levenburg-Marquadt algorithm*** and secondly, the ***orthogonal descent algorithm***. Both processes are iterative processes and usually require a significant number of iterations in order to converge satisfactorily. The damped least squares method is essentially a non-linear least squares algorithm whose speed of convergence is determined by a damping parameter which is automatically selected in the computer-based algorithm. Orthogonal descent relies on computation of the merit function's 'direction' of steepest descent with respect to all variables (five in our case). This 'direction' is effectively a linear combination of all variables. Having 'set off' in this direction a merit function minimum along this specific path is reached. The process is then repeated on an iterative basis.

As a rule of thumb, the orthogonal descent method is most appropriate to initiate the optimisation process. Damped least squares is preferred ultimately to refine the optimisation. Both minimisation processes, as outlined, are exceptionally demanding on computational resources. Unfortunately, this is not the whole story. A combination of these two methods is an efficient way of identifying a ***local*** minimum in the merit function with respect to all variables. However, in a real system there may be a large number of minima, so there is no guarantee that the local minimum that has been identified is also the global minimum. This scenario may be understood by imagining the most simple situation where there are only two variables to be optimised. In this case, the merit function may be pictured as a 3D map, with the value of the merit function assigned to the vertical axis. As such, the merit function may be viewed as a topographic landscape with many minima. When viewed from a specific location, it is not instantly clear whether an individual minimum represents the global minimum. This problem may, to a degree, be offset by an analytical understanding of the system in question. In the case of our system, our trial solution was an analytical solution derived from the thin lens approximation. In this instance, therefore, we may be reasonably confident that our original trial was close to the final solution. Therefore we may state with some certainty that the iterative solution obtained represents the global minimum.

In more complex systems we cannot necessarily be so confident that we are close to the global minimum. As a consequence, Optic Studio provides two further optimisation tools specifically to search for the global minimum, ***Hammer Optimisation*** and ***Global Optimisation***. The search for the global minimum is a decidedly a non-trivial process. To understand the issues involved, we might search for the global minimum by initiating a local minimum optimisation starting from a gridded array of points within the possible variable parameter solution space. In our case, we have five variable parameters and we might assign to each parameter 10 possible starting values across some reasonable bound. For our five variables, this would correspond to 10^5 or 100 000 possible starting points for the optimisation process.

Hammer optimisation works by introducing and element of randomness to the optimisation process, in the hope of 'shaking' the current solution out of a shallow local minimum into a deeper hollow. If, as a metaphor one can imagine the merit function represented as a 3D landscape model and the current solution as a small marble or ball bearing, then shaking the entire model would have the tendency to drive the marble into the deepest depression. The success of this procedure is, to a significant degree, a matter of chance. However, its

Table 18.6 Optimised prescription.

Surface	Type	Comment	Radius	Thickness	Material	Semi-Dia.	Conic
0	Standard	Object	Infinity	Infinity			0
1	Standard	Lens 1[a)]	121.25 V	9.0	N-BK7	22.54	0
2	Standard		−81.78 V	1.5		22.21	0
3	Standard	Lens 2	−81.29 V	5.0	SF2	21.83	0
4	Standard		−281.88 V	200.0 V		21.63	0
5	Standard	Image	Infinity	–		4.97	0

a) Surface 1 is denoted as the stop or entrance pupil.

Table 18.7 Merit function following optimisation process.

#	Type	Comment	Surf 1	Surf 2		Target	Weight	Value	Contrib.
1	MNCG	Min centre thick glass	1	4		5.0	0.0001	5.0	0%
2	MNEG	Min edge thick glass	1	4		2.5	0.0001	2.5	0%
3	MNEA	Min edge thick air	1	4		1.5	0.0001	1.5	0%
4	MNCA	Min centre thick air	1	4		0.5	0.0001	0.5	0%
#	Type	Comment		λ#		Target	Weight	Value	Contrib.
5	EFFL	Focal Length		2		200.0	1	200.0	0.0002%
#	Type	Comment	Zern.#	λ#	Field#	Target	Weight	Value	Contrib.
6	ZERN	Defocus	4	1	1	0	1	−0.042	3.01%
7	ZERN	Defocus	4	2	1	0	1	0.186	58.51%
8	ZERN	Defocus	4	3	1	0	1	−0.15	38.48%
9	ZERN	Coma	8	2	2	0	1	−0.0007	0.0007%
10	ZERN	Spherab	11	2	1	0	1	−0.0004	0.0003%

virtue is that it is relatively rapid. By contrast, global optimisation is a more thorough process, searching for the global minimum in a more systematic way. By necessity, as previously outlined, this process proceeds by initiating local optimisation at a very large number of starting points across the solution space. As such, the global optimisation process is extremely time consuming even on powerful computing platforms.

In the event, our simple design does not require the application of global optimisation; only local optimisation is effected. Table 18.6 shows the revised system prescription. Only relatively small adjustments have been made to the four lens curvatures. The finite lens element thicknesses account for the significant difference in the back focal distance. Otherwise, our initial analysis produced a solution that is close to the final optimised design. Table 18.7 shows the tabulated merit function illustrating the reduction in the key aberrations.

As all the glass and air thickness requirements have not been breached, these make no contribution to the merit function. It is clear that both coma and spherical aberration have been reduced to a negligible value. The bulk of the merit function contributions arise from the three defocus terms. This is an expression of the non-zero secondary colour that is present in a doublet lens system. However, the merit function does not provide a complete picture of system performance. We have omitted both astigmatism and field curvature from the picture. This is because in a classical Fraunhofer doublet we do not have enough variables to control

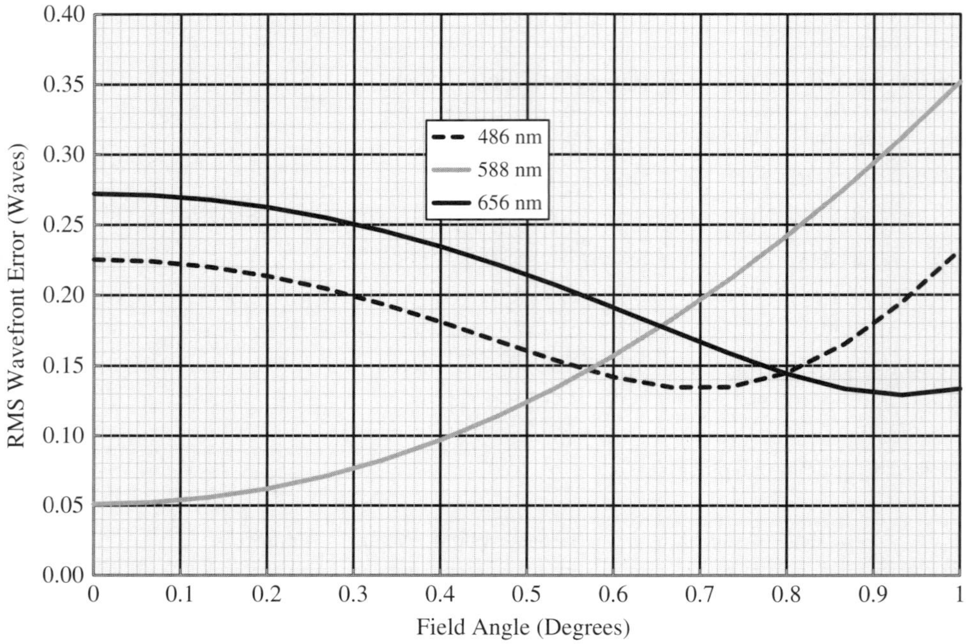

Figure 18.7 Doublet wavefront error vs field angle (after optimisation).

them. Therefore, at the edge of the field, we must accept the increased wavefront error that results from field curvature and astigmatism.

To summarise the system performance, as before, the wavefront error is traced as a function of field angle for all the wavelengths in Figure 18.7. Clearly, the performance has been substantially improved, with the wavefront error increasing with field angle, for reasons previously outlined.

18.3.5 Tolerancing

18.3.5.1 Background

We have now established a basic design with the detailed lens prescription established. However, we must convert this prescription into detailed manufacturing drawings for all the individual elements. As well as supplying the basic parameters, such as surface radii and element thicknesses, we must provide the manufacturer with a tolerance for each parameter specified on the drawing. For example, we have established a thickness of 9.0 mm for the first lens element and we might ascribe a tolerance of ±0.1 mm to this parameter. That is to say, a thickness of anywhere between 8.9 and 9.1 mm would be acceptable in this case. At first sight, the best strategy might be to restrict the tolerance to the very smallest possible value. However, the purpose of the tolerancing exercise is to ***optimise*** performance, **not** to ***maximise*** it. Unnecessarily tight tolerances will add cost and manufacturing difficulty (time) to the process. The overall objective of the tolerancing process is to establish the reasonable bounds of each parameter such that the performance requirements are (just) met.

Much of the approach we have described hitherto represents an extension of the classical design process, albeit effected with orders of magnitude greater speed and efficiency. However, there is no place in the traditional design process for the rigorous examination of tolerances. Historically, this aspect of the design process was covered by instinct gained through many years of practical experience. Inevitably, the lack of a rigorous approach was, to an extent, compensated through design conservatism, leading to a sub-optimal design in which performance and manufacturability were not adequately balanced.

18.3.5.2 Tolerance Editor

To initiate the tolerancing process, we must attempt to assert the bounds within which system parameters, such as lens thickness might reasonably be expected to lie. This must be applied to all potentially variable parameters within the lens data editor. This information is captured in another spreadsheet referred to as the tolerance editor. Each line within the tolerance editor is captured by a specific tolerance operand which quantifies the uncertainty in a parameter pertaining to a specific surface or group of surfaces.

One may split the tolerance operands into three broad categories. First, there is a group of parameters that describe the uncertainties in the material properties. For example, for optical glasses, the tolerance in the refractive index, Abbe number, and the impact of stress-induced birefringence should be estimated. Second, a large number of tolerance parameters relate to uncertainties in the manufacturing process. This might include shape (form) errors inherent in the fabrication of optical surface and errors in the element thickness. Finally, there are errors relating to the final assembly. In many respects, these mirror the manufacturing errors, with errors in spacing mapping onto errors in component thickness and tilt errors mapping onto the wedge errors of individual components. Table 18.8 shows a selective list of tolerancing operands in Optic Studio.

Each line in the tolerance data editor, as well as including the tolerance operand and other useful parameters, includes the range of values ascribed to the specific parameter. Naturally, the surface or range of surfaces to which the operand applies is also included. Since there may be many different operands relating to one individual surface, the tolerance data editor tends to be rather longer than the lens data editor.

For illustration, Table 18.9 shows a portion of the tolerance editor for our simple system. The first line of the spreadsheet introduces an operator not hitherto described. This is the so-called ***compensator*** operator, COMP which is not strictly a tolerancing operator, per se. As we will see a little later, the tolerancing process ascribes random values to each tolerancing parameter to simulate manufacturing and assembly imperfections.

Table 18.8 Selective list of tolerancing operands in Optic Studio.

Operand	Category	Comment
TIND	Material	Tolerance in Refractive Index
TABB	Material	Tolerance in Abbe Number
TRAD	Manufacturing	Tolerance in lens/mirror radius
TCUR	Manufacturing	Tolerance in lens/mirror curvature
TFRN	Manufacturing	Tolerance of surface curvature expressed in fringes
TCON	Manufacturing	Tolerance of conic constant
TIRR	Manufacturing	Simple model of surface form error (in fringes) allocating half to astigmatism and half to spherical aberration
TEZI	Manufacturing	More sophisticated model of surface form error, allowing the user to define it form in terms of (standard) Zernike polynomials
TSDX	Manufacturing	Tolerance of surface decentre in X
TSDY	Manufacturing	Tolerance of surface decentre in Y
TSTX	Manufacturing	Tolerance of surface tilt (wedge) in X
TSTY	Manufacturing	Tolerance of surface tilt (wedge) in Y
TTHI	Manuf. & Align.	Tolerance on thickness (glass) or spacing (air)
TEDX	Alignment	Tolerance on element decentre in X
TEDY	Alignment	Tolerance on element decentre in Y
TETX	Alignment	Tolerance on element tilt in X
TETY	Alignment	Tolerance on element tilt in Y
TETZ	Alignment	Tolerance on element rotation in Z

Table 18.9 Portion of tolerance editor.

Type	Surf. 1	Surf. 2	Nominal	Min	Max	Comment
COMP	4	0	189.01	−5	5	Focus Compensator
TWAV			0.6328			Test Wavelength (μm)
TRAD	1		111.632	−0.2	0.2	Radius tolerance
TRAD	2		−85.329	−0.2	0.2	Radius tolerance
TTHI	1	2	9.0	−0.2	0.2	Element thickness tolerance
TEDX	1	2	0	−0.2	0.2	Element decentre (X)
TEDY	1	2	0	−0.2	0.2	Element decentre (Y)
TETX	1	2	0	−0.2	0.2	Element tilt (X)
TETY	1	2	0	−0.2	0.2	Element tilt (Y)
TSTX	1		0	−0.2	0.2	Surface tilt (X)
TSTY	1		0	−0.2	0.2	Surface tilt (Y)
TIND	1		1.5168	−0.001	0.001	Index variation
TIND	3		1.6477	−0.001	0.001	Index variation
TABB	1		64.167	−0.642	0.642	Abbe number tolerance
TABB	3		33.848	−0.338	0.338	Abbe number tolerance

For an optimised system, this inevitably degrades system performance. However, for most optical systems there is some post assembly adjustment that can, to some degree, counteract these imperfections. For instance, a camera lens is designed to have some manual (or automatic) adjustment of focus. Thus any errors that lead to defocus may be compensated by adjustment of the relative location of the output focal plane. In this instance, the compensator operand permits one to move the focus by up to ±5 mm

18.3.5.3 Sensitivity Analysis

Having adequately described the individual tolerances in the tolerance editor, the first part of the tolerancing exercise is to understand the sensitivity of system performance with respect to small perturbations in system parameters, such as lens thickness. This process is referred to as ***sensitivity analysis***. We must, however, be able to codify system performance in a single quantity. This quantity could legitimately be the merit function used in the design process, or a proxy for it. For example, we might wish to exchange a relatively complex merit function for a simple metric of image quality based on solely on wavefront error or spot size. Most typically, the metric is based either on image quality, in the form of wavefront error, spot size, or on alignment in the form of ***boresight error***. Boresight error marks the deviation in the expected image centroid position produced by small errors in manufacturing and alignment. For the doublet optimisation we are using an average rms wavefront error aggregated across all fields and wavelengths as the figure of merit. In the subsequent analysis we are guided by the initial, unperturbed value of this figure of merit, known as the ***nominal value***.

Whichever metric is adopted, the purpose of the initial sensitivity analysis is simply to calculate the change in performance produced by a small perturbation in some system parameter, such as lens thickness. To initiate this ***sensitivity analysis***, we must provide an 'initial guess' as to the reasonable bounds that each parameter, such as thickness, might cover. This 'initial guess', of course, is based upon experience and we will return to this topic a little later. Ultimately, the sensitivity analysis simply calculates the effect on the figure of merit produced by this small perturbation, calculated as a deviation from the nominal value. Usually, the calculation is bipolar, so that if the nominal thickness of a component is 9.0 mm and the tolerance is ±0.2 mm, sensitivity calculations will be made for both 8.8 and 9.2 mm thickness values.

Table 18.10 Worst offenders in tolerance sensitivity analysis.

Type	Surface 1	Surface 2	Delta	FoM	Nominal	Change
TSDX	3		0.2	1.022	0.215	0.808
TSDY	3		−0.2	1.022	0.215	0.808
TSDY	3		0.2	1.021	0.215	0.806
TSDY	3		−0.2	1.021	0.215	0.806
TETX	3	4	0.2	0.949	0.215	0.735
TETX	3	4	−0.2	0.949	0.215	0.735
TETY	3	4	−0.2	0.949	0.215	0.734
TETY	3	4	0.2	0.949	0.215	0.734

The value of the sensitivity analysis is that it provides the designer with a comparison that identifies the most critical parameters affecting system performance. In Optic Studio, this is captured by setting out the five 'worst offenders', i.e. those tolerancing operands that have the largest negative impact upon performance. Table 18.10 shows the eight worst offenders for our simple system, using the default tolerance values. The default tolerances are, in this instance, somewhat loose. If the tolerance performance is inadequate, it is these tolerances we might have to tighten. This could be accommodated (in terms of cost and complexity) by relaxing other tolerances. To help guide us to the definition of reasonable and useful tolerances, an ***inverse sensitivity analysis*** can be performed. Here, the software seeks to elucidate the tolerance leading to a specific reduction in performance, rather than the other way round. Table 18.10 demonstrates that we are most concerned about surface decentres and element tilts, particularly those that relate to surfaces 3 and 4 (the diverging flint lens).

Before our understanding of the most critical tolerance can be translated into adjustments in the key tolerance parameters, we must conduct a full simulation of the impact of all the individual tolerances on the system performance as a whole. This more systematic system level modelling is a stochastic or Monte-Carlo simulation involving randomised perturbations of all tolerance operands based on their ascribed tolerances.

18.3.5.4 Monte-Carlo Simulation

In the Monte-Carlo simulation, each tolerance operand is randomly perturbed according to some favoured probability distribution. Most commonly, a Gaussian probability distribution is assumed and the maximum and minimum values ascribed to the tolerance operand are, in this instance, assumed to represent twice the standard deviation. That is to say, if the tolerance for an element thickness is set at ±0.1 mm, then this is modelled by a gaussian distribution whose mean is the nominal value and whose standard deviation is 0.05 mm. This is similar to the definition of extended uncertainty for type A uncertainty, where the extended uncertainty is defined as, for example, '2 sigma' or twice the standard deviation. Probability distributions other than the Gaussian are available to the user in Optic Studio.

To enact the Monte-Carlo simulation, a specific number of random trials must be selected, for example 200. This number should be sufficiently large to form a statistically significant conclusion about the probability distribution of the system performance metric. For each trial, every operand must be accorded a random value in line with the selected probability distribution and tolerance value. The random operand value is defined for each trial by generating a random number, r, with a value between 0 and 1. Having defined the mean (nominal value), μ, and the standard deviation, σ (from the tolerance), for each operand, the trial value, x, for the operand is generated by the standard error function erfc().

$$2r - 1 = erfc[(x - \mu)/(\sqrt{2}\sigma)] \tag{18.3}$$

For a specific trial, once all the random perturbations have been generated, then the figure of merit is calculated. This is then repeated for the requisite number of cycles. Subsequently a full statistical presentation

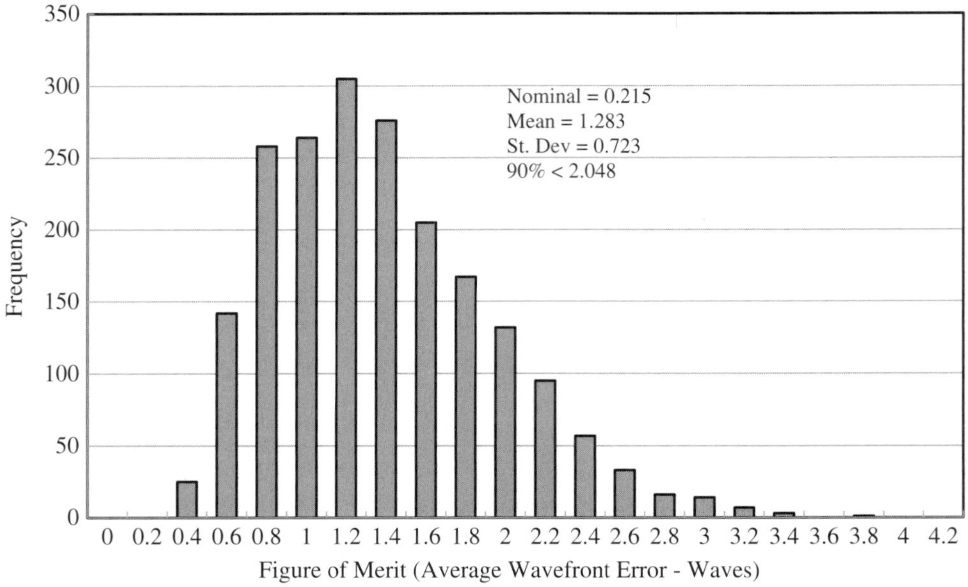

Figure 18.8 Monte Carlo simulation of tolerancing for simple doublet.

can be made of the random trials, providing the mean and standard deviation of the performance metric. Figure 18.8 shows a bar chart of the system performance (average wavefront error) for our system, following application of the basic default tolerance value. Also shown is the nominal (untoleranced) performance metric, revealing some degradation in average performance as a result of the tolerancing perturbations.

Of course, one needs to define a pass/fail criterion for the toleranced performance based on the statistical results. For example, one might be satisfied if the requirement for the average wavefront error lay within two standard deviations of the statistical average. Alternatively, one might require that the probability of satisfying the requirement is greater than some value, e.g. 90%. To illustrate how this might work in our simple case, we set the average wavefront error requirement to one wave rms. It is clear that both from the two sigma criterion and the 90% probability criterion that the tolerances need to be tightened in this instance. Therefore we must seek to refine our tolerance model to identify more appropriate tolerances.

18.3.5.5 Refining the Tolerancing Model

Our initial exercise provides a statistical analysis based on some initial default tolerances. The Monte-Carlo model previously described serves to appraise this initial choice of tolerances on a simple pass/fail basis. Clearly, if the results of the system model are unfavourable, then the tolerances must be tightened. However, if, conversely, the results are very favourable, the process does not end at that point. If the system requirements are met by some margin, then the tolerances prescribed are likely to be too tight and relaxing them will tend to reduce cost and manufacturing complexity without compromising system performance. The sensitivity analysis previously defined will help to select the most critical operands for influencing system performance. Where the tolerances need to be tightened to deliver the required performance, attention should be focused on the most sensitive parameters. Where one can afford to relax tolerances, attention should be turned to those that are most demanding in terms of cost and difficulty. As such, the tolerancing process is iterative, with a number of Monte-Carlo simulations being performed and relaxing or tightening the tolerances according to the results. The endpoint is reached when the statistical requirement is just met. This scheme is illustrated diagrammatically in Figure 18.9.

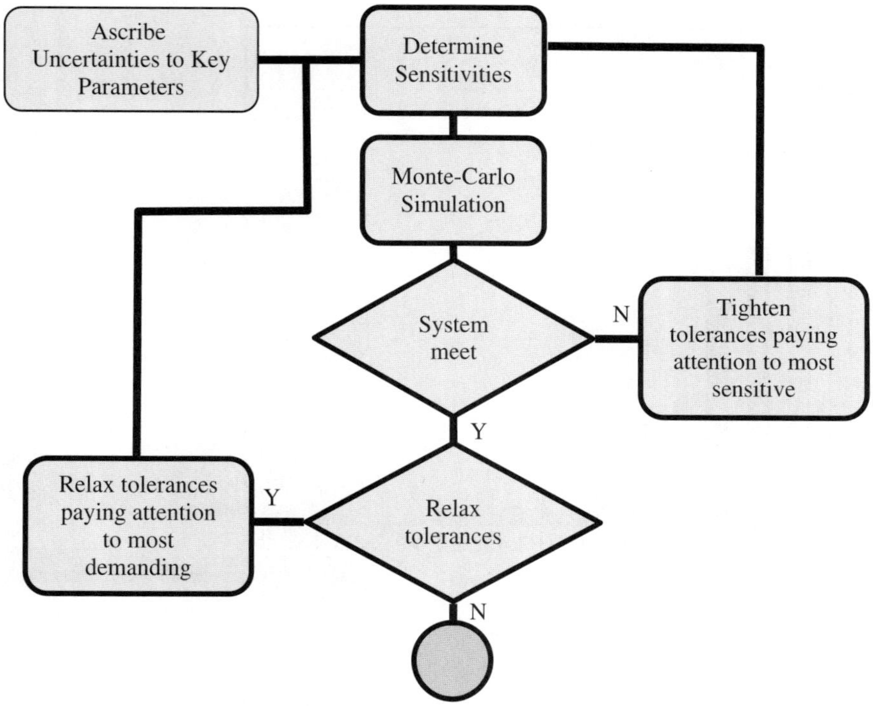

Figure 18.9 Tolerancing process.

In the simple example of our doublet, tolerances were tightened all round by a factor of two. For example, the thickness and decentre tolerances were changed from ±0.2 to ±0.1 mm. This is perhaps rather an oversimplified representation of the process, as, in practice, we would only be looking to tighten those tolerance operands that produce the greatest effect. Nonetheless, it provides a basic insight into the process. The results of this exercise, in the form of a histogram of the revised Monte-Carlo simulation is shown in Figure 18.10. This demonstrates that more than 90% of our trials produced an average wavefront error of less than one wave rms.

18.3.5.6 Default Tolerances

In initiating the tolerancing process, we need to define some initial tolerances. It is important that these initial selections and subsequent adjustments are somehow grounded in practical realities. To this end, it is useful to sketch out some benchmark tolerances for key parameters, such as lens thickness, as arbitrated by their (ascending) degree of difficulty – commercial, precision, and high precision. Furthermore, in refining tolerances as part of the final modelling exercise, it is imperative that we understand the burden that might be placed upon the manufacturing and alignment process as a result of tightening tolerances.

Table 18.11a sets out reasonable tolerances for material properties based on three standard grades expressing the manufacturing difficulty, namely commercial, precision and high precision. In terms of the discussion presented in this chapter, the variability in the refractive index and dispersion are the most relevant. Otherwise, the remaining parameters are discussed more fully in Chapter 9 which deals specifically with optical materials.

Similarly, Table 18.11b shows the equivalent tolerances for the manufacturing process. These are equivalent to the surface tolerances that we considered in the discussion of tolerance modelling. As such, these tolerances affect the relative relationship between surfaces in a single element, rather than the element tolerances, which affect the component as a whole.

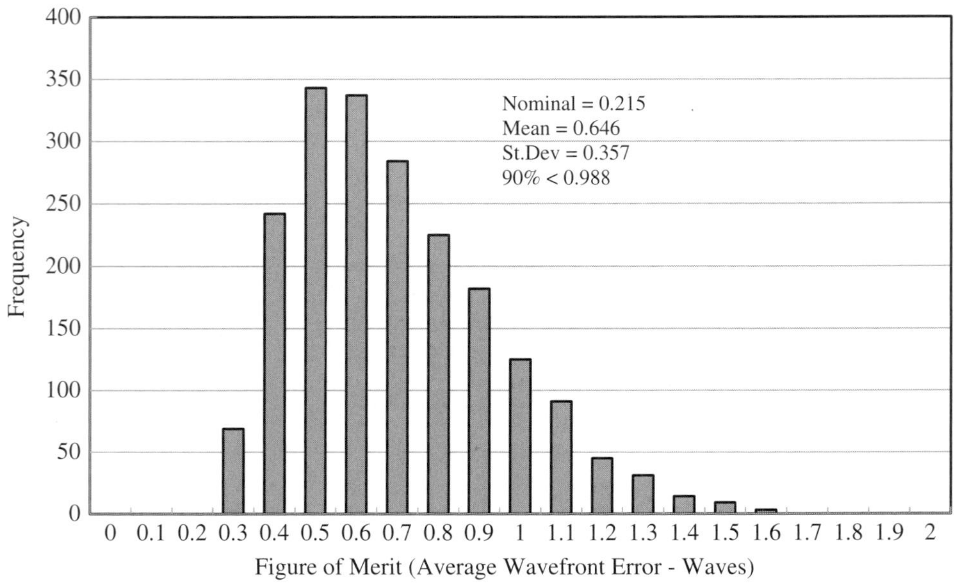

Figure 18.10 Revised Monte Carlo simulation of tolerancing for simple doublet.

Table 18.11 (a) Tolerances for material properties, (b) Tolerances for element manufacture, (c) Tolerances for alignment.

Parameter	Commercial	Precision	High precision
(a)			
Refractive index departure from nominal	±0.001	±0.0005	±0.0002
Dispersion departure from nominal	±0.8%	±0.5%	±0.2%
Index homogeneity	$\pm 1\times 10^{-4}$	$\pm 5\times 10^{-6}$	$\pm 1\times 10^{-6}$
Stress birefringence	$20\,\mathrm{nm\,cm^{-1}}$	$10\,\mathrm{nm\,cm^{-1}}$	$4\,\mathrm{nm\,cm^{-1}}$
Bubbles and inclusions (>50 μm) area per 100 $\mathrm{cm^3}$	0.5 $\mathrm{mm^2}$	0.1 $\mathrm{mm^2}$	0.03 $\mathrm{mm^2}$
Striae	Normal (fine striae)	Grade A (fine striae)	Precision (no detectable striae)
(b)			
Lens diameter	±100 μm	±25 μm	±6 μm
Lens thickness	±200 μm	±50 μm	±10 μm
Radius of curvature	±1%	±0.1%	±0.02%
Surface sag	±20 μm	±2 μm	±0.5 μm
Wedge	6 arcmin	1 arcmin	15 arcsec
Surface irregularity	λ (p. to v.)	λ/4 (p. to v.)	λ/20 (p. to v.)
Surface roughness[a)]	5 nm	2 nm	0.5 nm
Scratch/dig	80/50	60/40	20/10
Other dimensions (e.g. prism size)	±200 μm	±50 μm	±10 μm
Other angles (e.g. facet angles)	6 arcmin	1 arcmin	15 arcsec
(c)			
Element separation	±200 μm	±25 μm	±6 μm
Element decentre	±200 μm	±100 μm	±25 μm

a) Refers to polishing operations; equivalent diamond machining roughness would be ×5–10 higher.

Finally, Table 18.11c sets out the tolerances for alignment, equivalent to the lens element tolerances in the tolerancing model.

The surface irregularity of each surface has a clear and direct impact on the image quality. There is a transparent and proportional relationship between individual surface form error and system wavefront error. However, improved surface regularity comes at a cost (literally). As we will see, when we cover component manufacturing in a later chapter, moving from a surface irregularity specification of $\lambda \times 10$ to $\lambda/20$ entails a cost increase of over an order of magnitude. Broadly, the cost, which is a reflection of the manufacturing difficulty, is inversely proportional to the square root of the surface figure. More generically, moving from commercial to high precision increases costs by a factor of 2–3.

18.3.5.7 Registration and Mechanical Tolerances

At this point, we will say a little more about the derivation of mechanical tolerances in optical components and assemblies. The topic will be considered further in the chapters covering component manufacturing and mounting. Indeed, component and assembly tolerancing cannot be fully understood without an appreciation of the manufacturing and integration process.

Mechanical tolerancing of component systems is concerned specifically with the geometrical relationship with one (optical) surface with respect to some other surface. There may be one surface in particular that is the ***mechanical reference surface***, defining co-ordinate system of the optical assembly. When an optical surface is introduced into an optical mount or manufacturing jig that surface establishes a clear geometrical relationship with the corresponding mating surface. That process is referred to as ***registration***.

To provide an illustration of the geometrical subtleties of the tolerancing process we will consider the geometry of a simple lens. In common with a significant proportion of optical systems, it shares a nominally axially symmetric geometry. The impact of mechanical perturbations is to break that symmetry. Hitherto, in our more abstracted treatment of classical optics, a lens consists of just two surfaces, most normally spherical. The optical axis of such a system is uniquely defined by a line joining the two centres. Hence, such a component will always be aligned if that axis coincides with the optical axis of the system. However, as a solid object, this simple lens must have at least one other surface. Typically, this comes in the form of cylindrical surface formed in the grinding of the edge of the lens. Of critical importance in the manufacturing process, is the alignment of this mechanical surface with respect to the axis formed by the two optical surfaces. The two axes could be both tilted and offset. This is illustrated in Figure 18.11.

From Figure 18.11, it is clear that the lens may be both decentred and tilted with respect to its mechanical registration (the ground edges). The effect of the tilt, $\Delta\theta$, is to produce a global tilt of the lens about its nominal centre. Hence, this global tilt has no effect on the passage of the central field chief ray. It merely impacts off-axis aberrations with the on-axis field points acquiring off-axis type aberrations, such as coma. On the other hand,

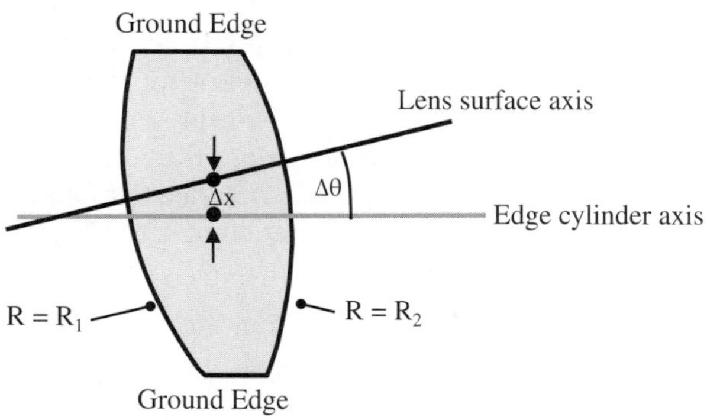

Figure 18.11 Mechanical tolerances in a simple spherical single lens.

the effective decentration, Δx, produced contributes to creating a wedge angle, $\Delta\phi$, in the lens. This wedge angle is given by:

$$\Delta\phi = \left[\frac{1}{R_1} - \frac{1}{R_2}\right]\Delta x \tag{18.4}$$

That is to say, for a biconvex lens with base radii of 100 mm, a 50 μm axial decentre will produce an effective wedge of 1 mrad or 3.4 arcminutes. Needless to say, both decentre and tilts must be considered for both the x and y axes. Of course, this analysis applies strictly to spherical surfaces. With a conical surface, each individual surface has its own unique axis of symmetry, whereas for spherical optics, two surfaces are required to define an axis of symmetry. This is why conical or aspheric surfaces have more degrees of freedom with respect to misalignment and are therefore more difficult to align.

This scenario marks out the manufacturing tolerance and, in the context of the tolerancing exercise, may be modelled by a surface tilt and decentre applied to a ***single surface***. Where a number of lenses are, for example, assembled in a lens tube, then mechanical and alignment errors introduce tilts and decentres in element with respect to the common tube axis. It is this process that is modelled by the element tolerancing operands.

This discussion illustrates the care that must be taken in modelling geometrical tolerances in a system. It is easy to be overzealous and to create too many operands. In the simple example of a singlet lens, we need just one set of surface operands and one set of element operands. Similarly, care must be taken where lenses are integrated into lens groups or other sub-assemblies within a system. If the alignment tolerance of a group of, say four, lenses within a system is to be modelled, then only three of the four lenses should be modelled as individual elements.

18.3.5.8 Sophisticated Modelling of Form Error

For the most basic analysis of component form error, the use of the TIRR feature provides a useful starting point. It treats form error as some random combination of astigmatism and spherical aberration. A more thorough treatment in Optic Studio is provided by the TEZI function. This adds a form error that is defined by a random summation of Zernike polynomial contributions between some minimum and maximum Zernike term number (Noll convention). It must be emphasised here that the contribution from each term is ascribed an equal weighting. In many respects, this is unrealistic and applies undue weighting to higher spatial frequency components of form error. All manufacturing processes tend to lead to the creation of surfaces whose form departure falls monotonically with spatial frequency. This is typically expressed as a ***Power Spectral Density, PSD*** curve, which is derived from a two-dimensional Fourier transform of the surface departure. It quantifies the power (i.e. the square of the modulus) of the form departure per unit spatial frequency interval.

The PSD of an optical surface is often modelled by a power law dependence:

$$PSD = A/f^{\beta}\; f \text{ is the spatial frequency} \tag{18.5}$$

Typically, the exponent β, taken on values between 2 and 3. Of course, the relationship set out in Eq. (18.5) is an ideal one. It is perhaps most descriptive of conventionally polished spherical or planar surfaces. Sub-aperture polishing of aspheric surfaces or diamond machining of optical surfaces create some anomalies at mid-spatial frequencies. Nonetheless, Eq. (18.5) forms a useful empirical basis for the more sophisticated tolerancing modelling of form error. It is possible to translate the Fourier representation of form error, as described in Eq. (18.5) by a Zernike-based format. In this representation, the form error contribution is dependent solely upon the (Born and Wolf) Zernike polynomial order, n. Indeed, it follows the same power law dependence as offered in Eq. (18.5), with the average rms contribution of each Zernike term represented as the square root of the PSD:

$$\sigma_{rms}(n) = A/n^{(\beta/2)} \tag{18.6}$$

Equation (18.5) may be implemented in OpticStudio® by designating a surface as 'Zernike Standard Sag'. In the tolerance data editor, each individual Zernike polynomial component may be ascribed its individual rms tolerance value. Values are ascribed, as per Eq. (18.6), ensuring that the RSS sum, as determined by the

Table 18.12 Cumulative contribution to form error by Zernike order.

Zernike order	Cumulative % of form error
2	84.1
3	86.8
4	93.6
5	93.9

coefficient A is equal to the desired rms form error tolerance. Equation (18.6) is a reasonable approximation that is adequate for tolerancing purposes. However, strictly speaking the overlap between the (Fourier derived) PSD and the Zernike representation is given by the following equation, as derived by Noll:

$$\sigma_{rms}^2(n) = A\frac{(\mathrm{n}+1)\Gamma(\mathrm{n}+1-\beta/2)}{\Gamma(\mathrm{n}+2+\beta/2)} \tag{18.7}$$

Data on the Zernike decomposition of real form error confirms the thesis that form error contribution declines with Zernike polynomial order. Table 18.12 shows the proportional of overall form error encompassed by Zernike terms up to a particular order. These data originate from a large number of diamond machined surfaces produced for the K-band multi-object spectrograph (KMOS) Integral Field Spectrometer deployed on the VLT (Very Large Telescope) facility at Paranal in Chile. It demonstrates that the majority of the form error is described by Zernike polynomials of up to the fourth order.

18.4 Non-Sequential Modelling

18.4.1 Introduction

As alluded to earlier, non-sequential modelling discards the assumption inherent to sequential modelling, that light progresses in a deterministic sequential fashion from one surface to the next. As such it is essentially a stochastic modelling tool. Any non-sequential model must start with the definition of source, as opposed to an object, as in sequential modelling. Such a source is defined by its radiometric properties which include both its spatial form (point source, uniform disc, etc.) and its angular characteristics (e.g. lambertian, etc.). Having defined the source in this way, the analysis then proceeds by the stochastic generation of a large number of individual rays. Each ray is uniquely described by its wavelength, point of origin and angle and is generated according to a probability distribution defined by the source characteristics. A large number of individual rays, often in excess of a million will be generated in an analysis. Another essential feature of non-sequential modelling is the substitution of a detector for the image in sequential modelling. Typically, in the model, a square pixelated detector is defined. Rays striking a particular 'pixel location' on the detector are recorded and this enables a map of the irradiance distribution over the detector area to be presented. Between the source(s) and detector(s) a number of surfaces will be described, both mechanical and optical.

From the source, an individual ray will be traced until it strikes a surface, which may be any of the surfaces listed in the model. When the ray strikes a surface, it will be treated in one of three possible ways. First, it may be either reflected or refracted. This process is deterministic and governed by the same rules pertaining to sequential modelling. Second, the ray may be absorbed by a surface, in which case further calculations for that ray are terminated. Often, but not always, ray tracing is terminated at a detector by describing it as absorbing. Finally, the ray may be scattered by at a surface according to some predefined distribution, e.g. Lambertian. As with the original generation of the ray, this process is stochastic.

18.4.2 Applications

Applications of non-sequential modelling fall into two broad categories. First, there is the simulation of illumination, as opposed to imaging systems. Here, the intent of the model is to simulate the primary function of the optical system. This might, for example, include the simulation of an automobile headlight system, optimising reflector geometry to produce uniform illumination. Alternatively, for example, the illumination stage of an optical microscope may be modelled, characterising a range of options incorporating ground glass screens or integrating spheres. Most often, the primary consideration is the delivery of uniform illumination at some plane or other surface.

The second area of interest is the modelling of straylight. Here we are not interested in the primary function of the optical system, but rather the characterisation of parasitic behaviour. Examples might include the modelling of scattering in a monochromator or spectrometer. In this type of application, particularly when dealing with weak optical signals, we are compelled to maximise the signal-to-noise ratio. The presence of background illumination not associated with the primary image both degrades image contrast and also enhances noise levels by adding background to the signal. This behaviour is troublesome in spectrometers where, for example, one is attempting to discriminate against a powerful source, such as a laser beam. This consideration also applies in imaging systems where scattering from powerful illumination sources (e.g. solar or lunar) may degrade image contrast. The design of surfaces to block or ***baffle*** straylight is an essential part of this aspect of the design process.

18.4.3 Establishing the Model

18.4.3.1 Background and Model Description

To describe the operation of the non-sequential model, a very simple application will be used to illustrate in the most basic way. Naturally, a full description of all the capabilities of the model falls outside the scope of this brief chapter. As before, we will illustrate the modelling process using the non-sequential modelling capability of Optic Studio. The system in question, in this instance, is an illumination system for an optical microscope. The illumination system consists of a point source located within an integrating sphere whose exit port forms the field stop and is conjugated with the object plane of the microscope. A condensing lens projects the exit port of the sphere onto the object plane. In addition, an aperture located before the condensing lens is used to define a f#1 system entrance pupil. The condensing lens then images the exit pupil at the correct location for the microscope objective. This is illustrated in Figure 18.12.

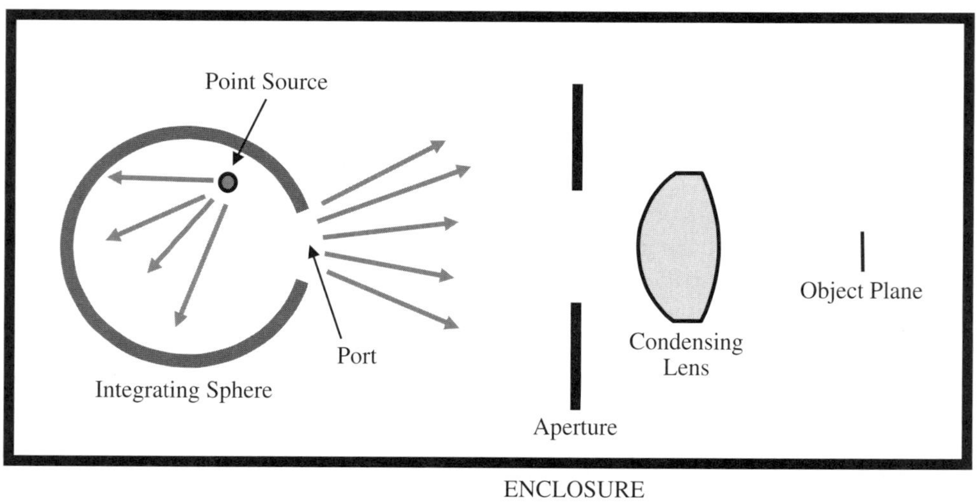

Figure 18.12 Model for microscope illumination system.

The integrating sphere has a diameter of 24 mm with an exit port of 10 mm diameter. An aspheric lens of approximately 9 mm focal lengths projects the exit port to produce an illumination pool approximately 0.5 mm in diameter at the input focal plane of the microscope. This gives a lateral magnification of 0.05. An 8.6 mm diameter stop placed about 30 mm from the lens defines the entrance pupil providing f#1 illumination with the pupil conjugated at the microscope objective aperture some 4 mm from the input focal plane.

In evaluating the design, we are interested in the uniformity of the illumination at the microscope objective plane. To illustrate the analysis of straylight, we will also evaluate the distribution of light at the object plane that lies outside the nominal area of illumination. To this end, the entrance pupil aperture is extended somewhat to act as a baffle. Furthermore, the whole assembly is enclosed in a 'light tight box'. In fact the 'light tight box' is to be modelled as a Lambertian scatterer with a hemispherical reflectance of 5%, equivalent to a generic black anodised or black coated opto-mechanical surface.

Of course, the model, as prescribed is very basic and illustrative. All components, especially the lens, will have to be mounted and attached to some common substrate. In practice, these component mounts would have to be modelled. Usually, these mounts would be designed to minimise scattering by using some proprietary black coating or, for example, making component mounts from black anodised aluminium. To model complex mechanical structures, the non-sequential model is able to import mechanically defined surfaces from CAD design files, e.g. .STEP or .IGES. Such complex surfaces are represented mathematically in the model as non-uniform rational B-spline (***NURBS***) surfaces.

18.4.3.2 Lens Data Editor

As with the sequential model, the essential system information is entered into the system in spreadsheet form in the lens data editor. Unlike in the sequential model, individual entries in the editor are described as ***objects*** rather than ***surfaces***. These objects, however, may be surfaces, but equally they may be 3D objects, such as lenses or solid objects such as cylinders. As such, the lens data editor provides a large number of different object types to place within the editor. The sequence in which these objects are placed within the editor is entirely unimportant, unlike for the sequential model. Broadly, there are four categories of objects. First, there are illumination sources, which are aimed at simulating a variety of sources, from point sources and LED sources. The lens data editor has provision for controlling the size and the flux distribution of these sources. In addition, the editor also provides a group of detector objects with various geometries, whose size and pixel resolution may be controlled in the editor. Thirdly, the editor provides a range of surface forms, including annuli, pipes, and discs. Finally, a number of 3D objects may be specified, including cylinders, lenses, and spheres.

Another distinctive feature of the non-sequential model is that the location of each object is defined by its absolute co-ordinate location with respect to some global co-ordinate system. This stands in contrast to the sequential model where the location of each surface is defined with respect to the previous surface. As such, for each object, the user must enter the x, y, z co-ordinate along with the object tilt about each axis.

The first task is to identify an active illumination source. In our case, the illumination source is a point source, simulating an LED or lamp filament. For all sources, there is some provision for specifying the angular distribution of the light source. In this instance, the angular distribution of the point source extends uniformly over some nominated half angle. Other sources include elliptical sources, whose geometry is captured by specifying the semi-major and semi-minor axis sizes. For the elliptical source and some other source types, the angular distribution may be defined with a range of mathematical relations incorporating trigonometric or Gaussian (as per Gaussian beam) relationships.

For all surfaces and surfaces that form part of solid objects, the model is able to specify the addition of a coating to the surface and to describe the scattering properties of that surface. The coating specification modifies the transmission and reflection properties of the surface. In this way, mirror surfaces, for example may be rendered highly reflective or only partially reflective (transmissive). Anti-reflection coatings may also be added to lens surface. Critically, we are able to specify the scattering properties of each surface, including the scattering model for the surface. At the most general, we may specify a user defined bi-directional reflection distribution function (BRDF) for the surface to describe the scattered distribution of the rays. This topic is described in more detail in Chapter 7. In addition, the model requests the level of scattering in the form of the total hemispherical scattering; the remainder of the rays are assumed to be reflected or transmitted, depending upon the medium. For example, we can estimate the level of ('small signal') scattering from a lens surface or mirror from the surface roughness. This is detailed in Chapter 7.

Our model only has one source, although, in principle, it is possible to specify any number of surfaces. After the source object, we must consider the integrating sphere. The integrating sphere surface is specified as a mirror surface in a similar manner to the material definition in a sequential surface. However, in this instance, a Lambertian scattering distribution has been specified. The coating reflectivity of the surface is defined to produce 100% scattering, a reasonable model for a Spectralon® integrating sphere. To allow the light to escape from the sphere, an aperture or exit port must also be provided in the model.

Before we can add the detector at the microscope object plane, a number of other objects must be inserted. First, there is the aperture defining the entrance pupil. This aperture is presented as a real object with a finite size. In practice, it is represented by an annulus, the inner diameter defining the aperture itself (8.6 mm) and the outer diameter describing the physical extent of the real aperture (50 mm). This latter point addresses in important point concerning the practical implementation of the pupil. If light emanating from the exit port of the integrating sphere is approximately Lambertian in angular distribution, then some of this light will inevitably skirt around the outer perimeter of the aperture stop. By scattering off other surfaces, this could contribute to straylight at the object plane of the microscope. The entrance pupil aperture is modelled as a Lambertian scatterer. However, it is acknowledged to be 'black'; the hemispherical reflection is modified to 5% by an appropriately specified coating.

The next object to add is the condensing lens. This is modelled as an aspheric lens object. As previously described, the whole lens is modelled as one entity. As well as providing the aspheric prescription for each surface and the lens material, the semi-diameter of each surface must also be entered. Having entered these data, the edge of the lens is automatically defined as a cylindrical or conical surface depending upon the surface semidiameters.

The lens object is a solid object with three surfaces, the two aspheric surfaces plus the (ground) edges. We may, if we wish, assign different properties to each surface. Each surface in turn may be designated transmissive, reflective, or absorbing. In addition, each surface may be modified according to the coating and scattering properties previously described. In practice, in this instance, we model the two aspheric surfaces as purely transmissive. Polished lens surfaces tend to contribute little to scattering and this scattering may be ignored in most applications. However, this does not apply where very low optical fluxes need to be detected in the presence of a high illumination source, e.g. in solar telescopes or high power laser diagnostics. By contrast, the lens edges are considered to be reflective with 100% scattering. This may be a little unrealistic, but it is a simple illustrative description of edge scattering which tends to exaggerate the overall magnitude of the effect. Of course, the lens edge could be blackened to ameliorate the problem. In addition, each surface may be provided with a coating, whose definition allows the modelling of anti-reflection coatings or bandpass coatings, etc. No coatings are modelled in this simple example.

In summary, the following objects are considered in the model:

- Point source
- Integrating Sphere
- Integrating Sphere Port
- Entrance pupil aperture
- Condensing lens
- Enclosure
- Detector

A highly edited version of the lens data editor is shown in Table 18.13.

Table 18.13 Non-sequential lens data editor (much condensed).

Obj.	Type	Comment	x	y	z	R1	R2	Width/Dia	Length	Matl.
0	Source Point	Source	0	7.2	8.4	—				
1	Sphere	I. Sphere	0	0	0	12				Mirror
2	Stand. Surf.	Port	0	0	12			10		
3	Annular Vol.	Aperture	0	0	172			8.6/50		Mirror
4	Stand. Lens	Condenser	0	0	199.6	7.27	−66.76	14	5	N-LAK3
5	Rect. Vol.	Image	0	0	−30			100	250	Mirror
6	Detector	Detector	0	0	211.6			0.64[a)]		Absorb

a) Detector is 0.64×0.64 mm and 101×101 pixels.

Those object rows shown shaded have additional information regarding scattering and coating properties; these are not shown here.

18.4.3.3 Wavelengths

As with the sequential model, a number of wavelengths may be specified. In this specific instance, for simplicity, only one wavelength is specified, 550 nm. When the established system is modelled, rays are launched by the model for all specified wavelengths according to a weighting parameter supplied by the user, which establishes the relative importance of each wavelength. Naturally, no fields or entrance pupil sizes are delineated, as the distribution of rays emanating from the source(s) is determined by the properties of the source(s).

18.4.3.4 Analysis

The analysis proceeds according to a Monte-Carlo ray tracing process. Rays are launched randomly from the source, but according to a spatial and angular probability distribution that fits the source characteristics. Rays are traced through the system until they are absorbed by some surface. Each time a ray strikes a detector at a particular point (pixel), this event is recorded and used to build up a picture of the irradiance distribution at the detector. The irradiance pattern at the object focal plane is shown in Figure 18.13. The false 'colour' plot reveals a broadly uniform disc in line with expectations. However, there is some speckle evident in the plot. In fact, this speckle is, in effect, 'shot noise'. The legend notes that some 2×10^6 rays struck the target. This seems rather a lot; however, it represents only some 200 rays per pixel. In effect, the detector is 'photon counting' and much of the variation is due to the impact of Poisson statistics at each detector pixel. In fact, the majority of the rays did not even reach the detector. A total of 2×10^9 rays were launched. Only 0.1% of the rays actually reached the target. It is thus clear that a simulation of this kind is extremely demanding on computer resources. Each of the 2×10^9 rays had to be traced over several segments, taking into account refraction, reflection, scattering, and the impact of any coating.

The analysis can also be used to characterise the straylight. Whilst the characterisation of low levels of straylight is not necessarily critical in this specific application, the analysis does, nevertheless, serve to illustrate the process. Figure 18.14 shows a section displaying the relative irradiance across the illuminated object plane.

The straylight levels around the illuminated area amount to a few parts in 10^5. If this were critical, a few modifications could be made. For instance, the maximum radius of the physical aperture could be extended from 50 mm or the edges of the lens could be black coated.

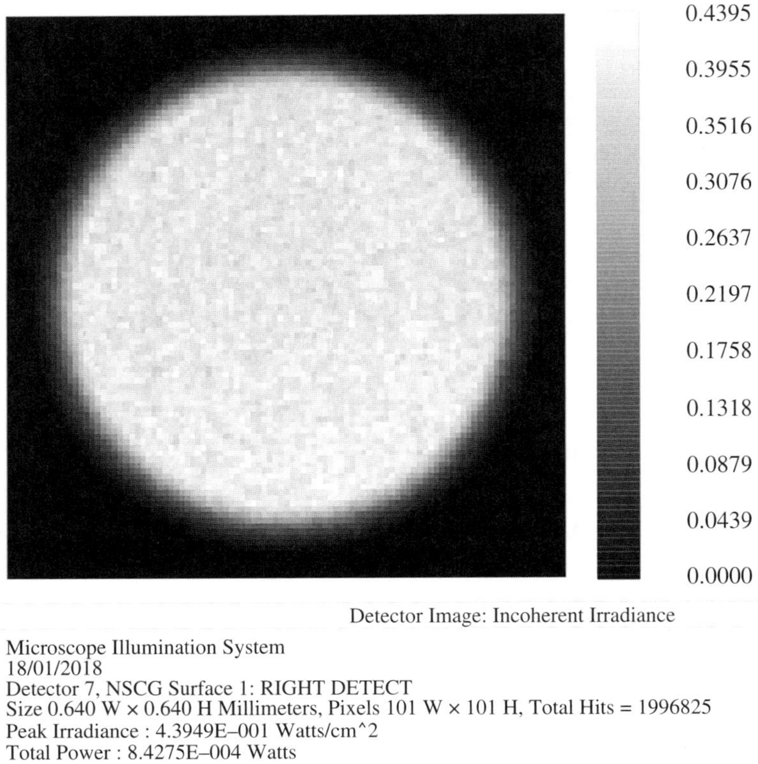

Figure 18.13 Microscope illumination – irradiance uniformity.

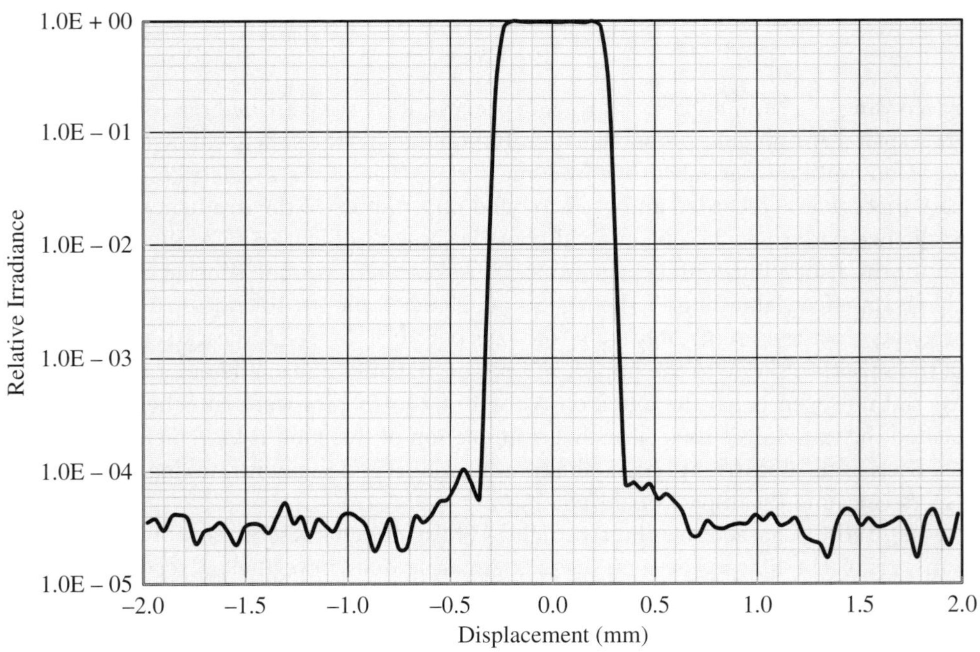

Figure 18.14 Relative irradiance across illuminated area.

18.4.4 Baffling

Baffling is an important topic in non-sequential modelling. Although the analysis of straylight in the previous example was a little artificial, it did introduce the subject of straylight control. If after the analysis presented, the level of straylight were unsatisfactory, then further modifications would have to be made to restrict the straylight contribution. This would generally involve the incorporation of additional structures designed to block the passage of straylight and to minimise the further generation of scattered light. Such structures are referred to as ***baffles***.

If one imagines an imaging system that is designed to convey light from object space to a detector located at the image plane, the sequential design is intended to accept a very specific bundle of rays, as defined by the detector field and the entrance pupil. This forms the system étendue, and we are naturally anxious to prevent light originating outside the system étendue from accessing the detector. The simplest example of a baffling structure is a lens tube. Not only does it provide mechanical integration for an axially symmetric system, it also baffles light from outside the tube structure. This is illustrated by a very basic refracting telescope consisting of an achromatic lens and a detector. The lens defines the entrance pupil and has some nominal aperture, e.g. f#8. Without the lens tube, then the image viewed at the detector would be polluted by light from outside the system étendue. This is illustrated in Figure 18.15.

The most important aspect of straylight analysis is consideration of the field of view open to the detector. In the example provided, the detector will, of course, view the system étendue. Outside this, the detector has a clear view of internal surface of the black tube. As illustrated, the straylight performance will be dictated by the light scattered from this surface. Depending on the internal surface coating the straylight performance may be adequate. However, there are additional measures we might take to further reduce straylight levels.

This is perhaps a rather basic example of direct 'contamination' of the signal by straylight from the external environment. Straylight analysis must explicitly address the scattering of light from the optical surfaces themselves. The scattering process from these surfaces has the potential, by definition, to transform rays lying outside the system étendue, enabling them to access the detector by sequential progression through the remaining surfaces. Generally, since the surface roughness of polished glass is low, surface scattering from lens surfaces, for the most part, may be neglected. However, the impact of bubbles and inclusions within the glass and the presence of dust or contamination on the lens surfaces cannot be ignored. All these produce scattering. Polished mirror surfaces produce somewhat more scattering that the equivalent lens surface. This is because an equivalent amount of surface departure in a mirror produces a greater OPD than would pertain to a transmissive component. A single lens surface with a refractive index of 1.5 would produce about one sixteenth of the scattering of a mirror surface with an equivalent roughness.

Generally, the amount of scattering produced by polished surfaces is relatively low. However, the effects of scattering may, nonetheless, be significant in the presence of a parasitic light source (e.g. laser or solar) with high flux. Some optical surfaces, however, may be produced by a machining process (***diamond machining***), for example diffraction gratings. This process produces optical surfaces with a considerably higher surface roughness than the comparable polishing process. Naturally, these surface produce substantially more scattering than the equivalent polished surface.

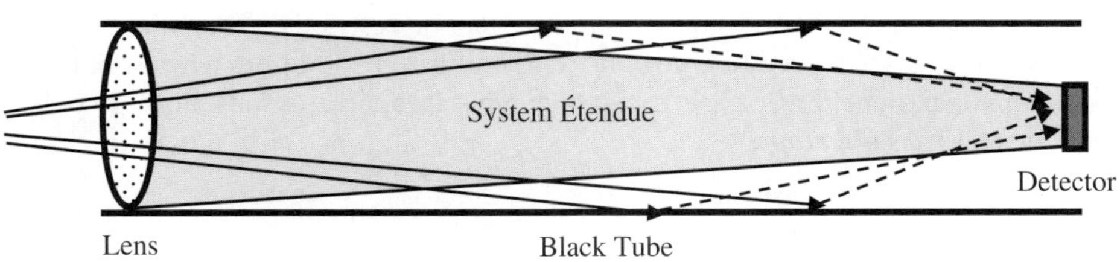

Figure 18.15 Baffling effect of lens tube.

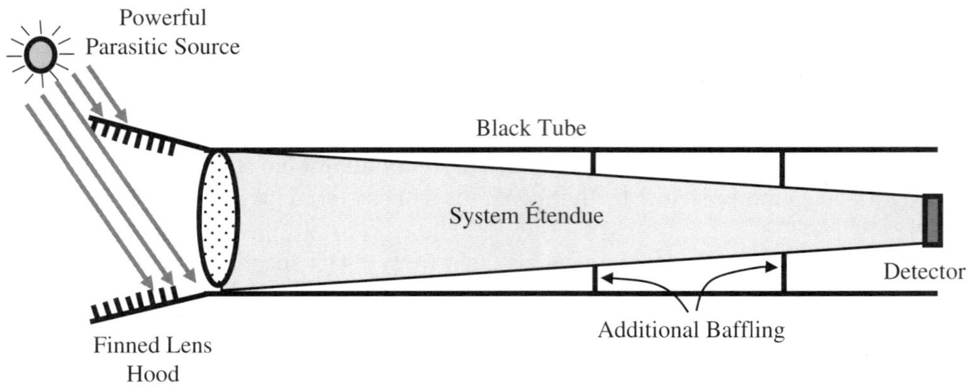

Figure 18.16 Lens hood and additional baffling.

Where scattering from optical surfaces introduces a significant amount of straylight, we must attempt to restrict the amount of light falling on them from outside the system étendue. A very simple example of this is the lens hood. This is effectively an extension to the lens tube that serves to shield the lens itself from direct illumination by the sun. This is illustrated in Figure 18.16.

In Figure 18.16 we have created a slightly more sophisticated solution for tackling straylight. Depiction of the lens hood itself is relatively straightforward. However, as will be noted, the lens hood benefits from the incorporation of ***internal fins*** which are effectively blackened annular plates affixed to the internal surface of the lens hood. The purpose of these is to further restrict the amount of light scattered from the internal surfaces of the lens hood. As a convenient and rather simplistic model, we have hitherto thought of the behaviour of (matt) blackened surfaces as low level Lambertian scatterers. Unfortunately, in practice, such surfaces produce markedly enhanced scattering or reflection at grazing angles of incidence. The purpose of the fins is to remedy this deficiency. Additionally, such a strategy is also useful for further reducing the scattering from the internal surfaces of a simple lens tube. Further baffling within the lens tube has been added that restricts the view of light scattered directly from the internal surface of the tube. Such baffling must not, of course, stray into the system étendue or vignetting would result. In complex systems, the provision of additional apertures, for the sole purpose of restricting straylight, is a common practice.

Choice of baffling material will depend upon the criticality of the application. For the most basic demands, black plastic or black anodised aluminium suffice. However, for more critical applications, particularly in the aerospace domain, there are proprietary black coatings with exceptionally low (e.g. <3%) hemispherical reflection over an extended spectral range. This is particularly true of the infrared spectral region, where the performance of traditional coatings, such as black anodising, is rather poor. Examples of such coatings include MetalVelvet™ and Martin Black™. More recently, very black coatings have emerged, based upon aligned carbon nanotubes.

Analysis of straylight in the infrared region is substantially complicated by the presence of thermal radiation particularly at 'thermal' wavelengths in excess of, say, 2.5 μm. In our treatment of detectors in Chapter 14, the value of cooling detectors to reduce the dark current was articulated. In addition, to take advantage of the enhanced sensitivity produced, background (thermal) radiation must also be restricted. This might involve the provision of optical sub-systems that are cooled to cryogenic temperatures. In addition, where reduction of straylight involves the provision of extra baffling or apertures, such physical apertures must also be cooled. These cooled apertures are referred to as ***cold stops***.

18.5 Afterword

Within the limited confines of a single chapter, only the briefest overview of modern optical design software has been provided. As such, it provides the reader, who might be unfamiliar with the topic, a useful introduction. In so doing, it remedies a deficiency inherent to most texts on optics where the topic is avoided entirely, or previous knowledge is assumed. What should be emphasised, though, is the depth and subtlety of analysis that may be gained through acquiring a thorough working familiarity of these design tools. For example, the range of different surface types and analytical tools available to the designer is truly prodigious and no attempt has been made to replicate that here. Training courses in the use of this software are provided and these provide a useful starting point for using the design software. That said, optical design is ultimately a practical discipline, and there is no substitute for the day-to-day practical use of such tools over an extended period to gain both experience and 'design intuition'.

The theme of this book is that one's experience and 'design intuition' is enhanced by an in-depth knowledge of the underlying principles of optics. Attempting blindly to exploit the capabilities of these powerful and flexible software tool in ignorance of the underlying principles is ultimately futile. That is not to say that the design process should always proceed mathematically and analytically from 'first principles'. The illustration within this chapter of the optimisation process illustrates the difficulty of optimising a complex system with many variables. A useful metaphor for this might be the game of chess. The range of possible options even a few moves into the game becomes so astronomical as to render a systematic deterministic analysis intractable. Therefore, the budding player is obliged to draw on the vast experience of those that have gone before by becoming familiar with libraries of opening moves and studying the overall strategy of the game. In the same way, the budding designer should be prepared to 'stand on the shoulders of giants' by studying, for example, libraries of lens designs – cameras, telephoto lenses, microscope objectives, etc. and, where necessary, to adapt them. It is never obligatory to 'reinvent the wheel'.

Further Reading

Fischer, R.E., Tadic-Galeb, B., and Yoder, P.R. (2008). *Optical System Design*, 2e. New York: McGraw-Hill. ISBN: 978-0-071-47248-7.

Geary, J.M. (2002). *Introduction to Lens Design: With Practical Zemax Examples*. Richmond, VA: Willman-Bell. ISBN: 978-0943396750.

Kidger, M.J. (2001). *Fundamental Optical Design*. Bellingham: SPIE. ISBN: 0-8194-3915-0.

Kidger, M.J. (2004). *Intermediate Optical Design*. Bellingham: SPIE. ISBN: 978-0-8194-5217-7.

Laikin, M. (2012). *Lens Design*, 4e. Boca Raton: CRC Press. ISBN: 978-1-4665-1702-8.

Noll, R.J. (1976). Zernike polynomials and atmospheric turbulence. *J. Opt. Soc. Am.* 66 (3): 207.

Rolt, S. (2012). A review of the KMOS IFU component metrology programme. *Proc. SPIE* 8450: 84500.

Rolt, S., Dubbeldam, C.M., Robertson, D.J. et al. (2008). Design for test and manufacture of complex multi-component optical instruments. *Proc. SPIE* 7102: 71020A.

Shannon, R.R. (1997). *The Art and Science of Optical Design*. Cambridge: Cambridge University Press. ISBN: 978-0521454148.

19

Mechanical and Thermo-Mechanical Modelling

19.1 Introduction

19.1.1 Background

In the previous chapter, we stressed the importance of considering the optical design as a compromise and a dialogue formed by the interaction between many different disciplines. The ideal design must fulfil all stated requirements at a reasonable cost employing manufacturing and assembly processes that are practicable. As such, a concurrent engineering approach is favoured, embracing a wide range of subjects that lie outside the narrow confines of optics. In the practical realisation of an optical design, mechanical engineering is one of the most salient disciplines that needs to be addressed. There is an exceptionally strong synergy between optics and mechanics. The formulation of an optical design is based upon the specification of a geometrical relationship between optical surfaces. This geometrical relationship can only be realised, in practice, through mechanical engineering.

A good mechanical design not only strives to replicate faithfully the geometrical layout of the optical design, but also must maintain that relationship over time and environmental stress. It is the ***stability*** of the geometrical relationships between all components that is of paramount importance. As a consequence, in the mechanical analysis of an optical system, it is the mechanical or thermo-mechanical stability of the system that is the focus of any analysis. An optical system is designed with a use environment in mind. This environment might embrace variations in temperature, humidity, exposure to chemicals or salt spray, in addition to dynamic loads, such as shock and vibration. Of course, many environments, particularly consumer environments are benign, with temperature and humidity contained within a narrow envelope and with limited exposure to dynamic loads. On the other hand, in other arenas, such as in aerospace and defence, the environment may be considerably more aggressive.

The focus of this chapter will be on the analysis of the impact of varying mechanical loads and thermal loads on system performance. Physical forces have a direct impact in producing deformation and particularly flexure and thus changing the geometrical relationship between optical surfaces. The same consideration applies to changes in temperature. Differential thermal expansion produces relative movement between components and subsystems, producing an impact upon system performance. Thus far, we have emphasised the impact of thermo-mechanical stress on the system alignment. However, we must also take into account the direct impact produced by thermo-mechanical distortion of individual surfaces. Whereas global changes in geometry impact alignment, with an indirect influence on image quality, surface distortion affects image quality through direct modulation of the wavefront error.

One impact of thermo-mechanical stress that is often neglected is the propensity to produce stress-induced polarisation effects. This is of some salience in critical scientific applications, particularly in instrumentation designed to monitor and analyse polarisation.

Optical Engineering Science, First Edition. Stephen Rolt.

Companion website: www.wiley.com/go/Rolt/opt-eng-sci

19.1.2 Tolerancing

From a practical point of view, the greatest impact of thermo-mechanical stress is on the tolerance analysis of the design. Having characterised component movement and distortion, this information must be fed back into the (optical) tolerance model to assess the impact upon performance. This may be accomplished in a generic way, by using the mechanical model to set reasonable limits on component movement, or to characterise additional contributions to form error through surface distortion. However, it is also possible to examine a specific environmental stress scenario in detail. A detailed mechanical model is capable of capturing the translations and rotations of each optical surface in addition to any distortion in that surface. This detailed quantitative information may be fed directly into the original optical model, thus 'closing the loop'. Changes in wavefront error, spot size, MTF, and image location, etc. may then be directly computed.

19.1.3 Athermal Design

Aggressive environments, such as those pertaining to military or aerospace applications require a particular focus on the mechanical design. In particular, a wide operational temperature range, for example, −50 to 70 °C poses an especial challenge. This temperature range is perhaps not untypical for some thermal camera applications. Therefore, there is a premium, for such an application, to develop an ***athermal design***. An athermal design is a design where the effects of thermal expansion are fully compensated by careful selection of materials. In particular, changes in the focal plane location with temperature are minimised. The designer has some freedom in offsetting thermal changes in material refractive index and the impact of thermal expansion. A notable and very specific example of an athermal design is a mirror-based system where the substrate material for the mirrors is identical to the material used in forming the optical bench. Changes in separation between components are entirely compensated by changes in the mirror radii or effective focal length.

19.1.4 Mechanical Models

This chapter will focus on some useful analytical tool to help understand the key issues involved. A brief introduction to computer-based ***Finite Element Analysis*** (FEA) models will also be given. Whilst an awareness of the capabilities of these models is useful, more detailed coverage is beyond the scope of this book. However, although this is an area that lies outside the traditional discipline of optics, the reader is very much encouraged to pursue this avenue. Training courses are available and they, together with the relevant experience, should be sought out. Opto-mechanics is a discipline that lies at the interface of optics, materials science, and mechanical engineering. As with many of these 'interfacial topics', it is often neglected, presenting the curious and motivated engineer with many opportunities.

That said, as with optical modelling, it is possible to sketch out a basic thermo-mechanical model, that helps to build a general picture of the mechanical design, before embarking on more detailed analysis.

19.2 Basic Elastic Theory

19.2.1 Introduction

Before discussing simple basic analytical modelling, or FEA modelling, we will briefly introduce the reader to basic elastic theory which underpins the characterisation of mechanical distortion in continuous media. Fundamentally, the theory asserts a simple linear relationship between material stress and strain. For many practical scenarios in optics, this is a perfectly reasonable assumption; stress levels are generally sufficiently low that this linear relationship is maintained. Of course, at higher stress levels, this linearity no longer holds. Furthermore, irreversible (plastic) deformation or brittle fracture may occur, under these conditions. However, for the most part, the assumption of linear behaviour is a useful generalisation.

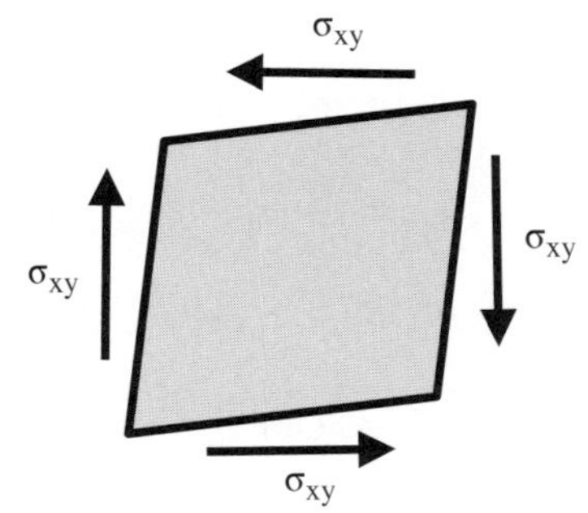

Figure 19.1 Uniform shear forces acting on an element.

19.2.2 Elastic Theory

In its simplest form elastic modelling represents stress and strain as simple scalar quantities, linked by a linear elastic constant, the ***elastic modulus*** or ***Young's modulus.*** However, in reality, both stress and strain are best described geometrically as a second order tensor. Starting with the strain tensor, we may consider a material as a network of nodes, or discrete points. Each node starts out with some unstrained location and ***small*** incremental shifts in each nodal position are described by the Cartesian vector – $u_x\mathbf{i}$, $u_y\mathbf{j}$, $u_z\mathbf{k}$. All these incremental displacements are themselves dependent upon the local co-ordinates, x,y,z. In describing (elastic) distortion, we are interested in displacements that vary with position, rather than an invariant vector displacement that could be sufficiently described by global displacement of the whole system. Similarly, we are not interested in a pattern of local displacements that can be defined by bodily rotation. As such, the local strain is defined by a set of partial derivatives of displacement with respect to spatial co-ordinate:

$$\varepsilon_{xx} = \frac{\partial u_x}{\partial x}\ \varepsilon_{yy} = \frac{\partial u_y}{\partial y}\ \varepsilon_{zz} = \frac{\partial u_z}{\partial z} \tag{19.1}$$

$$\varepsilon_{xy} = \left(\frac{\partial u_x}{\partial y} + \frac{\partial u_y}{\partial x}\right)\ \varepsilon_{xz} = \left(\frac{\partial u_x}{\partial z} + \frac{\partial u_z}{\partial x}\right)\ \varepsilon_{yz} = \left(\frac{\partial u_y}{\partial z} + \frac{\partial u_z}{\partial y}\right) \tag{19.2}$$

Equation (19.1) summarises the three tensile or compressive strains, ε_{xx}, ε_{yy}, ε_{zz}, whereas Eq. (19.2) sets out the shear strains, ε_{xy}, ε_{xz}, ε_{yz}. Care should be taken with Eq. (19.2), as alternative formulations exist, with the shear strain defined as half that set out in Eq. (19.2). By virtue of Eqs. (19.1) and (19.2) there is a set of six independent strains. Similarly, there are six stress components, three describing tensile stress and three describing shear stress.

If a continuous solid were to be cut by an imaginary plane, each part of the solid may be thought to impose an equal and opposite force on the other. Any force that is perpendicular to that plane is related to a tensile or compressive stress. It is important to recognise that uniform tensile or compressive stress exerts no net force on any element within the solid. For each stress component, the force imposed at one end of the element is balanced by the tensile force at the other.

Any force that lies in this imaginary plane is related to a shear stress. For a plane cut with its normal oriented along the x axis, then shear forces may be arrayed along the y or z axes. As with tensile stress, uniform shear stress imposes no net force upon a solid element. Furthermore, in addition, no net couple is produced. This is because the couple produced by the shear force acting in the y direction in the yz plane is precisely opposed by the shear force operating in the x direction in the xz plane. These two shear stresses are fundamentally identical. This is illustrated in Figure 19.1

The essence of elastic theory is that a simple linear relationship exists between the stress and strain tensor. In principle, this linear relationship might be defined by thirty-six independent material constants linking the six stress and six strain components. In practice, symmetry substantially reduces this figure. Specifically, in the case of isotropic materials, the single most important class, the relationship between stress and strain is defined by only two independent parameters. The relationship for isotropic materials is represented in matrix form as follows:

$$\varepsilon_{xx} = (\sigma_{xx} - \nu\sigma_{yy} - \nu\sigma_{zz})/E + \alpha\Delta T \tag{19.3a}$$

$$\varepsilon_{yy} = (-\nu\sigma_{xx} + \sigma_{yy} - \nu\sigma_{zz})/E + \alpha\Delta T \tag{19.3b}$$

$$\varepsilon_{zz} = (-\nu\sigma_{xx} - \nu\sigma_{yy} + \sigma_{zz})/E + \alpha\Delta T \tag{19.3c}$$

$$\varepsilon_{xy} = 2(1+\nu)\sigma_{xy}/E \tag{19.3d}$$

$$\varepsilon_{xz} = 2(1+\nu)\sigma_{xz}/E \tag{19.3e}$$

$$\varepsilon_{yz} = 2(1+\nu)\sigma_{yz}/E \tag{19.3f}$$

In Eqs. (19.3a–f), we have presented the relationship between stress and strain taking full account of thermal expansion, through the coefficient of thermal expansion (CTE), α, and the temperature excursion ΔT. The total strain comprises both a thermal component of strain and an elastic component of strain. In an isotropic material, the two sole parameters defined are the elastic or Young's modulus, E, and the Poisson's ratio, ν. The elastic modulus is equivalent to the uniaxial tensile stress required to produce unit strain, assuming linearity is preserved under such extreme conditions. As this uniaxial stress is applied, Poisson's ratio is the proportional contraction produced along the two orthogonal directions, as referenced to the elongation produced in the direction of the applied stress. These two parameters also define the relationship between the shear stresses and strains. This is because, as a fundamental geometrical property of the stress and strain tensor one can always, by rotation, find a co-ordinate representation where all shear stresses/strains are zero. In this co-ordinate frame, only tensile stresses or strains are present. The axes represented are the ***principal axes***, with the three ***principal stresses*** and three ***principal strains***. For an isotropic material, the principal axes for stress and strain must, necessarily, be identical.

In many simple scenarios, the behaviour of a material, particularly at high stress, is viewed in terms of a simple uniaxial tensile stress. This is particularly the case when analysing non-elastic (i.e. plastic) behaviour. Of course, FEA modelling of complex structures will inevitably reveal a correspondingly complex distribution of triaxial stress. It is desirable to characterise this triaxial stress to yield a single parameter that adequately describes the extremity of the local stress. One such representation is the so-called ***von-Mises stress***, σ_v. It is derived from the local principal stresses, σ_1, σ_2, σ_3.

$$\sigma_v^2 = [(\sigma_1 - \sigma_2)^2 + (\sigma_1 - \sigma_3)^2 + (\sigma_2 - \sigma_3)^2]/2 \tag{19.4}$$

Von-Mises stress is useful for predicting the limits of elastic behaviour. However, it is clear from Eq. (19.4) that the von-Mises stress is zero for purely hydrostatic stress (all principal stresses equal) ***whether that stress is compressive or tensile***. Clearly, this symmetry is not recognised in fracture mechanics, where tensile stress is substantially more destructive than compressive stress. This is because fracture mechanics is largely driven by defects, such as cracks, rather than the properties of the bulk material. Of course, cracks have a tendency to be 'opened' by tensile stress, in contrast to their behaviour under the imposition of compressive loads. Therefore, in the consideration of brittle fracture, it is the maximum principal stress that is of the most moment.

As outlined, just two independent constants define the elastic properties of an isotropic material. Rotation of the co-ordinate frame away from the principal axes will, where the principal stresses are unequal, reveal shear stresses and strains. This is why the shear modulus, G, the linear constant defining the relationship between the shear stress and the shear strain, is itself to be expressed solely in terms of the elastic modulus and the Poissons's ratio:

$$G = E/(2(1+\nu)) \tag{19.5}$$

19.3 Basic Analysis of Mechanical Distortion

19.3.1 Introduction

Initially, we will focus on the analysis of mechanical effects on alignment. All opto-mechanical designs feature some kind of common mechanical structure that integrates individual components or subsystems. It is these structures that effectively delineate a common optical axis. A common geometry for such a platform is the optical bench. This is a nominally planar structure whose thickness is small compared to its lateral dimensions. Such structures are particularly prone to flexure both from mechanical and thermo-mechanical effects. Another common platform is the lens tube, which serves to integrate circularly symmetric components. Within reason, due to their geometry, these structures tend to be more rigid than an optical bench. However, in terms of integrating and aligning large numbers of optical subsystems over an extended region, a planar optical bench structure is often preferred.

19.3.2 Optical Bench Distortion

19.3.2.1 Definition of the Problem

Analysis of optical bench distortion proceeds by the classical mechanical engineering discipline of plate or beam theory. In its one dimensional form, describing a beam, this analysis is referred to as Euler-Bernoulli beam theory. Its two dimensional counterpart is referred to as the Kirchoff-Love theory of plates. Neither of these theories is suitable for the modelling of large stresses and deflections. However, they are eminently suited to the modelling of small deflections (compared to plate thickness) that are typical of the majority of opto-mechanical applications. As with much of the basic optical analysis, the techniques described here form a useful precursor to more detailed modelling. Figure 19.2 shows an optical bench flexing under a load applied at its centre whilst being supported at the edges.

Beam theory is applicable for geometries where the beam thickness is small compared to the length and where any deflection is substantially less than the thickness. The assumption of beam theory is the centre line of the beam or plate is unstressed. As such, in the example illustrated in Figure 19.2, those areas above the centre line are in compressive stress, whilst those areas below are in tensile stress. For simplicity, Figure 19.2 views the problem in one dimension. Application of an external stress produces bending in the plate or beam. We may quantify the bending stress at any given location by a quantity called the bending moment. The ***bending moment*** is the couple produced by all the external (perpendicular) loads that is opposed by the couple produced by the tensile and compressive forces engendered across the section of the beam. For a simple homogeneous beam, it is simple to calculate this bending couple for a given amount of bending. The bending itself is quantified by the second derivative of the perpendicular displacement.

Firstly, we might like to derive the local couple that is produced in bending a thin structure, as shown in Figure 19.2. Distance from the nominal centreline is described by the parameter, a, and, at any location within the cross section, the local tensile force is clearly proportional to a. That is to say, the stress at the centre line is zero and is tensile or compressive on either side. To make the analysis general, we do not assume that the

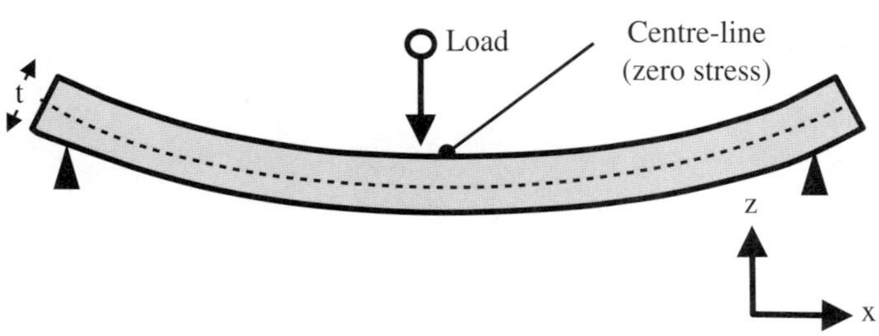

Figure 19.2 Flexure of optical bench under load.

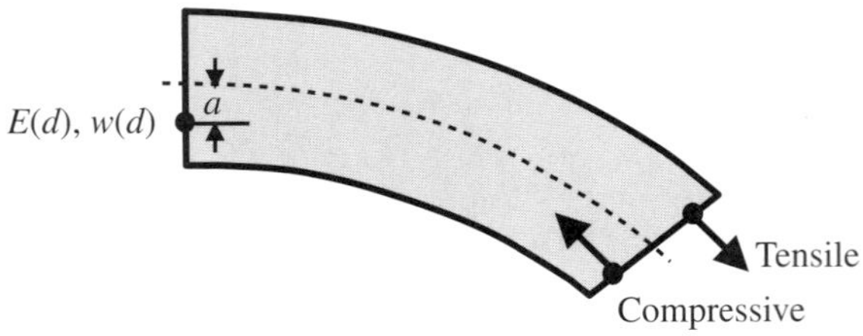

Figure 19.3 Flexure in a beam element.

structure is homogenous across its thickness; it is assumed to be homogenous along its length. As we progress through the thickness, the local elastic modulus, E, may be described as $E(a)$, a function of a. Furthermore, the width of the structure, $w(a)$, may not be uniform across the thickness. The problem is defined in a little more detail in Figure 19.3.

The local curvature of the beam, $C(x)$ is described by the second derivative of the displacement in y, $y(x)$.

$$C(x) = \frac{\partial^2 z}{\partial x^2} \tag{19.6}$$

The local cross-sectional force at a point displaced by a from the centre line is given by:

$$F(a) = C(x) \times E(a) \times w(a) \times a \times da \tag{19.7}$$

And the couple, $\tau(d)$ is given by:

$$\tau(a) = C(x) \times E(a) \times w(a) \times a^2 \times da \tag{19.8}$$

Finally, the total couple, which we define as the ***bending moment***, M(x), is established merely by integrating Eq. (19.8) with respect to a:

$$M(x) = C(x) \times \int_{a_1}^{a_2} E(a)w(a)a^2 da \tag{19.9}$$

Some care must be exercised in denominating the centre line. This is not necessarily the geometrical centre of the beam or plate. In practice, the centre is defined as centre, as weighted by local width and elastic modulus. In other words, the following condition must apply:

$$\int_{a_1}^{a_2} E(a)w(a)a\, da = 0 \tag{19.10}$$

Where the beam is of uniform composition (not width), the torque is given by:

$$M(x) = C(x) \times E \times \int_{a_1}^{a_2} w(a)a^2 da = C(x) \times E \times I \text{ where } I = \int_{a_1}^{a_2} w(a)a^2 da \tag{19.11}$$

The quantity, I, is referred to ***as the second moment of area***. To make expression more general, we define a parameter, κ_B, the ***bending stiffness***:

$$\kappa_B = \int_{a_1}^{a_2} E(a)w(a)a^2\, da \text{ and } M(x) = \kappa_B \frac{\partial^2 z}{\partial x^2} \tag{19.12}$$

For a homogenous beam of width, w, and thickness, t, the bending stiffness is easy to derive:

$$\kappa_B = \frac{Ewt^3}{12} \tag{19.13}$$

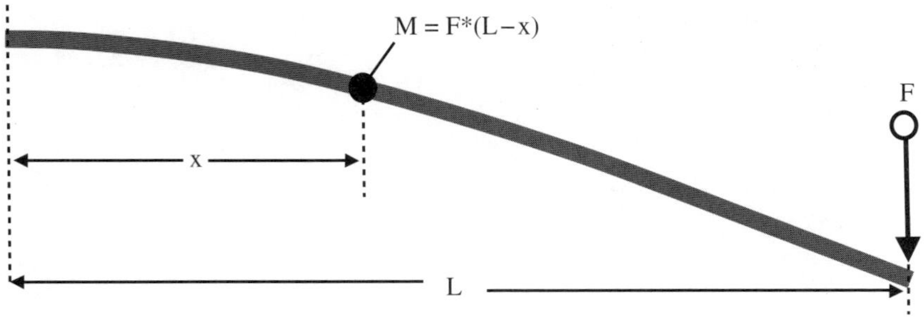

Figure 19.4 Force and bending moment in a cantilever.

19.3.2.2 Application of External Forces

Application of an external force to a beam introduces a bending moment. Initially, the simplest case to describe is that of the cantilever beam. A beam of length, L, is clamped horizontally at one end and a vertical force, F is applied to the opposite end. This is illustrated in Figure 19.4. At any location along the beam, the bending moment acting on the beam is proportional to the force and the displacement from its point of action.

The bending moment is as given in Figure 19.4 and the flexure in the beam is therefore given by:

$$\kappa_B \frac{\partial^2 z}{\partial x^2} = F(L - x) \tag{19.14}$$

And:

$$z = \frac{FL^3}{\kappa_B}\left[\frac{1}{2}\left(\frac{x}{L}\right)^2 - \frac{1}{6}\left(\frac{x}{L}\right)^3\right] \tag{19.15}$$

Equation (19.15) gives the deflection at any point along the beam. The deflection at the end of the beam is simply given by:

$$z(L) = \frac{FL^3}{3\kappa_B} \tag{19.16}$$

The analysis pertaining to Eq. (19.16) assumes the load is offered at a single point. This is clearly not the case in practice as, in reality, any load will be distributed in some way. Nonetheless, there are occasions where this idealised analysis is a convenient approach, providing a workable approximation. For a more generalised framing of the problem, we must assume that the load is characterised as a pressure, P – a force per unit area distributed in some way across the beam.

We now consider what might happen to a single element of length along the beam. For a bending moment that is constant along the length of the beam, the couples generated by the tensile and compressive forces at each end of the element are equal and opposite. In this situation, there is no net couple engendered by those means in that element. However, if there is a gradient in the bending moment, these forces will generate a net couple on the element. For static equilibrium to be maintained, an opposing shear couple must therefore act on the element.

This is illustrated in Figure 19.5 showing the shear force, F_S, providing a couple equal to $F_S \times \Delta x$ that balances the differential bending moment. This observation may be expressed more formally:

$$F_S = \frac{\partial M}{\partial x} \tag{19.17}$$

However, provided the shear force, F_S, is constant along the length of the beam, there will be no net force acting on the beam. On the other hand, if there is a gradient in that shear force, then there will be a net force acting on that individual element. This force is proportional to the product of the gradient in the shear force and the length of the element, Δx and, for static equilibrium to prevail, is equal and opposite to any externally applied force. The external pressure, $P(x)$ applies a force that is proportional to the sectional area

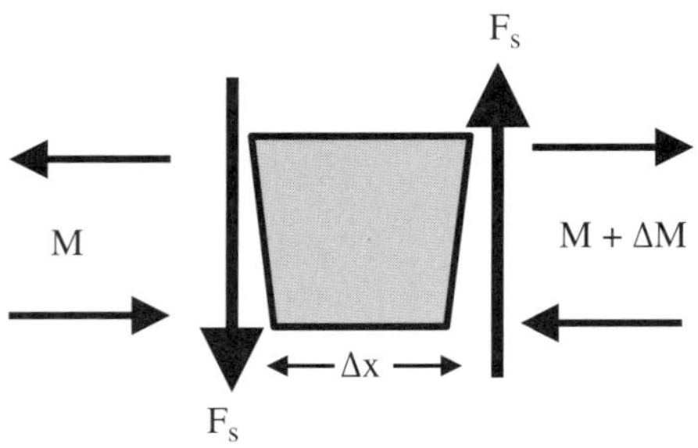

Figure 19.5 Forces acting on single beam element.

of the element or the product of its width, w, and its length, Δx. Therefore, the equilibrium condition is given by:

$$\frac{\partial^2 M}{\partial x^2} = P(x)w \tag{19.18}$$

Finally, from Eq. (19.12), we can express the beam deflection in terms of the externally applied pressure, P:

$$\frac{\partial^4 z}{\partial x^4} = \frac{P(x)}{\kappa_B} w \tag{19.19}$$

As discussed previously, there are many situations where we might wish to consider the application of a force at a point, rather than distributed over a wide area in the form of an external pressure. Technically, this produces a discontinuity in the profile of the beam. Nonetheless, it does provide a useful and tractable estimate of the beam deflection. In this instance, the behaviour of the beam is expressed in terms of the locally applied force, F, and a discontinuity of the third derivative of the beam deflection:

$$\frac{\partial^3 z(x+\Delta x)}{\partial x^3} - \frac{\partial^3 z(x)}{\partial x^3} = \frac{F}{\kappa_B} \tag{19.20}$$

Equation (19.19) is applied to one dimension only. If we wish to apply this to two dimensions (x and y), this must be modified slightly.

$$\left(\frac{\partial^2}{\partial x^2} + \frac{\partial^2}{\partial y^2}\right)\left(\frac{\partial^2 z}{\partial x^2} + \frac{\partial^2 z}{\partial y^2}\right) = \frac{P(x)w(1-\nu^2)}{\kappa_B} \quad \nu \text{ is Poisson's ratio} \tag{19.21}$$

Armed with Eqs. (19.19)–(19.21), we may attempt to analyse some real scenarios. However, these differential equations do not quite establish the whole picture. Before we proceed further we must attempt to define the ***boundary conditions***. That is to say, Eq. (19.21) quantifies the behaviour of the higher order derivatives of the deflection with respect to position. It does not establish the lower order derivatives, such as the local gradients, etc. This can only be done if we understand how these quantities are fixed at specific locations in the beam. These are the so called boundary conditions.

19.3.2.3 Establishing Boundary Conditions

For simplicity, if one views a one-dimensional beam problem, such as defined in Eq. (19.19), the fourth order differential equation requires the provision of four further constraints to define the solution absolutely. In fact, all boundary conditions are set at the edges of the beam. The balancing of forces on a single beam element, as shown in Figure 19.5 suggest that the forces and couples acting on one end of the element are, to degree, balanced by those acting on the other. However, where the final element on the end of the beam is considered, the forces and couples acting on one end of the element cannot be opposed. Therefore, the necessary condition for the free end of a beam is that the bending moment, M, and the shear force, F_s, must both vanish. Therefore, both the second and third derivatives in the deflection must also vanish:

$$\frac{\partial^2 z}{\partial x^2} = 0 \text{ and } \frac{\partial^3 z}{\partial x^3} = 0 \tag{19.22}$$

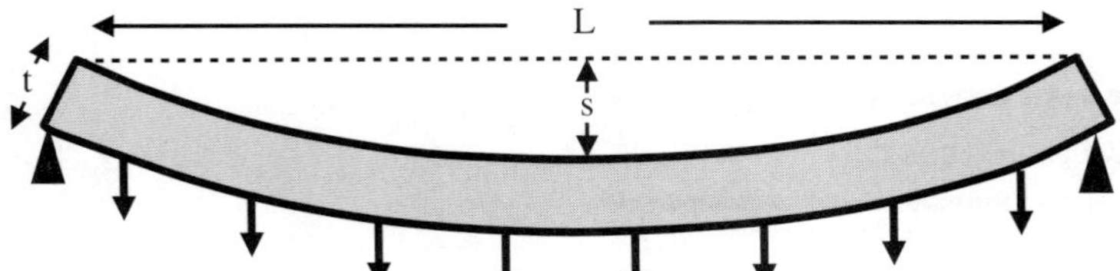

Figure 19.6 Beam deflection due to self-weight.

19.3.2.4 Modelling of Deflection under Self-Loading

We now consider the situation of a uniform beam supported at either end and deflecting under the influence of its own weight. The pressure load imposed upon the beam is given by the product of its mass per unit area, M_A, and the acceleration due to gravity, g. We wish to determine the sag, s, at the central point. Figure 19.6 illustrates the scenario.

$$\frac{\partial^4 z}{\partial x^4} = \frac{M_A g}{\kappa_B} w \tag{19.23}$$

We need now to establish the boundary conditions. The problem is defined by its symmetry and, accordingly, we set the origin of the x axis to coincide with the centre of the beam. At both free ends we may assume that the bending moment and the second derivative is zero. As far as the shear force is concerned, there is a discontinuity around the location of the supports. Therefore, it is not legitimate to apply a zero shear force at the supports. Nevertheless, we may apply the principle of symmetry and set the shear force at the two supports to be equal. By convention, we set the vertical deflection to be zero at the origin. We now have a complete set of defining boundary conditions:

$$z(0) = 0; z''(-L/2) = 0; z''(L/2) = 0; z'''(-L/2) = z'''(L/2) \tag{19.24}$$

The general solution to Eq. (19.23) is a quartic polynomial with constants of integration applied to the linear to cubic terms. In fact symmetry dictates that we may ignore the linear and cubic terms. If we assume that the gravity vector, in this instance is directed towards negative z, then the solution is given by:

$$z = \frac{M_A g w L^4}{24\kappa_B}\left[\frac{3}{2}\left(\frac{x}{L}\right)^2 - \left(\frac{x}{L}\right)^4\right] \tag{19.25}$$

The sag, s, at the centre is then given by:

$$s = \frac{5 M_A g w L^4}{384\kappa_B} \tag{19.26}$$

Worked Example 19.1 ***Deflection of Optical Table***

A typical large laboratory optical table consists of two flat steel plates bonded to a steel honeycomb core. In this exercise we will calculate the self-load deflection of a table 4 m long by 1.5 m wide and constructed from a stainless steel honeycomb core with a thickness of 300 mm and surrounded by two stainless steel plates 5 mm thick. The table is supported at the extreme ends of the long (4 m) dimension and, in this instance, we may analyse the problem as a one dimensional problem.

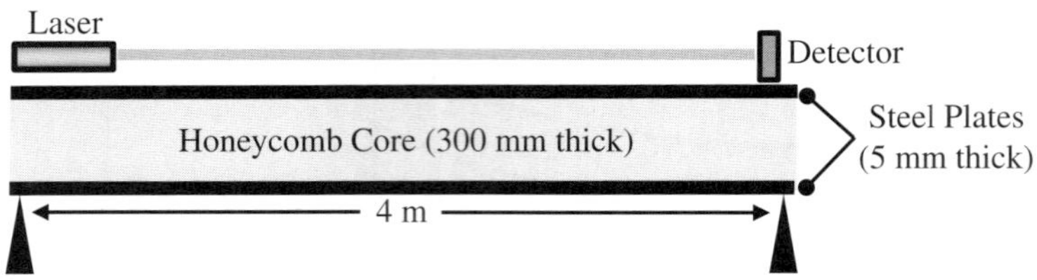

The stainless steel plates have a density of 7750 kgm^{-3} and an elastic modulus of 2×10^{11} Nm^{-2}. The honeycomb core is modelled in the same way as the 'skins', but using an effective 'fill factor' of 0.04. That is to say, the effective density of the core is 310 kgm^{-3} and the elastic modulus is 8×10^9 Nm^{-2}. Having calculated the self-deflection, we wish to determine its impact on optical alignment. A laser is mounted at one extreme end of the table in such a way that the beam is parallel to the surface of the table. The height of the beam above the table is 50 mm, at this point. A detector is mounted at the opposite (extreme) end of the table. At what height above the table does the beam strike the detector?

As the table is symmetrical, the central axis runs down the centre of the core and, from Eq. (19.8), the contribution to the bending stiffness from the uniform core is given by:

$$\kappa_B(core) = \frac{Ewt^3}{12} = \frac{(8 \times 10^9) \times (1.50.3)^3}{12} = 2.7 \times 10^7 \mathrm{Nm}^2$$

Contribution to the stiffness by each skin is determined by the integral given in Eq. (19.7). If the separation of the skins (305 mm at their centreline) is given by, *s*, and their individual thickness by, *t*, their contribution to the bending stiffness is given by:

$$\kappa_B(skins) = \frac{Ews^2t}{2} = \frac{(2 \times 10^{11}) \times (1.5 \times 0.305^2) \times 0.005}{2} = 7.0 \times 10^7 \mathrm{Nm}^2$$

Therefore the total bending stiffness is equal to $(2.7 \times 10^7) + (7.0 \times 10^7) = 9.7 \times 10^7$ Nm^2.

The mass per unit area, M_A, is simply equal to the aggregate of the respective thicknesses and densities:

$$M_A = (310 \times 0.31) + (2 \times 7750 \times 0.005) = 170.5\ \mathrm{Kgm}^{-2}.$$

Substituting all values into Eq. (19.21) gives:

$$s = \frac{5M_A gwL^4}{384\kappa_B} = \frac{5 \times 170.5 \times 9.81 \times 1.5 \times 4^4}{384 \times (9.7 \times 10^7)} = 8.6x10^{-5}m$$

Therefore, the self-weight deflection is 8.6×10^{-5} m or 86 μm.

Since the laser and target is symmetrically disposed about the centre of the optical table both are at the same height. However, the laser, which is situated at one end, is significantly tilted with respect to the horizontal. The gradient at the end is simply given by differentiating Eq. (19.20) with respect to *x*:

$$\frac{dz}{dx} = \frac{M_A gwL^4}{24\kappa_B}\left[\frac{3x}{L^2} - \frac{4x^3}{L^4}\right]$$ and the value at the end ($x = -L/2$) is $$\frac{dz}{dx} = \frac{M_A gwL^4}{24\kappa_B}\left[\frac{3}{2L} - \frac{1}{2L}\right] = \frac{M_A gwL^3}{24\kappa_B}$$

The laser beam will then 'lose height' with respect to the table and the total height lost, Δh, is given by the product of the gradient and the table length:

$$\Delta h = \frac{M_A gwL^4}{24\kappa_B}$$ This is simply 16/5 times the self-deflection or 276 μm.

The height of the beam above the table at the detector is 50–0.276 mm or 49.724 mm.

19.3.2.5 Modelling of Deflection Under 'Point' Load

Application of a point load is marked by a discontinuity in the third derivative, as exemplified by Eq. (19.15). In the absence of any distributed load, the fourth differential must be zero elsewhere. To give this analysis some substance, we return to the previous worked example. We now wish to calculate the external force distributed

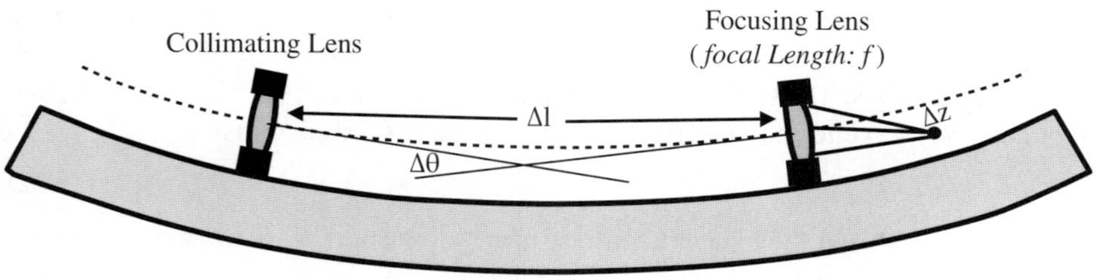

Figure 19.7 Generalised illustration of optical bench distortion.

across the centre that is required to produce the same deflection as the self-loading. If we assume symmetry, then the third derivative immediately either side of the centre should be equal and opposite. Therefore, if the force applied is F, the following applies:

$$\frac{\partial^3 z(0)}{\partial x^3} = \pm \frac{F}{2\kappa_B} \tag{19.27}$$

The table is still supported at either end, so, as in the previous exercise, the second derivative vanishes at the ends ($x = \pm L/2$). If we wish to calculate the impact of the external load alone, we may assume the table itself to be massless. In this case, the fourth derivative will be zero for all x. Finally, we can place the z origin at the centre and, by virtue of symmetry, we may assume the gradient is zero at the centre. Therefore, the general solution is in the form of a cubic equation, but with only second order and third order terms. Because of the discontinuity at the centre, different solutions apply either side of the discontinuity:

$$z = -\frac{FL^3}{12\kappa_B}\left[\left(\frac{x}{L}\right)^3 + \frac{3}{2}\left(\frac{x}{L}\right)^2\right] \ (\mathit{for}\, x < 0);\ z = \frac{FL^3}{12\kappa_B}\left[\left(\frac{x}{L}\right)^3 - \frac{3}{2}\left(\frac{x}{L}\right)^2\right] \ (\mathit{for}\, x > 0) \tag{19.28}$$

The deflection, s is simply given by:

$$s = \frac{FL^3}{24\kappa_B} \tag{19.29}$$

We are now interested in the force, F, required to produce a deflection of 86 μm. From our previous computations, we know that the bending stiffness, κ_B is equal to 9.7×10^7 Nm^2. And the length, L, is 4 m:

$$8.6x10^{-5} = \frac{F \times 4^3}{24 \times (9.7 \times 10^7)} \quad \mathrm{F} = 3{,}130\ \mathrm{N}$$

The force is 3130 N, equivalent to a loading by a mass of about 319 kg.

19.3.2.6 Impact of Optical Bench Distortion

The prime impact of the mechanical distortions that we have attempted to model is the introduction of alignment or boresight errors. That is to say, the effective optical axis of the system follows a curved as opposed to a straight path. A simplified exposition of the general problem is shown in Figure 19.7.

For any limited section of the optical bench, the distortion may be expressed as the local curvature, $C(x)$. This curvature is the inverse of the local radius and, in terms of the preceding discussion, is equal to the second derivative of the height. Thus, for a linear problem, as highlighted by the previous optical table exercise, the self-load curvature as a function of position for a table of length, L, is given by:

$$C(x) = \frac{M_A g w L^2}{8\kappa_B}\left[1 - 4\left(\frac{x}{L}\right)^2\right] \tag{19.30}$$

In the case of a centrally imposed external force, F, the curvature is given by:

$$C(x) = -\frac{FL}{4\kappa_B}\left[2\left(\frac{x}{L}\right) + 1\right] \ (\text{for x} < 0);\ C(x) = \frac{FL}{4\kappa_B}\left[2\left(\frac{x}{L}\right) - 1\right] \ (\text{for x} > 0) \tag{19.31}$$

The impact of any distortion may be visualised from the illustration in Figure 19.7. Imposed bending of the optical axis means that the chief ray launched from some other subsystem integrated onto the optical bench may not be parallel to the local optical axis. The extent of this angular divergence, $\Delta\theta$, may be approximated as the product of the local curvature, $C(x)$ and the distance, Δl between the two subsystems.

$$\Delta\theta = C(x)\Delta l \tag{19.32}$$

The previous exercise gave us some sense of the likely magnitude of any distortion. If we imagine an optical assembly with two subsystems separated by 1 m and arranged symmetrically about the centre of the bench, then the angular divergence may be computed as:

$$\Delta\theta = 52\ \mu\text{rad or } 10.6\text{ arcseconds (self–loading)};\ \Delta\theta = 32\ \mu\text{rad or } 6.7\text{ arcseconds (point loading)}$$

In themselves, these tilts are very small. The impact these might have in introducing additional off-axis aberrations into the system is likely to be negligible. Inevitability, therefore, we should be pre-occupied with the boresight errors introduced by these substrate distortions, rather than the impact on image quality.

In the light of this discussion, the most fruitful approach is to undertake a paraxial analysis of the system and to characterise any movement in the image position. In the specific example, as illustrated in Figure 19.6, the focusing lens sees an angular shift in the chief ray that is equal to $C(x) \times \Delta l$. For a lens of focal length, f, this produces a lateral shift in the focal spot of $C(x) \times \Delta l \times f$. However, we must not ignore any curvature of the optical bench between the lens and its focus. Therefore, the lateral image shift, Δz, is given by:

$$\Delta z = C(x)\left[\Delta l f + \frac{f^2}{2}\right] \tag{19.33}$$

For a lens with a focal length of 150 mm and Δl equal to 1 m, then, for the two distortion scenarios, the focal point shifts are given by:

$$\Delta z = 8.3\ \mu\text{m (self–loading)};\ \Delta z = 5.2\ \mu\text{m (point loading)}$$

Relative to the pixel size of a detector, these shifts are not insignificant. The assumptions and exercises presented here are relatively elementary. However, using these basic tools, it is possible for the reader to extend this analysis to slightly more ambitious scenarios. Since all practical problems in opto-mechanics are inherently 'small signal' – all deflections are small compared to substrate thickness, then the assumptions underlying plate theory are inherently valid. In obtaining a basic, initial understanding of an optical design, the engineer must be prepared to make some imaginative simplifications to the problem, in order to render it tractable. Thereafter, should this analysis highlight potential problem areas, or should the structure geometry be too complex, then the engineer must proceed to detailed Finite Element modelling.

19.3.3 Simple Distortion of Optical Components

19.3.3.1 Introduction

In this subsection we will consider the analysis of distortion in optical components, presenting the most basic scenarios. For example, we may be interested in the mounting of mirror components and to understand the impact of self-loading. This specific scenario is amongst a suite of tractable problems where uniform loading is imposed upon a geometrically simple structure. Other common examples might include optical windows that are deployed in vacuum or pressurised systems.

It is clear that, for mirrors and lenses, the impact of self-loading becomes more critical for larger component sizes. The most typical problem that is presented is the loading of a uniform disc, such as a mirror blank. Initially, we will consider the situation where the optic is oriented with the surface normal parallel to the gravity vector and the optic supported at the rim. The latter constraint cannot be circumvented in the case of a transmissive (lens) component. This is an important consideration in the design of telescope optics, as this limitation imposes severe mechanical constraints (lens thickness) on the design. In the case of mirror components, is it possible to offer distributed support, as will be discussed later.

19.3.3.2 Self-Weight Deflection

As a most general rule of thumb, the thickness of mirror or lens blanks is defined by some ratio of the diameter to thickness. This is typically of the order of a factor of six for mirror blanks. To make the analysis a little more generic, we assume that the disc of diameter, D and uniform thickness, t, is subject to a uniform force per unit area, P. Furthermore, we may assume that the problem is defined by its radial symmetry, so we may cast Eq. (19.16) in radial coordinates

$$\frac{\partial^4 z}{\partial r^4} + \frac{2}{r}\frac{\partial^3 z}{\partial r^3} - \frac{1}{r^2}\frac{\partial^2 z}{\partial r^2} + \frac{1}{r^3}\frac{\partial z}{\partial r} = \frac{12P(1-\nu^2)}{Et^3} \tag{19.34}$$

As a general solution to this equation, we will propose a quartic equation of the type generated for the optical table exercise. Symmetry dictates that only even powers of r are acceptable and, as a boundary condition, we assume that the bending moment vanishes at the edge ($r = D/2$). An alternative condition is the so called 'clamped' condition, whereby an external mechanical clamp maintains the gradient of the displacement at zero at the edges. In practice, the assumption of 'free edges' is more realistic. However, plate analysis for the two-dimensional problem produces a slightly amended version of the free edge boundary condition:

$$\frac{\partial^2 z}{\partial r^2} + \nu\frac{1}{r}\frac{\partial z}{\partial r} = 0 \tag{19.35}$$

This gives the following solution:

$$z(r) = \frac{3PD^4(1-\nu^2)}{16Et^3}\left[\left(\frac{r}{D}\right)^4 - \frac{3+\nu}{2(1+\nu)}\left(\frac{r}{D}\right)^2\right] \tag{19.36}$$

First, we will consider the impact of this distortion on the wavefront error produced by a mirror. It must be remembered that, for a mirror surface, the imposed wavefront error, $\Phi(r)$ is double the change in the surface displacement. Inspecting Eq. (19.31) it is clear that the additional wavefront error may be presented as two distinct contributions. First, the quadratic term will produce pure defocus (Zernike 4), whereas the quartic term is associated with spherical aberration (Zernike 11). Assuming that the diameter, D, is equivalent to the aperture of the optic, then the rms contributions to the wavefront error may be computed as follows:

$$\Phi(Zernike4) = \frac{\sqrt{3}\times 10\times PD^4(5+\upsilon)(1-\nu)}{256Et^3} \tag{19.37a}$$

$$\Phi(Zernike11) = \frac{PD^4(1-\upsilon^2)}{\sqrt{5}\times 256Et^3} \tag{19.37b}$$

Naturally, for self-loading, the pressure P may be expressed in terms of the mass per unit area, which is related to the density of the material, ρ, and the thickness, t. This gives the two contributions to the wavefront error of the mirror as:

$$\Phi(Zernike4) = \frac{\sqrt{3}\times 10\times \rho\times g\times D^2\alpha^2(1-\nu^2)}{256E} \tag{19.38a}$$

$$\Phi(Zernike11) = \frac{\rho\times g\times D^2\alpha^2(1-\nu^2)}{\sqrt{5}\times 256E} \tag{19.38b}$$

α is the ratio of the diameter to thickness.

It is now possible to characterise the wavefront error produced for a typical mirror material (fused silica) as a function of mirror size. Of primary concern is the spherical aberration contribution; the defocus term may be compensated by focus adjustment. For this exercise, we will delineate the size by the mirror diameter, D, and express the wavefront error in terms of the mirror aspect ratio (ratio of diameter to thickness), r, as expressed in Eqs. (19.38a) and (19.38b). In this example, we will choose fused silica as a representative substrate material. Its modulus of elasticity is 7.25×10^{10} Nm^{-2} and it has a Poisson's ratio of 0.17. The density

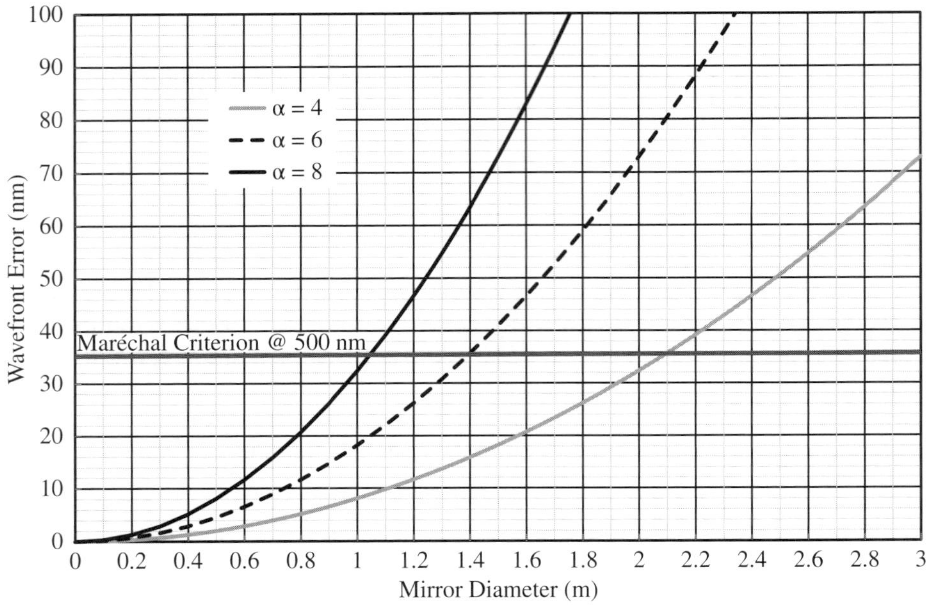

Figure 19.8 Self-deflection induced aberration in fused silica mirror.

is 2200 kgm^{-3}. Figure 19.8 shows a plot of the spherical aberration produced by a fused silica mirror supported at the edges, as a function of mirror diameter. For comparison, the Maréchal criterion for diffraction limited performance at 500 nm is shown. This analysis suggests that once the mirror diameter approaches 1m and greater, peripheral support becomes inadequate. As will be seen a little later, alternative strategies must be adopted.

Of course, all this analysis is applied to a uniform structure. For large mirrors, the practice is to provide a lighter 'honeycomb' substrate, by removing substrate material through milling and creating a ***lightweight*** structure. As with the sandwich structure of the optical table seen earlier, this can create a stiffer structure by reducing the density, but not reducing the bending stiffness in proportion. However, the benefits of lightweighting lie mainly in weight reduction of the mirror itself and greatly reducing the mass and complexity of the associated support structure. Naturally, this consideration applies to terrestrial applications. For space applications, the benefits of lightweighting are rather more obvious.

19.3.3.3 Vacuum or Pressure Flexure

In this section, we will consider the impact of flexure of a vacuum window. More specifically, we are interested in the ***defocus*** that would be produced in a collimated beam entering a vacuum system; we are not concerned, in this exercise, with spherical aberration. The scenario is sketched out in Figure 19.9, with the deformation greatly exaggerated.

According to Figure 19.9, the deformed plane window acts as a lens. The focusing effect has its origin in two distinct mechanisms. First, if we assume that both surfaces of the window adopt the same radius of curvature in the central region, R, then the finite thickness of the lens produces some focusing power. If the refractive index of the window is n, then the effective focal length of the distorted window according to this mechanism, f_w, is given by:

$$\frac{1}{f_w} = \frac{(n-1)^2 t}{nR^2} \tag{19.39}$$

In addition to the focusing power produced by the glass window itself, the curved interface between the air and the vacuum also introduces focusing power. If the focal length contribution of this mechanism is denoted

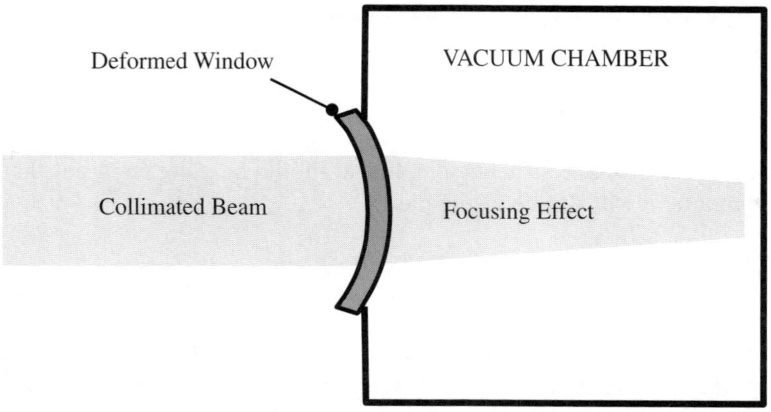

Figure 19.9 Impact of vacuum window deformation.

by f_v and the refractive index of air is n_{air}, then the value is given by:

$$\frac{1}{f_v} = \frac{(n_{air} - 1)}{R} \tag{19.40}$$

The deformation radius of the window, towards its centre is given by Eq. (19.36):

$$R = \frac{16Et^3}{3D^2P_A(3 + \upsilon)(1 - \upsilon)} \tag{19.41}$$

P_A is atmospheric pressure, 1.01×10^5 Nm^{-2}.

We now illustrate this analysis by a concrete example. A vacuum window of fused silica, 25 mm thick, is supported at a diameter of 340 mm. The edges may be assumed to be unstressed (no bending moment). Its modulus of elasticity is 7.25×10^{10} Nm^{-2} and it has a Poisson's ratio of 0.17. We are required to calculate the bending radius of the window at the centre. In addition, assuming the refractive index of the silica at 633.8 nm is 1.457 and the refractive index of air at the same wavelength is 1.000277, we wish to determine the focal power of the distorted window. Finally, the rms wavefront error produced by this distortion on a 80 mm diameter collimated beam must be evaluated.

The bend radius of the window is given by Eq. (19.41):

$$R = \frac{16 \times (7.25 \times 10^{10}) \times 0.025^3}{3 \times 0.34^2 \times (1.01 \times 10^5) \times 3.17 \times 0.83} = 196.7\ m$$

The focusing effect of the window is given by:

$$\frac{1}{f_w} = \frac{(n-1)^2 t}{nR^2}\ \frac{1}{f_w} = \frac{(0.457)^2 \times 0.025}{1.457 \times 196.7^2} = 9.26 \times 10^{-8} m^{-1}$$

The focusing effect of the vacuum is given by:

$$\frac{1}{f_v} = \frac{(n_{air} - 1)}{R} = \frac{0.000277}{196.7} = 1.408\text{x}10^{-6} m^{-1}$$

It is clear that the focusing effect of both mechanisms is extremely small, with the vacuum effect preponderating. Overall, the nett focusing effect is the sum of the two effects:

$$\frac{1}{f} = \frac{1}{f_w} + \frac{1}{f_v} = 9.26 \times 10^{-8} + 1.408 \times 10^{-6} = 1.5 \times 10^{-6} m^{-1}$$

This is equivalent to a focal length of nearly 700 km! Clearly, the effect in this instance is small. The root mean square defocusing for a beam of radius, a, is given by:

$$\frac{1}{f}\Delta\Phi = \frac{a^2}{\sqrt{48}f} = 0.347\ nm$$

As expected, the impact on defocusing, in this instance, is negligible. It also should be quite apparent that the impact on the spherical aberration component will also be negligible.

19.3.4 Effects of Component Mounting

19.3.4.1 General

The previous analysis focused on the useful, but narrow, scenario of simple flexure in optical components and support structures. Of course, in many practical applications, we are concerned with the impact of localised stresses produced by component mounting. In these cases, we must have recourse to more generic 'rules of thumb', or more detailed FEA. Very commonly, lens components are constrained by imposing forces at the edge of the optic. This might, for example, be imposed by a retaining ring and/or mounting shoulders or other features machined into a lens barrel. Naturally, this process results in the significant concentration of stresses around the periphery of the component. Although these stresses will clearly influence the form of the lens, they may be sufficient to cause complete mechanical failure.

In much of the previous analysis, we have been solely concerned with the deflection of surfaces. At no time have we actually analysed the internal stresses. For the most part, the impact of the internal stresses per se, is of secondary importance in the optical performance. Therefore, in considering the impact of mechanical stress on image quality and alignment, it is the change in surface geometry that is of primary importance.

There is one very specific exception to this general rule. Wherever stress is induced in a component where light is transmitted, we must be concerned about stress-induced birefringence. The impact of stress within a nominally isotropic material, such as glass, is to remove the inherent symmetry and to create a birefringent material. The refractive index change between the 'ordinary' and 'extraordinary' axes is, according to Eq. (8.49), proportional to the difference in stress and the stress optic coefficient, C:

$$\Delta n = C(\sigma_1 - \sigma_2)$$

Typical values of C for vitreous glassy materials are of the order of $2.5 \times 10^{-12}\ m^2\ N^{-1}$. In the previous vacuum window problem, the peak local stress amounted to about 50 MPa. This would produce a refractive index change of the order of one part in 10^5. This is not altogether insignificant. These levels of birefringence may be significant where one is designing an instrument for sensitive measurements of polarisation. However, in this specific example, the impact of stress-induced birefringence is muted by the symmetry of the stress induced about the window centre line. As such, any birefringence induced in the first half of the window will be cancelled by that experienced in the second half of the window. These sympathetic considerations do not necessarily apply for stresses produced in mounting components. It is important, in critical designs, that such stresses are analysed, by Finite Element techniques, if necessary.

The most significant stress, from an optical perspective, is the internal stress that is 'locked into' the window during the thermal processing cycle. That stress does produce significant birefringence and this must be carefully analysed in critical applications. Otherwise, the most significant optical stresses are those that are locally produced by large constraining forces inherent in mechanical mounting of components.

19.3.4.2 Degrees of Freedom in Mounting

Without any constraints, a solid body has six degrees of freedom in motion – three translational and three rotational. An ideal mounting solutions allows for a minimum of six constraints. Whilst additional mounting constraints would seem to confer greater solidity, there is always a tendency for additional constraints to produce geometrical conflicts which can only be resolved by elastic deformation of the part. Therefore an optimum design should, in fact, use the minimum number of constraints consistent with the geometry.

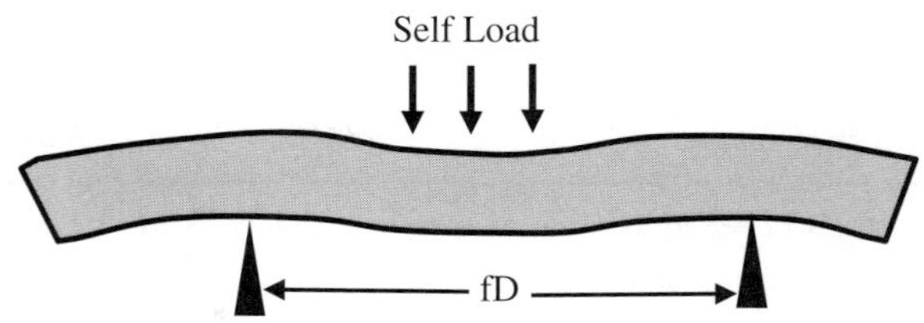

Figure 19.10 Mirror supported by a ring mount.

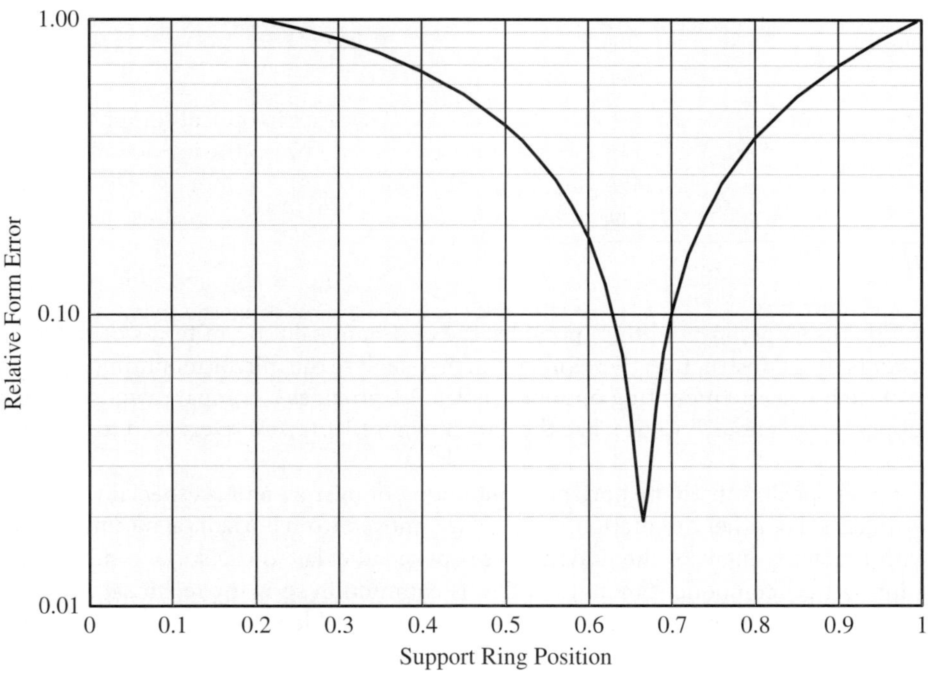

Figure 19.11 Impact of support ring position on mirror deflection.

19.3.4.3 Modelling of Mounting Deformation in Mirrors

Our previous analysis of deformation in mirrors dealt with the simple case of a mirror supported at its edge. We take this treatment a little further and apply it to more realistic mounting solutions. Supporting a mirror at its edge is clearly a sub-optimal solution in terms of the self-load distortion induced. We begin by examining a mirror that is supported by a ring located at some fraction, f, of its radius. Most particularly, we seek a value for f that minimises the rms distortion. The scenario is illustrated in Figure 19.10.

Varying the diameter of the ring enables optimisation of the support and significant reduction in the rms error attributable to self-loading. Figure 19.11 graphically shows how the rms distortion varies with mounting ring radius, expressed as a fraction of the part diameter. As Figure 19.11 shows, there is an optimum ratio, corresponding to about 67% of the mirror diameter.

Optimisation of the mounting not only enables the form error to be reduced for a given mirror thickness, but also facilitates the design of thinner, lighter, mirror substrates for a given performance requirement.

Although such a mounting strategy would seem to provide a highly optimised solution an additional problem is introduced with the use of a ring mounting structure. As outlined in the previous sub-section, one has to be extremely careful not to over-constrain the mounting solution. In this instance, the shape of the mounting ring would have to be a perfect fit to the form of the corresponding mating surface on the mirror. Any deviation in this fit will be accommodated by distortion in the mirror. Therefore, the preferred solution for mirror mounting is by bonding the reverse surface of the mirror at a limited number of discrete points.

Table 19.1 Relative mirror distortion for different mounting strategies.

Support mode	Relative distortion
Support over entire circumference	1
Clamped at circumference	0.226
3 point support at circumference	1.638
Ring at 68.1% of circumference	0.0338
3 point support at 68.1% radius	0.401
6 point support at 68.1% radius	0.0495
3 point support at 65% radius	0.381
3 point support at 70% radius	0.424
Support pillar at centre	1.457
Support along diameter	1.139

In this case, mounting at 3 points (3 point mounting) provides the optimum solution in terms of reducing distortion-inducing constraints. It is constructive, therefore, to further analyse the mirror mounting problem by considering a number of such discrete mounting options. Table 19.1 provides a comparative listing of a number of competitive support strategies. All values, like the data in Figure 19.11, are referenced to the edge support scenario.

The preceding analysis applies specifically to the horizontal mounting of mirrors and is especially relevant to large astronomical instruments. For other orientations, not so far removed from horizontal mounting, calculations may proceed with the component of the gravity vector perpendicular to the surface substituted in the analysis. However, for vertical component mounting, that is common in so many technical and consumer applications, the preceding constraints are inadequate to define the problem. Fundamentally, vertical mounting produces a distortion that is not rotationally symmetric. In fact, the primary aberration produced is asymmetric in nature. Furthermore, it must be realised, intuitively, that a simple plane surface engenders no distortion perpendicular to its surface when oriented vertically. Unlike the horizontally oriented mirror, any distortion produced is, in some way, dependent upon the original form of the mirror. An empirical formula exists to estimate the rms distortion produced in a vertically oriented mirror. Naturally, this depends upon the mode of mounting. We will consider two different mounting strategies, whereby the mirror is supported in a 'V block' and alternatively, the mirror is supported by a belt. These strategies are illustrated in Figures 19.12a and 19.12b.

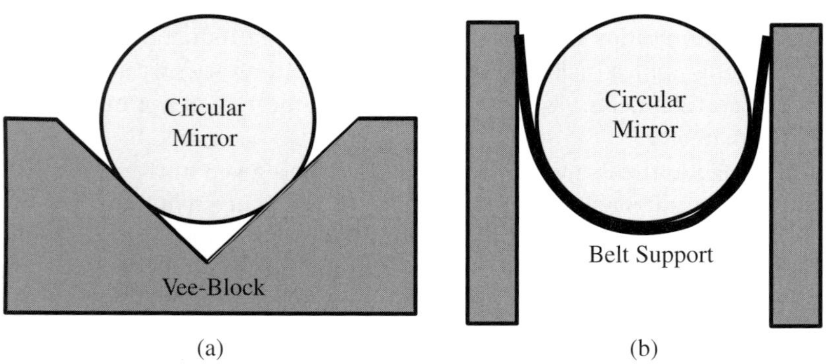

Figure 19.12 (a) Mirror vee block support (b) Mirror belt support.

First, we define a mechanical shape factor, f, for the mirror, an indication of the physical rigidity of the mirror. Specifically, this shape factor takes into account the base radius, R, of the mirror, as well as its diameter, D, and thickness:

$$s = \frac{D^2}{8Rt} \tag{19.42}$$

Equation (19.42) suggests, as expected, that a higher base radius confers greater resistance to distortion. The rms distortion induced by loading is then given by the following empirical expression:

$$\Phi = 100a_2 \frac{\rho g D^2 s}{2E} \tag{19.43}$$

$a_2 = 0.11$ for vee-block support; $a_2 = 0.031$ for belt support

As with the horizontal mounting scenario, the fundamental scaling with the square of the diameter is preserved. However, dependence upon the thickness of the part is less acute. To gauge the scale of the distortion produced, it is useful to introduce a realistic laboratory scenario with a fairly large bench mounted mirror. The mirror is fabricated from fused silica and is 300 mm in diameter, with a thickness of 50 mm; it has a base radius of 3000 mm. In this instance, the mirror is to be supported in a vee-block arrangement.

Firstly, we need to calculate the shape factor:

$$s = \frac{D^2}{8Rt} = \frac{300^2}{8 \times 3000 \times 50} = 0.075$$

The rms distortion is given by:

$$\Phi = 100a_2 \frac{\rho g D^2 s}{2E} = 100 \times 0.11 \frac{2200 \times 9.81 \times 0.3^2 \times 0.075}{2 \times (7.5 \times 10^{10})} = 1.06 \times 10^{-8}$$

The distortion thus amounts to about 11 nm rms.

Of course, it must be emphasised that the full evaluation of any design should, if necessary, be detailed by FEA. Nonetheless, analysis of this type is useful for a preliminary sketch and in assessing the order of any distortion effects.

19.3.4.4 Modelling of Mounting Stresses in Lens Components

The mounting of circularly symmetric lens components is generally accomplished by mounting them in a lens barrel and constraining them at the edges. The geometry is illustrated in Figure 19.13.

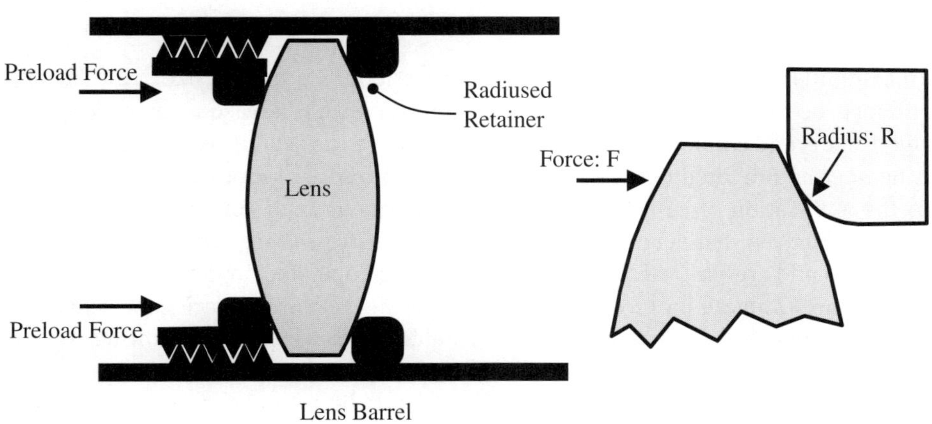

Figure 19.13 Lens mounting in a lens barrel.

Table 19.2 Allowable stresses for some optical materials.

Material	Allowable stress (MPa)
Sapphire	170
Silicon	55
BK7	40
Silica	35
Zinc Sulfide	29
SF5	27
F2	26
Zinc Selenide	17
Calcium Fluoride	17

An approximate magnitude of the mounting stresses at the edge of the lens may be computed from an empirical formula. The stress, σ, is dependent upon the applied mounting force, F, the radial position of the contacting retainer, a, and the radius of curvature of the retainer, R.

$$\sigma = 0.4\sqrt{\frac{FE}{2\pi Ra}} \tag{19.44}$$

E is the retainer elastic modulus

The stress may be controlled by selecting the curvature of the retainer and adjusting the preload, F. Since the preload is applied by the rotation of a threaded component, the preload force may be directly related to the torque imposed upon the threaded component. This may be measured. In terms of judging the maximum imposed stress, this is governed, from a mechanical perspective, by fracture mechanics. Lens materials do not possess ductile strength, so their failure is governed by catastrophic crack propagation. As such, we are interested, primarily, in the maximum crack length, a_c, left by the lens ***grinding*** process. The concept of fracture toughness was touched on in Chapter 9. Given a maximum crack length of a_c, left and a fracture toughness of K_c, the critical stress, σ_c, at which failure is expected to occur is:

$$\sigma_0 = \frac{K_c}{\sqrt{\pi a_c}} \tag{19.45}$$

As a rule of thumb, the maximum crack length for a polished surface is about 3 times the size of the smallest size of grit used in the grinding process. A crack length of a few microns represents a reasonable working estimate. If we take this rough estimate, and employ a safety factor of 10, then we can set out some recommended maximum stress levels using the fracture toughness data from Table 9.6. These are set out in Table 19.2.

The above data refer to the application of compressive stress by the retainer. Of course, ultimately, it is the tensile stresses produced in the glass than lead to failure. These tensile stresses are produced around the periphery of the load bearing area and are, empirically, somewhere between one-sixth and one-quarter of the compressive stress. With the above evaluation of local stress levels within the glass, we are concerned only about mechanical effects. Where there are concerns about stress-induced birefringence, mounting stresses should be kept below about 4 MPa – much lower than the values listed above.

We now turn to a specific example, where a 50 mm diameter BK7® lens is to be mounted in a lens barrel using a retainer whose base radius is 1 mm. The diameter of the aluminium retaining ring is 45 mm and its elastic modulus is 6.9×10^{10} Nm^{-2}. The preload force may be calculated from Eq. (19.44). The maximum allowable stress for BK7 is 40 MPa. From Eq. (19.38), we have:

$$F = 25\pi\frac{Ra\sigma^2}{2E} = 78.5 \times \frac{10^{-3} \times (2.25 \times 10^{-2}) \times (4\text{x}10^7)^2}{2 \times (6.9\text{x}10^{10})} = 20.5\,N$$

The required preload is thus 20.5 N. In practice this preload is realised by applying a torque to the threaded mount. The relationship between the preload and the applied torque depends upon the nature of the contact between the two sets of mating threads. As such, the correspondence between the two can be somewhat variable. Nonetheless, with that caveat in mind, a useful empirical relationship between the preload, F, and the applied torque, Γ, is given in Eq. (19.46).

$$\Gamma = 0.2D_pF \ (D_p \text{ is the thread pitch diameter}) \tag{19.46}$$

In the above example, assuming a 55 mm thread pitch diameter, the required torque amounts to 0.23 Nm.

19.4 Basic Analysis of Thermo-Mechanical Distortion

19.4.1 Introduction

As with the simple mechanical modelling presented in the previous section, much may be understood about the thermal stability of an optical design through some simple and basic analysis. Two factors drive the thermal stability of an optical design. First, and most obviously, thermal expansion changes the geometrical relationship between optical components, for example changing the location of the focal plane relative to the detector. Second, the temperature dependence of material refractive indices changes the focusing power of individual optical components. However, for the most part, we will focus on the impact of thermal expansion.

The most obvious impact of thermal expansion is simple unconstrained dimensional change. Typical thermal expansion coefficients are of the order of 10 ppm and this has the potential to change the axial separation of optical components and the location of the focal plane. For example a 1 m focal length lens integrated onto an aluminium breadboard could expect to see an axial focal shift of about 0.26 mm for a temperature excursion of 10 °C, ***considering expansion of the breadboard alone.*** The propensity for thermal expansion in a material is described by its ***coefficient of thermal expansion*** (CTE), α. This CTE represents the proportional increase in the ***unstressed*** length of a material as a function of temperature. As a working approximation, over a restricted temperature range, the CTE is broadly constant for some materials. However, more generally, it should be accepted that the CTE tends to vary with temperature to a degree. More formally, the CTE is expressed as follows:

$$\alpha(T) = \frac{1}{L(T)}\frac{dL(T)}{dT} \ L(T) \text{ is the unstressed length and } T \text{ the temperature} \tag{19.47}$$

Table 19.3 gives the thermal expansion for a range of materials of practical interest around 20 °C. Naturally, this includes both optical materials and materials favoured in the design of support structures.

In practice, thermal expansion is rarely unconstrained. As such, consideration of thermal expansion must be integrated into the overall mechanical model. In particular, it is often the case that contiguous juxtaposition of different materials (i.e. with differing thermal expansion) creates conflicts. As a consequence of these constraints and conflicts, thermally induced stresses are produced in a system. These inevitably lead to mechanical distortion and, in extreme cases, to material failure. It is perhaps instructive to understand the stresses generated where the freedom of a material to expand is entirely removed. The strain, ε, in a material may be broken down into two components, the thermal strain, ε_T, and the elastic strain, ε_E. This may be written as:

$$\varepsilon = \varepsilon_T + \varepsilon_E = \alpha\Delta T + \varepsilon_E \tag{19.48}$$

Table 19.3 Thermal expansion for some useful materials.

Material	CTE (ppm°C^{-1})	Material	CTE (ppm°C^{-1})
Aluminium	23	ZnSe	7.1
Brass	19	ZnS	6.9
Copper	17	GaAs	5.7
Steel	11	Sapphire	5.3
Invar	1[a)]	Kovar	5
CaF_2	18.9	Silicon Carbide	2.8
F2	8.2	Silicon	2.6
SF5	7.9	Fused Silica	0.5
BK7®	7.1	Zerodur	~0

a) Expansion for invar is variable around ambient temperature.

From Eq. (19.48), if we constrain a material so that it cannot expand, then the elastic strain induced is given by:

$$\varepsilon_E = -\alpha\Delta T \text{ and } \sigma_E = -\alpha E\Delta T \tag{19.49}$$

Taking the example of calcium fluoride with a CTE of 18.8 ppm and an elastic modulus E of 7.6×10^{10} Nm^{-2}, we find that a stress of 1.4 MPa°C^{-1} of temperature excursion is induced. Calcium fluoride is well known for its sensitivity to thermal shock.

19.4.2 Thermal Distortion of Optical Benches

An important practical case for thermo-mechanical distortion is the analysis of optical benches. In some respects, this analysis is analogous to the simple mechanical plate theory previously discussed. Although comprehensive analysis of real optical systems defies the simple analysis presented here, some imaginative simplification of a more complicated problem can yield some useful insights.

We might imagine that an optical system is built upon on some kind of thin planar structure, as per plate theory. However, in this instance, we introduce some asymmetry into the structure. This might, for example, be the creation of a heterogeneous bonded structure with plates of different materials bonded together. The simplest example is a 'bimetallic strip' structure where two materials of differing thermal expansion coefficients are bonded together. This is illustrated in Figure 19.14.

If the optical bench is heated and material 2, as shown in Figure 19.14, has a tendency to expand more, then this will be accommodated by the distortion shown. That is to say, the 'strip' will tend to bend, with the high expansion material adopting the larger bend radius. In so doing, each strip seeks to minimise the amount of elastic stress along its centreline as indicated by Eq. (19.49). However, with the two strip joined, there is a conflict in that this action produces bending in the beam which, in itself produces stress. In fact, an equilibrium is attained whereby the overall strain energy is minimised, offsetting the strain arising from

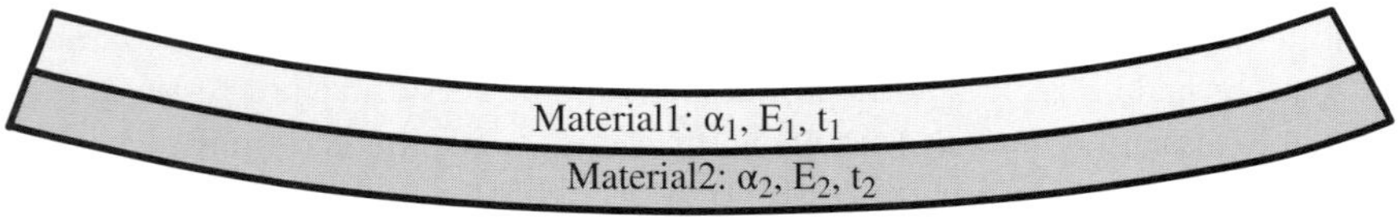

Figure 19.14 Composite optical bench.

bending against that due to the centreline thermal strain. A simple elastic model may be used to predict the bending. We define the bending in terms of the curvature, C:

$$C = \frac{1}{R} = \frac{\partial^2 z}{\partial x^2} \tag{19.50}$$

If one assumes that the strip is unstrained when bonded at temperature, T_0, and that the average temperature of the first material is T_1 and the second material T_2, then the strip curvature is given as follows:

$$C = \frac{6t_1 t_2(t_1 + t_2)[\alpha_2(T_2 - T_0) - \alpha_1(T_1 - T_0)]}{(E_1/E_2)t_1^4 + 4t_1^3 t_2 + 6t_1^2 t_2^2 + 4t_1 t_2^3 + (E_2/E_1)t_2^4} \tag{19.51}$$

Equation (19.51) shows that any bending in the optical bench is dependent not only on material inhomogeneity, but also on any temperature difference through the thickness of the bench. Such an eventuality could be brought about by uneven thermal loading of the bench. Of course, the natural lesson to take from this would be the avoidance of all material and thermal inhomogeneity.

We now turn to a (slightly idealised) example involving a simple micro-optic device designed to couple light into an optical fibre. The optical fibre is integrated into a silicon 'V-groove' structure. Onto the same structure is attached a semiconductor laser chip plus a focusing lens. The lens has a focal length of 3 mm, and its task is to image the laser output onto the input facet of the single mode fibre, whose characteristic mode size is 5.0 μm. We assume that the mode size of the laser is 2.0 μm, and the lens provides a magnification of 2.5 times, matching the mode size and optimising the coupling. Mechanically, we may think of the optical bench consisting of a 1 mm thick strip of silicon cemented to a 2 mm strip of aluminium underneath. The system is undistorted and perfectly aligned at a temperature of 20 °C. Subsequently, the entire device is warmed to 50 °C. Given the thermal expansion of silicon and aluminium (2.6 ppm and 23 ppm respectively) and their elastic moduli (1.5×10^{11} Nm^{-2} and 6.9×10^{10} Nm^{-2} respectively) we may calculate the curvature of the optical bench. In addition, given the layout of the system, the movement in the focused spot at the fibre may be computed and thence the loss in optical coupling produced by the temperature excursion may be deduced.

First, we need to calculate the bending produced by the temperature excursion using Eq. (19.51):

$$C = \frac{6t_1 t_2(t_1 + t_2)[\alpha_2(T_2 - T_0) - \alpha_1(T_1 - T_0)]}{(E_1/E_2)t_1^4 + 4t_1^3 t_2 + 6t_1^2 t_2^2 + 4t_1 t_2^3 + (E_2/E_1)t_2^4}$$

$$= \frac{6 \times 1 \times 2 \times (1+2)[(2.3 \times 10^{-5}) \times 30 - (2.6 \times 10^{-6} \times 30)]}{(15/6.9) \times 1 + 4 \times 1 \times 2 + 6 \times 1 \times 2^2 + 4 \times 1 \times 2^3 + (6.9/15) \times 2^4}$$

$$C = 3\text{x}10^{-4}\ \text{mm}^{-1}$$

The curvature of the bench is 3×10^{-4} mm^{-1}, equivalent to a bend radius of 3300 mm. The impact of this distortion is to produce misalignment of the laser and the fibre. We know that the focal length of the lens is 3 mm and that the magnification is 2.5. It is straightforward to calculate the object and image distances: $\frac{1}{u} + \frac{1}{v} = \frac{1}{f} = \frac{1}{3}$ and $\frac{v}{u} = M = 2.5$; u = 4.2 mm and v = 10.5 mm.

The scenario might look thus:

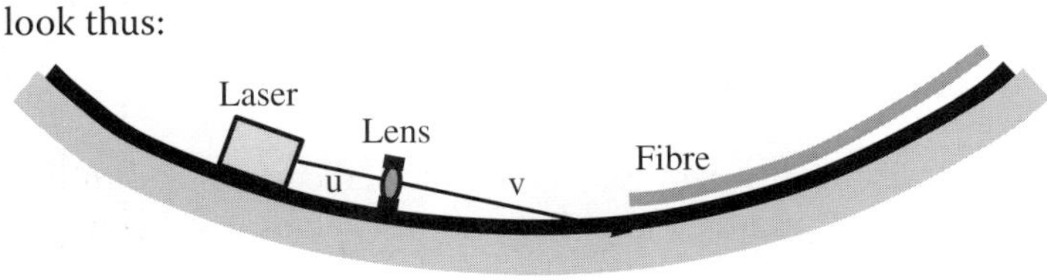

The displacement, Δz of the imaged beam at the fibre is given by the object and image distances, u and v, and the bench curvature, C, and may be calculated by projecting the chief ray through the centre of the lens:

$$\Delta z = \frac{C}{2}(u^2 + uv - v^2) = \frac{3\text{x}10^{-4}}{2}(4.2^2 + 4.2 \times 10.5 - 10.5^2) = -0.0073\text{mm}$$

Thus, the beam has moved by 7.3 μm with respect to the centre of the fibre. We were told that the characteristic mode size of the fibre is 5.0 μm. From Chapter 13, Eq. (13.58), we know that the coupling coefficient, C_{coup}, is given by:

$$C_{coup} = e^{-\frac{\Delta z^2}{w_0^2}}$$

Δz is 7.3 μm and w_0 is 5.0 μm and therefore the coupling is 11.9%. This is a substantial reduction in the optical coupling and illustrates the impact of a seemingly modest thermal stress. In practice, the mechanical construction of an optical package is rather more complex than presented here. Nonetheless, used with some imagination, analysis of this type may be used to provide some kind of feel for the impact of thermo-mechanical distortions. Of course, in practice, detailed analysis requires the use of finite element modelling.

19.4.3 Impact of Focal Shift and Athermal Design

As well as the impact of optical bench distortion, we need to consider the impact of focal shift produced by changes in the length of the optical path. A design in which we are able to eliminate any thermal shift in the output focal plane is referred to as an ***athermal design***. As outlined previously, there are three factors we need to consider in evaluating this problem. First, expansion of the optical bench or substrate, upon which the components are mounted, will produce a change in the axial separation of those components. Second, a proportional change in the radii of optical surfaces will be produced by thermal expansion; this will change the focusing power of those elements. Taken together, those two effects suggest a fairly obvious implementation of an athermal design. Should the thermal expansion of the optical bench material match that of the components that populate it, then the two effects are entirely complementary and an athermal design will result. The third factor to consider is the variation in refractive index with temperature of any lens material. As indicated in Chapter 9, this temperature dependence in the refractive index may be either positive or negative depending upon the material. Typically, it amounts to a few tens ppm°C^{-1}

Opto-mechanical substrates are often composite in nature, so that estimating their thermal expansion is a non-trivial problem. In general, accurate evaluation of their thermal behaviour requires detailed finite element modelling. However, as with many previous problems, useful insights may be gained by simpler analytical techniques. What we are concerned with here is a composite, perhaps laminar structure, involving layers of different materials. Each layer in the laminar structure is constrained to expand by the same amount. As each material has a different thermal expansion, clearly the thermal strain invested in each layer cannot be cancelled out entirely. Those layers with the lowest expansion will experience tensile stress, being stretched by the other more thermally expansive materials. The opposite applies to those layers with the highest thermal expansion; they will be under compressive stress. Overall, the system will seek to minimise the total strain energy. If we consider a sandwich structure with n layers, with the thickness, CTE, elastic modulus and Poisson's ratio of the ith layer represented as t_i, α_i, E_i, and υ_i respectively, then the aggregate CTE, $\alpha_{i,}$ is given by:

$$\alpha_0 = \sum_1^n \frac{E_i t_i}{(1-\upsilon_i)}\alpha_i / \sum_1^n \frac{E_i t_i}{(1-\upsilon_i)} \tag{19.52}$$

In the case of our simple optical bench with the 1 mm thick silicon bonded to the 2 mm aluminium, the aggregate thermal expansion is 12.9 ppmK^{-1}. That is to say, as well as undergoing bending, the bench also stretches.

Armed with this specific example we can now examine the impact of thermal expansion on the location of the focused image. We assume, in this example, that the 3 mm focal length lens is made from BK7 We now turn to the impact of the thermal property of the lens material. Although, as we discussed previously, two factors are at play in determining the change in focusing power – thermal expansion and refractive index variation – we may subsume them into one parameter, Γ, the optical power coefficient. Further details of this are presented in Chapter 9 Eq. (9.13) and, if we represent the temperature coefficient of refractive index as β and the CTE as α, then the optical power coefficient is given by:

$$\Gamma = -\frac{\Delta f}{f} = \frac{\beta}{n-1} - \alpha \tag{19.53}$$

In this case, the optical power coefficient for BK7 is equal to −2.7 ppmK^{-1}. To evaluate the overall impact in our optical bench problem, we may analyse the shift in the paraxial focus using simple matrix analysis:

$$\begin{bmatrix} A & B \\ C & D \end{bmatrix} = \begin{bmatrix} 1 & v(1+\alpha_0\Delta T) \\ 0 & 1 \end{bmatrix} \begin{bmatrix} 1 & 0 \\ (1/f)(1+\alpha_0\Delta T) & 1 \end{bmatrix} \begin{bmatrix} 1 & u(1+\alpha_0\Delta T) \\ 0 & 1 \end{bmatrix}$$

In particular, we are interested in the change in the matrix element, B. In fact, the focal shift is given by ΔB/D. For $u = 4.2$ mm, $v = 10.5$ mm, $f = 3$ mm, $\Gamma = -2.7$ ppmK^{-1}, and $\alpha = 12.9$ ppm K^{-1}, as previously advised, then the focal shift amounts to 23.6 μm. This is not an insignificant shift and could be reduced by use of an alternative lens material. In this instance, the BK7 power coefficient serves to further re-inforce the impact of optical bench thermal expansion. In fact, for a system of arbitrary complexity, the trade-off between thermal expansion and optical power may be expressed more succinctly. If we arbitrarily retain the same object, image, and optics locations, we may understand any focal shift entirely in terms of an effective change in the focal power. The essential principal of an athermal design is to ensure the optical power coefficient and the substrate thermal expansion are identical.

When analysing a more complex system, with multiple components, the effective contribution of each individual component to the defocus wavefront error is additive. Assuming all components have an identical optical power coefficient and the substrate is homogenous, then the change in focal power is given by the effective system focal power multiplied by the product of the difference of the optical power coefficient and the expansion coefficient and the temperature excursion. In other words:

$$\Delta\left(\frac{1}{f_{system}}\right) = \left(\frac{1}{f_{system}}\right)(\Gamma - \alpha)\Delta T \tag{19.54}$$

19.4.4 Differential Expansion of a Component Stack

As part of a mechanical design, lenses, and optical components are mounted to some common platform. Different lens components and mounts may possess differing coefficients of thermal expansion. As a result, the vertex of lens components will experience lateral displacement relative to some common optical axis. For example, consider two lenses mounted on a common optical bench, each with a common axial height of 50 mm. If one of the mounts is made from stainless steel (CTE 11 ppm) and the other from aluminium (CTE 23 ppm), then a temperature excursion of 5 °C will produce a relative shift of 3 μm. This may be sufficient to produce some noticeable misalignment. However, it is generally the case that such direct manifestations of lateral misalignments are smaller than those brought about by substrate bending. One may understand this point with reference to the previous example of the silicon and aluminium micro-optical substrate. If this were populated with components with an axial height of about 4 mm, then a 30 °C excursion would produce a lateral shift of about 1.5 μm. This compares with a bending induced shift of about 7 μm.

19.4.5 Impact of Mounting and Bonding

19.4.5.1 Bonding

It is frequently the case, in the mounting and bonding of optical components, that materials are severely constrained in their freedom to expand. For example, in the fabrication of achromatic doublets, two different glasses, e.g. BK7 and SF5, are bonded together. In practice, for many such materials the differential expansion is small and the effect is small. However, where the temperature excursions are large the resultant thermal stresses may be sufficient to produce delamination in the bond or catastrophic brittle failure of glass. Cryogenic environments often feature in the application of infrared instruments, principally to reduce thermal and background noise. Such environments pose especial challenges. Naturally, system assembly is likely to be undertaken at ambient temperatures, leading to a temperature excursion of over 100 K when the system is deployed in the application environment. Severe interfacial stresses develop in bonded components and this is further militated by embrittlement of the bonding compound itself. Of course, military applications require deployment in uncontrolled environments, ranging from −40 to 80 °C. It must be remembered that the thermal environment includes not only the ambient conditions, but also any thermal load produced by local heat dissipation – power supplies, solar loading, etc.

Adhesive bonding of optical components is a common mechanical operation. The thermo-mechanical shear stress that is developed in the bond depends upon the shear modulus of the adhesive, G, the thickness of the bond, t and the characteristic length of the bond, l, as well as the thermal expansion mismatch. To a degree, long bond lines tend to exacerbate the impact of any expansion mismatch, with the shear stress rising to a maximum at the free ends of the bond. Thicker bonds tend to ameliorate the shear stress, τ. It is possible to model the shear stress induced in a bond when two thin layers of material of different thermal expansions are co-joined. If the two material thicknesses are t_1 and t_2, their expansion coefficients and elastic moduli α_1, α_2 and E_1 and E_2, then the interfacial shear stress is given by:

$$\tau = \frac{(\alpha_1 - \alpha_2)tanh(l/l_0)G}{(t/l_0)} \Delta T \textit{ is the temperature excursion} \tag{19.55}$$

And

$$l_0 = 1/\sqrt{(G/t)((1/E_1t_1) + (1/E_2t_2))}$$

The parameter l_0 is a characteristic length associated with the joint and is of the order of the thickness of the material stack. In general, it will be small compared to the length and, therefore, $\tan(l/l_0)$ will normally be close to one. In this case the following approximation applies:

$$\tau \approx \frac{(\alpha_1 - \alpha_2)G}{(t/l_0)} \Delta T \tag{19.56}$$

All this analysis assumes perfectly elastic behaviour in all the materials. A specific problem often encountered in this type of modelling, which also applies to finite element modelling, is the impact of any discontinuities – sharp corners, etc. In these circumstances, purely elastic analysis leads to the generation of singularities – point localities with seemingly infinite stress. Indeed, the whole study of fracture mechanics is predicated upon notion that amplification of stress is produced by sharp features, such as cracks. In the event, this is ultimately resolved by non-elastic behaviour, such as creep or plastic deformation. Such plastic deformation is inevitable in bonded joints and is somewhat difficult to model rigorously.

Depending upon the application, compliant, low-shear modulus bonding materials are favoured, such as silicone resins. Otherwise, more rigid compounds, such as epoxy resins may be used. Flexibility tends to be favoured where substantial changes in the operating environment are to be expected. The rigidity of such materials is occasionally expressed by the glass transition temperature – broadly the temperature at which the transition from a rubbery to a glassy state occurs. For epoxy adhesives, this temperature is rather high, around 80 °C. At the other extreme, silicone resins have a sub-zero glass transition temperature. In between these two extremes are the acrylate and cyanoacrylate compounds. Epoxies and acrylates are generally binary compounds with hardening generated by the admixture of two components. Curing is effected either thermally or by ultraviolet or short wavelength irradiation. Cyanoacrylates are cured upon contact with airborne moisture.

19.4.5.2 Mounting

The mounting of glassy components in metallic or plastic mounts inevitably produces unresolved thermal expansion mismatch. Generally, although not exclusively, metallic materials tend to have a higher CTE than

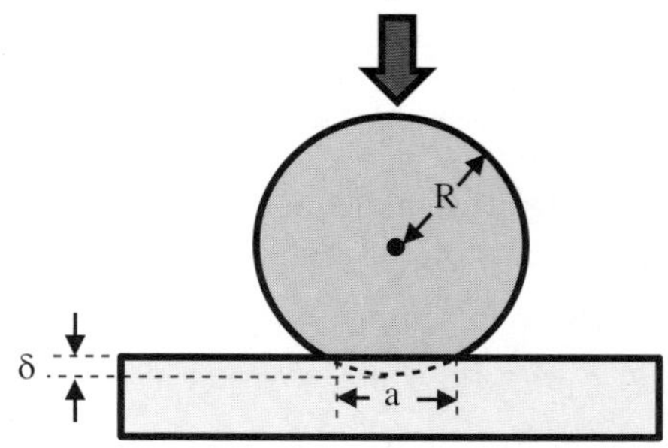

Figure 19.15 Compliance of radiused retainer.

glass materials. For plastic mounts, the difference is more marked. In the mounting of lenses in lens barrels, the impact of thermal expansion is to modulate the pre-load stress on the lenses significantly. In examining the impact of these relatively complex geometries, the scope of simple analytical modelling is rather limited. However, with some simplifying assumptions, it is possible to define the problem to understand the impact on a simple radiused retainer. With any relative movement of the retainer, the preload stress will change according to the elastic compliance of this arrangement. We therefore need to understand the elastic compliance of a ring of sectional radius, R, and hoop radius, r, in contact with a plane surface. The problem is illustrated in Figure 19.15.

The effect of the preload force, F, is to depress the ring by a distance δ into the plane surface. In so doing it produces an annular contact area with a width of a. The elastic indentation distance, δ, is proportional to the applied force, F, according to the following relation:

$$\delta = (4/3)F/Kr \tag{19.57}$$

The effective, composite elastic modulus of the system is given by K:

$$K = [(1 - \nu_1^2)/2E_1 + (1 - \nu_2^2)/2E_2]^{-1} \tag{19.58}$$

Of course, the width of the contact area, a, may be derived from simple geometry. From Eq. (19.57), it is possible to calculate the compliance for any given preload force, F. Quantitatively, the compliance, C, is the inverse of the effective spring constant, or the differential of the force with respect to displacement:

$$\frac{d\delta}{dF} = 4/3Kr \tag{19.59}$$

The striking feature of Eq. (19.59) is that the compliance is independent of the sectional radius of the retaining ring. To make this analysis more meaningful, we will return to our example of the 50 mm diameter BK7 lens mounted in a lens barrel, using a ring with a hoop radius of 22.5 mm. We assume that, at the mounting radius, the lens thickness is 6 mm. The whole system is assembled at 20 °C with the preload force of 20.5 N, as previously calculated. We wish to know how this force might change when the lens barrel is warmed to 50 °C.

Like the retaining ring, the lens barrel is made from aluminium. Thus far, in the analysis, we have only derived the depression from contact at one ring interface. For simplicity, we will assume that the total compliance of the ring is double that presented by a single interface. On the other hand, a rigorous analysis would have to calculate the depression at the other interface, using the appropriate material properties – probably aluminium on aluminium, as opposed to aluminium on BK7. Furthermore, in this analysis, we are assuming that the contact on the other side of the lens is absolutely non-compliant – i.e. it is 'hard mounted'.

First, we must calculate the compliance from Eq. (19.59). The elastic moduli of BK7 and aluminium are 8.2×10^{10} Nm^{-2} and 6.9×10^{10} Nm^{-2} respectively and the respective Poisson's ratios, 0.21 and 0.334. This gives the composite modulus, K, as:

$$K = [(1 - 0.21^2)/(2 \times 82) + (1 - 0.33^2)/(2 \times 69)]^{-1} = 81\text{GPa or } 8.1 \times 10^{10}\ \text{Nm}^{-2}$$

We must remember, from our previous discussion, that the overall compliance is double that of the single interface compliance in Eq. (19.52):

$$\frac{d\delta}{dF} = 8/3Kr = 8/(3 \times (8.1 \times 10^{10}) \times (2.25\mathrm{x}10^{-2})) = 1.46 \times 10^{-9}\mathrm{mN}^{-1}$$

The relative movement of the two material stacks is proportional to the difference in the coefficients of thermal expansion (23 and 7.1 ppm) multiplied by the thickness of the 'stack' (6 mm) and temperature excursion, and given by:

$$\delta = (\alpha_1 - \alpha_2)t\Delta T = (1.59 \times 10^{-5}) \times (6x10^{-3}) \times 30 = 2.87 \times 10^{-6}\mathrm{m}$$

From this movement, the change in force is equal to 1950 N, which is clearly excessive. In fact, in this instance, as the temperature is increased, the retaining ring will retreat from the glass and the lens would be unconstrained in its mount. Therefore, in this application, some additional compliance would need to be introduced. Selection of more compliant materials or adjusting the size of the retaining ring might assist. Otherwise, extra compliance might be introduced, for example, by incorporating sprung or 'wave washers' into the mount.

In other mounting arrangements, we might replace the retaining ring with the minimum effective number of geometrical constraints. The problem arises with the retaining ring because it is seeking to mate two surfaces over the entire annulus. In practice, the mount will not be flat with respect to the lens surface and this mismatch can only be resolved by distortion. The implication of this is that the bending or distortion would introduce extra compliance into the mounting arrangement. As such, the real change in preload force with temperature is likely to be smaller than calculated in the previous example. Adherence to the calculated values would demand that the relative flatness of the mating surfaces is less than the relative expansion of about 3 μm.

Minimum constraint might involve '***three point mounting***' where the two surfaces are forced together at three points only. Of course, in practice, the three points are not really points at all, but might be modelled by the contact of spheres with a planar surface. As with the retaining ring analysis, we can model the indentation displacement, δ, as a function of the applied force, F, the sphere radius, R, and the composite modulus K:

$$\delta = \sqrt[3]{8RF/3K} \tag{19.60}$$

The relationship is no longer linear and, unlike the annular geometry, the indentation is dependent upon the sphere radius. In addition, we may also calculate the radius of the indentation, a, from simple geometry:

$$a = \sqrt{2R\delta} \tag{19.61}$$

Thence, the average compressive stress, σ_c, is given by:

$$\sigma_c = (1/4\pi)\sqrt[3]{3KF^2/R^4} \tag{19.62}$$

If we need to calculate the compliance, this is calculated by the differentiation of Eq. (19.62):

$$\frac{d\delta}{dF} = (1/3)\sqrt[3]{8R/3KF^2} \tag{19.63}$$

At this point we might like to illustrate the analysis with an example. A mirror is retained against three stainless steel spherical bearings of diameter 10 mm. The mirror is 12 mm thick and its substrate material is BK7 with an elastic modulus of 8.2×10^{10} Nm^{-2} and a Poisson's ratio of 0.21. The elastic modulus of the stainless-steel bearing is 2×10^{11} Nm^{-2} with a Poisson's ratio of 0.27. If the maximum allowable compressive stress is 40 MPa, what is the total preload force on the mirror?

First, we need to calculate K, which is given by:

$$K = [(1 - 0.27^2)/(2 \times 200) + (1 - 0.33^2)/(2 \times 82)]^{-1} = 123\mathrm{GPa}$$

Thence, the preload force for each load point is given by substituting K into Eq. (19.62):
$4\text{x}10^7 = (1/4\pi)\sqrt[3]{3 * 1.23\text{x}10^{11} * F^2/0.005^4}$ and F = 463 N.

Thus, the total preload force is equal to 3 × 463 or 1389 N.

We might also care to assess the compliance of the mount from Eq. (19.63). Again, we need to assess the impact of deformation at both ends of each sphere. For simplicity of computation, once more we ascribe a compliance of double that suggested directly in Eq. (19.63). As such, the compliance of each mount is given by:

$$\frac{d\delta}{dF} = (2/3)\sqrt[3]{8R/3KF^2} = 0.67 \times \sqrt[3]{8 \times 0.005/3 \times (1.23 \times 10^{11}) \times 463^2} = 5.31 \times 10^{-7}\text{mN}^{-1}$$

The total compliance is one-third of this value (taking into account the three mounting points) and is equal to 1.7×10^{-7} mN^{-1}. If we imagine, once more, the BK7 substrate constrained in an aluminium mount and we prescribe the preload force at 20 °C, then we can use this analysis to estimate the change in the preload force as the mount is heated to 50 °C. The differential expansion related displacement is equal to $0.012 \times (2.3 \times 10^{-5} - 7.1 \times 10^{-6}) \times 30 = 5.7 \times 10^{-6}$ m or 5.7 μm. From the calculated compliance, the change in force produced amounts to 32 N. Unlike in the previous ring-mounted example, the component will still be well secured. Indeed, it is quite possible that the initial loading force may be reduced.

19.5 Finite Element Analysis

19.5.1 Introduction

As with optical modelling, the previous basic thermo-mechanical analysis is an attempt to attain an instinctive insight into the mechanical behaviour of optical assemblies. It also provides a basic sketch of the magnitude of some thermo-mechanical effects prior to detailed modelling. However, for the detailed analysis of complex assemblies, there is no substitute for detailed ***Finite Element Analysis (FEA)***. In the context of this discussion FEA is derived from a continuum mechanics description of the deterministic linear elastic behaviour of continuous solid media. Although non-linear behaviour and plastic (irreversible) deformation may be addressed, these topics lie beyond the scope of this discussion.

In describing purely elastic behaviour, continuum mechanics offers a series of linear partial differential equations that entirely encapsulate system behaviour. The only description that is relevant to this particular discussion is that of statics, as opposed to dynamics. That is to say, each point in the solid continuum is static and subject to no net force. As with the analysis of partial differential equations in general, the exact form of any solution is critically dependent upon the ***boundary conditions***, as much as the equations themselves. These conditions encapsulate the expected system behaviour at the interface of two dissimilar materials, or at the edge of the solution space. For example, it might dictate that certain stress elements vanish at free edges.

In some simple geometries, the suite of differential equations offers straightforward analytical solutions. Regrettably this is by no means a common scenario and, for complex optical assemblies, no tractable solution is offered. Therefore the partial differential equations are represented as (or approximated by) a set of finite difference equations. Here the infinitesimals of differential calculus are replaced by spatial nodes with finite separation. Provided that the separation of these spatial nodes is small, then the set of linear equations produced is a reasonable approximation to the original differential equations. In effect, the transformation is carried out by virtue of a Taylor Series approximation. Transformation to this set of finite difference equations produces an exceptionally large number of coupled linear equations. As such, implementation of FEA has only become possible with the development of extremely powerful computational resources. In implementation of FEA, there is always a tension between accuracy, which is enhanced by reducing spatial node separation and computational time which increases with the number of nodes deployed.

19.5.2 Underlying Mechanics

19.5.2.1 Definition of Static Equilibrium

Perhaps more usefully, in formulating the underlying differential or finite difference equations, the stress components may be expressed in terms of the strain components by inverting the matrix in Eq. (19.3). This is shown in Eqs. (19.64a) to (19.64f).

$$\sigma_{xx} = E((1-\nu)\varepsilon_{xx} + \nu\varepsilon_{yy} + \nu\varepsilon_{zz})/[(1+\nu)(1-2\nu)] + E\alpha\Delta T/(1-2\nu) \tag{19.64a}$$

$$\sigma_{yy} = E(\nu\varepsilon_{xx} + (1-\nu)\varepsilon_{yy} + \nu\varepsilon_{zz})/[(1+\nu)(1-2\nu)] + E\alpha\Delta T/(1-2\nu) \tag{19.64b}$$

$$\sigma_{zz} = E(\nu\varepsilon_{xx} + \nu\varepsilon_{yy} + (1-\nu)\varepsilon_{zz})/[(1+\nu)(1-2\nu)] + E\alpha\Delta T/(1-2\nu) \tag{19.64c}$$

$$\sigma_{xy} = E\,\varepsilon_{xy}/[2(1+\nu)] \tag{19.64d}$$

$$\sigma_{xz} = E\,\varepsilon_{xz}/[2(1+\nu)] \tag{19.64e}$$

$$\sigma_{yz} = E\,\varepsilon_{yz}/[2(1+\nu)] \tag{19.64f}$$

Static equilibrium of each element within the continuous solid demands that there must be no net force imposed on any element. Expressed more formally, this condition sets all three components of force per unit volume to zero, as follows:

$$\frac{\partial\sigma_{xx}}{\partial x} + \frac{\partial\sigma_{xy}}{\partial y} + \frac{\partial\sigma_{xz}}{\partial z} = 0 \tag{19.65a}$$

$$\frac{\partial\sigma_{xy}}{\partial x} + \frac{\partial\sigma_{yy}}{\partial y} + \frac{\partial\sigma_{yz}}{\partial z} = 0 \tag{19.65b}$$

$$\frac{\partial\sigma_{xz}}{\partial x} + \frac{\partial\sigma_{yz}}{\partial y} + \frac{\partial\sigma_{zz}}{\partial z} = 0 \tag{19.65c}$$

We can then substitute the above expressions to give the full set of differential equations for static equilibrium in a material ***with no imposed internal forces***.

$$(1-\nu)\frac{\partial^2 u_x}{\partial x^2} + \nu\frac{\partial^2 u_y}{\partial x\partial y} + \nu\frac{\partial^2 u_z}{\partial x\partial z} + \left(\frac{1-2\nu}{2}\right)\left(\frac{\partial^2 u_x}{\partial x\partial y} + \frac{\partial^2 u_y}{\partial y^2} + \frac{\partial^2 u_x}{\partial x\partial z} + \frac{\partial^2 u_z}{\partial z^2}\right) + \alpha(1+\nu)\frac{\partial\Delta T}{\partial x} = 0 \tag{19.66a}$$

$$(1-\nu)\frac{\partial^2 u_x}{\partial x^2} + \nu\frac{\partial^2 u_y}{\partial x\partial y} + \nu\frac{\partial^2 u_z}{\partial x\partial z} + \left(\frac{1-2\nu}{2}\right)\left(\frac{\partial^2 u_x}{\partial x\partial y} + \frac{\partial^2 u_y}{\partial y^2} + \frac{\partial^2 u_x}{\partial x\partial z} + \frac{\partial^2 u_z}{\partial z^2}\right) + \alpha(1+\nu)\frac{\partial\Delta T}{\partial x} = 0 \tag{19.66b}$$

$$(1-\nu)\frac{\partial^2 u_x}{\partial x^2} + \nu\frac{\partial^2 u_y}{\partial x\partial y} + \nu\frac{\partial^2 u_z}{\partial x\partial z} + \left(\frac{1-2\nu}{2}\right)\left(\frac{\partial^2 u_x}{\partial x\partial y} + \frac{\partial^2 u_y}{\partial y^2} + \frac{\partial^2 u_x}{\partial x\partial z} + \frac{\partial^2 u_z}{\partial z^2}\right) + \alpha(1+\nu)\frac{\partial\Delta T}{\partial x} = 0 \tag{19.66c}$$

The most common internal force, for practice purposes, acting on a continuous medium is that of weight. The force per unit volume acting is simply the product of the density, ρ and the acceleration due to gravity, g. If the gravitation vector is deemed to be directed in the opposite direction to the positive z axis, then, in this instance, Eq. (19.66c) may be modified to give:

$$(1-\nu)\frac{\partial^2 u_x}{\partial x^2} + \nu\frac{\partial^2 u_y}{\partial x\partial y} + \nu\frac{\partial^2 u_z}{\partial x\partial z} + \left(\frac{1-2\nu}{2}\right)\left(\frac{\partial^2 u_x}{\partial x\partial y} + \frac{\partial^2 u_y}{\partial y^2} + \frac{\partial^2 u_x}{\partial x\partial z} + \frac{\partial^2 u_z}{\partial z^2}\right) + \alpha(1+\nu)\frac{\partial\Delta T}{\partial x} = \frac{(1+\nu)(1-2\nu)\rho g}{E} \tag{19.67}$$

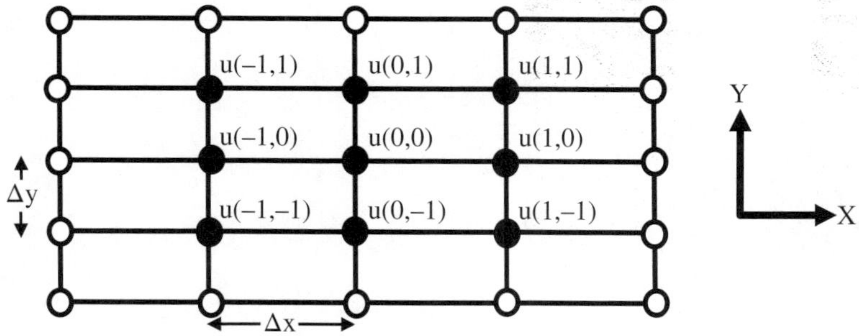

Figure 19.16 Simple rectangular mesh.

19.5.2.2 Boundary Conditions

In formulating the FEA analysis, presentation of equations, such as Eqs. (19.66a–c), are not sufficient to define a problem. Whilst they completely capture what is happening inside a solid, the definition of what happens at interfaces, both external and internal, is critical to the complete framing of a problem. One straightforward principle encapsulates the definition of all boundary conditions. Individual stress elements must be continuous across any material and at the boundaries. For specific components, any discontinuity in the stress at an interface would lead to the imposition of a finite force upon an infinitesimal element. This consideration applies specifically to any tensile stress that is perpendicular to the interface. In addition, any shear stress acting in the plane of the interface, must also be continuous. Therefore, at boundaries involving the external environment (i.e. air), then all free surfaces must have no component of shear stress acting in the plane of the boundary. Furthermore, at an external (free) surface, the normal component of tensile stress should (at least notionally) be zero. Otherwise, the impact of externally applied forces must be taken into consideration. For example, application of atmospheric pressure will produce a compressive stress at the interface equal to that imposed pressure.

Some care must be taken in the definition of external forces. Static equilibrium should pervade, with no net force or couple imposed. As such, any coherent definition of the problem should only invoke elastic distortion in a system. We must therefore suitably constrain the position and rotational state of the system in our FEA modelling.

19.5.3 FEA Meshing

Perhaps the most important element in the definition of an FEA model is the splitting of a spatially continuous mechanical system into a discrete array of three dimensional points. This process is referred to as ***meshing***. By so doing, the set of differential equations, as per Eqs. (19.66a–c), may be converted into a large set of linear simultaneous equations. It must be remembered, however, that this conversion is an approximation, based on a Taylor series expansion.

To illustrate the meshing process in a most simple form, we will consider a system as represented locally in just two dimensions, x and y. The mesh that we will create is perhaps the simplest imaginable – a rectangular grid array in x and y. The spacing of the grid is Δx in x and Δy in y. Within that grid, we will consider only nine points, clustered around a central point. For that central point we will derive expressions approximating various derivatives of some displacement component, u. This simple mesh is illustrated in Figure 19.16.

The first and second order derivatives may be represented thus:

$$\frac{\partial u}{\partial x} \approx [u(1,0) - u(-1,0)]/2\Delta x; \ \frac{\partial u}{\partial y} \approx [u(0,1) - u(0,-1)]/2\Delta y \tag{19.68a}$$

$$\frac{\partial^2 u}{\partial x^2} \approx \frac{[u(1,0) - 2u(0,0) + u(-1,0)]}{\Delta x^2} \ \frac{\partial^2 u}{\partial y^2} \approx \frac{[u(0,1) - 2u(0,0) + u(0,-1)]}{\Delta y^2} \tag{19.68b}$$

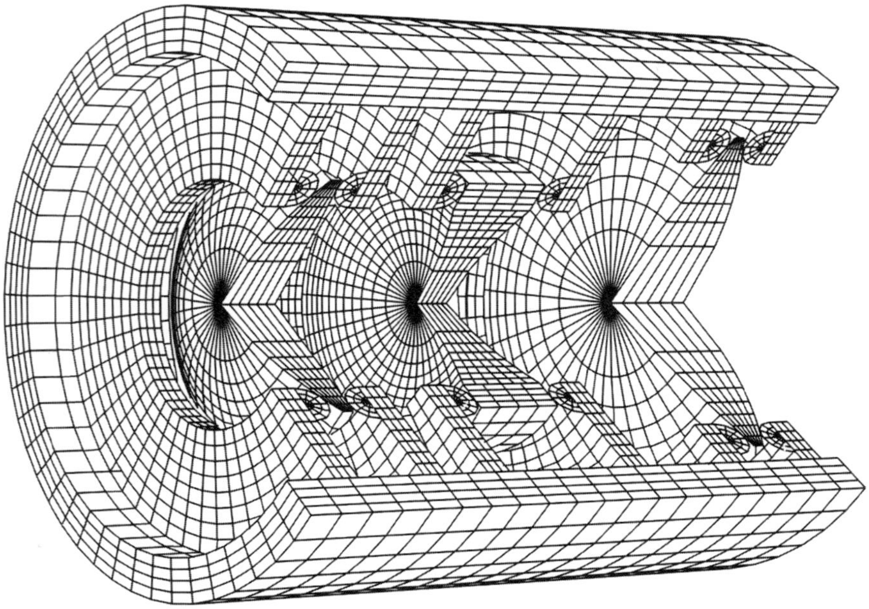

Figure 19.17 Meshing structure for simple barrel-mounted lens.

$$\frac{\partial^2 u}{\partial x \partial y} \approx \frac{[u(1,1) + u(-1,-1) - u(1,-1) - u(-1,1)]}{4\Delta x \Delta y} \tag{19.68c}$$

Of course, it should be understood that expressions are approximations only; higher order terms in the Taylor series approximation are effectively ignored. The practical validity of Eqs. (19.68a–c) is fundamentally dependent upon the choice of Δx and Δy. Broadly, Δx and Δy should be chosen such that any change in the stress, strain or displacement is small. Although the mechanics of solving the finite difference equations are fully under control of the FEA software, construction of the mesh geometry and choice of the mesh interval is determined by the user. Clearly, choice of a small mesh size promotes accuracy. However, in a complex system, reducing the mesh size substantially increases the number of mesh points and the number and complexity of calculations to be performed.

Resolution of this conflict demands considerable flexibility in setting up the mesh for a real system. Most importantly, in any practical simulation, the mesh will never match the simple neat and uniform structure displayed in Figure 19.16. There will often be large uniform areas where the stress varies very slowly with position. In these areas, a sparse mesh is quite justified. On the other hand, there will be areas, such as corners, material interfaces and localised areas where external force is applied, where the stress varies very rapidly. In these areas, a much denser mesh must be applied. In addition, the meshing must follow the local geometry, to match the symmetry of system components – e.g. circular or cylindrical, etc. That is to say, the underlying mesh symmetry will not always be rectangular or cubic. An example of a (*hypothetical*) FEA mesh is shown in Figure 19.17, showing the simulation of a simple Cooke Triplet mounted in a lens barrel structure, broadly reflecting the simple design established in Chapter 15. The illustration clearly demonstrates the non-uniformity of the meshing process.

Whilst the meshing ***process*** is under the control of the user, there are tools in the FEA package to assist the user in filling in a complex mesh structure. Therefore, there is no requirement to locate each mesh point individually. These tools are particularly useful in defining the meshing structure around boundaries or interfaces. Of course, all material properties, elastic modulus, Poisson's ratio, and CTE will have been notified and interfaces between the different materials defined.

Having defined the simulation mesh, it is clear that the system of linear equations will not appear exactly as in Eqs. (19.66a–c). Nevertheless, the principle is the same. The various partial differentials at a specific mesh point will be expressed in terms of a linear combination of the value of that mesh point and those of its nearest neighbours. Boundary conditions will be expressed according to a similar principle. The decision on which of the neighbouring points to employ and determination of the relevant weighting will be determined by the FEA program itself. In addition, the program also automatically handles the application of boundary conditions which will have been specified by the user.

At the end of the process, a very large array of coupled simultaneous linear equations will be produced. Solution of these equations requires exceptional computational power and, naturally, FEA modelling has developed alongside expanding processing power. The approach by which such solutions are effected is beyond the scope of this text.

As with optical modelling, some understanding of the underlying principles is useful. One example of this is the development of meshing around geometrical or material discontinuities, e.g. notches and corners. Provision of ever fining meshing around such features may not produce a convergent solution. This is because, in elastic theory, such discontinuities often produce a singularity. That is to say, the solution will suggest that the stress tends towards infinity at such locations. This behaviour (stress concentration) is the primary concern of fracture mechanics. In practice, such singularities tend to be resolved by non-elastic behaviour, e.g. irreversible, plastic deformation around the discontinuity. As with optical modelling, FEA should ***never*** be applied 'by rote', but should be underpinned by solid understanding.

19.5.4 Some FEA Models

A number of sophisticated FEA packages exist. The briefest of outlines has been provided as to how they operate. However, effective use of these packages represents a significant investment of time. Introduction to such software tools is best provided by specific training courses dedicated to that particular package. Many of these FEA tools form part of a wider analysis package, embracing 'Multi-Physics' analysis, including such elements as the analysis of fluid mechanics and heat transfer. Examples of such packages include NASTRAN™, PATRAN™, and ABAQUS™. From an optical standpoint, our interest in these models is restricted to the computation of surface deformations and tilts and the characterisation of stress in birefringent materials. As such, the FEA modelling directly impacts the tolerancing process. Detailed, complex analysis adds precision to the tolerancing process. Otherwise, our ignorance must be compensated by the provision of overgenerous tolerances leading to unnecessary manufacturing difficulties and expense.

Ideally the information derived from these FEA models should be fed back into the optical model. That is to say, knowledge of the precise deformation of mirror surfaces should be fed back into the original optical (e.g. Optic Studio ™) model and the impact on wavefront error and image quality determined. One such software package does exist. SigFit ™, from Sigmadyne Inc. acts as a direct interface between, for example, NASTRAN ™ and Optic Studio ™. From the FEA simulation, SigFit™ generates data files for direct input into the optical model. In this way, the impact of thermal or mechanical stress on alignment, image quality, or stress-induced birefringence may be directly characterised.

Further Reading

Ahmad, A. (2017). *Handbook of Opto-Mechanical Engineering*, 2e. Boca Raton: CRC Press. ISBN: 978-1-498-76148-2.

Budynas, R., Young, W., and Sadegh, A. (2012). *Roark's Formulas for Stress and Strain*, 8e. New York: McGraw-Hill. ISBN: 978-0-071-74247-4.

Doyle, K.B., Genberg, V.L., and Michels, G.J. (2012). *Integrated Opto-Mechanical Analysis*, 2e. Bellingham: SPIE. ISBN: 978-0-819-49248-7.

Friedman, E. and Miller, J.L. (2003). *Photonics Rules of Thumb*, 2e. New York: McGraw-Hill. ISBN: 0-07-138519-3.

Schwarz, U.D. (2003). A generalized analytical model for the elastic deformation of an adhesive contact between a sphere and a flat surface. *J. Colloid Interface Sci.* 261: 99.

Schwertz, K., *Useful Estimations and Rules of Thumb for Optomechanics*, MSc Dissertation, University of Arizona (2010)

Vukobratovich, D. and Yoder, P.R. (2018). *Fundamentals of Optomechanics*. Boca Raton: CRC Press. ISBN: 978-1-498-77074-3.

Yoder, P.R. (2006). *Opto-Mechanical Systems Design*, 3e. Boca Raton: CRC Press. ISBN: 978-1-57444-699-9.

20

Optical Component Manufacture

20.1 Introduction

20.1.1 Context

This chapter is not intended as a detailed introduction to the practice of optical component manufacture. It is more intended to guide the engineer whose role is in the specification and procurement of components. Above all, the purpose of this chapter is to assist the designer in formulating optical designs that are reasonable and practicable. To this end, the designer should have a thorough grounding in the manufacturing processes and technologies and a clear understanding of the boundaries of what is practicable. As was articulated in Chapter 18, the design process is, to a large degree, a process of negotiation between all stakeholders who have a role in bring a concept into fruition. In understanding the unique challenges faced by the manufacturer, the designer facilitates this process.

To a significant degree, optical component manufacture is a highly specialised activity, requiring significant skill and equipment resource to implement. As such, the creation of custom designs is necessarily a time consuming and costly activity. That said, there are a range of useful ***commercial off the shelf*** **(*COTS*)** components available to the designer. However, as a general rule, these are only available in sizes up to an equivalent (physical) diameter of 50 mm. Beyond this size, the available choices become rather more restricted.

This chapter will focus on the manufacture of individual optical components. The creation of optical materials and the provision of optical coatings have been detailed in previous chapters. In addition, the mounting and assembly of individual components will be left to the next chapter.

20.1.2 Manufacturing Processes

The manufacturing processes we are concerned with here involve the production or figuring of optical surfaces to a specific design shape. These processes are, almost exclusively, ***subtractive*** in nature. That is to say, they involve the removal of material either by grinding, machining, or polishing to create the final form. What is unique to optical component manufacture is the precision demanded of these subtractive processes. A reference sphere or flat used in precision metrology might have a form requirement of better than $\lambda/50$ rms, equivalent to 10 nm at 500 nm. Such precision cannot be attained without the close integration of metrology into the manufacturing process. In this instance, feedback from interferometric measurements is essential to deliver this level of form accuracy.

Three processes dominate optical component manufacturing – grinding, polishing, and machining. Grinding is fundamentally an abrasive process, involving the abrasive removal of material using abrasive particles with a specified size distribution. Larger particle sizes generate higher material removal rates, but produce rougher surface finishes. Furthermore, grinding is fundamentally a destructive process, producing sub-surface damage in the form of a network of small cracks. From the perspective of fracture mechanics, this weakens the material and polishing serves to remove or ameliorate this damage, as well as figuring the surface.

Optical Engineering Science, First Edition. Stephen Rolt.

Companion website: www.wiley.com/go/Rolt/opt-eng-sci

Polishing itself differs from grinding, in that it is not an abrasive process. That is to say, it is not merely an abrasive process with very small (10s–100s nm scale) particles. It is thought that polishing is fundamentally a chemical process. Historically, jeweller's rouge, very fine iron oxide, was the polishing medium of choice; zirconia has also been used. More recently, optical polishing has been dominated by the use of ceria (cerium oxide) and it is almost the universal material of choice. The polishing process itself is characterised by slow material removal rates, but it generates very smooth surface finish. Surface roughness values of a fraction of a nanometre are possible, although this depends significantly on the substrate material. Hard, amorphous materials, i.e. glasses tend to generate better finishes, whereas crystalline materials, such as calcium, fluoride can be a little troublesome, producing somewhat inferior surface finishes.

Direct machining processes have found increasing application in more recent years. For the most part, this is based upon ***single point diamond turning*** and related diamond machining techniques. A diamond machine is essentially a highly stable precision lathe that uses a small diamond tool to remove material in a similar manner to a conventional lathe. However, material removal rates are much slower and the surface precision is of the order of 10s–100s of nanometres.

The most obvious part of the manufacturing process is the figuring of the optical surfaces themselves. However, this is only part of the picture. As has been emphasised in the discussion on tolerancing and mechanical modelling, the optical surfaces themselves cannot be considered in isolation and their relative positioning fidelity is conferred by the referencing them to mechanical surfaces. For example, these mechanical surfaces might include the edges of lenses or other mounting surfaces. In the creation of these surfaces, it is important to maintain the precision of their geometrical relationship with respect to the optical surfaces. As mechanical surfaces, their form accuracy and surface roughness are less critical than for the corresponding optical surfaces. Generally, these surfaces are ground and not polished. However, as stated earlier, ground surfaces have some sub-surface damage, rendering them more vulnerable in terms of fracture mechanics. Therefore, in some critical applications, polishing may also be specified for certain mechanical surfaces.

Another process that should not be neglected is that of bonding, for example in achromatic doublets. Specialist optical adhesives have been developed for a variety of applications. For 'line of sight' applications, such as in the bonding of doublets, then their transmission properties and stability over time are of immense importance, as well as their thermo-mechanical properties. Having selected a bonding compound with the appropriate properties, the procedure for aligning the component during the bonding process must be considered carefully. For example, in bonding two singlet lenses together, optimum alignment must be maintained throughout the bonding process. This requires both the facility to adjust the alignment and to monitor it.

20.2 Conventional Figuring of Optical Surfaces

20.2.1 Introduction

The process of figuring an optical surface starts with selection of the material. For lens components, particular concern is attached to material quality – the presence of bubbles and striae, refractive index uniformity, and stress-induced birefringence. Glasses, as amorphous rather than polycrystalline materials, are dimensionally stable and are amenable to grinding and polishing. Polishing and grinding rates are isotropic and not impacted by crystal geometry. As such, where dealing with exotic materials, such as calcium fluoride, silicon, or zinc selenide, especial care must be taken with the geometry of the polishing process. Glass materials are formally classified for their 'grindability', being sub-divided into six classes from HG1 to HG6, with the highest class (i.e. HG6) promoting the most rapid removal of material.

Material selection is based upon the demands of the application. That is to say, for high-end applications, individual blanks must be inspected and graded according to quality. Naturally, higher quality material attracts a premium. A blank is generated by (diamond) sawing, such that the piece is slightly larger than demanded by the application, allowing for subtractive (grinding) processes. Thereafter the generation of the optical surface itself may proceed.

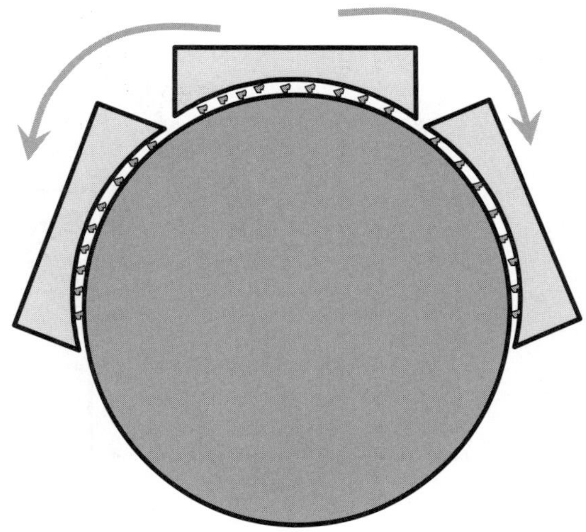

Figure 20.1 The generation of spherical surfaces by grinding.

The vast majority of optical surfaces generated are spherical or planar in form. Notwithstanding this, some aspherical surfaces, particularly conic sections, such as parabolas, find critical niche applications. However, optical figuring work is predominantly concerned with the generation of spherical or planar surfaces. This is important, since the generation of a spherical surface represents the default condition in the grinding of optical surfaces. In general, the grinding process involves the rubbing of two surfaces separated by a layer of abrasive particles. One of these surfaces is the component or 'workpiece' and the other is the tool. The important point is that this process has a natural tendency to general spherical surfaces. This is because a spherical surface is the only form whereby the two surfaces will fit together regardless of orientation. Any asperities generated on either surface would have a tendency to be preferentially abraded on account of their prominence. Thus, any departure from spherical form will be preferentially eroded producing two spherical surfaces. This illustrated schematically in Figure 20.1.

Notwithstanding the very high form accuracies required in generating optical surfaces, this principle of preferential (spherical) shaping greatly facilitates the process. That is to say, the fabrication of spherical surfaces is 'relatively easy'. Coarse grinding of the basic shape is followed by fine grinding whose purpose is to attenuate the layer of sub-surface damage produced by the rough grinding. This is followed by the polishing process which further attenuates the sub-surface damage and generates the final shape. At this point, it is possible to use optical techniques, such as interferometry to measure the surface form. Metrology is an essential part of the process; generation of highly accurate surface form cannot be assured without the process feedback provided by measurement.

20.2.2 Grinding Process

The grinding process typically uses rigid tools to generate the basic shape. These tools often use diamond abrasive in some form. A typical set up is shown in Figure 20.2.

Figure 20.2 shows the grinding process for a single piece. Grinding is effected by a cup shaped tool mounted on a rotating spindle. The (diamond) abrasive is attached to the 'lip' of the cup. The workpiece itself is rotating about a central axis. In addition, the axis of the rotating spindle is itself free to rotate about a point that lies on the axis of rotation of the workpiece. This set up preserves the spherical symmetry of the grinding process.

The set-up, as shown is for the grinding of a single, perhaps moderately large, piece. For reasons of economy, especially when cutting smaller pieces, it is preferable to generate multiple parts in one batch. This is accomplished by the process of ***'blocking'***. In the blocking process, individual blanks are mounted into recesses in the specially machined spherical block, using wax, or pitch, as shown in Figure 20.3. The spherical block broadly

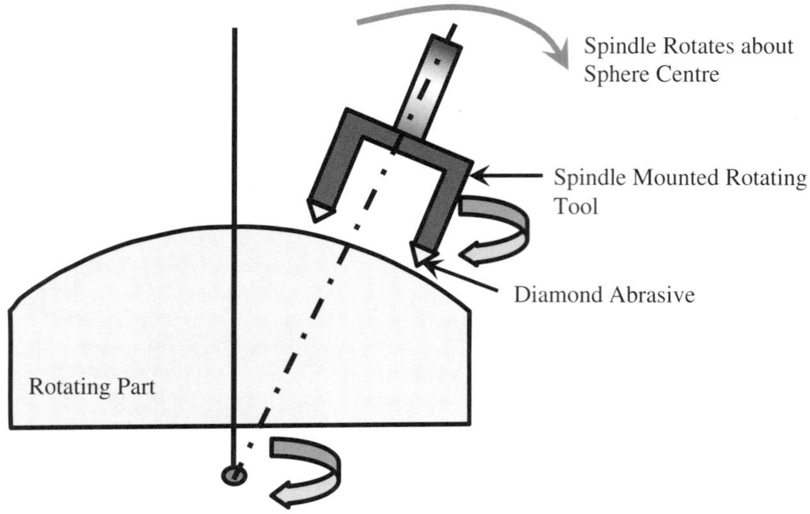

Figure 20.2 Typical grinding process for single piece.

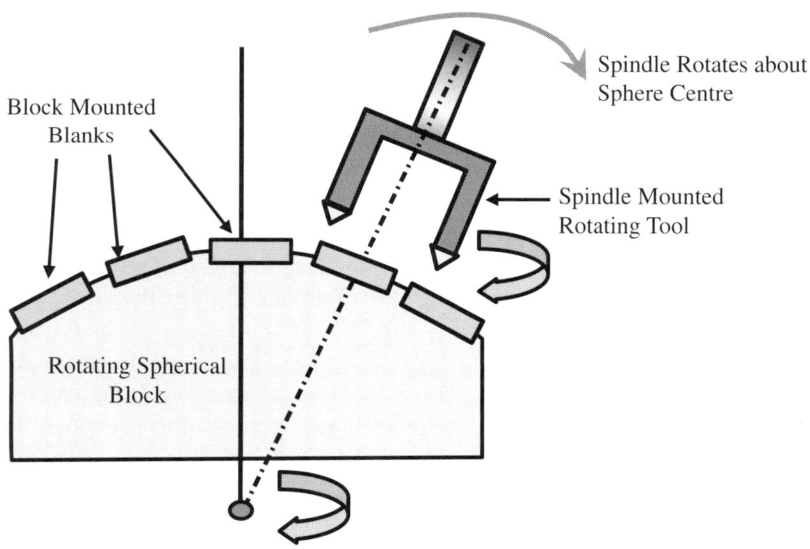

Figure 20.3 Blocking process.

follows the shape of the individual spherical surfaces. Naturally, this process will only work for the batch production of spherical surfaces having the same radius of curvature. The radius of the block is nominally that of the base radius of the sphere to be cut.

The discussion, thus far, has focused on the generation of spherical surfaces. In principle, the grinding of planar surfaces follows a similar overall principle. The set-up, in this case, involves the use of a rotating tool similar to that shown in Figures 20.2 and 20.3. However, the workpiece is mounted on a lathe bed provided with a linear axis (axes). Grinding takes place in a 'flycutting' operation, with the tool mounted perpendicularly to the lathe bed and workpiece. The workpiece is then traversed or rastered with respect to the rotating tool. Not only does this facilitate the creation of simple plane surfaces, but also the fabrication of faceted surfaces (e.g. prisms) with a high degree of precision in their relative orientation.

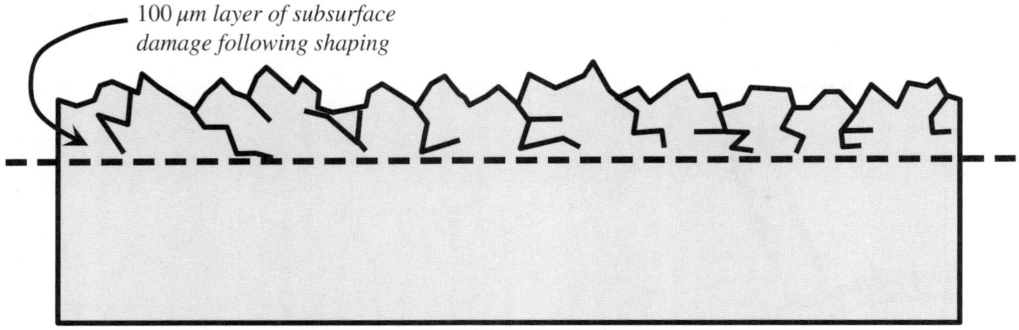

Figure 20.4 Subsurface damage following grinding.

In any precision machining process, workpiece mounting is a major concern. The workpiece must be mounted in such a way as to allow access to all parts of the surface without the risk of collision. However, most importantly, the component must be held securely without being unduly stressed. Any significant mounting stress has the potential to produce distortion in the ground surface when the workpiece is released. In addition, the mounting technique should permit the ready release of the component following grinding. Some form of bonding process on an obverse surface allows access to all parts of the optical surface. Wax is widely used for securing smaller components; heating facilitates attachment and removal. For larger parts, pitch, with its viscoelastic properties, promotes stress relief.

20.2.3 Fine Grinding

As outlined previously, the grinding process described leaves residual sub-surface damage in the glass optic. The damage takes the form of a network of sub-surface cracks whose presentation would otherwise degrade optical performance after polishing. In addition, the network of cracks accentuates the propensity for catastrophic crack propagation and material failure. This sub-surface cracking is illustrated in Figure 20.4.

This damaged layer must be removed. Typically, the thickness of this layer is a few tens of microns. Removal is accomplished by loose abrasive grinding, with the abrasive mixed with a liquid to form a loose slurry. Of course, this process, as a grinding process, also has a tendency to produce damage. However, the thickness of the damaged layer is directly related (in proportion) to the size of the abrasive particles. By grinding with a succession of abrasive slurries with diminishing particle size, the subsurface is attenuated sufficiently to allow polishing. In theory, it is possible to remove all the subsurface damage using a polishing process alone. However, the removal of 100 μm of subsurface damage would take unacceptably long.

20.2.4 Polishing

For the time being, we will restrict the description of the polishing process to the generation of 'conventional' surfaces. That is to say, we will initially consider only the creation of spherical or planar surfaces. In the vast majority of applications, ceria (CeO_2) is the polishing medium of choice. It should be emphasised that the polishing process cannot be described in terms of simple abrasive removal of material. It should rather be thought of as a pseudo chemical process and it is often described as ***Chemical-Mechanical Planarization*** (CMP).

For the generation of spherical surfaces, the geometry is, in some respects, similar to the grinding process. The major difference is that the polishing tools tend to be conformal, rather than rigid. Polishing compound is applied as a slurry to the surface of a conformal tool, often described as a bonnet, and the rotating tool is moved across the surface of the workpiece in a pattern similar to that of the grinding process. The significant point about the conformal tool is that it adopts the (spherical) shape of the workpiece. Historically, pitch was

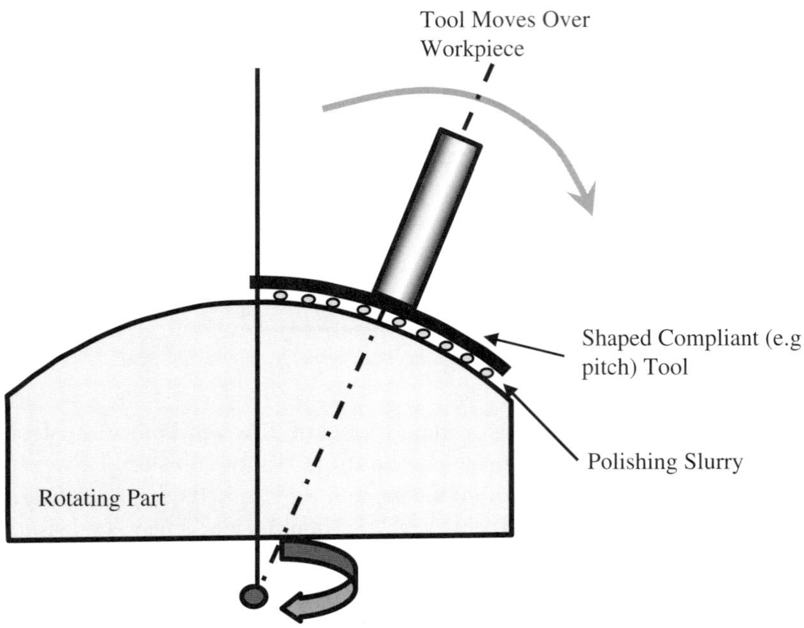

Figure 20.5 Polishing process (for spherical components).

the preferred material for the polishing bonnet. However, most contemporary polishing machines employ PTFE (polytetrafluoroethylene) or polyurethane as the bonnet material.

As indicated, a typical set-up substantially replicates the grinding process, except with a rotating bonnet replacing the cup grinding tool. The set-up is shown in Figure 20.5. Once again, the polishing bonnet rotates on a spindle and is impressed upon the rotating workpiece. As with the grinding process, the spindle axis itself is configured to rotate about a common spherical axis. Figure 20.5 shows the polishing process for a single workpiece. However, mass production may be facilitated, as for the grinding process, by 'blocking' multiple pieces.

One factor that the designer needs to be particularly aware of is the impact of edge effects. Depending upon the precise set up, uniform, controlled material removal cannot be guaranteed in regions close to the edge of the blank. In effect, the edge of the blank marks a geometrical discontinuity. Therefore, the designer needs to specify a ***clear aperture***, which might, as a guide, be 85–90% of the physical surface aperture. It is only within this clear aperture that any optical requirements, such as form error or surface finish specifications apply. To establish a reasonable clear aperture, with respect to the physical aperture, dialogue with the manufacturer is essential.

The previous discussion applies to the generation of spherical surfaces. The process for polishing plane surfaces is a little more elaborate. Plane surfaces are generally polished in a continuous process on a large, flat, rotating lap. The workpiece(s) itself is held within a rotating holder, called a septum, and is impressed upon the rotating lap. During the polishing process, the septum rotates in synchrony with the lap. The lap itself is compliant, as with the generality of polishing tools, but must itself be kept flat as the polishing process proceeds. This is effected by a large rotating glass blank impressed upon the lap. Typically, the diameter of the blank is approximately equal to the radius of the lap. The process is illustrated in Figure 20.6. The compensator compensates for uneven lap wear as the polishing process evolves. Optimisation of the compensation process depends upon fine adjustment of the radial position of the compensator. As with the polishing process in general, optimisation of process parameters is substantially dependent upon the feedback provided by dimensional metrology, especially interferometric measurement of surface form.

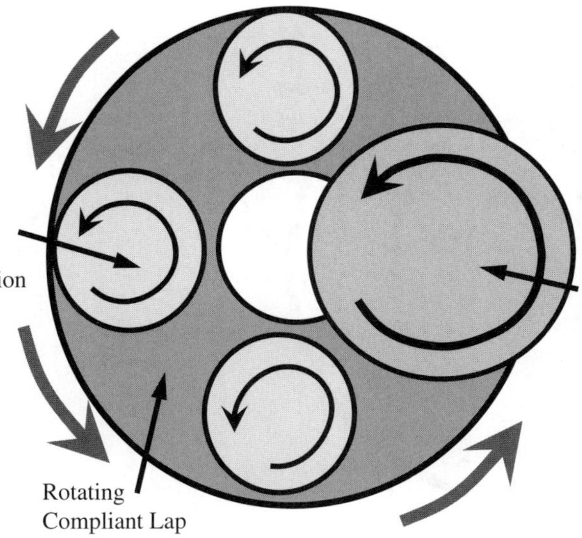

Figure 20.6 Continuous lap polishing of flats.

20.2.5 Metrology

The generation of optical surfaces with a form error of a few nanometres cannot be assured without the feedback of metrology. Testing the form of polished surfaces relies on interferometry. Typically, the shape of a surface will be tested with respect to a specially manufactured reference surface known as a ***test plate***. A Fizeau geometry is most widely used for this test. Any departure between the shape of the workpiece and the test plate will be recognised by the appearance of fringes. If the process has generated a perfectly spherical surface, but with a base radius different to that of the test plate, then a series of circular fringes will be produced. Otherwise, any departure from an ideal spherical shape will be characterised by some degree of irregularity in these nominally circular fringes. As such, in reporting the measurement of surface form error, results are usually presented in the form of separate figures for radial departure and irregularity, as expressed in fringes. In the example illustrated in Figure 20.7, an interferogram with three fringes of power and one fringe of irregularity is presented.

As discussed in Chapter 16, there is a historical tendency to evaluate interferograms on the basis of their visual appearance. That is to say, interferograms are characterised in terms of their peak to valley departure, as opposed to the (computer) analytically derived rms measure. Of course, it must be remembered that each fringe represents half a wavelength of form departure.

The important point to remember about the test interferogram illustrated in Figure 20.7 is that it does not simply represent a post-manufacturing verification test. Interferometric tests, such as illustrated above are an integral part of the manufacturing process, particularly for high specification components. That is to say, the character of form departure revealed by a test plate interferogram also informs the adjustments that need to be made in further polishing steps. This principle is illustrated in Figure 20.8 which shows a greatly simplified process flow for optical surface shaping.

Naturally, as Figure 20.8 illustrates, the achievement of the highest form accuracy places the highest demands on manufacturing resource. As indicated earlier, the specification of form error tends to be quoted in terms of peak-to-valley rather than rms form error. To illustrate the manufacturing premium inherent in high form specification, Figure 20.9 shows a plot of relative cost against surface form specification expressed in peak to valley form for a given size and geometry. An empirical observation can be made from Figure 20.9 in that the manufacturing difficulty, i.e. cost, increases as the inverse square root of the form error.

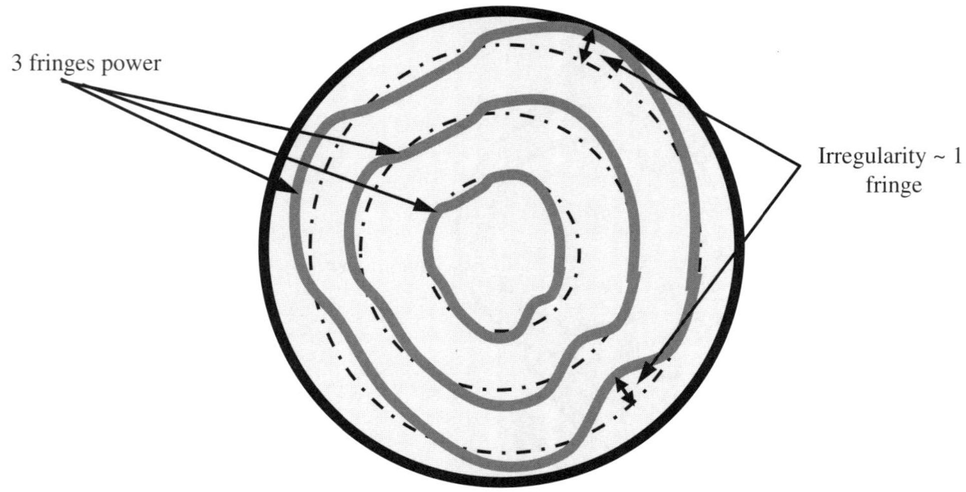

Figure 20.7 Test plate interferogram.

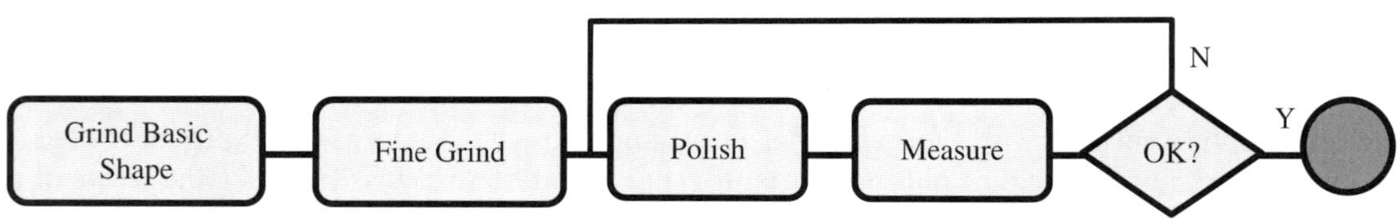

Figure 20.8 Simplified process flow for grinding and polishing.

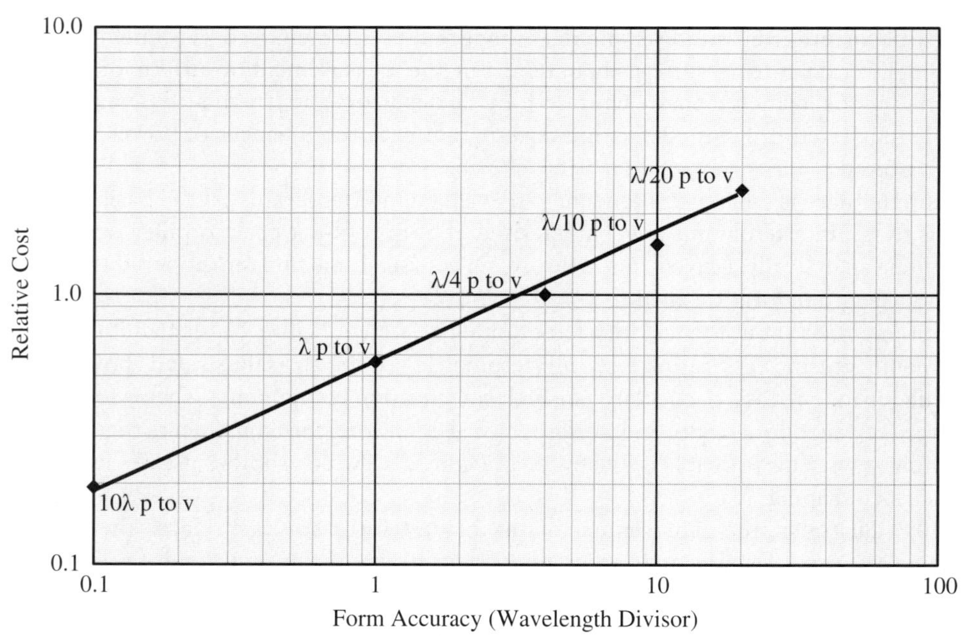

Figure 20.9 Relative cost vs form accuracy.

In terms of the form error requirement for individual surfaces or components, it must be understood that the specification for mirror surfaces is much more demanding than that for lens surfaces. Each mirror surface makes a system wavefront error contribution that is double its form error – for a double pass. However, for refraction at a single lens surface the wavefront error contribution is only half the form error (for $n = 1.5$), i.e. only one quarter of that of a mirror surface.

20.3 Specialist Shaping and Polishing Techniques

20.3.1 Introduction

Conventional polishing techniques are remarkably efficient at generating spherical surfaces to an exceptionally high precision. However, the generation of aspheric surfaces is considerably more problematical. Polishing or surface generation is only a part of the overall problem. Accurate generation of aspheric surfaces is even more dependent upon precision metrology than is the case for spherical surfaces. Metrology for aspheric surfaces is fundamentally more difficult than for spheres or flats. As detailed in Chapter 16, testing relies on the generation of special test geometries or the provision of computer generated holograms. To a degree, the range of aspheric shapes that may be characterised is limited. True freeform surfaces, those lacking any symmetry, as opposed to off axis conics, etc., are exceptionally difficult to characterise. To a degree, the majority of ***precision*** optical surfaces do not fall into this latter category and off-axis conics or off-axis (even) aspheres with tangible symmetry do make up a significant proportion of aspheric surfaces manufactured. However, freeform surface generation and metrology is an expanding topic and there is increasing interest in the wide range of applications that beckon.

The majority of aspheric surfaces are generated by a controlled polishing process. Historically, this controlled polishing was a hand polishing process requiring much skill and patience. However, this has been superseded by computer controlled polishing techniques. Where the degree of aspheric departure is small (e.g. tens of microns), then the surface may be finished from a spherical surface that has been ground in the conventional manner. Otherwise, the rough shape must first be formed by some form of computer-controlled grinding process. For the most part, such processes are used to produce high value parts in low volume. For instance, the lack of spherical symmetry means that the standard blocking procedure cannot be used to facilitate mass production. However, there are some circumstances, for specific materials, where aspheric lenses can be moulded. This includes not only polymers and resins, but also some specialist low melting point glasses. In this scenario, a high-value precision mould might be generated by a combination of machining and polishing and used to replicate large numbers of aspheric components.

20.3.2 Computer-Controlled Sub-Aperture Polishing

Conventional polishing seeks to generate spherical surfaces by removing material evenly across the entire component aperture. By contrast, sub-aperture polishing seeks to remove material preferentially in a local area by using a compliant polishing bonnet that is smaller than the workpiece that is being polished. The process is illustrated in Figure 20.10. Compliance and size are important, because lack of sphericity means that the tool cannot fit the workpiece over the whole of the geometry. Therefore, shape mismatch must be accommodated in the process by optimising the size of the tool and ensuring that its compliance is adequate.

In the sub-aperture process, the tool is moved under control over the surface of the workpiece, e.g. in a raster pattern. Control of the process is achieved by adjusting the contact pressure at any locality and the dwell time. By intuition, the material removal rate is increased by 'polishing harder' or applying more pressure. This empirical insight is formalised in ***Preston's Law*** which proposes that the material removal rate is proportional to the local pressure $P(t)$ applied to the work piece and the (rotary) velocity of the workpiece, $V(t)$.

$$\frac{dh}{dt} = C_P P(t)V(t) \tag{20.1}$$

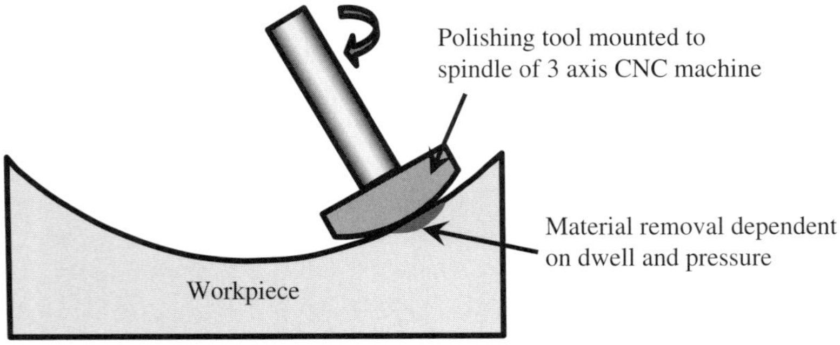

Figure 20.10 Subaperture polishing process.

The constant of proportionality, C_P is the Preston constant. Values for this constant are material and process dependent, but typically lie around 10^{-12} Pa^{-1}. However, it must be remembered that, since the workpiece is non-spherical, then the fit of the tool with the workpiece is variable across different areas of the workpiece. As a consequence, the pressure applied will not only vary across the surface of the tool, but will also depend upon the area of the workpiece being polished. Part of the computer polishing process is the generation of a removal function, a spatial model of the variable material removal rate as a function of position. This model may be used to inform the polishing process which is controlled by varying the dwell time over any workpiece location and the pressure applied. This will provide a tolerable 'first pass' recipe to polishing to the final shape. However, there are limits as to how deterministic this process can be made. Inevitably, tool wear will change the shape of the polishing bonnet which itself affects the removal function. The final shape can only be attained by an iterative process, employing precision metrology in between polishing stages, as per Figure 20.8.

When combined with precision metrology, computer-assisted sub-aperture polishing is capable of producing highly accurate aspheric surfaces. However, the lack of spherical symmetry in the process promotes the generation of ***mid-spatial frequency form errors***. The variability in the removal function across the part and the tool can, to a degree, be compensated by adjusting the computer-based polishing recipe. However, notwithstanding this, small residual mid-spatial frequency errors will remain. These errors are not present in conventional polishing, by virtue of the 'happy accident' entailed in a spherical processing geometry; any such asperities will have a tendency to be preferentially removed. Diligence will, of course, usually reduce these mid-spatial frequency form errors to an acceptable level. The important point, as discussed later, is that the spatial frequency distribution characteristics of form error are fundamentally different for sub-aperture polishing when compared to conventional polishing. The presence of enhanced mid-spatial frequency errors contributes very specifically to system image quality and needs to be accounted for in any optical modelling undertaken by the designer.

20.3.3 Magneto-rheological Polishing

Magneto-rheological polishing or finishing is a specialised technique for the controlled finishing of optical surfaces. The technique uses a specially formulated magneto-rheological fluid as a polishing slurry. A magneto-rheological fluid is a suspension of ferromagnetic particles in a fluid accompanied by polishing compounds. When this slurry encounters a high magnetic field, alignment of the magnetic particles causes its viscosity to increase substantially. This increase in viscosity is naturally accompanied by an increase in the Preston coefficient and hence the polishing rate. The overall process is shown in Figure 20.11.

In magneto-rheological polishing, the ferromagnetic polishing slurry is directed at a rotating wheel. The wheel carries around a film of fluid which encounters the workpiece at some point in its rotation. The workpiece itself is mounted on a rotating spindle and encounters the polishing slurry at a point on the rotating wheel where a magnetic field is applied. In this region, the fluid becomes viscous and the polishing efficiency is increased substantially. Material is removed from the workpiece at a controllable rate that depends upon the

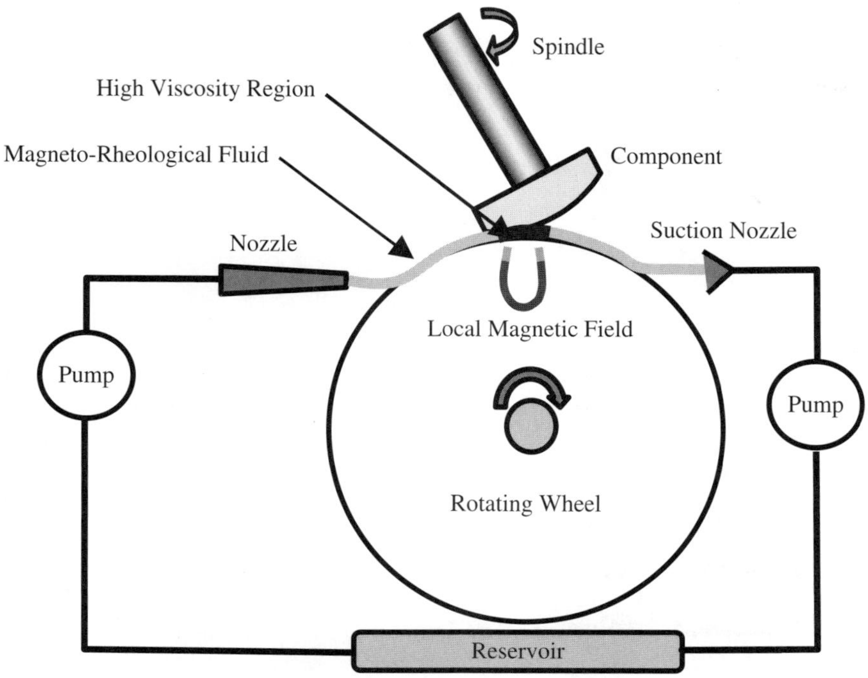

Figure 20.11 Magneto-rheological polishing.

applied magnetic field, as well as local pressure and workpiece velocity. In effect, the magnetic field functions as an additional useful process parameter in shaping the surface. As material is removed from a part of the workpiece, shaping can take place by controlling the dwell time and applied field under computer control.

20.3.4 Ion Beam Figuring

Ion beam figuring is a highly specialised technique for precise material removal. It has been used in niche applications for the precise figuring of unorthodox surfaces. It was famously used in the final figuring of the Keck Observatory primary mirror segments. Ion beam figuring is essentially a controlled sputtering process. The workpiece must be loaded into a vacuum chamber and exposed to a beam of (e.g. gallium) ions. Collision of these ions with the workpiece results in the sputtering or removal of material. Naturally, the removal rate may be adjusted by controlling the ion beam current as the ion beam is moved across the surface of the workpiece. Unlike other sub-aperture polishing techniques, the process is inherently deterministic. As such, exceptionally precise figuring is possible. However, removal rates are very low and facilities are few, unlike the other specialised techniques described here. The process is illustrated schematically in Figure 20.12.

20.4 Diamond Machining

20.4.1 Introduction

Diamond machining is a widely used technique for the direct machining of precision optical surfaces. As a process, it is fundamentally different from conventional grinding and polishing techniques and so is considered separately here. In many respects, it is similar to a conventional machining process, except for the precision and surface finish obtained. Diamond machining is characterised by the use of simple (radiused) tool geometries and relatively slow (compared to general machining) material removal rates. Naturally, the

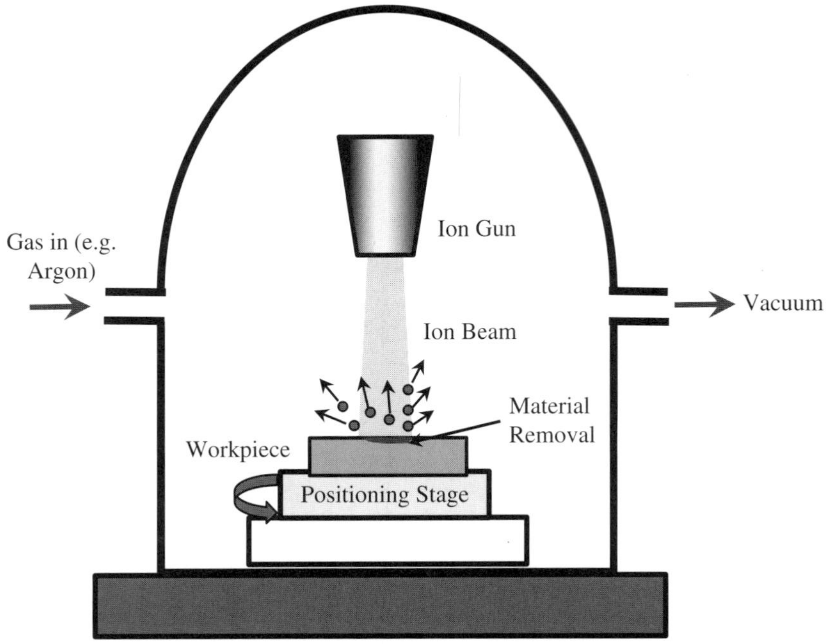

Figure 20.12 Ion beam figuring.

machining process is under computer control and, unlike conventional polishing processes, there is no fundamental restriction in the surface geometries that may be generated.

The use of small cut depths and feedrates allows the direct machining of surfaces with a specular finish, without the need for post-polishing steps. For many applications, the resulting surface texture is perfectly adequate. However, with a minimum practicable surface roughness of 2–3 nm rms, the surface finish is significantly inferior to that generated by polishing processes. For many visible applications, the scattering produced by such a surface finish is not acceptable. Therefore, there is a tendency for diamond machining to find niche applications in infrared instruments where scattering is substantially reduced on account of the longer wavelength.

Diamond machining excels in the generation of 'difficult' surface forms, particularly ***freeform*** shapes which lack any inherent defining symmetry. In principle, any surface that can be defined mathematically may be machined. In addition, surfaces having discontinuities, such as those with contiguous, separate facets, are amenable to diamond machining. This latter group might include, for instance, diffraction gratings, or diffractive optics in general.

For the most part, diamond machining is used to shape metals, either directly for use as mirror surfaces or indirectly in the fabrication of lens moulds. The range of materials that may be satisfactorily machined is necessarily restricted. Aluminium alloys are particularly favoured for diamond machining, in addition to copper, brass, gold, zinc, and tin. However, significant difficulty is experienced in machining iron and nickel alloys, as well as alloys containing titanium, molybdenum, and beryllium. These elements have a marked tendency to promote rapid chemically based wear of the diamond tool, where the carbon from the diamond tool is abstracted to form the metal carbide. Some crystalline materials can be machined, such as germanium, silicon, zinc selenide, zinc sulfide, and lithium niobate. Of course, these materials have wide application in the infrared. In addition, a wide range of optical polymers may be machined. These include acrylics, zeonex, polycarbonate, and optical resins, such as CR9 and CR39.

The utility of diamond machining lies in its ability to generate complex optical quality surfaces in a single processing step. The precision of the process is sufficient to replicate surfaces to a form accuracy of 10s–100s nm.

A computer-controlled machine tool, in general, relies on a number of rotary and linear stages to provide precise movement of the cutting tool with respect to the workpiece. Clearly, in the case of a diamond machining centre, the precision must be around two orders of magnitude superior to that of a conventional machine tool.

In the light of this, diamond machine tools are designed to be exceptionally rigid and stable. To this end, motion stages, such as spindles, are designed to move on air bearings and to be stable and robust. Positioning, to sub-nanometre resolution is achieved through the use of interferometrically derived optical encoders. Most importantly, in the design, there is a full appreciation of the impact of temperature drift on positioning stability. As with the thermomechanical modelling discussion from the previous chapter, differential expansion between the workpiece and tool material stacks may lead to relative motion of far greater than the 10s of nanometers precision desired. Therefore, diamond machining centres are operated in a carefully temperature controlled environment, stabilised typically to ±0.1 °C.

20.4.2 Basic Construction of a Diamond Machine Tool

Figure 20.13 shows the example of a five-axis diamond machine. It has three (X, Y, Z) linear axes controlling the relative position of the workpiece and tool, a fast rotating spindle (C axis) and an additional rotational axis (B axis).

The machine illustrated has five axes. Neither tool nor workpiece are shown in the illustration but both tool or workpiece could be deployed either on the spindle, or the B axis unit, depending upon the machining configuration. Compared to conventional machining tools, the geometry of a diamond tool is relatively simple.

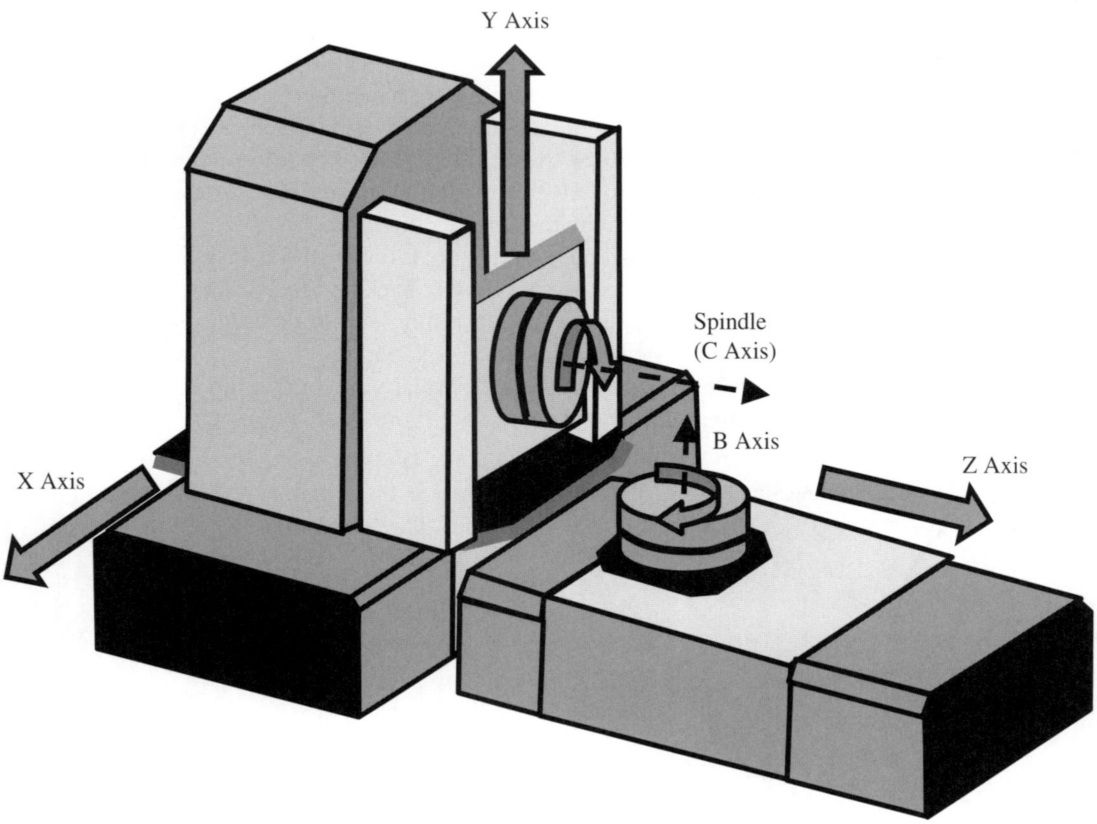

Figure 20.13 Five axis diamond machining tool.

Most commonly used are small radiused tools, with a chisel like edge presented along a circular arc with a radius of a fraction of a millimetre or a few millimetres.

The ***five-axis machine*** depicted allows for great flexibility in the relative positioning of tool and workpiece. As such, it can be used for the machining of extremely complex freeform surfaces. Another common configuration is the ***three-axis machine***. In this case, the B-axis and the Y-axis movement is dispensed with. The geometry for a three-axis machine is thus not dissimilar to that of a conventional lathe.

As illustrated in Figure 20.13, the construction of the machine is exceptionally robust. As a consequence, the machine is exceptionally stable and the machine head is extremely stiff. Of course, in order to maintain relative positioning fidelity of tool and workpiece to tens of nanometres, system compliance must be kept to an absolute minimum. The flatness of the machine slides is such that the positional uncertainty (runout) over the whole travel (~500 mm) is of the order of 100 nm. Over smaller intervals, the precision is much greater than this. The length of travel along the slides is recorded interferometrically to a ***precision*** (not accuracy) of 10s of picometres.

20.4.3 Machining Configurations

20.4.3.1 Single Point Diamond Turning

The process of ***single point diamond turning*** uses a three-axis machine configuration, as illustrated in Figure 20.14. The workpiece is mounted on the spindle and the diamond tool can be moved independently along the X and Z axes. In many respects, the process is similar to a conventional turning process, albeit with higher precision. As the workpiece rotates rapidly on the spindle, the tool is translated, under control, along an arc in the XZ plane. Precise delineation of the arc represents the control input to the machine. As is usual in computer numerical control (CNC) machining, definition of the tool path must account for geometrical variables, such as the diamond tool radius (assuming a radius tool).

It is easy to see how, in a conventional set up, it might be possible to machine shapes that are rotationally symmetrical about the spindle axis. That is to say, it is relatively straightforward to machine spherical surfaces and on-axis aspheres or conics. According to Figure 20.14, the workpiece rotates rapidly on the spindle and the tool progresses radially across the workpiece by controlled movement in the x direction, as shown. Generation of the correct sag is obtained by precise movement of the tool in the z direction.

Implicit in this narrative is the assumption that movement of the tool is much slower than the rotary motion of the spindle mounted workpiece. However, moving the tool rapidly back and forth along the spindle axis during a single rotation cycle, in a controlled way, allows the machining of more complex shapes. As such, non-rotationally symmetric parts may be produced. For instance, it is possible to create astigmatic shapes or other forms that might be defined by Zernike polynomials. There are, of course, fundamental limits to the rapidity of this additional tool movement and therefore the additional non-symmetric sag that can be

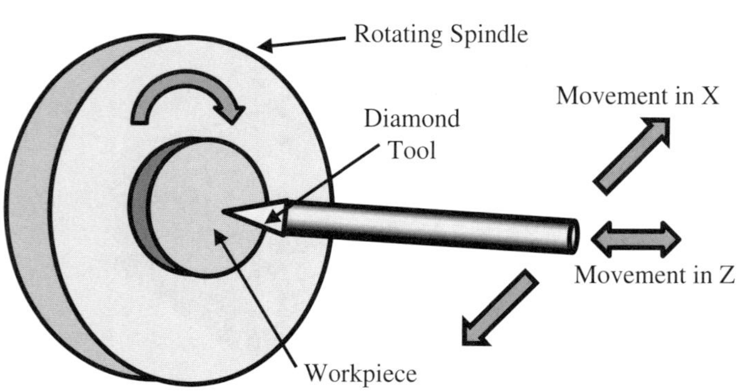

Figure 20.14 Single point diamond turning process.

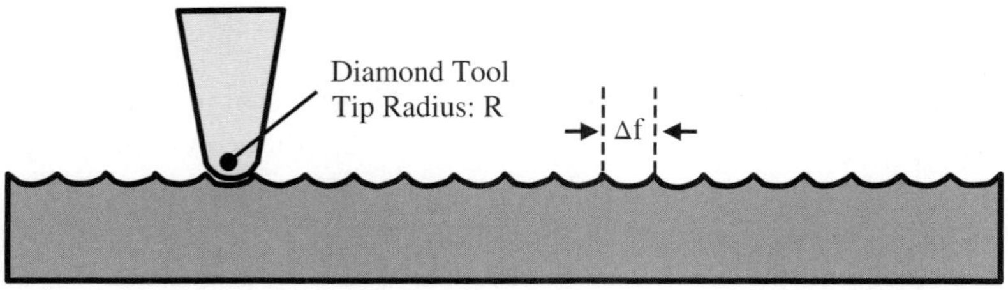

Figure 20.15 Surface texture generated during single point diamond machining.

machined. Two different techniques exist for generating this additional movement. The first is called ***slow slide servo***, where the spindle is rotated slowly and the extra motion of the tool is produced by z movement of the slide in the usual form. By contrast, in ***fast tool servo***, the additional motion is applied by a fast piezo-electric pusher, able to generate small displacements at frequencies of several kHz. Whilst allowing faster machining speeds than slow slide servo, the amount of additional sag that can be generated is considerably smaller.

Much of the diamond machining process is based on a deterministic and geometrically repetitive cutting process. As such, the surface texture of a diamond machining process replicates these intrinsic geometrical structures. The surface roughness inherent in diamond machining, as indicated earlier, is rather larger than that for polishing. A large part of this surface roughness is in the form of grating like structures that follow the repetitive application of, for example, a radiused tool. Figure 20.15 illustrates this in the context of a single point diamond turning process. The tool itself has a radius of R and, for each rotation of the spindle, the tool is fed by Δf in the x direction.

The geometrical surface roughness produced, σ_{rms}, is simply given by:

$$\sigma_{rms} = \frac{(\Delta f)^2}{12\sqrt{5}R} \tag{20.2}$$

For a typical feed of 15 μm and a 1 mm radius tool, the geometrical surface roughness produced is 8 nm rms. At a spindle speed of 1000 rpm, this process would machine a circular part 100 mm in diameter in approximately four minutes. Reduction of surface roughness requires either a smaller feed or a larger tool radius. Increasing the tool radius can, to a degree reduce the machining precision of the process. On the other hand, reducing the tool feed inevitably slows down the machining process. The optimisation of machining parameters is inevitably a matter of compromise. In modelling the scattering produced by diamond machined optics, due account must be taken of such structured surface texture. This would have a tendency to produce grating type effects with strong scattering along specific directions.

The preceding analysis assumes that the material removal is a simple process dependent only upon the tool shape. This is true to a degree. However, the workpiece material does play a role. In polycrystalline materials, such as metal alloys, the cutting process is dependent upon the local crystal structure. For this reason, in general, fine grained, or amorphous materials are preferred in machining applications. Hence, local, random variations in cutting behaviour contribute to the overall surface roughness. In a suitable material, such as aluminium alloy 6062, this is a matter of a few nanometers.

20.4.3.2 Raster Flycutting

Although the use of slow slide servo and fast tool servo can be used to increase the flexibility of single point diamond turning, there is a limit to the range of geometries that may be produced with these techniques. The flexibility of five axis machines may be fully exploited in a configuration known as raster flycutting. The set-up is shown in Figure 20.16.

In this instance, the cutting tool is mounted on the rotating spindle. As shown in Figure 20.16, when the tool is at the bottom of its arc, then a scallop of material will be removed from the workpiece. In a five axis machine, as depicted in Figure 20.13, the component can be moved in three dimensions with respect to the rotating tool.

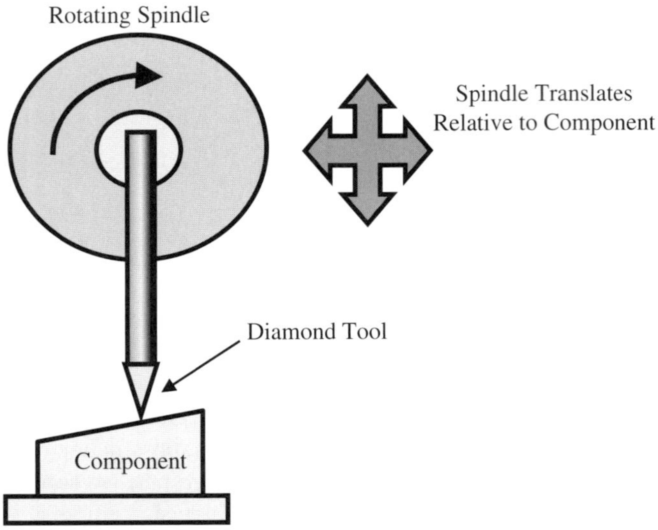

Figure 20.16 Raster flycutting.

Making due allowance for the tool geometry, the shape that is generated broadly follows the surface defined by the three axis movement of the tool relative to the workpiece. For instance, the path followed in x and z (the horizontal axis) will generally be described by a raster scan. The contour height of the surface is then simply generated by programming the tool height in y. Within reason, raster flycutting can generate any surface from that can be described mathematically. However, as is evident from Figure 20.16, the material removal is an intermittent process, occurring only at the bottom of the tool's arc. Therefore the greater flexibility invested in raster flycutting comes at the price of a slower machining process. In addition, the surface roughness generated is rather larger.

20.4.4 Fixturing and Stability

During the machining process, the workpiece must be held firmly. However, any forces applied to the workpiece have the propensity to generate elastic distortion. If the fixturing process has been poorly designed, significant distortion will be produced at the optical surface itself. Whilst the machining process will, in situ, generate the desired surface, when the part is removed from its jig, the clamping distortion will 'unwind' leaving a 'negative imprint' of the original distortion in the final workpiece. Of course, as described earlier, this effect has been historically exploited in the (conventional) polishing of the Schmidt corrector plate where a vacuum distorted plate is polished flat and the desired surface results after removal of the vacuum. In fact, vacuum is often used (in 'vacuum chucks') to attach a workpiece to a spindle. A particular issue in fixturing for diamond machining is the problem of 'over constraint'. If the registration between the workpiece and the spindle surface does not follow the minimum (e.g. three point) constraint then the workpiece will accommodate the extra constraints through distortion. In other words, the same considerations apply to the mounting of a workpiece on a spindle as to the mounting of a mirror in an optical assembly. For example, if the mounting of the workpiece is via the mating of two nominally plane surfaces, then any departure in the flatness of the two surfaces will produce distortion in the workpiece.

Thermal stability is another important issue. As indicated, diamond machining is a relatively slow process; a simple part may take several minutes (or indeed hours) to machine. Even with careful design, relative movement of tool and workpiece may be as much as 1 μm per °C. Thus the logic of controlling the temperature of the machine environment (to $<\pm 0.1\,°C$) is quite compelling. For example, if a simple flat is being turned as per the set up in Figure 20.14, then temperature drift will produce a conical shape error in the final workpiece.

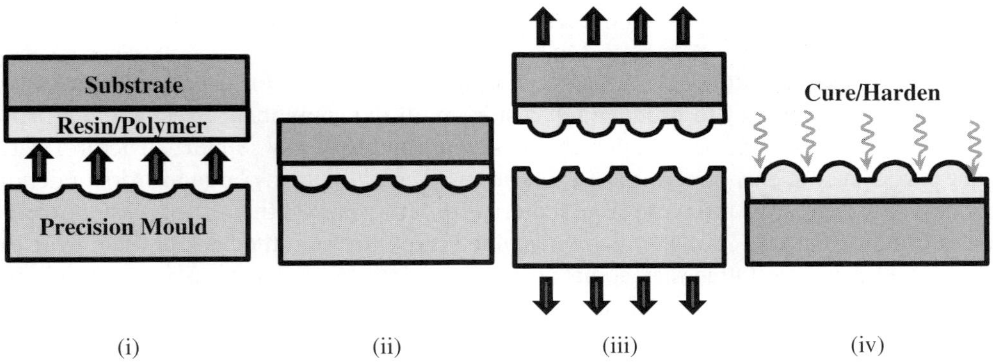

Figure 20.17 Replication of micro-optics. *(a) Mould application (b) Pressing (c) Withdrawal (d) Hardening.*

20.4.5 Moulding and Replication

As a direct fabrication process, diamond machining is a serial process for producing precision parts. In itself, it is not suitable to direct application in volume manufacture. Nevertheless, the technique is widely used for the production of precision moulds for optics fabrication. One example is the production of spectacle lenses from resins, such as CR39. Polymer aspheric lenses may also be produced in this way. One specific example of a moulding process is illustrated in Figure 20.17. Diamond machining may be used for the fabrication of moulds for micro-lens arrays and micro-optics in general. The process follows the steps used in the replication of diffraction gratings. A layer of resin is applied to a transparent substrate and the precision mould impresses the pattern upon the resin. After the mould has been withdrawn, the resin is cured, forming the hardened lenses.

Diamond machining of ceramic moulds can be used to facilitate the manufacture of glass aspheres. The range of glass materials that can be used in the production of moulded optics is necessarily restricted. Examples include P-SK57 and are characterised by a low softening temperature. A small, generally spherical, glass blank is inserted into a two part mould. The mould and glass are heated and pressure is applied to both parts of the mould, carefully controlling the separation of the two parts of the mould and thus the lens thickness. Subsequently, the mould is cooled and thence retracted, creating the moulded part. Careful design must account for any shrinkage or refractive index change as part of the thermal process.

20.5 Edging and Bonding

20.5.1 Introduction

Thus far, we have dealt purely with the shaping of the functioning optical surfaces. As any optical surface needs to be mounted in some geometrically predictable way, the generation of mechanical surfaces bearing a known relation to the optical surfaces is essential. For the most part, it is customary that these surfaces be ground. The finish is specified by the final ANSI (American National Standards Institute) grit size, G, used in the finishing process, e.g. 400. For the finer grit sizes used in this processing, an approximately inverse relationship exists between the standard grit size, G, and the average particle size, d, in microns:

$$d = 5000/(G - 143.5) \tag{20.3}$$

In some critical applications, it is desirable that the edges be polished rather than ground; this reduces the propensity for catastrophic crack propagation should the component become unduly stressed. The generation of mechanical surfaces naturally produces edges at the intersection with optical surfaces. Sharp edges are exceptionally vulnerable to mechanical damage. Therefore, it is customary to specify small bevels (usually 45°) at the intersection of these surfaces.

For smaller optical components, gravitational loading is of little consequence. However, for larger components, their self-loading has a tendency to produce self-distortion in the component itself, and, perhaps more significantly, places unacceptable demands upon the mass and rigidity of any support structure. Naturally, this issue is felt particularly keenly in aerospace applications. Therefore, many large mirrors undergo a 'lightweighting' process which involves the removal of material from the back of the mirror without compromising its structural rigidity. Pockets' of material are machined (ground) from the back of the mirror to create a stiff, but light ribbed structure. Often the geometry is similar to the honeycomb pattern prevalent in the design of lightweight optical tables; the principle is the same.

20.5.2 Edging of Lenses

The most important consideration in the edging of lenses is to ensure that the optical axis of the lens surface is aligned with the axis defining the cylindrical edges of the lens. The optic axis is defined by the line joining the centres of the two lens surfaces (assuming spherical geometry). This can be accomplished in two ways, either by relying on mechanical registration or by controlling the alignment optically. Figure 20.18 illustrates a common mechanical approach for edging a lens. The two lens surfaces are sandwiched in a '***Bell Chuck***' which, by default, aligns the optical axis to that of the chuck.

The edge of the rotating lens is ground by a diamond edging tool which is itself rotating. An additional tool is provided to cut the bevel, as indicated in Figure 20.18. In addition, the lens can also be centred in a chuck using an optical alignment technique. For example, by directing laser light through a lens mounted on a rotating chuck, the decentre is visible as the chuck rotates. Any decentre will then be manifested in the form of wobble produced in the focused laser beam. The position of the lens may then be adjusted to null out the wobble and the lens fixed in position by wax or pitch or similar. The arrangement is illustrated in Figure 20.19.

If the lens is off centre, then the focus spot follows the optical axis as it rotates about the spindle axis, creating a circle with a diameter equal to twice the decentre, Δy. Of course, there will be a residual centration error, where the optical axis does not coincide with the mechanical centre of the lens. A centration error of 0.01 mm

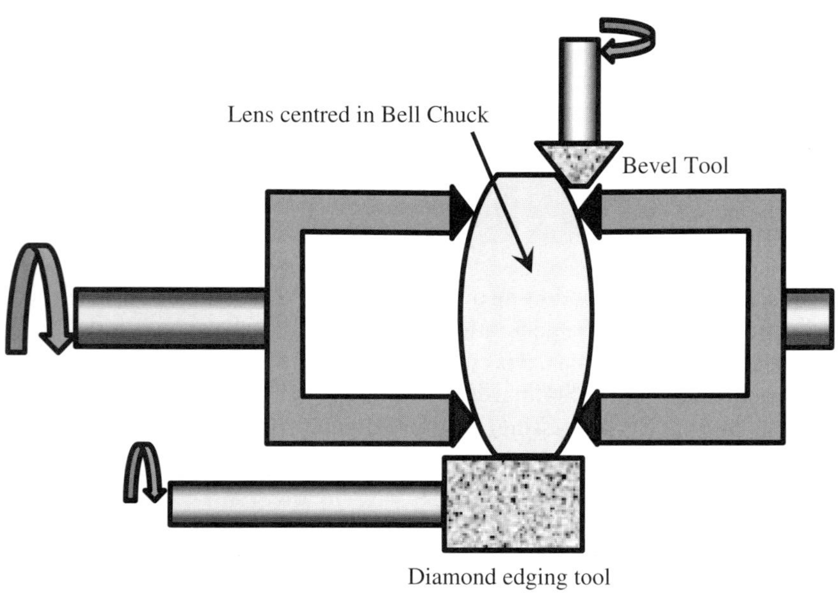

Figure 20.18 Lens edging in a bell chuck.

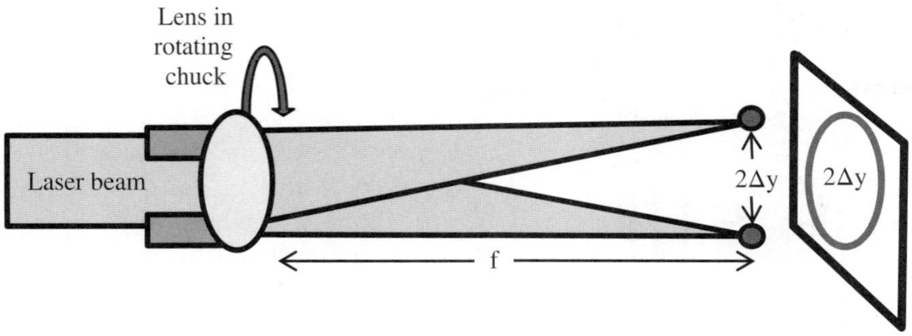

Figure 20.19 Lens centring in a chuck.

represents high precision. A centration error may equally be interpreted as an error in the angle between the two surfaces, i.e. a wedge error. Equally, centration error may also be interpreted also as a 'runout error' where the edge thickness of the lens varies with (spindle) rotation angle. This runout error may be monitored with a 'dial gauge', as the spindle is rotated. Centration errors tend to be correlated with uncertainties in the edge thickness of the lens, so that wedge errors tend to reduce as the lens size becomes larger. Such uncertainties in edge thickness might be of the order of 5 μm in precision applications.

If θ is the wedge angle and Δy is the centration error, the relationship between the two for a lens of focal length, f is given by:

$$\theta = \Delta y/[(n-1)f] \tag{20.4}$$

20.5.3 Bonding

Bonding of optical components, particularly the cementing of doublets and triplets, is a common manufacturing process. Where the adhesive compound is in the 'line of sight', as for the achromatic doublet, care must be exercised in the selection of the appropriate formulation. Not only must the adhesive be transparent over the useful wavelength range, it must also not degrade in its operating environment. The operating environment naturally includes any temperature excursions or temperature cycling and the ingress of moisture. In addition, the operating environment might also include exposure to ultraviolet radiation, potentially promoting discolouration of the adhesive. Historically, Canada balsam was used in the cementing of lasers, but ultraviolet-curing adhesives have been favoured for very many years.

The process for bonding doublet lenses is illustrated in Figure 20.20. Of critical importance is the alignment of the two optical axes. Misalignment may be monitored optically and, as indicated in Figure 20.20, a means of alignment adjustment provided. As indicated, the adhesive is an ultraviolet-curing adhesive. As soon as satisfactory alignment has been achieved, the glue line is irradiated and cured, cementing the two components permanently in their aligned state.

A bonding and alignment process is used in the production of micro-optics, particularly in the manufacture of packaged laser devices, such as fibre pigtailed lasers and packaged laser modulators or waveguide devices. The equivalent of the centring adjustment illustrated in Figure 20.20 aims to maximise fibre or waveguide coupling from one device to another. Alignment precision in these applications is submicron. As such, precision adjustment relies on piezo-electric nano-positioning stages. Thermally or ultraviolet-curing adhesives are used for bonding and active alignment using the nano-positioning stages is maintained throughout the curing process.

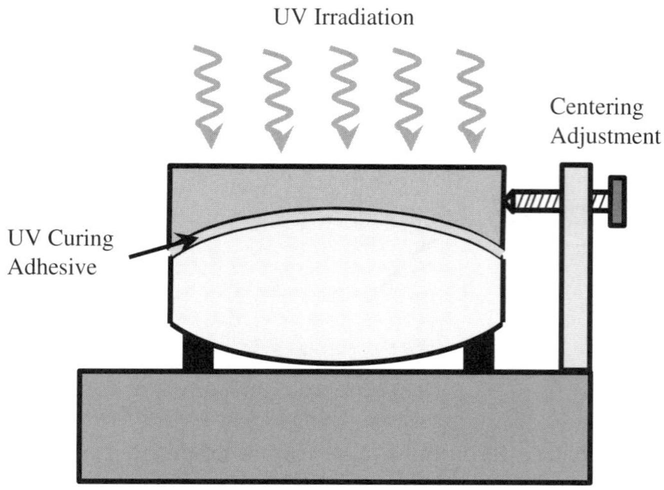

Figure 20.20 Bonding of doublets.

20.6 Form Error and Surface Roughness

There is a tendency to regard the form error and surface roughness as two distinct and separate attributes of a manufactured surface. In reality, they are arbitrarily assigned portions of a continuous PSD (power spectral density) spectrum. Splitting these two separate contributions so clearly expresses our desire to unambiguously separate image quality degradation from the phenomenon of scattering. In reality, particularly for high-quality surfaces, the PSD and scattering profiles produce a continuous point spread function (PSF) distribution that declines with displacement from the nominal image position. This is not wholly surprising. As established in Chapter 18, the PSD of manufactured surfaces tends to follow an inverse power law dependence with an exponent of between two and three. By virtue of the analysis of Fraunhofer diffraction, this is manifested at the image plane by an irradiance distribution following the same power law dependence as the original surface PSD.

Nevertheless, notwithstanding this, it is useful to establish some useful boundaries for distinguishing surface roughness and form error. If we establish scattering as a purely far field diffractive phenomenon, then the spatial frequency period, Δ, and the nominal propagation distance, L, should define a Fresnel number, F, that is much less than one, e.g. 0.1. With this in mind, we may establish the critical spatial period as:

$$\Delta = \sqrt{0.1\lambda L} \tag{20.5}$$

In practice, for a wavelength of 500 nm and a propagation distance of 1 m, this amount to a spatial period of about 0.2 mm. This figure might be applied more generally as a boundary between form error and surface roughness. Of course, what is described as mid-frequency errors are, by the above definition, form errors. What sets them apart is determined exclusively by the idiosyncrasies of the manufacturing process. As exemplified by sub-aperture polishing, the extent of this range must be related in some way to the size of the aperture or the sub-aperture. Thus, conventionally, the mid-spatial frequencies extend from the 0.2 mm surface roughness boundary to some fraction (e.g. 10%) of the total workpiece aperture. Thus for a 50 mm diameter optic, we might ascribe the different ranges in the following manner:

$$0.2\ \mu\text{m} < \textit{Roughness} < 0.2\ \text{mm} < \textit{Mid-Spatial Frequency} < 5\ \text{mm} < \textit{Form Error} < 50\ \text{mm}.$$

Figure 20.21 shows a PSD plot for a polished and a diamond machined component, illustrating the monotonic decay in the PSD with spatial frequency. Often, where extracting form and roughness data from a linear measurement, the PSD is expressed as the square of the amplitude per 1D spatial frequency interval. In this case, the dimensions of the (1D) PSD are length cubed. For the more informative 2D data analysis, the PSD

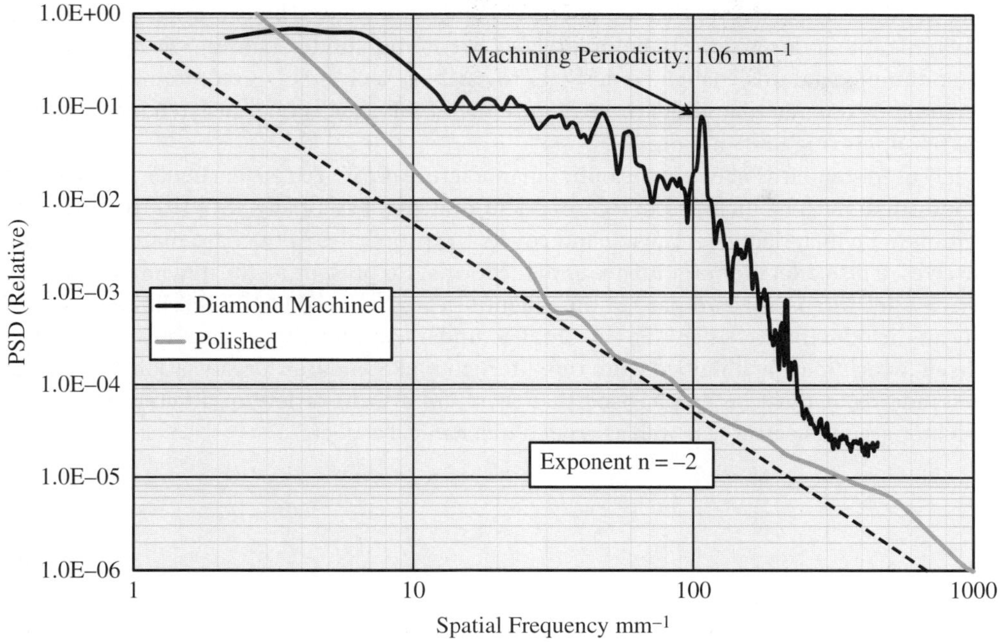

Figure 20.21 PSD spectra for polished and diamond machined components.

component is captured within a 2D (x, y) spatial frequency interval. In this case the dimensions of PSD are equivalent to the fourth power in length.

A very useful approximation to a PSD spectrum for polished glass may be made for a power coefficient of −(11/3). This is a workable approximation for a conventional polishing process. The significance of this particular exponent is that the analysis of random sinusoidal form error components ***approximately*** follows the well-known Kolmogorov theory for modelling refractive index variations in atmospheric turbulence. This theory is widely applied to propagation of light through the atmosphere and adaptive-optics techniques for ameliorating any image degradation due to atmospheric turbulence. In this case, it is possible to calculate the rms form error between two spatial frequency bounds, f_{low} and f_{high} as a proportion of the total, global form error.

$$\frac{\sigma}{\sigma_0} = \frac{0.806}{D^{(5/6)}}\sqrt{\left[\frac{1}{f_{low}^{5/3}} - \frac{1}{f_{high}^{5/3}}\right]} \tag{20.6}$$

D is the part diameter

However, a 'health warning' should be added to this analysis. The association with the Kolmogorov type of analysis is promoted as a convenient and tractable approximation that may be related to a large volume of analytical studies in the scientific literature. The spatial frequency dependence of PSD in polished surfaces is often not dissimilar to that of the Kolmogorov power law. However, this is matter of empirical convenience, rather than the representation of a coherent analytical theory.

20.7 Standards and Drawings

20.7.1 Introduction

Standards are an important part of the manufacturing process. It is essential that all stakeholders have a common understanding of design and performance metrics that underpin the ultimate functionality of the component or system. Whilst it is inevitable that some creative judgement must be exercised in those few areas where ambiguity remain, standards serve to enforce a level of common understanding within the optics community. As such, common standards enable the designer to establish a clear set of requirements that are unambiguously understood by the manufacturer. These requirements will usually be translated into a formal drawing accompanied by supporting information.

One of the most widely used standards in use in the optics manufacture is ***ISO 10110***. Particularly, it sets out and standardises the information to be included in a component drawing. Essentially, ISO 10110 represents a checklist of information to be included in a drawing and covers the properties of the base material and the optical surfaces themselves. It also offers some guidance as to the format for presenting the information. In this section we will present a brief and useful summary of the salient features of the ISO 10110. The description, as presented, is purely an introductory overview. For a complete understanding, the reader must refer to the standard document itself, which may be obtained from the International Standards Organisation. Of course, in the wider field of standards, it must be emphasised that there are a wide variety of different standards pertaining to optical materials and coatings, as well as related mechanical aspects.

20.7.2 ISO 10110

20.7.2.1 Background

The ISO standard is split into 13 sub-sections each describing a particular requirement category. The first section, ***ISO 10110 Part 1*** is introductory, setting out the specific differences between optical and mechanical drawings. Thereafter, the next three sections concentrate on a description of the material properties and the remaining nine sections focus on the optical surfaces themselves. Table 20.1 summarises the content of the standard.

20.7.2.2 Material Properties

Material properties are set out in three specific categories, each category defined by a descriptor consisting of a digit (either 0, 1, or 2) followed by a forward slash (i.e. '0/', '1/', '2/') and thence the codified information pertinent to that category. This information is presented in the manufacturing drawing in tabular form.

Table 20.1 ISO 10110 summary.

Section	Type	Description	Symbol
Part 1	General	Drawing conventions	
Part 2	Material	Stress birefringence	0/
Part 3	Material	Bubbles and inclusions	1/
Part 4	Material	Inhomogeneity and striae	2/
Part 5	Surface	Form tolerance	3/
Part 6	Surface	Centring (wedge angle) tolerances	4/
Part 7	Surface	Cosmetic surface quality	5/
Part 8	Surface	Surface texture	(λ) ▽‾‾
Part 9	Surface	Coating description	
Part 10	General	Drawing format	
Part 11	General	Default tolerances	
Part 12	Surface	Aspheric shape definition	
Part 14	General	Transmitted wavefront error	
Part 17	Surface	Laser damage threshold	6/
Part 19	Surface	Complex surface definition	

Table 20.2 Index inhomogeneity classification.

Inhomogeneity class (A)		Striae class (B)	
Class	Index homogeneity (ppm)	Class	Percentage area > 30 nm OPD
0	±50	1	≤10
1	±20	2	≤5
2	±5	3	≤2
3	±2	4	≤1
4	±1	5	~0 (Extremely low)
5	±0.5		

ISO 10110 Part 2 describes the requirement for stress induced birefringence. In the descriptor format it is presented as '0/A', where A is the difference in optical path length between the two polarisations expressed in nanometres per centimetre of propagation distance. For example, 0/20 represents a maximum stress induced birefringence of 20 nm cm^{-1}, equivalent to an effective refractive index difference of 2×10^{-6}. A stress induced birefringence of 20 nm cm^{-1} is consistent with general or commercial applications. At the other extreme, a value of 2 nm cm^{-1} is generally reserved for precision applications, particularly those involving the measurement of polarisation.

The next section, ***ISO 10110 Part 3*** deals with bubbles and inclusions within a solid glass matrix. Both types of imperfection are treated together and the standard expresses the maximum allowable number of bubbles and inclusions, N, with a size up to a maximum allowable, A (in mm). Although glass manufacturers tend to express material quality in terms of the number of bubbles and inclusions per unit volume, ***ISO 10110- Part 3*** refers to the number in the specific component. The requirement is expressed in the form, '1/N × A'. For example, 1/5 × 0.1 means a maximum of five bubbles or inclusions to a maximum size of 0.1 mm.

Material uniformity is covered by ***ISO 10110 Part 4***. Two types of non-uniformity are described and categorised in terms of two separate classes, A and B. The first class sets the maximum permissible refractive index variation across the whole part. The second class refers to the presence of striae or filamentary strands of non-uniformity produced during the glass mixing process. Imperfect mixing produces index inhomogeneities with length scales from a fraction of a millimetre to a few millimetres in the form of long filamentary strands. Determination of the striae class rests on the percentage of the part area seeing an optical path variation of greater than 30 nm. Table 20.1 sets out the classifications for both classes.

The format for specifying inhomogeneity is '2/A; B', where A and B are the inhomogeneity and striae classes, as indicated in Table 20.2. For example, specification of an index homogeneity of better than 2 ppm and a striae area of ≤2% of the part area would carry a legend of '2/3; 3'.

20.7.2.3 Surface Properties

Specification of surface properties cover the definition of form error, wedge angle, and centration, cosmetic surface quality and surface texture (roughness). In addition, the description of optical coatings and their propensity for laser induced damage is included, as well as the definition of aspheric surfaces.

Surface form error, and its definition is covered in ***ISO 10110 Part 5***. As indicated in Section 20.2.5, the custom is to express form error in fringes, most usually as a peak to valley figure. This format, as discussed elsewhere in this text, is very much a traditional approach inherited from the (human) visual inspection of fringe patterns. To a large degree, it does not anticipate the computer-based analysis of interferogram data, where rms form error would be a more useful single entry parameter. However, whilst ISO-11010 Part 5 does make provision for the expression of form error as an rms figure, this is not the default.

Three figures are called for in the form error description – the power error in fringes, the irregularity in fringes, and the rotationally symmetric error in fringes. Naturally, all figures must be referred to an analysis wavelength, e.g. 589 or 633 nm. The power error effectively defines the focusing error and may, alternatively, be expressed as an error in the base radius of the surface, rather than in fringes of sag. The second term refers to the non-rotationally symmetric form error as expressed in fringes. The last term describes the symmetrical form error, excluding form error that contributes to the power. For example, any form error described by a rotationally symmetric Zernike term (other than the second order defocus term) would be included under this heading. The format for form error description is '3/A(B, C)' where A is the power error, B is the asymmetric irregularity and C the symmetric irregularity. If the power error is expressed as a radius error, the A is replaced with a dash '-'. For example, if the power error is 1 fringe (p to v), the irregularity 0.5 fringes (p to v) and the symmetric irregularity 0.25 fringes, then this would be expressed as '3/1(0.5, 0.25)'. Of course, one might prefer to express the form error as an rms figures. As a rough rule of thumb, the rms figure is to be obtained from the p to v figure by dividing by a factor of 3.5.

Centration of an optical surface with respect to a specified mechanical reference is captured by an angular orientation error or wedge angle. The reference surface is specified in the drawing. According to ISO ***11010 Part 6***, wedge angle error is to be specified in the format '4/A', where A is the wedge angle error, often specified in arcminutes. For example, specification of a wedge angle error of 5 arcminutes would be expressed as '4/5'.

The specification of cosmetic surface quality is an attempt to capture the imperfections produced by an abrasive shaping process. Following the shaping of a surface by grinding, it is impossible altogether to eliminate all the surface scratches and pits that the process has a natural tendency to generate. Of minor concern is the additional scattering produced, over and above that produced by the general texture (roughness) of the surface. The primary concern is that where the optical surface lies close to an image conjugate, then these imperfections become very visible. In ***ISO 10110 Part 7***, the surface quality is usually defined by three classes of defects, pits, scratches, and edge chips. Both the linear dimension (in mm) and maximum number of each feature are to be specified. Pits and scratches occur over the general area of the optical surface. Edge chips are generally specified for faceted components, such as prisms or light pipes, particularly in the absence of any protective bevel. The format for presenting the information is '5/N1 × A1; LN2 × A2; EA3', where N1 and N2 are the maximum allowable number of pits and scratches. The size of the digs and edge chips are represented by A1 and A3 respectively, giving the maximum effective side length (in mm) of a nominally square feature. Scratches are defined by the maximum allowable width, A2, given in mm. The maximum number of edge chips and the maximum scratch length are provided separately.

It is interesting to compare the ISO definitions for surface quality with that embraced in the still widely used ***MIL-O-13830A*** specification. This standard generally comes under the heading of the 'scratch/dig' specification – S/D, where the maximal permitted scratch and dig sizes are specified. Unfortunately, the scratch width does not equate to a specific dimension, but is an arbitrary designation assigned by reference to a standard simple. The dig specification is the dimension of the pit in tens of microns. A specification of 80/50 is considered to be commercial quality whereas for critical (e.g. reticle) applications, 10/5 is appropriate. Returning to the ISO standard, the equivalent dig size is easy to derive from the MIL-O-13830A dig number by multiplying by 0.01 mm. Although, by contrast, the MIL-O-13830A scratch dimension is not explicitly given, useful comparison may be effected by considering the scratch number to be the width in microns. That is to say the scratch number should be multiplied by 0.001 mm to convert to the ISO designation. Therefore, in the ISO scheme, a designation of '5/5 × 0.8; L1 × 0.05; E0.8' would represent commercial quality. For critical applications, a designation of '5/5 × 0.05; 1 × 0.001, E0.5' would be more appropriate.

In the designation of cosmetic surface quality, there is also provision for describing defects in any surface coating. This is effectively described in the same manner as surface digs, with a maximum allowable number, N and size, A. The entry for coating defects is preceded by the letter 'C'.

Surface texture is a term describing the high spatial frequency content of the surface, i.e. roughness. It must be emphasised that the description in ***ISO-10110 Part 8*** is based on a one-dimensional measurement of the surface, rather than a 2D representation of the surface. The measurement can be reported as a standard Rq, RMS (roughness) measurement or a PSD spectrum. In the case of the PSD measurement, the exponent N is

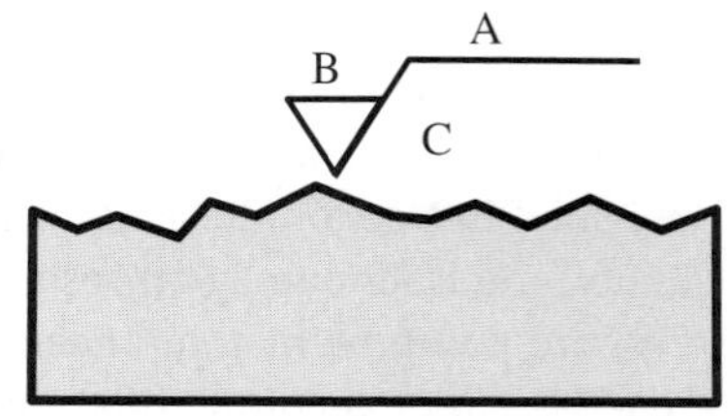

Figure 20.22 Designation for surface texture.

listed along with the proportional constant A.

$$PSD = Af^{-N} \tag{20.7}$$

The units for all roughness measurements are microns. Again, it must be emphasised that the reported values refer to a 1D measurement. For comparison, the exponent, N, in a 2D measurement will be equal to the 1D exponent plus one. That is to say, a 1D exponent of 1.5 is equivalent to a 2D exponent of 2.5. Measurements of roughness (Rq and RMS) are entirely meaningless without a cut-off spatial frequency being reported. In practice, this is reported via the scan length, S, of the measurement. In standard measurements of surface roughness, the scan length is equivalent to five times the wavelength of the cut-off filter used in the measurement analysis. The texture is reported in the drawing itself, ***with the scan length given in millimetres***, not microns.

The drawing contains a designation for the type of surface (G for ground and P for polished) as well as the type of measurement, the measurement value, and the scan length. This is illustrated in Figure 20.22.

In Figure 20.22, A describes the surface type, B the type of measurement together with any data and C the scan length.

As prescribed in ***ISO 10110 Part 9***, thin film coatings are specified by a λ symbol within a bold circle. Details of the coating, antireflection, bandpass, dichroic, etc. are laid out as specified in ISO. If no specific wavelength for the coating is specified, the default wavelength is assumed to be 546.07 nm (a prominent atomic spectral line emitted by a mercury lamp). For some ground surfaces, a blackened surface may be specified to reduce scattering. This is specified by the inclusion of a thick dashed line next to the surface in question.

Not surprisingly, spherical surfaces are assumed to be specified unless specifically indicated otherwise. ***ISO 10110 part 12*** sets out how aspheric surfaces are to be specified. The form of the aspheric surface definition is very much as set out in previous chapters covering both polynomial terms and surface definitions. However, unless otherwise indicated, any Zernike representation is assumed to consist of Zernike Fringe Polynomials rather than the Standard Polynomials. The reader should note this. Additionally, definition of more complex freeform surfaces, encompassing discontinuous surfaces, spline surfaces (NURBS – Non-uniform rational basis spline) is provided for in ***ISO 10110 Part 19***.

Finally, ***ISO 10110 Part 17 (formerly ISO 110110 Part 13)*** covers the susceptibility of the surface coating to high power laser damage. This is particularly relevant for optics used in high power laser materials processing or high-power laser systems used in research applications. In this case, the information is preceded in the drawing by the descriptor, '6/', followed by the laser damage threshold in J cm^{-2} for pulsed lasers or W cm^{-2} for continuous wave (cw) lasers. The laser wavelength is also included in the description.

20.7.2.4 General Information

The recommended tabular format of the part drawing is described in ***ISO10110 Part 1*** and ***ISO 10110 Part 10***. In the generalised block format, an engineering sketch will be set out at the top of the drawing. This will contain the basic mechanical information, prescribing the physical dimensions, definition of the surfaces, particularly reference surfaces and information about surface texture. Underneath the sketch, three columns are to be provided. The leftmost column contains information about the left-hand surface; the column on the extreme right defines the right-hand surface. The central column provides information about the material.

The material will be specified (i.e. named) at the top of the central column, together with the refractive index at some nominated wavelength. Thereafter, the information specified in Section 20.7.2.2 is to be provided.

For the columns defining the properties of the surface, the surface radius is generally specified first, together with a tolerance figure, if power error is to be defined in this way. Subsequently, the clear aperture might be presented for each surface, followed by information about protective chamfers and surface coatings as described in Section 20.7.2.3. Thereafter, details of form error requirements, wedge angle, cosmetic surface quality, and laser damage thresholds follow in due course, as described in Section 20.7.2.3.

Where component tolerances or details of protective chamfers are not explicitly provided, ***ISO 10110 Part 11*** offers recommendations as to 'default tolerances'. Dimensional tolerances (e.g. part diameter, thickness) scale with component size, as does the suggested width of any protective chamfer. Centring (angular) tolerances decline with component size. Recommendations for default material and surface quality are also provided in ISO 10110 Part 11. For details, the reader is referred to the standard itself.

The overall performance of a component in terms of its transmitted (or reflected) wavefront error is described according to ***ISO 10110 Part 14***. In presentation, this is similar to ISO 10110 Part 5, which covers surface form, except it applies to the transmitted wavefront error.

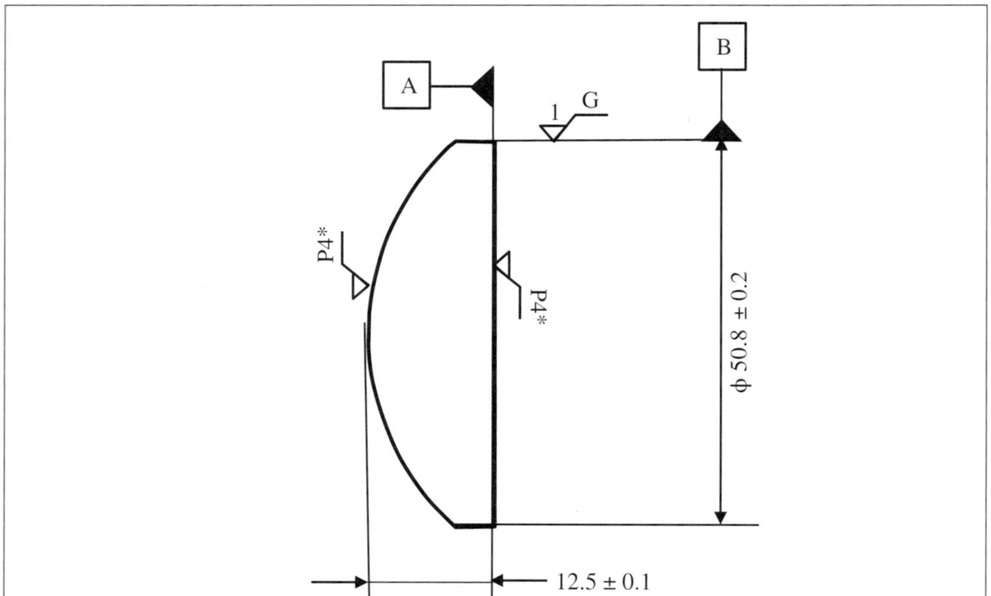

Left Surface:	**Material Specification:**	**Right Surface:**
R: 38.6 ± 0.4 CX	**Schott N-BK7**	**R:** ∞
ϕ_0: 45 (*Clear Aperture*)	**n** (694 nm) 1.5132 ± 0.0004	**ϕ_0:** 45 (*Clear Aperture*)
Prot. Chamfer: 0.5 – 0.75	**ν -**	**Prot. Chamfer:** 0.5 – 0.75
(λ) AR .694	0/10	(λ) AR .694
3/ – (0.5, 0.25)	**1**/2 x 0.1	**3**/1 (0.5, 0.25)
4/3.0	**2**/2; 3	**4**/-
5/5 x 0.1; C 5 x 0.1; L 1 x 0.002; E 1 x 0.2		**5**/ 5 x 0.1; C 5 x 0.1; L 1 x 0.002; E 1 x 0.2
6/ 1Jcm^{-2}; 694 nm; 10		**6**/ 1Jcm^{-2}; 694 nm; 10

Figure 20.23 Example drawing. **P4 designates a polished surface whose quality equates to an rms surface roughness of approximately 1 nm.*

20.7.3 Example Drawing

Figure 20.23 shows a sample drawing of a plano-convex lens, illustrating the features to be included. The format is as broadly prescribed in Part 10 of the standard.

Further Reading

Aikens, D., DeGroote, J.E., and Youngworth, R.N., *Specification and Control of Mid-Spatial Frequency Wavefront Errors in Optical Systems*, Frontiers in Optics 2008/Laser Science XXIV/Plasmonics and Metamaterials/Optical Fabrication and Testing, paper **OTuA1**, OSA Technical Digest. Optical Society of America, 2008.

Asadchikov, V.E., Duparré, A., Jakobs, S. et al. (1999). Comparative study of the roughness of optical surfaces and thin films by use of x-ray scattering and atomic force microscopy. *Appl. Opt.* 38 (4): 684.

Bass, M. and Mahajan, V.N. (2010). *Handbook of Optics*, 3e. New York: McGraw Hill. ISBN: 978-0-07-149889-0.

Duparré, A., Ferre-Borrull, J., Gliech, S. et al. (2002). Surface characterization techniques for determining the root-mean-square roughness and power spectral densities of optical components. *Appl. Opt.* 41 (1): 154.

Friedman, E. and Miller, J.L. (2003). *Photonics Rules of Thumb*, 2e. New York: McGraw-Hill. ISBN: 0-07-138519-3.

Harris, D.C. (2011). History of magnetorheological finishing. *Proc. SPIE* 8016, 0N.

ISO 10110-1:2006 (2006). *Preparation of Drawings for Optical Elements and Systems. General*. Geneva: International Standards Organisation.

ISO 10110-2:1996 (1996). *Preparation of Drawings for Optical Elements and Systems. Material Imperfections. Stress Birefringence*. Geneva: International Standards Organisation.

ISO 10110-3:1996 (1996). *Preparation of Drawings for Optical Elements and Systems. Material Imperfections. Bubbles and Inclusions*. Geneva: International Standards Organisation.

ISO 10110-4:1997 (1997). *Preparation of Drawings for Optical Elements and Systems. Material Imperfections. Inhomogeneity and Striae*. Geneva: International Standards Organisation.

ISO 10110-5:2015 (2015). *Preparation of Drawings for Optical Elements and Systems. Surface Form Tolerances*. Geneva: International Standards Organisation.

ISO 10110-6:2015 (2015). *Optics and Photonics. Preparation of drawings for optical elements and systems. Centring tolerances*. Geneva: International Standards Organisation.

ISO 10110-7:2017 (2017). *Preparation of Drawings for Optical Elements and Systems. Surface Imperfections*. Geneva: International Standards Organisation.

ISO 10110-8:2010 (2010). *Preparation of Drawings for Optical Elements and Systems. Surface Texture; Roughness and Waviness*. Geneva: International Standards Organisation.

ISO 10110-9:2016 (2016). *Preparation of Drawings for Optical Elements and Systems. Surface Treatment and Coating*. Geneva: International Standards Organisation.

ISO 10110-10:2004 (2004). *Preparation of Drawings for Optical Elements and Systems. Table Representing Data of Optical Elements and Cemented Assemblies*. Geneva: International Standards Organisation.

ISO 10110-11:2016 (2016). *Preparation of Drawings for Optical Elements and Systems. Non-toleranced Data*. Geneva: International Standards Organisation.

ISO 10110-12:2007 (2007). *Preparation of Drawings for Optical Elements and Systems. Aspheric Surfaces*. Geneva: International Standards Organisation.

ISO 10110-14:2007 (2007). *Preparation of Drawings for Optical Elements and Systems. Wavefront Deformation Tolerance*. Geneva: International Standards Organisation.

ISO 10110-17:2004 (2004). *Preparation of Drawings for Optical Elements and Systems. Laser Irradiation Damage Threshold*. Geneva: International Standards Organisation.

ISO 10110-19:2015 (2015). *Preparation of Drawings for Optical Elements and Systems. General Description of Surfaces and Components*. Geneva: International Standards Organisation.

Jiao, X., Zhu, J., Fan, Q. et al. (2015). Mechanistic study of continuous polishing. *High Power Laser Sci. Eng.* 3: e16.

Macalara, D. (2001). *Handbook of Optical Engineering*. Boca Raton: CRC Press. ISBN: 978-0-824-79960-1.

Symmons, A., Huddleston, J., and Knowles, D. (2016). Design for manufacturability and optical performance trade-offs using precision glass molded aspheric lenses. *Proc. SPIE* 9949: 09.

Yoder, P.R. (2006). *Opto-Mechanical Systems Design*, 3e. Boca Raton: CRC Press. ISBN: 978-1-57444-699-9.

21

System Integration and Alignment

21.1 Introduction

21.1.1 Background

The previous chapter considered the creation of optical components very much as entities in isolation. However, in order to create a functioning optical system, these components must be integrated. Moreover, the designated geometrical relationship between these components must be preserved to some appointed degree of precision. The first part of this exercise involves the design of a mechanical assembly that will serve to constrain the parts to an adequate precision. On a practical level, a process must then be established such that, following assembly, the required geometry is preserved. This is the process of alignment. It may be either passive or active. In the former case, the designer relies on the inherent fidelity of the mechanical design to ensure the required geometrical registration of all components. Active alignment, on the other hand, requires the provision of limited mechanical adjustment in some components and the ability to actively monitor some system performance attribute, e.g. wavefront error or boresight error and thence to correct it.

Design of the mechanical assembly is naturally underpinned by the type of modelling exercises outlined in Chapter 19. Since the system must perform to the desired requirement within some specified environment, due regard must be paid to thermal and mechanical loads, particularly in the operating environment.

21.1.2 Mechanical Constraint

Individual components are geometrically registered within a system by constraining their mechanical surfaces against matching or mating surfaces in the mechanical assembly. This registration may be maintained either by bonding or by the use of preload forces. This constraint entails the generation of mechanical forces that inevitably induce elastic deformation in the component. The generation of these forces is self-evident in the case of mounting under preload. For adhesive bonding, elastic forces will be generated as a consequence of shrinkage in the adhesive matrix during the curing process and through environmental (temperature) fluctuations.

Elastic compliance within a system is essential to maintaining these mating forces under varying environmental conditions. In particular, changing environmental temperatures serve to modify these mating forces due to the release or amplification of elastic stress through differential expansion. On the one hand, it is important that the forces do not interfere with the performance of the component, either by distorting the functional optical surfaces, inducing significant stress-induced birefringence, or by causing mechanical failure of the part. On the other hand, where a preload force is used to constrain components, it is important that changing environmental conditions do not lead to the removal of this preload force altogether. Indeed, a minimum preload must be maintained under all conditions to ensure that the component does not move under shifting gravitational orientation or reasonable shock or vibration levels.

On a more fundamental level, an optical component may, in continuum mechanics, be considered as a solid body. As such, its motion may be encapsulated in 6 degrees of freedom, three rotational and three translational.

Optical Engineering Science, First Edition. Stephen Rolt.

Companion website: www.wiley.com/go/Rolt/opt-eng-sci

Additional movement that might be ascribed to the individual particles within the matrix of the solid body may be described by distortion. In the context of mechanical mounting, this suggests that six constraints are required to define the position of a solid body. Furthermore, any additional constraints would have the propensity to generate distortion in the object, as constraint cannot be accommodated by rotation or translation. A component mounted in this way is said to be ***overconstrained***. On a practical level, this problem tends to be more salient for larger components, such as large mirrors; distortion, if not controlled, has the propensity to create significant wavefront distortion. Great attention, therefore, is paid to optimising the mounting of such large components. For simpler components, such as lenses, mounting might be effected by a linear contact zone, such as a ring or annulus. This does not, of course, represent a mathematically optimum mounting arrangement. It is inevitable that the two mating surfaces will, to some degree, be mismatched in terms of their form along the contact line. It is inevitable, then, that some distortion will occur when the pieces are forced into contact. Nonetheless, for smaller components, where the contact line is substantially outside the clear aperture, any distortion can be reduced to an acceptable level.

In many applications, the alignment of the system is assured to an adequate level through the mechanical design itself; no further adjustment is necessary for the system requirements to be met. However, in precision applications, such 'passive alignment' will not always be adequate to ensure the system is properly aligned. Therefore, provision for alignment adjustment must be allowed for in the mechanical design. Identification of which alignment degrees of freedom need to be applied to which specific components is carried out as part of the optical tolerance modelling.

21.1.3 Mounting Geometries

Many optical systems are defined by their axial symmetry. Indeed, as established in the earliest chapters, the analysis of Gauss-Seidel aberrations is predicated upon such symmetry. Therefore, a cylindrical symmetry inevitably characterises the geometry of such systems. Many such systems therefore consist of circular lens type components integrated into a ***lens barrel*** assembly. Such assemblies can be very simple, consisting of a uniform cylinder with components held against radiused projections and secured by threaded retaining rings. This simple geometry depends upon the assumption that all components have the same diameter. However, for most systems, this is not the case. Therefore, most lens barrel arrangements exhibit a more complex geometry. Whilst retaining the fundamental rotational symmetry, a typical lens barrel will possess a more complicated geometry, incorporating a range of projections and threaded inserts of varying inner and outer diameter.

If secured by retaining rings under preload, then mechanical registration is dictated by the spherical lens surfaces themselves. In the example included at the end of Chapter 19, the wedge angle was defined with respect to the two optical surfaces, rather than the ground edges. When the two spherical surfaces are thus mated, they are, to a degree, self-centred, although friction can frustrate this process. When mounted in this way, care must be taken to ensure that the components are solidly mounted under all environmental conditions. For the most part, lens barrel materials (generally metallic) exhibit higher thermal expansion than the optical (lens) material, although this is not true for optical polymers. Therefore, as the environmental temperature increases, then any preload force on a lens has a tendency to be released. Conversely, cooling increases any mounting stress. As a consequence, care must be taken to incorporate the necessary mounting compliance to forestall these effects. If necessary, additional compliance in the form of compliant mountings (e.g. spring washers) must be introduced.

Alternatively, components may be mounted individually on a planar optical bench structure. Unlike the lens barrel format, this arrangement facilitates the adjustment and alignment of individual components. As such, the arrangement is widely adopted in experimental and prototype configurations and for scientific instruments. On the other hand, this configuration is perhaps less robust and compatible with volume production methods. Independent mounting of each component, in principle allows for the maximum possible adjustment, corresponding to 6 degrees of freedom. As will be seen presently, a wide variety of mounting configurations are available for this purpose – ***kinematic mounts***, gimbal mounts, and flexure mounts, etc. In many cases, only a restricted degree of adjustment will be required, e.g. a simple tip-tilt mount that provides two degrees of angular adjustment.

Larger components are inevitably more sensitive to mounting stresses produced by the effects of self-loading and thermo-mechanical distortion. Therefore, great care must be taken in the mounting of such components, particularly in the distribution of any mounting support. Indeed, such considerations set a fundamental limit to the physical size of transmissive components. Under the assumption that the clear aperture can in no way be obscured, one is compelled to use the limited region lying outside the clear aperture for support. For example, in refractive telescope optics, as the aperture is increased, the thickness must be increased disproportionately to avoid sensitivity to gravitational distortion. Ultimately, when a certain size is reached, this consideration becomes inimical to the creation of a reasonable design.

This limitation does not apply to mirror-based optics where support on the obverse may be effected without interfering with the clear aperture. By distributing the support carefully, distortion may be minimised; some analysis of three-point mounting was provided in Chapter 19. In any mounting scheme, great care must be taken to minimally constrain the system. For very large mirrors, support may be distributed over many more supports to further reduce distortion. However, any support linkage must be so constructed to provide only six constraints. For example, mounting may be accomplished by the connection of the mirror substrate to a fixed base plate by means of six linkages. However, these linkages must be free to articulate, either by connection to swivel joints or flexures, so that each linkage only provides one constraint – that of its scalar length. More complex systems, with more linkages, employing a ***whiffletree*** or ***Hindle mount*** configuration may be employed.

21.2 Component Mounting

21.2.1 Lens Barrel Mounting

A wide variety of standard hardware is available for the mounting of commercial, stock lenses. These come in the form of '***lens tubes***' in standard diameters of up to 50 mm. The lens tubes will usually come with an internal thread, thus allowing the retention of the components by threaded retainers. Components may be positioned axially with the desired separation. In addition, adaptors may be used to concatenate lens tubes of different sizes thus enabling the incorporation of different part diameters. This approach is suited to the research and development environment, but lacks the flexibility for commercial design.

For a commercial product where the use of specially designed 'custom' optics, as opposed to commercial off the shelf (COTS) optics, is justified on a cost basis, then it is likely that a bespoke mounting solution is also justified. In this case, the lens barrel must cope with a range of different lens geometries. Broadly, the lens tube will consist of a cylindrical structure provided with internal threads. Into this structure will be integrated a potentially complex hierarchy of threaded inserts designed to hold the individual lenses. These inserts will often have both internal and external threads and the component itself retained by the threaded retaining ring.

As alluded to in Chapter 19, components may be retained within this structure with a retaining ring. Contact is maintained towards the edge circumference of each lens by means of radiused rings. As the mechanical modelling demonstrates, greater retention compliance is assured by reducing the effective radius of these rings. In many cases 'burnished' edges are employed for retention. In this case, these 'sharp edges' are assumed to have a nominal radius of 0.1 mm. However the components are held, sufficient compliance must be provided to ensure that the components are securely held over the whole temperature range without incurring undue stress. In addition, retention may be further promoted by the application of optical cement. Holes and channels may be machined into the barrel to facilitate application of the adhesive. In any case, the threaded inserts themselves are often fixed with a thread-locking compound for further stability. Figure 21.1 is a stylised illustration of a lens barrel mount. It is broadly illustrative of a double Gauss camera lens.

In the context of Figure 21.1, the overall objective is to ensure that all spherical surfaces are aligned with their centres lying on a common axis. Designing the assembly with the spherical surfaces as the mechanical

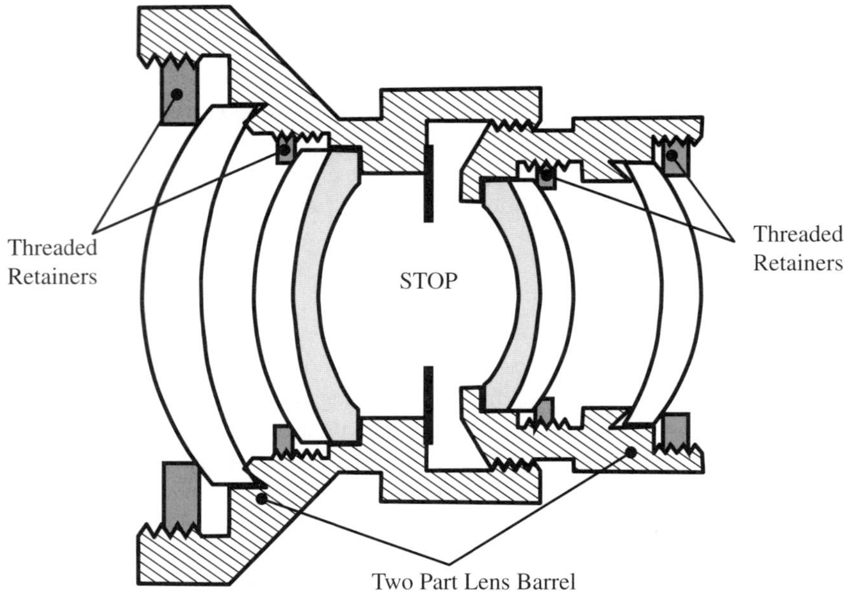

Figure 21.1 Schematic diagram of lens barrel mounting.

reference substantially facilitates this. This is of course conditional on the axis defined by the (ground) edges of the lenses being sufficiently co-incident with that defined by the optical surfaces. If this is not the case (to within the appropriate tolerance) the lens will not fit into its allotted recess in the desired orientation.

In terms of the misalignment of a lens within the lens barrel, there are two separate components to consider. First, the lens may be decentred with respect to the barrel axis, but the lens axis is parallel to this mechanical axis. In this case, boresight error is produced. Rotation of the lens about its axis will have a propensity to produce accompanying rotation of the final image about some centre. Second, the axis of the lens may be tilted with respect to the mechanical axis without any accompanying decentre. In this case, no boresight error, as described previously, will be produced. However, this tilt will produce off-axis aberrations, such as coma for the nominal central field position. Overall, such axial errors, whilst not producing lateral displacement in the image, will produce enhanced aberrations.

For precision applications, active alignment may need to be carried out during the assembly process. As with precision grinding of mechanical surfaces on individual components, alignment may be tested by rotation of the lens barrel. This can be done by illuminating the optic with a laser beam and viewing the back-reflection from either the lens surfaces themselves or from a separate external reference mirror. Any decentre that is present at any surface will produce an angle of incidence that varies slightly during the rotation of the lens barrel. This rotation may be viewed at an image plane where the laser beam is focused. Centration of the laser spot on a pixelated detector provides a very sensitive means for measuring 'decentration wobble'. Separation of the reflected and incident beams is accomplished by a standard arrangement using a quarter waveplate and beamsplitter combination; the laser is assumed to be linearly polarised. A sketch of the set-up is shown in Figure 21.2.

A microscope objective is a classical precision optical sub-system that is barrel mounted. For a high magnification objective, alignment adjustment may be required to meet the image quality and other requirements. As described in Chapter 15, such an objective usually comprises a hyperhemisphere as the first element, followed by a combination, perhaps, of two doublet groups. Typically, it is desirable to provide adjustment for spherical aberration and coma correction, the two most prominent aberrations. System wavefront error may be monitored and analysed using an interferometer arrangement. Spherical aberration is minimised by adjusting the distance between the hyperhemisphere and the succeeding lens group. Adjustment is accomplished

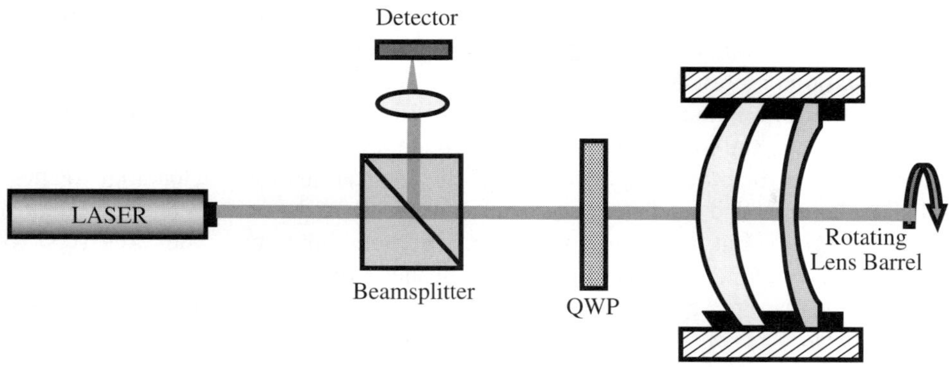

Figure 21.2 Active lens centring.

by the judicious insertion of lens spacers into the barrel assembly. This must, of course, take into account the presence of the cover slip, if any. To adjust for coma at the central field position, one of the lens groupings within the barrel is decentred by means of adjustable centring screws. In this way, the image quality is actively optimised. Following this adjustment process, the adjustment may be locked in some way, e.g. by the application of adhesive.

In addition to the image quality of the objective, there is an extra first order parameter that needs to be adjusted. The focal plane needs to be at some specific axial location with respect to the lens barrel mechanical reference. This is to ensure that when any standard microscope objective is interchanged within a system, the focal point does not change. This condition is known as ***parfocality***. By rotating a threaded outer sleeve, the location of the objective focus with respect to the standard mechanical reference may be corrected.

21.2.2 Optical Bench Mounting

21.2.2.1 General

Where space constraints apply, or where the design requirements mandate the insertion of mirror components, then the optical path must be folded in some way. Therefore, it is not possible for all components to be arranged co-axially. Most commonly, the folding of the optical axis produces a co-planar arrangement with the optical axis typically arrayed in a horizontal plane. Integration of components is facilitated by arranging individual component mounts on a common planar surface referred to as an optical bench. In the case of a typical instrument, such as a monochromator or spectrometer, the components are arrayed on a solid base-plate. For laboratory applications, there is the ubiquitous honeycomb optical table, consisting of a lightweight honeycomb core with flat metal skins attached either side to form a sandwich. The skins are usually provided with an array of tapped mounting holes for flexible clamping of optical component mounts.

21.2.2.2 Kinematic Mounts

Components or subsystems are mounted in separate holders and then attached to the bench. This type of configuration allows for the individual alignment of components, as required. As indicated earlier, the ideal goal of a mount is to provide optimum constraint. For a solid body, there should be six and no more than six geometrical constraints. A mount that fulfils this condition is referred as a ***kinematic mount***. By making some of these constraints adjustable, the mount can be used for alignment, for example, in a tip-tilt stage.

Ideally, kinematic mounting offers the optimum of six constraints through ***point contact*** of the mating surfaces. In this ideal scenario, perfectly reproducible registration of one surface with respect to another is offered. This property of ideal kinematic registration is referred to as ***kinematic determinacy***. However, in practice, the impact of mating forces between the surfaces is to create elastic distortion at the 'so-called' contact points, producing an area contact. It is inevitable that this process is accompanied by surface friction leading to the non-deterministic registration of the two parts, particularly as the mount is adjusted. This problem may be ameliorated by the use of very hard (minimal deformation) and smooth contact materials. Friction may be further reduced by the incorporation of lubricants.

In understanding the application of a kinematic mount it is useful to illustrate this with a discussion of some common kinematic elements and the constraint provided. Throughout this discussion, the operation of some mating force is assumed; this might either be gravitational or the application of spring loading. The first element to consider is a ball (sphere) loaded against a cone. This provides three degrees of constraint, fixing the position of the centre of the sphere, but in no way constraining rotation about any of the three axes. To be strict, the geometry described is not kinematic, as contact is established over a line in this instance. A true kinematic representation would be that of a ball contacting three regularly spaced inclined planes, rather like a set of plane surfaces forming three facets of a regular tetrahedron. Secondly, we might consider a ball loaded against a V-groove. Here the sphere contacts only two points, imposing just two constraints. The sphere is now free to move along the axis of the V-groove. Finally, a sphere in contact with a plane surface provides only one constraint. There is now freedom of translational movement in the two axes that define the plane. Of course, there are a number of different kinematic elements that may be used to provide this single point constraint. Other examples include cylinder upon cylinder contact, which produces just one constraint.

The three elements, as sketched in Figure 21.3 may be used as a basis for a kinematic mount if integrated together as part of a solid structure. Taken together, the three elements provide the ideal total of six constraints. For example, this type of arrangement may be used as a kinematic tip-tilt platform. Such a platform is usually understood to be horizontally oriented. A baseplate, which may be attached to the optical bench incorporates the three features (3 plane, V-groove, and Plane) sketched in Figure 21.3. To the upper platform are attached three corresponding spheres that mate with the three features in the baseplate, providing stable registration. By substituting two of the spheres with the rounded ends of micrometres or precision screws, an adjustable platform is created. In this example, gravitational loading clearly assists the location of the upper platform. However, in practice, spring loading is used to supplement the mating forces. Where this arrangement is used in a vertically oriented mirror mount, for example, some form of spring loading is essential. The principle is illustrated in Figure 21.4.

In the example shown in Figure 21.4, the kinematic mount is realised as a tip tilt stage. As illustrated, the mounting plate shows three mating spheres in position in the corners. In the assembled stage, two of the spheres have been replaced by micrometres. These micrometres would be arranged at opposite corners giving independent rotation (tip tilt) about orthogonal axes. Such simple tip-tilt mounts find widespread use, as this type of (fine) adjustment is the most useful practical alignment adjustment. Addition of decentring about the two axes orthogonal to the optical axis may be warranted on some occasions. Other mounting geometries may be realised based upon the kinematic principle.

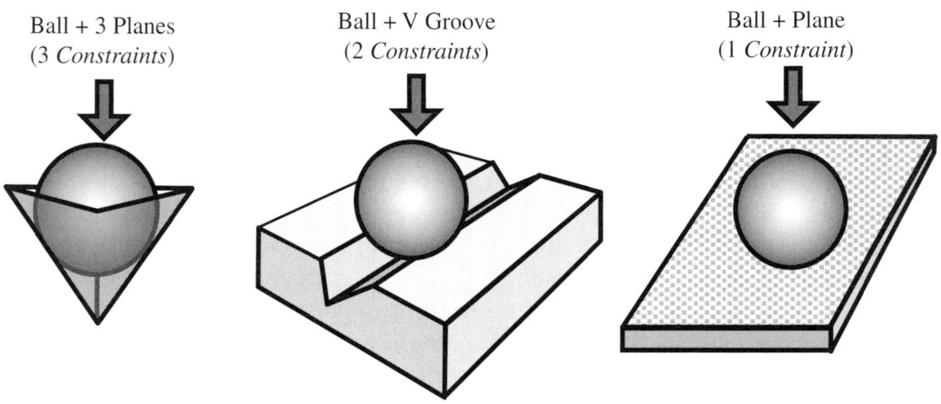

Figure 21.3 Kinematic constraints.

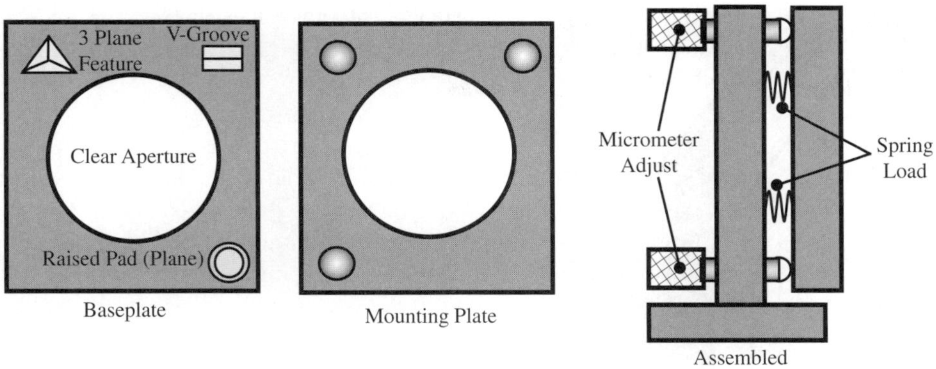

Figure 21.4 Kinematic mount example.

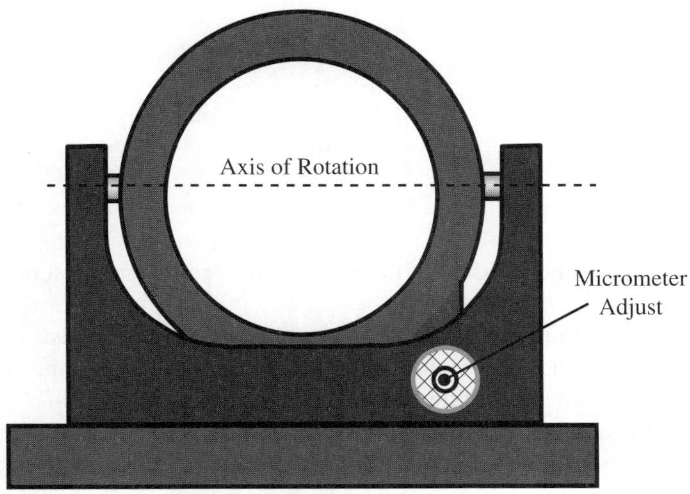

Figure 21.5 Gimbal mechanism.

Of course, such mounts are intended for fine adjustment. The range of angles over which the adjustment may be effected is naturally restricted to a few degrees or so.

21.2.2.3 Gimbal Mounts

A ***Gimbal mount*** allows for the rotation of a component about its centre along one, two or three axes. However, it is not strictly a kinematic mount, as movement is usually effected by rotation about a bearing whose axis is aligned to that of the component's centre. As such, the contact between mating parts is distributed, rather than at single points. Although a Gimbal mount can be used for fine adjustment by incorporating micrometres or fine adjustment screws, it can also, in principle, allow for rotation over a full 360°. Figure 21.5 shows an example of a gimbal mechanism with one axis of movement illustrated; most mounts have at least two axes of rotation.

21.2.2.4 Flexure Mounts

As previously outlined, the ideal kinematic mounting principle is dependent upon the establishment of six ***points*** of contact. Practical implementations tend to fall short of this ideal, and are often dubbed as 'semi-kinematic'. As the mount is adjusted, any re-adjustment in relative position of the two mating surfaces is affected by irreproducible frictional contact forces. This problem becomes more acute as the contact forces

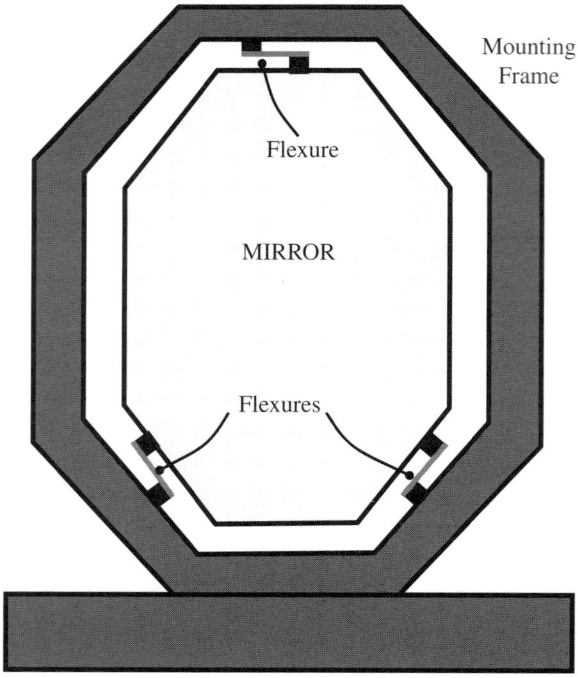

Figure 21.6 Mirror mount with flexures.

(e.g. weight) increase and the establishment of genuine kinematic adjustment is naturally more troublesome for larger components.

By contrast, a ***flexure mount*** is specifically designed to introduce solid connections between two mating surfaces. Adjustment is facilitated by the incorporation of directional compliance into each connection. In particular, flexure mounts are able to accommodate the effects of differential thermal expansion between the component and its mount. Most particularly, in the absence of any sliding contact, the positioning is substantially deterministic. The connections that are introduced are generally in the form or cantilever or similar flexures. That is to say, they have one or more axes where the linkage is stiff and one or more axes where they are compliant. The nominally stiff axes provide geometrical constraint, so there should be six of these for optimal mounting. Any small residual forces attributable to the 'compliant' axes will have a natural tendency to produce a low level of distortion.

Figure 21.6 illustrates the principal of a flexure mount used to secure a large mirror in a holder. Three individual cantilever flexures hold the mirror within the mount. Each flexure, as a cantilever, is flexible along one axis and thus provides two constraints.

In the example shown in Figure 21.6, the (three) individual flexures are bonded to the mirror with adhesive. The design is relatively straightforward, with each flexure implemented as a cantilever flexure. As far as possible, the compliant axis for each of the flexures is aligned to the centre of gravity of the mirror. Any relative expansion of the mounting frame and the mirror is then accommodated by the compliance afforded by the flexures. Furthermore, assuming the compliance of each flexure is matched, then any differential expansion of mirror and mount will leave the central position of the mirror unchanged.

Flexure elements may also be incorporated into adjustable mounts. Materials favoured for use in flexures are naturally similar to those used in spring applications, such as phosphor bronze and special steels. There must be the minimum of hysteresis and little irreversible (non-elastic) deformation. In practice, however, all materials, to a degree, suffer from creep. ***Creep*** describes time dependent strain behaviour in response to an applied load. Creep behaviour tends be most prominent in materials with a high homologous temperature (the ratio of the absolute environmental temperature and that of the melting point of the material). Ultimately, significant creep will lead to non-deterministic placement of the optic, which is not desirable.

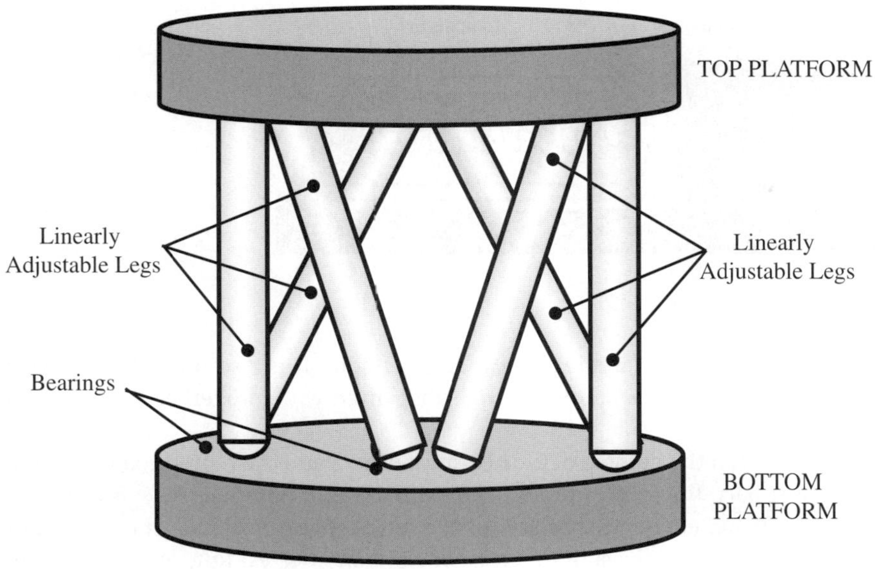

Figure 21.7 Example of a hexapod mount.

21.2.2.5 Hexapod Mounting

The kinematic principle may be extended to define the relationship of two platforms solely by the provision of six connecting rods between them. The scalar length of each connecting rod is defined absolutely and provides the six necessary constraints. However, all six rods are free to rotate or swivel at both ends with respect to the two platforms. For example, this could be accomplished by the provision of ball joints at both ends of the rod. With this additional freedom, the fixed length of the six rods is sufficient to constrain the relative positions and orientations of the two platforms absolutely. Furthermore, by adjusting the length of each of the connecting rod, it is then possible to provide relative movement with 6 degrees of freedom. Such an assembly is known as a ***hexapod mount***. Figure 21.7 shows a typical embodiment of a hexapod mount, with the two mating surfaces implemented as annular platforms. Connection of each leg to the platforms at either end may be accomplished either by a ball joint or a universal joint to allow the necessary freedom of articulation.

Of course, it is clear from the geometry of the hexapod mount illustrated in Figure 21.7, that there is no simple relationship linking connector extension with motion along an individual axis or rotation about a specific axis. However, by extending connectors in a co-ordinated fashion under computer control using a tailor-made algorithm, it is possible to provide controlled movement over 6 degrees of freedom. Each connecting rod is implemented in the form of a linear actuator. For large extensions and movement ranges, the linear actuator might comprise a leadscrew and motor drive. By contrast small scale but precise movement may be afforded by adding piezo-electric pushers to an otherwise rigid rod.

21.2.2.6 Linear Stages

In the design of kinematic mounting schemes, the implementation of angular movement, such as tip and tilt, is generally straightforward to implement. On the other hand, there are a wealth of optical systems, both in research and commercial applications, where the generation of single axis motion is required. This requirement is most frequently implemented in the form of a linear stage where a solid platform is guided along a linear path by some form of contact bearing. In the majority of applications, particularly where substantial travel is required, the movement itself is facilitated via a rotating leadscrew that drives a nut physically attached to the platform. The use of bearing surfaces and multiple contact points marks a significant departure from the constrictions of a kinematic design. As such, the linear stage suffers from a number of limitations that are important for the designer to understand. The general layout of a linear stage is shown in Figure 21.8. In this

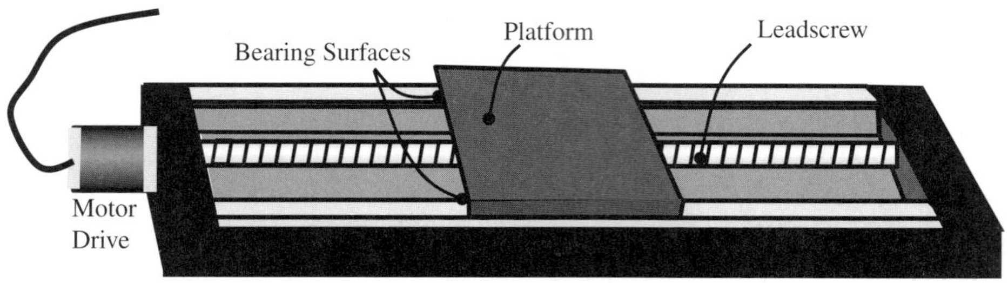

Figure 21.8 General layout of a linear stage.

particular instance, the stage is driven by a motor and leadscrew arrangement. It is also possible to drive a linear stage directly, using linear motors. Otherwise, motor drive is dispensed with altogether and replaced by manual adjustment.

The fidelity of the linear motion rests upon the flatness of the bearing surfaces and the reproducibility of the contact at those surfaces. For the most part, the linear motion in the direction of nominal travel is the most reproducible. Indeed, incorporation of linear encoders (precision fixed reticles) into the linear stage provides feedback control of motion along that axis to a precision of a few tens of nanometres. Rotary encoders can be used to monitor the position of the motor, which, to a degree is correlated to the stage position, allowing for leadscrew errors. Unfortunately, the correspondence between leadscrew rotation and stage position is not entirely deterministic. In particular, the force required to move the platform along the stage produces some variable compliance and slippage in the leadscrew mechanism, leading to the phenomenon of ***backlash***. That is to say, whenever motion in any direction is reversed, this non-reproducible backlash must be 'unwound' before the stage will progress. Furthermore, to obviate leadscrew seizure, some finite amount of 'play' (small in precision screws), must be introduced between leadscrew and nut. This further amplifies backlash. Backlash may be ameliorated by spring loading the nut/leadscrew mechanism. However, linear encoders (at a cost) are preferred for precision applications.

Ultimately, high precision is relatively easy to achieve in the direction of travel. However, it is deviations along the perpendicular axes that are of principal concern. As might be expected, flatness deviation of the mating surfaces causes the platform to deviate from its nominal straight-line path by several microns or tens of microns. One might understand this as a 'run out' error, producing unexpected excursions that are perpendicular to the direction of travel.

In practice, any lateral run out error is not the greatest concern. In general, the chief issue is the angular deviation of the platform as it progresses along the stage. Description of these angular deviations follows a convention borrowed from the aerospace industry. If we define the axis of the leadscrew as the z axis and the vertical axis (normal to the platform in Figure 21.8) as the y axis, then rotation about the x axis is referred to as ***pitch***, rotation about the y axis as ***yaw*** and rotation about the z axis as ***roll***. All these motions may be translated into positional errors since, in practice, the optical axis will be offset from the platform. For example, 50 μrad is a reasonable specification for angular deviation in a precision linear stage. If we take an example of a system where the optical axis lies 200 mm above the platform, then the 50 μrad pitch, roll or yaw translates into a 10 μm deviation. These positional uncertainties arising from the combination of axial offsets and angular deviation are referred to as ***Abbe errors***. It is always good practice, therefore, to minimise these errors, as far as possible, by reducing the offset of the optical axis from the platform to a minimum.

Amelioration of these positioning errors rests on the quality of the bearing surfaces. A number of different bearing types exist and, not surprisingly, choice of bearing is dictated by a compromise that encompasses cost, positioning uncertainty and maximum load. Loading is an important factor and any lack of stiffness in the bearing will produce significant excursions as the load is traversed.

At the most basic, the ***dovetail slide*** brings trapezoidal angled surfaces into direct sliding contact. No (ball or roller) bearing mediates the contact. In some designs, a thin shim of low friction material, such as PTFE

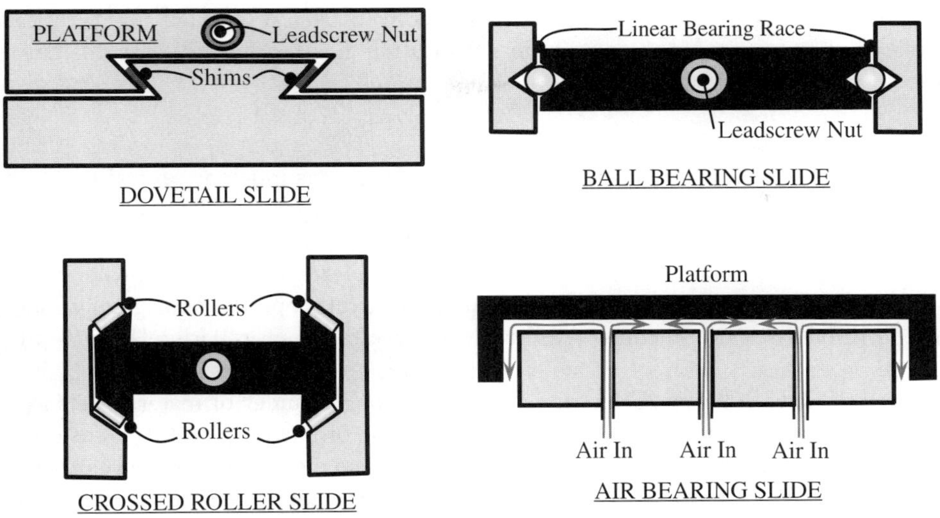

Figure 21.9 Types of linear slide.

(polytetrafluoroethylene) or lead is introduced between the two surfaces. The design is low cost, stiff with a high loading capacity, but its precision is limited. The next stage of refinement introduces linear ball bearing slides at either edge of the platform. ***Ball bearing slides*** provide added positional stability at the expense of stiffness with a moderate increase in cost. As such, they are useful in applications which call for modest loading at a reasonable degree of precision. In a ***crossed roller bearing***, each set of linear bearing slides is replaced by two lines of roller bearings arranged perpendicularly to each other. Roller bearings, in any case, feature a higher loading capacity than comparable ball bearing sets, due to their increased contact area. Here, the stiffness is further enhanced by arranging the two roller sets in a crossed configuration and augmenting the stiffness about the two independent axes. A linear slide with crossed roller bearings may be used in precision applications with relatively high loading. Naturally, this increased performance is attended by higher cost. Ultimately, the highest accuracy and load bearing capability is provided by the use of ***air bearings***. Essentially, this technology uses compressed air as a lubricant by injecting it into a very small gap between mating surfaces through micro-nozzles. Figure 21.9 summarises the geometry of the different bearing types.

Implicit in many of the linear slide designs is the incorporation of a centrally located leadscrew with either manual adjustment or (rotary) motor control. However, particularly popular with air bearing slides is the incorporation of linear motors. In effect, these motors have the stator magnets 'unwound' along a linear track, effectively converting rotary traction into linear traction.

In terms of leadscrew driven applications, incorporation of ***stepper motors*** is a popular choice. These motors allow for incremental or 'quantised' rotary location of the motor, by sequential input of electrical pulses. The disadvantage of this approach is that there is no feedback as to the real location of the rotary shaft. Occasionally, for a variety of reasons, the motor may not respond to the input activation. Therefore, ***rotary encoders*** are often incorporated on the motor shaft. These devices are essentially radially patterned reticles with alternately reflective and transmissive radial patterns. Interrogation of this pattern with an LED (light emitting diode) detector combination translates real incremental shaft rotation into a series of electrical pulses. Although this arrangement can be used to verify stepper motor positioning, most commonly it is used with DC motors as part of a servo-controlled loop. There is, however, a further refinement that may be added. The rotary encoder is only providing an indirect indication of the linear position of the stage; the rotary position of the shaft is, at best, only a proxy for the linear slide position. A variety of effects, such as leadscrew errors and backlash may conspire to limit the deterministic correlation of shaft rotation and linear slide position. Therefore, to further increase the positioning precision, a ***linear encoder*** may be used. As with the comparison of rotary and linear motors, the linear encoder is a rotary encoder that has effectively been unravelled along a linear scale. Finally, further increase in accuracy may be conferred by substituting the linear encoder for a distance measuring interferometer.

21.2.2.7 Micropositioning and Piezo-Stages

The linear stages previously described allow for substantial linear movement, as much as several hundred millimetres or more. However, the presence of long bearing surfaces inevitably leads to lateral positioning errors, as previously described. For small displacements, of the order of tens of microns, a motor driven stage may be replaced with a piezo-electric actuator. These devices rely on the piezo-electric effect, wherein an electric field applied to a piezoelectric crystal generates a small strain. This useful property was originally noted for crystalline quartz, although modern devices incorporate ferroelectric crystals, such as Barium Titanate and Lead Zirconate Titanate.

A common actuator geometry is that of a piezoelectric rod, where a long cylinder of material acts as a 'pusher'. An applied electric field then creates strain along the axis of the rod producing a useful extension of the actuator. Alternatively, the transducer may act to produce a lateral shearing movement. The development of shear strain (or tensile strain) depends upon the relative orientations of the applied electric field and the principal axes of the crystal. Such piezo-electric actuators offer highly linear displacement as a function of applied voltage and may be integrated into the structure of a mechanical stage. For example, where integrated into a flexure mount, piezo-actuators may be used to deliver precision, deterministic movement in more than one axis.

The movement directly afforded by piezo-actuators is necessarily limited to a few tens of microns. However, this can be extended by incorporating a piezoelectric element into a flexure element. This effectively 'amplifies' the strain produced by the crystal, at the expense of reduced actuator stiffness. However, the 'inchworm' device allows for movement of several millimetres through the ingeniously devised co-ordinated action of three separate piezo-electric actuators. Figure 21.10 illustrates the principle. The inchworm drive is designed to move a cylindrical pusher along its axis. Only one of the piezoelectric elements is used to provide the axial movement. The other two elements are in the form of annuli and designed merely to grip the cylindrical pusher at either end. As such, in the de-activated state, it has sufficient clearance to accommodate the pusher, but grips it firmly when activated. The axial separation of the two grippers is determined by the controllable length of the main axial actuator. In an eight-cycle activation sequence involving the three different actuators, the central pusher may be progressively moved in either direction; the reader might like to deduce this sequence, for either direction of movement.

21.2.3 Mounting of Large Components and Isostatic Mounting

We discussed the principles of kinematic mounting in the context of mounting and adjusting smaller optical components. Provision of a deterministic static location of a large component or platform is referred to as ***isostatic mounting***. As with kinematic mounting, the provision of six constraints and no more offers a

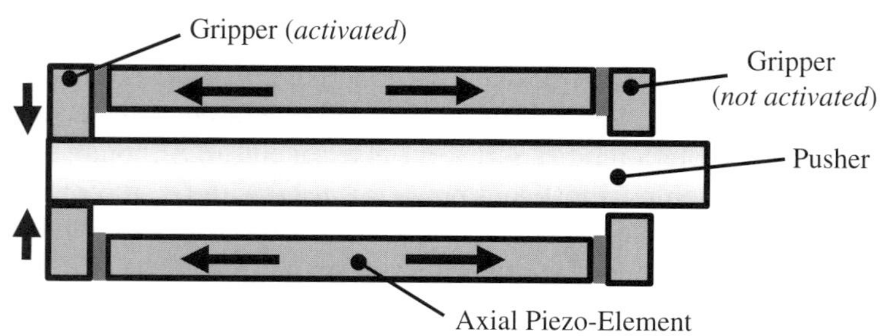

Figure 21.10 Inchworm piezoelectric drive.

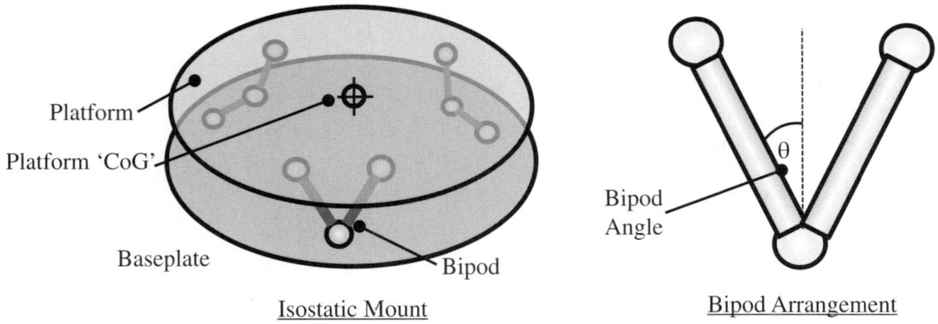

Figure 21.11 Isostatic mounting arrangement.

distortion free solution. The particular concern of isostatic mounting is to provide reproducible location of a physical platform in the presence of substantial dimensional changes, most notably due to differential thermal expansion. One example might be the mounting of a platform on a baseplate where significant differential expansion is anticipated. This scenario might be particularly relevant in cryogenic applications where relative thermal strains as high as 0.5% may be anticipated. In the presence of uniform expansion, it is possible to design a mount in such a way as to minimise the relative movement of the platform and baseplate 'centres of gravity'. Furthermore, such geometrical adaptation should be smooth and continuous without any slippage or sticking. The simplest mounting regimes follow the principle of hexapod mounting where the optimum six constraints are offered by joining baseplate and platform with six legs that are free to articulate at either end. These six legs are often mounting in pairs, known as ***bipods***. A typical isostatic mounting arrangement is shown in Figure 21.11, connecting two platforms with six legs arranged in three bipod pairs.

If the bipod terminations are arranged in such a way as to have their centres coincide with the baseplate and platform 'centres of gravity', then there will be no relative ***in plane*** movement of the two centres in the event of isotropic expansion. It has to be emphasised that this consideration applies only to in-plane movement. If now we imagine the baseplate and platform to have a coefficient of thermal expansion coefficient of α_b and α_p respectively, then for a bipod angle of θ (Figure 21.10), the bipod expansion, α_{bp}, for zero relative axial movement, is given by:

$$\alpha_{bp} = (\alpha_{p\_}\alpha_b)sin^2\theta \tag{21.1}$$

In practice, for the axial movement to be negligible, the bipod thermal expansion should be small. In many cases, it is customary to make the bipods from low thermal expansion materials, such as invar.

As outlined previously, these legs must be able to articulate such that the only significant constraint is the physical length of each leg. The most obvious way of achieving this is to incorporate either ball joints or universal joints at each end of the leg. However, the bearings themselves can be troublesome, with a tendency to exhibit unpredictable behaviour, such as slipping or sticking. Therefore, as an alternative, these bearings can be replaced with linkages that are designed to maximise stiffness along the axis of the flexure, but minimise stiff along the two perpendicular orientations. Figure 21.12 shows an embodiment of such a linkage. These linkages are designed to flex near their ends by incorporating two necked regions where the linkage diameter has been substantially reduced.

In Chapter 19, we described the modelling of self-deflection in mirrors, and location of the optimum mounting points. In particular, an estimate of mirror deflection was outlined for mounting on six evenly spaced points along the 68% radius circle. This arrangement is extremely efficient, and for mirror diameters up to 1–2 m ***and with reasonable thickness***, produces negligible wavefront distortion due to gravitational flexure. However, the combination of larger mirror sizes and the natural tendency to prefer thinner substrates presents the designer with particular challenges.

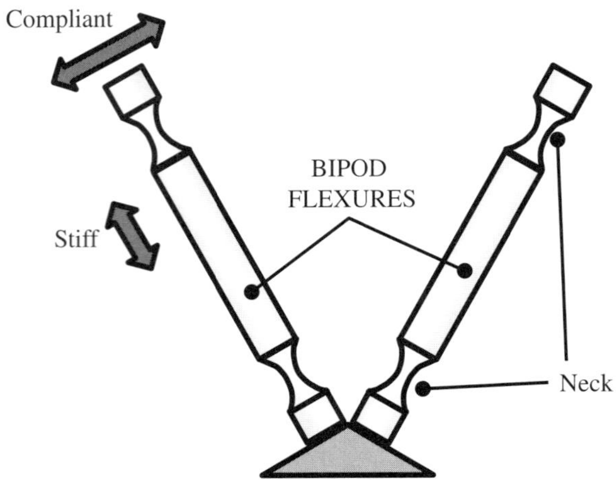

Figure 21.12 Flexure linkages.

For larger mirrors, a simple and obvious solution might be to spread the load over a larger number of points. The problem is that adding extra supports over constrains the mounting and the most minute geometrical imperfection in the mounting or the effect of differential expansion will cause significant distortion. It is inevitable that this distortion will be larger than any self-loading effects one is trying to ameliorate. Therefore, if extra supports are to be provided, then it must be done in such a way as not to introduce additional constraints. Of course, self-loading effects are particularly significant in astronomical applications where the gravitational vector is likely to change substantially with respect to the mirror axis as the optical axis of the telescope is manoeuvred. In this scenario, it is clearly impossible to ameliorate the impact of wavefront distortion by optical (as opposed to mechanical) design as it is inherently variable.

Thus, for supporting a large mirror there is a clear need to distribute the load more evenly (i.e. across more points) without over constraining the physical mount. Rather than directly linking N points on the back of the mirror, a network of linkages is created terminating in N points. However, the network is so contrived as to offer exactly six geometrical constraints, notwithstanding the number of physical mounting locations on the back of the mirror. The establishment of these six constraints is dependent upon the arrangement of bearings and pivots within the network and, in particular, the rotational degrees of freedom they offer. This is the principle of the ***Hindle mount*** and the so-called ***whiffletree*** arrangement. Historically, the whiffletree mount takes its name from the arrangement of articulated linkages used to distribute the load in horse drawn ploughing. Instead of linking the two surfaces directly by linkages connecting them, an additional layer (or layers) of points is provided, for example in the form of separate plates.

One embodiment of this is the Hindle mount uses three triangular plates each having three connections to the mirror. At first sight, with the nine linkages, the mount may seem over constrained. However, the freedom of movement of the intermediate mounting layer reduces the number of constraints to six. More specifically, the three plates are each connected to the baseplate by an arm, which is free to pivot in one orientation. Thus, taken together, the three plates 'consume' three degrees of freedom, leaving a total of six degrees of freedom for the mirror mounting itself. The scheme is illustrated in Figure 21.13.

The nine linkages, as indicated in Figure 21.13, are bonded onto the underside of the mount. The three attachment points on the mounting plate allow for tip-tilt adjustment of the mirror. For larger mirrors, more complex schemes have been used involving a more extensive distribution of the load. Broadly, these arrangements use a rather more extensive collection of nominally triangular plates each implemented as a framework known as a ***whiffletree***. As an example, the mounting arrangement for the hexagonal segments of the Keck telescope primary mirror involve the use of 12 whiffletree frameworks allowing for support distribution over 36 separate linkage points. As with the basic Hindle mount, the whole structure is supported at three points

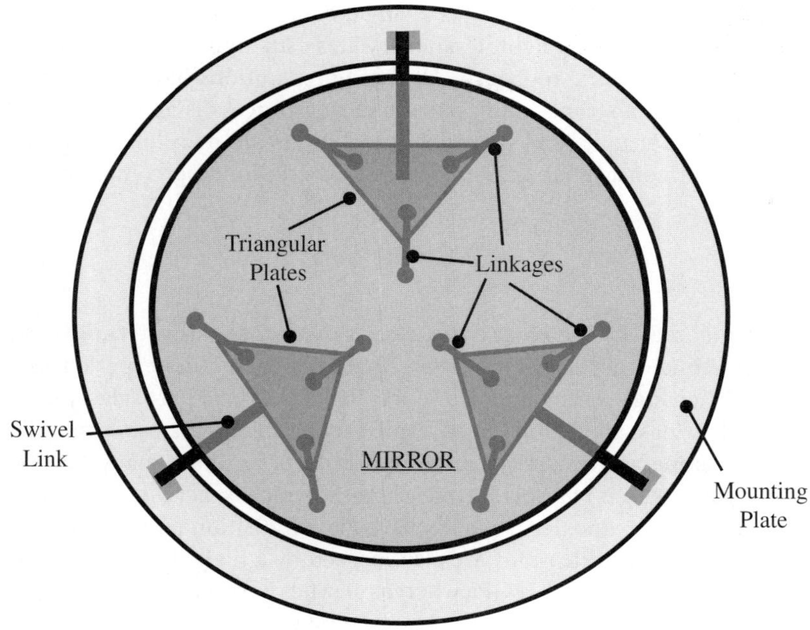

Figure 21.13 Hindle mount.

on the underlying baseplate with 18 intermediate flex pivot points. Analysis of the structure of linkages suggests that it offers a total of 60 constraints with 54 degrees of freedom applied to the 18 intermediate pivots. Again, this provides the necessary six degrees of freedom.

It will be noted that, as with the Hindle mount, three baseplate attachments underlie the tip tilt support. In the case of the Keck primary mirror, each attachment point was served with an actuator that provided alignment adjustment. Of course, if more attachment points and actuators are provided, then the actuators may be used to provide controllable distortion of the mirror surface. For example, the ***James Webb Space Telescope*** segmented primary mirror uses an additional central support actuator specifically designed to provide limited radius adjustment for each segment. In the case of the Keck mirror, previously highlighted, adjustable spring loading was incorporated into the whiffletree network, specifically to provide some controlled distortion to adjust for small low frequency form errors.

21.3 Optical Bonding

21.3.1 Introduction

Bonding of optical surfaces is a very common process in optical assembly. This discussion will revolve around the heterogeneous process of applying a separate adhesive layer between the two surfaces in question. This is fundamentally a low temperature process compatible with optical assembly. Of course, other higher temperature processes, such as welding and sintering are available but are not generally applicable in delicate optical assembly. It is possible to join directly two highly polished and clean glass surfaces directly by a ***molecular adhesion*** process. However, this is very much a niche application.

For the most part, in modern applications, the bonding process is based upon organic adhesive formulations drawn from a restricted range of material families. These include ***epoxies***, ***urethanes***, ***silicones***, ***acrylics***, and ***cyanoacrylates***. A significant number of these compounds are in binary form. That is to say the two components are dispensed in (viscous) liquid form and harden or ***cure*** by chemical reaction upon admixture. Whether the adhesive is in binary form or presented as a single component, the cure process may be accelerated thermally or by exposure to ultraviolet light. Cyanoacrylates ('superglue') are unusual in that their curing is initiated by exposure to atmospheric moisture. All these preparations have their niche applications. Epoxies are naturally hard and form a strong bond, but the curing process is generally more protracted. Acrylics, by contrast, are softer but readily lend themselves to UV curing in volume applications. Silicone based adhesives are rubbery in consistency and are applied where bonding compliance is demanded.

21.3.2 Material Properties

The 'hardness' of adhesives is most readily marked by the ***glass transition temperature***. This term is used loosely to describe the temperature of the second order phase transition where the compound experiences a rapid change in its specific heat capacity. The change is accompanied by a transition from a hard, glassy consistency to a soft rubbery one. Generally, a low glass transition temperature is consistent with a soft material where, by contrast, a higher glass transition temperature marks a harder material. For example, many silicone preparations have significantly sub-zero glass transition temperatures, whereas, typically, acrylics have glass transition temperatures of a few tens of degrees (C). Epoxies generally have glass transition temperatures of 85–100 °C and higher. Not surprisingly, higher glass transitions are also marked by a high elastic modulus. Epoxies generally have a high elastic modulus, of the order 3–4 GPa whereas acrylics have a rather low modulus, 2 GPa being typical. For silicone preparations, the elastic modulus is very much lower, and quoted in MPa rather than GPa.

Compared to glasses and metals, the thermal expansion coefficient of adhesives is high. Epoxies, in their native state, typically have a coefficient of thermal expansion of 50 ppm, whereas for acrylics, the figure is even higher, at around 70–100 ppm. However, many adhesive preparations incorporate additives that modify both there thermal and mechanical properties. Minute glass or silica beads are often added, for example, to epoxy formulations with the effect of substantially reducing the naturally high thermal expansion coefficient of the matrix (epoxy) material. Other additives, such as graphite, or silver may be incorporated to enhance thermal or electrical conductivity.

A complicating factor in the modelling of adhesives is that they exhibit marked plasticity. That is to say, they have a propensity to develop irreversible (non-elastic) deformation. Finite element analysis is substantially, although not exclusively, based upon linear elastic behaviour. For the most part, non-elastic behaviour in adhesives is modelled as Newtonian ***creep***, informed by a linear viscoelastic model whereby an imposed stress produces not only a proportional strain but a time dependent (first differential) component that is proportional to the applied stress. This behaviour is captured by the Maxwell model of viscoelastic behaviour, one of a number of such models and illustrated here in Eq. (21.2).

$$\frac{d\varepsilon}{dt} = \frac{\sigma}{\eta} + \frac{1}{E}\frac{d\sigma}{dt} \tag{21.2}$$

The first term on the RHS of Eq. (21.2) is effectively a viscous term and describes the creep behaviour and the second term the standard elastic deformation. That viscoelastic behaviour is particularly marked in adhesives and, more generally in polymers, is a consequence of their high homologous temperature. The homologous temperature is the ratio of the absolute temperature of the environment to that of the material melting or softening point. Furthermore, the effective creep viscosity, η, as revealed in Eq. (21.2), is a marked function of temperature, increasing very rapidly as the temperature approaches the material softening point. In understanding the impact on real adhesive bonds, the impact of uneven stress distribution must be appreciated. In particular, there is a tendency where (e.g. thermal) stresses develop in a bondline, then they are substantially exaggerated towards the edges of the bondline. Therefore, it is in these regions where creep flow or even delamination is likely to occur. Unfortunately, on a practical level, data regarding the creep behaviour of commercial preparations is somewhat sparse. Therefore, the engineer is reliant on the performance of laboratory tests to elucidate these properties.

21.3.3 Adhesive Curing

As outlined earlier, adhesive hardening or curing is initiated by application of heat or ultraviolet radiation or atmospheric moisture in the case of cyanoacrylate adhesives. The curing process results in the polymerisation and cross-linking of the adhesive elements to form a hard matrix.

Thermally setting adhesives, such as epoxies are generally available as binary mixtures, consisting of an (epoxy) resin and hardener. The curing process is initiated once the two components are admixed. However, it is common practice for the adhesive components to be pre-mixed and available as a single preparation. To avoid curing during storage, the pre-mixed adhesive must be stored at very low temperature ($<-40\,°C$) to slow down the polymerisation reaction. In any case, it is common practice to refrigerate industrial adhesives to enhance their shelf life. For thermally cured adhesives, the cure temperature affects the rate of cure; higher temperatures correspond to a faster cure. Indeed, as with chemical reactions in general, the reaction rate follows an Arrhenius type dependence with the cure rate doubling for a specified temperature increment. Cure temperature not only affects the cure rate, but also the glass transition temperature of the cured resin. A higher cure temperature equates to a higher glass transition temperature.

Ultraviolet curing adhesives, typically acrylics, contain a photosensitive component known as a photo-initiator. In effect, this is a photo-sensitive catalyst that initiates the polymerisation on irradiation. For the most part, these preparations are active in the ultraviolet region and are generally activated by irradiation with a mercury lamp at 365 nm. However, it is possible to incorporate photo-initiators that are sensitive to longer (visible) wavelengths, particularly in the 'blue' region of the visible spectrum. UV curing adhesives are widely used on account of their rapid cure cycle, typically of the order of a minute or less. Naturally, the bondline must be optically accessible.

Adhesive curing is accompanied by volume shrinkage. Cure shrinkage is an important attribute of an adhesive system, especially for optical applications, particularly those that impact system alignment. The concern is that as a bond is cured, the adhesive preparation experiences shrinkage as it changes from a viscous liquid to a hard, amorphous glassy material. As such, any 'unresolved' strain associated with shrinkage is translated into elastic stress capable of moving components during critical alignment procedures. Therefore, for many optical applications, preparations have been devised with especially low cure shrinkage, of the order of a fraction of a percent. In terms of modelling cure shrinkage, in many respects, its effects resemble those of thermal contraction. The important element of the cure shrinkage is the ***unresolved cure shrinkage***. Up to a certain point in the cure cycle, the cure shrinkage is accommodated by the rheology or flow of the adhesive preparation. Thereafter, as the material hardens, further contraction cannot be accommodated and is accompanied by the development of internal stress with the potential for producing component misalignment.

21.3.4 Applications

Many applications involve the bonding of optical assemblies and, in particular, glass-to-metal interfaces are common. Epoxies provide high bonding strength (as measured by the failure stress in shear) for these dissimilar materials. Shear strength is generally specified by manufacturers for both glass to glass and metal to glass bonds. As previously indicated, differential thermal expansion can place significant stress on such joints which, to a degree, may be ameliorated by using more compliant (i.e. rubbery) formulations that may have lower bond strength. As such, there is a clear trade-off between the absolute shear strength of a joint and its vulnerability to environmental degradation. Chapter 19 provides some additional details about the modelling of stresses in adhesive joints. Naturally, this consideration is most salient for those applications, such as military environments where large swings in ambient conditions are to be expected.

Modelling of stresses in an adhesive bond indicates the pre-eminence of bond thickness in determining the stresses in the joint. Therefore, in any manufacturing process, it is critical that the application of adhesive and therefore its thickness is tightly controlled. For example, adhesive can be 'printed' in a process akin to inkjet printing in order to deliver precisely controlled quantities. Another example of how bond thickness may be controlled is the incorporation of (glass) microspheres of some fixed diameter, which determines the offset of the two surfaces.

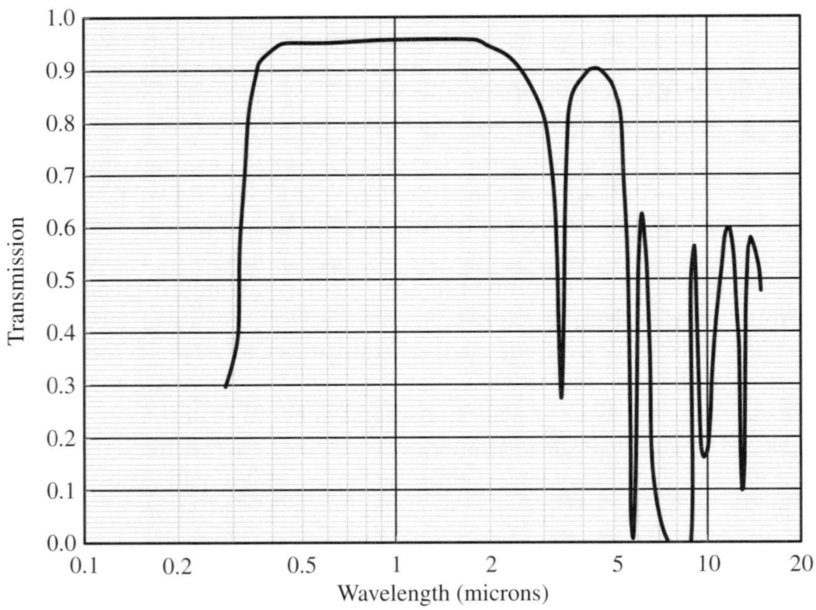

Figure 21.14 Transmission spectrum for acrylic adhesive (Norland NOA 61). Source: Data Courtesy of Norland Products Incorporated.

There are some applications, such as the bonding of lens doublets and triplets where the adhesive is in the 'line of sight'. Other examples include the bonding of beamsplitter cubes. In these cases, their transmissive properties are of direct importance. As organic formulations, these applications are necessarily restricted to visible and the near infrared, perhaps extending to around 350 nm in the ultraviolet. Most organic compounds exhibit strong electronic absorption features that render them opaque in the ultraviolet. Furthermore, in common with other organic materials, adhesives can degrade photochemically on exposure to UV radiation. In the infrared, there are characteristic vibrational features, such those arising from C—H bonds, that produce localised absorption. Infrared absorption properties can be modified by addition of fluorinated compounds, substituting C—F bonds for C—H bonds and shifting the vibrational spectrum further into the infrared. Figure 21.14 shows the transmission spectrum for a typical acrylic adhesive.

Bonding of metals and glass surfaces is an especially common process. This might include the bonding of prisms onto platforms or the direct assembly of lenses in lens tubes. Earlier in this chapter, we discussed the use of flexures in the mounting of mirrors and other large components; these metal elements are invariably bonded to the optic by means of an adhesive preparation.

Adhesives are widely used in the bonding of semiconductor lasers, optoelectronic devices, and fibres into component packages. Essentially these applications depend upon the efficient coupling of light into waveguides or optical fibres. As such, the sub-micron alignment of critical components is mandated. For example, a semiconductor laser chip may be bonded into a package using a silver loaded epoxy for thermal conduction. In addition, a single mode fibre is cemented into a silicon V-groove bonded to the package. Light is coupled into the fibre by a lens bonded into a metal annulus. At this point, the mounted lens is to be bonded into the package with the thermally cured adhesive. However, the alignment must be effected whilst the curing process is underway. To achieve this, the lens is attached to a gripping arm which is positioned by a piezo-electric stage with the appropriate number of degrees of freedom. Optimum alignment is attained when the fibre throughput is maximised, and a servo loop is used to maintain this condition as the curing process proceeds. This is illustrated in Figure 21.15.

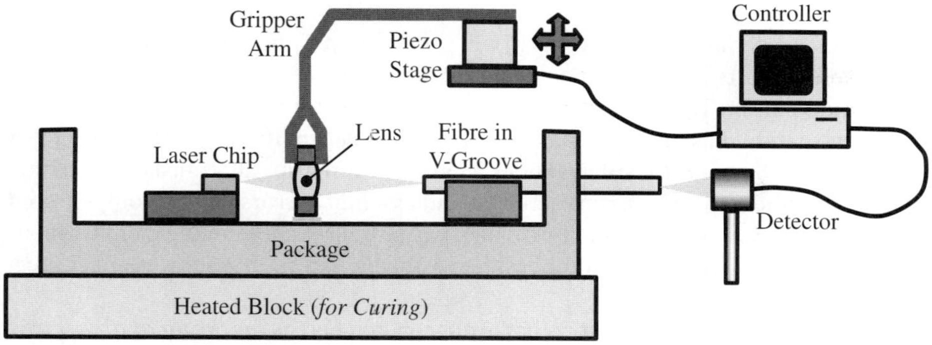

Figure 21.15 Opto-electronic component bonding and alignment.

As well as the obvious function of bonding optical components into assemblies, adhesive compounds may be used to provide sealing against environmental (e.g. moisture) ingress and for the locking in of alignment adjustments through thread sealing, etc. Adhesives are widely used for bonding in aerospace or research applications, where they are required to operate in a low pressure or reduced pressure environment. As organic compounds with appreciable vapour pressure, they are prone to outgassing. This process of outgassing leads to deposition elsewhere in the system and potential contamination. Therefore the use of low outgassing, 'space qualified' adhesive materials is called for in these applications. Outgassing is measured by virtue of the degree of mass loss during vacuum baking. Silicone preparations have a specific issue with regard to vacuum and aerospace applications. Silicone compounds are notoriously surface mobile and have a very strong tendency to strong surface diffusion, migrating along connected surfaces.

21.3.5 Summary of Adhesive Types and Applications

Table 21.1 summarises the main groups of adhesive materials, together with their properties and range of applications.

21.4 Alignment

21.4.1 Introduction

Implicit in any discussion of the alignment of an optical system is the question as to whether active alignment is required at all. In many cases, the fidelity of the component manufacture and assembly process is sufficient

Table 21.1 Summary of adhesive properties and applications.

Family	Cure process	Tg	Elastic modulus	Shear strength	Application
Epoxy resin	Thermal, some photo-curing	80–140 °C	3–4 GPa *(Higher for filled material)*	5–30 MPa	Strong glass to metal or metal to metal bonds
Acrylic resin	Thermal and photo-curing	0–100 °C	1.2–2.5 GPa	2–10 MPa	Bonding of transmissive components
Silicone resin	Thermal and photo-curing	−130–0 °C	<1.0 GPa	<5 MPa	Sealing and compliant mounting. High temperature use
Cyano-acrylate	Atmospheric moisture	20–120 °C	1.0–1.5 GPa	2–5 MPa	Temporary 'tacking' of components

to permit all system requirements to be fulfilled. That is to say, the mechanical tolerances within the system are sufficient to guarantee satisfactory alignment. This scenario will generally apply in the case of volume production. However, where the requirements are exceptionally demanding, it is often impossible to meet the requirements without further intervention. Therefore, provision must be made for some kind of active alignment process. This will involve the careful design of mechanical adjustment capability within the mechanical mounting arrangements of the system. Specifically, sufficient degrees of freedom (tip, tilt, translate, etc.) must be added to provide confidence that the impact of manufacturing and assembly errors can be compensated. This, of course, does not imply that every mount should be provided with 6 degrees of freedom of adjustment. What is required is ***sufficient*** adjustment capability to compensate reasonable errors. The adjustment strategy is determined from the optical tolerance modelling.

This section introduces the topic of ***optical alignment***. What is not covered here is ***mechanical alignment***. Mechanical alignment may be thought of as the placement of optical components or subsystems without any feedback associated with the ***optical*** performance characteristics or behaviour of the system. Such mechanical alignment may be entirely passive and rely on the inherent reliability of the manufacturing process to ensure that mounted components are in the correct position to the prescribed tolerance. Otherwise, mechanical alignment rests on the ability to measure the position of components and sub-systems in three dimensions, using ***co-ordinate measurement machines*** (CMM) and ***laser trackers***, etc. Thereafter, the position and orientation of components must be adjusted accordingly. In practice, mechanical alignment often precedes the more sensitive optical alignment process.

In practical terms, the assembly of an optical system must be based upon an ***alignment plan***. Generally, this will be a sequential process with the alignment of individual components or subsystems to some central axis. As the alignment process proceeds, then more components may be added in turn and aligned. This process is relatively straightforward for components with a clearly marked axial symmetry. However, alignment is complicated by the introduction of components lacking spherical symmetry, especially off-axis aspheres. Here, the alignment process, to a large extent, revolves around the elimination of specific types of aberration and relies upon the deployment of interferometric techniques.

21.4.2 Alignment and Boresight Error

The impact of optical misalignment can be characterised as boresight error or in terms of its impact on wavefront error. ***Boresight error*** occurs where the image location at the output focal plane is laterally shifted with respect to the intended position due to tilts and decentres within the system. This error is, in effect, an angular deviation of the chief ray and may be wholly understood in terms of the paraxial properties of the system. By contrast, the impact on wavefront error can only be interpreted in terms of the creation of off-axis aberrations by component decentring.

Much of practical laboratory and systems alignment is aimed at combatting boresight errors. The invention of the laser heralded a revolution in practical and convenient optical alignment. In effect, a laser beam can serve as a proxy for the chief ray. As such, general centration of system components may be evaluated by directing a laser beam along the path of the chief ray for the central field position. As outlined previously, a laser beam may be used to align reflective or refractive surfaces to some rotational axis. When the surface is rotated about this axis, any misalignment will be revealed as a cyclic deviation in the reflected beam.

For accurate alignment of reflective surfaces, an autocollimator may be used. This instrument functions as a telescope, producing a collimated beam from an illuminated object. Traditionally, this illuminated object was in the form of a reticle imprinted with cross hairs. Any plane mirror produces an image that is located in the same plane as the original object, but displaced laterally according to the angular offset of the reflected beam. Any small tilt in the mirror could be measured or offset by visual observation of the imaged cross-hairs. However, these 'ocular' instruments have been largely superseded by the use of digital image processing techniques which enable the achievement of far higher levels of precision.

Even greater sensitivity in angular alignment measurements can be attained through the use of interferometry. Imaging techniques, such as autocollimation, discard any information relating to the phase of the collimated beam. Although, in many respects not as direct or convenient as the use of imaging techniques, such as autocollimation, interferometry is, in principle, capable of measuring tilts across a wavefront to nanometre precision.

21.4.3 Alignment and Off-Axis Aberrations

Of course, the alignment process not only corrects the orientation of the chief ray, but also serves to minimise off-axis aberrations. An example of this, which we encountered earlier, is the centration adjustment of a microscope objective lens in order to minimise coma. In that instance, the procedure is relatively straightforward to grasp, with one adjustment (centration) and one parameter to optimise (coma). However, the principle may be extended to more complex systems, as will be revealed later.

Once the alignment process is complete, any adjustments must be 'frozen in'. Provision, therefore, must be made for 'locking' any adjustments made, for example, by the use of adhesives and thread sealing compounds or the use of locking screws.

21.4.4 Autocollimation and Alignment

Perhaps the simplest 'rough and ready' technique for quick alignment of reflective surfaces is the simple technique of laser retroreflection. This is particularly appropriate for the laboratory environment. We might wish to align a surface perpendicularly to some central axis which is somehow mechanically defined on an optical bench. To effect this, the laser beam is aligned with respect to some mechanical reference on the bench. This could be by virtue of simple alignment of the laser using pinhole targets or more precisely, using a quadrant or pixelated detector. Having set the laser beam as the 'proxy' for the optical axis of the system, the mirror or reflecting component simply needs to be tilted such that the laser beam returns along its original path. This illustrated in Figure 21.16.

The illustration in Figure 21.16 opens the possibility of co-alignment of two separate surfaces. In this example, the aligned laser may be traversed laterally on a linear stage to align the second mirror with respect to the first. However, in contemplating this, we must account for the characteristics of the linear stage. As outlined earlier in this chapter, the linear stage produces finite pitch roll and yaw error, so there is some tilt uncertainty that arises from the traverse between the two mirrors. For a typical stage, this is of the order of tens of micro-radians or more.

As indicated, the laser is a ubiquitous laboratory tool used in general purpose alignment tasks. Traditionally, the helium neon laser (more latterly the visible diode laser) has been used in this role and is even incorporated into instruments, particularly high power laser systems, for routine alignment. However, for more

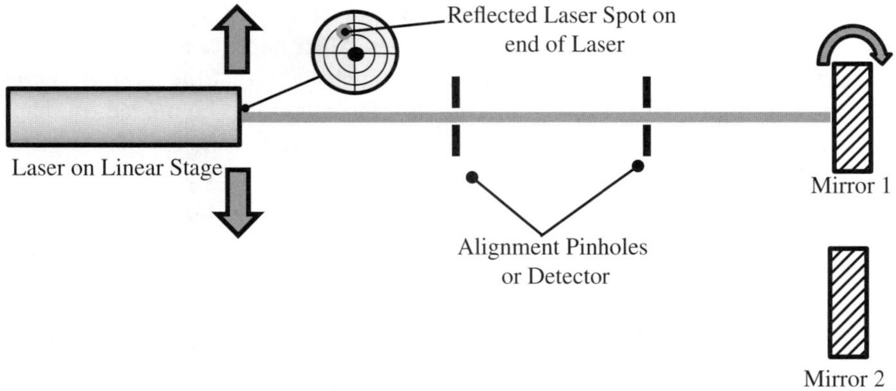

Figure 21.16 Simple laboratory alignment process.

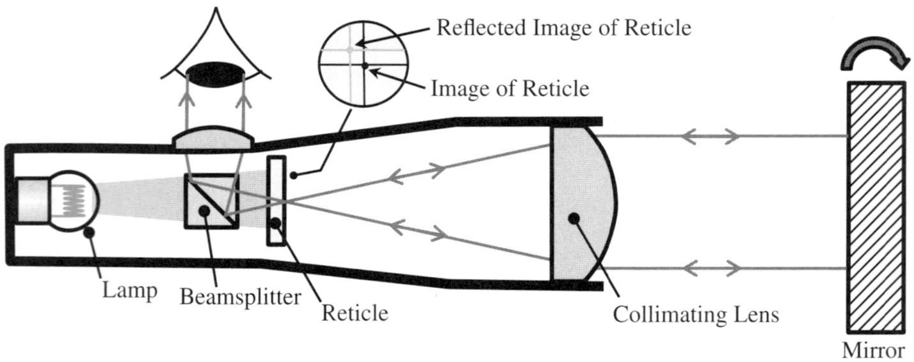

Figure 21.17 Principle of autocollimator.

accurate alignment, at the micro-radian level, an autocollimator or similar instrument is called for. Figure 21.17 illustrates the autocollimator in its traditional format, facilitating direct viewing by eye.

Figure 21.17 shows a traditional arrangement with adjustment proceeding visually by the alignment of two imaged sets of cross hairs. Contemporary arrangements use a focused spot or pattern of focused spots and attempts to centroid these features or spots on a pixelated detector. Spot centroiding has a precision of better than 0.1 times the pixel size, so centroiding to a few hundred nanometres or so is better. This affords an angular resolution of microradians or better.

It is instructive, here, to compare the performance of the autocollimator with that of an interferometer. If we substitute the autocollimator in Figure 21.17 for the output of a (collimated) Twyman-Green interferometer, the resolution afforded is even higher. If we imagine a mirror with a diameter of 100 mm and an interferometer with a resolution of 5 nm rms (in tip tilt), then, from analysis of the Zernike tilt contribution, this resolution corresponds to an angle of 0.2 μrad. The technique is so sensitive that 'pitch' and 'yaw' drifts are seen due to small changes in the thermal or mechanical environment. This also applies to density changes in the ambient air. In terms of understanding the evolution of air density, air in an internal or external environment is thought to be ***stratified***. That is to say, the thermal and therefore density profile shows a clear and systematic variation in height. Any change in this stratification will be seen as a small pitch angle in the resulting interferogram.

Accurate co-alignment of two mirrors may, in principle, proceed as in Figure 21.16, with the interferometer (or autocollimator) mounted on a linear stage. However, the pitch, roll, and yaw of the stage will dominate the uncertainty in this procedure. This problem may be alleviated by use of a reference flat to maintain alignment across any spatial interval between the mirror surfaces. Firstly, the interferometer is aligned to the first mirror and the reference flat inserted between the interferometer and mirror. At this point, the reference flat is aligned to the interferometer, by adjusting its tilt to eliminate any fringes. Most importantly, the reference flat is larger than the aperture of the interferometer and it is positioned in such a way that the interferometer aperture rests at one edge, e.g. the left, of the flat. At this point, the interferometer can be moved (perhaps on a linear stage) to the opposite diameter, i.e. the right, of the reference flat and then realigned. If necessary, at this point, the reference flat can be moved again, such that the interferometer aperture is returned to the original (left) side of the flat. Subsequently the process may be repeated. In this way, the interferometer may be 'walked' from the first mirror to the second, whilst retaining its original alignment. This procedure is illustrated in Figure 21.18.

The flatness of the reference mirror may typically be of the order of $\lambda/10$ or $\lambda/20$ peak to valley, corresponding to $\lambda/35$ or $\lambda/70$ rms. Where reference flat and interferometer are moved over multiple cycles, then, effectively, the flatness of the flat is being 'stitched together' across several diameters. This process does, to some extent, magnify the very small uncertainties associated with the optic's original flatness. In practice, thermal drift limits the accuracy of this process.

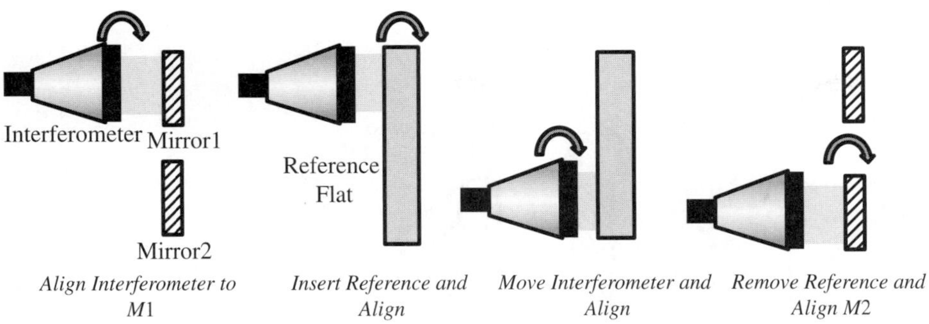

Figure 21.18 Use of interferometer or autocollimator in co-alignment of plane surfaces.

21.4.5 Alignment and Spot Centroiding

As we have seen, there are many circumstances in which a laser beam represents a very convenient proxy for the system chief ray. In particular, reflections can be used to align a surface normal to some rotational axis. Even when used as a visual tool, as illustrated in Figure 21.16, it is extremely useful. However, full exploitation of the laser beams geometrical attributes requires some form of ***spot centroiding.*** Spot centroiding may be carried out by a quadrant detector, as illustrated in Figure 12.22. However, with the availability of low cost CCD (charge coupled device) or CMOS (complementary metal oxide semiconductor) pixelated detectors, quadrant detectors have, to an extent, been replaced with low cost cameras in this specific application. The process of spot centroiding is illustrated in Figure 21.19.

As an illustration of how laser spot centroiding might work, we present one of the simplest centroiding algorithms that computes the centre of gravity of the spot. If the (known) x, y co-ordinate of the ith pixel is $\mathbf{x_i}$, $\mathbf{y_i}$, and the signal at that pixel is **Si**, then the centroid may be simply calculated from:

$$\langle x\rangle = \frac{\sum S_i x_i}{\sum S_i} \quad \langle y\rangle = \frac{\sum S_i y_i}{\sum S_i} \tag{21.3}$$

Of course, more sophisticated and efficient algorithms exist that take due account of the profile of the laser beam and the presence of any background illumination, dark current, and noise, etc. Taking these factors into account, the precision of centroiding is of the order of 0.05–0.1 times the width of a pixel. For a 2 μm pixel size, the spot centration precision is of the order of 100–200 nm, perhaps better. Converting this into angles, for a camera focal length of 50 mm, this centration precision is consistent with an angular uncertainty of 2–4 μrad or 0.4–0.8 arcseconds.

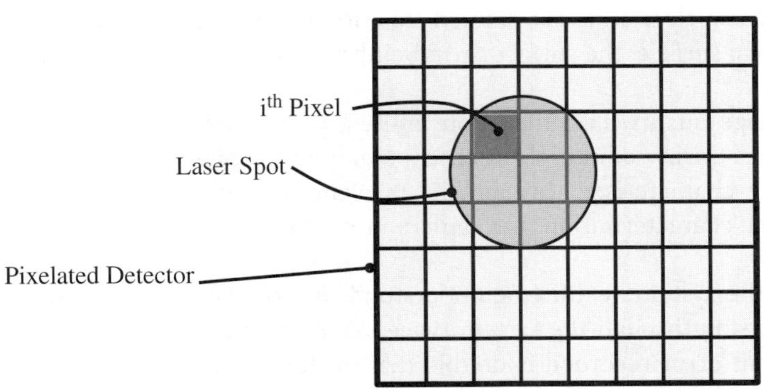

Figure 21.19 Spot centroiding.

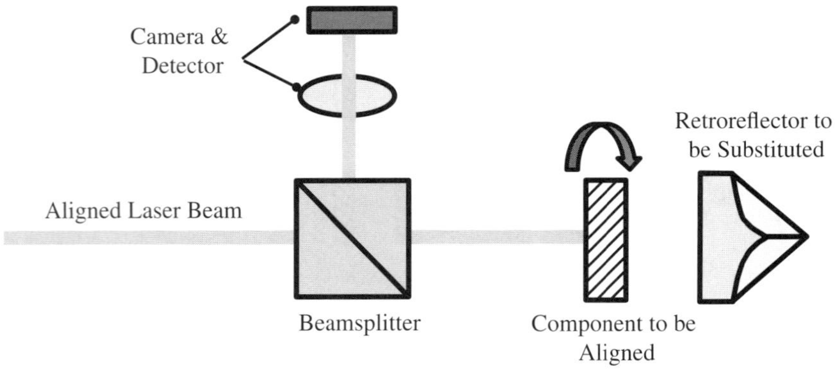

Figure 21.20 Alignment process with beamsplitter.

This spot centroiding process may be used to align a laser beam precisely to some mechanical reference. In other words, the camera may be moved along some linear path, defined by some referencing straight edge that constrains the location of the camera in its physical mount. Having precisely defined a reference line in this way, components may be axially aligned to it, as previously described. Arrangements for doing this are many and varied. Useful tools in experimental set ups include beamsplitters and retroreflectors (corner cubes). As cited earlier in this text, corner cubes have the extremely useful property of returning the incident beam on its original path. This property may be exploited to align an optical surface (normal) to the incoming aligned (to some reference) laser beam. Furthermore, the return beam may be viewed without hindrance by incorporating a beamsplitter into the optical set up. The basic arrangement is shown in Figure 21.20.

In the set up illustrated, initial incorporation of the retroreflector defines the course of the aligned laser beam, as viewed at the detector. When the component to be aligned is substituted, the alignment process merely has to replicate the original centroid location seen with the retroreflector in place. Of course, this arrangement may then be combined with axial rotation of the component with respect to some reference axis, facilitating the precise alignment or centration of components, as previously described.

21.4.6 Alignment and Off-Axis Aberrations

Much of the preceding discussion has focused on the impact of boresight error and alignment through replication of the chief ray. Tilts and decentres naturally tend to degrade image quality through the introduction of off-axis aberrations, such as coma. For the most part, the sensitivity of axially symmetric components to tilt and decentre is relatively modest. However, the sensitivity is particularly marked where off-axis aspheric or conic parts are introduced into a system, for example an off-axis parabola. In the case, for example, of a parabolic mirror, a single mirror surface has a clear and well-defined axis which represents an alignment constraint. This axis is absent from a spherical surface, the surface form being preserved in any angular orientation.

Alignment of systems requiring high image quality, i.e. diffraction limited, particularly in the systems incorporating non-spherical surfaces, requires the use of interferometric techniques to diagnose and remedy off-axis aberrations. System aberrations are characterised through the measurement of wavefront error. Figure 21.21 shows an example set up for the characterisation of a camera or collimator system in a double pass arrangement.

In characterising the wavefront error of the system, it must be understood that the above arrangement is a double pass arrangement, with light passing through the system twice. As such, since wavefront error is additive through the system, the wavefront error recorded is double that of the system in isolation. Of course, the reference sphere is a precision optic, with a peak to peak form error of $\lambda/20$ or less and does not contribute significantly to the measured wavefront error. Most importantly the wavefront error, as measured,

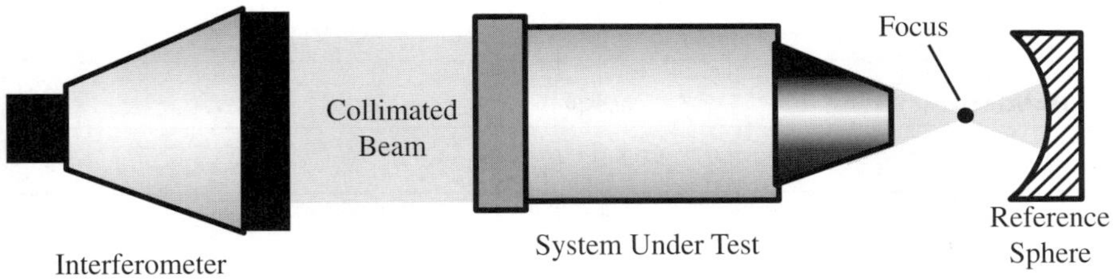

Figure 21.21 Double pass interferometry.

is decomposed into Zernike polynomials. As intimated in Chapter 18, the tolerancing analysis element optical modelling process will provide alignment sensitivities for all components of the wavefront error. That is to say, each Zernike polynomial contribution of the wavefront error has a known dependence on the decentre or tilt of each optical element. This information can be used to characterise and minimise the off-axis aberrations in a systematic way, using the sensitivities derived from the optical model.

21.5 Cleanroom Assembly

21.5.1 Introduction

Cleanliness is critical to any optical assembly process. If particularly serious, particulate contamination of optical surfaces leads to undesired scattering affecting image contrast. However, particulate contamination is especially salient for those surfaces that are closest to the focal plane. These include reticle components, and, most pertinently, pixelated detectors. For detectors with pixel sizes of less than 2 μm or so, then even a submicron particle can cause significant problems. In addition, components subject to high power laser flux may be prone to catastrophic damage through absorption of laser radiation by absorbing particles present on the surface.

All internal spaces contain small particles from a variety of sources. Inorganic and organic particulates from soil erosion and combustion products combine with fibres from clothes, carpets, and fabrics, as well as human detritus. In terms of numbers, smaller particles tend to predominate, with a large number of particles with a size below 3 μm. There is much discussion in the literature about the size distribution; one popular model of size distributions is the lognormal distribution. In any case, in practical terms, the preponderance of small particles poses an especially serious problem for optical assembly. With a large number of particles present in the air volume, these will deposit on critical surfaces, or be attracted electrostatically to vulnerable surfaces.

The solution to this problem encompasses two strands. First of all, the generation rate of fresh contamination must be restricted. This is accomplished by carrying out all assembly work in a special environment known as a ***cleanroom***. All materials used in the construction of a cleanroom are designed to minimise the shedding of particles. This consideration also applies very strictly to the introduction of materials from outside the cleanroom environment. In particular, personnel operating in a cleanroom are required to wear special clothing that minimises the shedding of fibres. In addition, the hair and face are generally covered to minimise the distribution of particles of human origin. Having minimised the generation rate, particles are removed by rapid air circulation through high efficiency particulate air filters (HEPA filters) which are capable of removing over 99.5% of particles with a size of greater than 0.3 μm.

21.5.2 Cleanrooms and Cleanroom Standards

The assembly of sensitive equipment therefore takes place in a cleanroom environment, with steps taken to minimise handling of optical surfaces. As previously advised, personnel must wear clothes that are compatible with the cleanroom environment, including gloves. Exposure of skin to the environment is to be avoided.

Table 21.2 Cleanroom particulate standards.

Standard designation		Particle size (microns)					
FS209E	**ISO1464**	**>0.1**	**>0.2**	**>0.5**	**>1.0**	**>2.0**	**>5.0**
	1	10	2				
	2	100	24	4	1		
1	3	1 000	237	35	8	2	
10	4	10 000	2 370	352	83	20	3
100	5	100 000	23 700	3 520	832	197	29
1 000	6	1 000 000	237 000	35 200	8 320	1 970	293
10 000	7	10 000 000	2 370 000	352 000	83 200	19 700	2 930
100 000	8		23 370 000	3 520 000	832 000	197 000	29 300
	9			35 200 000	8 320 000	1 970 000	293 000

Two formal standards define the quality of the cleanroom airspace. The performance metrics describe the maximum allowable concentration of particles per unit volume greater than a specific size. The original FS209E standard is being replaced by the relevant ISO standard (ISO14644). For each standard designation, maximum particle counts per cubic metre, for each particle size are set out in Table 21.2.

21.5.3 Particle Deposition and Surface Cleanliness

One might consider that particles present in an air volume will tend to settle out under gravity. However, the process is not as straightforward as this. Gravitational settling velocities of micron sized particles are of the order of a few tens of microns per second. Natural and artificial engendered turbulence will tend to ameliorate the gravitational settling process. Particles then deposit onto surfaces by a process akin to convective mass transport. As a consequence, for the smallest particles, it is not the vertical facing surfaces that are most vulnerable; all surfaces are exposed to particle deposition, especially those surfaces exposed to a high air flow rate.

Of course, the rate at which particles settle on surfaces is proportional to their concentration. The distribution and magnitude of particle size distributions on contaminated surfaces are well characterised by the standard IEST-STD-CC1246E. The cleanliness level, L, characterises the particle size distribution in the following fashion:

$$logN = -0.926[log^2(x) - log^2(L)] + 0.03197 \tag{21.4}$$

N is the number of particles per 0.1 m², whose effective diameter in microns is greater than x.

To put Eq. (21.4) into some context, a cleanliness level of 200 would be regarded as moderately clean, whereas levels above 500 would be visibly dirty. For demanding applications, such as those that pertain to high power laser beam lines, then a cleanliness level of 10–20 might be demanded. Cleanliness levels according to this standard are illustrated graphically in Figure 21.22.

It is possible to equate the cleanliness levels described in Figure 22.22 to scattering levels. The simplest model relates the scattering cross section of each particle to its physical area. Whilst this is certainly not realistic for the smallest particles, it does provide some measure of the scattering power of surfaces at specific cleanliness levels. This is illustrated graphically in Figure 21.23.

As previously indicated, particles are deposited from the environment onto the surface, although not by gravitational settlement. An empirical formula describes the rate of this settling process as a function of

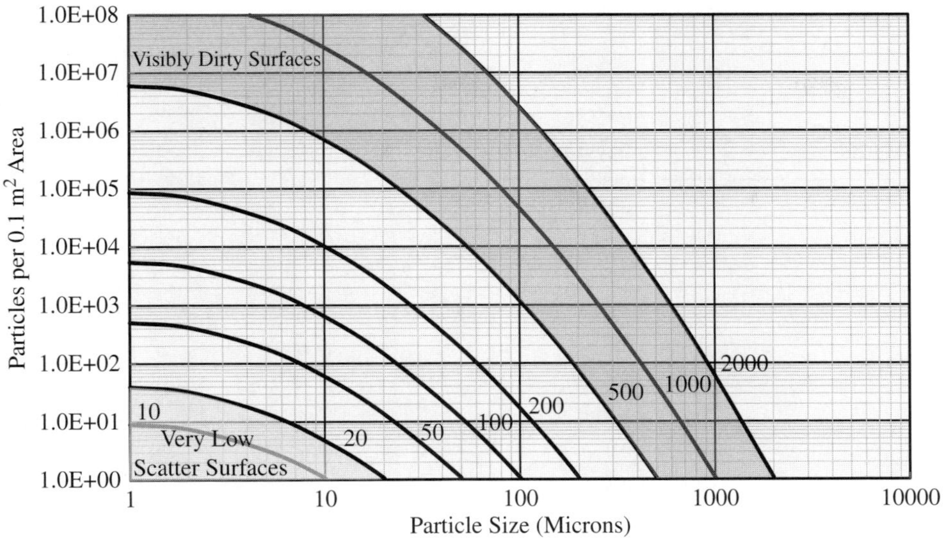

Figure 21.22 Cleanliness levels according to IEST-STD-CC1246D.

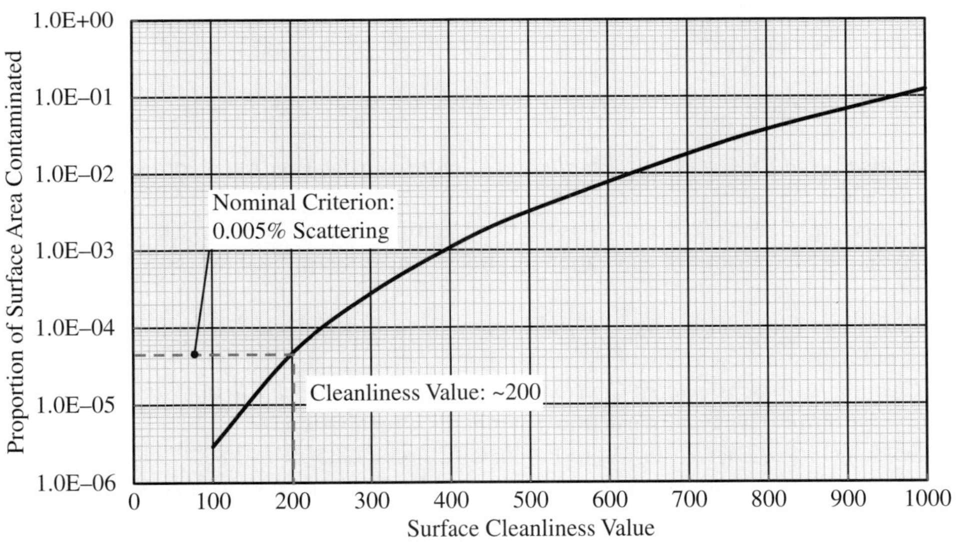

Figure 21.23 Impact of surface contamination on scattering.

particle size. The particle fallout rate, P, is expressed as mm^2 particulate covering per square metre per day. Effectively, this is the ppm surface coverage per day.

$$P = 0.069 \times 10^{(0.72N-2.16)} \tag{21.5}$$

N is the ISO Cleanroom Designation.

Equation (21.5) is an empirical formula taken from a number of observations. The fact that it is sublinear (i.e. $0.72 < 1$) with respect to particle concentration suggests that in the real cleanroom environment the particle size distribution function changes with cleanroom characteristics, not just the overall concentration. Based on Eq. (21.5) it is possible to calculate the number of days exposure in a given environment that would degrade a nominally clean surface to level 200. This is set out in Table 21.3.

Table 21.3 Days of cleanroom exposure required to produce cleanliness level L = 200.

Standard designation		
FS209E	**ISO14644**	**Days to L = 200**
	1	20 000
	2	3 800
1	3	725
10	4	138
100	5	26
1 000	6	5
10 000	7	1
100 000	8	0.2
	9	0.03

The table is not intended as a numerical guide, but merely a means of articulating the importance and significance of cleanroom practice in optical assembly.

Further Reading

ECSS-Q-ST-70-01C (2008). *Space Product Assurance: Cleanliness & Contamination Control*. Noordwijk: European Co-operation for Space Standardisation.

Freudling, M., Klammer, J., Lousberg, G. et al. (2016). New isostatic mounting concept for a space born three mirror anastigmat (TMA) on the Meteosat third generation infrared sounder instrument (MTG-IRS). *Proc. SPIE* 9912: 1F.

IEST-STD-CC1246E (2013). *Product Cleanliness Levels – Applications, Requirements, and Determination*. Schaumberg, IL: Institute of Environmental Sciences and Technology.

ISO 14644-1:2015 (2015). *Cleanrooms and Associated Controlled Environments – Part 1: Classification of Air Cleanliness by Particle Concentration*. Geneva: International Standards Organisation.

Macalara, D. (2001). *Handbook of Optical Engineering*. Boca Raton: CRC Press. ISBN: 978-0-824-79960-1.

Park, S.-J., Heo, G.-Y., and Jin, F.-L. (2015). Thermal and cure shrinkage behaviors of epoxy resins cured by thermal cationic catalysts. *Macromol. Res.* 23 (2): 156.

Vukobratovich, D. and Richard, R.M. (1988). Flexure mounts for high resolution optical elements. *Proc. SPIE* 0959: 18.

Williams, E.C., Baffes, C., Mast, T. et al. Advancement of the segment support system for the thirty meter telescope primary Mirror, *Proc. SPIE* 7018, 27 (2009).

Xu, Z. (2014). *Fundamentals of Air Cleaning Technology and its Application in Cleanrooms*. Berlin: Springer. ISBN: 978-3-642-39373-0.

Yoder, P.R. (2006). *Opto-Mechanical Systems Design*, 3e. Boca Raton: CRC Press. ISBN: 978-1-57444-699-9.

22

Optical Test and Verification

22.1 Introduction

22.1.1 General

In this book we have laid out an extensive narrative covering a detailed understanding of the optical principles underlying optical system design, before progressing to the more practical aspects of design, manufacture, and assembly. As this point, it would be tempting to think that, now we are in possession of an aligned and apparently working system, our work is now complete. However, our task is only complete when we can transfer responsibility for the assembled optical system to the end user. Before we can do this, we are obliged to demonstrate to the end user that the system does indeed conform to the originally stated requirements over the specified environmental conditions.

Unfortunately, this critical aspect is not attributed its due prominence in most treatments of the broad subject of optics. Therefore, at this point, we introduce the somewhat neglected topic of ***test and verification***.

22.1.2 Verification

During the design process, a clear set of requirements will have been agreed between the different stakeholders and documented. Of course, dependent upon the level of technical risk, it should be understood that, as a consequence of the uncertainties inherent in the development process, a number of these requirements may have to be modified with the agreement of all stakeholders. Nonetheless, eventually, a list of all requirements must be assembled and ordered, and a process of verification for each requirement clearly articulated. The practice is to capture this in a formal document, described variously as a ***verification matrix***, ***verification cross reference matrix,*** or ***requirements traceability matrix***. It is by no means assured that each requirement listed is to be accompanied by a physical test. Whilst it is absolutely clear that a performance attribute, such as wavefront error (WFE) must be tested, this is not the case for all requirements. For example, where an operating wavelength range is specified, verification may be covered by a formal statement to the effect that all subsequent performance tests cover the stated range. A similar consideration might apply to environmental specifications. Moreover, there may be other requirements that can be satisfactorily verified through recourse to modelling and analysis rather than physical testing. Whatever the preferred route to verification, this must be clearly outlined in the verification matrix.

The process of testing that is designed to assure the end user that the system conforms to the listed requirements is referred to as ***acceptance testing***. More specifically the suite of tests that are mapped against the verification matrix is often referred to as the ***factory acceptance tests*** or ***FATs***.

22.1.3 Systems, Subsystems, and Components

Whilst verification is ultimately aimed at establishing conformance to system-level requirements, testing also proceeds at the subsystem level or even component level. As indicated earlier, part of the overarching design

Optical Engineering Science, First Edition. Stephen Rolt.

Companion website: www.wiley.com/go/Rolt/opt-eng-sci

process at the system level is to partition specific requirements to individual subsystems or even components. Care, of course must be taken to establish the interfaces between these subsystems and that their conformance to the requirements is also verified. For example, in an imaging spectrometer, it may be necessary to measure the image quality (modulation transfer function [MTF], Strehl Ratio, WFE) for the camera sub-system alone, as well as for the end-to-end system. At the component level, for example, we might wish to specify surface roughness or cosmetic surface quality. The surface roughness for each component would have to be verified by means of a non-contact (e.g. confocal length gauge) probe or contact probe, whereas the cosmetic surface quality would be verified by inspection.

A clear distinction in practice emerges between high-volume and low-volume applications. At one extreme, we have consumer products, such as mobile phone cameras with production volumes amounting to many millions. By contrast, large astronomical or aerospace projects invest huge resources into the construction of one large system, which, of course, must work at the end of the project cycle. In the former case, the engineer has the benefit of a series of prototype developments with respectable volumes. Most particularly, product refinement is substantially promoted by the large quantity of useful process statistics derived from factory testing. Unfortunately, this benefit is not available in large high value projects where, as a consequence, the attendant development risks are much higher. In this latter case, limited sub-system testing, or breadboard testing is carried out. Otherwise, in some aerospace applications, there is some scope for providing tests based on the provision of pseudo-prototypes with limited functionality, e.g. ***engineering test models***. However, it is clear that for a system, such as a large astronomical telescope, end-to-end system testing must inevitably be confined to a single unit.

A distinction is often made between functional and performance testing. Strictly speaking, a ***functional test*** simply verifies conformance to a particular requirement and no more, whereas a ***performance test*** established how well (or fast, etc.) that requirement is met. That is to say, the anticipated result of a functional test is pass/fail, whereas a performance test demands numerical data to be presented. However, in practice, considering such attributes as WFE and MTF, there is, in many optical requirements, no clear distinction between the two.

22.1.4 Environmental Testing

We will presently consider the broad categories of verification testing that might be contemplated for an optical system. However, before we embark on this description, it is important to consider the environmental conditions that are presented to the system. Any test or verification programme must take due account of this. At first sight, it is the ***operational environment*** alone that needs to be considered. That is to say, the environment that the system is presented with when it is actually used. However, there are two other broad categories of environmental exposure that also need to be considered. Both these relate to situations in which the system is not operational. First, there is the **storage environment;** we need to have a clear understanding of the temperature and humidity ranges experienced, as well as extraneous factors such as the propensity for mould growth, etc. Perhaps more important is the so-called ***transport environment***, relating to the transport of optical systems by land, air, or sea. Naturally, in this environment, the most salient of exposures relates to mechanical shock and vibration experienced, for example, in road haulage. Another example is that of air freight, where exposure to temperature extremes and reduced external pressure must also be contemplated. For space applications, survival of the launch episode (acceleration and vibration) as well as the obvious vacuum environment and exposure to ionising radiation must be accounted for.

To attempt to unravel the great risks and uncertainties inherent in the exposure to these environments, standards have been agreed defining maximum anticipated exposure in these environments. These standards help define environmental tests that simulate exposure to these conditions. A variety of 'shake, rattle, and roll' tests expose the system, sub-modules or components thereof to vibrational stress and shock with clearly defined parameters. Other environmental tests expose the system to thermal shock and temperature and humidity cycling. Whilst these tests cannot be described as functional optical tests, they do form a very important part of the verification process for an optical system. This is especially true for systems designed to perform in demanding environments, such as those that pertain to military and aerospace applications.

22.1.5 Optical Performance Tests

One may conveniently divide the suite of optical tests into a number of broad categories. First, there are what we might refer to as the ***geometrical tests***. These might encompass the determination of the cardinal points of the system, its focal length, distortion, and the alignment of the system as described by the geometry of the chief path, for example. These tests would necessarily be followed by ***image quality tests***, quantifying WFE, MTF, Strehl Ratio, etc. Thereafter, we must consider the ***radiometric tests*** where the system flux, radiance, irradiance, and throughput might be characterised. Within this category, we might also include those tests related to establishing the polarisation state of the system. In addition, we might include the measurement of spectral characteristics such as spectral distribution and resolution. Finally, there are a series of ***material and component tests*** that are designed to validate attributes, such as surface roughness and refractive index uniformity, and so on.

For all the tests described, if the system itself is designed to make measurements, then the uncertainty of all relevant parameters needs to be estimated. For example, the camera plate scale, or focal length may be used to measure the angle subtended by distant objects. In this instance, the uncertainty in the plate scale and distortion test measurements must be presented. Similar considerations apply to radiometric characterisation, if the test measurements form part of the instrument calibration process.

22.2 Facilities

The performance of verification tests is fundamentally dependent upon a stable environment. This is particularly true of sensitive tests, such as those involving interferometry. Therefore, provision of adequate facilities is essential for conducting optical tests.

To a large extent, with the exception of the environmental testing process, we proceed under the premise that the tests are carried out under broadly ambient conditions. However, it should be understood that where systems are to be deployed in unusual environments, e.g. underwater or in space, then the testing process must reflect this. For example, for optical payloads deployed in space, then the system testing must be carried out under vacuum conditions. Furthermore, the thermal environment may be substantially different to the terrestrial environment. For example, background signal and noise conditions may demand that the system operates under cryogenic conditions. Therefore, in this eventuality, the facility should be equipped with a ***thermo-vacuum chamber***, to replicate the operating environment. These facilities, naturally are large and costly and are restricted to large facilities for use on substantial projects.

For all optical measurements, a controlled environment is essential. Ideally, the environmental temperature should be constrained to within $\pm 1.0\,^{\circ}\mathrm{C}$. Where sensitive geometrical and interferometric measurements are made, changes in temperature cause movement or distortion in the underlying mechanical support structure of an optical system. Furthermore, changes in ambient temperature lead to variations in air density creating fluctuations in the optical path over the pupil. Air movement produced by circulation creates further fluctuations in air density. For example, for a path length of 2 m, a change in temperature of $1.0\,^{\circ}\mathrm{C}$ corresponds to a change in the optical path of about 2 μm, or 4 waves at 500 nm.

Most typically, in any test laboratory environment, the ambient air is '***stratified***', exhibiting some vertical temperature gradient. Any changes in this gradient will lead to changes in the apparent tilt of a collimated beam. In any case, a poorly controlled environment leads to significant stochastic temporal shifts in the apparent pitch and yaw of a collimated beam. Therefore, the test environment must be carefully controlled and, in any case, temperature and humidity, etc. should be logged during any measurement process.

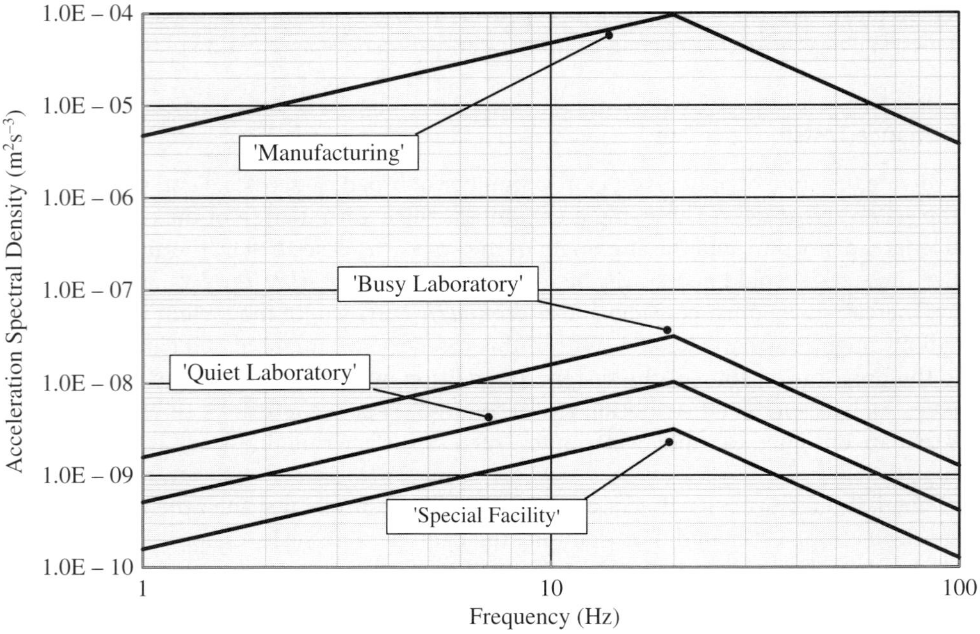

Figure 22.1 Background vibration levels in some environments.

Most interferometric measurements require a high degree of mechanical stability. Very small changes in optical path length, of a fraction of a wavelength lead to substantial loss in fringe contrast or visibility. Ideally, any site or facility should not be impacted to any significant degree by vibration from obvious sources, such as road traffic. It is also customary to locate sensitive test facilities at ground floor level on solid flooring. Vibration tends to be transmitted readily through raised or upper floors. Furthermore, to a significant degree, strategies to ameliorate vibration often involve the substantial addition of inertia, i.e. mass, to a test system, and this consideration tends to militate against deployment on upper floors.

Naturally, the greatest immunity to vibration is achieved by locating facilities deep underground, e.g. in abandoned mines or in specially prepared facilities. Aside from seismic activity, the principal sources of vibration are human in origin. Typically, the peak vibration amplitude occurs at around a few tens of Hertz. Figure 22.1 shows a plot of the effective background vibration in selected environments. The amplitude of the random (i.e. dephased) vibration is expressed as a power or acceleration spectral density (ASD) against frequency. The concept of ASD will be discussed in more detail presently when we examine environmental tests. To provide a broad perspective of naturally occurring vibration levels, 10^{-8} m^2 s^{-3} is representative of a 'quiet' laboratory environment, whereas 10^{-7} m^2 s^{-3} might describe a more 'busy' environment. Otherwise, a manufacturing environment may range from 10^{-4} to 10^{-3} m^2 s^{-3}.

The data shown have collated data from a variety of sources and fit the ASD plot to an empirical formula:

$$ASD = A_0 f \ (1\ \text{Hz} < \text{f} < 20\ \text{Hz});\ ASD = A_1 f^{-2}\ (20\ \text{Hz} < \text{f} < 100\ \text{Hz}) \tag{22.1}$$

where optical path lengths are substantial, i.e. several metres, it is very unlikely that interferometric measurements will be viable without additional measures. Long path interferometry is especially challenging and requires the adoption of underground facilities in 'quiet locations'. ***Vibration criterion curves*** do provide a useful estimate of the requirements for critical applications. These curves express the required environment in terms of rms velocity arising from random vibration over a one-third octave bandwidth. These criteria apply to vibration over a 1–80 Hz bandwidth. For the most critical applications, perhaps pertaining to long path interferometry, the rms velocity should not exceed 3 μm s^{-1}. Otherwise, a useful laboratory environment for interferometry might be described by a limit of 6–12 μm s^{-1}. For comparison, taking the quiet laboratory noise

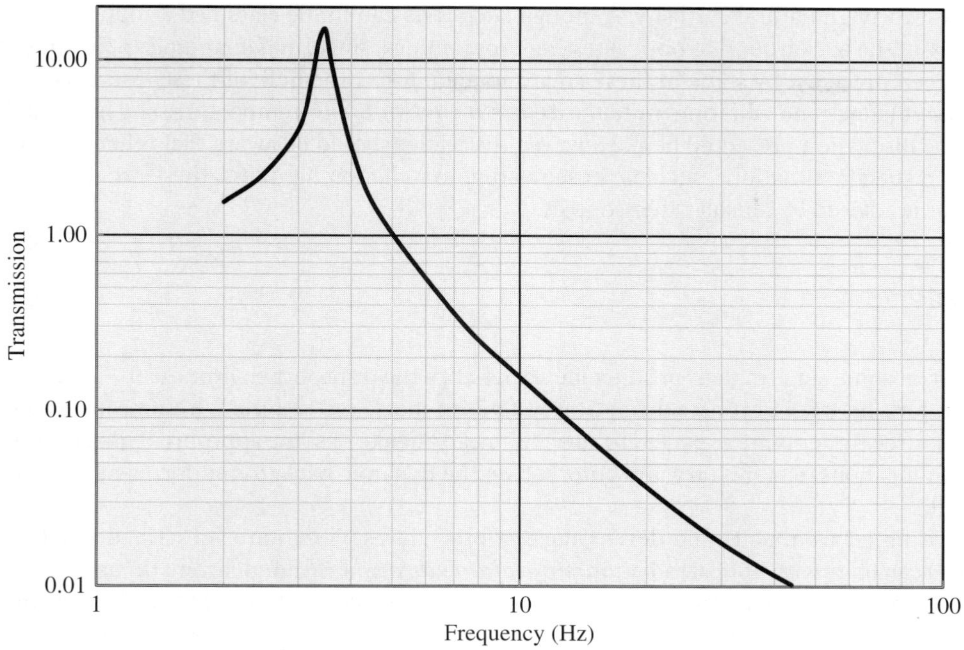

Figure 22.2 Vibration transmission for a typical passive isolation system.

level of 10^{-6} m^2 s^{-3} and integrating from 1 to 80 Hz, this gives an rms velocity of 400 μm s^{-1}, a factor of about 40 higher than desired. Therefore, it is clear that some form of ***vibrational isolation*** must be provided.

Vibrational isolation may either be ***passive*** or ***active***. In the former case, vibrational isolation relies upon the provision of a large mass which is floated upon some damped elastic mounting. The large mass, which functions as an optical table, is often in the form of a laminated honeycomb structure, as introduced in earlier chapters. Most commonly, the elastic mounting is achieved through the use of pneumatically inflated, damped mounts. Active isolation uses electromagnetic actuators to actively oppose any vibration sensed by accelerometers located on the optical table. Either way, the process works well at high frequencies, but less well at lower frequencies. This is adequate to ameliorate floor vibrations within the 'troublesome' range of 20–100 Hz. Furthermore, the optical table itself is designed to ensure that its resonance frequencies are well in excess of this range, so that any residual vibration that is transmitted does not lead to significant flexure of the bench. For an optical system entirely constrained to a bench, only distortion or flexure of the bench will contribute to changes in the optical path length. In general, optical tables are designed to have their fundamental resonance at a frequency higher than 100 Hz. A plot of the degree of vibrational isolation (in dB) against frequency is shown in Figure 22.2 for a representative system.

As with the system integration, for critical systems, especially in aerospace applications, any testing must be carried out under clean conditions. Hence clean room facilities must be provided and any equipment should be compatible with operating in that environment.

22.3 Environmental Testing

22.3.1 Introduction

Environmental testing is a very substantial topic in itself. The very brief description provided here is intended merely as an introduction and to make the reader aware of the significance of environmental testing and to stimulate further interest through natural curiosity. Although such tests cannot be classified as optical tests, they will often form part of the background to optical testing programmes, since, in demanding applications, optical tests will have been preceded by some of these environmental tests. The bulk of these tests relate to the ***dynamic environment*** (shock and vibration) or to the ***ambient environment*** (temperature and humidity) and form the basis of the discussion presented here. However, the reader should be aware that other types of test may be warranted in specific situations, such as those relating to radiation hardness; these are not dealt with here and the reader is advised to consult other sources.

22.3.2 Dynamical Tests

22.3.2.1 Vibration

All optical systems must, at some stage in their product life, experience the transport environment. Transport by road and handling by fork-lift trucks exposes the optical systems to shock and vibration. In some instances, e.g. large telescope structures, systems are expected to survive an earthquake of some appropriate magnitude. For some demanding applications, e.g. military and automotive, the dynamic background forms part of the use environment as well.

In terms of characterising environments and developing test procedures, vibration is described as either sinusoidal or 'random'. By analogy with optical radiation, ***sinusoidal vibration*** is 'monochromatic' and coherent and is described by its frequency and its amplitude (usually presented as acceleration). ***Random vibration***, by contrast, is 'incoherent' noise. It is characteristic of acceleration noise produced by a multiplicity of individual sources, such as in the transport environment. Analytically, random vibration is described by a broadband frequency distribution whose individual components are assigned a random phase. Summation of these individual random components, therefore, proceeds through a summation of the square of the amplitude. This process is exactly the same process as the summation of surface roughness as described by the power spectral density (PSD). By analogy with PSD, random vibration is quantified by the squared acceleration per unit frequency bandwidth. Therefore, the physical dimensions of random vibration are $m^2\,s^{-3}$. If the random vibration as a function of frequency is denominated as $\alpha(f)$, then the total (linear) acceleration, a, resulting from contributions between frequencies, f_1 and f_2 is given by:

$$a = \sqrt{\int_{f1}^{f2} \alpha(f)df} \tag{22.2}$$

The quantity, a, is effectively the root mean square acceleration. Sometimes the quantity, $\alpha(f)$, is referred to as the ***acceleration spectral density*** and represented in terms of the acceleration due to gravity as G^2/Hz, whose dimensions, as argued previously are $m^2\,s^{-3}$. Generally, for practical purposes, a random vibration spectrum, for example in the transport environment, is characterised in a range between a few tens of Hertz to a few thousand Hertz. A typical vibration spectrum shows a maximum spectral density at some mid frequency, perhaps around 20–50 Hz, tailing off at either extreme.

The most common approach to simulate the transport environment, for example, is to split the frequency spectrum into three regions, low, mid, and high. The low frequency region, from some frequency, f_0 to f_1 is characterised by a monotonically increasing acceleration spectral density modelled by a positive power law dependence on frequency. In the mid frequency range, between f_1 and f_2, the acceleration spectral density is constant. Finally, in the high frequency region, between f_2 and f_3, the acceleration spectral density declines monotonically and is described by a power law frequency dependence with a negative exponent. Such empirical definitions form the basis of useful standards upon which environment testing is based. A representative such description of a transport environment is illustrated in Figure 22.3.

Testing is carried out on a large platform that is disturbed by electromagnetic actuation. In effect it may be thought of as a very large and solid 'loudspeaker'. Vibration may be applied in any of three axes, for a duration specified in the standards derived test procedure. As intimated, this vibrational load may be applied either as

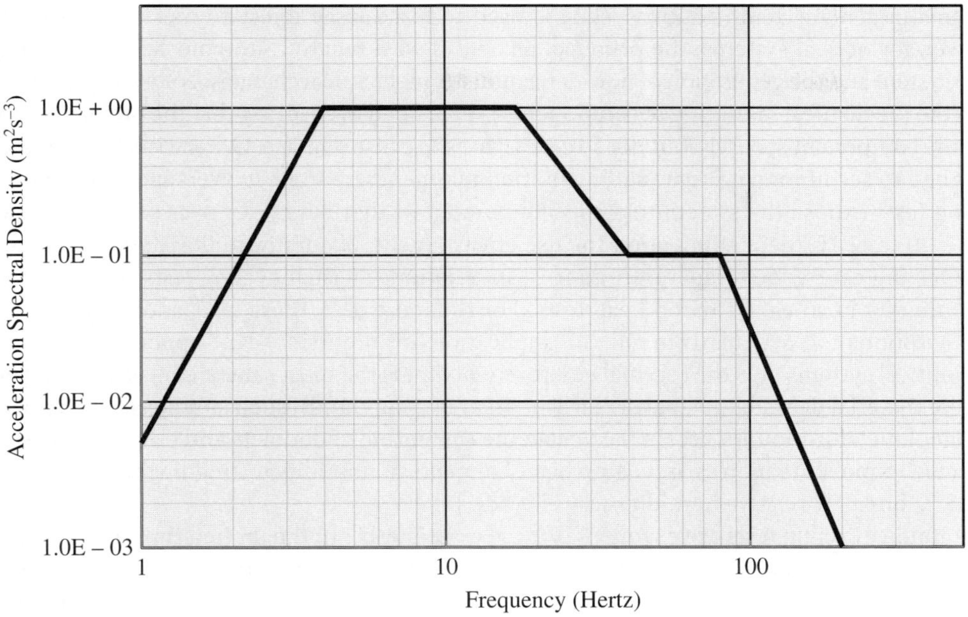

Figure 22.3 Typical transport environment vibrational load.

a sinusoidal vibration or as a random vibrational load. In the former case, the single frequency stimulation is swept over the relevant range, typically from a few tens of Hertz to a few kilohertz.

22.3.2.2 Mechanical Shock

Mechanical shock is represented by a brief episode of intense acceleration. Broadly, the intent of any test against mechanical shock is to simulate the effect of the system or module being dropped from some height. As suggested, the acceleration levels anticipated are very high, in the range of 500–1500 g in typical applications. However, the duration of the shock is only a millisecond or so. Classically, most tests prescribe a temporal profile for the acceleration that is characterised by a half sine wave. Typical test conditions are represented by a 500 g acceleration with a (half sine wave) duration of 1 ms or a 1500 g acceleration with a duration of 0.5 ms. As such, these conditions are consistent with sudden deceleration from a velocity of 5–7.5 ms^{-1}, equivalent to a drop from 1.25 to 2.5 m.

The test is often implemented as a drop test with the system under test mounted upon a platform located at the end of a swinging pendulum type arm. The arm is released under gravity and describes an arc until the platform strikes a buffer bringing it to a rapid halt. The buffer is constructed from some form of shock absorbent material, such as sorbothane, and designed to produce the desired deceleration profile.

Whilst shock testing as previously described, characterises most sudden load scenarios, there is another suite of tests described as '***bump tests***'. As the name suggests, a 'bump' refers to a lower frequency excursion, characterised by lower peak acceleration levels than pertain to shock loads, but with a longer duration, typically several milliseconds.

22.3.3 Thermal Environment

22.3.3.1 Temperature and Humidity Cycling

The thermal environment has a significant effect on the performance of an optical system. Most notably, for a system that is not athermal, significant change in the location of the focal plane is to be anticipated. However, this effect is largely predictable and not the chief concern of an environmental testing programme.

For mechanical systems in general, temperature cycling is used to test susceptibility to mechanical failure from fatigues. However, for optical systems, the principal anxiety is with regard to the mechanical stability and robustness of the system and the generation of non-deterministic mechanical changes. For example, much effort in the design of the mechanical mounting of components is expended in ensuring that the preload forces are sufficient to hold the component securely, but not so excessive as to cause damage. Otherwise, flexures and mating surfaces designed to accommodate some sliding motion may generate some irreversible behaviours.

Temperature cycling tests expose the system to a specific number of thermal cycles over some temperature range, e.g. from −40 to 85 °C, depending upon the use environment. Naturally, military and aerospace applications demand testing over wide temperature ranges. In addition, thermal cycling tests often feature the introduction of humidity in so-called 'damp heat' tests. For the most part, these tests address concerns over use in tropical environments, particularly in military applications or in 'outside plant' applications. In the majority of cases, for optical systems, it is the thermal environment that is the most salient concern. However, moisture can accelerate material degradation, particularly in organic compounds, such as adhesives, and also can promote corrosion of metallic mounts, etc. In particular, the cementing of doublets and other optical elements may be vulnerable to moisture ingress and damp heat. Elements of humidity testing might include, for example, a 'soak' at 85 °C and 85% relative humidity, as well as cycling.

Temperature and humidity cycling tests are examples of 'accelerated tests'. In the limited time available for testing, the tests are required to simulate environmental exposure of a system over many years of operating life. As such, temperature cycling might take place between extremes of −55 to 125 °C. The cycling process is characterised by a linear ramp between the two extremes, e.g. lasting for 20 minutes and then a dwell at each extreme lasting for an equivalent period. In this particular instance, a full cycle would last for two hours. An example thermal cycle is shown in Figure 22.4. This is for the deep cycle testing of the NIRSPEC Integral Field Unit to be deployed on the James Webb Space Telescope (JWST). As the instrument itself is designed for a cryogenic environment, but experiences cycling to ambient temperature over its operational and storage life, the test cycles are very deep.

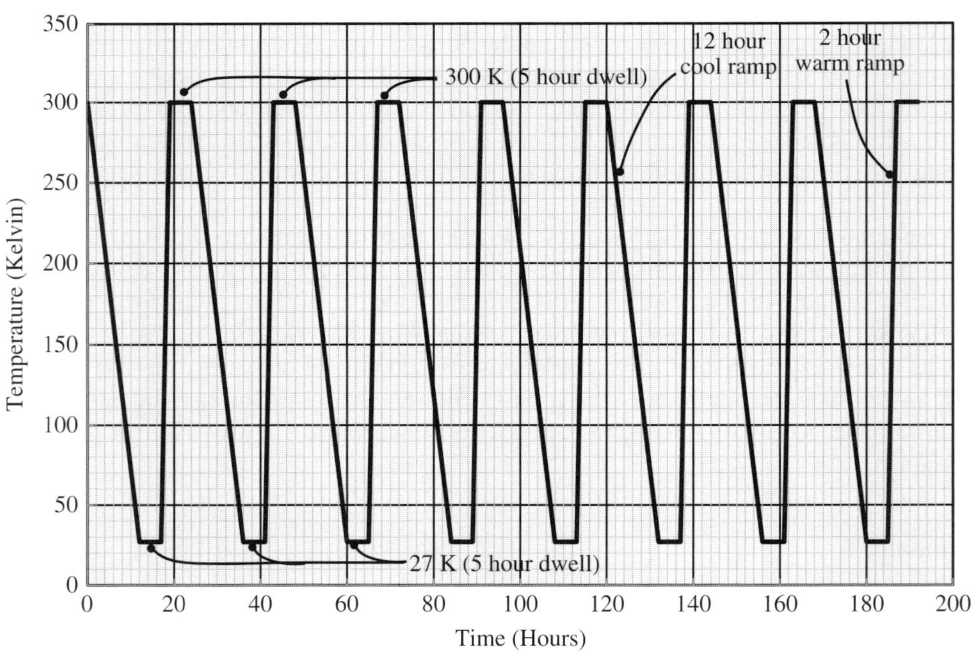

Figure 22.4 Temperature cycling profile for NIRSPEC integral field unit (IFU) test on JWST.

22.3.3.2 Thermal Shock

Thermal shock is characterised by a sudden change in the environmental temperature, potentially leading to catastrophic failure in a component thus exposed, usually by brittle fracture. Rapid cooling of a solid material leads to the establishment of very high temperature gradients and, in consequence, large thermal strains which, for non-uniform heating are translated into substantial internal stresses. Materials that are vulnerable are those with large thermal expansion coefficients, low thermal conductivity and low fracture toughness. The significance of thermal shock is not so much the rapidity of the temperature change in the environment, but rather the speed at which any temperature change is transferred to the component in question. Typically, vulnerability is engendered by rapid heat transfer processes, such as liquid to solid heat transfer or heat transfer by vapour condensation. With this in mind, one may propose a figure of merit, Γ, for thermal shock resistance, given that the stress induced is proportional to the thermal expansion coefficient, α, and the elastic modulus, E, but inversely proportional to the thermal conductivity, k, for a given heat transfer rate.

$$\Gamma = \frac{kK_C}{E\alpha} \quad K_C \textit{ is the fracture toughness of the material} \tag{22.3}$$

From an optical standpoint, the primary concern is the vulnerability of glass materials. There are particular concerns in the bonding of multi-element lenses, e.g. doublets, with the propensity for thermal shock to initiate delamination along the bond line. As indicated earlier, thermal shock is characterised by high heat transfer rates and test procedures are based upon the transfer of components between liquid baths at different temperatures. For example, this might involve the transfer between water baths maintained at 1 and 95 °C. In the case of cryogenic applications, cryogenic fluids, such as liquid nitrogen, are used in the liquid baths.

22.4 Geometrical Testing

22.4.1 Introduction

Geometrical testing validates the first order paraxial attributes of an optical system, confirming the location of the cardinal points and measuring the system focal length, magnification, etc. In these tests, we are not concerned about the image quality, but rather the dimensional characteristics of the system. Although recognised as a Gauss-Seidel aberration, the measurement of distortion falls within the compass of geometrical measurement.

The availability of low cost pixelated detectors and image processing tools has facilitated the rapid and accurate dimensional characterisation of optical images. A precision artefact can be used as an object target, providing a precise geometrical definition of an array of points constituting the input field. Thereafter, the image sensor located at the system focal plane helps to locate precisely the conjugated image points with the help of image processing (spot centration) software. The correspondence between object and image geometry enables computation of system focal length and distortion, etc.

22.4.2 Focal Length and Cardinal Point Determination

For a camera lens operating at the infinite conjugate, the most convenient method for determining the system focal length is the measurement of the magnification of a standard ***reticle*** or illuminated ***pinhole mask*** as projected by a precision collimator of known focal length, f_0. The collimated beam is focused by the camera under test producing an image at the detector whose geometry may be precisely evaluated by centroiding and associated image processing techniques. In essence, the measurement determines the magnification of the combined system and thus that of the lens under test. The arrangement is illustrated in Figure 22.5.

A ***pinhole mask*** is a precision array of pinholes made by a standard semiconductor type lithographic process, for example, using patterned chromium on silica. As such, high dimensional precision is assured. Otherwise, the pinhole mask could also be replaced by an array of illuminated optical fibres. As indicated in

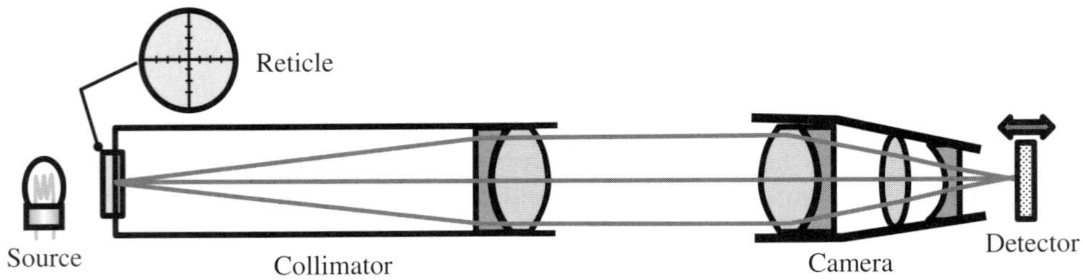

Figure 22.5 Focal length determination with precision collimator.

Figure 22.5, the axial position of the detector is adjustable and its location amended to provide optimum focusing of the pattern. In the case, for example, of the pinhole pattern, focusing could be attained by a formal algorithm that seeks the geometrical location at which the pinhole sizes are minimised. At this optimum focal location, the geometrical size of the imaged pattern can then be measured by spot centroiding, or other image processing technique. If the size of a feature on the precision reticle is h_0 and the measured size of the image is h_1, the magnification, M and the camera focal length, f, is given by:

$$M = h_1/h_0 \text{ and } f = Mf_0 \tag{22.4}$$

By mechanically referencing the position of the detector, the arrangement shown in Figure 22.5 gives the location of the second focal point of the camera and given the focal length previously determined, the location of the second principal point may be derived. By reversing the camera lens, the first focal and principal points may be located. Assuming both object and image points are located in the same refractive medium, then the nodal point locations are equivalent to those of the corresponding principal point locations. Otherwise, the location of the nodal points would have to be calculated from a knowledge of the refractive indices pertaining to the object and image spaces.

Of course, as the technique measures magnification, it may also be used to measure distortion, which is characterised by field varying magnification. However, in all these determinations, we are reliant upon the accuracy invested in the collimator as a precision instrument. The focal length of this instrument must be accurately known, as must its contribution to distortion, if we are to measure distortion. Calibration of such a precision instrument inevitably requires the accurate measurement of angles, a topic that we will return to shortly.

Another technique for the determination of focal length and cardinal point location is the principle of the ***nodal slide***. This technique is based upon the simultaneous determination of the location of the nodal point and its corresponding focal point. As argued earlier, for a system where the object and image media are identical, the nodal point corresponds to the principal point, enabling determination of the system focal length. Location of the nodal point is assisted by its fundamental definition, in that the orientation of object and image rays are identical for this pair of conjugate points. As a consequence, if the system is rotated about the second nodal point, then rays emerging from this point are undeviated. Where the object is located at the infinite conjugate, this means that the image location is unaffected by rotation of the system about the second nodal point. The principle is illustrated schematically in Figure 22.6, reverting to our original description of an optical system as a black box described wholly by the cardinal point locations.

In the nodal slide arrangement, the lens under investigation is mounted on a linear stage which itself is mounted on a rotary stage. As far as possible, the optical axis of the camera should be aligned laterally such that its optic axis intersects the centre of rotation of the turntable. The camera is then illuminated by light from a point object located at the infinite conjugate, e.g. from a collimator. Traditionally, the output from the camera would have been viewed by a microscope lens and the image position recorded using a travelling microscope arrangement. However, viewing the image with a digital camera allows very accurate monitoring of any lateral image movement. The digital camera is, itself, mounted on its own linear stage and its axial position adjusted to provide the optimum focus. At any given linear stage location, the rotary stage is adjusted

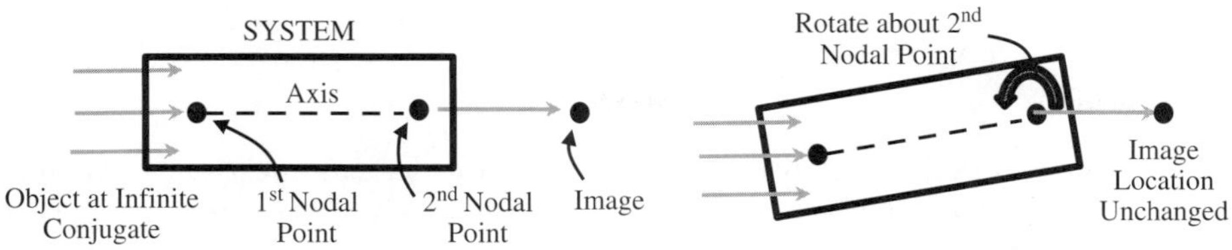

Figure 22.6 Nodal point location.

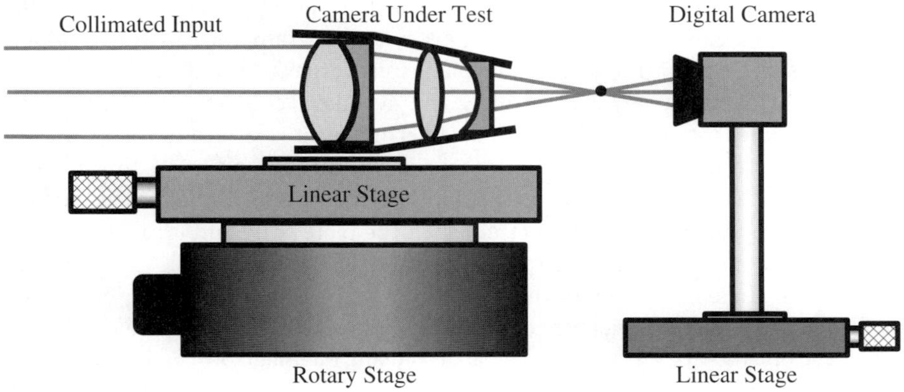

Figure 22.7 Nodal slide arrangement.

and the (linear) drift of image position as a function of (rotary) stage angle is calculated. The position of the linear stage is then adjusted and the measurements repeated. A plot of the rotary drift of the image against linear stage position gives the nodal point as the intercept of this plot.

The camera under test is then removed and replaced by an illuminated pinhole. As before, the linear position of the digital camera is optimised to obtain the best focus. In addition, the previous procedure adopted for the camera lens is adopted for the pinhole, thus co-locating the pinhole and digital camera focus with the centre of the turntable. The difference in the linear position of the digital camera in these two scenarios gives the test camera focal length, assuming principal and nodal points to be equivalent. Furthermore, by referencing the focus of the digital camera on some convenient mechanical feature, such as a lens mount or, if possible, the final lens vertex, the absolute positions of the nodal and focal points and the back focal length are also provided. In this way, one set of cardinal points (second) is derived and, by inverting the test lens and repeating the procedure the other set of cardinal points (first) may also be gleaned. The nodal slide arrangement is illustrated in Figure 22.7.

The process described above can be automated to a significant degree and, with the large number of individual measurements available for analysis, the precision is high. However, as with all image processing techniques the analysis only uses the amplitude of the optical wave; all phase information is discarded. Where the highest precision is demanded, then interferometric techniques may be used to measure the focal length and the location of any cardinal points. Using image processing to identify the location of the optimum focus involves determining a minimum image spot size with respect to axial position. Broadly, this can be viewed as locating the position of a local minimum in a locally quadratic profile that describes the dependence of spot size on axial location. By contrast, location of a focal spot using interferometry relies on plotting the (signed) Zernike defocus contribution as a function of axial position and calculating the axial position at which this coefficient vanishes. It is clear that this process is inherently more precise than the former. To illustrate the precision afforded by this process, one can assign some uncertainty, $\Delta\Phi$, to the determination of the rms defocus (Zernike 4) WFE; typically, this will be a few nanometres. If the numerical aperture of the system is, NA, then the defocus uncertainty, Δf, that is compatible with an rms wavefront uncertainty of $\Delta\Phi$, is given by:

$$\Delta f = \sqrt{48}\frac{\Delta\Phi}{NA^2} \tag{22.5}$$

For example, for an f#4 system ($NA = 0.125$) and a $\Delta\Phi$ equal to 5 nm rms, the defocus uncertainty amounts to about 2 μm. An inteferometric approach quickly establishes the back focal length of a system. The focus of a Twyman-Green interferometer is brought to the focus of a (camera) lens. The assumption here is of a typical scenario with a camera lens designed for operating at the infinite conjugate. A plane mirror placed at the output of the camera lens establishes a double pass set up with the interferometer (mounted on a linear stage) brought to the lens focal point to produce a null interferogram. This establishes the lens focal point. By translating the interferometer focus to the final lens vertex and observing the 'cat's eye' interferogram, the camera back focal length may be established with precision. In this arrangement, the plane mirror may be tilted on a rotary stage and the interferometer moved laterally to null out any tilt (Zernike 2 and Zernike 3) contributions. This lateral movement may be measured using a linear stage and the location determined to interferometric precision. It goes without saying, that a poorly controlled thermal environment significantly compromises this process, as random drift in these Zernike 2 and Zernike 3 components substantially compromises the measurements. Assuming the tilt of the plane mirror can be established precisely, then the plate scale and focal length may be derived to an equivalent precision. The principle is illustrated schematically in Figure 22.8.

It is possible to conceive of another arrangement whereby the focal length and cardinal points may be derived solely by axial adjustment of all elements. The arrangement depicted in Figure 22.9 is in many ways redolent of that in Figure 22.5 where the magnification of the system is measured using an image processing approach. Unlike the arrangement shown in Figure 22.8, where all measurements are made at the same conjugate ratio, the principle of the axial measurement is dependent upon the variation of the object and image locations.

As in Figure 22.8, the arrangement is a double pass configuration. However, the plane mirror of Figure 22.8 is substituted for a reference sphere whose axial location may be varied by movement of a linear stage. In the same way, the position of the object, i.e. the interferometer focus, is similarly adjustable by movement of a

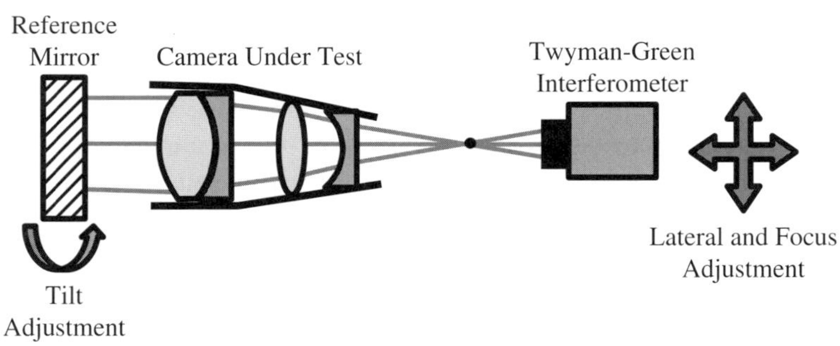

Figure 22.8 Interferometric measurement of focal length.

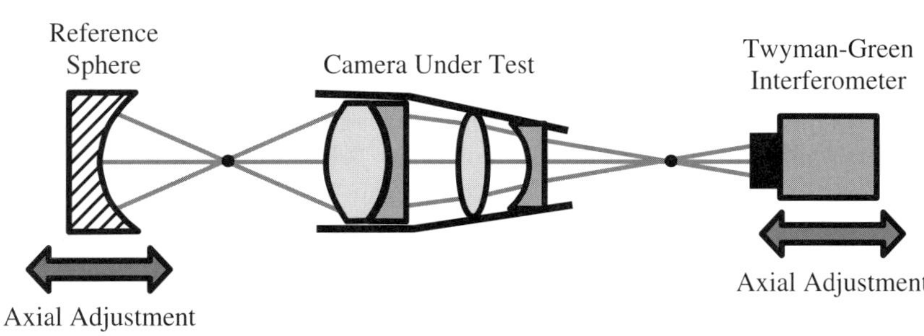

Figure 22.9 Interferometric measurement of focal length with axial adjustment.

linear stage. For referencing purposes, the interferometer focus is set to the vertex of the first lens to establish its axial position, thus enabling the determination of the back focal length. Subsequently, the interferometer position is set to some axial location and the position of the reference sphere adjusted until the defocus Zernike is nulled. In practice, this 'null position' is determined by plotting the Zernike 4 component of the WFE against the recorded axial position of the reference sphere. The focus position is given by the calculated intercept derived from a linear fit where the Zernike 4 component is nulled out. In this way, the reference sphere focus position is plotted against interferometer focus position for a range of positions. Before computation of the focal length and the cardinal point locations, the location of the reference sphere location may be fixed in the same way as the interferometer focus. For example, the vertex of the first lens may be used as the reference point and located in a similar manner as the last vertex and the interferometer focus.

In this way a series of data points may be derived, relating the position of the interferometer focus, x_1 and the position of the reference sphere centre, x_2, with respect to their corresponding reference points. If we assume that the interferometer reference point (the last lens vertex) is separated from the second principal plane by Δ_1, and the sphere reference point is separated from the second principal plane by Δ_2, then using Newton's equation, we may determine all parameters by fitting the following relationship:

$$(x_1 - \Delta_1)(x_2 - \Delta_2) = f^2 f \textit{ is the focal length} \tag{22.6}$$

The test arrangement is shown in Figure 22.9.

22.4.3 Measurement of Distortion

Measurement of distortion proceeds as per the set up shown in Figure 22.5. A precision target forms the object, for example an array of illuminated pinholes. It is to be expected that such targets are to be produced using precision lithographic techniques, for example as chrome on quartz masks. In this instance, the image is characterised by using image processing techniques, i.e. spot centroiding to locate the imaged pinholes in two-dimensional space. The two dimensional aspect is important in some instances, as distortion is not always manifested as a scalar phenomenon, particularly in off-axis systems. That is to say, distortion may produce a skew effect, even in central field locations, with a square replicated as a rhombus or parallelogram. As a consequence, distortion needs to be characterised as a vector quantity which describes any departure from uniform ***scalar*** magnification in vectorial form.

Distortion is a measure of the relationship between object field angle and transverse image location. As such, the measurement of distortion, as described, assumes the conversion of precise dimensional features of the target into similarly precise angles. This conversion is afforded by a calibrated collimator, as previously indicated. However, the collimator itself may contribute distortion to the system and its focal length must also be known accurately. Ultimately, therefore, there is a need to provide precise characterisation of angles in optical systems to effect this calibration.

22.4.4 Measurement of Angles and Displacements

22.4.4.1 General

In many respects, the measurement of linear displacement is straightforward compared to the measurement of angle. The use of mechanical stages equipped with precision ***linear encoders*** provides a robust and highly accurate means for measuring linear displacement. In effect, these precision optical encoders are implemented as precision linear reticle patterns. Most usually, these are in the form of glass strips onto which a transmissive periodic structure has been imprinted, for example, as a chrome on quartz pattern with a sinusoidally varying density. These patterns are then interrogated optically, e.g. by means of a light-emitting diode (LED) and photodiode combination. In this case, unambiguous derivation of the linear displacement is dependent upon the provision of two sinusoidal patterns, one in 'quadrature' with respect to the other. That is to say, one scale gives the sine of the phase and the other the cosine, yielding unambiguously the 'phase' of the displacement.

In this simple example, the encoder needs to be 'referenced' to some 'home displacement'. This is provided by an additional feature in the reticle pattern. Thereafter, when the linear slide is displaced, the system counts the whole number of 'wavelengths' moved plus the fractional wavelength provided by the phase. In this instance, the 'wavelength' is the period of the encoder, which might be several microns or tens of microns. The precision of this process is very high, with submicron resolution. Accuracy is dependent upon the fidelity of the replication process and also the temperature stability (thermal expansion) of the environment. Figure 22.10 illustrates the operation of the linear encoder.

Linear encoders are widely used in laboratory equipment and machine tools. Even greater precision may be conferred by substituting a length interferometer for the linear encoder. For measurement of angles, a rotary encoder may be used. In essence, the principle of operation is the same as for a linear encoder, except the reticle pattern is arranged around the circumference of a circle, rather than along a straight line. Whereas a linear encoder is incorporated into a linear stage, a rotary encoder is integrated into a rotary stage or platform. A rotary stage arrangement specifically designed for the measurement of angles is referred to as a ***goniometer***, derived from the Greek word for angle, gonia. As applied to optical measurements, a goniometer features two 'arms' one fixed and one that is permitted to rotate about a fixed axis. These two arms effectively define the optical axes of two path elements of an optical system prior and following deviation by a mirror or prism. A typical arrangement is shown in Figure 22.11.

The arrangement shown in Figure 22.11 is just one embodiment of a goniometer. In this instance, the turntable permits rotation about a full 360°. In other examples, a 'cradle' arrangement is adopted whereby

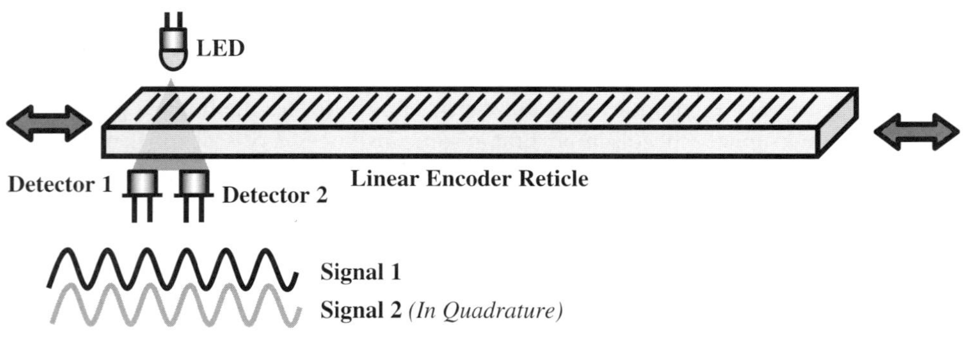

Figure 22.10 Schematic of linear encoder.

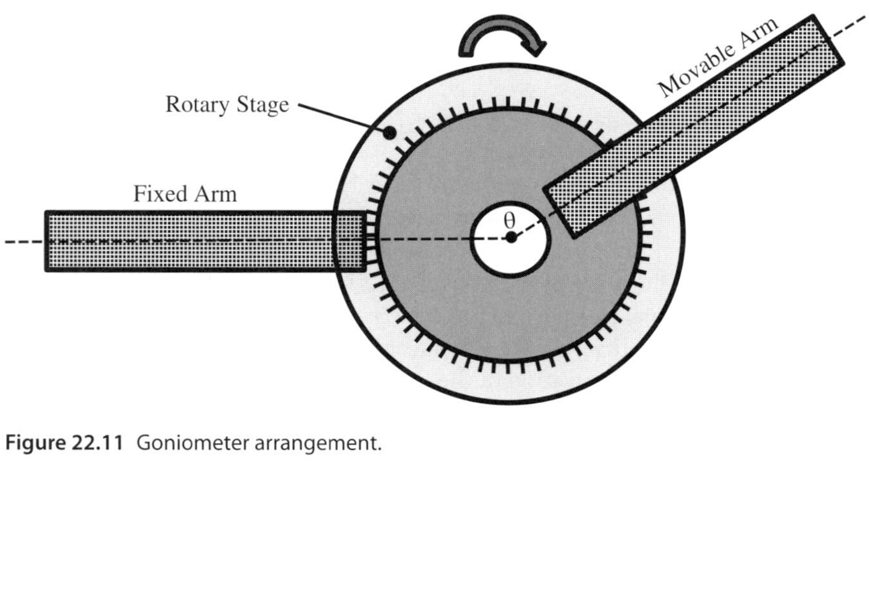

Figure 22.11 Goniometer arrangement.

a limited rotation encompasses some portion of the full arc. Traditionally, as implied in Figure 22.11, measurement of the angle was facilitated by a graduated scale, perhaps subdivided by means of a Vernier scale. Of course, contemporary systems use precision encoders for the measurement of angles.

22.4.4.2 Calibration

Calibration of linear displacement measurement is quite straightforward in that it can be directly supported by wavelength sub-standards through interferometry. One such set of sub-standards are so-called ***gauge blocks***. Gauge blocks are polished blocks of metal, typically hardened steel, whose thickness has been precisely calibrated using interferometric techniques. Thickness vary between a millimetre or so and about a 100 mm. For the calibration of longer lengths, gauge rods may be used. These are rods of low expansion material, e.g. invar, with a reference feature, such as a polished sphere at either end. They are precisely calibrated to standard lengths, such as 1 m.

Calibration of angles is a little more difficult, but can be effected through a thorough familiarity with the fundamental principles of geometry. This process generally requires the fabrication and test of precision geometrical artefacts for angle calibration. Ultimately, as with length measurement, the angular precision is informed in some way by the uncertainty in phase measurement between one wave and a reference wave. Removing this phase information does compromise accuracy. As such, the interferometric measurement of the tilt of a collimated beam is limited by the precision of determining the tilt component of the WFE. Assuming a WFE uncertainty of 5 nm rms, across a 100 mm diameter pupil, then a precision of 0.2 μrad or about 40 mas is possible. At this level, however, drift due to air currents and small thermal movements will be apparent, especially in the absence of an adequately controlled environment.

As intimated earlier, calibration is often effected by the generation and characterisation of precision artefacts. There are many such schemes. As an example, one may describe the generation of a reference prism whose three faces have a nominal internal angle of 60°. By generating two such equivalent prisms it is possible to characterise these angles to an interferometric precision. In essence, the arrangement measures the difference between two specific angles on the two different prisms to interferometric precision. This requires that all prism angles are ***close*** to 60°; they do not have to be exactly 60°. However, any difference should be sufficiently small as to be amenable to interferometric measurement through extraction of the relevant Zernike tilt term. By measurement of these differences and the knowledge that the internal angles sum to 360° permits precise determination of all angles. Generation of such artefacts, then allows the calibration of goniometers and rotary encoders, etc. The general scheme for this is illustrated in Figure 22.12.

As illustrated in Figure 22.12, the two prisms are placed on top of each other and broadly aligned and then clamped such that their relative tilts with respect to rotation about the vertical axis may be readily characterised by the interferometer. If the diameter of the interferometer analysis pupil is d and the ***difference***

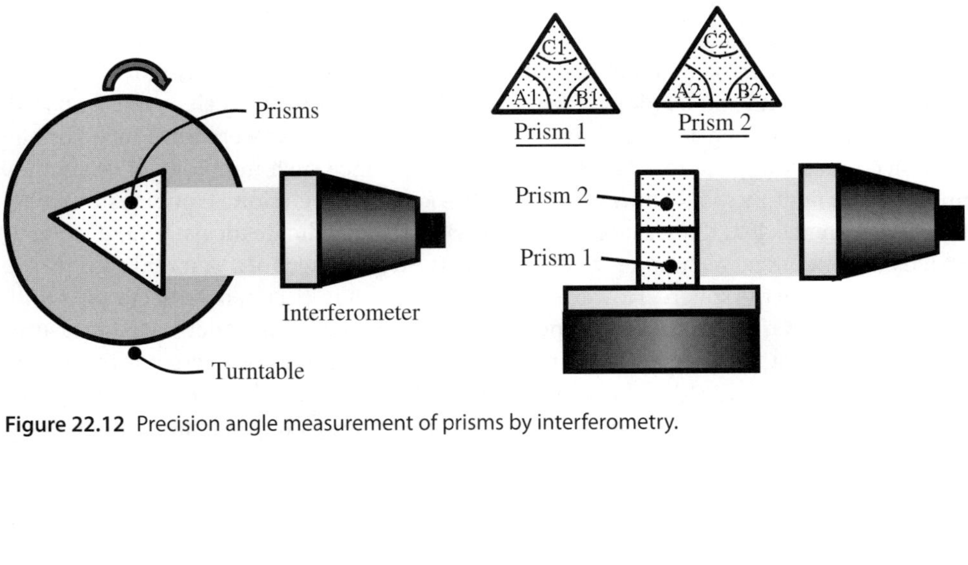

Figure 22.12 Precision angle measurement of prisms by interferometry.

(between the two prisms) in the rms value of the Zernike polynomial describing tilt about the vertical axis is $\Delta Z2$, then the relative angular tilt, $\Delta\varphi$, of the two prism faces is given by:

$$\Delta\varphi = \frac{\Delta Z2}{4d} \tag{22.7}$$

With the prisms clamped in place, the turntable is rotated, accessing two more faces. The difference in the relative tilts measured in these two arrangements will give the difference in internal angles for a specific pair of angles. By repeating the turntable rotation, two additional pairs of angles may be characterised in this way. Finally, by rotating the top prism to a new (aligned) position and repeating the previous steps, a further three pairs may be measured. Ultimately, a total of nine angle pairs may be characterised:

$$\Delta_1 = A_1 - A_2; \Delta_2 = B_1 - B_2; \Delta_3 = C_1 - C_2; \Delta_4 = A_1 - B_2; \Delta_5 = B_1 - C_2; \Delta_6 = C_1 - A_2;$$
$$\Delta_7 = A_1 - C_2; \Delta_8 = B_1 - A_2; \Delta_9 = C_1 - B_2; A_1 + B_1 + C_1 = 180^\circ; A_1 + B_1 + C_1 = 180^\circ \tag{22.8}$$

From Eq. (22.5) all six angles may be computed. Indeed, there is data redundancy (eleven relations for six unknowns) to allow the estimation and statistical reduction of uncertainty. This approach may be replicated for other regular solids.

22.4.4.3 Co-ordinate Measurement Machines

Co-ordinate measurement machines (CMMs) allow for 3D measurement of component positions. Such machines are widely used in the geometrical characterisation of optical and opto-mechanical systems and also in initial alignment. The important point is that they may be used to determine the 3D co-ordinates of a surface or feature with respect some specific Cartesian reference frame. Most such machines are tactile, relying on physical contact with a specific surface to determine its location. For example, a common arrangement uses a contact probe mounted on a set of orthogonally mounted linear stages, i.e. an XYZ stage. The contact probe consists of a hard, e.g. ruby, sphere, where contact with a surface is sensed by pressure or by virtue of microscopic displacement. The XYZ stage positions at which any surface contact occurs are recorded, and the particulars of the test surface (plane, sphere, cone, cylinder, etc.) computed after making due allowance for the geometry (e.g. spherical) of the contact probe. The accuracy of these machines is of the order of 2–5 μm.

Such coordinate measuring machines that employ XYZ stages are fixed and generally require any subsystem to be measured to be brought to the machine. However, there are portable CMMs that are based upon an articulated arm. In many respect, these machines function like a robotic arm. However, instead of the arm's several joints being driven by motors, each joint is 'passive', but supplied with a rotary encoder to establish the rotational state of each joint. The 'hand' of the robotic arm is replaced with a contact probe, e.g. a ruby ball, and, by calibrating the system using a precision artefact, the position of the contact probe may be established absolutely. This arrangement has the natural advantage of portability which is especially useful in the alignment and test of optical systems. The accuracy is somewhat reduced at 10 μm or so.

Tactile CMMs have the natural disadvantage associated with any contact method. The use of hard contact probes, such as ruby, carries a significant risk of inducing surface damage. This is a particularly salient issue for optical surfaces, especially for those with soft coatings, such as aluminium. There are a variety of laser based, non-contact methods that allow for 3 dimensional characterisation of objects. These were introduced briefly in Chapter 12, with reference to laser applications. Laser triangulation is once such non-contact technique. Another 3D measurement instrument introduced was the ***laser tracker***. This technique combines the measurement of angle with interferometric length measurement to provide 3D-coordinate measurement capability. However, it is not strictly a non-contact technique. It does require the deployment of a 'tracker target' in the form of a corner cube mounted in a tactile sphere. However, its principal asset is its enhanced accuracy. Quite naturally, the interferometric measurement of scalar distance is highly accurate; the encoder-based measurement of angle does compromise accuracy somewhat. Nevertheless, this difficulty may be overcome by making measurements with the laser tracker positioned at different locations. This approach, known as ***multilateration***, entirely removes the dependency of the measurement upon the more uncertain angle measurements.

Allied to the laser tacker is the technique known as ***laser radar***. As the name suggests, like its radio-frequency or microwave counterpart, it relies on the measurement of time of flight to build up a picture of surrounding objects. However, the return signal to the instrument is in the form of light scattered directly from the surface of interest, rather than retro-reflected from an object in indirect contact with the surface. As such, it is truly a non-contact measurement technique. However, the scattering process is fundamentally weaker than retro-reflection and, thus, accuracy is compromised. Indeed, the distance measurement function is directly based on a time of flight measurement, rather than an interferometric measurement of phase.

The use of CMMs plays an important part in the characterisation and alignment of optical systems, particularly larger systems.

22.5 Image Quality Testing

22.5.1 Introduction

With the exception of the characterisation of distortion, geometrical testing is concerned with the elucidation of the paraxial properties of a system. However, image quality testing seeks to quantify the departure from ideal behaviour. The notion of image quality suggests an interest in the preservation of detail and contrast in an image with respect to the original object. As such, the most basic tests seek to access this information directly, by the measurement of contrast degradation in real images of test objects. However, this type of process only has access to the spatial variation of image amplitude; any phase information is discarded. By contrast, interferometry exploits the phase information available to provide a more comprehensive and detailed picture of system aberration. However, analysis is required to convert this information into a useful assessment of image degradation; that is to say, interferometry does not provide a ***direct*** measure of image quality.

22.5.2 Direct Measurement of Image Quality

Direct measurement of image quality seeks to characterise the performance of an optical system through presentation of a standard illuminated object and the analysis of the resulting image. A typical example of a standard illuminated object is the USAF 1951 resolution target, which consists of a series of parallel bars of varying separation. The semi-quantitative analysis of such a pattern seeks to establish the minimum bar spacing that can be 'resolved' by an optical system. The theoretical background to image quality measurement and analysis was introduced in Chapter 6. Here, we were presented with the definition of the ***modulation transfer function*** (MTF) which expresses the contrast attenuation produced by an optical system as a function of object spatial frequency. This helps provide a more formally quantitative measure of image resolution. It is measured by presenting a standard illuminated pattern at the object and measuring the contrast ratio obtained at the image. In former times, this latter task was performed using a ***microdensitometer***, which recorded the image contrast through measurement of the transmission of exposed photographic film. Of course, the contemporary process is much more straightforward, where the pattern from a digital image may be analysed more directly.

MTF is an especially useful measure of image quality as it is multiplicative through a system. It is particularly popular in the characterisation of imaging lenses. Furthermore, as previously introduced, not only can it be used to describe the performance of the optical imaging system itself, but also the detector as well. For instance, traditionally, the performance of photographic film could be described with reference to its MTF. In addition, as related in Chapter 14, the resolution of pixelated detectors may also be used be characterised as a spatial frequency dependent MTF.

The most obvious approach for measurement of MTF, is the presentation of a sinusoidally varying object pattern followed by the direct measurement of contrast degradation at that particular spatial frequency. For example, this could be achieved by introduction of a number of replicated patterns, e.g. chrome on quartz, whose transmission varies sinusoidally with displacement. Alternatively, this could be achieved through a

spatial light modulator (liquid crystal display), whose spatially varying contrast is programmable. However, more generally, MTF is measured indirectly. A single target is presented at the object plane and its image recorded. Most usually, these patterns are remarkably simple, for example, in the form of a pinhole or slit or a scanning knife edge. Of course, in such a simple object, there is invested the whole range of the frequency spectrum. As such, computation of the resulting MTF simply involves the Fourier analysis of the input object and the resulting image. If the normalised Fourier transform (as a function of frequency) of the object pattern is $F_0(f)$ and that of the image, $F_1(f)$, then the MTF is given by:

$$MTF = F_1(f)/F_0(f) \tag{22.9}$$

22.5.3 Interferometry

This short section is not intended to provide a detailed description of interferometry and associated experimental arrangements; this is discussed in more detail in Chapter 16. In this context we are particularly concerned with its application in the testing of image quality. Interferometry, ultimately, provides the richest source of information about a system's image quality. Although the measurement of WFE does not translate directly to image quality, as it is perceived in terms of spatial resolution, such information may be derived from analysis. The convenience of computer-controlled instrumentation and analysis enables the WFE to be decomposed into polynomial representations, such as the Zernike polynomial series. In this format, the WFE data is directly related to the characterisation of system aberrations and represents a very powerful tool for understanding design, manufacturing and alignment imperfections.

From the analysis of the system WFE, the Huygens point spread function may be derived. This yields other useful image quality metrics, such as the Strehl ratio and the point spread function (PSF), full width half maximum, and other similar measures.

Although in many respects, a 'gold standard' for instrument characterisation, there are practical difficulties associated with its sensitivity to vibrations, as previously advised. This is perhaps more of a challenge for routine production test scenarios in 'noisy', i.e. manufacturing environments. As such, interferometry is more favoured in critical 'high value' applications. Of course, as outlined in Chapter 16, there are special instrument configurations, such as the Shack-Hartmann sensor and so-called 'vibration free interferometry' that address vibration sensitivity in noisy environments. However, these techniques tend be reserved for more specialist applications.

22.6 Radiometric Tests

22.6.1 Introduction

Radiometric tests are concerned with the absolute or relative levels of illumination, particularly as a function of wavelength. Not only are they concerned with the measurement of (spectral) radiance, irradiance and intensity, but also with the elucidation of optical system transmission and polarisation state. More particularly than is the case for other verification tests, the performance of radiometric tests is dependent upon calibration. The maintenance and transference of calibration standards that define levels of spectral radiance or radiance is central to radiometric measurements. In many respects, radiometric measurements represent the most challenging area in the suite of verification tests. As such, the levels of accuracy expected from these measurements are much lower than those expected from dimensional characterisation. For example, for the measurement of spectral irradiance, one might expect an accuracy of no more than a few percent, depending upon the spectral region of interest.

The maintenance of the primary radiometric standards is the province of the relevant National Measurement Institute (NMI). For example, as discussed in Chapter 7, highly controlled, high temperature blackbody sources may be used as a calibrated source of spectral radiance. However, such sources require a great deal

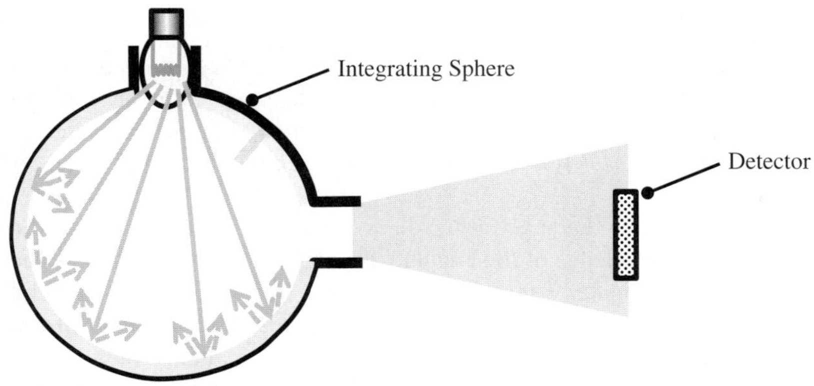

Figure 22.13 Detector flat fielding.

of nurturing and maintenance and are restricted to the NMIs. Therefore, to facilitate practical radiometric measurements, these primary calibration standards must be transferred. Most usually, the transfer is accomplished through the cross calibration of detectors, either directly from the primary standard itself, or indirectly through a secondary radiometric standard, such as a calibrated filament emission lamp (FEL).

22.6.2 Detector Characterisation

22.6.2.1 General

Photodiodes are useful transfer standards for radiometry. They are inherently stable, linear, and sensitive. In order to perform useful radiometric measurements their sensitivity, e.g. in amps per watt, must be calibrated. Calibration is carried out using a standard FEL lamp whose output spectral radiance has been calibrated at an NMI. Ultimately the calibration is traceable to a fundamental black body source. More details of the arrangements are included in Chapter 7.

For longer wavelength applications, pyroelectric detectors may be substituted for photodiodes. As thermal sensors, their wavelength sensitivity is purely dependent upon the absorptivity of their ('black') coating. As such, their sensitivity variation with wavelength is inherently smaller when compared with other sensor types. Furthermore, by virtue of substitution radiometry (see Chapter 7), their output may be related directly to the incident flux in watts. This increased flexibility, unfortunately, comes at the cost of reduced sensitivity, when compared to photodiode sensors.

22.6.2.2 Pixelated Detector Flat Fielding

In many image characterisation applications, we may exploit the versatility of the pixelated detector. For example, we may wish to locate the centroid of an imaged spot to sub-pixel accuracy. This process has been described previously and one simple algorithm involves a simple 'centre of gravity' calculation. However, this approach is predicated upon the implicit assumption that each pixel has precisely the same sensitivity and that the relative signals measured for each pixel faithfully replicates the flux incident upon that pixel. Unfortunately, process variation leads to the pixels in an array possessing a range of sensitivity. Indeed, some pixels may not be functional at all; these pixels are referred to as ***dead pixels***. Therefore, for critical metrology related applications, this sensitivity variation much be calibrated in some way. The procedure by which this is accomplished is referred to as ***flat fielding***. This is achieved by illuminating the detector with a spatially uniform pool of light. For critical applications, this uniform illumination is derived from an integrating sphere. The arrangement is illustrated in Figure 22.13.

22.6.3 Measurement of Spectral Irradiance and Radiance

Measurement of spectral irradiance requires the use of an apertured and calibrated photo-detector and a characterised bandpass filter. The basic arrangement is set out in Figure 22.14. The photodetector will have been calibrated, as previously advised. Using the lamp calibration curve, the flux incident upon the detector may be derived from the detector signal. Finally, the spectral irradiance is calculated by dividing by the filter bandwidth and the detector area.

Of course, the arrangement, as depicted measures the spectral irradiance of a beam. Calculation of the corresponding radiance value is derived from a knowledge of the angular size of the beam. For example, the collimated beam might represent the far field conjugate of some apertured field. In this instance, the solid angle is simply given by the physical area of the apertured field divided by the square of the collimator focal length. The spectral irradiance is simply divided by this solid angle to yield the spectral radiance value.

Due care must be taken to ensure the temperature stability of the detector. Detector sensitivity is a function of temperature and detector calibration only applies at the temperature at which it has been characterised. Figure 22.15 shows a plot of temperature sensitivity versus wavelength for a silicon detector. The temperature sensitivity is expressed as parts per million change per °C.

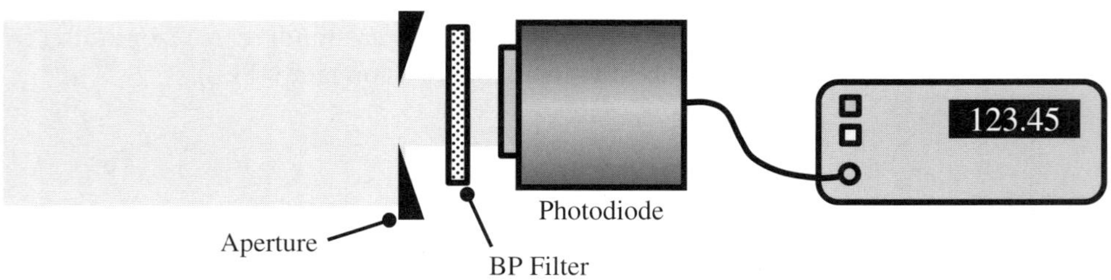

Figure 22.14 Measurement of spectral irradiance.

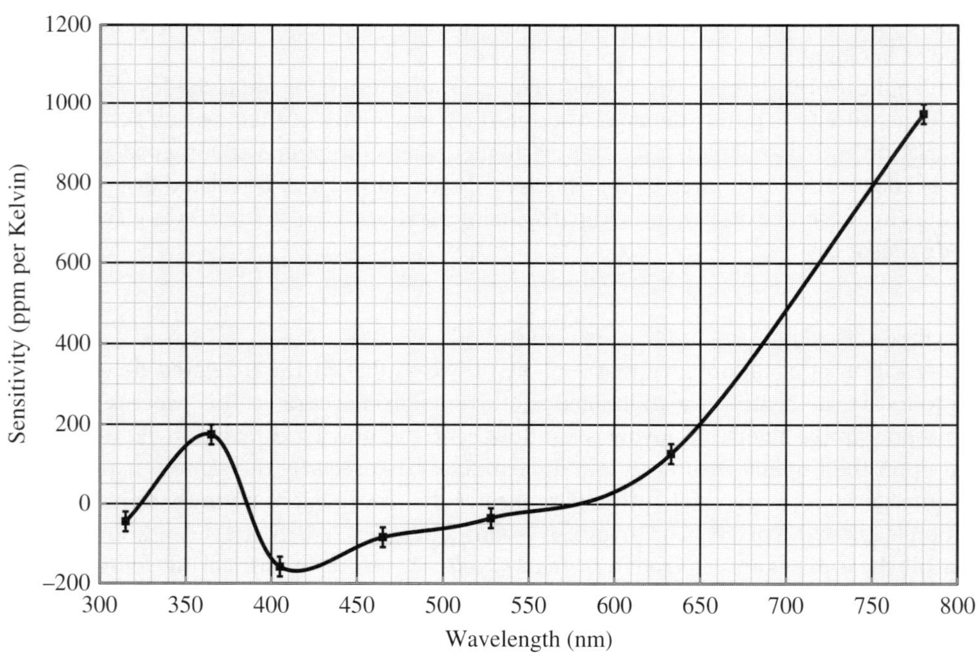

Figure 22.15 Silicon photodiode temperature sensitivity vs. wavelength.

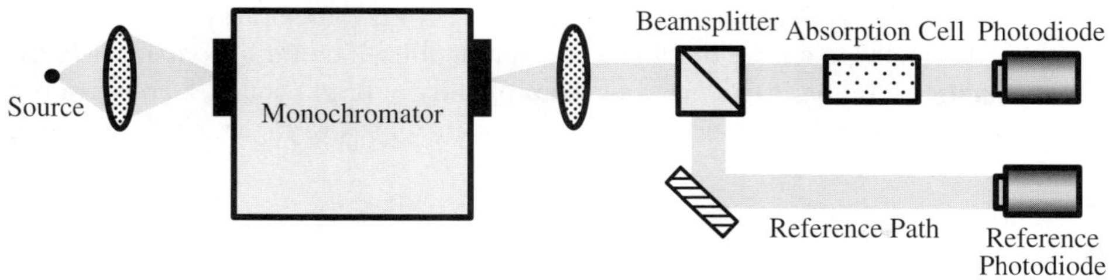

Figure 22.16 Layout of spectrophotometer.

22.6.4 Characterisation of Spectrally Dependent Flux

The arrangement previously described gives a 'radiometric' snapshot of radiance or irradiance at specific wavelengths. Such an instrument, providing radiometric information in a multiplicity of discrete bands may be described as a ***multi-channel radiometer***. Obtaining a full spectrum requires the introduction of a dispersive module, such as a monochromator. In order to measure the absolute radiance or irradiance across the spectrum, the combined instrument must be calibrated. This calibration process proceeds in the same manner as for the simple detector calibration process. That is to say, a calibrated FEL provides a known spectral irradiance to the system. In this case, the entire instrument is calibrated by scanning the system through its wavelength range. Such an instrument is known as a ***spectroradiometer***.

By contrast, unlike the spectroradiometer, a ***spectrophotometer*** is designed to measure relative flux as a function of wavelength. In particular, it is used to measure the absorption or reflectance of materials for assay and chemical analysis and for colour characterisation. Characterisation of absorption is derived from two sets of measurements. First a spectrum of the light transmitted through the material or compound in question is obtained, followed by a reference measurement, that replicates the original measurement, but with the absorber removed. The absorption is then computed from the ratio of the two measurements.

However, the sequential method just described presents a problem. In this type of instrument, the monochromator or dispersive sub-system will be illuminated by some continuum source, such as a filament or arc lamp. The output of such sources will tend to fluctuate by a percent or perhaps more over a measurement cycle. Therefore, the absorption measurement and the reference measurement should be contemporaneous, otherwise significant additional uncertainties will be introduced. As such a spectrophotometer provides, via a beamsplitter, a reference path that facilitates the simultaneous recording of the reference and measurement paths. This setup is shown in Figure 22.16.

22.6.5 Straylight and Low Light Levels

Much of the verification of optical performance is concerned with the behaviour of light that is restricted to the system's as designed étendue. However, as revealed in Chapter 18, even for imaging systems, we must also consider light that is scattered or reflected from both optical and mechanical surfaces. For illumination systems, the system étendue is even less clearly defined. From the design perspective, this difficulty is dealt with through the stochastic analysis of the non-sequential model. However, where this analysis supports a design goal, for example, in an imaging system, we desire to restrict the amount of straylight to such an extent that it does not interfere with the image contrast. To this end, mechanical surfaces are coated (black) and baffles added to restrict the amount of straylight. However, as with other performance requirements, we need to verify that any countermeasures are effective.

In many cases, the levels of straylight will be orders of magnitude lower than those pertaining to any imaging function. As such, evaluation of straylight must pay particular attention to the anticipated signal-to-noise ratios. The distribution of straylight may be analysed by low noise charge coupled device (CCD), or other pixelated detector type. However, where signal levels are especially low, the light may have to be 'chopped' and the signal recovered by a lock-in amplifier, as described in Chapter 14. In any experimental system looking at low level straylight, care must be exercised to ensure that the test equipment itself does not contribute to the burden of scattering and stray light generation, particularly in the proximity of bright sources. For example, it should be understood that a mirror surface has a much greater propensity for scattering when compared to the equivalent lens surface.

22.6.6 Polarisation Measurements

The characterisation of polarisation relies of the measurement of flux as mediated by a polarisation analyser. The purpose of the analyser is to admit one linear polarisation state whilst rejecting the orthogonal polarisation. Such an analyser might take the form of a highly efficient crystal polariser, such as a Wollaston prism, affording an extinction ratio of up to 10^6 between the two linear polarisations. However, depending upon the application and the wavelength range, other, perhaps less efficient polarisation devices might be employed, such as multi-layer polarising beamsplitters, wire grid polarisers and polarising film. Polarising film, of course, is convenient and low cost, but not efficient; wire grid polarisers tend to be adopted for infrared work.

In addition to the relative magnitudes of the two polarisation components, we are also interested in their phase. This information is derived through the insertion of 'waveplates' which provide an additional controlled phase difference between the two polarisation directions. This, in turn, has a deterministic effect upon the post analyser flux recorded by the photodetector(s), enabling the relative phase of the two components to be derived.

The characterisation of polarisation provides information about scattering or reflection from surfaces or the birefringent properties of materials. ***Ellipsometry*** is specifically concerned with the measurement of the change in polarisation state following reflection from a surface. More particularly, it relies on measurement of the amplitude ratio of the two polarisation components, **s**, and **p**. Moreover, the characterisation is not a purely scalar process as the relative change in phase between the two components is also recorded. A combination of analyser and waveplate enables the derivation of both relative phase and amplitude. This technique is ubiquitous in the characterisation of thin films, as the effective complex refractive index, $\mathbf{n} + \mathbf{ik}$, may be derived from such measurements.

The characterisation of polarisation is necessarily a very diverse topic. However, we will illustrate it here through the measurement of birefringence. This technique may, for example, be used to measure stress-induced birefringence in solid materials. The characterisation of birefringence proceeds through the measurement of the phase difference between the two principal axes produced by propagation through some thickness of material. It should be particularly noted that the orientation of the two axes is unknown at the outset. The basic arrangement is shown in Figure 22.17. A beam, e.g. tunable laser beam, is polarised in some orientation by a polarising crystal and then passes through a thickness of the material of interest. Finally, an analyser is placed after the specimen and a photodetector monitors the flux passing through it. The important point to note is that ***both*** polariser and analyser are rotated during the course of measurements.

We now select the system x axis as the major axis of the specimen's birefringence which produces a phase delay of δ. The linear polarisation produced by the polariser subtends an angle, θ, with respect to this x axis. Similarly, the angle of the analyser is defined as φ. Using the Jones Matrix formulation, the polarisation state arriving at the detector may be set out as follows:

$$M = \begin{bmatrix} cos^2\varphi & -cos\varphi sin\varphi \\ -cos\varphi sin\varphi & sin^2\varphi \end{bmatrix} \begin{bmatrix} 1 & 0 \\ 0 & e^{i\delta} \end{bmatrix} \begin{bmatrix} cos^2\theta & -cos\theta sin\theta \\ -cos\theta sin\theta & sin^2\theta \end{bmatrix} \tag{22.10}$$

From Eq. (22.10), we may deduce the total flux as a function of the three variables:

$$\Phi = \frac{1}{2}[cos^2(\theta + \varphi) + cos^2(\theta - \varphi) + sin2\theta sin2\varphi cos\delta] \tag{22.11}$$

Equation (22.11) can then be used to analyse the measurement flux as a function of polariser and analyser angles, thus yielding the specimen birefringence.

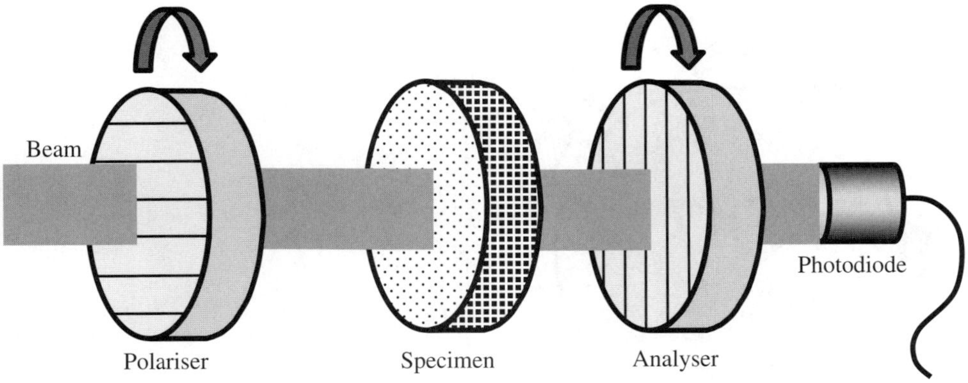

Figure 22.17 Measurement of birefringence and stress-induced birefringence.

22.7 Material and Component Testing

22.7.1 Introduction

Verification testing proceeds at different stages during the product cycle. Much of what has been described hitherto relates to testing on complete systems or sub-systems following integration. By contrast, material and component testing is the province of the manufacturer and opposed to the system integrator. In terms of material properties, from an optical perspective, we are primarily interested in the refractive properties of a material and its uniformity. Of course, there are many mechanical and thermo-mechanical parameters of interest; however, we will restrict our discussion here to the optical properties. Similarly, we restrict our coverage of component testing to a characterisation of optical surfaces, considering the measurement of surface roughness and the characterisation of surface cosmetic quality. Measurement of surface form error and geometrical parameters such as lens focal length and wedge will not be considered here as its characterisation follows upon the same lines as equivalent tests performed at the system level.

22.7.2 Material Properties

22.7.2.1 Measurement of Refractive Index

The measurement of refractive index of a glass is done on a batch to batch basis and across the batch to characterise the likely variation of index within the batch. The favoured method for measurement of refractive index is the so-called ***minimum deviation method***. A prism is fabricated and the minimum deviation angle measured using a goniometer arrangement. This very traditional arrangement is illustrated in Figure 22.18.

An autocollimator or interferometer provides a parallel beam which is deviated by the prism and retroreflected by a plane mirror situated on the moveable arm of the goniometer. Assuming the use of a precision, calibrated goniometer, this angle may be measured very accurately. The prism itself, as indicated may be rotated to establish the minimum deviation condition. The minimum deviation condition refers to the relative orientation of the prism and which the angular deflection of the prism is minimised. In practice, this is established in the symmetrical position where the relevant angles with respect to the surface normal are equal for both illuminated prism facets. If the prism apex angle is Δ and the deflection angle θ, then the refractive index is given by the following relationship.

$$n = \frac{sin(\theta/2)}{tan(\Delta/2)} + cos(\theta/2) \quad (22.12)$$

The prism angle, Δ, may be determined in a similar manner using Fresnel reflection from the facets and measuring the angle with the goniometer. The arrangement above is used for the characterisation of a new

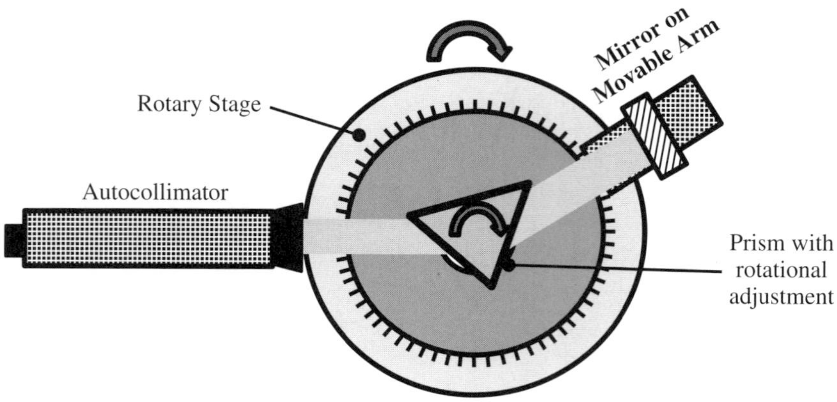

Figure 22.18 Measurement of refractive index through minimum deviation.

formulation rather than as a routine measurement. For the measurement to record the index of the base material, then the measurement should be performed in vacuum. Any measurements made in air must reference ambient conditions, i.e. temperature and pressure, etc. Otherwise, the preference must be for measurement under vacuum conditions and to derive the index relative to air from standard refractive data for air. Any measurement of refractive index must account for the material temperature. For a thorough characterisation of a material, it is customary to use a thermo-vacuum chamber. Thence measurements may encompass a range of temperatures for cryogenic to substantially elevated.

The preceding arrangement is too cumbersome for routine measurements in a manufacturing environment. The refractive index of manufacturing samples is usually measured with respect to some accurately characterised material artefact. One example of this is where the artefact is in the form of a V-block. This V-block is designed to accommodate small sample prisms fabricated from a manufactured batch of glass. Thereafter, a small angular deviation is measured, as per Eq. (22.12). In this case, the value of n set out in the formula refers to the ratio of the indices of the two materials. The temperature of the material must be restricted and controlled to some standard value.

These measurements will yield values across a range of wavelengths, sufficient to derive the Abbe number, etc. for the material. Index variability across a batch is derived from replicated measurements of a standard number of samples across the batch. Finally, measurement of striae is accomplished through interferometric measurements across a slab of material and characterising the localised index variations through analysis of phase contrast.

22.7.2.2 Bubbles and Inclusions

The presence of bubbles and inclusions within a batch of glass is established by their propensity for scattering light. A sample block of glass material is illuminated from the side and viewed against a black background. As such, the process for determining the density of bubbles and inclusions is based upon a (subjective) inspection process, as opposed to a deterministic measurement. The inspection process uses a viewing microscope and the density of bubbles or inclusions of a certain size is determined by counting the number of such features detected within a block of a certain volume.

22.7.3 Surface Properties

22.7.3.1 Measurement of Surface Roughness

Surface roughness can either be measured by a contact or non-contact method. The contact method involves drawing a hard (e.g. sapphire, diamond, etc.) stylus across the surface of interest. Such an instrument is referred to as a ***stylus profilometer***. The stylus is attached to a rigid arm, free to rotate about a precision pivot. Movement of the arm is detected at the other side of the pivot by precision length interferometry. Inevitably, the radius of the stylus limits the sensitivity of instrument at high spatial frequency. Access to higher frequencies, naturally, becomes accessible with the deployment of stylus tips with very small radii. However, this increases the surface load upon the specimen, increasing the likelihood of damage, i.e. scratching. Such stylus measurements may be replicated by non-contact optical instruments, such the Confocal Length Gauge and the White Light Interferometer. Details of these are to be found in Chapter 16.

The Confocal Gauge, like the Stylus Profilometer, samples along a linear track, whereas the White Light Interferometer collects data over an area. Detailed analysis of surface roughness data presents the information as a PSD spectrum. The concept of the surface roughness or form spectrum was introduced in Chapter 20 and forms part of the detailed specification of surface quality. However, where a single surface roughness number is to be presented, then the data presented must be analysed and digested in some way. To this end, the profile of the surface is fitted to some form (e.g. straight line or plane) and the residual filtered to remove low spatial frequency components. The spatial frequency filtering process is characterised by a specific 'cut-off' spatial wavelength, usually selected to be one-fifth of the length of the profilometer trace. This effectively acts as a high pass filter. As such, the measurement data is then restricted to a specific and well defined spatial frequency range. Thereafter, the surface roughness may be presented as an rms value. For analysis of linear tracks, the rms surface roughness is denominated as an $\boldsymbol{R_q}$ value, whereas the corresponding area value is designated as $\boldsymbol{S_q}$. The rms data presented in this way is relevant, optically, to the scattering process. However, roughness information is occasionally presented as a peak, as opposed to rms value and designated as the $\boldsymbol{R_a}$ and $\boldsymbol{S_a}$ values respectively. Presentation of roughness information in this way is generally associated with a mechanical as opposed to an optical specification.

As discussed, presentation of specific surface roughness values requires the introduction of a high pass spatial frequency filter, removing low spatial frequency components. At the high spatial frequency end, the data are filtered by the resolution of the measurement instrument itself. For the stylus profilometer, this is determined by the radius of the stylus tip. In the case of the non-contact optical probes, the lateral resolution is defined by the optical resolution. Either way, the high spatial frequency cut-off corresponds to spatial wavelengths of the order of a micron or a few microns. Therefore, to characterise particularly smooth surfaces, such as those used in X-Ray mirrors, higher spatial resolution is required. This may be obtained by exceptionally high-resolution instruments, such as ***atomic force microscopes***.

22.7.3.2 Measurement of Cosmetic Surface Quality

The concept of cosmetic surface quality was introduced in Chapter 20. Cosmetic surface quality is an indication of the thoroughness or effectiveness of the optical polishing process to remove abrasively generated scratches or digs earlier in the manufacturing process. The intent of the surface quality inspection process is to quantify the number of pits or digs over a certain specific size and the combined length of scratches over a certain size. In many respects, the inspection process replicates that of the assessment of bubbles and inclusions in glass blanks. Oblique illumination of the surface is used, which is viewed against a dark background. Ideally, the size of the features should be quantified during the inspection. However, the width of visible scratches, in particular can be very small and only quantifiable using high resolution microscopy. Unfortunately, such an arrangement is not highly productive and therefore is not appropriate in a production environment As a consequence, in practice, the process relies upon the feature size being gauged by an operator by virtue of its prominence when viewed against a dark background. To help the operator, a series of standard features, scratches, and digs, etc., of known size are provided for comparison. However, inevitably, the process relies upon the subjective judgement of the operator. The earlier military specification, MIL-PRF-13830B, was built upon the description of scratches in terms of standard calibration samples, rather than specific scratch dimensions. The relevant ISO standard (ISO-10110) attempts to resolve this by using definitions that are tied to specific dimensions. However, traditional inspection methods render implementation troublesome.

As with other optical measurements, the advent of digital imaging and powerful image processing tools has changed the picture significantly. Using a broadly similar arrangement to that of the traditional inspection process, a high resolution digital image of light scattered from the sample's surface is gleaned. Image processing enables the denomination of scratch and dig features in a more deterministic fashion. Chapter 20 presents more details of the cosmetic surface quality standards themselves.

Further Reading

Ahmad, A. (2017). *Handbook of Opto-Mechanical Engineering*, 2e. Boca Raton: CRC Press. ISBN: 978-1-498-76148-2.

Aikens, D.M. (2010). Meaningful surface roughness and quality tolerances. *Proc. SPIE* 7652: 17.

Gordon, C.G. (1999). Generic vibration criteria for vibration-sensitive equipment. *Proc. SPIE* 3786 12pp.

Macalara, D. (2001). *Handbook of Optical Engineering*. Boca Raton: CRC Press. ISBN: 978-0-824-79960-1.

Turchette, Q. and Turner, T. (2011). Developing a more useful surface quality metric for laser optics. *Proc. SPIE* 7921: 13.

Vukobratovich, D. and Yoder, P.R. (2018). *Fundamentals of Optomechanics*. Bota Racon: CRC Press. ISBN: 978-1-498-77074-3.

Yoder, P.R. (2006). *Opto-Mechanical Systems Design*, 3e. Boca Raton: CRC Press. ISBN: 978-1-57444-699-9.

Index

Optical Engineering Science, First Edition. Stephen Rolt.

Companion website: www.wiley.com/go/Rolt/opt-eng-sci

g